Linear Algebra
and Analytic Geometry

Linear Algebra and Analytic Geometry

HEINRICH W. BRINKMANN
EUGENE A. KLOTZ
Swarthmore College, Pennsylvania

 ADDISON-WESLEY PUBLISHING COMPANY
Reading, Massachusetts · Menlo Park, California
London · Don Mills, Ontario

This book is in the
ADDISON-WESLEY SERIES IN MATHEMATICS

Consulting Editor: LYNN H. LOOMIS

Copyright © 1971 by Addison-Wesley Publishing Company, Inc.
Philippines copyright 1971 by Addison-Wesley Publishing Company, Inc.

All rights reserved. No part of this publication may be reproduced, stored in a retrieval system, or transmitted, in any form or by any means, electronic, mechanical, photocopying, recording, or otherwise, without the prior written permission of the publisher. Printed in the United States of America. Published simultaneously in Canada. Library of Congress Catalog Card No. 79–132056.

For Betty and Debby

Prelude

Although we trust that some of our method of presentation is new, the mathematics was discovered by others. In this respect the writing of textbooks is usually a matter of assembly, rather than of creation. But in this we can sing along with Kipling:

> When 'Omer smote 'is bloomin' lyre,
>   He'd 'eard men sing by land an' sea;
> An' what he thought 'e might require,
>   'E went an' took—the same as me!
>
> The market-girls an' fishermen,
>   The shepherds an' the sailors, too,
> They 'eard old songs turn up again,
>   But kep' it quiet—same as you!
>
> They knew 'e stole; 'e knew they knowed.
>   They didn't tell, nor make a fuss,
> But winked at 'Omer down the road,
>   An' 'e winked back—the same as us!
>
> (*Introduction to the Barrack-Room Ballads in 'The Seven Seas'*)
>
> RUDYARD KIPLING

Preface

PURPOSE

This book is intended for courses in linear algebra and in n-dimensional analytic geometry. A number of different courses of varying lengths can be taught from the text because of the generous amount of supplementary material. Our basic aim has been to create a text that can be used to advantage by students with a wide range of abilities and interests, a book that would aid the student outside the classroom while freeing the instructor to concentrate on the specific needs of his particular class. The prerequisite for profitably using this is a year's course in single-variable calculus, or the equivalent mathematical maturity.

APPROACH

To meet these goals, we have:
1. included informal discussions providing intuitive and geometrical motivation;
2. written out many examples in detail;
3. carefully spelled out computational procedures;
4. kept notation to a minimum so that the instructor can add his own if he desires;
5. suppressed extraneous terminology, although the standard terms are defined, at least for reference.

We have tried to cut through a simple but cluttered subject in order to emphasize the central ideas, and have attempted to separate the superficial from the crucial. The expository style is relaxed and informal, but the mathematics itself is done with care and precision.

Our aims are also met through the variety of supplemental material that is presented. This includes not only more advanced mathematical topics, but also discussions of computational techniques (and the computer), subjects useful in the sciences, and some applications. Our direct applications are mainly in the biological and social sciences, since students in these fields often lack immediate access to relevant applications; students in the physical sciences and engineering usually do not have this problem.

USING THE BOOK

We have tried to make this text as flexible as the material will allow. We indicate how and where topics can be omitted so that short courses can be satisfactorily taught from parts of the book. In particular, starred sections may be omitted at first reading with minimal loss of continuity. Moreover, the chapters on determinants and on lines, planes, and flats contain summaries so that those who wish to may pass by quickly.

Although it would be a rare course which would cover all the material, we expect the book will serve the student well for future reference.

The text has taken several years to evolve, and in a number of incarnations it has been used by a variety of students. In particular, it has been used for semester linear algebra courses, and for half-semester courses in analytic geometry.

CONTENTS

There is an introduction, consisting of background information for the student to read on his own. This includes a discussion of the why and wherefore of linear algebra, the axiomatic method, and mathematical models. We explain to the student why he should be interested in developing a modest facility for making simple proofs. To further this we often ask students to fill in missing steps in proofs, and even ask for entire proofs, in simple or redundant situations, or in honors material. (In addition to removing some of the clutter and thereby exposing the basic outlines of the proof, we hope that this will also serve to engage the student more actively with the material).

Throughout the text, we officially work with abstract vector spaces over the field of real numbers (except when complex eigenvalues force us to consider complex spaces). However, there are supplementary comments and honors projects extending the material to vector spaces over the complex numbers (or even over arbitrary fields for the able and curious). Other aspects of our presentation should be readily visible from our Table of Contents. For example we introduce matrices before linear transformations in connection with a fundamental computational procedure.

There are a number of advanced topics which we do not even mention (quotient spaces, elementary divisors, rational canonical form, the tensor product). We have, however, included a discussion of the Jordan form because of its importance, although, due to its complexity, we have made it an optional topic. Dual spaces are included (optionally), but not emphasized; we try to give a concrete approach to serve as a firm foundation for later work.

We have presented some rather innovative material in our study of analytic geometry (again, see the Table of Contents), along with the standard facts about euclidean spaces and quadratic forms. Although a number of these topics are of interest to scientists, statisticians, and engineers, as well as future mathematicians, some of the material is not readily available in elementary texts. Starred sections may be omitted at first reading with minimal loss of continuity.

CUPM

The topics are presented in a spirit in keeping with the CUPM recommendations.

Swarthmore, Pennsylvania H.W.B.
February 1971 E.A.K.

Acknowledgments

We wish to thank:

—our Swarthmore students for their patience and consideration in the face of innumerable dittoed revisions.

—all our Swarthmore colleagues for their cooperation and advice.

—Philip Carruth for using and commenting upon a preliminary version.

—Mrs. Dorothy Blythe, both for her speedy and accurate typing, and for her warm cheerfulness throughout.

—Mrs. Ann DeRose and Miss Laura Denton for typing help when the chips were down.

—the Addison-Wesley staff for their patience and cooperation.

Contents

To the Reader xv

Introduction

1 What is linear algebra? 2
2 Why linear algebra? 3
3 The axiomatic method 4
4 Mathematical models 9
5 The relationship between mathematics and science 12

Chapter 0 Preliminaries

1 Geometrical vectors 16
   Supplement 1. Equivalence relations 18
   Supplement 2. Groups 20
2 Vectors and N-tuples 22
3 Real functions 24
4 Functions in general 27
5 Composition of functions. Inverse functions 31
6 Fields 33

Chapter 1 Vector Spaces

1 In which the central characters are introduced 38
   Supplement. Vector spaces over other fields 44

Contents

    2 Elementary properties of vector spaces 45
    3 Subspaces 48
         Supplement. The arithmetic of subspaces. 53
    4 Bases 56
         Supplement. The deletion lemma, proof of 4.6 65
    5 Isomorphic vector spaces 66

Chapter 2 Matrices

    1 Bases for subspaces 73
    2 The basis algorithm 82
         Supplement. Flow diagrams; a computer improvement . . 87
   3* Matrix arithmetic 90
         Supplement. Partitioned matrices 96
    4 Square matrices; transposition 104
   5* Some applications of matrix arithmetic 121

Chapter 3 Linear Transformations

    1 What they are 145
    2 Arithmetic properties 153
    3 Representing linear transformations. 160
    4 The dimension theorem; invertibility 170
         Supplement 1. Some consequences of the dimension theorem . . 177
         Supplement 2. Almost diagonal matrices 181
   5* Dual spaces 182

Chapter 4 Systems of Linear Equations

    1 An elementary approach 187
    2 Linear equations, matrices, and linear transformations . . . 201
   3* Matrix equivalence; another solution procedure. . . . 213
   4* Change of basis 219

Chapter 5 Determinants

    1 Definitions and the like 232
    2 Calculations 241
    3 Products and transpose 248
    4 Cofactors 254
   5* Proof of existence and uniqueness 262

Chapter 6 The Eigenvalue Problem; Similar Matrices

    1 Eigenvalues 267
         Supplement 1. Gershgorin's theorem 275

2	Distinct eigenvalues	279
	Supplement. The power method	284
3	The characteristic polynomial	289
4*	Similarity	293
5*	An introduction to Jordan form	299
6*	Jordan form proofs; nilpotent matrices	307

Chapter 7 Linear Planes, and Other Flats

1	Points and Lines	315
2	Parametric equations and affine equations	329
3	Solutions to systems of linear equations	336

Chapter 8 Euclidean Spaces

1	Distance, angles, and dot product	345
2	Orthogonality	352
	Supplement 1. Inner product spaces	358
	Supplement 2. Legendre polynomials	360
	Supplement 3. Complex inner product spaces	361
3*	Orthogonal complements and flats again	362
4*	Distance from points to subspaces; least squares	370
5	Orthogonal linear transformations	379
	Supplement 1. Unitary mappings	386
	Supplement 2. Orthogonal matrices expressed in terms of parameters	387
6*	Rigid motions, especially in 2- and 3-space	388
	Supplement. Rotation in $\mathbb{R}_3$; rotations and the cross product	402
7*	Rigid motions in $\mathbb{R}_n$	406

Chapter 9 Symmetric Matrices and Quadratic Forms

1	Symmetric matrices and transformations	415
	Supplement 1. Hermitian matrices	421
	Supplement 2. The Jacobi method	422
2	Quadratic forms	425
3*	The general 2nd degree equation	431
4	Positive definite matrices	443
5*	Nonorthogonal diagonalization of quadratic forms	450

Chapter 10* Matrix Calculus

1	Vector norms	456
2	Matrix norms	466
	Supplement 1. The spectral norm	471
	Supplement 2. The row and column norms	476

3	Infinite series of matrices	478
	Supplement. e^{A+B}	489
4	The complex numbers	491
	Supplement. Quaternions again	496
5	Analogies between $\mathbb{C}$ and $\mathbb{R}_{n \times n}$	499
	Answers to Selected Exercises	511
	Index	531

To the Reader

The linear algebra material which you are about to study may have a different mathematical flavor from much of your previous mathematics. Because of this, we have provided an Introduction, an attempt to explain not only where we are going, but why the trip is worthwhile. You are invited to consult this material at the outset, or at any time throughout your course of study when interest or energy flags.

There are answers to selected problems in the back of the book. Some exercises are marked with an H for "honors," and should be undertaken at your own risk. Titles such as "theorem," "lemma," "corollary," "proposition," etc., have been dispensed with. Items to be proved are simply numbered (beginning afresh each chapter: m.n means "theorem n of section m").

There are two abbreviations which are part of standard mathematical vocabulary, and of which we now remind you.

"iff" stands for "if and only if"

Thus, to prove the assertion "statement 1 iff statement 2" means that we must prove both that if statement 1 is true, then so is statement 2, and also that if statement 2 is true, then so is statement 1. Second, the notation

$\{x \text{ in } X \,|\, \text{statement about } x\}$

is to be read "the set of all x in X for which the statement about x is true."

The expressions "it is clear that," "obviously," and "it can be easily seen" stand for the longer phrase "you should be able to get this without too much trouble, and we think it more important to work it out yourself rather than for us to lead you by the hand." This is standard mathematical style and is not meant to be condescending. For further information the reader is urged to consult H. Pétard, "A Brief Dictionary of Phrases Used in Mathematical Writing," *American Math. Monthly* **73,** February 1966, pages 196–197.

Finally, there is a considerable amount of supplementary material (both in supplements and in starred sections) which is designed to introduce you to mathematics important in applications or in computations. We hope that you will find this and other aspects of our book useful for future reference.

Bon voyage!

Moral: to the vector belongs the spoils†

Introduction

Mathematicians have a long and glorious tradition of refusing to discuss the practical value of mathematics.‡ Indeed future mathematicians, physical scientists, and engineers often become so brainwashed that they cease asking "Why learn this topic?" Some of our newer clientele—social scientists, biologists, doctors, and merchant chiefs (lawyers may be added in due time)—might still wonder why such an abstract subject as linear algebra should engage their time and attention. This introduction is an unabashed attempt to draw forth the best energies of those not planning to become mathematicians, by pointing out some of the potential value not only of the subject itself, but also of our method of presenting it.

In this introduction we first make a brave attempt at indicating the contents and flavor of linear algebra, and then list some reasons why it might be of potential worth to a nonmathematician. The rest (which is most) of the introduction is spent in elucidating one of the less obvious items on the list: we discuss the axiomatic method in mathematics and then its application to science via the mathematical model. After this we make a few remarks on the relationship between mathematics and science. Our introduction even includes references at the end.

It is our hope that one of the conclusions which nonmathematicians will draw from this material is that the usual form of presenting mathematics in texts

---

† *The Dot and the Line, a Romance in Lower Mathematics*, by Norton Juster, Random House, 1963.
‡ There is a story that a student, having worked his way through his first theorem, asked Euclid what good it would do him. Euclid told his slave to give the student a quarter, if he had to profit from his learning.

and courses not only is suited for the apprenticeship of future mathematicians—it can be of great value for anyone to learn to work with axiom systems (and this includes both understanding the proofs of theorems, and also developing modest skills at actually constructing proofs). To encourage this and to clarify the structure of some theorems, we often leave selected steps in proofs (or even entire proofs) up to the reader.

This introduction contains little real mathematics (there is only one theorem), but future mathematicians may find it at least of cultural value.

1. WHAT IS LINEAR ALGEBRA?

One problem inherent in the learning of modern mathematics is that it is often impossible to describe where you're going until you get there. We shall try, nonetheless.

Linear algebra is a branch of modern algebra which has its origins, in part, in the solution of systems of linear equations. As is often the case, this algebra has a geometrical counterpart. In the two-dimensional situation, the basic geometrical entity here is the straight line, which gives rise to the linear part of the name.

Roughly speaking, linear algebra is concerned with abstract systems called "vector spaces". The basic constituents of a vector space are (1) a set of objects called "vectors" which can be added (and subtracted) in pleasing although unsurprising ways, and (2) a "field of scalars", which we shall for now consider to be the real numbers, with which the vectors can be "scalar-multiplied", again in a very sensible fashion.

One example of a vector space is the collection of all geometrical vectors in the plane (those movable arrows you've probably encountered in other mathematics and science courses). Another example (which you may know how to identify with the first) is the set of all pairs (a, b) of real numbers, with componentwise addition and scalar multiplication. Working our way up through triples, quadruples, quintuples, etc., we see that yet another example is the set of all "n-tuples" $(a_1, \ldots, a_n)$ of real numbers (fixed positive integer n), under the same kind of addition and scalar multiplication. Yet another example is the set of all real-valued functions defined on some interval of real numbers, under the usual definition of function addition and scalar multiplication.

You can ferret out a vector space in almost any context in which some sort of natural "addition" and "scalar multiplication" transpires (and such contexts occur more often than one might suppose). You can also usually concoct a vector space whenever you need to add and scalar-multiply disparate quantities [e.g., a apples $+$ b bananas can be conveniently represented by the ordered pair (a, b)].

The field of scalars need not be restricted to the real numbers, although we shall officially do so for simplicity. There are contexts in which physical scientists, at least, need to use the complex numbers as the field of scalars (we include honors projects for those who wish to keep up with this possibility, and other possibilities which we now discuss). Users of computers will be delighted to discover that the rational numbers (the only real numbers with which a computer operates) also form an appropriate field of scalars. There are even more bizarre mathematical confections (which turn out to have practical uses, of course) known as "fields" which are suitable as fields of scalars for vector spaces. The adventuresome will find a discussion of fields in Chapter 0, which is chiefly concerned with setting up important examples of vector spaces, and with reviewing basic notions about functions. Chapter 1 contains the main discussion of vector spaces themselves.

The other basic objects of study in linear algebra are those functions from one vector space to another which preserve the algebraic structure of the vector spaces. They are called "linear transformations" and represent an extremely important class of functions. They can often be used to approximate other functions and are usually far simpler to deal with.

Matrices (of whose existence you may already be aware) can be considered as linear transformations, among other things, and to a large extent the converse is also true: linear transformations can often be viewed as matrices. In fact, matrix algebra is one of the important topics within linear algebra.

In Chapter 2 we introduce the basic operations on matrices. Matrices then serve as an important example of linear transformations, which we study in Chapter 3. Chapter 4 deals with systems of linear equations from the point of view of matrices and linear transformations—their natural setting.

Chapter 5 is concerned with the determinant, a function defined on square matrices (you may also consider the determinant as an "oriented volume" if you prefer a geometric approach). Chapter 6 has to do with finding those vectors which are merely stretched by a given linear transformation (this is actually an important and useful quest).

Thus endeth the first part of the book. Most of the rest has to do with applying linear algebra to analytic geometry. The geometrical nature of this material makes it overtly more natural, and obviates any need to discuss it further at this time.

Linear algebra has a different mathematical flavor from much of your previous mathematics. In contrast with elementary calculus, for example, the proofs in linear algebra tend to be rather easy. Moreover, linear algebra is a curious blend of notions at once more abstract and more concrete than those one finds in calculus.

2. WHY LINEAR ALGEBRA?

0) It is an interesting subject in itself. (We won't count this since there is no disputing about tastes. However, most mathematics is created for pure interest. Since it is conceivable that you will never use this material, you might as well try to get something out of the course, if only enjoyment.)

1) Linear algebra discusses some important problems, e.g., the ubiquitous

problem of the solution to systems of linear equations. Actual procedures for solutions can be described in linear algebra terms (and they even work, unless a very large number of equations and unknowns are present, in which case it is necessary to introduce some approximation notions as well).

2) Linear algebra is a very basic language. In almost all areas of science and mathematics, its vocabulary has proved useful in illuminating fundamental structure, and its notation is often magnificent for dealing with otherwise cluttered and complicated situations.

3) In addition to serving as a basic language, linear algebra has more sophisticated applications in the various sciences—physical, social, and biological—and in engineering, business, and so forth. (You can easily convince yourself of the applicability of one branch of linear algebra by looking up "Matrix Algebra for _____" in *Books in Print*.) In fact, everything in this book is of use to *some* nonmathematicians (although not everything is used—at least currently—by *all* nonmathematicians). Those areas not immediately useful to you may well become important later, and they should at least provide mathematical insight into the subject as a whole.

4) Linear algebra is widely used in areas of mathematics which are of great importance to the nonmathematician. For example, in probability theory, statistics, the analytical geometry of n dimensions, several-variable calculus, and differential equations.

5) It is important for computational purposes, since it is usually much easier to make computations in a linear than in a nonlinear situation. Consequently, scientists often assume that the world is linear, at least as a first approximation.

6) Linear algebra provides an excellent opportunity to observe and experience the axiomatic method. We next explain what this is, and then discuss its relevance to the nonmathematician.

3. THE AXIOMATIC METHOD

The axiomatic method is essentially a way of presenting mathematics, but the pedagogical device is so intertwined with the subject itself that some might say that it *is* modern mathematics.

To begin our discussion of the axiomatic method, we shall consider the notion of a "field". (This is given a fuller treatment in Chapter 0, Section 6, and although it is an optional topic for the purposes of this book, it is well suited for the current discussion.) Roughly speaking, a "field" is a mathematical system in which one can add, subtract, multiply, and divide (except by an element called "0"), in a decent and reasonable fashion (whatever that means). Examples of fields are the rational numbers, the real numbers, the complex numbers, and certain finite "number systems" (some of which are discussed in Chapter 0). In fact, the notion of "field" represents a distillation of the mathematical properties common to some of our usual number systems.

To describe an object axiomatically, one postulates the existence of a set (or sets), together with operations on the set(s), and perhaps also relations which may hold among certain elements of the set(s); one then lists "axioms" which are rules governing the behavior of the operations and relations.

Thus, a "field" is any set F which is equipped with two operations called "addition" and "multiplication" (subtraction and division may be derived from these). The axioms (listed completely in Chapter 0) say such sensible things as:

There exists an element "0" in F for which $a + 0 = a$, for any a in F.

(As an example of a relation in an axiom system, an "ordered field" is a field F with a relation "less than", written "$<$", which satisfies all the usual properties of "$<$"; for example, "For any a, b, c, in F, if $a < b$, then $a + c < b + c$".)

By an *axiom system* we shall mean a collection of postulated sets, operations, and relations, together with axioms governing their behavior. After presenting an axiom system, the rest of the axiomatic method consists of deriving properties of the system (i.e., theorems) using only the axioms or previously proved theorems. For example, one property which holds in any field F is

If z is an element in F and if $a + z = a$ for some a in F, then z is the element "0" alluded to above (see Chapter 0, Section 6).

Axiom systems are used to present the basic properties common to a number of mathematical constructs, as we have seen with "fields". One can also use an axiom system to describe precisely and concisely a specific mathematical entity. For example, the real numbers may be described by means of the axiom system for "ordered field", together with one additional axiom (which essentially involves the calculus notion of a limit).

Most of you have already encountered an axiom system in euclidean geometry ("synthetic", not "analytic" geometry). However, the axiomatic method may have disappeared from view in your study of analytic geometry and calculus. It is still there—just hidden. The reason for this is that the axiom system used in these latter subjects consists of just the axioms for the real number system; all other concepts are *defined* from this. For example, in the analytic geometry of the plane, a **point** is an ordered pair of real numbers (x, y); a *line* is the set of all points (x, y) which satisfy a nontrivial equation $ax + by = c$; and so on.

In linear algebra the axiomatic method again comes to the surface, and is worth considering overtly.

A mathematician does not just sit down, make out a list of operations, relations, and axioms, and say "Well, now, what can we prove from this batch of basic assumptions?"—not if he expects to do interesting mathematics. Rather, an axiom system is usually an extraction of basic properties that occur in several important areas of mathematics (or science). Thus, we have seen that a "field" is given by an axiom system which represents the essence of the rational, real, and complex numbers, and certain quaint, finite number systems which are to be mentioned in Chapter 0. In like manner, the protagonist of this book, "vector space", is defined by a set of properties common to geometrical vectors, the set of functions defined on an interval, ordered quadruples of real numbers, and the set of all polynomials.

Virtues

There are several virtues in presenting mathematics by means of the axiomatic method. The first virtue (which also appears below as the first vice) is the generality which one obtains. According to H. Poincaré,

> Mathematicians do not study objects but the relations between objects. Matter does not engage their attention. They are interested in form alone.

By stripping off all connotations and working with abstract sets, together with operations and relations on these sets, and the axioms governing their behavior, one becomes truly free to study form and relations independent of the origins. Thus, the axiomatic method is an important aid to the understanding of mathematics.

In addition, the generality afforded by the axiomatic method can make the mathematics itself simpler. For example, some properties of matrix operations are quite involved to prove directly. However, when matrices are considered more abstractly as linear transformations, these properties are seen to be more transparent and easier to prove.

The axiomatic method is also an efficient learning device. If one studies "fields" in the abstract, then one learns properties common to the rational, real, and complex numbers, the finite "number systems" mentioned above, and other examples too numerous to enumerate.

Perhaps it should be pointed out that many areas of mathematics which at first appear esoteric have a way of becoming useful. For example, finite fields are now being used in the design of error-control codes for digital communication systems. The engineer who has studied fields in the abstract has no intellectual retooling to do in order to meet this situation; generality *can* be useful.

This example suggests another virtue: the axiomatic method is eminently applicable. If you encounter a situation in your work (in mathematics, science, business, engineering) which seems reminiscent of some familiar axiom system, all you need to do is to check that you have creatures around which satisfy the axioms of the system. Then each known theorem derived from the axiom system becomes a valid fact in your context. We will speak more of this in the discussion below of mathematical models.

A final pedagogical virtue of the axiomatic method is that it permits a student to know exactly where he stands. At any point of the course he knows precisely what he can assume (viz, the axioms and theorems thus far proved), and he can easily identify the ingredients of a valid proof (viz, the axioms and theorems thus far proved).

Vices, and Remarks on Overcoming Them

For many students, the main difficulty in learning by the axiomatic method is that they allow the generality to become disconcerting. Confronted with an axiom system which begins, for example,

> "a vector space $\mathcal{V}$ over the field of real numbers consists of two sets $\mathcal{V}$ and $\mathbb{R}$ ..."

the beginning student may feel discouraged if he cannot get the general notion of "vector space" into his head right from the outset. It may seem that "vector space" should be a creature for which he feels the same affinity as for geometrical vectors, pairs of real numbers, polynomials, and other (more or less) tangible mathematical notions. Actually, it usually takes some time to come to grips with such an abstract concept, and to obtain any comfort with the idea of vector space as an entity in its own right. We now suggest some helpful steps.

What to do when confronted with an unfamiliar axiom system:

1) Carefully read through the list of relations, operations, and axioms.
2) Take a couple of familiar special cases and go through the system, checking that these do indeed satisfy the axioms. (We are assuming that the axiom system is an extraction of properties common to several different areas, such as "vector spaces", and not a description of a single entity such as "real numbers".)
3) Memorize the axioms (and later check from time to time that you do recall the system perfectly).
4) Check a few more examples.
5) Try to invent some examples on your own (recognizing failures can be as instructive as complete successes).
6) Go through the proofs of a few simple theorems.

By this time you should be well on your way to a satisfactory working relationship with the system; and if it is not an out-and-out comradeship, at least it can be a dignified armed truce.

After this attempt at basic assimilation, you always have the problem of attempting to understand definitions, as well as the statements and proofs of theorems. Difficulties sometimes arise here in apprehending things in full generality. If this is the case, it is often beneficial to first study a definition, theorem, or proof in the light of a specific example. To this effect, after we define "vector space" in Chapter 1, we list some examples in increasing levels of abstraction.

It is often possible to appreciate mathematics at different levels of abstraction. In our particular case, it is not really necessary to understand "abstract vector space" to learn a great deal from the material in this book. One can take for "vector space" a specific example such as "triples of numbers", and obtain great profit from this interpretation of the material.

You may find some solace in the fact that the first sophisticated axiom system one encounters (and masters) is usually the hardest; skills can be developed here. There is an old saying that one of the great values in studying mathematics is the training in abstract thinking it affords. This applies in particular to working with axiom systems. We shall point out some practical values of this in our discussion of mathematical models.

Another difficulty which is sometimes encountered is that an axiom system can be rather complicated: lots of sets, operations, relations, and axioms, seemingly entangled in a positively perverse manner. The axiom system which we use to describe a "field" might appear to be of this sort, at first glance. For, in addition to one set and two operations, there are eleven axioms (which, we hasten to add, are all both natural and simple; moreover, two sets of five axioms are quite similar).

More to the point, the notion of "vector space $\mathcal{V}$ over the field of real numbers $\mathbb{R}$" consists of two sets $\mathcal{V}$ and $\mathbb{R}$ with four operations involved. The first set $\mathcal{V}$ satisfies the five axioms encountered twice in the axiomatic description of a field; the second set $\mathbb{R}$ is a field, the real number system; and there remains the extra operation of scalar multiplication and its axioms linking together the two sets.

All this may sound rather overwhelming, but there are a number of mitigating factors, so that there is really not too much to cope with, after all. First of all, for simplicity, we assume that you know (more or less) what the real numbers are, so we refrain from listing the axioms for the real number system (this clears the air tremendously). Second, the axioms and the situation are very natural: you are already quite aware of several important examples of vector spaces over the reals. Moreover, our plan of attack is to first isolate the basic properties of vector spaces from these examples so that you will be well prepared for the formal axiom system. We have already alluded to the various levels of abstraction which we present to help you reach a full appreciation of the concept. Finally, we even draw a picture.

There is another possible aid. Since five of the axioms occur essentially twice in the definition of "field" and once again in the set $\mathcal{V}$ which is the other part of a "vector space over the reals", some people may find it beneficial to dignify a system described by these five axioms with a name. This is, in fact, a "commutative group", a notion which is of great importance in its own right. As with fields, it is unnecessary to take a detour and study commutative groups before vector spaces, or even to deal with the term. However, we do include it in an optional manner for those who would find it a useful unifying concept.

Consistency

> Mathematical certainty is, after all, something insufferable. Twice two makes four seems to me simply a piece of insolence. Twice two makes four is a pert coxcomb who stands with arms akimbo barring your path and spitting. I admit that twice two makes four is an excellent thing, but if we are to give everything its due, twice two makes five is sometimes a very charming thing, too.
>
> Dostoevski, *Notes from the Underground*

A particularly unfortunate choice of axioms can lead to conclusions which are simultaneously both true and false. That is, it may be possible to deduce a certain theorem from a given system of axioms, and also to deduce from these same axioms the negation of the theorem. In this situation the axiom system is said to be **inconsistent**.

An inconsistent system is not only unfortunate; it is a complete disaster. For there is a logical principle which states that if it is possible to deduce one conclusion which is both true and false, then *everything* is both true and false. As a vivid illustration, we shall give an adaptation of a famous proof by Bertrand Russell that if $2 + 2 = 5$, then he is the Pope:

Theorem. If $2 \times 2 = 5$ (and all the usual laws of arithmetic hold), then you will get a grade of zero on the final examination for this course.

Proof. Since also $2 \times 2 = 4$, some sophisticated arithmetic leads to the conclusion that $0 = 1$. Now let G denote your grade on the final exam (a very large, positive number, we would hope). Then, by substitution and some basic multiplicative properties, $G = 1G = 0G = 0$, so $G = 0$. Q.E.D. ∎

We offer this material on consistency not only for its relevance both here and in the discussion to follow, but also to point out one of the reasons behind the mathematician's seemingly fanatical obsession with precise reasoning: he must avoid inconsistencies. Walt Whitman may have been able to get away with contradicting himself, but poets are about the only ones left who can afford to do so.

4. MATHEMATICAL MODELS

We are all aware of the long and distinguished relationship between mathematics and the physical sciences (and engineering). Recently, there has also been a surprising increase in both the amount and the sophistication of the mathematics used in such areas as the social and biological sciences, medicine, government, and industry. While some of this may be mere fashion, misuse, or that academic infirmity, obscurantism, it nonetheless requires both courage and naiveté to declare "Well, at least mathematics will never intrude in *my* particular bailiwick". (As an example—which may not be the wave of the future—one finds such titles as Nicolas Rashevsky's "Outline of a Mathematical Approach to History" in his *Mathematical Biology of Social Behavior*, University of Chicago Press, 1959.)

Some of this increased usage is routine: new formulas to use and new uses for old formulas. However, such simple applications hardly justify taking a course such as this. (Nonmathematicians: Would your mentors in your prospective field consign you to this course if instead they could supply you with a handy list of formulas which would cover your every mathematical need?) One of the main uses of mathematics lies in working with mathematical models, a construct closely connected to the axiomatic method.

A typical activity of the scientist consists in making assumptions as to the behavior of a particular aspect of the real world. Sometimes a collection of assumptions will lend itself to rational analysis: treating the assumptions as hypotheses, one can draw conclusions. Under these circumstances, we would say we have a "mathematical model" of this particular aspect of reality.

As a simple example, a biologist might conjecture that the growth of a strain of bacteria in his Petri dish is roughly proportional to the number present at any time. This might be considered a verbal model of the growth of the bacteria. If we let $N(t)$ denote the number of bacteria present at any time t, then we can make a mathematical statement of this verbal assumption†:

$$\frac{dN(t)}{dt} = kN(t).$$

This differential equation would then be a mathematical model of the growth of the bacteria culture.

Mathematical models might be better understood in relation to other methods of describing reality, such as the purely verbal model or the physical model (more or less as in "model airplane"). Of course, not every collection of assumptions regarding the real world lends itself to mathematical treatment. As Patrick Suppes observes, "the kind of theory which mainly consists of insightful remarks and heuristic slogans will not be amenable to this treatment" [4, p. 172]. The mathematical model is therefore only one possible aspect of the scientific method, and one which arises only under felicitous circumstances.

In the mathematical model, the scientist is dealing with certain sets abstracted from nature (e.g., the idealized bacteria population mentioned above). Moreover, he has relations and operations on his sets (e.g., growth—the number $N(t)$ present at any time t) and he makes assumptions as to the behavior of the relations and operations on these sets [for example, $dN(t)/dt = kN(t)$]. Thus, our scientist is essentially working with an axiom system! The only difference between the mathematical model and an axiom system is that the model may be given in a more informal fashion. For example, the mathematician prefers to work with "undefined

---

† Since bacteria come only in whole numbers, your first mathematical statement might more properly be the difference equation $\Delta N(t)/\Delta t = kN(t)$. However, differential equations are usually much simpler to handle, so you might wish to replace this by the indicated differential equation. You have simplified the model (or made a model of the model, if you wish).

terms" and might therefore wish to use "element of a set" rather than the biologist's "bacterium", and so forth.

After extracting a mathematical model of the real world, the scientist attempts to draw conclusions from his assumptions ("theorems", in the axiomatic method). For example, from the mathematical model for bacterial growth one can conclude that $N(t) = ce^{kt}$, which may or may not correspond to the reality of the situation.

If the conclusions derived from the model correspond fairly well to some appropriate segment of reality, the scientist sometimes confuses the issue by dubbing these conclusions "laws of nature". Actually, they are only "laws" for the mathematical model. For example, the "law of exponential growth" $N(t) = ce^{kt}$, applies exactly only to the abstract model and not to any particular Petri dish (especially over a long range of time).

What are the advantages in the use of mathematical models? A good model lays bare some basic aspects and interactions of a portion of nature, and in a concise and tractable form. Moreover, from such a model one can derive conclusions which are far-reaching and/or surprising (the existence of such conclusions is, in fact, the main criterion of the "goodness" of the model).

Predictions from a mathematical model often lead to experiments. Even if these invalidate the model to some extent, the form of the model is such that corrections are often easy to make. For example, the outcome of the experiments may make it apparent that an obvious modification is necessary in some particular axiom. Thus, a mathematical model can be convenient to work with and to alter.

A poor mathematical model can also be of value. To quote a social scientist:†

> The construction of a logically consistent and complete model provides an invaluable check against unclear thought and poorly formulated concepts. Often controversies arise and achieve mammoth proportions in both journals and debate, where the differences hinge on implicit assumptions well hidden in the wording. The better one understands the concept of a model and is able to use it as an aid to thought, the more quickly differences in fundamental points of view can be located. If for no other reason, the application of mathematical methods to the study of human behavior has great value in that bad simple mathematical models can be spotted in far less time than bad verbal models The same poor assumptions and conclusions clothed in 400 pages of words require much more time to locate the basic structure and with it the errors, fallacies, and omissions.

Even an inconsistent model can be of great value in that such a model shows that your basic assumptions (axioms) cannot simultaneously hold in the real world. One celebrated and striking example of the construction of an inconsistent model is "Arrow's Paradox". There are various methods which a society may use

---

† Shubik, M., ed., *Game Theory and Related Approaches to Social Behavior*. New York: Wiley, 1964.

to pass from the individual preferences of its members to a decision of the society at large (simple majority rule, dictatorial rule, etc.). Kenneth Arrow devised a very reasonable set of criteria which one would expect to hold in any "fair" way of reaching a societal decision from the preferences of its members (e.g., "there is no individual with the property that whenever he prefers alternative x to alternative y, then so does society, regardless of the preferences of the other individuals"). Using these criteria as axioms, Arrow showed that the criteria are in fact inconsistent. This rather disquieting conclusion indicates that it is necessary to reconsider carefully what one deems "fair" criteria for reaching societal decisions. For a good discussion of Arrow's Paradox, see [6].

The main case against the use of mathematical models seems to be the feeling that nature is too subtle and complex, perhaps especially so in some of the newer areas of mathematicization in the social sciences and biology. The possibility of capturing the subtle nuances of reality in a few precise axioms is therefore very slim indeed. Consequently, to do justice to some disciplines, it is necessary to devise models of such complexity as to be mathematically intractable: after a decent set of hypotheses, there is little possibility of drawing interesting conclusions.† However, it makes little sense to foreclose prematurely on a potentially useful tool.

Thus, the mathematical model is essentially the same as an axiom system. The main purpose of writing this section is to point out that valuable skills in dealing with mathematical models may be obtained from working with the axiom systems, such as the one for "vector space" to be encountered in this book.

5. THE RELATIONSHIP BETWEEN MATHEMATICS AND SCIENCE

In olden times, much of mathematics received its impetus from the real world (and there are great mathematicians who would argue that this is still a necessary source of inspiration [7]). However, in more recent times, it has occasionally been the case that mathematicians have already investigated the axiom systems which scientists have later developed as models. For example, the appropriate material in group theory was available when needed in the development of quantum mechanics, as was other material in group theory for the study of kinship systems in primitive societies [11]. Nonetheless, this is not always the case; the future scientist, like his predecessors, may well have to do some mathematical deductions and should be experienced accordingly.

It is worthwhile noting that the same axiom system can occur in many different mathematical models. A few entertaining examples:

The mathematical formulation of the birth-and-death process, originally

---

† There may be an unwarranted hesitancy against simplifying assumptions in some of the disciplines new to mathematics. Thus, physicists may casually make simplifications of an order of magnitude that might leave a social scientist aghast.

devised to investigate changes in living populations, can be applied to populations of inanimate objects such as telephone poles or vehicles.

The epidemic model, developed primarily to describe the spread of infectious diseases, can be applied to other infectious phenomena such as the wish for higher education or the demand for a new product.

Input–output analysis, evolved in the first place to study industrial interdependence, can be adapted to the study of demographic flows ... [10, page v].

We close with a remark on one value of introducing mathematics into some of its newer contexts, a value which seems particularly relevant today:

"The projection of mathematics into any argument, even though it turns out irrelevant, always has a purifying influence, because it is not possible to calculate and do algebra in hot blood. At this time when the world seems in danger of splitting because of lack of calm discussion, anything which contributes to such pacification is welcome..." [Remarks by G. A. Bernard on the article "An essay on the mathematical theory of freedom" by Denis and André Gabor in *Journal of the Royal Statistical Society* (Series A), **117**, I, 1954, pp. 31–72.]

REFERENCES

1. Bailey, N. T. J.: *The Mathematical Approach to Biology and Medicine*. New York: Wiley, 1967.
2. Bellman, R., and P. Brock: "On the Concept of a Problem and Problem-Solving," *Am. Math. Monthly* **67**, 1960, pp. 119–134.
3. Bourbaki, N.: "The Architecture of Mathematics" (tr. A. Dresden), *Am. Math. Monthly* **57**, 1950, pp. 221–231.
4. Freudenthal, H. (organizer), *The Concept and the Role of the Model in Mathematics and Natural and Social Sciences*. New York: Gordon and Breach, 1961.
5. Kemeny, J. G., and J. L. Snell: *Mathematical Models in the Social Sciences*. Ginn, Boston: 1962.
6. Luce, R. D., and H. Raiffa: *Games and Decisions*. New York: Wiley, 1957.
7. Neumann, J. von: "The Mathematician," from *Collected Works*, **1**, pp. 1–9, Oxford: Pergamon, 1961.
8. Noble, B.: "Applications of Undergraduate Mathematics in Engineering," *MAA*, Macmillan, 1968, pp. 6–7.
9. Shubik, M., ed., *Game Theory and Related Approaches to Social Behavior*. New York: Wiley, 1964.
10. Stone, R.: *Mathematics in the Social Sciences and Other Essays*. Cambridge: M.I.T. Press, 1966.

11. White, H. C.: *An Anatomy of Kinship*. Englewood Cliffs, N. J.: Prentice-Hall, 1963.
13. Wigner, E.: "The Unreasonable Effectiveness of Mathematics in the Natural Sciences," *Comm. Pure & Appl. Math.*, **XIII,** 1960, pp. 1–14.
13. Wilder, R. L.: "The Role of the Axiomatic Method," *Am. Math. Monthly* **74,** 1967, pp. 115–128.

This is a vector. It is about a foot long and looks like a cross between an arrow and a flatworm with a bit of snake thrown in.†

CHAPTER 0 **Preliminaries**

This chapter is primarily concerned with things you may already know about, geometrical vectors and real-valued functions. However, our emphasis may well be new to you. We shall be setting up these topics so that they will conveniently serve as examples for later work. It should be worthwhile covering the first three sections before beginning Chapter 1. The rest of the material may be interspersed as needed.

We do introduce one notion that is likely to be new to you, that of "commutative group". However, we are not interested in this as a subject itself, but rather as a convenient means of organizing a number of important notions which often occur together.

The chapter closes with an optional section on fields for those who wish to appreciate the full generality of linear algebra.

Our notation for vectors (and later for matrices) will be boldface letters **V**, **W**, etc. This makes clear the distinction between vectors and scalars (for which we shall use lightface italic type $a, b, \ldots, z$), especially when it comes to the vector **0** and the scalar 0. To make a handwritten distinction, some people use notations such as $\vec{V}$ for vectors, although others feel that such a notation is not worth the effort, since in a handwriting context, the distinctions are usually pretty obvious anyway.

---

† Elizabeth Yount, "A Child's Garden of Vectors" in *The Journal of Biological Psychology* (which upside down is *The Worm Runner's Digest*) **VIII**, April 1966.

1. GEOMETRICAL VECTORS

We shall be intuitive in much of this section. Let us imagine ourselves on the real line, or in the euclidean plane, or in 3-space, or in general in euclidean n-space (whatever that is; we shall define and discuss euclidean n-space later; for now just use your intuition). As you may well know, it is customary to define a **vector** as a directed line segment; that is, a line segment with a favored end. This end is called the *head* of the vector (the other end, the *tail*). One usually draws vectors thus:

———▷,   or ———▶,   etc.

We consider two vectors to be **equal** if
1) they lie on parallel lines,
2) their heads are pointing in the same direction, and
3) their underlying line segments have the same length.

Thus,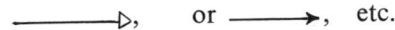

are all considered equal as vectors.† One may say that two vectors are equal if one can be obtained from the other by a parallel displacement. We define addition of vectors according to the following prescription: given vectors **V** and **W**, place the head of **V** on the tail of **W** by means of our allowed parallel displacement (if needed). Then **V + W** is the vector consisting of the line segment between the tail of **V** and the head of **W**, and whose head is at the head of **W** (Fig. 1):

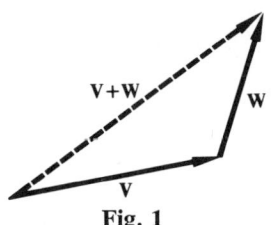

Fig. 1

This addition is sometimes called the "parallelogram law" since **V + W** is one (directed) diagonal of the parallelogram made from two copies of **V** and **W** (Fig. 2):

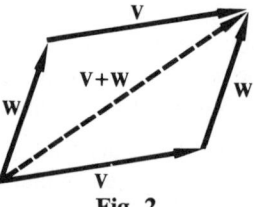

Fig. 2

---

† If this notion of equality causes you any concern, recall that one considers two fractions a/b and c/d as **equal** if $ad = bc$ and not necessarily if a/b and c/d are the *same*. Thus, $\frac{1}{2} = \frac{2}{4}$ as fractions even though $\frac{1}{2}$ and $\frac{2}{4}$ are not the same object. Thus, "equality" need not mean identity. A full discussion of the situation is given in Supplement 1 to this section.

By going around opposite sides of the parallelogram one observes that

$$V + W = W + V.$$

We distinguish the vector whose underlying line segment has zero length (and for which choosing a direction is either impossible or immaterial, depending upon your philosophic outlook), and call this vector the *zero vector*, written **0**. Note that

$$0 + V = V = V + 0$$

for any vector **V**. Moreover, for any **V** there is a unique vector which we call **−V** for which

$$V + (-V) = 0$$

(namely, **−V** is the vector whose line segment is the same as **V**'s, but whose head is the tail of **V**). Let $\mathcal{U}$ be the set of all vectors in euclidean n-space. One can observe the following properties of vector addition:

1) For any **V**, **W** in $\mathcal{U}$, **V + W** is an element of $\mathcal{U}$. (Closure)
2) For any **U**, **V**, **W** in $\mathcal{U}$, one has **U + (V + W) = (U + V) + W**. (Associativity)
3) There is an element called **0** in $\mathcal{U}$ for which $0 + V = V = V + 0$ for any **V** in $\mathcal{U}$. (Existence of an identity)
4) For any **V** in $\mathcal{U}$ there is an element called **−V** in $\mathcal{U}$ for which **V + (−V) = 0 = (−V) + V**. (Existence of an inverse)
5) For any **V**, **W** in $\mathcal{U}$, **V + W = W + V**. (Commutativity)

A system satisfying these conditions is called a **commutative group**. (See Supplement 2 for further discussion.) Note that the five conditions may be easily remembered by name (although some people might find it easier to remember the names by the five conditions). Associativity is evident if you draw a picture.

There are several other operations often associated with vectors. At this time we shall consider only *scalar multiplication*. For a real number (i.e., scalar) a and a vector **V**, we define a**V** to be the vector whose line segment has length

$$|a| \times \text{(length of } V\text{)},$$

where $|a|$ is the absolute value of a, and whose direction is:

the same as that of **V** if $a \geq 0$;
opposite to that of **V** (i.e., the same as that of **−V** if $a < 0$.

In particular, for the scalar 0, 0**V** = **0** (the zero vector) for any **V**. Thus if

$$V \text{ is } \longrightarrow,$$

then

$$2V \text{ is } \longmapsto,$$

$$-\tfrac{1}{2}V \text{ is } \leftarrow,$$

etc. Scalar multiplication satisfies the following properties: For a, b real numbers, **V**, **W** vectors in $\mathcal{V}$,

i) $a\mathbf{V}$ is in $\mathcal{V}$. (Closure)
ii) $1\mathbf{V} = \mathbf{V}$. (Identity)
iii) $a(b\mathbf{V}) = (ab)\mathbf{V}$. (A kind of associativity)
iv) $(a + b)\mathbf{V} = a\mathbf{V} + b\mathbf{V}.$
v) $a(\mathbf{V} + \mathbf{W}) = a\mathbf{V} + a\mathbf{W}.$ $\Big\}$ (Distributivity conditions)

EXERCISES

1. Draw pictures in the plane which illustrate each of the commutative group properties for vectors.
2. Draw pictures in the plane which illustrate each of the scalar-multiplication properties for vectors.
3. Show that $(\mathbf{W} - \mathbf{V}) + \mathbf{V} = \mathbf{W}$. Show that $-\mathbf{V} = (-1)\mathbf{V}$. Draw figures to illustrate this.
4. Note that $\mathbf{V} + \frac{1}{2}(\mathbf{W} - \mathbf{V}) = \frac{1}{2}\mathbf{V} + \frac{1}{2}\mathbf{W}$ and draw a figure to illustrate this. Show that this "proves" that the diagonals of a parallelogram bisect each other.

REFERENCE

Those who would enjoy learning more about geometrical vectors (and those who think they know everything) should consult Banesh Hoffman, *About Vectors*. Englewood Cliffs, N.J.: Prentice-Hall, 1966.

Supplement 1. Equivalence Relations

It may be somewhat disconcerting to *define* two vectors to be equal under certain circumstances—and come to think of it, by what right can we consider $\frac{1}{2}$ and $\frac{2}{4}$ as equal? Herein we hope to clarify the situation and to introduce the reader to a very far-reaching (and very elementary) notion.

The crux of any difficulties which may arise lies in distinguishing between vector and fractional "equality" and the slightly more common usage of "equality" as meaning "the same as". For, in the former case, we really mean "we may treat these vectors (or fractions) as if they were the same for the purposes which we have in mind". We then compound the felony by using the symbol "=" for both uses of the word. It would, of course, make more sense to use a different word for our fractional/vector equality, "equivalent", for example, and even a different symbol, "≡".

Given an arbitrary set S and a relation $\equiv$ which holds between certain pairs of elements of the set, we say that $\equiv$ is an **equivalence relation** if

i) $a \equiv a$ for a in S (the relation is reflexive).
ii) If $a \equiv b$, then $b \equiv a$ for any a, b in S (it is symmetric).
iii) If $a \equiv b$ and $b \equiv c$, then $a \equiv c$ for any a, b, c in S (it is transitive).

For example, let S be the set of all directed line segments in the plane, and for a, b in S let $a \equiv b$ if a can be obtained from b by parallel displacement. The three conditions certainly hold.

As you have noticed, given an equivalence relation, for convenience one is sometimes tempted to use the word "equal" for "equivalent" and also to conserve on symbol usage. It is nonetheless a good idea to clarify exactly what is going on.

The theory of equivalence relations has a certain charm in that there is really only one theorem to prove (and a converse—Exercise 5).

1.1 *The Partition theorem.* Given an equivalence relation $\equiv$ on a set S, there exists a partition of S into disjoint subsets $S_1, S_2, \ldots$ such that $a \equiv b$ iff a and b belong to the same subset S_i.

By "partition of S into disjoint subsets $S_1, S_2, \ldots$" we mean that every element of S belongs in some S_i and no element belongs to two different S_i.

We shall call the **equivalence class** of a, S_a, the set of all x for which $x \equiv a$. Thus,

$$S_a = \{x \text{ in } S \mid x \equiv a\}.$$

Sketch of Proof of 1.1 (You are asked to complete this in Exercise 4). Take the set of equivalence classes of the various elements of S to be the subsets we're looking for. Then every element of S is in one of these sets (why$_1$?) and if any element belongs to two of these subsets S_1 and S_2, then S_1 and S_2 are the same (why$_2$?), and we are done. ∎

The theorem shows that, given an equivalence relation on a set S, we split up the set into nonoverlapping parts (Fig. 3). For many purposes, we may then choose some particularly convenient element from the **equivalence class** with which to work (e.g., "the vector with tail at the origin").

We shall encounter many other examples of equivalence relations.

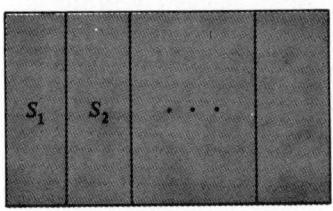

Fig. 3

EXERCISES

1. Show that fractional equality is an equivalence relation. What is the equivalence class of $\frac{1}{2}$? (List elements.)
2. Let S be the set of all triangles in the plane and let "$\equiv$" denote the relation of congruence of triangles. Show that $\equiv$ is an equivalence relation.
3. Let S be the set of all integers and define $a \equiv b$ iff $a - b$ is divisible by 3. How many equivalence classes are there?
4. Answer the why's in the proof of the theorem.
5. Prove this converse to the theorem: Given any set S and a partition of S into disjoint subsets $S_1, S_2, \ldots$, then there exists an equivalence relation on S relative to which the subsets $S_1, S_2, \ldots$ are the equivalence classes.
H6. Why can't you prove the reflexive condition from the conditions of symmetry and transitivity? Or can you?

Supplement 2. Groups

"Commutative group" will be our first decent axiom system. It involves one set G together with an operation defined on pairs of elements from G. This operation is a pasting together of two elements of G to get another element of G, and may be thought of as a generalization of addition and multiplication. We denote the operation by "$\circ$" and write the element which it determines from the elements a and b in G as $a \circ b$. There are five axioms governing the action of "$\circ$".

Formally, a **commutative group** is any set of objects G together with an operation "$\circ$" defined on pairs of elements of G and satisfying

1) Given any two elements a, b in G, $a \circ b$ is an element of G. (Closure)
2) For any a, b, c in G, $a \circ (b \circ c) = (a \circ b) \circ c$. (Associativity)
3) There exists an element e in G with the property that $e \circ a = a = a \circ e$ for any a in G. (Existence of identity)
4) For any a in G there exists an element a^{-1} for which $a \circ a^{-1} = e = a^{-1} \circ a$. (Existence of inverse)
5) For any a, b in G, $a \circ b = b \circ a$. (Commutativity)

Thus all geometrical vectors in the plane (or in 3-space) form a commutative group with operation $\circ$ interpreted as vector addition. The identity e is then the zero vector, and the inverse of an element $\mathbf{V}$ in the group is the vector $-\mathbf{V}$.

Similarly, the set of all integers (positive, negative, and zero) forms a commutative group under the operation of addition (with identity 0 and inverse the usual additive inverse of an integer).

Moreover, the set of all *nonzero* real numbers forms a commutative group under the operation of *multiplication* (note that the identity is 1 and the inverse is the multiplicative inverse). Commutative groups are also called **abelian groups** after the mathematician N. H. Abel.

For a taste of axiomatics we proceed to draw two modest consequences from the axioms. Other theorems are to be found in Exercise 7 through 9.

1.2 The identity e of a commutative group is unique. That is, if e' is another element satisfying Axiom 3,

$$e' \circ x = x = x \circ e' \text{ for all } x \text{ in } G,$$

then $e' = e$.

Proof. For any e' satisfying our hypothesis,

$$e' \circ e = e = e \circ e'.$$

By Axiom 3, $e \circ e' = e'$. Hence $e = e'$. ∎

1.3 If z is any element in a commutative group G which has the property that $z \circ x = x$ for some one element x in G, then $z = e$.

Proof. Since x^{-1} exists (Axiom 4), we can operate with it on both sides of the equation $z \circ x = x$ to obtain

$$(z \circ x) \circ x^{-1} = x \circ x^{-1}. \tag{1}$$

We proceed to show that this simply says that $z = e$. Now the lefthand side of (1) is $z \circ (x \circ x^{-1})$ by Axiom 2, which in turn is $z \circ e$ by Axiom 4, which is z by Axiom 3. On the other hand (or side), $x \circ x^{-1} = e$ by Axiom 4 again. Equating our simplifications of the two sides of (1), we obtain $z = e$. ∎

If we do not demand that the operation "∘" satisfy the commutative law (Axiom 5), we obtain the more general notion of a (not necessarily commutative) "group". Thus, a **group** is a set G together with an operation "∘" which satisfies all the above axioms except (possibly) Axiom 5. The axioms for a group are therefore: (1) closure, (2) associativity, (3) existence of identity, (4) existence of an inverse for each element. Any commutative group is, of course, a group.

Groups are of fundamental importance in many areas of mathematics (and its applications; e.g., the physicists' "eightfold way" involves a group, which we shall later point out). At this time it would be too much of a digression to develop examples of noncommutative groups, especially since we shall encounter many natural examples of such groups later in the book.

Note that 1.2 and 1.3 hold in arbitrary groups, since Axiom 5 was not used in their proof.

EXERCISES

1. What commutative group axioms are violated if we consider the set of all real numbers under multiplication?

2. Which commutative group axioms are violated by (a) the set of nonzero integers relative to the operation of multiplication? (b) the set of positive integers relative to addition? (c) the set of positive rationals relative to multiplication? (d) the set of negative rationals relative to multiplication?

3. Show that the set of integers $\{1, -1\}$ forms a group with respect to multiplication.

4. Let G be the set of all equivalence classes in Exercise 3, Supplement 1. With notation from that supplement, define

$$S_a \circ S_b = S_{a+b}.$$

Show that G is a commutative group under this operation. (Be sure you prove that if a' is in the same equivalence class as a, so that $S_{a'} = S_a$, and if b' is in the same equivalence class as b, so that $S_{b'} = S_b$, then $S_{a'} \circ S_{b'} = S_a \circ S_b$. This says that the operation is *well defined*.) What is e? How many elements does G have? Write out a "multiplication table" for G.

5. Think of some other examples of commutative groups. Be sure to specify the set involved and the interpretation of "$\circ$" and "e".

6. Think of some other examples of commutative groups with a "scalar multiplication" satisfying conditions (i) through (v) of Section 1. Specify what the "scalars" are and what the "multiplication" is. Check all five axioms.

7. Prove that the identity e of a commutative group is unique. That is, if $x \circ a = a = a \circ x$ for all a in G, then $x = e$.

8. For all elements x and y in a commutative group, prove that if $x \circ y = e$, then $y = x^{-1}$. (Thus, the inverse of an element is unique.)

9. Prove that for any element x in a commutative group, $(x^{-1})^{-1} = x$.

10. Which of the results in Exercises 7 through 9 hold in any noncommutative group? How would you prove the results?

2. VECTORS AND N-TUPLES

For many uses of vectors, it is convenient to introduce the machinery of analytic geometry. To do this, let us imagine our euclidean n-space as now equipped with n coordinate axes meeting at an origin, so that each point in the euclidean space may be described by an n-tuple of real numbers $(v_1, \ldots, v_n)$.

We shall use $\mathbb{R}$ to denote the real numbers system; we shall call the set of all (ordered) n-tuples of real numbers $\mathbb{R}_n$. That is

$$\mathbb{R}_n = \{(v_1, v_2, \ldots, v_n) \mid \text{each } v_i \text{ is a real number (i.e., } v_i \text{ in } \mathbb{R})\}.$$

For example, $\mathbb{R}_2$ is the set of pairs (v_1, v_2) where v_1 and v_2 are real.

We then obtain an extremely important one-to-one correspondence between vectors and n-tuples of real numbers as follows:

Given a vector V, place the tail of V at the origin by means of parallel displacement. Then the head of the vector V lies at a point with coordinates $(v_1, v_2, \ldots, v_n)$. We set up the correspondence $V \leftrightarrow (v_1, v_2, \ldots, v_n)$.

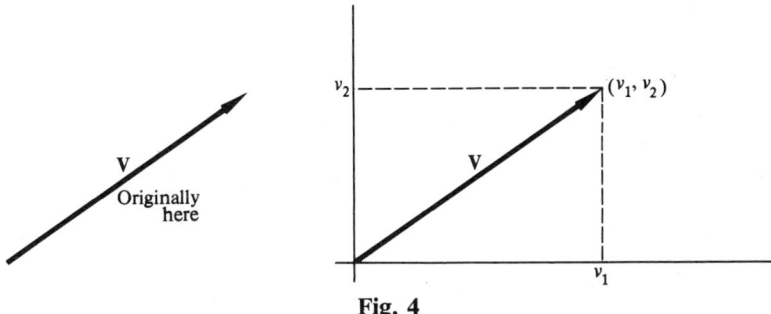

Fig. 4

This correspondence is clearly one-to-one: to each vector there corresponds exactly one n-tuple, and to each n-tuple there corresponds exactly one vector (namely, the vector whose head is at the point described by the n-tuple and whose tail is at the origin). The above correspondence has an important dividend:

2.1 If, in the above correspondence, $\mathbf{V} \leftrightarrow (v_1, v_2, \ldots, v_n)$ and if a is any scalar, then $a\mathbf{V} \leftrightarrow (av_1, \ldots, av_n)$. Moreover, if $\mathbf{W} \leftrightarrow (w_1, \ldots, w_n)$, then

$$\mathbf{V} + \mathbf{W} \leftrightarrow (v_1 + w_1, \ldots, v_n + w_n).$$

Note the value of this! We now have our choice of adding vectors geometrically or adding n-tuples of numbers componentwise, and of multiplying vectors by scalars geometrically or multiplying an n-tuple by a scalar componentwise.

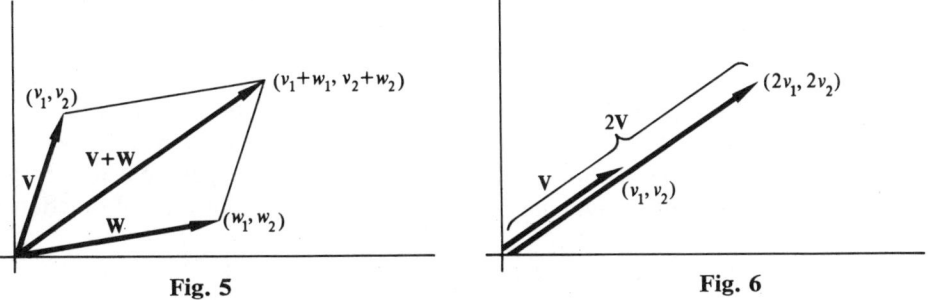

Fig. 5   Fig. 6

Because of 2.1 we can treat the plane and $\mathbb{R}_2$ as if they were algebraically the same, and similarly we can identify 3-space with $\mathbb{R}_3$, and geometrical n-space with $\mathbb{R}_n$. For a good portion of what lies ahead this informal identification is quite satisfactory. We shall reexamine this relationship between geometrical spaces and $\mathbb{R}_n$ in Chapter 7.

EXERCISES

1. Using analytic geometry, prove that at least in the euclidean plane (i.e., for $n = 2$) the correspondence between vectors and n-tuples has the stated arithmetic properties

(for addition of vectors, this amounts to showing that if a parallelogram has one vertex at the origin and the two neighboring vertices are at (v_1, v_2) and (w_1, w_2) respectively, then the remaining vertex is at $(v_1 + w_1, v_2 + w_2)$.

2. Prove directly that the set of all n-tuples of real numbers (fixed n, of course) satisfy the commutative group conditions under componentwise addition:

$$(v_1, \ldots, v_n) + (w_1, \ldots, w_n) = (v_1 + w_1, \ldots, v_n + w_n);$$

and that the scalar-multiplication conditions are satisfied under componentwise multiplication

$$a(v_1, \ldots, v_n) = (av_1, \ldots, av_n).$$

State precisely why this isn't very surprising.

3. The vectors $(1, 0, 0)$, $(0, 1, 0)$, and $(0, 0, 1)$ are often called **i**, **j**, and **k** respectively (especially by physicists). Show that $x\mathbf{i} + y\mathbf{j} + z\mathbf{k} = (x, y, z)$ for any real numbers x, y, z.

4. If $\mathbf{V} = (1, -2, 3, 0)$ and $\mathbf{W} = (2, \frac{1}{2}, -1, 3)$, find $\mathbf{V} + \mathbf{W}$, $\mathbf{V} - \mathbf{W}$, and $3\mathbf{V} + 2\mathbf{W}$.

5. Draw a picture in the euclidean plane which shows that if the tails of vectors are brought to any point other than the origin, the resulting one-to-one correspondence between pairs of real numbers and the heads of vectors in this position does *not* preserve addition and scalar multiplication.

3. REAL FUNCTIONS

In this section we shall remind you of some well-known properties of functions, and point out a striking analogy with properties of vectors discussed in Section 1.

Let us consider the collections of all real-valued functions defined on some set S of real numbers, say S in the interval $a \leq x \leq b$ (although the domain S makes absolutely no difference in our discussion). Thus, our collection consists of such creatures as $\sin(x^3 + 2)$, $e^{5x} + 12 \log |\arcsin x + 3|$, the function z for which $z(x) = 0$ for all x, all the pathological examples which are dragged out in a first-year calculus course, and many, many more. Let us call our collection of functions $\mathcal{F}$. Thus

$$\mathcal{F} = \{\text{functions } f \mid f(x) \text{ exists for all } x \text{ in } S\}.$$

In terms of the usual definition of function equality, we consider two functions in $\mathcal{F}$ equal if they agree on each element of S: two functions f, g in $\mathcal{F}$ are **equal**, written $f = g$, if $f(x) = g(x)$ for all x in S.

Recall also that we can add two functions in $\mathcal{F}$ and get another function in $\mathcal{F}$: given f, g in $\mathcal{F}$, the **sum** of f and g, written $f + g$, is that function whose value at x in S is $f(x) + g(x)$. That is, $(f + g)(x) = f(x) + g(x)$ for all x in S. Thus, we specify the function $f + g$ by giving its value for each x in S.

To connect with some of our previous work, we consider

3.1 *Functions as a commutative group.* The set $\mathcal{F}$ of all real-valued functions defined on some set S of real numbers forms a commutative group with respect to function addition. That is,

1. For any f, g in $\mathcal{F}$, $f + g$ is an element of $\mathcal{F}$. (Closure)
2. For any f, g, h in $\mathcal{F}$, $f + (g + h) = (f + g) + h$. (Associativity)
3. There is an element z in $\mathcal{F}$ for which $z + f = f = f + z$ for any f in $\mathcal{F}$ [that is, the function defined by $z(x) = 0$ for all x in S]. (Existence of an identity)
4. For an f in $\mathcal{F}$ there is an element called $-f$ for which $f + (-f) = z = (-f) + f$ [that is, the function defined by $(-f)(x) = -f(x)$ for all x in S]. (Existence of inverse)
5. For any f, g in $\mathcal{F}$, $f + g = g + f$. (Commutativity)

Start of Proof. (1) $f + g$ is a function (if you are hazy on the definition of function, we give it in the next section) and it is defined on S. Hence it is in $\mathcal{F}$. Now (2) through (5) all follow the same pattern: just show that the appropriate property holds at each x in S and use the definition of function equality. For example:

(5) To see that $f + g = g + f$, it suffices to show this amazing fact for each x in S by our definition of function equality again. Now $(f + g)(x) = f(x) + g(x)$, while $(g + f)(x) = g(x) + f(x)$; and since addition of real numbers is commutative.

$$f(x) + g(x) = g(x) + f(x).$$

Thus $(f + g)(x) = (g + f)(x)$ for all x in S. Thus $f + g = g + f$.

One can also multiply functions by scalars: if f is in $\mathcal{F}$ and a is a real number, then af is the function defined by $(af)(x) = af(x)$ for all x in f.

To complete our analogy with Section 1 we have

3.2 Scalar multiplication of functions satisfies the following properties:

i) For any f in $\mathcal{F}$ and real number a, af is in $\mathcal{F}$. (Closure)
ii) $1f = f$ for any f in $\mathcal{F}$. (Identity)
iii) $a(bf) = (ab)f$ for any f in $\mathcal{F}$ and real numbers a and b. (Associativity)
iv) $(a + b)f = af + bf$
v) $a(f + g) = af + ag$ } (Distributivity)

End of Proof. (v)

$$[a(f + g)](x) \stackrel{\text{defn. scalar mult.}}{=} a[(f + g)(x)]$$
$$\stackrel{\text{defn. func. add.}}{=} a[f(x) + g(x)] \stackrel{\text{distr. real nos.}}{=} af(x) + ag(x)$$
$$\stackrel{\text{defn. scalar mult.}}{=} (af)(x) + (ag)(x) \stackrel{\text{defn. func. add.}}{=} (af + ag)(x).$$

Since $a(f + g)$ and $af + ag$ agree everywhere on S, they are equal. ∎

Buried somewhere in your first-year calculus course, you will find statements to the effect that:

if f and g are continuous at x_0, then so is $f + g$; if f and g are differentiable at x_0, then so is $f + g$ (and $(f + g)' = f' + g'$); if f is continuous at x_0, then so is af for any real a; if f is differentiable at x_0, then so is af (and $(af)' = af'$).

In the exercises, there are games to be played with these notions.

We close this discussion with some remarks on polynomials. There are two approaches which one can take to polynomials:

1) They are functions. If
$$p(x) = a_0 + a_1 x + \cdots + a_n x^n,$$
this means that p is the function whose value at any x_0 is $a_0 + a_1 x_0 + \cdots + a_n x_0^n$.

2) They are formal symbols which may be added:
$$(a_0 + a_1 x + \cdots + a_n x^n) + (b_0 + b_1 x + \cdots + b_m x^m)$$
$$= (a_0 + b_0) + (a_1 + b_1)x + \cdots,$$
multiplied by a scalar
$$c(a_0 + a_1 x + \cdots + a_n x^n) = ca_0 + ca_1 x + \cdots + ca_n x^n,$$
and otherwise mutilated (e.g., multiplied). Moreover, two polynomials are equal iff they are of the same degree and all coefficients are the same:
$$a_0 + a_1 x + \cdots + a_n x^n = b_0 + b_1 x + \cdots + b_m x^m$$
iff $m = n$ and $a_i = b_i$ for $i = 1, \ldots, n$. (Any apparent mysticism in the "formal symbol" approach can be easily dispelled.)

Fortunately, the two approaches to polynomials agree in many important situations. For example, polynomials with real (or complex) coefficients may be treated either as formal symbols or as functions (provided the domain is large enough—and an interval $a \leq x \leq b$ is more than enough. See Exercise 12.) A situation where there is a difference is discussed in Section 6 (Exercise 6).

We shall denote by $\mathbb{R}[x]$ the set of all polynomials with real coefficients and by $\mathbb{C}[x]$ the set of all polynomials with complex coefficients. In either case the set of polynomials forms a commutative group under polynomial addition and satisfies the conditions of 3.2 for scalar multiplication (Exercise 3).

EXERCISES

1. Complete the proof of 3.1.
2. Complete the proof of 3.2.

In the following six exercises, show that (a) the given set of functions forms a commutative group (i.e., satisfies 3.1) with respect to function addition, and (b) the set of functions satisfies the scalar-multiplication conditions of 3.2. Either write out the gory details or state clearly why it is unnecessary to do so, or evolve some general principles to handle the situation.

3. a) $\mathbb{R}[x]$, the set of all polynomials with real coefficients.
   b) $\mathbb{R}_n[x]$, the set of all polynomials of degree less than n.
4. C = the set of all functions which are continuous for each x, $0 \le x \le 1$.
5. C' = the set of all functions which have continuous derivatives for each x, $0 \le x \le 1$.
6. The set of all functions defined for $0 \le x \le 1$ for which $f(\frac{1}{2}) = 0$.
7. The set of all continuous functions defined for $0 \le x \le 1$ for which $\int_0^1 f = 0$.
8. The set of all functions which are differentiable for $0 \le x \le 1$ and for which $f'(x) + 3f(x) = 0$. (Do you know any such? all such?—it doesn't matter here.)

In the following three exercises, find all the commutative group and scalar-multiplication conditions *not* satisfied by the given set.

9. The set of all functions defined for $0 \le x \le 1$ for which $f(\frac{1}{2}) = 2$.
10. The set of all functions f defined for $0 \le x \le 1$ for which $f(x) \le 1$ for all x, $0 \le x \le 1$.
11. All continuous functions f defined for $0 \le x \le 1$ for which $\int_2^1 f = 92$.
12. If S is the open interval $\{x \mid a < x < b\}$, show that the set of all polynomial functions defined on S behaves the same as does the set of all formal polynomials when it comes to addition, scalar multiplication, and equality. [*Suggestions*: If

$$p(x) = a_0 + a_1 x + \cdots + a_n x^n \quad \text{and} \quad q(x) = b_0 + b_1 x + \cdots + b_n x^n$$

are two polynomial functions, one can use induction to show that the function $p(x) + q(x)$ looks the way it is supposed to, and similarly for scalar multiplication. Show that the question of equality boils down to proving that if

$$p(x) = a_0 + a_1 x + \cdots + a_n x^n = 0 \quad \text{on} \quad S,$$

then each $a_i = 0$, $i = 0, 1, \ldots, n$. One rather heavy-handed method of doing this involves noticing that, if $p(x) = 0$ on S, then so does $p'(x), p''(x), \ldots$.]

4. FUNCTIONS IN GENERAL

This material is not really needed until Section 5 of Chapter 1, and may be omitted until then, if you wish. We assume that this is partially review, so that you can at least construct your own examples.

It will be necessary for us to consider functions in a rather general context. In fact, given two arbitrary sets of objects S and $\mathcal{C}$, a **function** f **from** S **into** $\mathcal{C}$ is any rule which assigns to each element s in S exactly one element $f(s)$ in $\mathcal{C}$. The set S

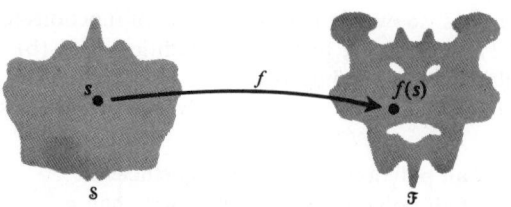

Fig. 7

is called the **domain** of the function, while the set $f(S)$ of all elements of the form $f(s)$, s in S, is called the **range** of the function:

$$f(S) = \{f(s) \mid s \text{ in } S\}.$$

A function may be conveniently envisioned as a box with a funnel and spout; drop in an element from S and out pops an element in $\mathcal{C}$ (the functional condition says that you do get an element of $\mathcal{C}$ each time—and only one element for each s).

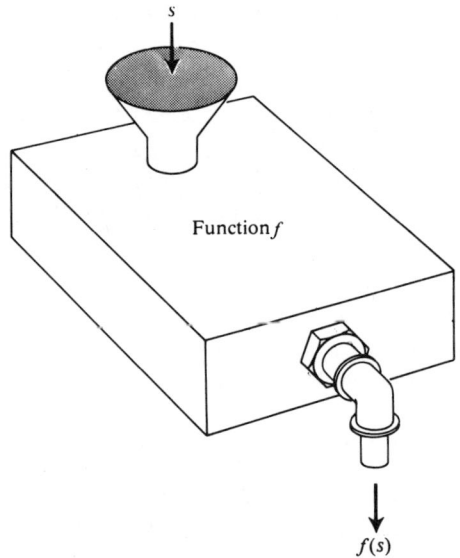

Fig. 8

You may have noticed that in our definition of function we used such words as "rule" and "assigns"—very functionlike words. If this strikes you as being improper, that is because it *is* improper. We shall give a complete exposé as to what functions really are at the end of this section. Our faulty definition has the advantage of being intuitively plausible: it feels good.

Note that it is possible to have several elements of S landing on the same element of $\mathcal{C}$. On the other hand, if this *never* happens, the function deserves a special name: a function f from a set S into $\mathcal{T}$ is **one-to-one** if $f(s_1) = f(s_2)$ only occurs if $s_1 = s_2$.

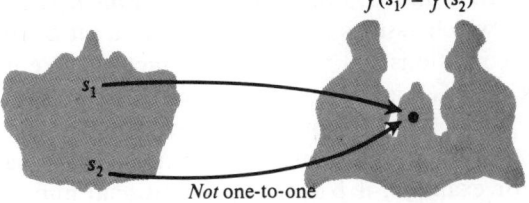

Fig. 9

If a function f from S into $\mathcal{T}$ has $\mathcal{T}$ as its range, we say that f is **onto** $\mathcal{T}$. Thus, f is a function from S *onto* $\mathcal{T}$ iff for every t in $\mathcal{T}$ there is at least one s in S for which $f(s) = t$. Note that a function from S *into* $\mathcal{T}$ is always *onto* its range. In our usage of the word "into" we also include the possibility of "onto".

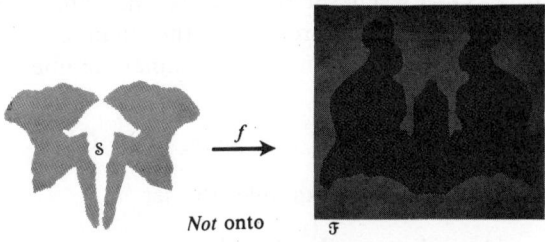

Fig. 10

A function blessed by being both onto and one-to-one is called a **one-to-one correspondence between** S and $\mathcal{T}$. Thus, if f is a one-to-one correspondence between S and $\mathcal{T}$, to each element of $\mathcal{T}$ there corresponds exactly one element of S and, conversely, to each element of S there corresponds exactly one element of $\mathcal{T}$.

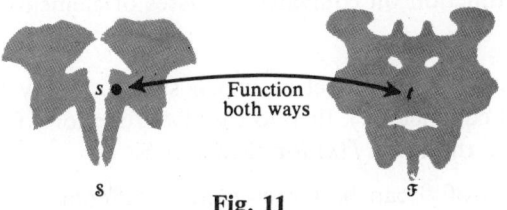

Fig. 11

Let f and g be functions from a set S into a set $\mathcal{T}$. We consider f and g to be **equal** if $f(s) = g(s)$ for all s in S.

There are several other schools of terminology in wide use, which we summarize below:

English		Francophile		Classical
onto	=	surjection	=	epimorphism
one-to-one into	=	injection	=	monomorphism
one-to-one correspondence	=	bijection	=	isomorphism

Some of these words also have useful adjectival forms. For example, if there is an isomorphism between S and $\mathcal{C}$, one says they are **isomorphic**.

Given a rule f which assigns to an element s of a set S some element $f(s)$ in a set $\mathcal{C}$, under what conditions can f fail to be a function? We need not bother with cases where the element $f(s)$ is ambiguously stated (for example, "$f(s)$ is that number whose square is s") which result from a rather elementary sort of sloppy thinking. A more interesting problem occurs when the elements of S can be represented in various forms. For example, if S is the set of rational numbers and $\mathcal{C}$ the set of integers, then the rule $f(m/n) = m$ is *not* a function since $m/n = (m \cdot r)/(n \cdot r)$ for any $r \neq 0$. The problem can be viewed in two slightly different ways, either (I) or (II) below.

I) The difficulty is with fractional equality. You can have $s_1 = s_2$ for two different elements of S (for example, $\frac{1}{2} = \frac{2}{4}$), and to define a function on S it is necessary that $f(s_1) = f(s_2)$ whenever $s_1 = s_2$. (In the language of Supplement 1 to Section 1, we have an *equivalence* relation rather than strict equality.)

II) The difficulty is with the meaning of rational numbers. The set you're interested in is not the set

$$S = \{m/n \mid m, n \text{ integers}, \ n \neq 0\},$$

but rather certain subsets of S, for example, the set

$$[\tfrac{1}{2}] = \{\ldots, -2/-4, -1/-2, \tfrac{1}{2}, \tfrac{2}{4}, \ldots\},$$

and it is *these* sets which comprise the rational numbers. Note that we abbreviate the statement "s_1 and s_2 are members of the subset $[\tfrac{1}{2}]$" by saying "$s_1 = s_2$". Thus, to specify a function on the rational numbers, you must have $f(s_1) = f(s_2)$ whenever s_1 and s_2 belong to the same subset, such as $[\tfrac{1}{2}]$, of S. (That is, we are actually trying to define a function on equivalence classes of elements of S, and not on S itself.)

Function principle. 1) If S is a set in which strict equality holds (that is, "$=$" means identity, not equivalence), then to specify a function f from S to a set $\mathcal{C}$, it is sufficient to indicate the value $f(s)$ for each s in S.

2) If the elements of S can be represented in different ways, then in order to define a function f on S, one must verify that $f(s_1) = f(s_2)$ whenever s_1 and s_2 are the same element written in different ways. (That is, if S is a set with an equivalence relation, then to specify a function f from the set of equivalence classes of S to a set $\mathcal{C}$, one must verify that $f(s) = f(s')$ whenever s and s' are in the same equivalence class.) The terminology is that you must show that f is **well defined**.

What Functions Really Are

It is possible to avoid the bootstrap aspect of our previous definition, although at the price of a bit of artificiality: a **function** f from a set S into a set $\mathcal{C}$ really consists of a collection of ordered pairs (s, t) (where s comes from S, and t from $\mathcal{T}$), with the property that if (s, t_1) and (s, t_2) are in the collection, then $t_1 = t_2$.

You should make sure that you understand how this does the trick: What is $f(s)$ here? the range of f? What assures you that to each element in the domain there corresponds an element in the range? and only one such element?

EXERCISES

1. Let $S = \{s_1, s_2\}$, and $\mathcal{T} = \{t_1, t_2\}$ be two sets containing two elements each.
   a) List all functions f from S into $\mathcal{T}$ by giving their effect on the elements of S (for example, $f(s_1) = t_1, f(s_2) = t_1$ completely specifies one such function).
   b) Which of the functions are one-to-one?
   c) Which are onto?
   d) Which are one-to-one correspondences?
2. Give an example of a function f from a set S onto itself which is not one-to-one. Is this possible if the set is finite?
3. Working with the definition of a function as a set of ordered pairs, (a) Answer the questions at the end of the section above. (b) What condition on the ordered pairs defines a one-to-one function? (c) an onto function? (d) a one-to-one correspondence? (e) Write out each of the functions in exercise 1(a) as a set of ordered pairs. (f) What does functional equality mean in terms of ordered pairs?

5. COMPOSITION OF FUNCTIONS. INVERSE FUNCTIONS

As was the case with the previous section, this might be mostly review. It is not needed until Chapter 3. The exercises in this section are informally inserted in the body of the text instead of at the end.

Given a function f from S into $\mathcal{T}$ and a function g from $\mathcal{T}$ into $\mathcal{U}$, it is very natural to paste them together and to obtain a function from S into $\mathcal{U}$. The **composition of f and g**, written gf, is the function from S into $\mathcal{U}$ defined by $(gf)(s) = g(f(s))$ for each s in S (Fig. 12). (Note that by gf we do *not* mean the product function $g \cdot f$ for which $(g \cdot f)(s) = g(s)f(s)$; we can't even discuss $g \cdot f$ unless both f and g map S into $\mathcal{T}$. Furthermore, $\mathcal{T}$ must be endowed with a multiplication.)

Of course, one has to check that gf is indeed a function.

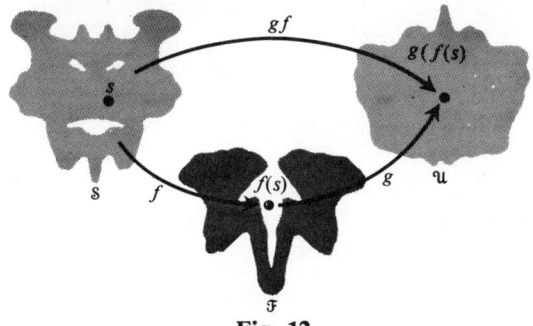

Fig. 12

Exercise 1. Do this (a) using the informal definition of function, and (b) using the ordered-pairs definition.

If one now has the misfortune to encounter another function h from $\mathcal{U}$ into a set $\mathcal{V}$, one could form the compositions $h(gf)$ and $(hg)f$ (Fig. 13). Luckily, these are the same.

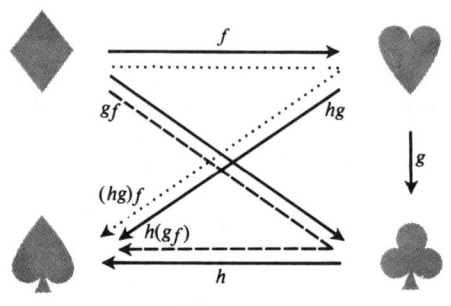

Fig. 13

5.1 Composition of function is associative. If f is a function from $\mathcal{S}$ into $\mathcal{C}$, g a function from $\mathcal{C}$ into $\mathcal{U}$, and h a function from $\mathcal{U}$ into $\mathcal{V}$, then the functions $h(gf)$ and $(hg)f$ from $\mathcal{S}$ into $\mathcal{V}$ are the same.

Proof. Just evaluate $h(gf)$ and $(hg)f$ at any s in $\mathcal{S}$. ∎

Exercise 2. Perform the above proof.

If f is a one-to-one function from a set $\mathcal{S}$ into a set $\mathcal{C}$, it is possible to define the **inverse function** f^{-1} from the range of f into $\mathcal{S}$:

f^{-1} is the function whose value at any $t = f(s)$ in the range of f is $f^{-1}(t) = s$. That is,

$$f^{-1}(t) = s \quad \text{iff} \quad f(s) = t.$$

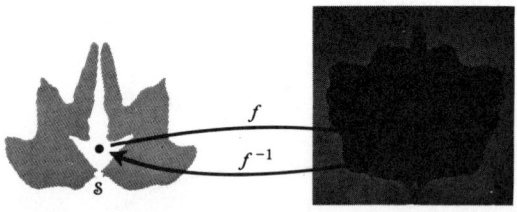

Fig. 14

Since f^{-1} has a specified value at each t in $f(\mathcal{S})$,—that is, $f^{-1}(t) = s$, where $f(s) = t$—it follows from the definition of functional equality that f^{-1} is unique. The fact that f is one-to-one makes f^{-1} a function.

Exercise 3. Show this (a) using the sloppy definition; (b) using ordered pairs.

Exercise 4. Show that f^{-1} is a one-to-one correspondence between $f(S)$ and S.

Exercise 5. Show also that the inverse of f^{-1} is f:
$$(f^{-1})^{-1} = f.$$

Exercise 6. One must be careful in computing the inverse of the composition fg, if both f and g are one-to-one. It turns out that $(fg)^{-1} = g^{-1}f^{-1}$, which is very likely different from $f^{-1}g^{-1}$. Prove these statements.

One very convenient function from any set S onto itself is the **identity function** I_S of S defined by $I_S(s) = s$ for all s in S. Inverses relate nicely here. If f is a one-to-one function from S into T, then $f^{-1}f = I_S$, while $ff^{-1} = I_f(S)$.

Exercise 7. Prove the preceding statement.

The set of all one-to-one correspondences from a set onto itself forms a group (Section 1, Supplement 2) with respect to the operation of composition of functions.

Exercise 8. Prove the preceding statement.

Exercise 9. This group is not commutative if S contains more than two elements.

Exercises 10. Write out the group elements for the functions on a set $S = \{a, b, c\}$ consisting of three elements and give a "multiplication table" for these functions.

The following relates function composition to the existence of an inverse.

5.2 If f and g are one-to-one functions from a set S onto itself and if $fg = I_S$, then $gf = I_S$ also. In fact $g = f^{-1}$.

Proof. We first show that $gf = I_S$, that is, that $gf(s) = s$ for any s in S. Given s let $s = g(t)$. (Why is there such a t?) Then

$$(gf)(s) \stackrel{\text{defn. } gf}{=} g(f(s)) \stackrel{\text{defn.}}{=}$$
$$g(f(g(t))) \stackrel{\text{defn. } fg}{=} g(fg(t)) \stackrel{\text{hyp.}}{=} g(I_S(t)) \stackrel{\text{defn. } I}{=} g(t) \stackrel{\text{defn. } t}{=} s.$$

Thus, $gf = I_S$.

To see that $g = f^{-1}$, note that f^{-1} exists (why?) and is defined by $f^{-1}(f(s)) = s$ for all $f(s)$. But also $g(f(s)) = s$ for all $f(s)$. Thus, g and f^{-1} agree everywhere on S (why?) Thus, they are the same function. ∎

Exercise 11. Answer the questions in the proof of 5.2.

Since the roles of f and g are symmetrical, we see that $f = g^{-1}$ as well, re-establishing the fact that $f = (f^{-1})^{-1}$.

6. FIELDS

This section is not absolutely necessary—we could make do with the remarks on fields at the beginning of the next chapter. This will serve, however, to point out the amazing generality of our subsequent linear algebra theory. Moreover, it provides a double dose of commutative groups for those who enjoy this notion.

Roughly speaking, a field is a system in which one can add, subtract, multiply, and divide according to rules very similar to those which obtain in the rational numbers, the real number system, and the complex number system. To be precise, a **field** is a set $\mathbb{F}$ together with two operations $+$ and $\cdot$ satisfying:

1A. For any x, y in $\mathbb{F}$, $x + y$ is an element of $\mathbb{F}$. (Closure of addition)

2A. For any x, y, z in $\mathbb{F}$, $x + (y + z) = (x + y) + z$. (Associativity of addition)

3A. There is an element called 0 in $\mathbb{F}$ for which $0 + x = x = x + 0$ for any x in $\mathbb{F}$. (Existence of additive identity)

4A. For any x in $\mathbb{F}$ there is an element called $-x$ in $\mathbb{F}$ for which $x + (-x) = 0$. (Existence of additive inverse)

5A. For any x, y in $\mathbb{F}$, $x + y = y + x$. (Commutativity of addition)

1M. For any x, y in $\mathbb{F}$, $x \cdot y$ is an element of $\mathbb{F}$. (Closure of multiplication)

2M. For any x, y, z in $\mathbb{F}$, $x \cdot (y \cdot z) = (x \cdot y) \cdot z$. (Associativity of multiplication)

3M. There is an element called 1 in $\mathbb{F}$ for which $1 \cdot x = x = x \cdot 1$ for all x in $\mathbb{F}$. Moreover, $1 \neq 0$. (Existence of multiplicative identity)

4M. For any x in $\mathbb{F}$ *different from* 0, there is an element called x^{-1} in $\mathbb{F}$ for which $x \cdot (x^{-1}) = 1 = (x^{-1}) \cdot x$. (Existence of multiplicative inverse for nonzero elements)

5M. For any x, y in $\mathbb{F}$, $x \cdot y = y \cdot x$. (Commutativity of multiplication)

D. For any x, y, z in $\mathbb{F}$, $x \cdot (y + z) = x \cdot y + x \cdot z$ and $(x + y) \cdot z = x \cdot z + y \cdot z$. (Distributivity)

Lest this seem a large potion of axioms to swallow, we point out what the reader has undoubtedly already noticed: 1A through 5A say that $\mathbb{F}$ is a commutative group under the operation $+$, 1M through 5M say that $\mathbb{F}$ is just about a commutative group under the operation $\cdot$, and D is a distributive law linking the operations of $+$ and $\cdot$. We shall clarify the second observation in 6.5.

Some examples of fields: the rational numbers; the real numbers; the complex numbers;† all numbers of the form $a + b\sqrt{2}$ where a and b are rational; the "numbers" $\bar{0}$ and $\bar{1}$ satisfying the following addition and multiplication tables:

+	$\bar{0}$	$\bar{1}$
$\bar{0}$	$\bar{0}$	$\bar{1}$
$\bar{1}$	$\bar{1}$	$\bar{0}$

$\cdot$	$\bar{0}$	$\bar{1}$
$\bar{0}$	$\bar{0}$	$\bar{0}$
$\bar{1}$	$\bar{0}$	$\bar{1}$

---
† Defined in Example 1 of Chapter 1, Section 1.

and the "numbers" $0'$, $1'$, $2'$, satisfying tables:

+	$0'$	$1'$	$2'$
$0'$	$0'$	$1'$	$2'$
$1'$	$1'$	$2'$	$0'$
$2'$	$2'$	$0'$	$1'$

$\cdot$	$0'$	$1'$	$2'$
$0'$	$0'$	$0'$	$0'$
$1'$	$0'$	$1'$	$2'$
$2'$	$0'$	$2'$	$1'$

Although fields may seem to be nothing more than entertaining mathematical confections, even those with only a finite number of elements have their practical uses (such as in constructing Latin squares for the statistical design of experiments and in designing error-control codes for digital communications systems).

We shall prove only a few essential theorems. The first three hold in any commutative group. Proofs of the first two are found in Supplement 2 to Section 1.

6.1 There is only one element 0 in any field which satisfies Axiom 3A.

6.2 If z is an element of a field $\mathbb{F}$ with the property that $z + x = x$ for some element x in $\mathbb{F}$, then $z = 0$.

6.3 If $x + y = 0$, then $y = -x$ (that is, the additive inverse of x is unique).

Proof.

$$-x \stackrel{3A}{=} (-x) + 0 \stackrel{\text{hyp.}}{=} (-x) + (x + y) \stackrel{2A}{=} ((-x) + x) + y$$
$$\stackrel{5A}{=} (x + (-x)) + y \stackrel{4A}{=} 0 + y \stackrel{3A}{=} y. \quad\blacksquare$$

6.4 For any x in $\mathbb{F}$, $0 \cdot x = 0$.

Proof. $0 \cdot x \stackrel{3A}{=} (0 + 0) \cdot x \stackrel{D}{=} 0 \cdot x + 0 \cdot x$. Hence $0 \cdot x = 0$, by 6.2. $\blacksquare$

6.5 The nonzero elements of $\mathbb{F}$ form a commutative group under the operation $\cdot$, with identity element 1.

Proof. Let $\mathbb{F}*$ denote the elements of $\mathbb{F}$ which are $\neq 0$. We first show closure; that is, x, y in $\mathbb{F}*$ implies that $x \cdot y$ is in $\mathbb{F}*$. We proceed by contradiction. If $x \cdot y$ is not in $\mathbb{F}*$, then $x \cdot y = 0$, by 1M. Since y is in $\mathbb{F}*$, $y \neq 0$. Hence, y^{-1} exists, by 4M. Then $0 \stackrel{6.4}{=} 0 \cdot y^{-1} \stackrel{\text{subst.}}{=} (x \cdot y) \cdot y^{-1} \stackrel{2M}{=} x \cdot (y \cdot y^{-1}) \stackrel{4M}{=} x \cdot 1 \stackrel{3M}{=} x$. Thus $x = 0$, contradicting our assumption that x is in $\mathbb{F}*$. The rest of the proof is now obvious. $\blacksquare$

Given any field $\mathbb{F}$, one can consider the set $\mathbb{F}[x]$ of all polynomials $a_0 + a_1 x + \cdots + a_n x^n$ with coefficients a_i in $\mathbb{F}$. However, the formal symbol approach (which is what we mean by $\mathbb{F}[x]$) can differ from the functional approach; see Exercise 6 at the end of this section.

To alert you to the possible difficulties involved in working over fields other than the reals, we now discuss the main distinctions.

Ordering

The real numbers are equipped with an ordering: Given any two real numbers a and b, one always has either $a < b$ or $b < a$ or $a = b$, and the relation "$<$" satisfies certain axioms which you may have encountered. On the other hand, it is impossible to equip the complex numbers with a similar ordering. For example, one cannot postulate that either $i < 0$ or $0 < i$ (certainly $i \neq 0$) without violating the axioms for "$<$".

Now, many pictures are dependent upon the natural ordering of the coordinate axes (and the attendant notion of "betweenness"). It is also unreasonable to attempt to interpret a coordinate axis in a picture as having points corresponding to complex numbers rather than real numbers. (Where would you squeeze in i?) Consequently, many pictures cannot be considered as valid depictions of events in vector spaces over the complex numbers instead of the reals $\mathbb{R}$. (Pictures do, of course, continue to give intuitive comfort and understanding; and a picture, together with the recognition of its limitations, may be better than no picture at all.)

There are fields other than the real numbers which can be ordered, for example, the field of rational numbers $\mathbb{Q}$; hence, the above objections do not apply to $\mathbb{Q}$. On the other hand, the two examples of finite fields given above cannot be ordered so as to preserve the basic properties of "$<$".

Calculus Notions

Limits, continuity, differentiation, and integration can all be developed for complex functions of a complex variable just as for real-valued functions of a real variable (with an occasional surprise). However, in terms of current curricula, it is unlikely that you have heretofore encountered this theory. Consequently, if you are rechecking the material to see what applies to vector spaces over the complex numbers, we suggest that you ignore those examples and exercises in which calculus notions are used.

Most other fields do not fare as well when it comes to calculus; e.g., the rationals $\mathbb{Q}$ are not closed with respect to taking limits (there are sequences of rational numbers which converge to nonrational numbers). You can't even speak intelligently of limits of sequences in finite fields.

Arithmetic Properties

Any nonconstant polynomial with real (or complex) coefficients has a complex root, while it may not have a *real* root. There are places in Chapter 6 where we make use of facts such as: there is no real x for which $x^2 + 1 = 0$.

Some fields are subject to certain types of arithmetic quirks to which both the complex numbers and rational numbers are immune. For example, in our two finite fields we find that

$$\bar{1} + \bar{1} = \bar{0}; \quad 1' + 2' = 0'.$$

EXERCISES

1. Convince yourself that all our examples of fields are actually fields.
2. a) Prove that there is just *one* element 1 in any field for which $1 \cdot x = x$ for all x in $\mathbb{F}$. (That is, the multiplicative identity is unique.)
   b) Prove that multiplicative inverses are unique.
3. Prove that division by 0 is impossible in any field. (Note that this is the same as multiplication by 0^{-1}. Show that 0^{-1} can't exist.)
4. Show that the set of all n tuples $(x_1, \ldots, x_n)$, where the x's are from some fixed field $\mathbb{F}$, form a commutative group under componentwise addition. Show that the set of all n-tuples satisfies the scalar-multiplication conditions under componentwise multiplication by elements of $\mathbb{F}$.
5. Why don't the integers form a field? Which axiom(s) are violated?
6. Prove that in the field with two elements, whose multiplication and addition tables are given in the examples, (a) the function $x^2 - x$ is different from the formal polynomial $x^2 - x$; (b) $a = -a$ for all a in the field.
7. Which of our theorems hold in any commutative group? Why?

Because they are kind hearted, some mathematicians have made little apartment houses for vectors to live in. They call them vector spaces.†

CHAPTER 1 # Vector Spaces

This chapter introduces the concept of "vector space" and attendant fundamental notions. The material is basic to the rest of the book, and in all likelihood basic to much of the mathematics you will later encounter (and, hopefully, of considerable importance to those who will later apply mathematics to other areas of endeavor). This chapter is also the most abstract of the book, although experience with the first three sections of Chapter 0 should bring things comfortably down to earth. We defend this abstract approach at the end of the final section of the chapter; also please refer to the Introduction if you are in need of further solace.

Some of this basic material does lend itself to immediate application. We shall point out some of these applications in Chapter 2, where the requisite computational techniques are developed.

1. IN WHICH THE CENTRAL CHARACTERS ARE INTRODUCED

We shall now begin our study of a very powerful and far-reaching generalization of the notion of geometrical vectors. You are invited to review the properties of vectors which we singled out in the first section of the previous chapter.

We first give an informal discussion of the notion of vector space. By a "vector space $\mathcal{V}$ over the field of real numbers $\mathbb{R}$", we shall mean a set $\mathcal{V}$ of objects called "vectors", and we shall demand that the set $\mathcal{V}$ have an operation of "vector addition" which satisfies axioms for closure, associativity, identity,

---

† *A Child's Garden of Vectors* by Elizabeth Yount, *Op. cit.*

inverses, and commutativity (that is, $\mathcal{V}$ must be a commutative group with respect to the operation of vector addition). Moreover, there must be a "scalar multiplication" of elements of $\mathcal{V}$ by elements of $\mathbb{R}$ (which we refer to as "scalars") satisfying appropriate conditions (we must have closure, the two possible kinds of distributivity, the only possible kind of associativity, and the real number 1 must scalar-multiply properly). Thus, the set of all vectors in euclidean n-space forms an example of a vector space over the reals. We shall consider other examples after we give the formal definition.

Although it will be our practice to adhere to the axiomatic method (see the Introduction), we shall make one exception: We do not give an axiomatic characterization of the real number system $\mathbb{R}$. Much of the material herein makes use only of the field properties of the real numbers (listed in Chapter 0, Section 6), and the other necessary axioms would take us into too great a digression. We are confident that an informal knowledge of the real numbers will not lead you into algebraic disasters. (An axiomatic description is to be found, for example, in Birkhoff and MacLane, *A Survey of Modern Algebra*, Macmillan, revised edition, 1953.)

To begin our axiomatic approach...

A **vector space $\mathcal{V}$ over the field $\mathbb{R}$ of real numbers** consists of a set $\mathcal{V}$ which is first of all a commutative group; that is, $\mathcal{V}$ is endowed with an operation "+" satisfying

1) For **V**, **W** in $\mathcal{V}$, **V** + **W** is an element of $\mathcal{V}$. (Closure)
2) For **U**, **V**, **W** in $\mathcal{V}$, **U** + (**V** + **W**) = (**U** + **V**) + **W**. (Associativity)
3) There is an element **0** in $\mathcal{V}$ for which **0** + **V** = **V** = **V** + **0** for every **V** in $\mathcal{V}$. (Existence of identity)
4) For any element **V** in $\mathcal{V}$ there is an element −**V** in $\mathcal{V}$ which satisfies **V** + (−**V**) = **0** = (−**V**) + **V**. (Existence of inverse)
5) For **V**, **W** in $\mathcal{V}$, **V** + **W** = **W** + **V**. (Commutativity)

Furthermore, there is a multiplication of elements of $\mathcal{V}$ by elements of $\mathbb{R}$ satisfying:

i) For a in $\mathbb{R}$, **V** in $\mathcal{V}$, a**V** is an element of $\mathcal{V}$. (Closure)
ii) For a, b in $\mathbb{R}$, **V** in $\mathcal{V}$, $(a + b)$**V** = a**V** + b**V**. (Distributivity)
iii) For a in $\mathbb{R}$, **V**, **W** in $\mathcal{V}$, a(**V** + **W**) = a**V** + a**W**. (Distributivity)
iv) For a, b in $\mathbb{R}$, **V** in $\mathcal{V}$, (ab)**V** = $a(b$**V**$)$. (Associativity)
v) If 1 is the multiplicative identity of $\mathbb{R}$, then 1**V** = **V** for any **V** in $\mathcal{V}$. (Scalar multiplication by 1)

Note that in order to specify a vector space $\mathcal{V}$ over the field $\mathbb{R}$, it is necessary to specify the set $\mathcal{V}$, the operation + in $\mathcal{V}$, and the scalar multiplication. One then has to check to see if conditions 1 through 5 and (i) through (v) are satisfied (see Fig. 1).

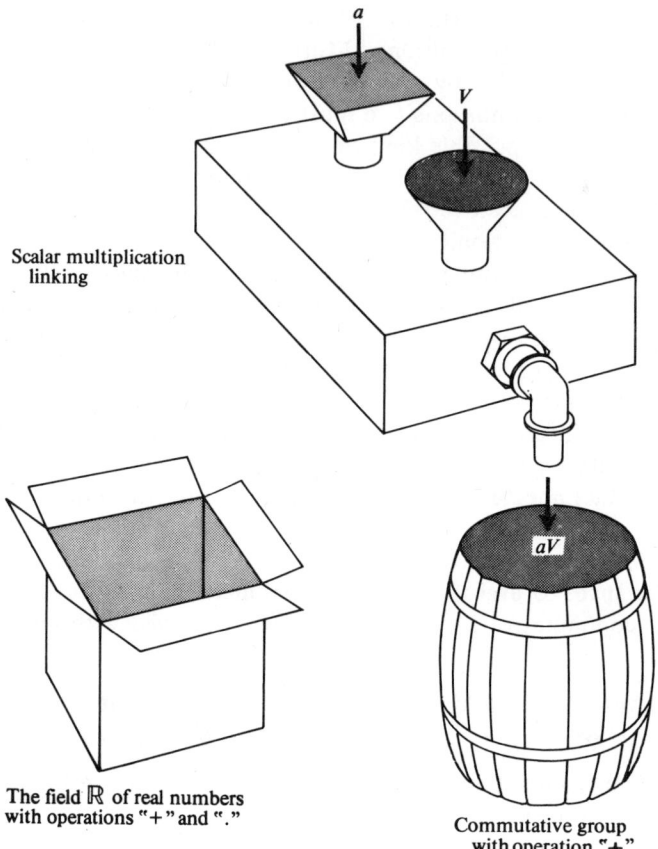

Fig. 1. A vector space $\mathcal{V}$ over $\mathbb{R}$.

The elements in $\mathcal{V}$ are called **vectors**, while we often refer to the field elements in $\mathbb{R}$ as **scalars.**

From the previous chapter you are already well acquainted with several extremely important examples of vector spaces:

The set of geometrical vectors in euclidean n-space forms a vector space over the field $\mathbb{R}$ of real numbers with respect to the usual addition and scalar multiplication of such vectors (Chapter 0, Section 1).

The set $\mathcal{F}$ of all real valued functions defined on a set S of real numbers forms a vector space over $\mathbb{R}$ with respect to the usual addition and scalar multiplication of functions (Chapter 0, Section 3).

The set $\mathbb{R}[x]$ of all polynomials with real coefficients forms a vector space over $\mathbb{R}$ with respect to polynomial addition and scalar multiplication (Chapter 0, Section 3, Exercise 3).

The set $\mathbb{R}_n$ of all n-tuples $(x_1, \ldots, x_n)$ of elements from the field $\mathbb{R}$ forms a vector space over $\mathbb{R}$ if vector addition is defined componentwise $(x_1, \ldots, x_n) + (y_1, \ldots, y_n) = (x_1 + y_1, \ldots, x_n + y_n)$, scalar multiplication is defined componentwise $a(x_1, \ldots, x_n) = (ax_1, \ldots, ax_n)$, and two n-tuples $(x_1, \ldots, x_n)$ and $(y_1, \ldots, y_n)$ are considered equal if and only if $x_1 = y_1, \ldots, x_n = y_n$. (See Chapter 0, Section 2, Exercise 2.) We have informally identified $\mathbb{R}_2$ with the euclidean plane and $\mathbb{R}_3$ with 3-space.

Some fresher examples:

Example 1. As you probably recall, the field $\mathbb{C}$ of **complex numbers** consists of all numbers of the form $a + ib$ where a and b are real numbers and $i(=\sqrt{-1})$ satisfies $i^2 = -1$. Two complex numbers $a + ib$ and $c + id$ are **equal** iff the real numbers $a = c$ and $b = d$. The following arithmetic holds:

Addition $\qquad (a + ib) + (c + id) = (a + c) + i(b + d)$

Multiplication by reals $\qquad c(a + ib) = ca + icb$

(Actually a special case
of) multiplication $\qquad (a + ib)(c + id) = (ac - bd) + i(ad + bc)$.

Conjugation $\qquad \overline{a + ib} = a - ib$.

The set $\mathbb{C}$ of all complex numbers forms a vector space over the field of real numbers with the usual addition and multiplication by reals.

Example 2. The set of all solutions (x, y) to the equation $ax + by = 0$, a, b, x, y in $\mathbb{R}$, forms a vector space over $\mathbb{R}$ if both addition and scalar multiplication are done componentwise. We check a few axioms:

1) If (x_1, y_1) and (x_2, y_2) are both solutions, then

$$0 = (ax_1 + by_1) + (ax_2 + by_2) \underset{\text{of addn.}}{\overset{\text{assoc. and commut.}}{=}} (ax_1 + ax_2) + (by_1 + by_2)$$

$$\overset{\text{distrib.}}{=} a(x_1 + x_2) + b(y_1 + y_2).$$

But this says that $(x_1 + x_2, y_1 + y_2)$ is a solution, so (1) holds.

3) $(x, y) + (0, 0) = (x + 0, y + 0) = (x, y)$.

4) $(x, y) + (-x, -y) = (x + (-x), y + (-y)) = (0, 0)$, so $(-x, -y) = -(x, y)$

5) $(x_1, y_1) + (x_2, y_2) \overset{\text{defn.}}{=} (x_1 + x_2, y_1 + y_2)$

$$\underset{\text{defn.}}{\overset{\text{commut. of real number addn.}}{=}} (x_2 + x_1, y_2 + y_1)$$

$$= (x_2, y_2) + (x_1, +y_1)$$

We leave the rest for you.

Example 3. The set of all (real-valued) functions f defined for $0 \leq x \leq 1$ and for which $\int_0^1 f(x)\, dx = 0$ forms a vector space over the reals $\mathbb{R}$ with respect to the usual definitions of addition and scalar multiplication of functions.

Let us pause for a moment and assess the situation. You have been presented with the definition of "vector space over the reals". You also have a number of specific examples of this notion, some of which can be arranged in a reasonable hierarchy of abstraction:

Some Vector Spaces over $\mathbb{R}$ (with respect to the usual operations)
Geometrical vectors in the plane, which we informally identify with $\mathbb{R}_2$ = all pairs of real numbers
Geometrical vectors in 3-space, which we informally identify with $\mathbb{R}_3$ = all triples of real numbers
$\mathbb{R}_n$ = all n-tuples of real numbers
$\mathbb{R}[x]$ = all polynomials with real coefficients
$\mathcal{F}$ = all real-valued functions defined on some interval

This, of course, is only a partial listing of vector spaces, and there is nothing sacred about it. It merely represents our attempt to indicate some vector spaces in increasing level of abstraction. You might wish to make your own list, or to incorporate some examples we have omitted.

In approaching the material to come, you may find it convenient first to interpret it at the highest level of abstraction you can apprehend, and then try to improve your level of abstract thinking. For example, if you feel comfortable on the list down as far as the vector space $\mathbb{R}[x]$ of polynomials with real coefficients, you might first approach each definition and theorem by (1) checking that it has something sensible to say about $\mathbb{R}[x]$ (and all vector spaces on the list preceding it), and (2) trying to understand what it says about the next vector space on the list, and so on.

All of the material in this book is valid when applied to the vector space $\mathbb{R}_2$ of pairs of real numbers (although some of the results are trivial in this simple situation). In fact, it would probably be better to end up with a thorough understanding of $\mathbb{R}_2$ than with a hazy acquaintance with some ephemeral creature called "vector space". (We do want you to get an appreciation of something more sophisticated than $\mathbb{R}_2$; in fact we hope you'll learn to contemplate the term "arbitrary vector space over $\mathbb{R}$" with some comfort and satisfaction.)

A large portion of the material in this book is applicable to vector spaces over fields other than the real numbers (this explains our insistence on the full phrase "vector space over the real numbers" rather than just "vector space"). For further information see the supplement at the end of this section.

EXERCISES

Many of the following exercises are concerned with showing that certain sets do form vector spaces. Your procedure should be to list the vector space axioms and give a reason why each one holds in the set. Allowable reasons would include "same as verification of this axiom in ———"; "this set is a subset of ——— which we know to be a vector space, so this property automatically holds"; "a well-known property of the reals". There is also some opportunity for more interesting and original work. Most of the steps are quite simple. The purpose of these exercises is (1) to acquaint you with the vector space axioms, and (2) to familiarize you with some further important examples of vector spaces. In Section 3 we shall develop a much more convenient procedure for testing whether certain sets are vector spaces.

1. Check all axioms in the examples which we did not discuss.
2. Show that the set $\mathcal{C}$ of all continuous functions defined for $0 \leq x \leq 1$ forms a vector space over the real numbers $\mathbb{R}$ with respect to the usual addition and scalar multiplication of functions.
3. Show that the set of all solutions (x, y, z) to the equations

    $$a_1 x + b_1 y + c_1 z = 0,$$
    $$a_2 x + b_2 y + c_2 z = 0,$$

    a's, b's, c's, x, y, z real

    forms a vector space over $\mathbb{R}$ relative to the usual addition and scalar multiplication of triples.
4. Show that the field of real numbers $\mathbb{R}$ forms a vector space over $\mathbb{R}$ with vector addition taken to be addition of reals and scalar multiplication the usual multiplication of real numbers.
5. Construct other examples of vector spaces over $\mathbb{R}$.
6. Show that all continuous functions f defined for $0 \leq x \leq 1$ for which $f(\tfrac{1}{2}) = 0$ form a vector space over the reals with respect to the usual definitions of addition and scalar multiplication.
7. Show that if we amend Exercise 6 to require $f(\tfrac{1}{2}) = 3$, we no longer obtain a vector space.
8. Given any vector space $\mathcal{V}$ over $\mathbb{R}$, show that the set $\{\mathbf{0}\}$ consisting of just the identity $\mathbf{0}$ of $\mathcal{V}$ forms a vector space over $\mathcal{V}$ with respect to its usual operations. (Assume for now that $c\mathbf{0} = \mathbf{0}$ for any scalar c; we prove this in the next section.)
9. (*Uses commutative groups*) Show that if $\mathcal{V}$ is any commutative group containing more than one element, we do *not* obtain a vector space if we define our scalar multiplication by $a\mathbf{V} = \mathbf{V}$ for all a in $\mathbb{R}$, $\mathbf{V}$ in $\mathcal{V}$. Which axioms do hold?
10. Prove that the set of all functions f which have second derivatives and which satisfy $f''(x) + 9f(x) = 0$ forms a vector space over $\mathbb{R}$ with respect to the usual definition of function addition and scalar multiplication.

Although vector spaces occur in many applications, there are a number of contexts in which one works with certain proper subsets of vector spaces. In the following two exercises we indicate two important subsets of $\mathbb{R}_n$ which are not vector spaces themselves. In each case (a) state which vector space axioms do hold, (b) determine necessary and sufficient conditions that the sum of two vectors in the set lies in the set, (c) determine necessary and sufficient conditions that a scalar multiple of an element in the set lies in the set.

11. There are many applications in which negative components of vectors in $\mathbb{R}_n$ are meaningless. The set of all vectors $\mathbf{V} = (v_1, \ldots, v_n)$ in $\mathbb{R}_n$ with all coordinates nonnegative is called the **nonnegative orthant** $\mathcal{N}$. That is,

    $$\mathcal{N} = \{\mathbf{V} \text{ in } \mathbb{R}_n \mid \mathbf{V} = (v_1, \ldots, v_n) \text{ and each } v_i \geq 0\}.$$

    Proceed as indicated just above.

12. A vector $\mathbf{p} = (p_1, \ldots, p_n)$ in $\mathbb{R}_n$ is a **probability vector** if
   i) each $p_j \geq 0$ and
   ii) $\sum_{i=1}^{n} p_i = p_1 + \cdots + p_n = 1$.

   These arise in situations where there are n possible courses of action, one of which must be chosen, and there is assigned to each possible course of action a probability p_j that it will be the one chosen. See, for example, Section 5 of Chapter 2. Proceed as indicated above Exercise 11.

Supplement. Vector Spaces Over Other Fields

There are contexts in mathematics and in its applications where one needs to deal with vector spaces over fields other than the real numbers. The field of complex numbers $\mathbb{C}$ is of special importance. (This field is discussed in Example 1; if you are interested in other fields, see Chapter 0, Section 6.) Almost all the contents of this book are valid for vector spaces over $\mathbb{C}$ (some indicated adjustments must occasionally be made). Through Chapter 7 almost everything is valid for vector spaces over any field.

If you wish to be efficient and learn linear algebra for vector spaces over the reals and over the complex numbers simultaneously (or for vector spaces over arbitrary fields), we include two honors projects where appropriate, asking you to go through the section and check that (except as noted) all the material in the section holds for vector spaces (1) over $\mathbb{C}$, and (2) over any field. We sometimes include additional exercises and remarks to help you appreciate the material from either of these more general approaches. Most of the possible difficulties involved in working over other fields are discussed at the end of Chapter 0, Section 6.

Although it involves great redundancy, we now give a definition for "vector space over the complex numbers":

A **vector space** $\mathcal{U}$ **over the field of complex numbers** $\mathbb{C}$ consists of a set $\mathcal{U}$ which is a commutative group under addition:

1) For $\mathbf{V}, \mathbf{W}$ in $\mathcal{U}$, $\mathbf{V} + \mathbf{W}$ is an element of $\mathcal{U}$.
2) For $\mathbf{U}, \mathbf{V}, \mathbf{W}$ in $\mathcal{U}$, $\mathbf{U} + (\mathbf{V} + \mathbf{W}) = (\mathbf{U} + \mathbf{V}) + \mathbf{W}$.
3) There is an element $\mathbf{0}$ in $\mathcal{U}$ for which $\mathbf{0} + \mathbf{V} = \mathbf{V} = \mathbf{V} + \mathbf{0}$ for every $\mathbf{V}$ in $\mathcal{U}$.
4) For any element $\mathbf{V}$ in $\mathcal{U}$, there is an element $-\mathbf{V}$ in $\mathcal{U}$ which satisfies $\mathbf{V} + (-\mathbf{V}) = \mathbf{0} = (-\mathbf{V}) + \mathbf{V}$.
5) For $\mathbf{V}, \mathbf{W}$ in $\mathcal{U}$, $\mathbf{V} + \mathbf{W} = \mathbf{W} + \mathbf{V}$.

Futhermore, there is a multiplication of elements of $\mathcal{U}$ by elements of $\mathbb{C}$ satisfying the following:

i) For a in $\mathbb{C}$, $\mathbf{V}$ in $\mathcal{U}$, $a\mathbf{V}$ is an element of $\mathcal{U}$.

ii) For a, b in $\mathbb{C}$, $\mathbf{V}$ in $\mathcal{U}$, $(a + b)\mathbf{V} = a\mathbf{V} + b\mathbf{V}$.
iii) For a in $\mathbb{C}$, $\mathbf{V}, \mathbf{W}$ in $\mathcal{U}$, $a(\mathbf{V} + \mathbf{W}) = a\mathbf{V} + a\mathbf{W}$.
iv) For a, b in $\mathbb{C}$, $\mathbf{V}$ in $\mathcal{U}$, $(ab)\mathbf{V} = a(b\mathbf{V})$.
v) If 1 is the multiplicative identity of $\mathbb{C}$, then $1\mathbf{V} = \mathbf{V}$ for any $\mathbf{V}$ in $\mathcal{U}$.

The definition of vector space over an arbitrary field $\mathbb{F}$ can be obtained from the above by simply crossing out "of complex numbers" and replacing $\mathbb{C}$ by $\mathbb{F}$.

An interesting example of a vector space over a field is obtained by taking $\mathcal{U} = \mathbb{R}$, the reals with vector addition the usual addition of real numbers, and field $\mathbb{F} = \mathbb{Q}$, the rationals, with scalar multiplication the usual multiplication of real numbers.

Honors projects. Tasks 1 and 2 are listed in the second paragraph of this supplement.

2. ELEMENTARY PROPERTIES OF VECTOR SPACES

We have now given a definition of "vector space over the reals" and we have encountered a rather rich supply of such creatures. After stating a formal axiom system, one usually has to derive some simple preliminary properties before one can proceed to deeper results. Although there are a number of these simple results which we must obtain, we are fortunate in that these results are all eminently reasonable and plausible; your previous experience with particular examples of vector spaces would lead you to expect them. Because of this, the style of this section will be rather different from our normal approach: we present a string of theorems and proofs unrelieved by examples (you can trivially construct your own) or explanations (none should be needed). Throughout the first four theorems some of you may be gripped with a sense of *déjà vu*.

2.1 *Uniqueness of zero.* There is only one vector in a vector space $\mathcal{U}$ which satisfies Axiom 3. That is,

$$\text{if } \mathbf{0} + \mathbf{V} = \mathbf{V} = \mathbf{V} + \mathbf{0} \quad \text{for all } \mathbf{V} \text{ in } \mathcal{U},$$

and

$$\text{if } \mathbf{0}' + \mathbf{V} = \mathbf{V} = \mathbf{V} + \mathbf{0}' \quad \text{for all } \mathbf{V} \text{ in } \mathcal{U},$$

then

$$\mathbf{0} = \mathbf{0}'.$$

Proof. Take $\mathbf{V} = \mathbf{0}'$ in the first equation to get $\mathbf{0} + \mathbf{0}' = \mathbf{0}'$. Take $\mathbf{V} = \mathbf{0}$ in the second equation to get $\mathbf{0} = \mathbf{0} + \mathbf{0}'$. Hence, $\mathbf{0}' = \mathbf{0}$. ∎

As a more convenient refinement of 2.1, the next result shows that you need only check a candidate for $\mathbf{0}$ on Axiom 3 for any single element rather than for every $\mathbf{V}$ in $\mathcal{U}$.

2.2 If $\mathbf{Z}$ is any element in a vector space $\mathcal{U}$ with the property that $\mathbf{Z} + \mathbf{X} = \mathbf{X}$ for any one element $\mathbf{X}$ of $\mathcal{U}$, then $\mathbf{Z} = \mathbf{0}$.

Proof. Add $-\mathbf{X}$ (which exists by Axiom 4) to both sides of the equation $\mathbf{Z} + \mathbf{X} = \mathbf{X}$:

$$(\mathbf{Z} + \mathbf{X}) + (-\mathbf{X}) = \mathbf{X} + (-\mathbf{X}) = \mathbf{0}.$$

Then

$$\mathbf{0} = (\mathbf{Z} + \mathbf{X}) + (-\mathbf{X}) \stackrel{\text{Ax.2}}{=} \mathbf{Z} + (\mathbf{X} + (-\mathbf{X})) \stackrel{\text{Ax.4}}{=} \mathbf{Z} + \mathbf{0} \stackrel{\text{Ax.3}}{=} \mathbf{Z}. \quad \blacksquare$$

2.3 Uniqueness of the additive inverse. If $\mathbf{X} + \mathbf{Y} = \mathbf{0}$, then $\mathbf{Y} = -\mathbf{X}$.

Proof. $-\mathbf{X} \stackrel{\text{Ax.3}}{=} (-\mathbf{X}) + \mathbf{0} = (-\mathbf{X}) + (\mathbf{X} + \mathbf{Y}) \stackrel{\text{Ax.2}}{=} ((-\mathbf{X}) + \mathbf{X}) + \mathbf{Y}$

$\stackrel{\text{Ax.5}}{=} (\mathbf{X} + (-\mathbf{X})) + \mathbf{Y} \stackrel{\text{Ax.4}}{=} \mathbf{0} + \mathbf{Y} \stackrel{\text{Ax.3}}{=} \mathbf{Y}. \quad \blacksquare$

2.4 Scalar multiplication by zero. For any $\mathbf{V}$ in $\mathcal{U}$ the element 0 in $\mathbb{R}$ satisfies $0\mathbf{V} = \mathbf{0}$.

Proof. $0\mathbf{V} \stackrel{\text{field property}}{=} (0 + 0)\mathbf{V} \stackrel{\text{Ax (ii)}}{=} 0\mathbf{V} + 0\mathbf{V}$. Hence $0\mathbf{V} = \mathbf{0}$ by 2.2. $\quad \blacksquare$

2.5 Scalar multiplication of the zero vector. For any c in $\mathbb{R}$ the vector $\mathbf{0}$ satisfies $c\mathbf{0} = \mathbf{0}$.

Proof. $c\mathbf{0} \stackrel{\text{Ax.3}}{=} c(\mathbf{0} + \mathbf{0}) \stackrel{\text{Ax(iii)}}{=} c\mathbf{0} + c\mathbf{0}$. Hence $c\mathbf{0} = \mathbf{0}$ by 2.2. $\quad \blacksquare$

2.6 For any $\mathbf{V}$ in a vector space $\mathcal{U}$, $-(-\mathbf{V}) = \mathbf{V}$.

Proof. Let $\mathbf{U} = -(-\mathbf{V})$, so by definition $(-\mathbf{V}) + \mathbf{U} = \mathbf{0}$. But also $(-\mathbf{V}) + \mathbf{V} \stackrel{\text{Ax.5}}{=} \mathbf{V} + (-\mathbf{V}) \stackrel{\text{Ax.4}}{=} \mathbf{0}$. Moreover, 2.3 guarantees the *uniqueness* of the additive inverse of $-\mathbf{V}$. Hence, $\mathbf{U} = \mathbf{V}$; that is, $-(-\mathbf{V}) = \mathbf{V}$. $\quad \blacksquare$

2.7 Multiplication by the scalar -1. For the element -1 in $\mathbb{R}$ and for any $\mathbf{V}$ in $\mathcal{U}$, $(-1)\mathbf{V} = -\mathbf{V}$.

Proof. $\mathbf{0} \stackrel{2.4}{=} 0\mathbf{V} = (1 + (-1))\mathbf{V} \stackrel{\text{(ii)}}{=} 1\mathbf{V} + (-1)\mathbf{V} \stackrel{\text{(v)}}{=} \mathbf{V} + (-1)\mathbf{V}$. Hence $(-1)\mathbf{V}$ is an additive inverse of $\mathbf{V}$. By 2.3, $(-1)\mathbf{V} = -\mathbf{V}$. $\quad \blacksquare$

From now on we shall write $\mathbf{U} - \mathbf{V}$ in place of $\mathbf{U} + (-\mathbf{V})$.

Axiom 2 says that both possible ways of grouping the sum of three elements into successive sums of two elements are the same. Thus, parentheses are unnecessary for writing sums of three elements. By means of mathematical induction, one can show that any two ways of grouping a sum of n elements are the same. That is,

parentheses are unnecessary for writing sums of n elements.

This is again a group-theoretic result. We omit the inductive proof, since it is both somewhat difficult and not particularly enlightening.

The following facts, valid in any vector space, are easily proved by induction. We ask you to do so in the exercises.

2.8 For vectors $V, V_1, \ldots, V_n$ in a vector space and scalars $a, a_1, \ldots, a_n$ in $\mathbb{R}$,

i) $a(V_1 + \cdots + V_n) = aV_1 + \cdots + aV_n$
ii) $(a_1 + \cdots + a_n)V = a_1V + \cdots + a_nV$
iii) $-a(V_1 + \cdots + V_n) = (-aV_1) + \cdots + (-aV_n)$
  (written $-aV_1 - aV_2 - \cdots - aV_n$)

Sample induction proof. To prove a result by induction, you must show two things:

I) The result is true for the integer $n = 1$ (or for some first integer n_0); and

II) If the result is true for an integer n ($n \geq n_0$), then it is also true for the integer $n + 1$. [Note that in proving it for $n + 1$, you are allowed to assume that it is true for n.] If you have demonstrated I and II, then it follows that the result holds for all positive integers. For

  by I, the result is true for $n = 1$ (or your first integer n_0);
  by II, since it is true for $n = 1$, it is true for $n + 1 = 2$;
  by II, since it is true for $n = 2$, it is true for $n + 1 = 3$;
  by II, since it is true for $n = 3$, it is true for $n + 1 = 4$;
  .
  .
  .

By one of the basic properties of the integers, it is true for all positive integers ($n \geq n_0$). We illustrate on (i):

I) For $n = 1$ this says $a(V_1) = aV_1$, which is trivially true since we interpret parentheses as just a means of grouping vectors (if you feel slighted, start with $n = 2$, for which our desired statement becomes an axiom).

II) Suppose, then, that for some integer n,

1) $a(V_1 + \cdots + V_n) = aV_1 + \cdots + aV_n$ for all a in $\mathbb{R}$, and $V_1, \ldots, V_n$ in $\mathcal{V}$.

We must show that if this holds then

2) $a(V_1 + \cdots + V_n + V_{n+1}) = aV_1 + \cdots + aV_n + aV_{n+1}$ for all a in $\mathbb{R}$, and $V_1, \ldots, V_{n+1}$ in $\mathcal{V}$.

To establish (2) we need only let $W = V_1 + \cdots + V_n$, so that we can rewrite

$$a(V_1 + \cdots + V_n + V_{n+1}) \stackrel{\text{defn.}}{=} a(W + V_{n+1}) \stackrel{\text{Ax. (iii)}}{=} aW + aV_{n+1}$$
$$\stackrel{\text{defn.}}{=} a(V_1 + \cdots + V_n) + aV_{n+1}$$
$$\stackrel{(1)}{=} aV_1 + \cdots + aV_n + aV_{n+1}.$$

The extremes of this string of equations being (2), we are done. ∎

2.9 Given vectors $V_1, \ldots, V_n$ in a vector space, and scalars $a_1, \ldots, a_n$, $b_1, \ldots, b_n$ in $\mathbb{R}$,

i) $(a_1V_1 + \cdots + a_nV_n) + (b_1V_1 + \cdots + b_nV_n)$
$$= (a_1 + b_1)V_1 + \cdots + (a_n + b_n)V_n;$$

ii) $(a_1V_1 + \cdots + a_nV_n) - (b_1V_1 + \cdots + b_nV_n)$
$$= (a_1 - b_1)V_1 + \cdots + (a_n - b_n)V_n.$$

The only possible surprising thing in this section is that arithmetic in an abstract vector space so closely parallels arithmetic in your well-known examples.

EXERCISES

1. a) Write all possible ways of grouping a sum of four elements V, W, X, Y into sums of fewer elements, keeping the order of the elements the same.
   b) Choose two interesting items on your list in (a) and show that they are equal, using only Axiom 2.
2. Prove that $-(U + V + W) = -U - V - W$ for any U, V, W in a vector space $\mathcal{V}$.
3. a) Prove that $a(U + V + W) = aU + aV + aW$ for a in $\mathbb{R}$, U, V, W in $\mathcal{V}$.
   b) Prove 2.8(ii).
4. a) Prove that $(U + V) + (X + Y) = (U + X) + (V + Y)$ for U, V, X, Y in $\mathcal{V}$.
   b) Prove 2.9(i).
5. a) Prove that $(-a)V = -(aV)$. b) Prove 2.8(iii). c) Prove 2.9(ii).
H6. (*Uses commutative groups*) Prove that parentheses are unnecessary in any commutative group in that, given any two sums of n elements which differ only in the placement of parentheses, these two sums are the same.
7. a) (*Uses commutative groups*) Explain why 2.1 is the same as Chapter 0, Supplement 2, 1.2; explain why 2.2 is the same as Chapter 0, Supplement 2, 1.3; explain why 2.3 is the same as Chapter 0, Supplement 2, Exercise 8. Are any other exercises there the same as theorems here?
   b) (*Uses Chapter 0, Section 6*) Find all theorems in Chapter 0, Section 6, which correspond to theorems of this section. Are they exactly the same? Explain.

Honors Projects

1. Go through the theorems and exercises of this section replacing "vector space over the reals" by "vector space over complex numbers", and verify that all the results are true.
2. Same for "vector spaces over an arbitrary field".

3. SUBSPACES

As you no doubt observed in Section 1, there are many examples of interesting vector spaces which are subsets of other vector spaces (the vector space $\mathcal{F}$ of functions defined on a given interval is one good parent space). We call such subsets "subspaces".

Let $\mathcal{V}$ be a vector space over $\mathbb{R}$. A nonempty subset $\mathcal{W}$ of $\mathcal{V}$ is called a **subspace** of $\mathcal{V}$ if $\mathcal{W}$ is itself a vector space (under the vector space operations of $\mathcal{V}$).

Examples. We can regard the vector space $\mathbb{R}[x]$ of polynomials as a subspace of the vector space $\mathcal{F}$ of functions defined on an interval. The set of all solutions (x, y) to $ax + by = 0$ of Section 1, Example 2, is a subspace of $\mathbb{R}_2$.

Since a subset inherits many properties from the containing space, it is particularly easy to test whether a subset is in fact a subspace:

3.1 The Subspace theorem. Let $\mathcal{W}$ be a nonempty subset of a vector space $\mathcal{V}$ over $\mathbb{R}$. Then $\mathcal{W}$ is a subspace of $\mathcal{V}$ iff

i) $\mathcal{W}$ is closed under vector addition (for $\mathbf{W}_1$ and $\mathbf{W}_2$ in $\mathcal{W}$, $\mathbf{W}_1 + \mathbf{W}_2$ is in $\mathcal{W}$); and

ii) $\mathcal{W}$ is closed under scalar multiplication (for $\mathbf{W}$ in $\mathcal{W}$ and c in $\mathbb{R}$, $c\mathbf{W}$ is in $\mathcal{W}$).

Proof. 1) If $\mathcal{W}$ is a subspace, then it is in particular a vector space, and thus satisfies all the vector space axioms including the two closure axioms.

2) Conversely, if a subset $\mathcal{W}$ satisfies (i) and (ii), then we must show that it is a vector space. The vector addition in $\mathcal{W}$ will surely satisfy the associative and commutative axioms, since these hold in all of $\mathcal{V}$, and we are assuming closure of addition. Hence, we need to demonstrate the existence of an identity and of additive inverses. Take $\mathbf{W}$ in $\mathcal{W}$. Then $0\mathbf{W} = \mathbf{0}$ (2.4) which must be in $\mathcal{W}$ because of closure of scalar multiplication. Thus, $\mathcal{W}$ contains $\mathbf{0}$. Moreover, $(-1)\mathbf{W} = -\mathbf{W}$ (2.7), which is in $\mathcal{W}$ for the same reason. Hence, all the vector addition axioms hold in $\mathcal{W}$. Further, $\mathcal{W}$ is closed under scalar multiplication, and all other scalar-multiplication axioms are true in $\mathcal{W}$ since they hold in all of $\mathcal{V}$. Hence, $\mathcal{W}$ is a subspace. ∎

Examples. Consider the subset $\mathcal{W}$ of $\mathbb{R}_2$ of all vectors whose coordinates are both the same,
$$\mathcal{W} = \{(x, x) \mid x \text{ in } \mathbb{R}\}.$$
It is closed under addition $((x, x) + (y, y) = (x + y, x + y))$ and scalar multiplication $(a(x, x) = (ax, ax))$, so $\mathcal{W}$ is a subspace of $\mathbb{R}_2$. In like manner, the subset $\mathcal{X} = \{(x, 0) \mid x \text{ in } \mathbb{R}\}$ is easily seen to be a subspace. However, the subset
$$\mathcal{Y} = \{(x, 1) \mid x \text{ in } \mathbb{R}\}$$
is not a subspace, since it is not closed under either scalar multiplication or vector addition (for example, $2(0, 1) = (0, 2)$, which is not in $\mathcal{Y}$).

The Subspace Theorem has a nice geometrical interpretation. Let us think of $\mathcal{V}$ as a collection of vectors with tails at the origin, and $\mathcal{W}$ a nonempty subset of $\mathcal{V}$. Then $\mathcal{W}$ is a subspace iff

i) Whenever two adjacent sides of a parallelogram are in $\mathcal{W}$, then the diagonal is in $\mathcal{W}$.

ii) For any vector $\mathbf{W} \neq \mathbf{0}$ in $\mathcal{W}$, all vectors on the line (through the origin) containing $\mathbf{W}$ lie in $\mathcal{W}$. For this last, we are interpreting the line through the origin containing $\mathbf{W}$ ($\neq \mathbf{0}$) as the set $\{c\mathbf{W} \mid c \text{ in } \mathbb{R}\}$. This is discussed further in Chapter 7.

We can use this geometrical approach to classify the subspaces of $\mathbb{R}_2$ (considered as the set of all vectors in the plane with tails at the origin). A moment's reflection should convince you that the only subsets of $\mathbb{R}_2$ satisfying (i) and (ii) are

the zero vector, any line through the origin,† $\mathbb{R}_2$ itself

In like manner, considering $\mathbb{R}_3$ as all vectors in 3-space with tails at the origin, we see that the only possible subspaces of $\mathbb{R}_3$ are

the zero vector, any line through the origin, any plane containing the origin, $\mathbb{R}_3$ itself.

We now work towards a means of describing subspaces of a vector space. We first introduce the basic notion of "linear combination".

Given a finite set $\mathbf{V}_1, \ldots, \mathbf{V}_n$ of vectors in a vector space $\mathcal{V}$ over $\mathbb{R}$, a **linear combination** of $\mathbf{V}_1, \ldots, \mathbf{V}_n$ is any expression of the form

$$c_1\mathbf{V}_1 + c_2\mathbf{V}_2 + \cdots + c_n\mathbf{V}_n, \qquad c_1, \ldots, c_n \text{ in } \mathbb{R}.$$

Thus, examples of linear combinations of $\mathbf{V}_1$, $\mathbf{V}_2$, $\mathbf{V}_3$ are

$$-\mathbf{V}_1 + 2\mathbf{V}_2 + \pi\mathbf{V}_3, \qquad \mathbf{V}_3 \; (= 0\mathbf{V}_1 + 0\mathbf{V}_2 + 1\mathbf{V}_3), \qquad \text{etc.}$$

By a **linear combination** of an infinite set of vectors we shall mean any linear combination of a finite subset of this set. Thus, we are specifically excluding infinite sums of vectors.

By the **span** of a set S of vectors in $\mathcal{V}$ we shall mean the collection of all linear combinations of elements of S. For notation we write "span S". Thus, if $S = \{\mathbf{V}_1, \ldots, \mathbf{V}_n\}$ is a finite set, then

$$\text{span } S = \{c_1\mathbf{V}_1 + \cdots + c_n\mathbf{V}_n \mid c_1, \ldots, c_n \text{ in } \mathbb{R}\}.$$

Example 1. Consider vectors $\mathbf{X} = (1, 2, 0)$ and $\mathbf{Y} = (0, 2, 2)$ in $\mathbb{R}_3$. For $S = \{\mathbf{X}, \mathbf{Y}\}$,

$$\text{span } S = \{a\mathbf{X} + b\mathbf{Y} = a(1, 2, 0) + b(0, 2, 2) = (a, 2a + 2b, 2b) \mid a, b \text{ in } \mathbb{R}\}.$$

It is easy to check that also

$$\text{span } S = \text{span } \{(1, 3, 1), (1, 4, 2), (1, 5, 3)\}.$$

In general, there is nothing sacrosanct about the vectors used in describing the span (Fig. 2).

---

† More precisely, all vectors with tails at the origin and heads on such a line.

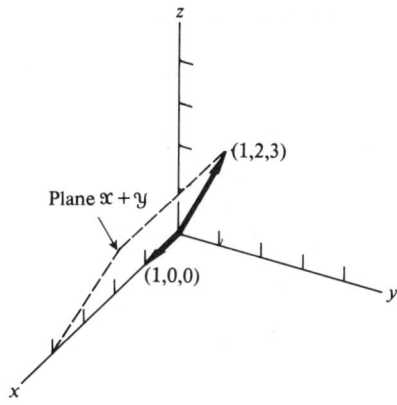

Fig. 2

The importance of the notion of the span of a set of vectors comes from the following:

3.2 If S is any nonempty† subset of a vector space $\mathcal{V}$, then span S is a subspace of $\mathcal{V}$.

Proof. Let us assume first that $S = \{V_1, \ldots, V_n\}$ is a finite set. By the subspace theorem, it suffices to show that any scalar multiple of a linear combination is a linear combination, as is the sum of two linear combinations. These follow at once from

$$c(a_1 V_1 + \cdots + a_n V_n) = c a_1 V_1 + \cdots + c a_n V_n$$

and

$$(a_1 V_1 + \cdots + a_n V_n) + (b_1 V_1 + \cdots + b_n V_n)$$
$$= (a_1 + b_1) V_1 + \cdots + (a_n + b_n) V_n$$

(see 2.8 and 2.9).

If S is an infinite set, the only difference is that the sum of two linear combinations of its elements may be of the form

$$(a_1 V_1 + \cdots + a_m V_m) + (b_1 W_1 + \cdots + b_n W_n)$$

which still works out to being a linear combination of elements of S (Exercise 9). ∎

Span S is often called the **subspace spanned** by S or the **subspace generated** by S. Note that span S contains the original set of vector S.

---

† If the subset S is empty (i.e., it contains no elements) then it is convenient to *define* span S to be the zero subspace. We shall see why in later footnotes. This is a mild mathematical sophistry and may be safely ignored by those uninterested in such simple entertainments.

The notion of span often allows us to describe subspaces in a very convenient fashion. In particular, it often happens that we can describe a subspace (which in general will contain an infinite number of vectors) in terms of a finite set of its elements.

Example 2. The subspace in the previous example can be written as span

$$\{(1, 2, 0), (0, 2, 2)\} = \{(a, 2a + 2b, 2b) \mid a, b \text{ in } \mathbb{R}\}.$$

This subspace can thus be described in terms of two vectors, even though it contains an infinite number of vectors.

EXERCISES

1. Go through the examples and exercises in Section 1 and list all vector spaces which are subspaces of other vector spaces there.
2. Prove that the subset $\mathcal{W} = \{\mathbf{0}\}$ consisting of just the zero vector forms a subspace of any vector space.
3. Let $\mathbb{R}_n[x]$ denote the set of all real polynomials of degree $< n$:

   $$\mathbb{R}_n[x] = \{a_0 + a_1 x + \cdots + a_m x^m \text{ in } \mathbb{R}[x] \mid m < n\}.$$

   Show that $\mathbb{R}_n[x]$ is a subspace of $\mathbb{R}[x]$.
4. For what subspaces of vectors in the euclidean plane are all vectors of length less than 100? Prove your assertion.
5. Considering the field $\mathbb{R}$ as a vector space over itself as in Exercise 4 of Section 1, find all subspaces of $\mathbb{R}$ over $\mathbb{R}$.
6. Let $\mathcal{V}$ be a vector space over $\mathbb{R}$ and choose any $\mathbf{W}$ in $\mathcal{V}$. Let $\mathcal{W}$ be the set consisting of $\mathbf{W}$ alone and define an addition in $\mathcal{W}$ by $\mathbf{W} + \mathbf{W} = \mathbf{W}$ and a scalar multiplication by $c \cdot \mathbf{W} = \mathbf{W}$ for all c in $\mathbb{R}$.
   a) Verify that $\mathcal{W}$ is a commutative group under the operation $+$ and that $\mathcal{W}$ is a vector space over $\mathbb{R}$ under the scalar multiplication.
   b) When is $\mathcal{W}$ a subspace of $\mathcal{V}$? Explain.
7. Let $\mathcal{V}$ be a vector space over $\mathbb{R}$, and $\mathcal{W}$ be a subspace of $\mathcal{V}$. If $\mathcal{X}$ is a subspace of the vector space $\mathcal{W}$ over $\mathbb{R}$ prove that $\mathcal{X}$ is also a subspace of $\mathcal{V}$.
8. Let $\mathcal{W} = \{(x, y, z) \text{ in } \mathbb{R}_3 \mid x = y\}$.
   a) Show that $\mathcal{W}$ is a subspace of $\mathbb{R}_3$.
   b) Identify $\mathcal{W}$ geometrically.
   c) Show that $\mathcal{X} = \{(x, y, z) \text{ in } \mathbb{R}_3 \mid x = y = z\}$ is a subspace of $\mathcal{W}$ and identify geometrically.
9. Prove 3.2 when S is infinite.
10. Let us use $\{a_n\}$ to denote an infinite sequence of reals:

    $$\{a_n\} = \{a_1, a_2, \ldots, a_n, \ldots\}.$$

Define $\{a_n\} + \{b_n\} = \{a_n + b_n\}$ and $c\{a_n\} = \{ca_n\}$. Prove:
a) With these definitions the set of all sequences forms a vector space S over $\mathbb{R}$.
b) The set of all *finite* sequences $\{a_1, a_2, \ldots, a_k, 0, \ldots, 0, \ldots\}$ is a subspace S' of S.
c) The set of all *convergent* sequences forms a subspace C of S.
d) The set of all sequences with limit 0 forms a subspace C_0 of C.
e) The set of all sequences $\{a_n\}$ for which the infinite series $\sum_1^\infty a_n$ converges forms a subspace C_1 of S. In fact, C_1 is a subspace of C_0. Also S' is a subspace of C_1. [*Résumé*: $S' \subset C_1 \subset C_0 \subset C \subset S$ ($X \subset Y$ means X is contained in Y).]
f) Show that all inclusions are proper (i.e., no equalities).

Supplement to Section 3. The Arithmetic of Subspaces

Given a vector space $\mathcal{V}$ over the reals $\mathbb{R}$ and two subspaces $\mathcal{X}$ and $\mathcal{Y}$ of $\mathcal{V}$, we define the *sum of $\mathcal{X}$ and $\mathcal{Y}$*, $\mathcal{X} + \mathcal{Y}$, to be the collection of all elements of $\mathcal{V}$ which can be written in the form $\mathbf{X} + \mathbf{Y}$, where $\mathbf{X}$ is in $\mathcal{X}$ and $\mathbf{Y}$ is in $\mathcal{Y}$. In symbols,

$$\mathcal{X} + \mathcal{Y} = \{\mathbf{X} + \mathbf{Y} \mid \mathbf{X} \text{ in } \mathcal{X}, \mathbf{Y} \text{ in } \mathcal{Y}\}.$$

3.3 The sum of two subspaces is a subspace: for any subspaces $\mathcal{X}$ and $\mathcal{Y}$ of a vector space $\mathcal{V}$ over $\mathbb{R}$, $\mathcal{X} + \mathcal{Y}$ is a subspace of $\mathcal{V}$. It contains both $\mathcal{X}$ and $\mathcal{Y}$.

Proof. $\mathcal{X} + \mathcal{Y}$ is a subset of $\mathcal{V}$. It is closed under addition: for $\mathbf{X}_1 + \mathbf{Y}_1$ and $\mathbf{X}_2 + \mathbf{Y}_2$ in $\mathcal{X} + \mathcal{Y}$, $\mathbf{X}_i$ in $\mathcal{X}$, $\mathbf{Y}_i$ in $\mathcal{Y}$, we have

$$(\mathbf{X}_1 + \mathbf{Y}_1) + (\mathbf{X}_2 + \mathbf{Y}_2) = (\mathbf{X}_1 + \mathbf{X}_2) + (\mathbf{Y}_1 + \mathbf{Y}_2), \tag{1}$$

after several uses of associativity and commutativity. Since $\mathbf{X}_1 + \mathbf{X}_2$ is in $\mathcal{X}$ (it is a subspace) and $\mathbf{Y}_1 + \mathbf{Y}_2$ is in $\mathcal{Y}$ (ditto), their sum is in $\mathcal{X} + \mathcal{Y}$, which is therefore closed under addition.

To show closure under scalar multiplication, just notice that $a(\mathbf{X} + \mathbf{Y}) = a\mathbf{X} + a\mathbf{Y}$, and proceed as above. By 3.1 we have that $\mathcal{X} + \mathcal{Y}$ is a subspace.

Since $\mathcal{X} + \mathcal{Y}$ contains everything of the form $\mathbf{X} = \mathbf{X} + \mathbf{0}$, $\mathbf{X}$ in $\mathcal{X}$ (why is $\mathbf{0}$ in $\mathcal{Y}$?), then $\mathcal{X} + \mathcal{Y}$ contains $\mathcal{X}$. Similarly, it contains $\mathcal{Y}$. ∎

Example 3. In $\mathbb{R}_3$ let
$$\mathcal{X} = \{(a, 0, 0) \mid a \text{ in } \mathbb{R}\}$$
and
$$\mathcal{Y} = \{(b, 2b, 3b) \mid b \text{ in } \mathbb{R}\}.$$

These are easily seen to be subspaces. Their sum $\mathcal{X} + \mathcal{Y}$ consists of all vectors of the form

$$\mathbf{X} + \mathbf{Y} = (a + b, 2b, 3b),$$

where $\mathbf{X} = (a, 0, 0)$ is in $\mathcal{X}$ and $\mathbf{Y} = (b, 2b, 3b)$ is in $\mathcal{Y}$. Thus,

$$\mathcal{X} + \mathcal{Y} = \{(a + b, 2b, 3b) \mid a, b \text{ in } \mathbb{R}\}.$$

We can describe $\mathcal{X} + \mathcal{Y}$ in terms of its counterpart in geometrical vectors. $\mathcal{X}$ corresponds to the subspace of vectors which (with tails at the origin) have heads on the x-axis, and $\mathcal{Y}$ to the subspace of vectors with heads on the line through 0 and the point $(1, 2, 3)$, say. The set of all sums of such vectors will clearly consist of all vectors with heads on the plane containing these two lines. Consequently, $\mathfrak{X} + \mathfrak{Y}$ will be the coordinates of all points in this plane. Of course we could also describe $\mathfrak{X} + \mathfrak{Y}$ as span $\{(1, 0, 0) (1, 2, 3)\}$.

With $\mathfrak{X}$ and $\mathfrak{Y}$ continuing to be subspaces of a vector space $\mathfrak{V}$ over $\mathbb{R}$, we next define the **intersection of** $\mathfrak{X}$ **and** $\mathfrak{Y}$, $\mathfrak{X} \cap \mathfrak{Y}$, to be all elements common to both $\mathfrak{X}$ and $\mathfrak{Y}$. That is,

$$\mathfrak{X} \cap \mathfrak{Y} = \{Z \text{ in } \mathfrak{V} \mid Z \text{ in } \mathfrak{X} \text{ and } Z \text{ in } \mathfrak{Y}\}.$$

Note that since **0** is in any subspace, by (3.1), the intersection of any two subspaces always contains **0**, and is therefore never empty.

3.4 The intersection of two subspaces is a subspace: For any subspaces $\mathfrak{X}$ and $\mathfrak{Y}$ of a vector space $\mathfrak{V}$ over $\mathbb{R}$, $\mathfrak{X} \cap \mathfrak{Y}$ is a subspace of $\mathfrak{V}$. It is contained in both $\mathfrak{X}$ and $\mathfrak{Y}$.

Proof. If Z_1 and Z_2 are in $\mathfrak{X} \cap \mathfrak{Y}$, then Z_1 and Z_2 are in $\mathfrak{X}$, so $Z_1 + Z_2$ is in $\mathfrak{X}$. Similarly, it is in $\mathfrak{Y}$. Hence $Z_1 + Z_2$ is in $\mathfrak{X} \cap \mathfrak{Y}$. Closure under scalar multiplication is similarly proved. Thus, $\mathfrak{X} \cap \mathfrak{Y}$ is a subspace. It is clearly contained in both $\mathfrak{X}$ and $\mathfrak{Y}$. ∎

Direct Sums

The following special type of sum of subspaces is often useful, and we shall encounter it later.

We say that the vector space $\mathfrak{V}$ over $\mathbb{R}$ is the **direct sum** of subspaces $\mathfrak{X}$ and $\mathfrak{Y}$ if

i) $\mathfrak{V}$ is the sum of $\mathfrak{X}$ and $\mathfrak{Y}$; that is, $\mathfrak{V} = \mathfrak{X} + \mathfrak{Y}$, and

ii) the only vector in both $\mathfrak{X}$ and $\mathfrak{Y}$ is **0**

$$\mathfrak{X} \cap \mathfrak{Y} = \{\mathbf{0}\}.$$

We write $\mathfrak{V} = \mathfrak{X} \oplus \mathfrak{Y}$.

Example 4. Let $\mathfrak{Z} = \mathfrak{X} + \mathfrak{Y}$, with $\mathfrak{X} = $ span $\{(1, 0, 0)\}$ and $\mathfrak{Y} = $ span $\{(1, 2, 3)\}$, as in Example 3. if Z is in $\mathfrak{X} \cap \mathfrak{Y}$, then $Z = a(1, 0, 0) = b(1, 2, 3)$, whence it easily follows that $Z = \mathbf{0}$ (why?). Thus,

$$\mathfrak{X} \cap \mathfrak{Y} = \{\mathbf{0}\} \quad \text{and} \quad \mathfrak{Z} = \mathfrak{X} \oplus \mathfrak{Y}.$$

On the other hand, we also have

$$\mathfrak{Z} = \underbrace{\text{span }\{(3, 4, 6)\}}_{\mathfrak{X}'} + \underbrace{\text{span }\{(2, 4, 6), (1, -2, -3)\}}_{\mathfrak{Y}'}.$$

The vector $(3, 4, 6)$ is in both $\mathfrak{X}'$ (clearly) and $\mathfrak{Y}'$, for $(3, 4, 6) = \frac{5}{4}(2, 4, 6) + \frac{1}{2}(1, -2, -3)$. Hence, $\mathfrak{X}' \cap \mathfrak{Y}' \neq \{\mathbf{0}\}$, and $\mathfrak{Z}$ is *not* the direct sum of $\mathfrak{X}'$ and $\mathfrak{Y}'$ even though it is their (regular) sum.

The following theorem is a key to the importance of direct sums.

3.5 *Direct sums and uniqueness of expression.* $\mathcal{U}$ is the direct sum of subspaces $\mathcal{X}$ and $\mathcal{Y}$ iff each $\mathbf{V}$ in $\mathcal{U}$ can be written in one and only one way as

$$\mathbf{V} = \mathbf{X} + \mathbf{Y}, \tag{2}$$

with $\mathbf{X}$ in $\mathcal{X}$, $\mathbf{Y}$ in $\mathcal{Y}$.

Proof.
1. Suppose $\mathcal{U} = \mathcal{X} \oplus \mathcal{Y}$ and

$$\mathbf{V} = \mathbf{X}_1 + \mathbf{X}_2$$
$$\mathbf{V} = \mathbf{X}_2 + \mathbf{Y}_2$$

with $\mathbf{X}_i$ in $\mathcal{X}$, $\mathbf{Y}_i$ in $\mathcal{Y}$, are two expressions for some $\mathbf{V}$ in $\mathcal{U}$. We must show $\mathbf{X}_1 = \mathbf{X}_2$ and $\mathbf{Y}_1 = \mathbf{Y}_2$. But $\mathbf{X}_1 + \mathbf{Y}_1 = \mathbf{V} = \mathbf{X}_2 + \mathbf{Y}_2$ so that $\mathbf{X}_1 - \mathbf{X}_2 = \mathbf{Y}_2 - \mathbf{Y}_1$. Since $\mathbf{X}_1 - \mathbf{X}_2$ is in $\mathcal{X}$ (why?) and $\mathbf{Y}_2 - \mathbf{Y}_1$ is in $\mathcal{Y}$, $\mathbf{X}_1 - \mathbf{X}_2 = \mathbf{Y}_2 - \mathbf{Y}_1$ is in both $\mathcal{X}$ and in $\mathcal{Y}$. Since $\mathcal{X} \cap \mathcal{Y} = \{0\}$, we have $\mathbf{X}_1 - \mathbf{X}_2 = \mathbf{Y}_2 - \mathbf{Y}_1 = 0$, so that $\mathbf{X}_1 = \mathbf{X}_2$ and $\mathbf{Y}_1 = \mathbf{Y}_2$, as desired.

2. Conversely, suppose that each $\mathbf{V}$ has a unique expression of the form (2). Since there is at least one expression (2) for each $\mathbf{V}$, it follows that $\mathcal{U} = \mathcal{X} + \mathcal{Y}$. Moreover, if $\mathbf{Z}$ were in $\mathcal{X} \cap \mathcal{Y}$ then

$$\mathbf{Z} = 1\mathbf{Z} + \mathbf{0},$$

with $\mathbf{Z}$ in $\mathcal{X}$, $\mathbf{0}$ in $\mathcal{Y}$, and

$$\mathbf{Z} = \mathbf{0} + 1\mathbf{Z},$$

with $\mathbf{0}$ in $\mathcal{X}$, $\mathbf{Z}$ in $\mathcal{Y}$, are two expressions for $\mathbf{Z}$. Since these expressions must be the same, we can equate components and obtain $\mathbf{Z} = \mathbf{0}$. Thus

$$\mathcal{X} \cap \mathcal{Y} = \{0\} \quad \text{and} \quad \mathcal{Z} = \mathcal{X} \oplus \mathcal{Y}. \quad \blacksquare$$

EXERCISES

In the following we will be working, as usual, in a vector space $\mathcal{U}$ over $\mathbb{R}$.

1. a) If $S = \{\mathbf{V}_1, \mathbf{V}_2\}$ prove that span $S = $ span $\{\mathbf{V}_1\} + $ span $\{\mathbf{V}_2\}$.
   b) State and prove a similar result for $S = \{\mathbf{V}_1, \ldots, \mathbf{V}_n\}$.

2. a) For subspaces $\mathcal{X}$ and $\mathcal{Y}$, prove that $\mathcal{X} + \mathcal{Y}$ is the "smallest" subspace containing both $\mathcal{X}$ and $\mathcal{Y}$, in that if $\mathcal{Z}$ is any other subspace containing $\mathcal{X}$ and $\mathcal{Y}$, then $\mathcal{X} + \mathcal{Y}$ is contained in $\mathcal{Z}$.
   b) Prove that $\mathcal{X} \cap \mathcal{Y}$ is the "largest" subspace contained in both $\mathcal{X}$ and $\mathcal{Y}$ in that if $\mathcal{Z}$ is any other subspace contained in both $\mathcal{X}$ and $\mathcal{Y}$ then $\mathcal{Z}$ is contained in $\mathcal{X} \cap \mathcal{Y}$.
   c) Show that span S is the smallest subspace containing all the elements of S.

3. a) Show that if $\mathcal{U} = \mathbb{R}_2$ and if

   $$\mathcal{X} = \text{span}\{(1, 0)\}, \quad \mathcal{Y} = \text{span}\{(0, 1)\}, \quad \mathcal{Y}' = \text{span}(1, 1),$$

   then $\mathcal{X} \oplus \mathcal{Y} = \mathcal{X} \oplus \mathcal{Y}'$, but $\mathcal{Y} \neq \mathcal{Y}'$.
   b) Draw a picture in the euclidean plane illustrating this.

4. The **union** of two sets S_1 and S_2, written $S_1 \cup S_2$, is the set consisting of all elements which are in *either* S_1 or S_2 (or both):

$$S_1 \cup S_2 = \{s \mid s \text{ is in } S_1 \text{ or } s \text{ is in } S_2\}.$$

Prove or disprove: The union of two subspaces is a subspace. Interpret geometrically in the euclidean plane.

5. Write $\mathbb{R}_3$ as a direct sum of two subspaces in four different ways.

Honors Project

1. Show that all theorems in this section hold for vector spaces over (a) the complex numbers $\mathbb{C}$, (b) any field.

2. Same for the supplement.

3. Are there any difficulties with any of the examples or exercises?

4. BASES

We have thus far defined "vector space", investigated some simple properties of vector spaces, devised a convenient check as to when a subset of a vector space is itself a vector space, and discussed the subspace generated by a set of vectors.

Given such a far-reaching notion as "vector space", the mathematician often attempts to characterize all creatures satisfying his definition in terms of a few elementary building blocks. Such a classification of vector spaces will be our concern in this section and the next. In this section we shall distinguish between those vector spaces which have a finite spanning set and those which do not. We shall investigate properties of minimal spanning sets.

What is the distinguishing feature of the plane (as a vector space) as opposed to three-space, function spaces, and other vector spaces? One basic difference is that in the plane there exists a set of two vectors, neither of which can be written in terms of the other, and whose span is the entire plane.

In fact, any two nonzero vectors which do not lie on the same line have this property. Given two vectors V_1 and V_2 in the plane not lying on the same line through the origin, there is a well-known basic construction for writing any other vector W as a linear combination of V_1 and V_2. Simply draw a line through the tip of W parallel to V_1 until it intersects the line of V_2. This will be at a point $a_2 V_2$ which is a multiple of V_2. Similarly, the line through W and parallel to V_2 will intersect V_1 at a point $a_1 V_1$. Then $W = a_1 V_1 + a_2 V_2$.

Note that this construction makes use of properties such as "parallelism" which we have not officially considered. Therefore this is not a valid proof in our current linear-algebra context.

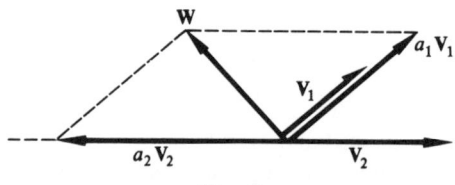

Fig. 3

In our new terminology, there is a set of two vectors in the plane, with neither a linear combination of the other, whose span is the entire vector space.

In like manner, a distinguishing feature of three-space is that there exists a set of three vectors, none expressible in terms of the others, and such that every other vector can be written in terms of these vectors. That is, there is a set of three vectors, none a linear combination of the others, and they span the entire space.

Similarly, in $\mathbb{R}_n$ there exists a set of n vectors, none of which can be written in terms of the others, and every other vector can be expressed as a sum of these. (For example, the set $e_1 = (1, 0, \ldots, 0)$, $e_2 = (0, 1, 0, \ldots, 0), \ldots, e_n = (0, \ldots, 0, 1)$ will do; see Exercise 3.) Thus, there is a set of n vectors, none a linear combination of the others, which spans the entire space.

We shall say that an ordered set of vectors $V_1, \ldots, V_n$ in a vector space $\mathcal{V}$ is a **basis** for $\mathcal{V}$ over $\mathbb{R}$ if

i) no one of the vectors $V_1, \ldots, V_n$ can be written as a linear combination of the other V_i; and

ii) span $\{V_1, \ldots, V_n\} = \mathcal{V}$.

Thus, $V_1, \ldots, V_n$ is a basis if

i) there is *no* expression of the form

$$V_j = c_1 V_1 + \cdots + c_{i-1} V_{j-1} + c_{j+1} V_{j+1} + \cdots + c_n V_n,$$

for any $j = 1, \ldots, n$ and $c_1, \ldots, c_n$ in $\mathbb{R}$.

ii) there *is* an expression of the form

$$V = c_1 V_1 + \cdots + c_n V_n$$

for any V in $\mathcal{V}$.

Example. The set $\{(2, 0), (1, -3)\}$ forms a basis for $\mathbb{R}_2$, for it is clear that neither vector can be written in terms of the other (why?), and any $V = (a, b)$ can be written

$$V = \left(\frac{a}{2} + \frac{b}{6}\right)(2, 0) + \left(-\frac{b}{3}\right)(1, -3)$$

The set $\{(2, 0), (1, -3), (-1, 6)\}$ cannot form a basis for $\mathbb{R}_2$ since, for example,

$$(-1, 6) = \tfrac{1}{2}(2, 0) + (-2)(1, -3),$$

violating (i).

The set $\{(1, 2, 3), (2, 2, 1)\}$ does not form a basis for $\mathbb{R}_3$ since its span is not all of $\mathbb{R}_3$. For example, $(1, 0, 0)$ cannot be written as a linear combination of these vectors (try it and see).

One can also speak of infinite bases. An infinite set S is a **basis** for $\mathcal{V}$ over $\mathbb{R}$ if

i) no element of S is a linear combination of other elements of S, and

ii) the span of S is $\mathcal{V}$.

For example, one basis for the vector space of polynomials $\mathbb{R}[x]$ is the set S of powers of x: $S = \{1, x, x^2, \ldots\}$.

One important property of a basis is the following:

4.1 A basis is a minimal spanning set. That is, if S is a set which spans the vector space $\mathcal{V}$ over $\mathbb{R}$ and if S is not a basis, then some element of S can be deleted and the resulting set still spans $\mathcal{V}$.

Proof. For convenience let us suppose $S = \{\mathbf{V}_1, \ldots, \mathbf{V}_n\}$ is finite. Since S spans $\mathcal{V}$ and S is not a basis, condition (i) must be violated. For simplicity of writing, suppose that $\mathbf{V}_1$ can be written in terms of $\mathbf{V}_2, \ldots, \mathbf{V}_n$:

$$\mathbf{V}_1 = c_2 \mathbf{V}_2 + \cdots + c_n \mathbf{V}_n$$

for some c's in $\mathbb{R}$. (Guess which element of S we shall delete.) Now any $\mathbf{V}$ in $\mathcal{V}$ can be written as

$$\mathbf{V} = a_1 \mathbf{V}_1 + a_2 \mathbf{V}_2 + \cdots + a_n \mathbf{V}_n$$

for some a's in $\mathbb{R}$ (from above) and thus

$$\mathbf{V} = a_1(c_2 \mathbf{V}_2 + \cdots + c_n \mathbf{V}_n) + a_2 \mathbf{V}_2 + \cdots + a_n \mathbf{V}_n$$
$$= (a_1 c_2 + a_2)\mathbf{V}_2 + \cdots + (a_1 c_n + a_n)\mathbf{V}_n.$$

Thus, $\mathbf{V}_2, \ldots, \mathbf{V}_n$ also spans $\mathcal{V}$, so $\mathbf{V}_1$ can be deleted. ∎

There is a more symmetrical and more convenient way of expressing the fact that "no vector in the set is a linear combination of other vectors in the set". This comes from the following fundamental notion.

A finite set $\{\mathbf{V}_1, \ldots, \mathbf{V}_n\}$ is **linearly independent** if the only linear combination of its elements yielding the zero vector comes from taking all the coefficients to be zero:

$$a_1 \mathbf{V}_1 + \cdots + a_n \mathbf{V}_n = \mathbf{0} \quad \text{only for} \quad a_1 = 0, \ldots, a_n = 0.$$

Examples. The set $\{(2, 0), (1, -3)\}$ is linearly independent since

$$a(2, 0) + b(1, -3) = 0 \quad \text{iff} \quad 2a + b = 0 \quad \text{and} \quad 0a - 3b = 0,$$

so that $b = 0$ and consequently $a = 0$.

The set $\{(2, 0), (1, -3), (-1, 6)\}$ is *not* linearly independent since

$$\tfrac{1}{2}(2, 0) + (-2)(1, -3) + (-1)(-1, 6) = 0.$$

A simple application of the definition shows (Exercise 8d) that

4.2 Subsets of linearly independent sets are linearly independent.

We say that an infinite set is **linearly independent** if every finite subset of it is linearly independent. (Once again we avoid infinite sums of vectors.)

We next tie in the notion of linear independence with our previous considerations.

4.3 If a set S contains more than one element then it is linearly independent iff none of its elements can be written as a linear combination of other elements of the set.

Proof. Again we restrict ourselves to finite sets $S = \{\mathbf{V}_1, \ldots, \mathbf{V}_n\}$ for convenience.
1. Suppose that S is linearly independent and that one of its elements, say $\mathbf{V}_1$, can be written in terms of the other elements,

$$\mathbf{V}_1 = a_2\mathbf{V}_2 + \cdots + a_n\mathbf{V}_n. \tag{1}$$

Then, adding $-\mathbf{V}_1$ to both sides, we have

$$\mathbf{0} = -\mathbf{V}_1 + a_2\mathbf{V}_2 + \cdots + a_n\mathbf{V}_n.$$

Since $-\mathbf{V}_1 = (-1)\mathbf{V}_1$, this means we have a linear combination equal to zero with not all zero coefficients, contradicting the linear independence. In sum, if S is linearly independent, it is impossible for a relation such as (1) to occur.
2. Suppose that no element of S can be written as a linear combination of other elements of S. If S were *not* linearly independent, then we would have some expression

$$a_1\mathbf{V}_1 + a_2\mathbf{V}_2 + \cdots + a_n\mathbf{V}_n = \mathbf{0}$$

with not all a's $= 0$. If $a_1 \neq 0$, say, then we could solve for $\mathbf{V}_1$,

$$\mathbf{V}_1 = (-a_1^{-1}a_2)\mathbf{V}_2 + \cdots + (-a_1^{-1}a_n\mathbf{V}_n).$$

Hence, it is impossible to have any of the a's nonzero, and S is linearly independent. ∎

When S contains a single vector, $S = \{\mathbf{V}\}$, none of its elements can be written as a linear combination of others of its elements.† However, there is an exceptional case when S can fail to be linearly independent: if $S = \{\mathbf{0}\}$, then it is not linearly independent since, for example, $1\mathbf{0} = \mathbf{0}$ is a linear combination of elements of S yielding $\mathbf{0}$ with not all zero coefficients

A set which is not linearly independent is called, *mirabile dictu*, **linearly dependent**. Since one notion is the negation of the other, it is unnecessary to use both; we shall restrict ourselves to using the word "independent". At least note that if a set is linearly dependent (and contains more than one element), then by 4.3, some element can be written as a linear combination of other elements. Thus, some element depends linearly on other elements, justifying the terminology.

Having obtained another convenient method for saying "no element is a linear combination of other elements", we return to our main discussion of bases.

4.4 A set S is a basis for a vector space $\mathcal{V}$ over $\mathbb{R}$ iff (i) it spans $\mathcal{V}$, and (ii) it is linearly independent.

---

† There is one way to finesse the situation and obtain 4.3 without any special case for $n = 1$. Just *define* $\mathbf{0}$ to be the span of the empty set (see Exercise 8c). This is not very important and can be safely ignored.

Proof. This follows immediately from 4.3 and the definition of basis, except for the case where $n = 1$, which we leave for you (Exercise 9a). ∎

Our next result shows that a linearly independent set can always serve as the starting point for a basis.

4.5 If $\mathcal{V}$ has a finite basis,† then any linearly independent set can be extended to a basis. That is, if S is a linearly independent set of vectors in a vector space $\mathcal{V}$ over $\mathbb{R}$, then there exists vectors which, together with the vectors in S form a basis for $\mathcal{V}$ over $\mathbb{R}$.

Proof. Suppose that $\mathcal{V}$ has the finite basis $\{\mathbf{V}_1, \ldots, \mathbf{V}_n\}$. For convenience (see Exercise 9b) also assume that the given set $S = \{\mathbf{W}_1, \ldots, \mathbf{W}_m\}$ is finite.

If S is a basis, we have nothing to prove. If it is not, we know by 4.4 that S must fail to span $\mathcal{V}$. If each basis element were in span S, then span $S = \mathcal{V}$ (prove₁), so we know this cannot be the case. Hence, there must be some one of $\mathbf{V}_1, \ldots, \mathbf{V}_n$, say $\mathbf{V}_1$, which is not in span S. *Claim:* $\{\mathbf{W}_1, \ldots, \mathbf{W}_m, \mathbf{V}_1\}$ is linearly independent. If not, there is some linear combination

$$a_1\mathbf{W}_1 + \cdots + a_m\mathbf{W}_m + b\mathbf{V}_1 = \mathbf{0}, \tag{2}$$

with not all coefficients zero. Now if $b \neq 0$, then we could solve for $\mathbf{V}_1$ as a linear combination of $\mathbf{W}_1, \ldots, \mathbf{W}_m$ and $\mathbf{V}_1$ would be in span S. Since this is contrary to assumption, we must have $b = 0$. But then (2) gives a linear combination of the $\mathbf{W}$'s with not all coefficients zero, contradicting linear independence. Since assuming that $\mathbf{W}_1, \ldots, \mathbf{W}_m, \mathbf{V}_1$ is *not* linearly independent leads to a contradiction, it must be linearly independent.

Let $S_1 = \{\mathbf{W}_1, \ldots, \mathbf{W}_m, \mathbf{V}_1\}$, which we now know to be linearly independent. If it is a basis, then it is our desired extension of S. If it is not, apply the argument of the previous paragraph to add on some other $\mathbf{V}_i$, say $\mathbf{V}_2$, to obtain a linearly independent set $S_2 = \{\mathbf{W}_1, \ldots, \mathbf{W}_m, \mathbf{V}_1, \mathbf{V}_2\}$. Continuing in this fashion one either obtains a linearly independent set which spans $\mathcal{V}$ (that is, a basis), or one can add on another basis vector without destroying linear independence. Since there are only a finite number of basis elements, this process must stop. At the very worst, one would obtain a linearly independent set

$$S_n = \{\mathbf{W}_1, \ldots, \mathbf{W}_m, \mathbf{V}_1, \ldots, \mathbf{V}_n\}$$

which must surely span $\mathcal{V}$ (prove₂) and hence be a basis. ∎

Thus, we can regard linearly independent sets as building blocks for bases. The next result is fundamental.

---

† This theorem is true even under the assumption that $\mathcal{V}$ doesn't have a finite basis, but the proof is immensely more difficult (see [1], for example). We will have no need of the infinite case.

4.6 If $\mathcal{U}$ has a finite basis $\{V_1, \ldots, V_n\}$ containing n elements, then any linearly independent set $\{W_1, \ldots, W_n\}$ in $\mathcal{U}$ containing n elements is also a basis for $\mathcal{U}$.

Proof of a special case. We shall attend to a simple but illuminating special case here, and give a complete proof in the supplement to this section.

For now, assume that the basis contains only two elements: let it be $\{V_1, V_2\}$. Given a linearly independent set $\{W_1, W_2\}$,

$$W_1 = a_1 V_1 + a_2 V_2$$

for some a's in $\mathbb{R}$, and since $W_1 \neq 0$ (Exercise 5), not both a's are 0. Suppose $a_1 \neq 0$. Then we can solve for V_1 in terms of W_1 and V_2. It is then easy to see that $\{W_1, V_2\}$ spans $\mathcal{U}$ (why$_1$?). Hence

$$W_2 = c_1 W + c_2 V_2,$$

and not both c_1 and c_2 are 0 (why$_2$?). If $c_2 = 0$ then $\{W_1, W_2\}$ is not linearly independent (why$_3$?). Hence $c_2 \neq 0$ and we can solve for V_2 in terms of W_1 and W_2. In fact $\{W_1, W_2\}$ spans $\mathcal{U}$ (why$_4$?). Hence $\{W_1, W_2\}$ is a basis for $\mathcal{U}$. ∎

The following easy but valuable result follows almost immediately from 4.6.

4.7 The number of elements in a basis is an upper bound for the number of elements in any linearly independent set. That is, if $\mathcal{U}$ has a basis containing n elements then any linearly independent set contains at most n elements.

Proof. Given any linearly independent set containing more than n elements, choose a subset containing $(n + 1)$ elements. By 4.2 we need only show that it is impossible to have such a linearly independent set containing $(n + 1)$ elements. Suppose $W_1, \ldots, W_{n+1}$ linearly independent; by 4.2, $W_1, \ldots, W_n$ would be linearly independent also and by 4.6 this would be a basis. Then $\{W_1, \ldots, W_{n+1}\}$ could not be linearly independent (why?). ∎

There is another very important consequence of 4.6:

4.8 The Basis Theorem. If $\mathcal{U}$ has one basis containing n elements, then any basis for $\mathcal{U}$ contains n elements.†

Proof. Let $\{V_1, \ldots, V_m\}$ be one basis for $\mathcal{U}$ and let $\{W_1, \ldots, W_n\}$ be another. If $m < n$, then 4.6 would attest that $\{W_1, \ldots, W_m\}$ is a basis. Then W_{m+1} could be written in terms of $W_1, \ldots, W_m$, and $\{W_1, \ldots, W_n\}$ would not be a basis. This contradiction shows that it is impossible to have $m < n$. Reversing roles shows the same for $n < m$. Hence, $m = n$. ∎

Thus, if $\mathcal{U}$ has one finite basis $\{V_1, \ldots, V_n\}$, the number n is quite intimately associated with $\mathcal{U}$. As we shall see in the next section, this number essentially completely describes $\mathcal{U}$.

---

† There is an infinite-dimensional extension of 4.8: any two bases for a vector space have the same cardinality. See [1] if you're both interested and stalwart.

If $\mathcal{V}$ has a finite basis $\{\mathbf{V}_1, \ldots, \mathbf{V}_n\}$ over $\mathbb{R}$, we call n the **dimension** of $\mathcal{V}$ over $\mathbb{R}$. For example the plane $\mathbb{R}_2$ has dimension 2. (It has a basis $\{(1, 0), (0, 1)\}$, and another basis $\{(1, 2), (3, 4)\}$. There is no unique basis; only the number of elements in a basis is unique.) In like manner, three-space $\mathbb{R}_3$ has dimension 3 and $\mathbb{R}_n$ has dimension n (see Exercise 3).

We define† the dimension of the zero vector space $\mathcal{V} = \{\mathbf{0}\}$ to be zero.

If $\mathcal{V}$ has no finite basis, we say that it is **infinite-dimensional** over $\mathbb{R}$. For example, the set of functions defined on some open interval is infinite-dimensional (see Exercise 15), as is the vector space $\mathbb{R}[x]$ of polynomials. In fact, many very important spaces are infinite-dimensional. This should not be held against the space, although one usually attempts to reduce computational problems to finite-dimensional subspaces. It turns out that every infinite-dimensional space has a basis, but this is one of those infinite-dimensional problems which are quite beyond the scope of this book (see [1] again).

We now see that the dimension of a subspace is related to the dimension of the containing space in the obvious fashion:

4.9 Let $\mathcal{V}$ be an n-dimensional vector space.‡ Then any subspace of $\mathcal{V}$ has dimension at most n. The only n-dimensional subspace of $\mathcal{V}$ is $\mathcal{V}$ itself.

Proof. For the subspace $\mathcal{W} = \{\mathbf{0}\}$, there is nothing to prove. If $\mathcal{W} \neq \{\mathbf{0}\}$, then it contains at least one element $\mathbf{W} \neq \mathbf{0}$ and hence contains at least one linearly independent set, namely $\{\mathbf{W}\}$. Now consider all possible linearly independent sets $\{\mathbf{W}_1, \mathbf{W}_2, \ldots, \mathbf{W}_k\}$ of vectors in $\mathcal{W}$. Then $k \leq n$, by 4.7. Hence there must be a linearly independent set $\{\mathbf{W}_1, \ldots, \mathbf{W}_m\}$ with maximum m, where $m \leq n$. Then $\{\mathbf{W}_1, \mathbf{W}_2, \ldots, \mathbf{W}_m\}$ is a basis of $\mathcal{W}$. For, if $\mathbf{W}$ is any vector in $\mathcal{W}$ then $\mathbf{W}_1, \ldots, \mathbf{W}_m, \mathbf{W}$ are not linearly independent, so that there is an expression with some nonzero coefficients

$$c_1 \mathbf{W}_1 + \cdots + c_m \mathbf{W}_m + c \mathbf{W} = \mathbf{0}$$

with $c \neq 0$, for otherwise $\mathbf{W}_1, \ldots, \mathbf{W}_m$ would be linearly dependent. Hence $\mathbf{W}$ is a linear combination of $\mathbf{W}_1, \ldots, \mathbf{W}_m$. Therefore dimension $\mathcal{W} \leq n$. If dimension $\mathcal{W} = n$, let $\mathbf{W}_1, \ldots, \mathbf{W}_n$ be a basis of $\mathcal{W}$. Then, by 4.6, this is also a basis for $\mathcal{V}$. Thus $\mathcal{V} = \mathcal{W}$. ∎

The following theorem gives us our final result of this section.

---

† Alternatively, note that the empty set is linearly independent: No linear combination of its element with nonzero coefficients yields zero, since it has no elements. On the other hand, by the convention of the footnote on p. 59, we are taking $\mathbf{0}$ to be the span of the empty set. Hence, $\mathcal{V} = \{\mathbf{0}\}$ has the empty set as its basis. Since the empty set contains zero elements, the dimension of $\mathcal{V}$ is thus zero. This approach has the advantage that it provides every vector space with a basis.

‡ We shall always understand n to be finite, so that "n-dimensional" implies finite-dimensional.

4.10 If $\mathcal{V}$ is n-dimensional and $\{\mathbf{W}_1, \ldots, \mathbf{W}_n\}$ is a subset of $\mathcal{V}$ containing n elements and spanning $\mathcal{V}$, then $\{\mathbf{W}_1, \ldots, \mathbf{W}_n\}$ is a basis for $\mathcal{V}$.

Proof. If $\mathcal{V} = \{\mathbf{0}\}$, then the assertion is either easy or irrelevant, depending upon whether you have or have not been following our footnotes; so assume $\mathcal{V} \neq \{\mathbf{0}\}$.

Proceeding by contradiction, suppose that $\mathbf{W}_1, \ldots, \mathbf{W}_n$ is *not* linearly independent. Then, by 4.1, we can delete some $\mathbf{W}_i$ and obtain a set of $(n-1)$ elements which also spans $\mathcal{V}$. If this new set is not linearly independent, we can likewise delete another $\mathbf{W}_j$ without affecting the span. Continuing in this fashion, we must obtain a set which we can call $\mathbf{W}_1, \ldots, \mathbf{W}_m$ for convenience, and which is both linearly independent and spans $\mathcal{V}$. (Otherwise, we could delete *all* the $\mathbf{W}$'s; at the last step this would mean that we had $\mathbf{W}_1$, say, spanning $\mathcal{V}$, and $\mathbf{W}_1$ was not linearly independent; hence $\mathbf{W}_1 = \mathbf{0}$ and $\mathcal{V} = \{\mathbf{0}\}$, contrary to current supposition.)

Thus, we have $\{\mathbf{W}_1, \ldots, \mathbf{W}_m\}$ which is linearly independent and spans $\mathcal{V}$. Thus, this is a basis for $\mathcal{V}$. But $m < n$ if we have made deletions. This contradicts 4.8. Hence deletions were impossible, the original $\{\mathbf{W}_1, \ldots, \mathbf{W}_n\}$ is linearly independent, and is hence a basis for $\mathcal{V}$. ∎

Combining this with 4.6, we see that

in an n-dimensional vector space, to show that a set of n vectors forms a basis, it suffices to show that

either the set is linearly independent *or* the set spans the space.

This can often be used with great convenience.

EXERCISES

1. a) Show that the set $\{(1, 0), (1, 1)\}$ is a basis for $\mathbb{R}_2$.
   b) Show that the vectors $\{(1, 0, 0, 0), (1, 1, 0, 0), (1, 1, 1, 0), (1, 1, 1, 1)\}$ form a basis for $\mathbb{R}_4$.
   c) What is a basis for $\mathbb{R}$ as a vector space over $\mathbb{R}$? What is the dimension of $\mathbb{R}$?

2. (a) Under what conditions can a set containing one vector fail to be linearly independent? (b) two vectors? (c) three vectors?

3. Prove that $\{\mathbf{e}_1 = (1, 0, \ldots, 0), \ldots, \mathbf{e}_n = (0, \ldots, 0, 1)\}$ is a basis for $\mathbb{R}_n$.

4. Either prove that the following sets of vectors are linearly independent, or write one of them as a linear combination of the others.
   a) $(1, 0, -2, 1), (2, 1, 3, -1), (4, 1, -1, 1)$.
   b) $(1, 0, -2, 1), (2, 1, 3, 5), (4, -1, -1, 1)$.
   c) What is the dimension of each subspace?

5. (a) Prove that no linearly independent set can contain zero. (b) prove that no basis can contain zero.

6. Prove that any linearly independent set is a basis for its span.
7. Prove that an n-dimensional vector space has subspaces of dimension $0, 1, \ldots, (n-1)$.
8. (a) Prove 4.1 if S is infinite. (b) Same for 4.3. (c) Prove 4.3 with $n = 1$, taking the span of the empty set to be $\mathbf{0}$, as in footnote 1. (d) Prove 4.5 for S infinite.
9. (a) Prove 4.4 with $n = 1$. (b) Prove 4.2. (c) Prove the indicated parts of 4.5. (d) Answer the questions in the proof of 4.6. (e) Answer the questions in the proof of 4.7.
10. a) Prove that a set S is linearly *dependent* iff it is possible to find distinct $\mathbf{V}_1, \ldots, \mathbf{V}_n$ in S and $c_1, \ldots, c_n$ all nonzero, for which
$$c_1 \mathbf{V}_1 + \cdots + c_n \mathbf{V}_n = \mathbf{0}.$$
b) Are subsets of linearly dependent sets linearly dependent? Prove or produce a counterexample.
c) Prove that, if S is a subspace, then it is linearly dependent.
11. Find a basis for $\mathbb{R}_n[x]$, the vector space of polynomials of degree $< n$.
12. If $\{\mathbf{V}_1, \mathbf{V}_2, \ldots, \mathbf{V}_n\}$ is a basis for $\mathcal{V}$, show that $\{\mathbf{V}_1 + c\mathbf{V}_2, \mathbf{V}_2, \ldots, \mathbf{V}_n\}$ is also a basis for $\mathcal{V}$ for any scalar c.
13. a) Prove that the powers of x are a basis for the vector space $\mathbb{R}[x]$ of polynomials.
b) Find another basis.
14. (For those who wish to consider polynomials as functions) Choose $x_1, \ldots, x_n$ to be distinct reals, $0 \le x_i \le 1$. For $i = 1, \ldots, n$ define
$$p_i(x) = a_i(x - x_1) \cdots (x - x_{i-1})(x - x_{i+1}) \cdots (x - x_n),$$
where a_i is chosen so that $p_i(x_i) = 1$ (note $p_i(x_j) = 0$ if $j \ne i$).
a) Show that $p_1(x), \ldots, p_n(x)$ are linearly independent. [*Suggestion:* If a linear combination is $\mathbf{0}$ (that is, the zero function), evaluate at $x_1, \ldots, x_n$.]
b) Prove that $\mathbb{R}[x]$ is infinite-dimensional.
c) Prove that the vector space $\mathcal{C}$ of continuous functions defined for $0 \le x \le 1$ is infinite-dimensional.
15. a) Let $\mathcal{F}$ be the vector space of all real-valued functions defined for $0 \le x \le 1$. For any a in that interval, define f_a to be the function
$$f_a(x) = \begin{cases} 0 & \text{for } x \ne a, \\ 1 & \text{for } x = a. \end{cases}$$
Prove that the set $\{f_a \mid 0 \le a \le 1\}$ is a linearly independent set of vectors in $\mathcal{F}$.
b) Prove that $\mathcal{F}$ is infinite-dimensional.
16. Prove that all of the sequence spaces in Section 3, Exercise 10, are infinite-dimensional.

The following five exercises make use of the supplement to Section 3.

17. a) If $\mathcal{X}$ and $\mathcal{Y}$ are subspaces of a vector space $\mathcal{V}$ over $\mathbb{R}$ and $\mathcal{V} = \mathcal{X} \oplus \mathcal{Y}$, prove that if α is a basis for $\mathcal{X}$ and β is a basis for $\mathcal{Y}$, then $\alpha \cup \beta$ is a basis for $\mathcal{V}$.
b) Is (a) true if merely $\mathcal{V} = \mathcal{X} + \mathcal{Y}$? Prove or disprove.
c) If $\alpha \cup \beta$ is a basis for $\mathcal{V} = \mathcal{X} \oplus \mathcal{Y}$ and α is a basis for $\mathcal{X}$, is β a basis for $\mathcal{Y}$? Prove or disprove.

18. Let $\mathfrak{X}$ and $\mathfrak{Y}$ be finite-dimensional subspaces of a vector space $\mathfrak{V}$. Show that $\mathfrak{X} + \mathfrak{Y}$ and $\mathfrak{X} \cap \mathfrak{Y}$ have finite dimensions.
19. Let $\mathfrak{X}$ and $\mathfrak{Y}$ be as in Exercise 18, and let their dimensions be n and m respectively. Let k be the dimension of $\mathfrak{X} \cap \mathfrak{Y}$. Prove that the dimension of $\mathfrak{X} + \mathfrak{Y}$ is $n + m - k$. (Note that Exercise 17(a) is a special case of this.) [Hint: Start with a basis of $\mathfrak{X} \cap \mathfrak{Y}$ and extend it to form a basis of $\mathfrak{X}$. Do the same for $\mathfrak{Y}$ and then proceed in a manner similar to that used in Exercise 17(a).]
20. Let $\mathfrak{X}$ and $\mathfrak{Y}$ be subspaces of dimensions 2 and 3 respectively, of $\mathbb{R}_4$
    a) Prove that the dimension of $\mathfrak{X} \cap \mathfrak{Y}$ is at least 1.
    b) What happens if the dimension of $\mathfrak{X} \cap \mathfrak{Y}$ is 2? Can it be 3?
21. Suppose that $\mathfrak{X}$ and $\mathfrak{Y}$ are subspaces and $\mathfrak{V} = \mathfrak{X} + \mathfrak{Y}$. Show that $\mathfrak{V}$ is their direct sum iff any set $\{X, Y\}$ with X in $\mathfrak{X}$, Y in $\mathfrak{Y}$, is either linearly independent or contains 0.

Supplement to Section 4

In this section we give a general proof of 4.6. Our proof will be based upon this stronger version of 4.1:

The Deletion Lemma. Suppose that $\{W_1, \ldots, W_k\}$ is linearly independent and $\{W_1, \ldots, W_k, V_1, \ldots, V_m\}$ is *not* linearly independent. Then one of the V_i can be deleted from the last set without affecting the span of $W_1, \ldots, W_k, V_1, \ldots, V_m$.

Proof. Since $\{W_1, \ldots, W_k, V_1, \ldots, V_m\}$ is not linearly independent, there are scalars, not all zero, for which

$$a_1 W_1 + \cdots + a_k W_k + b_1 V_1 + \cdots + b_m V_m = 0.$$

If all the b's are 0, then

$$a_1 W_1 + \cdots + a_k W_k = 0,$$

with not all a's $= 0$. This would contradict the independence of the W's. We conclude that some $b_i \neq 0$. One could then write the corresponding V_i in terms of the W's and the remaining V's. Hence, this V_i can be deleted and the span would not be changed. ∎

We now prove our main theorem.

4.6 If $\mathfrak{V}$ has a finite basis $\{V_1, \ldots, V_n\}$, then any linearly independent set $\{W_1, \ldots, W_n\}$ containing n elements is also a basis.

Proof. Since $S_0 = \{V_1, \ldots, V_n\}$ spans $\mathfrak{V}$, so does $\{W_1, V_1, \ldots, V_n\}$ Moreover, since W_1 is a linear combination of $V_1, \ldots, V_n$, the set is not linearly independent. By the Deletion Lemma, we can delete some one of $V_1, \ldots, V_n$ without affecting the span. Suppose we can delete V_1. (If not, renumber the V_i's so that this is the case; since we shall ultimately replace all V's by W's, the numbering is immaterial.) Then the span of $S_1 = \{W_1, V_2, \ldots, V_n\}$ is $\mathfrak{V}$, so that W_2 is a linear combination of $W_1, V_2, \ldots, V_n$. Hence, the set $\{W_1, W_2, V_2, \ldots, V_n\}$ is linearly dependent, and it surely spans $\mathcal{U}$. Apply the Deletion Lemma to see that another $\mathbf{V}_i$ can be deleted. Suppose it is $\mathbf{V}_2$. Then the span of $S_2 = \{\mathbf{W}_1, \mathbf{W}_2, \mathbf{V}_3, \ldots, \mathbf{V}_n\}$ is $\mathcal{U}$.

Continue in this fashion, adding on a $\mathbf{W}$ and deleting a $\mathbf{V}$, until one obtains a set $S_{n-1} = \{\mathbf{W}_1, \ldots, \mathbf{W}_{n-1}, \mathbf{V}_n\}$ which spans $\mathcal{U}$. Then $\{\mathbf{W}_1, \ldots, \mathbf{W}_n, \mathbf{V}_n\}$ also spans $\mathcal{U}$, and since $\mathbf{W}_n$ is in the span of the previous set, it is not linearly independent. Applying the Deletion Lemma one last time, we see that $\mathbf{V}_n$ can be deleted and the resulting set

$$S_n = \{\mathbf{W}_1, \ldots, \mathbf{W}_n\}$$

also spans $\mathcal{U}$. Since we know that S_n is linearly independent also, it is a basis. ∎

Honors Project

1. Show that all the theorems in the section hold for vector spaces over (a) the complex numbers, and (b) any field.
2. Same for the supplement.
3. Check the exercises and examples. What happens if we replace the real numbers in Exercise 4a by their corresponding elements from the field with two elements ($\bar{2} = \bar{1} + \bar{1} = \bar{0}$, $\bar{3} = \bar{1} + \bar{1} + \bar{1} = \bar{1}$, etc.)?

If you are working with different fields, it is important to make clear which one you mean when discussing linear independence, dimension and the like. The following exercises illustrate this.

4. Let $\mathbb{C}$ be the field of complex numbers considered as a vector space over $\mathbb{R}$.
    a) Find a basis for $\mathbb{C}$ over $\mathbb{R}$ and note the dimension.
    b) Now consider $\mathbb{C}$ as a vector space over $\mathbb{C}$, with vector addition its usual addition and the same for scalar multiplication. What happens to your linearly independent set in (a)? Demonstrate. What is the dimension of $\mathbb{C}$ over $\mathbb{C}$?
5. The following shows that the situation can be even worse. (a) Consider the real number $\mathbb{R}$ as a vector space over the rationals $\mathbb{Q}$. Let S be the set of logarithms of prime numbers $2, 3, 5, 7, \ldots$:

$$S = \{\log p \mid p \text{ prime}\}.$$

Prove that S is linearly independent over $\mathbb{Q}$. (b) Discuss the dimension of $\mathbb{R}$ over $\mathbb{Q}$ and of $\mathbb{R}$ over $\mathbb{R}$.

REFERENCE

The infinite-dimensional extensions mentioned in this section may be found in Chapter 9 of the following book; the discussion is far from easy. [1] Jacobson, Nathan, *Lectures in Abstract Algebra*, Vol. II. New York: Van Nostrand, 1953.

5. ISOMORPHIC VECTOR SPACES

In the last section we saw that vector spaces could be categorized as being either infinite-dimensional or finite-dimensional. In this section we see that a finite-dimensional vector space is essentially completely determined by its dimension.

In fact, an n-dimensional space is almost the same as $\mathbb{R}_n$. Thus, an abstract vector space can be very nicely classified according to its dimension. Later in this section we shall discuss why one wishes to consider abstract vector spaces at all, and justify the existence of this chapter.

The following two theorems should be compared with the discussion of geometrical vectors and n-tuples which was presented in Chapter 0, Section 2.

5.1 Let $\mathcal{U}$ be an n-dimensional vector space with basis $\{\mathbf{V}_1, \ldots, \mathbf{V}_n\}$. Then every $\mathbf{V}$ in $\mathcal{U}$ can be written in one and only one way as a linear combination of the basis elements

$$\mathbf{V} = a_1 \mathbf{V}_1 + \cdots + a_n \mathbf{V}_n,$$

with $a_1, \ldots, a_n$ depending uniquely on $\mathbf{V}$. Consequently, the correspondence

$$\mathbf{V}(= a_1 \mathbf{V}_1 + \cdots + a_n \mathbf{V}_n) \leftrightarrow (a_1, \ldots, a_n)$$

is a one-to-one correspondence between elements of $\mathcal{U}$ and the n-tuples of $\mathbb{R}_n$.

Proof. Since $\{\mathbf{V}_1, \ldots, \mathbf{V}_n\}$ is a basis, we know that each $\mathbf{V}$ has some expression of the form

$$\mathbf{V} = a_1 \mathbf{V}_1 + \cdots + a_n \mathbf{V}_n.$$

Suppose also that

$$\mathbf{V} = b_1 \mathbf{V}_1 + \cdots + b_n \mathbf{V}_n.$$

Subtracting the second expression for $\mathbf{V}$ from the first, we obtain

$$0 = \mathbf{V} - \mathbf{V} = (a_1 \mathbf{V}_1 + \cdots + a_n \mathbf{V}_n) - (b_1 \mathbf{V}_1 + \cdots + b_n \mathbf{V}_n)$$
$$= (a_1 - b_1) \mathbf{V}_1 + \cdots + (a_n - b_n) \mathbf{V}_n.$$

Since we are working with a basis, the coefficients in such an expression must all be zero,

$$(a_1 - b_1) = 0, \ldots, (a_n - b_n) = 0;$$

so $a_1 = b_1, \ldots, a_n = b_n$ and the two expressions for $\mathbf{V}$ are exactly the same.

We have seen that to each $\mathbf{V}$ in $\mathcal{U}$ there corresponds a unique n-tuple $(a_1, \ldots, a_n)$. Conversely, given an n-tuple $(b_1, \ldots, b_n)$, there corresponds a unique element of $\mathcal{U}$, namely $b_1 \mathbf{V}_1 + \cdots + b_n \mathbf{V}_n$. Thus, we have a one-to-one correspondence between elements of $\mathcal{U}$ and those of $\mathbb{R}_n$. ∎

Example 1. Let us contrast a basis $\{(1, 0), (1, 1)\}$ for $\mathbb{R}_2$ with the behavior of the set of elements

$$S = \{(1, 0), (1, 1), (-1, 0)\},$$

which is *not* a basis even though it spans $\mathbb{R}_2$. For example, we can write the vector $(2, 3)$ in terms of the basis as

$$(2, 3) = (-1)(1, 0) + 3(1, 1),$$

and this expression is unique; for if also
$$(2, 3) = a(1, 0) + b(1, 1), \quad a, b \text{ in } \mathbb{R},$$
then
$$(2, 3) = (a, 0) + (b, b) = (a + b, b),$$
so that $a + b = 2$, $b = 3$, whence $b = 3$, $a = -1$, and uniqueness is established.

On the other hand, using S we can write
$$(2, 3) = -1(1, 0) + 3(1, 1) + 0(-1, 0).$$
or
$$= 5(1, 0) + 3(1, 1) + 6(-1, 0)$$
or
$$= 1(1, 0) + 3(1, 1) + 2(-1, 0),$$
and an infinite number of other ways as well.

5.2 Let $\mathcal{V}$ be an n-dimensional vector space with basis $\{V_1, \ldots, V_n\}$. Then the correspondence
$$V = a_1 V_1 + \cdots + a_n V_n \leftrightarrow (a_1, \ldots, a_n)$$
between $\mathcal{V}$ and $\mathbb{R}_n$

i) preserves scalar multiplication (that is, if $V \leftrightarrow (a_1, \ldots, a_n)$, then $cV \leftrightarrow (ca_1, \ldots, ca_n)$); and

ii) preserves vector addition (that is, If $V \leftrightarrow (a_1, \ldots, a_n)$ and $W \leftrightarrow (b_1, \ldots, b_n)$, then $V + W \leftrightarrow (a_1 + b_1, \ldots, a_n + b_n)$).

Proof. (i) If $V = a_1 V_1 + \cdots + a_n V_n$, then
$$cV = (ca_1)V_1 + \cdots + (ca_n)V_n.$$
Hence $cV \leftrightarrow (ca_1, \ldots, ca_n)$, establishing (i).

We leave the equally difficult proof of (ii) for the reader. ∎

Note carefully what 5.1 and 5.2 say: *when vectors in $\mathcal{V}$ are written in terms of a fixed basis $\{V_1, \ldots, V_n\}$, equality of vectors, addition of vectors, and scalar multiplication of vectors all behave exactly as if one were working with n-tuples in $\mathbb{R}_n$*
(with the n-tuple corresponding to a vector being given by the coefficients in its expression in terms of the basis).

Example 2. Let us take $\mathcal{V} = \mathbb{R}_2$, but instead of the usual basis, let us take the basis $\{(1, 0), (1, 1)\}$ of Example 1. Now any V in $\mathcal{V}$ is of the form $V = (a, b)$, a, b in $\mathbb{R}$. We can thus write
$$V = (a, b) = (a - b)(1, 0) + b(1, 1).$$
Hence, in terms of this basis the correspondence we have been discussing would be $V \leftrightarrow (a - b, b)$. This means that rather than writing vectors in $\mathbb{R}_2$ in the usual form (a, b), which is the way we express them in terms of the basis $\{(1, 0), (0, 1)\}$, we could

just as well write these vectors in the form $(a - b, b)$, corresponding to the basis $\{(1, 0), (1, 1)\}$.

We have established the fact that, *after a basis has been chosen*, any n-dimensional vector space behaves essentially the same as $\mathbb{R}_n$.

In general, we say that two vector spaces $\mathcal{V}$ and $\mathcal{W}$ are **isomorphic**† if there is a one-to-one correspondence $\mathbf{V} \leftrightarrow \mathbf{W}$ between the spaces which

   i) preserves scalar multiplication: If $\mathbf{V} \leftrightarrow \mathbf{W}$, then $c\mathbf{V} \leftrightarrow c\mathbf{W}$ for any scalar c.

   ii) preserves vector addition: If $\mathbf{V}_1 \leftrightarrow \mathbf{W}_1$ and $\mathbf{V}_2 \leftrightarrow \mathbf{W}_2$, then $\mathbf{V}_1 + \mathbf{V}_2 \leftrightarrow \mathbf{W}_1 + \mathbf{W}_2$.

We can therefore summarize 5.1 and 5.2 by saying

   5.3 Any n-dimensional vector space is isomorphic to $\mathbb{R}_n$.

Using a very modest generalization of the above approach we obtain

   5.4 Any two n-dimensional vector spaces are isomorphic.

Start of proof. Let $\mathcal{V}$ and $\mathcal{W}$ have bases $\{\mathbf{V}_1, \ldots, \mathbf{V}_n\}$ and $\{\mathbf{W}_1, \ldots, \mathbf{W}_n\}$, respectively. It is a straightforward review of 5.1 and 5.2 to show that the correspondence

$$a_1\mathbf{V}_1 + \cdots + a_n\mathbf{V}_n \leftrightarrow a_1\mathbf{W}_1 + \cdots + a_n\mathbf{W}_n$$

makes $\mathcal{V}$ and $\mathcal{W}$ isomorphic (Exercise 7b). ∎

If isomorphic vector spaces are really essentially the same, we should have a converse to 5.4. We do, in fact:

   5.5 If two finite-dimensional vector spaces are isomorphic, then they have the same dimension.

Proof. Let $\mathcal{V}$ and $\mathcal{W}$ be isomorphic under the correspondence $\mathbf{V} \leftrightarrow \mathbf{W}$, and suppose that $\{\mathbf{V}_1, \ldots, \mathbf{V}_n\}$ is a basis for $\mathcal{V}$. We show that if $\mathbf{V}_1 \leftrightarrow \mathbf{W}_1, \ldots, \mathbf{V}_n \leftrightarrow \mathbf{W}_n$, then $\{\mathbf{W}_1, \ldots, \mathbf{W}_n\}$ is a basis for $\mathcal{W}$. The set spans $\mathcal{W}$: Given $\mathbf{W}$ in $\mathcal{W}$, there is a $\mathbf{V}$ in $\mathcal{V}$ for which $\mathbf{V} \leftrightarrow \mathbf{W}$. Since

$$\mathbf{V} = a_1\mathbf{V}_1 + \cdots + a_n\mathbf{V}_n,$$

repeated use of properties (i) and (ii) and the correspondence shows that

$$\mathbf{W} = a_1\mathbf{W}_1 + \cdots + a_n\mathbf{W}_n,$$

so the set spans $\mathcal{W}$. No element of the set can be written in terms of other elements of the set: if, for example,

$$\mathbf{W}_1 = a_2\mathbf{W}_2 + \cdots + a_n\mathbf{W}_n,$$

---

† From the Greek "of the same form."

then we would have both $\mathbf{W}_1$ and $a_2\mathbf{W}_2 + \cdots + a_n\mathbf{W}_n$ corresponding to the same element of $\mathcal{U}$. But the former corresponds to $\mathbf{V}_1$ and the latter to
$$a_2\mathbf{V}_2 + \cdots + a_n\mathbf{V}_n.$$
Hence, we would have
$$\mathbf{V}_1 = a_2\mathbf{V}_2 + \cdots + a_n\mathbf{V}_n.$$
Since the $\mathbf{V}_i$'s are a basis this is impossible, and hence $\mathbf{W}_1$ cannot be written in terms of the other $\mathbf{W}$'s. The same reasoning applies to $\mathbf{W}_2, \ldots, \mathbf{W}_n$. ∎

We shall later have occasion to shift our attention to the correspondence involved in the definition of "isomorphic". A function $\mathbf{f}$ from $\mathcal{U}$ onto $\mathcal{W}$ which is a one-to-one correspondence and which

i) preserves scalar multiplication, $\mathbf{f}(c\mathbf{V}) = c\mathbf{f}(\mathbf{V})$
(that is, if $\mathbf{V} \leftrightarrow \mathbf{W}$, then $c\mathbf{V} \leftrightarrow c\mathbf{W}$), and

ii) preserves vector addition, $\mathbf{f}(\mathbf{V}_1 + \mathbf{V}_2) = \mathbf{f}(\mathbf{V}_1) + \mathbf{f}(\mathbf{V}_2)$,
(that is, if $\mathbf{V}_1 \leftrightarrow \mathbf{W}_1$ and $\mathbf{V}_2 \leftrightarrow \mathbf{W}_2$, then $\mathbf{V}_1 + \mathbf{V}_2 \leftrightarrow \mathbf{W}_1 + \mathbf{W}_2$),

is called an **isomorphism from the vector space** $\mathcal{U}$ **onto the vector space** $\mathcal{W}$.

If two vector spaces are isomorphic, then there is an isomorphism from one vector space to the other. Conversely, if there is an isomorphism of one space onto the other, then the two spaces are isomorphic. Notice that the correspondence between geometrical vectors and $\mathbb{R}_n$ discussed in Chapter 0, Section 2, is an isomorphism.

Example 3. In Example 2 we obtained a correspondence between $\mathbb{R}_2$ and itself by writing vectors $\mathbf{V} = (a, b)$ in terms of the basis $\{(1, 0), (1, 1)\}$. The actual correspondence was $\bar{\mathbf{V}} \leftrightarrow (a - b, b)$. We now see that the function $(a, b) \to (a - b, b)$ is an isomorphism of $\mathbb{R}_2$ onto itself.

An isomorphism preserves all the vector space properties we have thus far encountered (see Exercise 8). In particular,

5.6 Under an isomorphism the image of a basis is a basis. That is, if $\{\mathbf{V}_1, \ldots, \mathbf{V}_n\}$ is a basis for $\mathcal{U}$ and if there is an isomorphism from $\mathcal{U}$ onto $\mathcal{W}$ for which $\mathbf{V}_1 \to \mathbf{W}_1, \ldots, \mathbf{V}_n \to \mathbf{W}_n$, then $\{\mathbf{W}_1, \ldots, \mathbf{W}_n\}$ is a basis for $\mathcal{W}$.

Proof. This can be easily deduced from the proof of 5.5 (Exercise 6c). ∎

Having observed that any n-dimensional vector space $\mathcal{U}$ over $\mathbb{R}$ is essentially the same as $\mathbb{R}_n$, the reader may be prepared to spend the rest of his time working in $\mathbb{R}_n$, which is apparently cleaner and easier to work in, and is certainly more familiar and concrete. In fact, why did we spend so much time developing things abstractly if we were just working with n-tuples all along? Before the reader asks for a refund, let us point out why the evidence overwhelmingly favors the abstract approach.

1. Even if one were to start out only in $\mathbb{R}_n$, linear algebra problems very often lead to the consideration of subspaces of $\mathbb{R}_n$. Although, as we know, any such subspace will be *isomorphic* to $\mathbb{R}_m$ for some $m \leq n$, such a subspace will not *be* $\mathbb{R}_m$. For example, in section 3 we considered the subspace

$$\{(a + b, 2b, 3b) \mid a, b \text{ in } \mathbb{R}\}$$

of $\mathbb{R}_3$. Although this subspace is readily seen to be isomorphic to $\mathbb{R}_2$, it is surely not $\mathbb{R}_2$, whose elements are of the form (a, b). Thus, one must be able to think of vector spaces as being comprised of creatures of a form other than $(v_1, \ldots, v_m)$.

2. Starting in $\mathbb{R}_n$, problems often force us to consider not only subspaces, but also bases for these subspaces, linear combinations of vectors, spans of sets, linearly independent sets, and in short, all the vector-space accoutrements we have developed. Now, the only contribution gained by restricting yourself to $\mathbb{R}_n$ is the comfort of knowing exactly where you are; but this is gained at the price of a notation that not only is cumbersome, but can be downright misleading. Except for some trivia in section 2, there is no proof in this chapter that is any easier to give for n-tuples than for abstract vectors.

3. A vector space may be presented in many forms, and some bases may be much more convenient for some purposes than others. Moreover, the correspondence between abstract vectors and n-tuples is very heavily dependent on the choice of basis. The situation is quite analogous to that in analytic geometry where we consider the euclidean plane as a geometrical entity and not just the set of ordered pairs of real numbers. For example, by choosing a new coordinate system in the euclidean plane, one may reduce the equation

$$7x^2 - 8xy + y^2 - 50x + 26y + 79 = 0$$

to the rather more comfortable $9X^2 - Y^2 = 9$; we shall work with higher-dimensional analogs of this in Chapter 9.

4. Finally, there are exceedingly important vector spaces which are not finite-dimensional, and thus cannot be considered as n-tuples (function spaces, polynomials, sequence spaces, and the like).

We readily concede that for computational purposes working with n-tuples is often absolutely necessary; but as in the above example from analytic geometry, if you do have the choice, don't choose your n-tuples too hastily.

EXERCISES

1. Under the isomorphism established in 5.1 and 5.2, find the vectors in $\mathbb{R}_n$ corresponding to the basis vectors $\mathbf{V}_1, \ldots, \mathbf{V}_n$.
2. a) Show that the subspace of $\mathbb{R}_3$ spanned by $\mathbf{i} = (1, 0, 0)$ and $\mathbf{j} = (0, 1, 0)$ is isomorphic to $\mathbb{R}_2$.
   b) Describe all subspaces of $\mathbb{R}_3$ that are isomorphic to $\mathbb{R}_2$.

3. Show that for any $0 \leq m \leq n$, there is a subspace of $\mathbb{R}_n$ that is isomorphic to $\mathbb{R}_m$.
4. (a) Find an isomorphism from the vector space $\mathbb{R}_n[x]$ of polynomials of degree less than n, over $\mathbb{R}$, onto the vector space of n-tuples $\mathbb{R}_n$ over $\mathbb{R}$. (b) Verify that you have an isomorphism. (c) Find another isomorphism.
5. a) Prove that if **f** is an isomorphism from $\mathcal{U}$ over $\mathbb{R}$ onto $\mathcal{W}$ over $\mathbb{R}$, then the inverse function $\mathbf{f}^{-1}$ (defined by $\mathbf{f}^{-1}(\mathbf{W}) = \mathbf{V}$ if $\mathbf{f}(\mathbf{V}) = \mathbf{W}$) is an isomorphism from $\mathcal{W}$ over $\mathbb{R}$ onto $\mathcal{U}$ over $\mathbb{R}$.
   b) Prove that if **f** is an isomorphism from the vector space $\mathcal{U}$ over $\mathbb{R}$ onto the vector space $\mathcal{W}$ over $\mathbb{R}$, and if **g** is an isomorphism from the vector space $\mathcal{W}$ over $\mathbb{R}$ onto the vector space $\mathcal{X}$ over $\mathbb{R}$, then their composition **gf** is an isomorphism from $\mathcal{U}$ over $\mathbb{R}$ onto $\mathcal{X}$ over $\mathbb{R}$ (see Section 5 of Chapter 0).
6. (a) Prove 5.2(ii). (b) Finish 5.4. (c) Prove 5.6.
7. (*Uses equivalence relations*) Prove that isomorphism is an equivalence relation among vector spaces over $\mathbb{R}$.
8. This exercise shows that an isomorphism preserves all the vector-space properties we have thus far encountered. Let **f** be an isomorphism from a vector space $\mathcal{U}$ onto a vector space $\mathcal{W}$. Show that **f** preserves

   a) the zero vector (that is, $\mathbf{f}(\mathbf{0}) = \mathbf{0}$) [*Suggestion:* What is $\mathbf{f}(\mathbf{0} + \mathbf{0})$?];
   b) inverses ($\mathbf{f}(-\mathbf{V}) = \mathbf{f}(\mathbf{V})$);
   c) subspaces (if $\mathcal{X}$ is a subspace of $\mathcal{U}$, then
   $$\mathbf{f}(\mathcal{X}) = \{\mathbf{f}(\mathbf{X}) \mid \mathbf{X} \text{ in } \mathcal{X}\}$$
   is a subspace of $\mathcal{W}$);
   d) linear combination (that is, if $\mathbf{V} = a_1\mathbf{V}_1 + \cdots + a_n\mathbf{V}_n$, then
   $$\mathbf{f}(\mathbf{V}) = a_1\mathbf{f}(\mathbf{V}_1) + \cdots + a_n\mathbf{f}(\mathbf{V}_n));$$
   e) spans ($\mathbf{f}(\text{span } S) = \text{span } \mathbf{f}(S)$ for any subset S of $\mathcal{U}$);
   f) linear independence (if S is linearly independent, then so is
   $$\mathbf{f}(S) = \{\mathbf{f}(\mathbf{V}) \mid \mathbf{V} \text{ in } S\});$$
   g) bases (i.e., if S is a basis for $\mathcal{U}$, then $\mathbf{f}(S)$ is a basis for $\mathcal{W}$);
   and, from the supplement to Section 3,
   h) sums of subspaces (if $\mathcal{X}_1$ and $\mathcal{X}_2$ are subspaces of $\mathcal{U}$, then
   $$\mathbf{f}(\mathcal{X}_1 + \mathcal{X}_2) = \mathbf{f}(\mathcal{X}_1) + \mathbf{f}(\mathcal{X}_2));$$
   i) intersections of subspaces (that is,
   $$\mathbf{f}(\mathcal{X}_1 \cap \mathcal{X}_2) = \mathbf{f}(\mathcal{X}_1) \cap \mathbf{f}(\mathcal{X}_2));$$
   j) direct sums (that is, $\mathbf{f}(\mathcal{X}_1 \oplus \mathcal{X}_2) = \mathbf{f}(\mathcal{X}_1) \oplus \mathbf{f}(\mathcal{X}_2)$).

Honors Project. Verify that with the obvious modifications the material in this section goes over to vector spaces over (a) $\mathbb{C}$, (b) any field.

If different fields are under consideration, it is important to use the phrase "isomorphism over ——" to keep things clear. For example, consider the conjugate mapping on the complex numbers defined by $f(z) = \bar{z}$.

9. a) Show that f is an isomorphism from $\mathbb{C}$ onto $\mathbb{C}$ as a vector space over $\mathbb{R}$.
   b) Show that f is *not* an isomorphism when $\mathbb{C}$ is considered a vector space over $\mathbb{C}$.

If Adam Smith and Karl Marx were young would-be economists these days, one suspects they would make sure they knew matrix algebra . . . †

CHAPTER 2 **Matrices**

In this chapter we shall first develop a computational approach to a problem left over from the previous chapter: given a set of vectors which span a subspace, how can you find a basis for the subspace? This problem is often important in applications, and the techniques we develop are also crucial for the solution of systems of linear equations, which we discuss in Chapter 4.

We introduce matrices first as a notational convenience, and then endow them with arithmetic properties. Some applications of these simple properties are discussed in Section 5. Matrices are of central importance for the next chapter, where they are viewed in a more sophisticated light.

1. BASES FOR SUBSPACES

Our objective is to develop techniques for finding a basis for a subspace $\mathcal{W}$ of a vector space $\mathcal{V}$ over $\mathbb{R}$. We shall, however, attack a simplified version of the problem and assume that the containing space $\mathcal{V}$ is $\mathbb{R}_n$ and that we know a spanning set for $\mathcal{W}$,

$$\mathcal{W} = \text{span } \{\mathbf{V}_1, \ldots, \mathbf{V}_m\},$$

where $\mathbf{V}_1, \ldots, \mathbf{V}_m$ are known vectors in $\mathbb{R}_n$. Our approach will depend upon the following simple facts:

1.1 A permutation‡ of $\mathbf{V}_1, \ldots, \mathbf{V}_m$ has the same span as $\mathbf{V}_1, \ldots, \mathbf{V}_m$.

---

† Harry Schwartz, *The New York Times*, Nov. 2, 1969.
‡ A *permutation* is just a one-to-one rearrangement of a finite set. For example, $\mathbf{V}_2, \mathbf{V}_1, \mathbf{V}_3$ is a permutation of $\mathbf{V}_1, \mathbf{V}_2, \mathbf{V}_3$.

1.2 Multiplying any vector in the set $V_1, \ldots, V_m$ by a *nonzero* scalar leaves the span unchanged.

1.3 Replacing any vector in the set $V_1, \ldots, V_m$ by that vector plus a scalar multiple of another vector in the set does not affect the span.

A moment's reflection should make 1.1 and 1.2 quite apparent. Since 1.3 might take two moments, we shall supply part of a proof:

Outline of proof of 1.3. For notational convenience, let us suppose that V_1 has been replaced by $V_1 + cV_2$. We must show span $\{V_1, V_2, \ldots, V_m\}$ = span $\{V_1 + cV_2, V_2, \ldots, V_m\}$. It's pretty clear that any linear combination of $V_1 + cV_2$, $V_2, \ldots, V_m$ is also a linear combination of $V_1, V_2, \ldots, V_m$. It remains to show that any linear combination of $V_1, V_2, \ldots, V_m$ is a linear combination of $V_1 + cV_2, V_2, \ldots, V_m$; that is, given that

$$V = a_1 V_1 + a_2 V_2 + \cdots + a_m V_m,$$

show that

$$V = b_1(V_1 + cV_2) + b_2 V_2 + \cdots + b_m V_m$$

also, for some choice of $b_1, \ldots, b_m$. We leave this to you. ∎

There are two other known facts which will be of occasional use:

1.4 If $V_1, \ldots, V_m$ are linearly independent, then they form a basis for their span.

1.5 Zero vectors may be deleted from $V_1, \ldots, V_m$ without altering the span.

For the time being we shall use only 1.1 through 1.3, with the following in mind:

OBJECTIVE. Given a set $\{V_1, \ldots, V_m\}$ of vectors in $\mathbb{R}_n$, by using the operations in 1.1 through 1.3 we wish to replace this set by a set $\{\bar{V}_1, \ldots, \bar{V}_k, \bar{V}_{k+1}, \ldots, \bar{V}_m\}$ where $\bar{V}_1, \ldots, \bar{V}_k$ are linearly independent and $\bar{V}_{k+1}, \ldots, \bar{V}_m$ are all **0**.

Since 1.1 through 1.3 state that the operations involved do not change the span, we see that the span of $\bar{V}_1, \ldots, \bar{V}_m$ is the same as the span of $V_1, \ldots, V_m$. By 1.5, we also note that the span of $\bar{V}_1, \ldots, \bar{V}_k$ is the same as the span of $V_1, \ldots, V_m$. We will have thus found (by 1.4) a basis for the span of $V_1, \ldots, V_m$, namely $\bar{V}_1, \ldots, \bar{V}_k$. Moreover, the dimension of the span of $V_1, \ldots, V_m$ must be k.

Example 1. Given vectors

$$V_1 = (1, 3, -2), \quad V_2 = (2, 4, 3), \quad \text{and} \quad V_3 = (-1, 1, -12)$$

in $\mathbb{R}_3$, using the operation in 1.3 we replace V_2 by $V_2 - 2V_1$; let

$$V_1' = V_1 = (1, 3, -2), \quad V_2' = V_2 - 2V_1 = (0, -2, 7), \quad V_3' = V_3 = (-1, 1, -12).$$

Next, replace V_3' by $V_3' + V_1'$; let

$$V_1'' = V_1' = (1, 3, -2), \qquad V_2'' = V_2' = (0, -2, 7), \qquad V_3'' = V_3' + V_1' = (0, 4, -14).$$

Finally, replace V_3'' by $V_3'' + 2V_2''$ to obtain

$$\bar{V}_1 = V_1'' = (1, 3, -2), \qquad \bar{V}_2 = V_2'' = (0, -2, 7), \qquad \bar{V}_3 = V_3'' + 2V_2'' = (0, 0, 0).$$

By 1.3 (applied three times) we see that the span of $\bar{V}_1, \bar{V}_2, \bar{V}_3$ is the same as that of V_1, V_2, V_3. By 1.5 we can dispense with $\bar{V}_3 = \mathbf{0}$. Moreover, we claim that $\bar{V}_1$ and $\bar{V}_2$ are linearly independent (note that $\bar{V}_1$ has a nonzero entry in the same position in which $\bar{V}_2$ has a zero entry). For, if $a\bar{V}_1 + b\bar{V}_2 = \mathbf{0}$, this says that

$$(a, 3a - 2b, -2a + 7b) = (0, 0, 0),$$

so that one quickly obtains $a = 0$ and $b = 0$. Hence,

$$\bar{V}_1 = (1, 3, -2) \quad \text{and} \quad \bar{V}_2 = (0, -2, 7)$$

form a basis for the subspace spanned by V_1, V_2, V_3. Its dimension is therefore 2.

Working a few examples of this sort would quickly convince you of the need for an efficient system of bookkeeping. To this effect, we now introduce the star of this chapter, the matrix. Rather than working with

$$V_1 = (1, 3, -2), \qquad V_2 = (2, 4, 3), \quad \text{and} \quad V_3 = (-1, 1, -12)$$

as separate entities, it is much more convenient to stack them on top of each other and dispense with commas, obtaining

$$\begin{pmatrix} V_1 \\ V_2 \\ V_3 \end{pmatrix} = \begin{pmatrix} 1 & 3 & -2 \\ 2 & 4 & 3 \\ -1 & 1 & -12 \end{pmatrix},$$

a fine example of a matrix.

In general, a **matrix** is any rectangular array of numbers† (real, complex, or even elements from any field; unless otherwise stated we shall consider our matrices to have real entries).

A general matrix is usually written

$$\mathbf{M} = \begin{pmatrix} a_{11} & a_{12} & \cdots & a_{1n} \\ a_{21} & a_{22} & \cdots & a_{2n} \\ \vdots & & & \vdots \\ a_{m1} & a_{m2} & \cdots & a_{mn} \end{pmatrix}, \qquad \text{with all } a_{ij} \text{ in } \mathbb{R}.$$

---

† Note that this really isn't a precise definition, since it uses words such as "rectangular," "array", etc., which have not been previously defined in our axiom system. It should be clear what is intended, nonetheless (cf. the casual definition of "function" in Chapter 0).

The one depicted here is an **m ✕ n** matrix, since it has m rows and n columns. We could regard **M** as either a stack of m vectors from $\mathbb{R}_n$,

$$\mathbf{M} = \begin{pmatrix} \mathbf{V}_1 \\ \mathbf{V}_2 \\ \vdots \\ \mathbf{V}_m \end{pmatrix}, \qquad \begin{array}{l} \mathbf{V}_1 = (a_{11}, a_{12}, \ldots, a_{1n}) \\ \mathbf{V}_2 = (a_{21}, a_{22}, \ldots, a_{2n}) \\ \vdots \\ \mathbf{V}_m = (a_{m1}, a_{m2}, \ldots, a_{mn}), \end{array}$$

or as a collection of n vectors from $\mathbb{R}_m$ written as columns:

$$\mathbf{M} = (\mathbf{W}_1 \, \mathbf{W}_2 \cdots \mathbf{W}_n),$$

$$\mathbf{W}_1 = \begin{pmatrix} a_{11} \\ a_{21} \\ \vdots \\ a_{m1} \end{pmatrix}, \; \mathbf{W}_2 = \begin{pmatrix} a_{12} \\ a_{22} \\ \vdots \\ a_{m2} \end{pmatrix}, \ldots, \mathbf{W}_n = \begin{pmatrix} a_{1n} \\ a_{2n} \\ \vdots \\ a_{mn} \end{pmatrix}.$$

We shall have occasion to adopt both points of view. For this section we shall take the first approach and consider a matrix as a stack of row vectors.

The **row space** of an $m \times n$ matrix

$$\mathbf{M} = \begin{pmatrix} \mathbf{V}_1 \\ \mathbf{V}_2 \\ \vdots \\ \mathbf{V}_m \end{pmatrix}$$

is the subspace of $\mathbb{R}_n$ spanned by the rows $\mathbf{V}_1, \ldots, \mathbf{V}_m$ of **M**. (We shall later consider the **column space** also, that is, the subspace of $\mathbb{R}_m$ spanned by the vectors which form the columns of **M**.)

The **row rank** of **M** is the dimension of its row space (with a similar definition for **column rank**; very surprisingly, we shall later see that these ranks are the same!).

Given a set of vectors $\mathbf{V}_1, \ldots, \mathbf{V}_m$ in $\mathbb{R}_n$, we can stack them up to form the matrix **M** whose row space is the span of $\mathbf{V}_1, \ldots, \mathbf{V}_m$. Our original quest can therefore be restated in matrix terms as follows:

MATRIX OBJECTIVE. Given an $m \times n$ matrix

$$\mathbf{M} = \begin{pmatrix} \mathbf{V}_1 \\ \vdots \\ \mathbf{V}_m \end{pmatrix},$$

find a means of replacing **M** by a matrix **M̄** which has the same row space as **M**, and with **M̄** of the form

$$\bar{\mathbf{M}} = \begin{pmatrix} \bar{\mathbf{v}}_1 \\ \vdots \\ \bar{\mathbf{v}}_k \\ \bar{\mathbf{v}}_{k+1} \\ \vdots \\ \bar{\mathbf{v}}_m \end{pmatrix},$$ with the first k rows of **M̄** linearly independent and the last $(m-k)$ rows all **0**.

We shall accomplish this objective by means of operations on the rows of the matrix **M**, which are just the operations involved in 1.1 through 1.3, translated into matrix terms. The matrix forms of 1.1 through 1.3 are

1.1′ If the rows of **M′** are a permutation of the rows of **M**, then **M** and **M′** have the same row space.

1.2′ If **M′** is a matrix obtained from **M** by multiplying one of the rows of **M** by a *nonzero* scalar, then the row spaces of **M** and **M′** are the same.

1.3′ If **M′** is obtained from **M** by adding a scalar multiple of one row of **M** to another row of **M**, then **M** and **M′** have the same row space.

There is nothing really to prove here; everything follows from the definition of row space and our previous results. The same is true of the following.

1.4′ If the rows of **M** are linearly independent, then they form a basis for the row space of **M**.

1.5′ Rows of **M** which consist entirely of zeros can be deleted and the row space of the resulting matrix will be the same as that of **M**.

We shall call the row operations mentioned in 1.1′–1.3′ the **elementary row operations**. These consist of permuting the rows, multiplying a row by a nonzero scalar, and adding a scalar multiple of one row to another row. From 1.1′–1.3′ we have

1.6 If **M′** is obtained from **M** by a sequence of elementary row operations, then the row space of the two matrices is the same.

From the definition of row rank and 1.6, we have

1.7 If **M′** is obtained from **M** by a sequence of elementary row operations, then the row ranks of the two matrices are the same.

Example 2. Let us redo example 1 in our matrix context. Stacking up V_1, V_2, V_3 we have

$$\begin{pmatrix} V_1 \\ V_2 \\ V_3 \end{pmatrix} = \begin{pmatrix} 1 & 3 & -2 \\ 2 & 4 & 3 \\ -1 & 1 & -12 \end{pmatrix}.$$

We proceed to redo the steps in example 1; each involves the elementary row operation of multiplying a row of the matrix by a constant and adding it to another row. We indicate this by placing the constant to the right of the multiplied row and directing an arrow from this row to the one to which it is added:

$$\begin{pmatrix} V_1 \\ V_2 \\ V_3 \end{pmatrix} = \begin{pmatrix} 1 & 3 & -2 \\ 2 & 4 & 3 \\ -1 & 1 & -12 \end{pmatrix} \begin{matrix} -2 \\ \downarrow \\ \end{matrix} \quad \text{becomes} \quad \begin{pmatrix} V'_1 \\ V'_2 \\ V'_3 \end{pmatrix} = \begin{pmatrix} 1 & 3 & -2 \\ 0 & -2 & 7 \\ -1 & 1 & -12 \end{pmatrix} \begin{matrix} 1 \\ \downarrow \\ \end{matrix} \quad \text{becomes}$$

$$\begin{pmatrix} V''_1 \\ V''_2 \\ V'_3 \end{pmatrix} = \begin{pmatrix} 1 & 3 & -2 \\ 0 & -2 & 7 \\ 0 & 4 & -14 \end{pmatrix} \begin{matrix} \\ 2 \\ \downarrow \end{matrix} \quad \text{becomes} \quad \begin{pmatrix} \bar{V}_1 \\ \bar{V}_2 \\ \bar{V}_3 \end{pmatrix} = \begin{pmatrix} 1 & 3 & -2 \\ 0 & -2 & 7 \\ 0 & 0 & 0 \end{pmatrix}.$$

As before, we find that

$$\bar{V}_1 = (1, 3, -2) \quad \text{and} \quad \bar{V}_2 = (0, -2, 7)$$

form a basis for the row space of

$$\begin{pmatrix} V_1 \\ V_2 \\ V_3 \end{pmatrix}.$$

It should be quite evident (even without our having proved this in Example 1) that the first two rows of

$$\begin{pmatrix} \bar{V}_1 \\ \bar{V}_2 \\ \bar{V}_3 \end{pmatrix}$$

are linearly independent (obviously no nontrivial linear combination of these rows can be **0**). Hence the row rank of this matrix is 2. By 1.7 then, 2 is also the row rank of

$$\begin{pmatrix} V_1 \\ V_2 \\ V_3 \end{pmatrix}.$$

It is often convenient to choose a particular nonzero entry of a matrix and use it to clean out the entire column in which it originally appears, by means of the elementary row operation of adding appropriate multiples of its row to the other rows. (This was done in the first two steps in the above example, using the upper left-hand element of the matrix.) The chosen nonzero element is often replaced

1 / Bases for subspaces

by a 1 by dividing its row by itself (perhaps before using it to clean out its column). This entire process is known as "pivoting".

Example 3. Continuing with the matrix in Example 2, we "pivot" on the -2 in the center:

$$\begin{pmatrix} 1 & 3 & -2 \\ 0 & -2 & 7 \\ 0 & 0 & 0 \end{pmatrix} \begin{Bmatrix} \text{divide row} \\ \text{by } -2 \end{Bmatrix} \begin{pmatrix} 1 & 3 & -2 \\ 0 & 1 & -\frac{7}{2} \\ 0 & 0 & 0 \end{pmatrix} \begin{matrix} \uparrow \\ -3 \\ {} \end{matrix} \quad \text{becomes} \quad \begin{pmatrix} 1 & 0 & \frac{17}{2} \\ 0 & 1 & -\frac{7}{2} \\ 0 & 0 & 0 \end{pmatrix}.$$

No steps are needed, since the bottom entry of the second column is already 0.

To describe the process more generally, we need addresses for the entries of a matrix. The **i, j-entry** of the matrix

$$\mathbf{M} = \begin{pmatrix} a_{11} & \cdots & a_{1j} & \cdots & a_{1n} \\ \vdots & & \vdots & & \vdots \\ a_{i1} & \cdots & a_{ij} & \cdots & a_{in} \\ \vdots & & \vdots & & \vdots \\ a_{m1} & \cdots & a_{mj} & \cdots & a_{mn} \end{pmatrix}$$

is the element a_{ij} in the ith row and jth column of the matrix. For example, if $\mathbf{M}$ is the 3×4 matrix

$$\mathbf{M} = \begin{pmatrix} 1 & 3 & -2 & 5 \\ 2 & 4 & 3 & -3 \\ -1 & 1 & -12 & 8 \end{pmatrix},$$

then the 1,2-entry is 3, the 2,1-entry is 2, the 2,3-entry is 3, etc.

To **pivot** on the i, j-entry a_{ij} of a matrix (which must be nonzero),

i) multiply the ith row by $1/a_{ij}$ (the i, j-entry of the resulting matrix will then be a 1), and

ii) add an appropriate multiple of the ith row to each of the remaining rows, so that the other entries in the jth column are all replaced by 0's.

Example 4. To pivot on the 1,1-entry of the following matrix,

$$\begin{pmatrix} -3 & -9 & 6 & -15 \\ 2 & 4 & 3 & -3 \\ -1 & 1 & -12 & 8 \end{pmatrix} \begin{matrix} \text{divide} \\ \text{by } -3 \end{matrix} \quad \text{to get} \quad \begin{pmatrix} 1 & 3 & -2 & 5 \\ 2 & 4 & 3 & -3 \\ -1 & 1 & -12 & 8 \end{pmatrix} \begin{matrix} -2 & 1 \\ \downarrow & \\ & \downarrow \end{matrix}$$

which becomes

$$\begin{pmatrix} 1 & 3 & -2 & 5 \\ 0 & -2 & 7 & -13 \\ 0 & 4 & -14 & 13 \end{pmatrix}.$$

We shall circle the pivot entry and draw an arrow from one matrix to another to denote pivoting. Thus, for the above example we would write

$$\begin{pmatrix} \boxed{-3} & -9 & 6 & -15 \\ 2 & 4 & 3 & -3 \\ -1 & 1 & -12 & 8 \end{pmatrix} \to \begin{pmatrix} 1 & 3 & -2 & 5 \\ 0 & -2 & 7 & -13 \\ 0 & 4 & -14 & 13 \end{pmatrix}.$$

A more detailed description of the pivot operation would be: to pivot on the i,j-entry a_{ij} ($\neq 0$) of a matrix

$$\mathbf{M} = \begin{pmatrix} \mathbf{V}_1 \\ \cdot \\ \cdot \\ \mathbf{V}_i \\ \cdot \\ \cdot \\ \mathbf{V}_m \end{pmatrix} \quad (\mathbf{V}\text{'s are the rows}),$$

i) multiply $\mathbf{V}_i$ by $1/a_{ij}$ so

$$\mathbf{M} = \begin{pmatrix} a_{11} & \cdots & a_{1j} & \cdots & a_{1n} \\ \cdot & & \cdot & & \cdot \\ \cdot & & \cdot & & \cdot \\ a_{i1} & \cdots & a_{ij} & \cdots & a_{in} \\ \cdot & & \cdot & & \cdot \\ \cdot & & \cdot & & \cdot \\ a_{m1} & \cdots & a_{mj} & \cdots & a_{mn} \end{pmatrix}$$

becomes

$$\begin{pmatrix} \mathbf{V}_1 \\ \cdot \\ \cdot \\ \left(\dfrac{1}{a_{ij}}\right)\mathbf{V}_i \\ \cdot \\ \cdot \\ \mathbf{V}_m \end{pmatrix} = \begin{pmatrix} a_{11} & \cdots & a_{1j} & \cdots & a_{1n} \\ \cdot & & \cdot & & \cdot \\ \cdot & & \cdot & & \cdot \\ a'_{i1} & \cdots & 1 & \cdots & a'_{in} \\ \cdot & & \cdot & & \cdot \\ \cdot & & \cdot & & \cdot \\ a_{m1} & \cdots & a_{mj} & \cdots & a_{mn} \end{pmatrix};$$

ii) to each row $k \neq i$, add $-a_{kj}$ times the new ith row to obtain

$$\begin{pmatrix} \mathbf{V}_1 - a_{1j}\left[\left(\dfrac{1}{a_{ij}}\right)\mathbf{V}_i\right] \\ \vdots \\ \left(\dfrac{1}{a_{ij}}\right)\mathbf{V}_i \\ \vdots \\ \mathbf{V}_m - a_{mj}\left[\left(\dfrac{1}{a_{ij}}\right)\mathbf{V}_i\right] \end{pmatrix} = \begin{pmatrix} a'_{11} & \cdots & 0 & \cdots & a'_{1n} \\ \vdots & & \vdots & & \vdots \\ a'_{i1} & \cdots & 1 & \cdots & a'_{in} \\ \vdots & & \vdots & & \vdots \\ a'_{m1} & \cdots & 0 & \cdots & a'_{mn} \end{pmatrix}.$$

Since pivoting is accomplished by a succession of elementary row operations, we obtain, from 1.6 and 1.7,

1.8 If $\mathbf{M}'$ is a matrix obtained from $\mathbf{M}$ by a sequence of pivot operations, then the row spaces of $\mathbf{M}'$ and $\mathbf{M}$ are the same. So are the row ranks.

Example 4, continued.

$$\mathbf{M} = \begin{pmatrix} -3 & -9 & 6 & -15 \\ 2 & 4 & 3 & -3 \\ -1 & 1 & -12 & 8 \end{pmatrix} \to \begin{pmatrix} 1 & 3 & -2 & 5 \\ 0 & -2 & 7 & -13 \\ 0 & 4 & -14 & 13 \end{pmatrix} \to$$

$$\begin{pmatrix} 1 & 0 & \frac{17}{2} & -\frac{29}{2} \\ 0 & 1 & -\frac{7}{2} & \frac{13}{2} \\ 0 & 0 & 0 & -13 \end{pmatrix} \to \begin{pmatrix} 1 & 0 & \frac{17}{2} & 0 \\ 0 & 1 & -\frac{7}{2} & 0 \\ 0 & 0 & 0 & 1 \end{pmatrix} = \overline{\mathbf{M}}.$$

The rows of this matrix are clearly linearly independent (you supply the argument). Hence, by 1.8, $\mathbf{M}$ is of row rank 3 and a basis for the row space of $\mathbf{M}$ consists of the rows of $\overline{\mathbf{M}}$.

Not all of the operations used in pivoting are absolutely necessary for finding bases for row spaces and row ranks (e.g., step (i) and some of the cleaning out of column entries). However, pivoting leads to forms which are easy to describe theoretically (and to computers), so we shall often work with it, leaving you to avoid such unnecessary steps as you wish in your homework computations.

Example 5. Let

$\mathbf{V}_1 = (-3, -9, 6, -13)$, $\mathbf{V}_2 = (2, 4, 3, -3)$, and $\mathbf{V}_3 = (-1, 1, -12, 21)$.

To find a basis (and dimension) for the subspace spanned by these vectors, we stack them up and pivot:

$$\begin{pmatrix} V_1 \\ V_2 \\ V_3 \end{pmatrix} = \begin{pmatrix} \boxed{-3} & -9 & 6 & -15 \\ 2 & 4 & 3 & -3 \\ -1 & 1 & -12 & 21 \end{pmatrix} \to \begin{pmatrix} 1 & 3 & -2 & 5 \\ 0 & \boxed{-2} & 7 & -13 \\ 0 & 4 & -14 & 26 \end{pmatrix} \to \begin{pmatrix} 1 & 0 & \frac{17}{2} & -\frac{29}{2} \\ 0 & 1 & -\frac{7}{2} & \frac{13}{2} \\ 0 & 0 & 0 & 0 \end{pmatrix}.$$

Hence, the span of V_1, V_2, V_3 is a 2-dimensional subspace of $\mathbb{R}_4$, with basis $(1, 0, \frac{17}{2}, -\frac{29}{2})$ and $(0, 1, -\frac{7}{2}, \frac{13}{2})$.

It should be just about apparent that we have at our disposal the means to accomplish our objectives. In matrix terms, to find a basis for the row space of a given matrix, you pivot on nonzero entries until every row of the resulting matrix either has a pivot entry or is entirely zero. The nonzero rows form the desired basis. In the next section we shall formalize the procedure and supply proofs. For now we abandon you to some exercises, which should render the following section quite transparent, if not entirely unnecessary.

EXERCISES

In each of problems 1 through 5, find a basis for (and hence the dimension of) the space spanned by the vectors V_i given in the space indicated.

1. In $\mathbb{R}_4$: $V_1 = (1, 0, -2, 1)$, $V_2 = (2, 1, 3, -1)$, $V_3 = (4, 1, -1, 1)$.
2. In $\mathbb{R}_4$: V_1, V_2, V_3 from Exercise 1 and $V_4 = (-1, 2, 1, 1)$.
3. In $\mathbb{R}_4$: V_1, V_2, V_3, V_4 from Exercise 2 and $V_5 = (1, 1, 1, 1)$.
4. In $\mathbb{R}_3$: $V_1 = (1, 2, 4)$, $V_2 = (0, 1, 1)$, $V_3 = (-2, 3, -1)$, and $V_4 = (1, -1, 1)$.
5. In $\mathbb{R}_6$: $V_1 = (1, -2, 3, 0, 5, 6)$,
   $V_2 = (-2, 3, -1, 4, 1, 1)$,
   $V_3 = (3, 0, 1, 2, -1, 1)$,
   $V_4 = (3, -1, 6, 6, 10, 14)$,
   $V_5 = (-1, 1, 2, 4, 6, 7)$.
6. Let V_1, V_2, V_3, V_4, V_5 be the vectors in $\mathbb{R}_4$ of problem 3. Let $\mathfrak{X}$ be spanned by V_1, V_2 and $\mathfrak{Y}$ be spanned by V_3, V_4, V_5. Find the dimensions of $\mathfrak{X}, \mathfrak{Y}, \mathfrak{X} + \mathfrak{Y}, \mathfrak{X} \cap \mathfrak{Y}$.

2. THE BASIS ALGORITHM

We shall now mold our informal technique for finding bases for subspaces into an "algorithm" (by which we mean a repetitive computational procedure). Since you already know how to formulate the basis problem in matrix terms, we shall state the algorithm in matrix form.

2.1 *The Basis Algorithm.* Given a matrix **M**, to find a basis for the row space of **M**, first pivot on a nonzero entry of the first column of **M** (if there is one, otherwise go on to the second column), then pivot on a nonzero entry of the second column, one which is not in the same row as your first pivot entry (if there is one; otherwise go on to the third column). Continuing from left to right, pivot on a nonzero entry in each column if there is one which is not in the same row as a previous pivot entry.

The nonzero rows of the resulting matrix $\bar{\mathbf{M}}$ form a basis for the row space of $\mathbf{M}$, and their number is the row rank of $\mathbf{M}$.

The algorithm has already been illustrated in Example 4 of Section 1. Another example will help to illustrate further what can happen:

Example 1.

$$\mathbf{M} = \begin{pmatrix} \boxed{-3} & -9 & 6 & -15 \\ 2 & 6 & 3 & -3 \\ -1 & -3 & -12 & 21 \end{pmatrix} \rightarrow \begin{pmatrix} 1 & 3 & -2 & 5 \\ 0 & 0 & \boxed{7} & -13 \\ 0 & 0 & -14 & 26 \end{pmatrix} \rightarrow \begin{pmatrix} 1 & 3 & 0 & \frac{9}{7} \\ 0 & 0 & 1 & -\frac{13}{7} \\ 0 & 0 & 0 & 0 \end{pmatrix}.$$

Hence $(1, 3, 0, \frac{9}{7})$ and $(0, 0, 1, -\frac{13}{7})$ form a basis of the row space, and the row rank is 2.

The same example done with a different choice of pivots gives:

Example 2.

$$\mathbf{M} = \begin{pmatrix} -3 & -9 & 6 & -15 \\ 2 & 6 & 3 & -3 \\ \boxed{-1} & -3 & -12 & 21 \end{pmatrix} \rightarrow \begin{pmatrix} 0 & 0 & 42 & -78 \\ 0 & 0 & \boxed{-21} & 39 \\ 1 & 3 & 12 & -21 \end{pmatrix} \rightarrow \begin{pmatrix} 0 & 0 & 0 & 0 \\ 0 & 0 & 1 & -\frac{13}{7} \\ 1 & 3 & 0 & \frac{9}{7} \end{pmatrix},$$

and the result is the same as before.

Proof of the basis algorithm. We saw in 1.8 that pivoting preserves row space and row rank, so it suffices to show that the nonzero rows of $\bar{\mathbf{M}}$ are linearly independent (why?). We first notice that each of these nonzero rows, row i, say, contains a pivot entry. For if we could get through each column without pivoting on row i, then it was possible to pivot in another row at each step, and this contributed a zero to row i at each step. Hence, if row i does not consist entirely of zeros, it contains a pivot entry, a 1 above and below which are all zeros.

The following picture should clarify the rest of this verbal argument. For convenience we have assumed that the first k rows are the nonzero ones, and that the pivot entries occur in a descending fashion from left to right (this can always be arranged by applications of 1.1'):

If $a_1\bar{V}_1 + a_2\bar{V}_2 + \cdots + a_k\bar{V}_k = 0$, then from the form of the matrix we can write the components of this vector as
$$(0 \cdots a_1 \square \ a_2 \square \ a_3 \square \ a_k \square \ ) = (0, 0, \ldots, 0).$$
Hence
$$a_1 = 0, a_2 = 0, \ldots, a_k = 0.$$
Hence the $\bar{V}_i$ are linearly independent. As observed at the start, this shows that the algorithm works as stated. ∎

Since we have allowed some latitude in the choice of the pivot entry in each column, we should, strictly speaking, call this "a" Basis Algorithm rather than "the" Basis Algorithm. The supplement to this section is concerned with a specific choice of entry which is desirable for computers. Another special choice is needed for Exercise 6. (See also Exercise 8.)

One often wishes to choose a basis from the given spanning set $V_1, \ldots, V_m$ itself (or from the rows of M, in our matrix formulation). The Basis Algorithm actually gives us such a basis! This is seen as follows:

2.2 *A basis for the row space as a subset of the given rows.* If the Basis Algorithm is applied to a matrix M, resulting in a matrix $\bar{M}$, then those rows of M in the same position† as the nonzero rows of $\bar{M}$ form a basis for the row space of M. That is, if

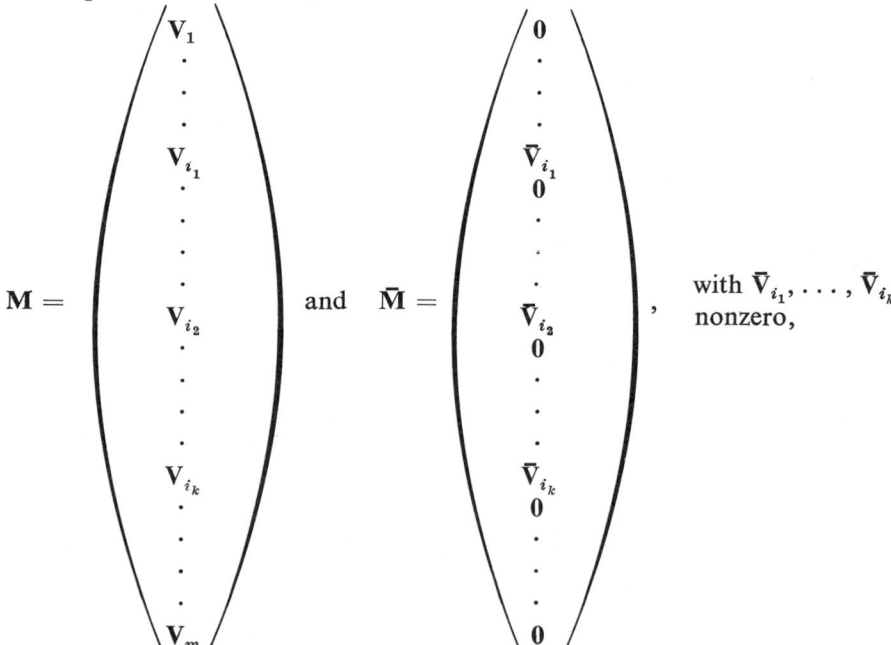

then $V_{i_1}, V_{i_2}, \ldots, V_{i_k}$ form a basis for the row space of M.

---

† Recall that the Basis Algorithm does not call for row interchanges. Even though such interchanges do not affect the space spanned by the rows, they very obviously *cannot* be applied in conjunction with this theorem.

Proof. Suppose, for convenience, that the first k rows are those pivoted upon, so that the last $(n - k)$ rows of $\bar{\mathbf{M}}$ are zero (cf. the proof of 2.1). By checking the row operations involved in pivoting, one sees that this means that the last $(n - k)$ rows of $\mathbf{M}$ can be written as linear combinations of the first k rows (you should supply more detail). Thus, the first k rows of $\mathbf{M}$ span the row space of $\mathbf{M}$. From the Basis Algorithm, the dimension of the row space of $\mathbf{M}$ is k. Hence, the first k rows of $\mathbf{M}$ form a basis (you should supply more detail), which is the statement of 2.2. ∎

Example 3. For the row space of the matrix in Example 1, we can conclude that
$$\mathbf{V}_1 = (-3, -9, 6, -15) \quad \text{and} \quad \mathbf{V}_2 = (2, 6, 3, -3)$$
form a basis. For the same matrix we can conclude, by Example 2, that
$$\mathbf{V}_2 = (2, 6, 3, -3) \quad \text{and} \quad \mathbf{V}_3 = (-1, -3, -12, 21)$$
form a basis also.

2.3 Computational Test for Linear Independence. Given vectors $\mathbf{V}_1, \ldots, \mathbf{V}_m$ in $\mathbb{R}_n$, apply the Basis Algorithm to the matrix

$$\mathbf{M} = \begin{pmatrix} \mathbf{V}_1 \\ \vdots \\ \mathbf{V}_m \end{pmatrix}$$

The vectors $\mathbf{V}_1, \ldots, \mathbf{V}_m$ are linearly independent iff the resulting matrix $\bar{\mathbf{M}}$ contains no zero rows.

Proof. See Exercise 10. ∎

Direct applications of this material to problems in dimensional analysis, radiative heat transfer, stoichiometry, and systems of differential equations may be found in Ben Nobel's *Applications of Undergraduate Mathematics to Engineering* (Macmillan, 1967).

Our algorithms can be nicely exhibited in the form of "flow diagrams". This is a method of organizing a repetitive procedure which was invented by John von Neumann; they are often used in writing up computer programs. Aside from its importance here, the flow diagram is a very useful pedagogical device: you really learn a process when you make up a flow diagram for it. (See the supplement to this section.)

EXERCISES

In each of problems 1–4 find a basis for (and hence the dimension of) the space spanned by the vectors $\mathbf{V}_i$.

1. In $\mathbb{R}_3$: $\mathbf{V}_1 = (1, 2, -1)$, $\mathbf{V}_2 = (3, 4, 1)$, $\mathbf{V}_3 = (-2, 3, -12)$, and $\mathbf{V}_4 = (5, -3, 21)$.
2. In $\mathbb{R}_5$: $\mathbf{V}_1 = (1, 2, 4, -1, 1)$, $\mathbf{V}_2 = (0, 1, 1, 2, 1)$, $\mathbf{V}_3 = (-2, 3, -1, 1, 1)$, and $\mathbf{V}_4 = (1, -1, 1, 1, 1)$.

3. The *row* space of the matrix

$$\begin{pmatrix} 0 & 1 & 2 & 3 \\ 3 & -3 & 0 & 0 \\ 4 & -4 & 2 & 1 \\ 5 & 0 & 10 & 15 \end{pmatrix}.$$

4. The *column* space of the matrix in problem 3.
5. In each of problems 1 through 4 find a basis from among the given V_i's.
6. Prove the following:

   Algorithm for extending linearly independent sets to bases. Given a linearly independent set of vectors $V_1, \ldots, V_m$ in $\mathbb{R}_n$, to extend this set to a basis for $\mathbb{R}_n$,
   i) form the matrix

   $$M = \begin{pmatrix} V_1 \\ \cdot \\ \cdot \\ \cdot \\ V_m \\ e_1 \\ \cdot \\ \cdot \\ \cdot \\ e_n \end{pmatrix},$$

   where e_i's are the usual basis for $\mathbb{R}_n$;
   ii) apply the Basis Algorithm to M, pivoting whenever possible on one of the first m rows (but not on a row previously pivoted upon, of course).

   Then the nonzero rows of M corresponding to nonzero rows of the resulting matrix $\overline{M}$, form the desired basis for $\mathbb{R}_n$. That is, this basis will be comprised of $V_1, \ldots, V_m$ plus $(n - m)$ of the e_i's.

7. a) Prove that the vectors $V_1 = (0, 1, -2, 0, 3)$, $V_2 = (1, -3, 2, 4, 1)$, and $V_3 = (-2, 4, 1, -2, 2)$ in $\mathbb{R}_5$ are linearly independent.
   b) Extend the set V_1, V_2, V_3 to obtain a basis for $\mathbb{R}_5$.
8. Find a basis for $\mathfrak{X} \cap \mathfrak{Y}$ in Exercise 6 of Section 1. [*Hint:* Begin by finding a basis of $\mathfrak{X} + \mathfrak{Y}$ which contains V_1 and V_2.]
9. Show that the Basis Algorithm can be modified by pivoting on elements *not* necessarily in consecutive columns. Give an exact statement of the resulting algorithm, and try it one some of the problems or examples in this section.
10. Prove 2.3.

H11. If $\tilde{V}_j = 0$, write out a precise expression for V_j as a sum of $V_{i_1}, \ldots, V_{i_k}$.

Honors Projects. Show that the results of these sections carry over to:

1. $\mathbb{C}_n$, the vector space of complex *n*-tuples;
2. $\mathbb{F}_n$, $\mathbb{F}$ an arbitrary field,

with some possibility for arithmetic quirks in the numerical examples.

Supplement: Flow Diagrams; A Computer Improvement

As far as we are concerned, a computer is able to perform the following operations:

1. Enter numbers in named storage places, called "memory cells". (This "input" operation is denoted in a flow diagram by a box shaped ▽.)
2. Replace the contents of one memory cell by the contents of another. (The flow diagram command for this is contained in a rectangular box.)
3. Perform the usual arithmetic operations of addition, subtraction, multiplication, and division on the contents of given memory cells (likewise indicated in a rectangle).
4. Check to see if the contents of two memory cells are the same. (The question "are they the same?" is contained in a diamond shaped box.)
5. Go from one operation to a next operation, either directly (following an arrow) or according to the answer to a yes–no question (following one of two arrows leading out of a diamond).
6. Print out the contents of prescribed memory cells in a prescribed array. (This "output" operation is in the same type of box as an input.)

For example, there follows (see Fig. 1, p. 88) a flow diagram for the Basis Algorithm (just follow the arrows).

The following flow diagram would, of course, have to be broken down into more detail to satisfy a computer. Pivoting would have to be explained by means of a "subroutine" which told exactly what to do with the entries of the matrix. The question "Is some element...?" would also have to be expressed in terms of matrix entries. (See Exercise 2.)

A Computer Improvement

A computer performs arithmetic operations to a specified number of "significant digits". For example, if you are working with 5 significant digits, then only 5 digits of any number are recorded by the computer. If arithmetic operations produce a number with more than 5 digits, this number is rounded off to a 5-digit number. Thus, to a computer working with 5 significant digits, the 7-digit number 1000.001 is stored as the 5-digit number 1000.0. We shall now see how this can lead to difficulties.

Suppose that we ask a computer working with 5 significant digits to perform the Basis Algorithm on the vectors

$$\mathbf{V}_1 = (.001, -1, -1), \qquad \mathbf{V}_2 = (1, .001, 0), \qquad \text{and} \qquad \mathbf{V}_3 = (1, 0, .001).$$

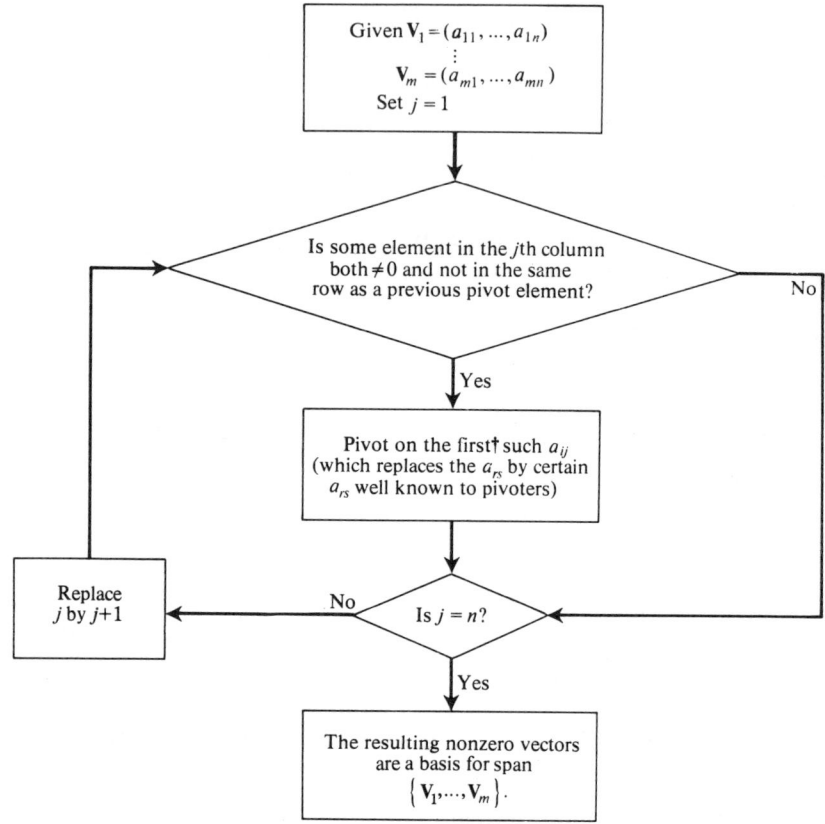

Fig. 1 (The Basis Algorithm)

Pivoting on the first available entry in each column, we would obtain

$$\begin{pmatrix} .001 & -1 & -1 \\ 1 & .001 & 0 \\ 1 & 0 & .001 \end{pmatrix} \to \begin{pmatrix} 1 & -1000 & -1000 \\ 1 & .001 & 0 \\ 1 & 0 & .001 \end{pmatrix}$$

$$\to \begin{pmatrix} 1 & -1000 & -1000 \\ 0 & 1000.001 & 1000 \\ 0 & 1000 & 1000.001 \end{pmatrix}$$

which to the computer is

$$\begin{pmatrix} 1 & -1000 & -1000 \\ 0 & 1000 & 1000 \\ 0 & 1000 & 1000 \end{pmatrix} \xrightarrow{\text{pivoting on the 2,2-entry}} \begin{pmatrix} 1 & -1000 & -1000 \\ 0 & 1 & 1 \\ 0 & 0 & 0 \end{pmatrix}.$$

---

† For simplicity we have suggested pivoting on the *first* nonzero candidate in a column. We shall shortly suggest that the one which is *largest in absolute value* should be chosen by a computer. When working by hand, just take the most convenient entry.

The computer would thus inform us that a basis for span $\{V_1, V_2, V_3\}$ is $\{V_1, V_2\}$. Now $\{V_1, V_2, V_3\}$ certainly *looks* linearly independent, and in fact it is; the computer is utterly wrong.

The problem here is that in pivoting on the small number .001, we first change it to a 1. This is accomplished by multiplying the first row by 100, which increases the other entries in the row in such a fashion that they subsequently obliterate the distinction between the remaining rows.

This problem could have been avoided here by choosing a larger pivot entry. A principle often used in computer programming is:

> **1.** Pivot on an entry of the column which is as large as possible in absolute value.

By following this maxim, the computer would work the above example correctly.

There are serious difficulties with error accumulation when working with computers. We could easily give examples of repetitive approximation procedures which it is better to carry out 10 times by hand than 10,000 times by computer. Moreover, error analysis and estimation can be quite difficult, and often requires sophisticated linear algebra in its discussion.

We close with a quote from L. Fox [2]:

> It is a bore to talk about checking, and most workers find it such a bore to listen, and to act on the advice, that checking is unfortunately rarely practiced. As a result, I estimate that some 50 per cent of all computer results have bigger errors than the author of the problem will know about or admit. In the desk-machine era ... the possibility of errors and the perpetual need for guarding against them were fully realized.

REFERENCES

[1] R. L. Ashehurst and N. Metropolis, "Error Estimation in Computer Calculations", Am. Math. Monthly **72**, Feb. 1965; Part II, pp. 47–58 (#10 in the Slaught Memorial Papers).

[2] L. Fox, *An Introduction to Numerical Linear Algebra*. Oxford: Clarendon Press, 1964.

[3] George E. Forsythe, "Today's Computational Methods of Linear Algebra", *SIAM Rev.*, **9**, 1967, pp. 489–515. A good expository article on computers in linear algebra.

EXERCISES

1. Show that by following principle (**1**), the computer would work the example correctly.
2. a) Make up a flow diagram for pivoting.
   b) Make a flow diagram for answering the question "Is some element ... ?" in the Basis Algorithm flow diagram.
   c) If possible, construct and test a computer program for the Basis Algorithm.

3. MATRIX ARITHMETIC

In this section we treat $m \times n$ matrices as entities unto themselves and introduce an addition and scalar multiplication (which is really just the componentwise addition and scalar multiplication of mn-tuples). We then define a multiplication of certain pairs of matrices which may seem a bit peculiar at first, but after Section 5 and the next chapter, you will see that it is a natural multiplication procedure.

Operating now just as if $m \times n$ matrices are simply an eccentric way of writing mn-tuples, we consider two $m \times n$ matrices

$$\mathbf{A} = \begin{pmatrix} a_{11} & a_{12} & \cdots & a_{1n} \\ a_{21} & a_{22} & \cdots & a_{2n} \\ \vdots & & & \vdots \\ a_{m1} & \cdots & & a_{mn} \end{pmatrix} \quad \text{and} \quad \mathbf{B} = \begin{pmatrix} b_{11} & b_{12} & \cdots & b_{1n} \\ b_{21} & b_{22} & \cdots & b_{2n} \\ \vdots & & & \vdots \\ b_{m1} & \cdots & & b_{mn} \end{pmatrix}$$

to be **equal**, and write $\mathbf{A} = \mathbf{B}$, if their entries are the same:

$$a_{ij} = b_{ij} \quad \text{for} \quad i = 1, \ldots, m \quad \text{and} \quad j = 1, \ldots, n.$$

Matrix addition and scalar multiplication are also done componentwise: the **sum** of two $m \times n$ matrices $\mathbf{A}$ and $\mathbf{B}$ is the $m \times n$ matrix whose entries are the sums of the corresponding entries of the summands:

$$\mathbf{A} + \mathbf{B} = \begin{pmatrix} a_{11} + b_{11} & a_{12} + b_{12} & \cdots & a_{1n} + b_{1n} \\ a_{21} + b_{21} & & & \\ \vdots & & & \vdots \\ a_{m1} + b_{m1} & \cdots & & a_{mn} + b_{mn} \end{pmatrix}.$$

If $\mathbf{A}$ and $\mathbf{B}$ are not both $m \times n$ matrices, then addition is not defined.† Similarly, if $\mathbf{A}$ is the above $m \times n$ matrix and c is in $\mathbb{R}$, then the **scalar product of A by c** is

$$c\mathbf{A} = \begin{pmatrix} ca_{11} & ca_{12} & \cdots & ca_{1n} \\ ca_{21} & & & \\ \vdots & & & \vdots \\ ca_{m1} & \cdots & & ca_{mn} \end{pmatrix}.$$

---

† As usual, subtraction may be defined in terms of addition: for $m \times n$ matrices $\mathbf{A}$ and $\mathbf{B}$, $\mathbf{A} - \mathbf{B}$ is that $m \times n$ matrix which, added to $\mathbf{B}$, yields $\mathbf{A}$: $(\mathbf{A} - \mathbf{B}) + \mathbf{B} = \mathbf{A}$. Of course it turns out that with $\mathbf{A}$ and $\mathbf{B}$ as above,

$$\mathbf{A} - \mathbf{B} = \begin{pmatrix} a_{11} - b_{11} & \cdots & a_{1n} - b_{1n} \\ \vdots & & \vdots \\ a_{m1} - b_{m1} & \cdots & a_{mn} - b_{mn} \end{pmatrix}.$$

Example 1.

$$\begin{pmatrix} 1 & 0 \\ -1 & 3 \\ 2 & 0 \end{pmatrix} + \begin{pmatrix} 2 & 1 \\ 1 & 0 \\ -3 & 2 \end{pmatrix} = \begin{pmatrix} 3 & 1 \\ 0 & 3 \\ -1 & 2 \end{pmatrix};$$

$$(-2)\begin{pmatrix} 1 & 0 \\ -1 & 3 \\ 2 & 0 \end{pmatrix} = \begin{pmatrix} -2 & 0 \\ 2 & -6 \\ -4 & 0 \end{pmatrix}$$

We denote the set of all $m \times n$ matrices with real entries by $\mathbb{R}_{m \times n}$.

Now matrix addition and scalar multiplication are performed (and equality of matrices holds) just as if $m \times n$ matrices were vectors in $\mathbb{R}_{mn}$, written in a slightly peculiar fashion. Thus, we obtain, for free,

3.1 $\mathbb{R}_{m \times n}$ **as a vector space.** The set $\mathbb{R}_{m \times n}$ of all $m \times n$ matrices with elements from $\mathbb{R}$ is a vector space over $\mathbb{R}$ with respect to the operations of addition and scalar multiplication. Moreover, this vector space $\mathbb{R}_{m \times n}$ is isomorphic to $\mathbb{R}_{mn}$.

By Chapter 1, 5.5:

3.2 The dimension of the set of all $m \times n$ matrices as a vector space over $\mathbb{R}$ is mn.

As a special case we have $\mathbb{R}_{1 \times m}$, whose elements $(v_1, \ldots, v_m)$ are scarcely distinguishable from those of $\mathbb{R}_m$. We also have the very similar space of "column vectors" $\mathbb{R}_{m \times 1}$ whose elements are of the form

$$\begin{pmatrix} v_1 \\ \cdot \\ \cdot \\ \cdot \\ v_m \end{pmatrix}.$$

Matrix multiplication is not a difficult process, but we shall define it in three stages in order to emphasize three important properties.

I. *The product of a row vector and a column vector.* Given a row vector

$$\mathbf{U} = (u_1, \ldots, u_n)$$

and a column vector

$$\mathbf{V} = \begin{pmatrix} v_1 \\ \cdot \\ \cdot \\ \cdot \\ v_n \end{pmatrix}$$

containing the same number of entries, we define their **product** to be the scalar $u_1 v_1 + \cdots + u_n v_n$,

$$\mathbf{UV} = (u_1, \ldots, u_n)\begin{pmatrix} v_1 \\ \cdot \\ \cdot \\ \cdot \\ v_n \end{pmatrix} = u_1v_1 + \cdots + u_nv_n.\text{†}$$

II. *The product of an $m \times n$ matrix and an $n \times 1$ column vector.* Given an $m \times n$ matrix $\mathbf{A}$ with rows $\mathbf{R}_1, \ldots, \mathbf{R}_m$ (each a $1 \times n$ row vector) and an $n \times 1$ column vector $\mathbf{V}$, we define

$$\mathbf{AV} = \begin{pmatrix} \mathbf{R}_1 \\ \cdot \\ \cdot \\ \cdot \\ \mathbf{R}_m \end{pmatrix} \mathbf{V} = \begin{pmatrix} \mathbf{R}_1\mathbf{V} \\ \cdot \\ \cdot \\ \cdot \\ \mathbf{R}_m\mathbf{V} \end{pmatrix}.$$

The $\mathbf{R}_i\mathbf{V}$ are defined in I. Thus, in terms of components, if

$$\mathbf{R}_1 = (a_{11}, \ldots, a_{1n}), \ldots, \mathbf{R}_m = (a_{m1}, \ldots, a_{mn}),$$

then

$$\mathbf{AV} = \begin{pmatrix} a_{11} & \cdots & a_{1n} \\ \cdot & & \cdot \\ \cdot & & \cdot \\ a_{m1} & \cdots & a_{mn} \end{pmatrix} \begin{pmatrix} v_1 \\ \cdot \\ \cdot \\ v_n \end{pmatrix} = \begin{pmatrix} a_{11}v_1 + \cdots + a_{1n}v_n \\ \cdot \\ \cdot \\ a_{m1}v_1 + \cdots + a_{mn}v_n \end{pmatrix} = \begin{pmatrix} \sum_{i=1}^{n} a_{1i}v_i \\ \cdot \\ \cdot \\ \sum_{i=1}^{n} a_{mi}v_i \end{pmatrix}.$$

For example,

$$\begin{pmatrix} 3 & 5 & 1 \\ -2 & 0 & 2 \\ 1 & 0 & 0 \\ -1 & 0 & 1 \end{pmatrix} \begin{pmatrix} 1 \\ -1 \\ 2 \end{pmatrix} = \begin{pmatrix} 3 - 5 + 2 \\ -2 + 0 + 4 \\ 1 + 0 + 0 \\ -1 + 0 + 2 \end{pmatrix} = \begin{pmatrix} 0 \\ 2 \\ 1 \\ 1 \end{pmatrix}.$$

III. *The product of an $m \times n$ matrix and an $n \times r$ matrix.* Given an $m \times n$ matrix $\mathbf{A}$ and an $n \times r$ matrix $\mathbf{B}$ with columns $\mathbf{C}_1, \ldots, \mathbf{C}_r$ (which are $n \times 1$ matrices), we define

$$\mathbf{AB} = \mathbf{A}(\mathbf{C}_1, \ldots, \mathbf{C}_r) = (\mathbf{AC}_1, \ldots, \mathbf{AC}_r).$$

The $\mathbf{AC}_i$ are defined by II. Thus, if

$$\mathbf{A} = \begin{pmatrix} \mathbf{R}_1 \\ \cdot \\ \cdot \\ \cdot \\ \mathbf{R}_m \end{pmatrix}$$

---

† This is the same as if $\mathbf{U}$ and $\mathbf{V}$ were both row vectors and we took their dot product $\mathbf{U} \cdot \mathbf{V}$. The dot product will make its official appearance in Chapter 8.

in terms of its rows, then

$$AB = \begin{pmatrix} R_1C_1 & R_1C_2 & \cdots & R_1C_r \\ R_2C_1 & R_2C_2 & \cdots & R_2C_r \\ \vdots & & & \vdots \\ R_mC_1 & R_mC_2 & \cdots & R_mC_r \end{pmatrix},$$

and in terms of components,

$$i\begin{pmatrix} a_{11} & \cdots & a_{1n} \\ \vdots & & \\ a_{i1} & \cdots & a_{in} \\ \vdots & & \\ a_{m1} & \cdots & a_{mn} \end{pmatrix} \begin{pmatrix} b_{11} & \cdots & \overset{j}{b_{1j}} & \cdots & b_{1r} \\ \vdots & & \vdots & & \vdots \\ b_{n1} & \cdots & b_{nj} & \cdots & b_{nr} \end{pmatrix}$$

$$= i\begin{pmatrix} \sum_{k=1}^{n} a_{1k}b_{k1} & \cdots & \overset{j}{\sum_{k=1}^{n} a_{1k}b_{kr}} \\ \vdots & & \vdots \\ \cdots & \boxed{\sum_{k=1}^{n} a_{ik}b_{kj}} & \cdots \\ \vdots & & \vdots \\ \sum_{k=1}^{n} a_{mk}b_{k1} & \cdots & \sum_{k=1}^{n} a_{mk}b_{kr} \end{pmatrix}.$$

Example 2.

$$\begin{pmatrix} 3 & 5 & 1 \\ -2 & 0 & 2 \\ 1 & 0 & 0 \\ -1 & 0 & 1 \end{pmatrix} \begin{pmatrix} 1 & 2 \\ -1 & 3 \\ 2 & 0 \end{pmatrix} = \begin{pmatrix} 0 & 21 \\ 2 & -4 \\ 1 & 2 \\ 1 & -2 \end{pmatrix}.$$

With a little practice (ample opportunity is afforded in the exercises), matrix multiplication becomes completely mechanical. When no one is looking, you might practice the basic gesture, which consists in moving the left hand horizontally from left to right, while the right hand tranverses a vertical line from top to bottom.

It is important to realize that

> unless the number of *columns* of **A** is the same as the number of *rows* of **B**, the product **AB** is *not* defined.

Note also that

$$(\text{an } m \times n \text{ matrix}) \times (\text{an } n \times r \text{ matrix}) = (\text{an } m \times r \text{ matrix});$$
$$(m \times n) \times (n \times r) = (m \times r).$$

The following theorem lists the basic properties of matrix multiplication:

3.3 Whenever the appropriate products are defined, matrix multiplication satisfies the following laws:

i) $A(cB) = c(AB) = (cA)B$ (Homogeneity)

ii) $A(B + C) = AB + AC$
$(A + B)C = AC + BC$ (Distributivity)

iii) $(AB)C = A(BC)$ (Associativity)

Proof. (i) Let

$$A = \begin{pmatrix} a_{11} & \cdots & a_{1n} \\ \vdots & & \vdots \\ a_{m1} & \cdots & a_{mn} \end{pmatrix} \quad \text{and} \quad B = \begin{pmatrix} b_{11} & \cdots & b_{1r} \\ \vdots & & \vdots \\ b_{n1} & \cdots & b_{nr} \end{pmatrix},$$

so that

$$cB = \begin{pmatrix} cb_{11} & \cdots & cb_{1r} \\ \vdots & & \vdots \\ cb_{n1} & \cdots & cb_{nr} \end{pmatrix}$$

and the i,j-entry of $A(cB)$ is

$$a_{i1}(cb_{1j}) + a_{i2}(cb_{2j}) + \cdots + a_{in}(cb_{nj}). \tag{1}$$

Since the i,j-entry of AB is $a_{i1}b_{1j} + \cdots + a_{in}b_{nj}$, we know that the i,j-entry of $c(AB)$ is

$$c(a_{i1}b_{1j} + a_{i2}b_{2j} + \cdots + a_{in}b_{nj}). \tag{2}$$

Since this is the same as (1), the matrices $A(cB)$ and $c(AB)$ agree entry-for-entry; hence they are the same. You use a similar argument to prove the second equality in (i) (Exercise 15a).

ii) With A and B as above, let

$$C = \begin{pmatrix} c_{11} & \cdots & c_{1r} \\ \vdots & & \vdots \\ c_{n1} & \cdots & c_{nr} \end{pmatrix}.$$

The i,j-entry of $\mathbf{A}(\mathbf{B} + \mathbf{C})$ can be rapidly calculated to be
$$a_{i1}(b_{1j} + c_{1j}) + a_{i2}(b_{2j} + c_{2j}) + \cdots + a_{in}(b_{nj} + c_{nj}), \tag{3}$$
while that of $\mathbf{AB}$ is
$$a_{i1}b_{1j} + a_{i2}b_{2j} + \cdots + a_{in}b_{nj}, \tag{4}$$
and that of $\mathbf{AC}$ is
$$a_{i1}c_{1j} + a_{i2}c_{2j} + \cdots + a_{in}c_{nj}. \tag{5}$$
Since (3) = (4) + (5), we have established the first distributive law. You do the second (Exercise 15b).

iii) Let us leave $\mathbf{A}$ and $\mathbf{B}$ as in (i), but change $\mathbf{C}$ into

$$\mathbf{C} = \begin{pmatrix} c_{11} & \cdots & c_{1s} \\ \cdot & & \cdot \\ \cdot & & \cdot \\ \cdot & & \cdot \\ c_{r1} & \cdots & c_{rs} \end{pmatrix},$$

so that we can perform the multiplication $(\mathbf{AB})\mathbf{C}$. The ith row of $\mathbf{AB}$ is

$$a_{i1}b_{11} + \cdots + a_{in}b_{n1}, a_{i1}b_{12} + \cdots + a_{in}b_{n2}, \ldots, a_{i1}b_{1r} + \cdots + a_{in}b_{nr},$$

so the i,j-entry of $(\mathbf{AB})\mathbf{C}$ is

$$(a_{i1}b_{11} + \cdots + a_{in}b_{n1})c_{1j} + (a_{i1}b_{12} + \cdots + a_{in}b_{n2})c_{2j}$$
$$+ \cdots + (a_{i1}b_{1r} + \cdots + a_{in}b_{nr})c_{rj}. \tag{6}$$

You calculate the i,j-entry of $\mathbf{A}(\mathbf{BC})$ by first calculating the jth column of $\mathbf{BC}$, and show that this is the same as (6) after a bit of rearranging (Exercise 15c). ∎

Starting out with a vector $\mathbf{V}$ in $\mathbb{R}_n$ (written as a column vector), if we premultiply by an $m \times n$ matrix $\mathbf{A}$, we obtain an $m \times 1$ vector $\mathbf{AV}$, which we can regard as being in $\mathbb{R}_m$ (written as a column). Given such a $\mathbf{V}$ and $\mathbf{A}$, matrix multiplication yields a unique vector $\mathbf{AV}$, so that the correspondence $\mathbf{V} \to \mathbf{AV}$ is a function from $\mathbb{R}_n$ into $\mathbb{R}_m$ (see Chapter 0, Section 4). Let us denote this function by $\mathbf{T}$ (for transformation), so that
$$\mathbf{T}(\mathbf{V}) = \mathbf{AV}$$
is a function from $\mathbb{R}_n$ into $\mathbb{R}_m$, where $\mathbf{A}$ is an $m \times n$ matrix, and $\mathbf{V}$ an $n \times 1$ matrix. The crucial properties of this function (readily obtainable from 3.2) are

(Homogeneity)
$$\mathbf{T}(c\mathbf{V}) = c\mathbf{T}(\mathbf{V}) \quad \text{for any } c \text{ in } \mathbb{R} \tag{7}$$
(that is, $\mathbf{A}(c\mathbf{V}) = c\mathbf{AV}$),

(Additivity)
$$\mathbf{T}(\mathbf{V}_1 + \mathbf{V}_2) = \mathbf{T}(\mathbf{V}_1) + \mathbf{T}(\mathbf{V}_2) \tag{8}$$
(that is, $\mathbf{A}(\mathbf{V}_1 + \mathbf{V}_2) = \mathbf{AV}_1 + \mathbf{AV}_2$).

Functions on vector spaces which satisfy these two conditions are called *linear transformations*. The study of this topic is one of the most central issues of this book. You encountered a special type of linear transformation in the previous chapter, the isomorphisms of Section 5. We shall leave the subject of linear transformations until the next chapter, where they receive star billing.

There are two other simple, special properties of matrix multiplication which are often useful.

3.4 Let $\mathbf{A} = (\mathbf{A}_1, \ldots, \mathbf{A}_n)$ be an $m \times n$ matrix written in terms of its columns. Then

i) $\mathbf{A}\mathbf{e}_i = \mathbf{A}_i$, where $\{\mathbf{e}_1, \ldots, \mathbf{e}_n\}$ is the usual basis for $\mathbb{R}_n$, and

ii) $\mathbf{A}\begin{pmatrix} v_1 \\ \vdots \\ v_n \end{pmatrix} = v_1 \mathbf{A}_1 + v_2 \mathbf{A}_2 + \cdots + v_n \mathbf{A}_n$ for any vector $\begin{pmatrix} v_1 \\ \vdots \\ v_n \end{pmatrix}$ in $\mathbb{R}_n$.

Proof. (i) Observe that for $\mathbf{W} = (w_1, \ldots, w_m)$, $\mathbf{W}\mathbf{e}_i = w_i$; then use II in the definition of matrix multiplication. You should write out the details (Exercise 17a).

ii) Write

$$\begin{pmatrix} v_1 \\ \vdots \\ v_n \end{pmatrix} = v_1 \mathbf{e}_1 + v_2 \mathbf{e}_2 + \cdots + v_n \mathbf{e}_n,$$

and use distributivity and (i). (See Exercise 17b.) ∎

Supplement to Section 3. Partitioned Matrices

As we shall discover, it is often useful to be able to slice up matrices into submatrices, e.g., when the submatrices have special meanings, or when you're working with a computer and your matrix is too large to feed in all at once. If you have two matrices partitioned into submatrices, say

$$\mathbf{A} = \begin{pmatrix} \mathbf{A}_{11} & \mathbf{A}_{12} & \cdots & \mathbf{A}_{1m} \\ \mathbf{A}_{21} & \mathbf{A}_{22} & \cdots & \mathbf{A}_{2m} \\ \vdots & & & \vdots \\ \mathbf{A}_{l1} & \mathbf{A}_{l2} & \cdots & \mathbf{A}_{lm} \end{pmatrix}, \quad \mathbf{A}_{ij} \text{ submatrices,}$$

and

$$\mathbf{B} = \begin{pmatrix} \mathbf{B}_{11} & \mathbf{B}_{12} & \cdots & \mathbf{B}_{1n} \\ \mathbf{B}_{21} & \mathbf{B}_{22} & \cdots & \mathbf{B}_{2n} \\ \vdots & & & \vdots \\ \mathbf{B}_{m1} & \mathbf{B}_{m2} & \cdots & \mathbf{B}_{mn} \end{pmatrix}, \quad \mathbf{B}_{rs} \text{ submatrices,}$$

it is often desirable to be able to multiply **AB** by performing appropriate matrix operations on the submatrices. It turns out that you can multiply these matrices just as if the entries were scalars (rather than matrices themselves), provided one can perform each of the matrix multiplications $\mathbf{A}_{ik}\mathbf{B}_{kj}$ and also add the appropriate sums

$$\mathbf{A}_{i1}\mathbf{B}_{1j} + \cdots + \mathbf{A}_{im}\mathbf{B}_{mj}.$$

3.5 *Multiplication of partitioned matrices.* The equation

$$\begin{pmatrix} \mathbf{A}_{11} & \cdots & \mathbf{A}_{1m} \\ \vdots & & \vdots \\ \mathbf{A}_{l1} & \cdots & \mathbf{A}_{lm} \end{pmatrix} \begin{pmatrix} \mathbf{B}_{11} & \cdots & \mathbf{B}_{1n} \\ \vdots & & \vdots \\ \mathbf{B}_{m1} & \cdots & \mathbf{B}_{mn} \end{pmatrix}$$

$$= \begin{pmatrix} \sum_{k=1}^{m}\mathbf{A}_{1k}\mathbf{B}_{k1} & \cdots & \sum_{k=1}^{m}\mathbf{A}_{1k}\mathbf{B}_{kn} \\ \vdots & & \vdots \\ \sum_{k=1}^{m}\mathbf{A}_{lk}\mathbf{B}_{k1} & \cdots & \sum_{k=1}^{m}\mathbf{A}_{lk}\mathbf{B}_{kn} \end{pmatrix} \quad (9)$$

holds whenever the expressions $\sum_{k=1}^{m}\mathbf{A}_{ik}\mathbf{B}_{kj}$ have meaning (for $i = 1, \ldots, l$ and $j = 1, \ldots, n$). That is, you can multiply two partitioned matrices just as if their entries were real numbers, provided you can form the appropriate sums of products of the submatrices.

Before giving a proof, let us note a numerical example,

Example 3. Let us redo Example 2, partitioned as indicated:

$$\begin{pmatrix} \overset{\mathbf{A}_{11}}{3 \quad 5} & \overset{\mathbf{A}_{12}}{1} \\ -2 \quad 0 & 2 \\ \hline \overset{\mathbf{A}_{21}}{1 \quad 0} & \overset{\mathbf{A}_{22}}{0} \\ -1 \quad 0 & 1 \end{pmatrix} \begin{pmatrix} \overset{\mathbf{B}_{11}}{1 \quad 2} \\ -1 \quad 3 \\ \hline \overset{\mathbf{B}_{21}}{2 \quad 0} \end{pmatrix} = \begin{pmatrix} 0 & 21 \\ 2 & -4 \\ 1 & 2 \\ 1 & -2 \end{pmatrix}$$

vs. $\begin{pmatrix} \mathbf{A}_{11} & \mathbf{A}_{12} \\ \mathbf{A}_{21} & \mathbf{A}_{22} \end{pmatrix} \begin{pmatrix} \mathbf{B}_{11} \\ \mathbf{B}_{21} \end{pmatrix} = \begin{pmatrix} \mathbf{A}_{11}\mathbf{B}_{11} + \mathbf{A}_{12}\mathbf{B}_{21} \\ \mathbf{A}_{21}\mathbf{B}_{11} + \mathbf{A}_{22}\mathbf{B}_{21} \end{pmatrix}$

Now

$$\mathbf{A}_{11}\mathbf{B}_{11} + \mathbf{A}_{12}\mathbf{B}_{21} = \begin{pmatrix} 3 & 5 \\ -2 & 0 \end{pmatrix}\begin{pmatrix} 1 & 2 \\ -1 & 3 \end{pmatrix} + \begin{pmatrix} 1 \\ 2 \end{pmatrix}(2, \; 0)$$

$$= \begin{pmatrix} -2 & 21 \\ -2 & -4 \end{pmatrix} + \begin{pmatrix} 2 & 0 \\ 4 & 0 \end{pmatrix} = \begin{pmatrix} 0 & 21 \\ 2 & -4 \end{pmatrix},$$

and
$$A_{21}B_{11} + A_{22}B_{21} = \begin{pmatrix} 1 & 0 \\ -1 & 0 \end{pmatrix}\begin{pmatrix} 1 & 2 \\ -1 & 3 \end{pmatrix} + \begin{pmatrix} 0 \\ 1 \end{pmatrix}(2, \ 0)$$

$$= \begin{pmatrix} 1 & 2 \\ -1 & -2 \end{pmatrix} + \begin{pmatrix} 0 & 0 \\ 2 & 0 \end{pmatrix} = \begin{pmatrix} 1 & 2 \\ 1 & -2 \end{pmatrix}.$$

Proof of **3.5.** Given any specific partitioning of a specific matrix, it is easy to see that 3.4 holds. Moreover, one can appreciate that (9) is a simple consequence of the definition of matrix multiplication: it essentially boils down to the fact that you can write $a_{i1}b_{1j} + \cdots + a_{im}b_{mj}$ as

$$(a_{i1}b_{1j} + \cdots + a_{ir_1}b_{r_1 j})$$
$$+ \ (a_{i,r_1+1}b_{r_1+1,j} + \cdots + a_{ir_2}b_{r_2 j}) + \cdots + (a_{i,r_m+1}b_{r_m+1,j} + \cdots + a_{im}b_{mj})$$

in a consistent fashion. Unfortunately, trying to prove (9) in general results in one of those maddening situations where you end up tripping over your own subscripts.

We shall content ourselves with showing that the 1,1-entry of **AB** ends up as it should. Using obvious notation, if $\sum_{k=1}^{m} \mathbf{A}_{1k}\mathbf{B}_{k1}$ is to have meaning, then we can write

$$\mathbf{A} = \begin{pmatrix} \mathbf{A}_{11} & \cdots & \mathbf{A}_{1m} \end{pmatrix} = \begin{pmatrix} a_{11}^{(1)} \cdots a_{1r_1}^{(1)} & a_{11}^{(2)} \cdots a_{1r_2}^{(2)} & \cdots & a_{11}^{(m)} \cdots a_{1r_m}^{(m)} \\ \mathbf{A}_{11} & \mathbf{A}_{12} & \cdots & \mathbf{A}_{1m} \end{pmatrix},$$

$$\mathbf{B} = \begin{pmatrix} \mathbf{B}_{11} \\ \vdots \\ \mathbf{B}_{m1} \end{pmatrix} = \begin{pmatrix} b_{11}^{(1)} \\ \vdots \\ b_{r_1 1}^{(1)} \\ \hline b_{r_2 1}^{(2)} \\ \vdots \\ \mathbf{B}_{21} \\ \hline \vdots \\ b_{11}^{(m)} \\ \vdots \\ b_{r_m 1}^{(m)} \end{pmatrix}$$

with $\mathbf{B}_{11}$, $\mathbf{B}_{21}$, $\mathbf{B}_{m1}$ labeling the respective blocks.

Hence, the 1,1-entry of $\sum_{k=1}^{m} \mathbf{A}_{1k}\mathbf{B}_{k1}$ will be

$$\sum_{k=1}^{r_1} a_{1k}^{(1)}b_{k1}^{(1)} + \sum_{k=1}^{r_2} a_{1k}^{(2)}b_{k1}^{(2)} + \cdots + \sum_{k=1}^{r_m} a_{1k}^{(m)}b_{k1}^{(m)},$$

which is also the 1,1-entry of $\mathbf{AB}$.

Perhaps you will be thoroughly convinced if you show the 1,2-entries are the same (Exercise 14).

The main question left is, given two matrices $\mathbf{A}(l \times m)$ and $\mathbf{B}(m \times n)$, how can you partition $\mathbf{A}$ and $\mathbf{B}$ to ensure that the resulting submatrices will be compatible in $\sum \mathbf{A}_{ik}\mathbf{B}_{kj}$? The answer (which you may prove in Exercise 18) is:

3.6 To multiply $\mathbf{AB}$ as in (9), the *rows* of $\mathbf{B}$ must be partitioned to conform to the *columns* of $\mathbf{A}$; the columns of $\mathbf{B}$ and rows of $\mathbf{A}$ may be partitioned in any manner.

$$\begin{pmatrix} \overbrace{\mathbf{A}_{11}}^{r_1} & \overbrace{\mathbf{A}_{12}}^{r_2} & \cdots & \overbrace{\mathbf{A}_{1m}}^{r_m} \end{pmatrix} \begin{matrix} r_1 \{ \\ r_2 \{ \\ \\ \\ r_m \{ \end{matrix} \begin{pmatrix} \mathbf{B}_{11} \\ \hline \mathbf{B}_{12} \\ \hline \vdots \\ \hline \mathbf{B}_{m1} \end{pmatrix}$$

(Slice in any way)

EXERCISES

1. a) Multiply $\begin{pmatrix} 1 & 2 & 3 \\ 1 & 4 & 9 \\ 1 & 1 & 1 \end{pmatrix} \begin{pmatrix} -\frac{5}{2} & \frac{1}{2} & 3 \\ 4 & -1 & -3 \\ -\frac{3}{2} & \frac{1}{2} & 1 \end{pmatrix}$.

   b) Multiply in the reverse direction.

2. For

$$\mathbf{A} = \begin{pmatrix} 0 & 1 & 2 \\ 3 & -3 & 0 \\ 4 & -4 & 2 \\ 5 & 0 & 10 \end{pmatrix}, \quad \mathbf{B} = \begin{pmatrix} 1 & 0 & 0 & 0 \\ 0 & \frac{1}{3} & 0 & 0 \\ 0 & 0 & 1 & 0 \\ 0 & 0 & 0 & 1 \end{pmatrix},$$

$$\mathbf{B}' = \begin{pmatrix} 1 & 0 & 0 \\ 0 & \frac{1}{3} & 0 \\ 0 & 0 & 1 \end{pmatrix}, \quad \mathbf{C} = \begin{pmatrix} 1 & 0 & 0 & 0 \\ 0 & 1 & 0 & 0 \\ 0 & -\frac{4}{3} & 1 & 0 \\ 0 & 0 & 0 & 1 \end{pmatrix}, \quad \mathbf{C}' = \begin{pmatrix} 1 & 0 & 0 \\ 0 & 1 & 0 \\ 0 & -\frac{4}{3} & 1 \end{pmatrix},$$

$$\mathbf{D} = \begin{pmatrix} 0 & 1 & 0 & 0 \\ 1 & 0 & 0 & 0 \\ 0 & 0 & 1 & 0 \\ 0 & 0 & 0 & 1 \end{pmatrix}, \quad \mathbf{D}' = \begin{pmatrix} 0 & 1 & 0 \\ 1 & 0 & 0 \\ 0 & 0 & 1 \end{pmatrix},$$

calculate
(a) **BA** (b) **AB'** (c) **CA**
(d) **AC'** (e) **DA** (f) **AD'**.

3. Let
$$e_1 = (1, 0, \ldots, 0), \quad e_2 = (0, 1, 0, \ldots, 0), \ldots, e_n = (0, \ldots, 0, 1)$$
be matrices in $\mathbb{R}_{1 \times n}$, and take **A** in $\mathbb{R}_{n \times m}$. Prove that $e_i \mathbf{A}$ is the matrix which consists of the ith row of **A**, $i = 1, \ldots, n$. That is, if

$$\mathbf{A} = \begin{pmatrix} A_1 \\ \vdots \\ A_n \end{pmatrix}$$

is a partitioning of **A** into its rows, then $e_i \mathbf{A} = A_i$.

4. (a) Under what conditions are both **AB** and **BA** defined? (b) When is $\mathbf{A}^2 = \mathbf{AA}$ defined? (c) When are both **AB** and **A + B** defined?

5. In the text we calculated **VW** for

$$\mathbf{V} = (v_1, \ldots, v_m) \quad \text{and} \quad \mathbf{W} = \begin{pmatrix} w_1 \\ \vdots \\ w_m \end{pmatrix}.$$

You calculate **WV**. Warm up on
$$\mathbf{V} = (1, 2, 0), \quad \mathbf{W} = \begin{pmatrix} 3 \\ -1 \\ 4 \end{pmatrix}.$$

6. Let

$$\mathbf{E}_{ij} = i \begin{pmatrix} & & j & & \\ & & \vdots & & \\ \cdots & & 1 & \cdots & \\ & & \vdots & & \end{pmatrix}$$

be the $m \times n$ matrix which has a 1 as its i,j-entry and 0's elsewhere. The collection of these matrices $\mathbf{E}_{ij}$ (where $i = 1, \ldots, m$, and $j = 1, \ldots, n$) are called the $m \times n$ **matrix units**.

a) Write out the 3×2 matrix units.
b) Write the matrix
$$\begin{pmatrix} 1 & 2 \\ -1 & 3 \\ 2 & 0 \end{pmatrix}$$
as a sum of these matrix units.
c) Show that these 3×2 matrix units form a basis for $\mathbb{R}_{3 \times 2}$.
d) Show that the $m \times n$ matrix units are a basis for $\mathbb{R}_{m \times n}$.

7. (a) Find an isomorphism between $\mathbb{R}_{m \times n}$ and $\mathbb{R}_{mn}$. (b) Find another.

8. Let

$$A = \begin{pmatrix} a_{11} & \cdots & a_{1n} \\ \vdots & & \vdots \\ a_{m1} & \cdots & a_{mn} \end{pmatrix}, \quad \mathbf{u} = \underbrace{(1, 1, \ldots, 1)}_{n}, \quad \mathbf{v} = \left.\begin{pmatrix} 1 \\ 1 \\ \vdots \\ 1 \end{pmatrix}\right\} n.$$

   a) Calculate **uA**. State verbally what this is.
   b) Same for **Av**. (c) Same for **uAv**
   d) If $m = n$ what is **uv**? **vu**?

9. Let

$$A = \begin{pmatrix} 1 & 1 \\ 2 & -1 \end{pmatrix}, \quad B = \begin{pmatrix} 1 & 1 \\ 2 & 2 \end{pmatrix}, \quad C = \begin{pmatrix} 1 & 1 \\ 2 & -1 \\ 1 & -3 \end{pmatrix},$$

$$D = \begin{pmatrix} 1 & 1 \\ 2 & 2 \\ 3 & 3 \end{pmatrix}, \quad E = \begin{pmatrix} 1 & 2 & 1 \\ 1 & -1 & -3 \end{pmatrix}, \quad F = \begin{pmatrix} 1 & 2 & 3 \\ 1 & 2 & 3 \end{pmatrix}.$$

   a) Calculate $A + B$; $C + D$; $E + F$.
   b) Calculate AB, BA, CE, and DF.
   c) What other products are possible among **A**, **C**, and **E**?
   d) Calculate these.

10. Let $\mathbf{J}_{m \times n}$ be the $m \times n$ matrix all of whose entries are 1. For example,

$$\mathbf{J}_{2 \times 3} = \begin{pmatrix} 1 & 1 & 1 \\ 1 & 1 & 1 \end{pmatrix}.$$

   Prove that:
   a) $\mathbf{J}_{2 \times 3} \mathbf{J}_{3 \times 4} = 3 \mathbf{J}_{2 \times 4}$.
   b) $\mathbf{J}_{m \times 3} \mathbf{J}_{3 \times n} = 3 \mathbf{J}_{m \times n}$.
   c) $\mathbf{J}_{m \times l} \mathbf{J}_{l \times n} = l \mathbf{J}_{m \times n}$.

   For **A** an arbitrary $m \times n$ matrix, what is:
   d) $\mathbf{J}_{l \times m} \mathbf{A}$?   e) $\mathbf{A} \mathbf{J}_{n \times r}$?   f) $\mathbf{J}_{l \times m} \mathbf{A} \mathbf{J}_{n \times r}$?

11. The following means of partially checking matrix multiplication may be of interest to those who do not have instantaneous access to a computer, and whose lack of arithmetic skill is not made up for by great luck in making compensating errors. Let

$$A = \begin{pmatrix} a_{11} & \cdots & a_{1m} \\ \vdots & & \vdots \\ a_{l1} & \cdots & a_{lm} \end{pmatrix}, \quad B = \begin{pmatrix} b_{11} & \cdots & b_{1n} \\ \vdots & & \vdots \\ b_{m1} & \cdots & b_{mn} \end{pmatrix},$$

$$\mathbf{u} = \underbrace{(1, 1, \ldots, 1)}_{l} \quad \text{and} \quad \mathbf{v} = \left.\begin{pmatrix} 1 \\ 1 \\ \vdots \\ 1 \end{pmatrix}\right\} n.$$

As you saw in Exercise 10, **uA** is the vector whose components are the column sums of the entries in **A**, while **Bv** has as components the row sums of the entries of **B**. Since matrix multiplication is associative,

$$(\mathbf{uA})(\mathbf{Bv}) = \mathbf{u}(\mathbf{AB})\mathbf{v}.$$

Thus, one way of partially checking your work in constructing **AB** is to compare (**uA**)(**Bv**), the product of the vector of column sums of **A** and the vector of row sums of **B**, with **u**(**AB**)**v**, the total sum of the entries of **AB**. Use this to check 1(a) and (b); 2(a) and (c).

12. Partitioned matrices afford a means for carrying out matrix multiplication and the check in Exercise 11 simultaneously.
    a) Show that

$$\left(\frac{\mathbf{A}}{\mathbf{uA}}\right)(\mathbf{B} \mid \mathbf{Bv}) = \left(\begin{array}{c|c} \mathbf{AB} & \mathbf{ABv} \\ \hline \mathbf{uAB} & \mathbf{uABv} \end{array}\right),$$

where **ABv** is the sum of the rows of **AB**, **uAB** the sum of the columns of **AB**, and **uABv** is the total sum of the entries in **AB**.

To use this as a check, given **A** and **B**, one augments them by **uA** and **Bv**, respectively, and calculates

$$\left(\begin{array}{ccc} a_{11} & \cdots & a_{1m} \\ \cdot & & \cdot \\ \cdot & & \cdot \\ a_{l1} & \cdots & a_{lm} \\ \hline \sum_{k=1}^{l} a_{k1} & \cdots & \sum_{k=1}^{l} a_{km} \end{array}\right) \left(\begin{array}{ccc|c} b_{11} & \cdots & b_{1n} & \sum_{k=1}^{n} b_{1k} \\ \cdot & & \cdot & \cdot \\ \cdot & & \cdot & \cdot \\ b_{m1} & \cdots & b_{mn} & \sum_{k=1}^{n} b_{mk} \end{array}\right)$$

Partition the resulting product as

$$\left(\begin{array}{ccc|c} c_{11} & \cdots & c_{1n} & d_1 \\ \cdot & & & \cdot \\ \cdot & \mathbf{C} & & \mathbf{d} \\ \cdot & & & \cdot \\ c_{l1} & \cdots & c_{ln} & d_l \\ \hline e_1 & \cdots & e_n & f \\ & \mathbf{e} & & \end{array}\right),$$

and check that the sum of the rows of **C** is **d**, the sum of the columns of **C** is **e**, and the sum total of the entries of **C** is f. If all these check, then either **C** = **AB** or your errors compensate beautifully!
    b) Illustrate this on 1(b), and
    c) on 2(d),   (d) on 2(e).

13. a) Make up a flow diagram for matrix addition. Thus, you are given two sets of mn numbers $\{a_{ij}\}$, $\{b_{ij}\}$, where $i = 1, \ldots, m$, and $j = 1, \ldots, n$; and you are to make up a flow diagram (as in the supplement to Section 1) which completely instructs a computer to give you back the matrix **A** + **B**.

b) Make up a flow diagram for matrix multiplication, given the lm numbers $\{a_{ij}\}$ and the mn numbers $\{b_{jk}\}$.

14. Prove that the 1,2-entry of (9) is what it should be.
15. Finish proving 3.2 (a) (i); (b) (ii); (c) (iii).
16. Prove (a) (7); (b) 8.
17. Finish proving 3.3, (a) (i); (b) (ii).
18. Prove 3.5.

Honors Projects. Show that this section goes over for matrices with entries from (1) $\mathbb{C}$ (we write $\mathbb{C}_{m \times n}$ for all such $m \times n$ matrices); (2) $\mathbb{F}$, an arbitrary field (we write $\mathbb{F}_{m \times n}$, of course).

It might be worthwhile for you to acquire some familiarity with the manipulation of complex matrices. We therefore present the following exercises (which contain "real life" matrices).

19. Some of the isospin matrices used by physicists are called

$$\tau_1 = \begin{pmatrix} 0 & 1 \\ 1 & 0 \end{pmatrix}, \quad \tau_2 = \begin{pmatrix} 0 & -i \\ i & 0 \end{pmatrix}, \quad \tau_+ = \begin{pmatrix} 0 & 1 \\ 0 & 0 \end{pmatrix}, \quad \text{and} \quad \tau_- = \begin{pmatrix} 0 & 0 \\ 1 & 0 \end{pmatrix}.$$

a) Write τ_+ as a linear combination of τ_1 and τ_2 (you'll need complex coefficients).
b) Same for τ_-.

20. The Pauli spin matrices are

$$I_2 = \begin{pmatrix} 1 & 0 \\ 0 & 1 \end{pmatrix}, \quad \sigma_1 = \begin{pmatrix} 0 & 1 \\ 1 & 0 \end{pmatrix}, \quad \sigma_2 = \begin{pmatrix} 0 & -i \\ i & 0 \end{pmatrix}, \quad \sigma_3 = \begin{pmatrix} 1 & 0 \\ 0 & -1 \end{pmatrix}.$$

Fill in the following "multiplication table".

	σ_1	σ_2	σ_3
σ_1			
σ_2			
σ_3			

21. Some of the SU_3 matrices involved in the physicist's "eightfold way" are

$$\lambda_1 = \begin{pmatrix} 0 & 1 & 0 \\ 1 & 0 & 0 \\ 0 & 0 & 0 \end{pmatrix} \quad \lambda_2 = \begin{pmatrix} 0 & -i & 0 \\ i & 0 & 0 \\ 0 & 0 & 0 \end{pmatrix} \quad \lambda_3 = \begin{pmatrix} 1 & 0 & 0 \\ 0 & -1 & 0 \\ 0 & 0 & 0 \end{pmatrix},$$

$$\lambda_4 = \begin{pmatrix} 0 & 0 & 1 \\ 0 & 0 & 0 \\ 1 & 0 & 0 \end{pmatrix}, \quad \lambda_5 = \begin{pmatrix} 0 & 0 & -1 \\ 0 & 0 & 0 \\ i & 0 & 0 \end{pmatrix}, \quad \lambda_6 = \begin{pmatrix} 0 & 0 & 0 \\ 0 & 0 & 1 \\ 0 & 1 & 0 \end{pmatrix},$$

$$\lambda_7 = \begin{pmatrix} 0 & 0 & +i \\ 0 & 0 & 0 \\ -i & 0 & 0 \end{pmatrix}. \quad \lambda_8 = \begin{pmatrix} \frac{1}{3} & 0 & 0 \\ 0 & \frac{1}{3} & 0 \\ 0 & 0 & -\frac{2}{3} \end{pmatrix}.$$

As examples calculate $\lambda_1 \lambda_2$; $\lambda_3 \lambda_5$; $(\lambda_2 + \lambda_3)\lambda_5$; $(\lambda_2 + \lambda_3)(\lambda_5 - \lambda_4)$.

4. SQUARE MATRICES; TRANSPOSITION

There are some special privileges attendant upon the space $\mathbb{R}_{n \times n}$ of $n \times n$ "square" matrices: you can both add and multiply everything in sight (and still remain in the space). To celebrate the closure of multiplication here, we shall denote $\mathbf{AA}$ by $\mathbf{A}^2$, $\mathbf{AA}^2$ by $\mathbf{A}^3$, etc. ($\mathbf{AA}^2$ and $\mathbf{A}^2\mathbf{A}$ are the same by the associative law.) Among other pleasant multiplicative features which hold in $\mathbb{R}_{n \times n}$, let us note the $n \times n$ *identity matrix*

$$\mathbf{I}_n = \begin{pmatrix} 1 & & & & 0\text{'s} \\ & 1 & & & \\ & & \cdot & & \\ & & & \cdot & \\ & & & & \cdot \\ 0\text{'s} & & & & 1 \end{pmatrix};$$

this is often written just $\mathbf{I}$. It is the unique matrix for which $\mathbf{AI} = \mathbf{A} = \mathbf{IA}$ for all $\mathbf{A}$ in $\mathbb{R}_{n \times n}$ (Exercise 3). Moreover, at least some matrices $\mathbf{A}$ in $\mathbb{R}_{n \times n}$ have inverses relative to this multiplicative identity: there may exist a matrix $\mathbf{B}$ in $\mathbb{R}_{n \times n}$ for which $\mathbf{BA} = \mathbf{I}$. We write $\mathbf{B} = \mathbf{A}^{-1}$.

Example 1. If

$$\mathbf{A} = \begin{pmatrix} 1 & 2 & 3 \\ 1 & 4 & 9 \\ 1 & 1 & 1 \end{pmatrix}, \quad \text{then } \mathbf{A}^{-1} = \begin{pmatrix} -\frac{5}{2} & \frac{1}{2} & 3 \\ 4 & -1 & -3 \\ -\frac{3}{2} & \frac{1}{2} & 1 \end{pmatrix},$$

as you can readily verify. In the next example we shall see that there are nonzero matrices without inverses.

To take stock of the basic properties of matrix arithmetic, we compare them with the field axioms (Chapter 0, Section 6), which hold for the real numbers, complex numbers, rational numbers, etc. These are listed on p. 105.

In summary, the basic laws of matrix addition obey the field axioms, while matrix multiplication has its differences when it comes to commutativity and inverses for nonzero elements.

Along with addition and multiplication, matrices are endowed with a scalar multiplication (which has no field counterpart) satisfying the usual vector-space axioms, together with the relation of **homogeneity** linking scalar and matrix multiplication:†

$$c(\mathbf{AB}) = (c\mathbf{A})\mathbf{B} = \mathbf{A}(c\mathbf{B}).$$

You have already encountered noncommuting matrices in the exercises for the last section. We present another in Example 2.

---

† A mathematical system which satisfies the ring axioms (see following footnote) and which is equipped with a scalar multiplication satisfying the vector space axioms and homogeneity is known as an **algebra**. $\mathbb{R}_{n \times n}$ is a fine example, as is $\mathbb{R}[x]$.

4 / Square matrices; transposition

Field Axioms		*Arithmetic Properties of $n \times n$ matrices*†
Name	*Statement*	
Closure of addition	For x, y in the set, $x + y$ is a unique member of the set.	
Associativity of addition	For any x, y, z in the set, $x + (y + z) = (x + y) + z$.	
Additive identity	There is an element 0 in the set for which $0 + x = x = x + 0$ for all x in the set.	Yes (These are properties of $\mathbb{R}_{n \times n}$ as a vector space.)
Additive inverse	For any x in the set, there is an element $-x$ for which $x + (-x) = 0 = (-x) + x$.	
Commutativity of addition	For x, y in the set, $x + y = y + x$	
Closure of multiplication	For any x, y in the set, xy is a unique member of the set.	Yes (Definition of multiplication.)
Associativity of multiplication	For any x, y, z in the set, $x(yz) = (xy)z$.	Yes
Multiplicative identity	There is an element $1 \neq 0$ in the set for which $1x = x = x1$.	Yes ($\mathbf{I}_n$, defined above)
Multiplicative inverse for nonzero elements	For any $x \neq 0$ in the set there is an element x^{-1} in the set for which $xx^{-1} = 1 = x^{-1}x$.	No, not in general
Commutativity of multiplication	For any x, y in the set, $xy = yx$.	No, not in general
Distributivity	For any x, y, z in the set $x(y + z) = xy + xz$ $(x + y)z = xz + yz$.	Yes

Example 2. If

$$\mathbf{A} = \begin{pmatrix} 0 & 0 \\ 1 & 0 \end{pmatrix} \quad \text{and} \quad \mathbf{B} = \begin{pmatrix} 0 & 0 \\ 0 & 1 \end{pmatrix},$$

then $\mathbf{AB} = \mathbf{0}$, while $\mathbf{BA} = \mathbf{A} \neq \mathbf{0}$. Since these matrices are **zero divisors** (nonzero, but multiply with something nonzero to yield zero), they cannot have multiplicative inverses. For, if $\mathbf{A}$, for example, did have an inverse $\mathbf{A}^{-1}$, then

$$\mathbf{A}^{-1}(\mathbf{AB}) = (\mathbf{A}^{-1}\mathbf{A})\mathbf{B} = \mathbf{IB} = \mathbf{B},$$

but it is also true that

$$\mathbf{A}^{-1}(\mathbf{AB}) = \mathbf{A}^{-1}(\mathbf{0}) = \mathbf{0}.$$

---

† A mathematical system which satisfies the same axioms here as does $\mathbb{R}_{n \times n}$ is called a *ring*.

Thus, if **A** did have an inverse, we would arrive at the nonsensical conclusion that
$$\mathbf{B} = \begin{pmatrix} 0 & 0 \\ 1 & 0 \end{pmatrix} = \mathbf{0} = \begin{pmatrix} 0 & 0 \\ 0 & 0 \end{pmatrix},$$
and hence that $1 = 0$.

We illustrate further some of the broader possibilities for behavior which one finds in $\mathbb{R}_{n \times n}$ as compared with the number systems you are more accustomed to:

Example 3. For
$$\mathbf{C} = \begin{pmatrix} 2 & 3 \\ -1 & -2 \end{pmatrix},$$
one finds that $\mathbf{C}^2 = \mathbf{I}$ but $\mathbf{C} \neq \pm \mathbf{I}$.† For
$$\mathbf{B} = \begin{pmatrix} 0 & 0 \\ 0 & 1 \end{pmatrix},$$
as above, one finds that $\mathbf{B}^2 = \mathbf{B}$ but $\mathbf{B} \neq \mathbf{0}$ or $\mathbf{I}$. (This also could not happen in a field; see Exercise 5.) For
$$\mathbf{A} = \begin{pmatrix} 0 & 0 \\ 1 & 0 \end{pmatrix}$$
as above, one has $\mathbf{A}^2 = \mathbf{0}$ but $\mathbf{A} \neq \mathbf{0}$.

Despite these endearing idiosyncrasies, $\mathbb{R}_{n \times n}$ is actually a rather decent place. One of the purposes of Chapter 10 is to show you that $\mathbb{R}_{n \times n}$ has much in common with the complex numbers. (A relationship between complex numbers and certain matrices in $\mathbb{R}_{2 \times 2}$ is given in Exercise 13.)

For many purposes the "good guys" in $\mathbb{R}_{n \times n}$ are those matrices **A** which have an inverse $\mathbf{A}^{-1}$. In a heroic break with tradition, we shall call such matrices **invertible** (much more common names are **regular** and **nonsingular**). Thus, **A** is invertible if there is a **B** in $\mathbb{R}_{n \times n}$ for which $\mathbf{BA} = \mathbf{I}$, and we write $\mathbf{B} = \mathbf{A}^{-1}$. A matrix inverse works on both sides (that is, $\mathbf{AB} = \mathbf{I}$), but one must take care of the ordering of products of inverses:

4.1 Let **A** and **B** be matrices in $\mathbb{R}_{n \times n}$.

i) If $\mathbf{AB} = \mathbf{I}$ then $\mathbf{BA} = \mathbf{I}$.
ii) If **A** has an inverse, then so does $\mathbf{A}^{-1}$; in fact $(\mathbf{A}^{-1})^{-1} = \mathbf{A}$.
iii) If **A** and **B** have inverses, then so does **AB**, and $(\mathbf{AB})^{-1} = \mathbf{B}^{-1}\mathbf{A}^{-1}$. (The product of the inverses is the inverse of the product *but taken in reverse order.*)

---

† $\mathbf{C}^2 = \mathbf{I}$ is the same as $\mathbf{C}^2 - \mathbf{I} = \mathbf{0}$, and
$$\mathbf{C}^2 - \mathbf{I} = (\mathbf{C} - \mathbf{I})(\mathbf{C} + \mathbf{I}).$$
In a *field*, if $(c-1)(c+1) = 0$, and if $c \neq +1$ say, then you could multiply by $(c-1)^{-1}$ and obtain $c + 1 = 0$ or $c = -1$. Thus, $c = \pm 1$. Here we cannot conclude this.

Proof. i) We defer proving (i) until Chapter 3, Section 6, since the best proof makes use of material developed at that time. Of course, the subsequent material does not depend on (i); we could introduce it now but it would necessitate a lengthy detour.

ii) Using (i) we see that $(A^{-1})A = I$, so by definition $(A^{-1})^{-1} = A$.

iii) $(B^{-1}A^{-1})(AB) \stackrel{\text{assoc.}}{=} B^{-1}(A^{-1}A)B = B^{-1}IB = B^{-1}B = I.$ ∎

You should calculate $(ABC)^{-1}$ (Exercise 6b).

We now turn our attention to a special collection of invertible matrices in $\mathbb{R}_{n \times n}$ called the "elementary matrices". We shall see in Chapter 4, Section 3, that any invertible matrix can be written as a product of these elementary matrices, a fact both entertaining and useful. For now, we show that the elementary matrices serve another function: If X is an elementary matrix in $\mathbb{R}_{n \times n}$ and if A is a matrix in $\mathbb{R}_{n \times m}$ (m arbitrary), then XA is a matrix related to A by one of the row operations used in pivoting! Our matrices come in three types:

I. The **row interchange matrix**

$$N_{ij} = \begin{matrix} \\ i \\ \\ j \\ \\ \end{matrix} \begin{pmatrix} 1 & & & & & \\ & \ddots & & & & \\ & & 0 & \cdots & 1 & \\ & & & 1 & & \\ & & \vdots & & \ddots & \\ & & 1 & \cdots & 0 & \\ & & & & & \ddots \\ & & & & & & 1 \end{pmatrix} \begin{matrix} i \\ \\ \\ j \\ \\ \end{matrix}$$

Left multiplication by N_{ij} interchanges the ith and jth rows of A:

$$\text{For } A = \begin{matrix} \\ i \\ \\ j \\ \\ \end{matrix} \begin{pmatrix} A_1 \\ \vdots \\ A_i \\ \vdots \\ A_j \\ \vdots \\ A_n \end{pmatrix}, \quad N_{ij}A = \begin{matrix} \\ i \\ \\ j \\ \\ \end{matrix} \begin{pmatrix} A_1 \\ \vdots \\ A_j \\ \vdots \\ A_i \\ \vdots \\ A_n \end{pmatrix}.$$

(We have not used this particular row operation as yet, but it will later prove important.)

II. The **row multiplier matrix**

$$\mathbf{M}_i(c) = i \begin{pmatrix} 1 & & & & \\ & \ddots & & & \\ & & c & & \\ & & & 1 & \\ & & & & \ddots \\ & & & & & 1 \end{pmatrix}$$

Left multiplication of a matrix $\mathbf{A}$ by $\mathbf{M}_i(c)$ multiplies the ith row of $\mathbf{A}$ by c:

$$\mathbf{M}_i(c)\mathbf{A} = \begin{pmatrix} \mathbf{A}_1 \\ \vdots \\ c\mathbf{A}_i \\ \vdots \\ \mathbf{A}_n \end{pmatrix} \quad \text{if } \mathbf{A} = \begin{pmatrix} \mathbf{A}_1 \\ \vdots \\ \mathbf{A}_i \\ \vdots \\ \mathbf{A}_n \end{pmatrix}.$$

(We must have $c \neq 0$ if $\mathbf{M}_i(c)$ is to be invertible.)

III. The **row addition matrix**

$$\mathbf{A}_{ij}(c) = i \begin{pmatrix} 1 & & & & & \\ & \ddots & & & & \\ & & 1 & \cdots & c & \\ & & & \ddots & & \\ & & & & 1 & \\ & & & & & \ddots \end{pmatrix}$$

Left multiplication of a matrix by $\mathbf{A}_{ij}(c)$ adds $c\mathbf{A}_j$ to the ith row:

$$\mathbf{A}_{ij}(c)\mathbf{A} = i \begin{pmatrix} \mathbf{A}_1 \\ \vdots \\ \mathbf{A}_i + c\mathbf{A}_j \\ \vdots \\ \mathbf{A}_j \\ \vdots \\ \mathbf{A}_n \end{pmatrix}.$$

Example 4. If

$$A = \begin{pmatrix} 0 & 1 & 2 \\ 3 & -3 & 0 \\ 4 & -4 & 2 \\ -5 & 0 & 10 \end{pmatrix},$$

then

$$M_2(\tfrac{1}{3})A = \begin{pmatrix} 1 & 0 & 0 & 0 \\ 0 & \tfrac{1}{3} & 0 & 0 \\ 0 & 0 & 1 & 0 \\ 0 & 0 & 0 & 1 \end{pmatrix} \begin{pmatrix} 0 & 1 & 2 \\ 3 & -3 & 0 \\ 4 & -4 & 2 \\ -5 & 0 & 10 \end{pmatrix} = \begin{pmatrix} 0 & 1 & 2 \\ 1 & -1 & 0 \\ 4 & -4 & 2 \\ -5 & 0 & 10 \end{pmatrix},$$

$$A_{32}(-\tfrac{4}{3})A = \begin{pmatrix} 1 & 0 & 0 & 0 \\ 0 & 1 & 0 & 0 \\ 0 & -\tfrac{4}{3} & 1 & 0 \\ 0 & 0 & 0 & 1 \end{pmatrix} \begin{pmatrix} 0 & 1 & 2 \\ 3 & -3 & 0 \\ 4 & -4 & 2 \\ -5 & 0 & 10 \end{pmatrix} = \begin{pmatrix} 0 & 1 & 2 \\ 3 & -3 & 0 \\ 0 & 0 & 2 \\ -5 & 0 & 10 \end{pmatrix},$$

$$N_{12}A = \begin{pmatrix} 0 & 1 & 0 & 0 \\ 1 & 0 & 0 & 0 \\ 0 & 0 & 1 & 0 \\ 0 & 0 & 0 & 1 \end{pmatrix} \begin{pmatrix} 0 & 1 & 2 \\ 3 & -3 & 0 \\ 4 & -4 & 2 \\ -5 & 0 & 10 \end{pmatrix} = \begin{pmatrix} 3 & -3 & 0 \\ 0 & 1 & 2 \\ 4 & -4 & 2 \\ -5 & 0 & 10 \end{pmatrix}.$$

It is easy to check that these matrices work as indicated (Exercise 7). It is also apparent that pivoting can be accomplished by multiplying by a succession of row multipliers and row addition matrices. That is,

4.2 Pivoting via matrix multiplication. Given an $m \times n$ matrix

$$A = i \begin{pmatrix} a_{11} & \cdots & & a_{1n} \\ \vdots & & & \\ & \cdots & a_{ij} & \cdots \\ & & & \vdots \\ a_{m1} & & \cdots & a_{mn} \end{pmatrix} \text{ with } i,j\text{-entry } a_{ij} \neq 0,$$

there exist elementary matrices $E_1, \ldots, E_k$ in $\mathbb{R}_{m \times m}$ for which $E_k E_{k-1} \cdots E_1 A$ is the same matrix as the one obtained from A by pivoting on its i,j-entry.[†]

[†] We should actually write $(E_k(E_{k-1}(\cdots(E_2(E_1 A))\cdots))$, but by the associativity of matrix multiplication, we can dispense with parentheses.

Proof. For nice notation, assume that $i = j = 1$, so we wish to pivot on a_{11} in

$$A = \begin{pmatrix} a_{11} & \cdots & a_{1n} \\ \vdots & & \vdots \\ a_{m1} & \cdots & a_{mn} \end{pmatrix}.$$

Then $\mathbf{M}_1(a_{11}^{-1})\mathbf{A}$ has first row $a_{11}^{-1}\mathbf{A}_1$, where $\mathbf{A}_1 = (a_{11}, \ldots, a_{1n})$ is the first row of $\mathbf{A}$. Then

$$\mathbf{A}_{21}(-a_{21})(\mathbf{M}_1(a_{11}^{-1})\mathbf{A}) \text{ adds } -a_{21}a_{11}^{-1}\mathbf{A}_1 \text{ to } \mathbf{A}_2 \text{ (the 2nd row of } \mathbf{A}),$$

yielding

$$\mathbf{A}_{21}(-a_{21})\mathbf{M}_1(a_{11}^{-1})\mathbf{A} = \begin{pmatrix} 1 & a'_{12} & \cdots & a'_{1n} \\ 0 & a'_{22} & \cdots & a'_{2n} \\ a_{31} & a_{32} & \cdots & a_{3n} \\ \vdots & \vdots & & \vdots \\ a_{m1} & a_{m2} & \cdots & a_{mn} \end{pmatrix}.$$

In like manner,

$$\mathbf{A}_{m1}(-a_{m1})\mathbf{A}_{m-1,1}(-a_{m-1,1}) \cdots \mathbf{A}_{21}(-a_{21})\mathbf{M}_1(a_{11}^{-1})\mathbf{A} = \begin{pmatrix} 1 & a'_{12} & \cdots & a'_{1n} \\ 0 & a'_{22} & \cdots & a'_{2n} \\ \vdots & \vdots & & \vdots \\ 0 & a'_{m2} & \cdots & a'_{mn} \end{pmatrix},$$

and we leave it for you to show that the a'_{rs} are the same as if we had pivoted on a_{11} (Exercise 10). ∎

Even though pivoting can be accomplished by means of matrix multiplication, this is *not* the way to go about it; it is better to do it in our usual straightforward fashion. It will be of theoretical use in Chapters 4 and 5, however.

4.3 The Basis Algorithm via matrix multiplication. If $\bar{\mathbf{A}}$ is the matrix obtained from $\mathbf{A}$ by applying the Basis Algorithm, then there is a matrix $\mathbf{P}$ for which $\bar{\mathbf{A}} = \mathbf{PA}$. Moreover, $\mathbf{P}$ is a product of elementary matrices.

Proof. See Exercise 11. ∎

The elementary matrices are all invertible, and one has (see Exercise 8):

$$\mathbf{M}_i(c)^{-1} = \mathbf{M}_i(c^{-1}), \qquad \mathbf{N}_{ij}^{-1} = \mathbf{N}_{ij}, \qquad \mathbf{A}_{ij}(c)^{-1} = \mathbf{A}_{ij}(-c). \tag{1}$$

You have no doubt already guessed that there are corresponding matrices for the elementary column operations, and that these operate by postmultiplying $\mathbf{A}$ rather than premultiplying.

I'. To interchange column j and column i, form $\mathbf{AN}_{ij}$, where $\mathbf{N}_{ij}$ is the row interchange matrix (Exercise 9).

II'. To multiply the ith column of $\mathbf{A}$ by c, form $\mathbf{AM}_i(c)$, where $\mathbf{M}_i(c)$ is the row multiplier matrix.

III'. To multiply column j of $\mathbf{A}$ by c and add it to column i, form $\mathbf{A}(\mathbf{A}_{ji}(c))$, where $\mathbf{A}_{ji}(c)$ is the row addition matrix.

Thus, row and column elementary matrices are the same; it is only a matter of which side you put them on.

Transposition†

There is a very reasonable operation which takes $m \times n$ matrices to $n \times m$ matrices, which we now introduce. Given the matrix

$$\mathbf{A} = \begin{pmatrix} a_{11} & a_{12} & \cdots & a_{1n} \\ a_{21} & a_{22} & \cdots & a_{2n} \\ \vdots & \vdots & & \vdots \\ a_{m1} & a_{m2} & \cdots & a_{mn} \end{pmatrix} \quad \text{in } \mathbb{R}_{m \times n},$$

the **transpose** of $\mathbf{A}$ is the matrix

$$\mathbf{A}^* = \begin{pmatrix} a_{11} & a_{21} & \cdots & a_{m1} \\ a_{12} & a_{22} & \cdots & a_{m2} \\ \vdots & \vdots & & \vdots \\ a_{1n} & a_{2n} & \cdots & a_{mn} \end{pmatrix} \quad \text{in } \mathbb{R}_{n \times m}. \tag{2}$$

That is, $\mathbf{A}^*$ is the $n \times m$ matrix whose i,j-entry is the j,i-entry a_{ji} of $\mathbf{A}$,

$$\mathbf{A}^* = i \begin{pmatrix} & j & \\ & \vdots & \\ \cdots & a_{ji} & \cdots \\ & \vdots & \end{pmatrix} \quad \text{if} \quad \mathbf{A} = j \begin{pmatrix} & i & \\ & \vdots & \\ \cdots & a_{ji} & \cdots \\ & \vdots & \end{pmatrix}.$$

Example 5. For

$$\mathbf{A} = \begin{pmatrix} 1 & 2 \\ 0 & 0 \\ -1 & 3 \\ 1 & 1 \end{pmatrix}, \quad \mathbf{A}^* = \begin{pmatrix} 1 & 0 & -1 & 1 \\ 2 & 0 & 3 & 1 \end{pmatrix}.$$

---

† In this discussion, matrices are no longer necessarily square.

Clearly, the columns of **A*** are the rows of **A**, and vice versa. That is, if **A** = (**A**$_1$, ..., **A**$_n$) is a partitioning of **A** into its column vectors, then

$$\mathbf{A}^* = \begin{pmatrix} \mathbf{A}_1^* \\ \vdots \\ \mathbf{A}_n^* \end{pmatrix}$$

is a partitioning of **A*** into its row vectors. In particular, if **V** = (v_1, ..., v_n) is in $\mathbb{R}_{1 \times n}$, then

$$\mathbf{V}^* = \begin{pmatrix} v_1 \\ \vdots \\ v_n \end{pmatrix}$$

is in $\mathbb{R}_{n \times 1}$, and conversely.

Other popular notations for the transpose of **A** include **A**′, **A**^{t}, **A**^{T}, etc.

4.4 Properties of the transpose operator. i) For **A**,**B** in $\mathbb{R}_{m \times n}$ and c in $\mathbb{R}$,

$$(\mathbf{A} + \mathbf{B})^* = \mathbf{A}^* + \mathbf{B}^* \qquad \text{(Additivity)}\dagger$$
$$(c\mathbf{A})^* = c\mathbf{A}^* \qquad \text{(Homogeneity)}\dagger$$

ii) For **A** in $\mathbb{R}_{l \times m}$ and **B** in $\mathbb{R}_{m \times n}$,

$$(\mathbf{AB})^* = \mathbf{B}^*\mathbf{A}^*;$$

thus the product of the transpose is the transpose of the product, *but taken in reverse order* (cf. the product of inverses).

iii) For **A** in $\mathbb{R}_{m \times n}$,

$$(\mathbf{A}^*)^* = \mathbf{A}.$$

iv) For **A** in $\mathbb{R}_{n \times n}$, **A** is invertible iff **A*** is. In fact,

$$(\mathbf{A}^*)^{-1} = (\mathbf{A}^{-1})^*.$$

Proof. (i) is Exercise 29a.

ii) For

$$\mathbf{A} = \begin{pmatrix} a_{11} & \cdots & a_{1m} \\ \vdots & & \vdots \\ a_{l1} & \cdots & a_{lm} \end{pmatrix} \quad \text{and} \quad \mathbf{B} = \begin{pmatrix} b_{11} & \cdots & b_{1n} \\ \vdots & & \vdots \\ b_{m1} & \cdots & b_{mn} \end{pmatrix},$$

---

† It readily follows that the mapping **A** → **A*** of $\mathbb{R}_{m \times n}$ into $\mathbb{R}_{n \times m}$ is a linear transformation, in fact, an isomorphism (see Exercise 28).

the i,j-entry of $\mathbf{AB}$ is
$$a_{i1}b_{1j} + a_{i2}b_{2j} + \cdots + a_{im}b_{mj}, \tag{3}$$
and this is also the j,i-entry of $(\mathbf{AB})^*$. Furthermore,

$$\mathbf{B}^* = \begin{pmatrix} b_{11} & \cdots & b_{m1} \\ \vdots & & \vdots \\ b_{1j} & \cdots & b_{mj} \\ \vdots & & \vdots \\ b_{1n} & \cdots & b_{mn} \end{pmatrix}$$

and $\quad \mathbf{A}^* = \begin{pmatrix} a_{11} & \cdots & a_{i1} & \cdots & a_{l1} \\ \vdots & & \vdots & & \vdots \\ a_{1m} & \cdots & a_{im} & \cdots & a_{lm} \end{pmatrix},$

so the j,i-entry of $\mathbf{B}^*\mathbf{A}^*$ is
$$b_{1j}a_{i1} + b_{2j}a_{i2} + \cdots + b_{mj}a_{im}. \tag{4}$$
Using commutativity of real-number multiplication, we see that (3) = (4), and the matrices in question have the same j,i-entries for any i and j. Hence the matrices are the same.

iii) See Exercise 29b.

iv) If $\mathbf{A}^{-1}$ exists, then
$$\mathbf{A}^*(\mathbf{A}^{-1})^* \stackrel{(ii)}{=} (\mathbf{A}^{-1}\mathbf{A})^* = \mathbf{I}^* = \mathbf{I}.$$
Complete the proof by showing that if $(\mathbf{A}^*)^{-1}$ exists, then so does $\mathbf{A}^{-1}$ (Exercise 29c). ∎

If $\mathbf{A}$ is an $n \times n$ matrix then so is $\mathbf{A}^*$. In fact, it might even turn out that $\mathbf{A}^* = \mathbf{A}$. If this is the case, we say that $\mathbf{A}$ is **symmetric**. Thus, $\mathbf{A}$ is symmetric if

i) $\mathbf{A}$ is square, and

ii) $a_{ij} = a_{ji}$ for all $i, j = 1, \ldots, n$.

Looking at $\mathbf{A}$ and $\mathbf{A}^*$ for square matrices,

$$\mathbf{A} = \begin{pmatrix} a_{11} & a_{12} & \cdots & a_{1n} \\ a_{21} & a_{22} & \cdots & a_{2n} \\ \vdots & \vdots & & \vdots \\ a_{n1} & a_{n2} & \cdots & a_{nn} \end{pmatrix} \quad \text{vs.} \quad \mathbf{A}^* = \begin{pmatrix} a_{11} & a_{21} & \cdots & a_{n1} \\ a_{12} & a_{22} & \cdots & a_{n2} \\ \vdots & \vdots & & \vdots \\ a_{1n} & a_{2n} & \cdots & a_{nn} \end{pmatrix},$$

we see that **A** is a symmetric matrix iff its elements are symmetrically placed with respect to the **main diagonal**

$$\begin{pmatrix} a_{11} & & & \\ & a_{22} & & \\ & & \ddots & \\ & & & a_{nn} \end{pmatrix}$$

that is, a reflection of **A** about the main diagonal does not change **A**.

Example 6. I_n is symmetric for any n, as is any matrix whose entries off the main diagonal are all 0. So is

$$\begin{pmatrix} 1 & 2 \\ 2 & 3 \end{pmatrix}.$$

By great good fortune (as we shall see in Chapter 9), many matrices encountered by scientists are symmetric.

It is important to realize that by taking transposes, *whatever you can do with row vectors you can do with column vectors* (and conversely), and *whatever you can do at one end of a matrix you can do at the other end.*

In the previous section we observed that if **A** is an $m \times n$ matrix, then left multiplication of column vectors yields a function from $\mathbb{R}_n$ into $\mathbb{R}_m$:

$$\mathbf{T(V)} = \mathbf{AV} = \begin{pmatrix} a_{11} & \cdots & a_{1n} \\ \vdots & & \vdots \\ a_{m1} & \cdots & a_{mn} \end{pmatrix} \begin{pmatrix} v_1 \\ \vdots \\ v_n \end{pmatrix} \quad (5)$$

is in $\mathbb{R}_m$ (considered as the space of *column* vectors). Similarly, if **W** is in $\mathbb{R}_m$ (considered as *row* vectors), then we can multiply **WA**. As you would expect,

$$\mathbf{T'(W)} = \mathbf{WA} = (w_1, \ldots, w_m) \begin{pmatrix} a_{11} & \cdots & a_{1n} \\ \vdots & & \vdots \\ a_{m1} & \cdots & a_{mn} \end{pmatrix} \quad (6)$$

is a function from $\mathbb{R}_m$ (row vectors) into $\mathbb{R}_n$ (row vectors). By the principle enunciated in the previous paragraph, it suffices to consider either of Eqs. (5) or (6). (Again, note that $(\mathbf{AV})^* = \mathbf{V}^*\mathbf{A}^*$, which is of the same form as Eq. (6).) We now explain our preference for Eq. (5).

Row vectors somehow seem more natural to write than column vectors, and they make a more visually pleasing printed page. However, Eq. (5) has the advantage that the function **T** occurs on the same side of the vector **V** as does the matrix

A. This may seem inconsequential, but the appeal of row vectors (together with certain facts about functions of functions) have led many card-carrying algebraists (and members of other small but powerful minority groups) to write the function symbol to the *right* of the vector **V**: **WT'** in Eq. (6), rather than **T'(W)**. (It is unclear whether any are sufficiently fanatical to write x sin for sin x.) We point this out in case the reader encounters this "right-handed" notation in other sources.

We shall use the more standard "left-handed" notation "**T(V)**" for the function **T** applied to the vector **V**, with which you have grown familiar in your previous mathematics. We also prefer to use row vectors because they look nicer, but shall often use them with a transpose *, so that they are really column vectors and conform to Eq. (5). Thus,

$$(v_1, \ldots, v_n)* \quad \text{in place of} \quad \begin{pmatrix} v_1 \\ \vdots \\ v_n \end{pmatrix}.$$

In general, you can expect us to abbreviate

$$\begin{pmatrix} a_{11} & \cdots & a_{1n} \\ \vdots & & \vdots \\ a_{m1} & \cdots & a_{mn} \end{pmatrix} \begin{pmatrix} v_1 \\ \vdots \\ v_n \end{pmatrix}$$

either by **AV** with

$$\mathbf{V} = \begin{pmatrix} v_1 \\ \vdots \\ v_n \end{pmatrix}$$

(usually written $\mathbf{V}* = (v_1, \ldots, v_n)$), or by **AV*** with

$$\mathbf{V} = (v_1, \ldots, v_n).$$

They have the same meaning.

EXERCISES

1. Let

$$\mathbf{A} = \begin{pmatrix} 0 & 2 & 0 & 0 & 0 \\ 0 & 0 & 2 & 0 & 0 \\ 0 & 0 & 0 & 2 & 0 \\ 0 & 0 & 0 & 0 & 2 \\ 0 & 0 & 0 & 0 & 0 \end{pmatrix}.$$

Calculate
   a) A^2 b) A^3 c) A^4
   d) A^5 e) A^n for any $n > 5$.

2. Let
$$B = \begin{pmatrix} 0 & 0 & 0 & 2 \\ 2 & 0 & 0 & 0 \\ 0 & 2 & 0 & 0 \\ 0 & 0 & 2 & 0 \end{pmatrix}.$$

Calculate
   a) B^2 b) B^3 c) B^4
   d) B^{-1} e) B^n, n any positive integer.

3. a) Prove that the $n \times n$ identity matrix I satisfies
$$AI = A = IA$$
for any A in $\mathbb{R}_{n \times n}$.
   b) Prove that I is the only matrix for which this is true. That is, $AX = A = XA$ for all A in $\mathbb{R}_{n \times n}$ implies that $X = I$.

4. We shall consider the matrix units
$$E_{ij} = i \begin{pmatrix} & & \overset{j}{\vdots} & \\ --&-&1&-- \\ & & \vdots & \end{pmatrix}$$
(0's elsewhere)

in the $n \times n$ context. Recall that these form a basis for $\mathbb{R}_{n \times n}$ as a vector space. They are sometimes (but not always!) convenient in multiplicative situations as well.
   a) Prove that one obtains the following multiplication table for matrix units
$$E_{ij}E_{kl} = \begin{cases} 0 & \text{if } j \neq k, \\ E_{il} & \text{if } j = k. \end{cases}$$

Note the particular instances of this:
$$E_{11}^2 = E_{11}, \qquad E_{11}E_{12} = E_{12}, \qquad E_{12}E_{11} = 0.$$

It is useful to use the **Kronecker delta** *delta* symbol in this context. By definition,
$$\delta_{ij} = \begin{cases} 0 & \text{if } i \neq j, \\ 1 & \text{if } i = j. \end{cases}$$
Thus, $E_{ij}E_{kl} = \delta_{jk}E_{il}$.
   b) One could develop matrix multiplication strictly in terms of the matrix units, but this would be a mistake. Prove this by writing out the matrix in Exercise 1 in terms of matrix units and performing the matrix multiplication by the above table.
   c) Prove that, except for the trivial case when $n = 1$, the matrix units are never invertible.

5. Some details in Example 3: a) Prove that in any field (axioms listed above 4.1), $a^2 = 0$ implies that $a = 0$.
   b) Prove that in any field, $b^2 = b$ implies that b is 0 or b is 1.
6. a) (HH) Prove that if $\mathbf{AB} = \mathbf{I}$ for $n \times n$ matrices then $\mathbf{BA} = \mathbf{I}$.
   b) Prove that
   $$(\mathbf{ABC})^{-1} = \mathbf{C}^{-1}\mathbf{B}^{-1}\mathbf{A}^{-1}.$$
7. a) Show that property I (see p. 107) holds for elementary matrices $\mathbf{N}_{ij}$;
   b) prove II; c) prove III.
8. Prove that the inverses of the elementary matrices are as indicated in Eq. (1).
9. a) Prove I'; b) II'; c) III'.
10. Finish the proof of 4.2.
11. Prove 4.3.
12. a) Find an inverse for the matrix
    $$\mathbf{A} = \begin{pmatrix} 0 & 1 \\ 1 & 0 \end{pmatrix}$$
    and verify that $\mathbf{AA}^{-1} = \mathbf{A}^{-1}\mathbf{A}$.
    b) By multiplying together the matrices
    $$\mathbf{A} = \begin{pmatrix} a_{11} & a_{12} \\ a_{21} & a_{22} \end{pmatrix} \quad \text{and} \quad \mathbf{B} = \begin{pmatrix} b_{11} & b_{12} \\ b_{21} & b_{22} \end{pmatrix} \quad \text{of } \mathbb{R}_{2 \times 2},$$
    obtain a set of conditions for a 2×2 matrix $\mathbf{A}$ to have an inverse.
    c) If $\mathbf{A}$ satisfies your conditions, write $\mathbf{A}^{-1}$ in terms of the entries of $\mathbf{A}$.
13. We shall obtain herein an alternate way of considering complex numbers. Let C be the set of all 2×2 real matrices of the form $\begin{pmatrix} a & -b \\ b & a \end{pmatrix}$.
    a) Show that the set C, with respect to matrix multiplication and matrix addition, satisfies all the field axioms (listed at the beginning of this section).
    b) Show that C is a two-dimensional vector space over $\mathbb{R}$.
    c) Show that each element of C can be written uniquely in the form $a\mathbf{I} + b\mathbf{i}$, where $\mathbf{I}$ is the 2×2 identity matrix and $\mathbf{i}$ is a matrix for which $\mathbf{i}^2 = -\mathbf{I}$. What specifically is $\mathbf{i}$? (Is it unique?)
    d) Prove that addition and multiplication is the same for elements $a + b\sqrt{-1}$ of $\mathbb{C}$ as for $a\mathbf{I} + b\mathbf{i}$ of C.
    e) What matrix operation corresponds to complex conjugation?
14. Prove by example that matrix multiplication is (usually) noncommutative (a) for 2×2 matrices, (b) for $n \times n$ matrices (where $n > 1$).
15. a) Write out the elementary matrices in terms of matrix units $\mathbf{E}_{ij}$ (see Exercise 4).
    b) Use this to do Exercise 7b.
16. a) Write out the matrix $\mathbf{P}_{12}$ for which $\mathbf{P}_{12}\mathbf{A}$ results in pivoting on the 1,2-entry of $\mathbf{A}$.
    b) Write out the matrix $\mathbf{P}_{ij}$.
17. Find the form of any matrix $\mathbf{A}$ in $\mathbb{R}_{2 \times 2}$ which commutes with all other matrices in $\mathbb{R}_{2 \times 2}$; i.e., for which $\mathbf{AB} = \mathbf{BA}$ for all $\mathbf{B}$ in $\mathbb{R}_{2 \times 2}$.

18. Let
$$\mathbf{A} = \begin{pmatrix} 1 & 2 & 3 \\ -2 & 1 & 4 \end{pmatrix}, \quad \mathbf{B} = \begin{pmatrix} 1 & \frac{3}{5} \\ -\frac{6}{5} & -\frac{9}{5} \\ \frac{4}{5} & 1 \end{pmatrix}.$$

Calculate **AB** and **BA**.

19. If
$$\mathbf{A} = \begin{pmatrix} a & b \\ c & d \end{pmatrix},$$
prove that $\mathbf{A}^2 - (a+d)\mathbf{A} + (ad-bc)\mathbf{I} = \mathbf{0}$.

20. Let
$$\mathbf{A} = \begin{pmatrix} a & b \\ c & d \end{pmatrix} \quad \text{and} \quad \mathbf{B} = \begin{pmatrix} d & -b \\ -c & a \end{pmatrix}.$$

a) Show that
$$\mathbf{AB} = \mathbf{BA} = (ad-bc)\mathbf{I}, \quad \mathbf{A} + \mathbf{B} = (a+d)\mathbf{I}.$$

b) Show that $\mathbf{A}^{-1}$ exists iff $ad - bc \neq 0$ and, in fact,
$$\mathbf{A}^{-1} = \frac{1}{ad-bc}\mathbf{B}.$$

c) Prove the result in Exercise 19 by using (a).

21. If
$$\mathbf{A} = \begin{pmatrix} 1 & h \\ 0 & 1 \end{pmatrix},$$
find $\mathbf{A}^2, \mathbf{A}^3, \ldots, \mathbf{A}^k$.

22. a) If
$$\mathbf{A} = \begin{pmatrix} 1 & h & 0 \\ 0 & 1 & h \\ 0 & 0 & 1 \end{pmatrix},$$
find $\mathbf{A}^k$.

b) Generalize this.

23. Let
$$\mathbf{E}_1 = \begin{pmatrix} 1 & 0 & 0 \\ -2 & 1 & 0 \\ 0 & 0 & 1 \end{pmatrix}, \quad \mathbf{E}_2 = \begin{pmatrix} 1 & 0 & 0 \\ 0 & 1 & 0 \\ 1 & 0 & 0 \end{pmatrix},$$

$$\mathbf{E}_3 = \begin{pmatrix} 1 & 0 & 0 \\ 0 & 1 & 0 \\ 0 & 2 & 1 \end{pmatrix}, \quad \mathbf{E}_4 = \begin{pmatrix} 1 & 0 & 0 \\ 0 & -\frac{1}{2} & 0 \\ 0 & 0 & 1 \end{pmatrix},$$

$$\mathbf{E}_5 = \begin{pmatrix} 1 & -3 & 0 \\ 0 & 1 & 0 \\ 0 & 0 & 1 \end{pmatrix}, \quad \mathbf{A} = \begin{pmatrix} 1 & 3 & -2 \\ 2 & 4 & 3 \\ -1 & 1 & -12 \end{pmatrix}.$$

Find: $\mathbf{E}_1\mathbf{A}$, $\mathbf{E}_2(\mathbf{E}_1\mathbf{A})$; $\mathbf{E}_3(\mathbf{E}_2\mathbf{E}_1\mathbf{A})$; $\mathbf{E}_4(\mathbf{E}_3\mathbf{E}_2\mathbf{E}_1\mathbf{A})$; $\mathbf{E}_5(\mathbf{E}_4\mathbf{E}_3\mathbf{E}_2\mathbf{E}_1\mathbf{A})$; $\mathbf{E}_5\mathbf{E}_4\mathbf{E}_3\mathbf{E}_2\mathbf{E}_1$; $(\mathbf{E}_5\mathbf{E}_4\mathbf{E}_3\mathbf{E}_2\mathbf{E}_1)\mathbf{A}$. Compare all this with Examples 2 and 3 in Section 1.

24. Take
$$A = \begin{pmatrix} -3 & -9 & 6 & -15 \\ 2 & 4 & 3 & -3 \\ -1 & 1 & -12 & 8 \end{pmatrix},$$
the matrix in Example 4 of Section 1. Write down the elementary matrices that correspond to the row operations in that example, and then work out the products analogous to those in the previous exercise.

25. Let
$$A = \begin{pmatrix} 2 & 0 & 1 \\ 3 & 1 & 2 \\ 0 & 1 & 1 \end{pmatrix}.$$
Show that $A^3 - 4A^2 + 3A = I$. Use this result to show that A is invertible and express A^{-1} in terms of powers of A.

26. Consider the set of matrices of the form
$$A = \begin{pmatrix} \alpha & \beta \\ -\bar{\beta} & \bar{\alpha} \end{pmatrix}$$
where α and β are complex numbers and $\bar{\alpha}$ and $\bar{\beta}$ are the complex conjugates of α and β respectively.
   a) Prove that these matrices form a vector space over $\mathbb{R}$ if addition and scalar multiplication are defined in the matrix way.
   b) Show that the matrices
$$e = \begin{pmatrix} 1 & 0 \\ 0 & 1 \end{pmatrix}, \quad i = \begin{pmatrix} 0 & 1 \\ -1 & 0 \end{pmatrix}, \quad j = \begin{pmatrix} 0 & i \\ i & 0 \end{pmatrix}, \quad k = \begin{pmatrix} i & 0 \\ 0 & -i \end{pmatrix}$$
   form a basis of this vector space.
   c) Prove that the product of any two matrices in the set is a matrix of the set. Calculate the multiplication table of the basis elements e, i, j, k.
   d) Prove that any nonzero matrix in the set has an inverse which is also in the set.

27. a) For which A in $\mathbb{R}_{n \times n}$ is $A^* = -A$? Give a condition in terms of the entries of A (such matrices are called **skew-symmetric**).
    b) Considering $\mathbb{C}$ as a subset of $\mathbb{R}_{2 \times 2}$, as in Exercise 13, for which A do $A^* = A$? $A^* = -A$? Comment.

28. Prove that the correspondence between $\mathbb{R}_{m \times n}$ and $\mathbb{R}_{n \times m}$ defined by $T(A) = A^*$ is (a) a linear transformation, (b) an isomorphism.

29. Prove 4.4(a) (i); (b) (iii); (c) (iv);

30. a) Show that the transpose of an elementary matrix is an elementary matrix. Develop specific formulas.
    b) Use this to prove the column operation properties I' through III'.

31. Prove that $(ABC)^* = C^*B^*A^*$.
32. Let

$$A = \begin{pmatrix} 2 & 0 & 1 \\ -3 & 1 & 2 \\ 1 & 0 & 1 \end{pmatrix}, \quad B = \begin{pmatrix} -1 & 1 & -2 \\ 0 & 3 & -1 \\ 2 & 1 & 0 \end{pmatrix}.$$

Calculate AB, A^*B^*, B^*A^*.

33. If

$$A = \begin{pmatrix} 2 & 2 & 1 \\ -1 & 0 & 2 \\ 4 & -5 & 2 \end{pmatrix},$$

find AA^*. Use the result to find A^{-1} in this case.

34. a) Let $e_1 = (1, 0, \ldots, 0)^*, \ldots, e_n = (0, \ldots, 0, 1)^*$ be the usual basis for $\mathbb{R}_n$, and let T be the function from $\mathbb{R}_n$ into $\mathbb{R}_m$ given by

$$T'(V) = AV, \quad V \text{ in } \mathbb{R}_n, \quad A \text{ fixed in } \mathbb{R}_{m \times n}.$$

Prove that $T(e_i)$ is the ith column of A, $i = 1, \ldots, n$.

b) Prove that if A and B are two matrices in $\mathbb{R}_{m \times n}$ and if $AV = BV$ for all V in $\mathbb{R}_n$, then $A = B$.

c) Let T be as in (a) and let T' be a function from $\mathbb{R}_n$ to $\mathbb{R}_m$ given by

$$T(V) = BV, \quad V \text{ in } \mathbb{R}_n, \quad B \text{ fixed in } \mathbb{R}_{m \times n}.$$

Prove that if T and T' are the same functions, then A and B are the same matrices. That is, if $T(V) = T'(V)$ for all V in $\mathbb{R}_n$, then $A = B$.

d) Part (c) did not require $T(V) = T'(V)$ for all V in $\mathbb{R}_n$. There is a fairly obvious set which will do. What do you basically need? Prove your assertion.

Honors Projects. Show that all this material holds for matrices over (1) $\mathbb{C}$, (2) any field $\mathbb{F}$ (with the possible exception of some fussy arithmetic; e.g., can you prove that the matrix of Example 1 is *not* invertible if its elements are replaced by the appropriate elements from the field with two elements?).

35. For $A = (a_{ij})$ in $\mathbb{C}_{n \times n}$, define

$$A\dagger = \bar{A}^*,$$

where $\bar{A} = (\bar{a}_{ij})$ is the matrix whose entries are the complex conjugates of the entries of A.

a) Prove that this "conjugate-transpose" operator $\dagger$ is a linear transformation of $\mathbb{C}_{n \times n}$ over $\mathbb{R}$ but not over $\mathbb{C}$, the difference being that $(cA)\dagger = \bar{c}A\dagger$ for c in $\mathbb{C}$.

b) Prove that (ii) of 4.4 holds with $\dagger$ in place of *.

c) Prove that (iii) of 4.4 holds with $\dagger$ in place of *.

d) Which matrices in $\mathbb{C}_{n \times n}$ are left fixed by $\dagger$?

5. SOME APPLICATIONS OF MATRIX ARITHMETIC

In this (optional) section we present some applications of the matrix notions thus far encountered to problems in the biological and social sciences. Other readers should not feel slighted; applications in other disciplines are in general easier to come by, but often involve more technical knowledge.† All these examples essentially require that we consider rectangular tables as matrices and handle the data in the tables by means of matrix arithmetic. We shall take up the following:

a) sociometric matrices,
b) the age distribution of a population,
c) matrices and game theory,
d) Markov Chains, with an application to social mobility, and
e) permutation matrices, with an application to the study of kinship systems.

The examples are independent as well as optional, and may be taken (or left) in any order. We shall refer to some of them in later exercises.

a. Matrix Multiplication and Sociometry

We discuss matrix techniques for analyzing interactions among people. With slight modification, these techniques can be used to consider the pecking order among chickens, athletic tournaments, and communications networks.

    Festinger *et al.* conducted a detailed sociological investigation of some married-student housing projects at M.I.T. One of their studies attempts to analyze who influenced whom within the separate courts (which were dwelling complexes of some 13 family units).‡ For each court they construct a matrix whose i,j-entry is a 1 if individual i had influence over individual j, and the i,j-entry was a 0 otherwise. Two examples (with blanks rather than 0's) are shown in Fig. 1. Thus, for example, in Tolman Court, 2 influenced 3, 11, and 13, and was influenced by 3 and 13. We shall call (1) the **sociometric matrix** for Tolman and Howe, respectively.

    Suppose that we wish to determine the secondary influences among the individuals. If i has been influenced by k and k by j, then it makes sense to consider that i has been secondarily influenced by j. Now

    i is influenced by k iff the sociometric matrix has a 1 in the ith row and kth column and k is influenced by j iff the sociometric matrix has a 1 in the kth row and jth column.

---

† A nice compendium is to be found in Ben Nobel, *Applications of Undergraduate Mathematics to Engineering*, Macmillan, 1967. Problems discussed there include framework calculations, electrical networks, and the determination of star positions from photographs.
‡ For present purposes the family is considered the basic unit, i.e., "individual" = family.

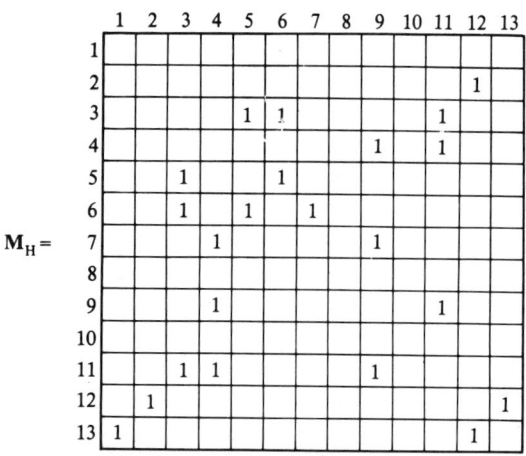

1 in *i*th row and *j*th column indicates that *i* was influenced by *j*. (Festinger, p. 139.)

Fig. 1

Therefore, i is secondarily influenced by j iff $\mathbf{M}^2$ has a 1 as its i,j-entry.

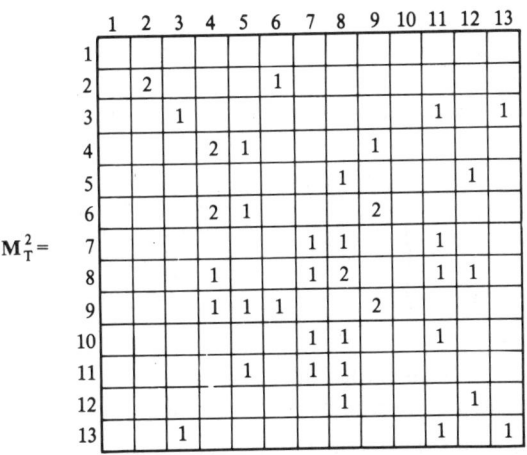

For example, the secondary influences in Tolman Court may be read off from this diagram:

$\mathbf{M}_T^2 =$

	1	2	3	4	5	6	7	8	9	10	11	12	13
1													
2		2				1							
3			1								1		1
4				2	1			1					
5						1					1		
6				2	1				2				
7							1	1		1			
8				1			1	2			1	1	
9				1	1	1			2				
10							1	1			1		
11					1		1	1					
12								1				1	
13				1							1		1

i,j-entry $= 1$ if i was secondarily influenced by j.
(Festinger, p. 141.)
Fig. 2

Thus, although 11 was only directly influenced by 6 (see $\mathbf{M}_T$), 11 was *secondarily* influenced by 5, 7, and 8. Note that the 2 in the 6,4-entry of $\mathbf{M}_T^2$ indicates that 6 (who seems quite influential) secondarily influenced 4 in two different ways. (Can you find these two chains of influence?)

What is the meaning of the i,i-entries of $\mathbf{M}_T^2$? According to the definition of matrix multiplication, the i,i-entry of $\mathbf{M}^2$ will be $a_{i1}a_{1i} + a_{i2}a_{2i} + \cdots + a_{ni}a_{ni}$ if

$$\mathbf{M} = \begin{pmatrix} a_{11} & \cdots & a_{1n} \\ \vdots & & \vdots \\ a_{n1} & \cdots & a_{nn} \end{pmatrix}.$$

Since all the a_{rs} are either 0 or 1, you obtain a contribution of 1 to the i,i-entry of $\mathbf{M}^2$ whenever both a_{ik} and a_{ki} are nonzero. Thus, the i,i-entry of $\mathbf{M}^2$ indicates the total number of instances of mutual influence which i holds with other members of the court. For example, 2 in Tolman had two such instances of reciprocal influence. These can easily be found in $\mathbf{M}_T$ by locating those entries in row 2 which have symmetrically located entries in column 2 (namely, 3 and 13).

In the back of your mind there may lurk the fear that we will ask you to compute the matrix $\mathbf{M}_H^2$. This is, in fact, correct, but $\mathbf{M}_H$ is fortunately not a typical 13 × 13 matrix. It is **sparse**: this means that it has a preponderance of zero entries. Large sparse matrices are not difficult to multiply. For example, since the first row of $\mathbf{M}_H$ is 0, the first row of $\mathbf{M}_H^2$ is 0. Since the second row of $\mathbf{M}_H$ is nonzero only in the 12th entry, the second row of $\mathbf{M}_H^2$ will have a nonzero entry only when this combines with some nonzero element of the 12th row of $\mathbf{M}$. Thus, row 2 of $\mathbf{M}_H^2$ will be nonzero only in its 2nd and 13th entries.

We finally illustrate this on row 3, one of the "tougher" rows of $\mathbf{M}_H$. Since the only nonzero entries of row 3 are entries 5, 6, and 11, the only relevant rows to consider in the multiplications are rows 5, 6, and 11, which have nonzero entries as indicated:

$$\text{Row 3}\begin{pmatrix} \overset{\text{Column}}{\underset{5\ \ \ 6\ \ 11}{1\ \ \ 1\ \ \ 1}} \\ \overline{M_H} \end{pmatrix} \begin{matrix} \text{Row 5} \\ \text{Row 6} \\ \text{Row 11} \end{matrix} \begin{pmatrix} \overset{\text{Column}}{\underset{3\ 4\ 5\ 6\ 7\ \ 9}{\ \ 1\ \ 1}} \\ 1\ \ 1\ \ 1 \\ 1\ 1\ \ \ \ 1 \end{pmatrix} = 3\begin{pmatrix} \overset{\text{Column}}{\underset{3\ 4\ 5\ 6\ 7\ 9}{3\ 1\ 1\ 1\ 1\ 1}} \end{pmatrix}.$$

Exercise 1. a) Calculate $\mathbf{M}_H^2$.
b) Determine the ways 7 is secondarily influenced by 11.
c) Determine the individuals with whom 3 is in mutual influence.

One could similarly consider tertiary influences, and find that the number of chains of influence of the form

i influenced j, $\quad j$ influenced k, $\quad k$ influenced l

would be the i,l-entry of $\mathbf{M}^3$, for example, as shown in Fig. 3.

Exercise 2. Calculate $\mathbf{M}_H^3$

One might expect the importance of these higher levels of influence to taper off rapidly. Thus, one measure of the extent to which individual i influenced j might be given by the sum total of the primary, secondary, and tertiary influence he imparts to j. This, of course, would be the i,j-entry of $\mathbf{M} + \mathbf{M}^2 + \mathbf{M}^3$ (you might wish to assign weights so that the importance of the higher levels is decreased).

Exercise 3. a) Calculate $\mathbf{M}_T + \mathbf{M}_T^2 + \mathbf{M}_T^3$.
b) Who was influenced the most by 8? Who influenced 8 the most?
c) Devise a simple measure of the total amount of influence exerted by an individual. According to your criterion, who is the most influential individual in Tolman Court?

$\mathbf{M}_T^3 =$

	1	2	3	4	5	6	7	8	9	10	11	12	13
1													
2		2		1		1	1			2		2	
3	2				1								
4			1			1	3			1	2		
5			2	1				1					
6			1			2	4			2	2		
7			2	1				2					
8			3	2	1		1	3			1		
9			1	1		3	4			2	1		
10			1	1	1			2					
11			2	1				2					
12			2	1				1					
13	2				1								

Fig. 3

A **clique** may be defined to be a collection of three or more individuals, each pair of which mutually influences each other, and to which no further individuals may be added and still retain this property. We shall now develop a matrix method for discovering cliques. Since we are considering only those relations of influence which are reciprocal, we delete those entries in the sociometric matrix **M** which do not correspond to relations of mutual influence, i.e., which are not symmetric with respect to the main diagonal. After so doing we get the **symmetric submatrix S** of **M**,

$$\mathbf{S} = \begin{pmatrix} s_{11} & \cdots & s_{1n} \\ \vdots & & \vdots \\ s_{n1} & \cdots & s_{nn} \end{pmatrix},$$

where

$$s_{ij} = \begin{cases} 0 & \text{if} \quad a_{ij} \neq a_{ji}, \\ 0 & \text{if} \quad a_{ij} = a_{ji} = 0, \\ 1 & \text{if} \quad a_{ij} = a_{ji} = 1, \end{cases} \quad \text{and} \quad \mathbf{M} = \begin{pmatrix} a_{11} & \cdots & a_{1n} \\ \vdots & & \vdots \\ a_{n1} & \cdots & a_{nn} \end{pmatrix}.$$

Note that **S** is symmetric, $\mathbf{S}^* = \mathbf{S}$. For example, the symmetric submatrix for Tolman Court is shown in Fig. 4. A convenient criterion for clique membership can be given in terms of $\mathbf{S}^3$:

Individual i is in a clique iff the i,i-entry of $\mathbf{S}^3$ is nonzero.

Exercise 4. a) Prove this. [*Suggestion:* Show that the i,i-entry of $\mathbf{S}^3$ is a sum of terms of the form $s_{ij}s_{jk}s_{ki}$. If there is such a nonzero creature, then

i influenced j, $\quad$ j influenced k, $\quad$ and $\quad$ k influenced i.

Show that the reverse relations also hold, so that i, j, and k are in a clique.]
b) Determine the meaning of the numerical value of the nonzero diagonal elements of $\mathbf{S}^3$ (you might wish to do Exercise 5 first).

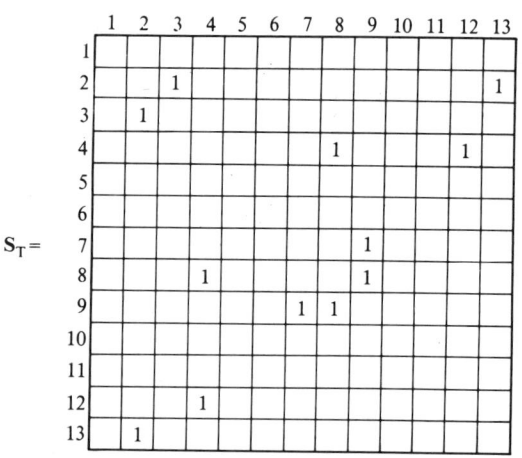

Fig. 4

The diagonal entries of $\mathbf{S}_T^3$ are all 0, so, alas, there are no cliques in Tolman. The reader will be delighted to learn that not only are there several cliques in Howe, but also that we leave it for him to find them.

Exercise 5. (a) Calculate the symmetric submatrix $\mathbf{S}_H$ for Howe. (b) Calculate $\mathbf{S}_H^3$. (c) List the members of the cliques in Howe. (d) Find any direct connections between the cliques.

Exercise 6. (a) Devise an insidious method for testing the influence of students in your math class upon one another. Determine all cliques. (b) Do the same for your dormitory, and (c) for your mathematics faculty. (d) Publish all these in your college paper. [*Suggestion:* Be prepared to transfer.]

REFERENCES

Festinger, L., S. Schachter, K. Back, *Social Pressures in Informal Groups*,† Harper & Rowe, 1950.

Luce, R. D., and A. D. Perry, "A Method of matrix analysis of group structure", *Psychometrika*, 14, 1949, pp. 95–116. (for the theory).

---

† Disclaimer: All groups in the above are sociological and are most unlikely to satisfy the axioms of Supplement 2 to Chapter 0, Section 1.

b. Age Distribution of a Population

We wish to study the future structure of a population, assuming we know the birth and survival rates of the various age groups which comprise it. The population could be human or animal (or even inanimate; there have been industrial applications to populations of telephone poles, boxcars, etc.). For some of the possible populations, mating may occur between members of different age groups, so for simplicity we shall restrict ourselves to considering only the female members of the population (leaving those more interested in telephone poles to make appropriate rewording).

Thus, suppose we have a population of females in $(m + 1)$ age groups 0 to 1 year old, 1 to 2 years old, ..., m to $(m + 1)$ years old, with essentially no survivors after $(m + 1)$ years. Suppose that the probability that a female alive in age group "k to $(k + 1)$ years" survives to age group "$k + 1$ to $k + 2$ years" is P_k, with $k = 0, \ldots, (m - 1)$. Suppose that the average number of daughters born to members of age group "k to $(k + 1)$ years" is F_k, with $k = 0, \ldots, m$. We shall make the rather strong assumption that these P_k and F_k do not vary with time (even if they do, this may yield a good first approximation).

Our work will center around the $(m + 1) \times (m + 1)$ matrix $\mathbf{M}$,

$$\mathbf{M} = \begin{pmatrix} F_0 & F_1 & \cdots & & F_m \\ P_0 & 0 & \cdots & & 0 \\ 0 & P_1 & 0 & \cdots & 0 \\ \vdots & & & & \vdots \\ \vdots & & & & \vdots \\ 0 & \cdots & 0 & P_{m-1} & 0 \end{pmatrix},$$

where F_k (≥ 0) is the average number of daughters born to members of age group "k to $(k + 1)$ years", and $(0 \leq) P_k (\leq 1)$ the probability that a member of age group "k to $(k + 1)$ years", survives to age group "$(k + 1)$ to $(k + 2)$ years".

The first two of the following examples are taken from Leslie, 1945, pages 200–201. He uses them in particular to illustrate the problems involved in setting up the material properly, a task which a lack of space and time forces us to ignore.

Example 1. Given a hypothetical population of beetles (hereafter called beetles$_1$) for which the females live approximately 3 years, give birth to approximately 6 female offspring apiece during their 3rd year and immediately die, we have three age groups,

0–1 year, 1–2 years, 2–3 years

and fertility figures

$$F_0 = 0, \quad F_1 = 0, \quad F_2 = 6.$$

Suppose that the probability of living from the age group "0–1 years" to the group

"1–2 years" is $P_0 = \frac{1}{2}$, and from "1–2 years" to "2–3 years" is $P_1 = \frac{1}{3}$. Then we have, for beetles$_1$,

$$\mathbf{M}_1 = \begin{pmatrix} 0 & 0 & 6 \\ \frac{1}{2} & 0 & 0 \\ 0 & \frac{1}{3} & 0 \end{pmatrix}.$$

Example 2. Let us suppose we have another beetle population, beetles$_2$, with the same figures as for beetles$_1$, except that some become nubile when 2 years old, and that the fertility figures are

$$F_0 = 0, \quad F_1 = 1, \quad F_2 = 3.$$

This leads to the matrix for beetles$_2$,

$$\mathbf{M}_2 = \begin{pmatrix} 0 & 1 & 3 \\ \frac{1}{2} & 0 & 0 \\ 0 & \frac{1}{3} & 0 \end{pmatrix}.$$

Example 3. Finally, let us endow beetles$_1$ with a possibility of two years of life beyond menopause, giving them survival probabilities of $P_2 = \frac{1}{5}$ and $P_3 = \frac{1}{10}$, respectively. We then have for beetles$_3$,

$$\mathbf{M}_3 = \begin{pmatrix} 0 & 0 & 6 & 0 & 0 \\ \frac{1}{2} & 0 & 0 & 0 & 0 \\ 0 & \frac{1}{3} & 0 & 0 & 0 \\ 0 & 0 & \frac{1}{5} & 0 & 0 \\ 0 & 0 & 0 & \frac{1}{10} & 0 \end{pmatrix}.$$

Returning for the moment to our general discussion, suppose you are given an initial population distribution of

n_0 individuals in age group 0–1 year,

n_1 individuals in age group 1–2 years,

$\cdot$

$\cdot$

$\cdot$

n_m individuals in age group m–$(m+1)$ years.

Let $\mathbf{n} = (n_0, \ldots, n_m)^*$ be the corresponding column vector. Similarly, let $\mathbf{n}' = (n'_0, \ldots, n'_m)^*$ be the vector of age distributions that you can expect a year from now. Then $\mathbf{n}' = \mathbf{Mn}$. That is,

$$\begin{pmatrix} n'_0 \\ \cdot \\ \cdot \\ \cdot \\ n'_m \end{pmatrix} = \begin{pmatrix} F_0 & \cdots\cdots\cdots & F_m \\ P_0 & 0 & \\ & \cdot & \\ & & \cdot \\ & P_{m-1} & 0 \end{pmatrix} \begin{pmatrix} n_0 \\ \cdot \\ \cdot \\ \cdot \\ n_m \end{pmatrix}.$$

Exercise 1a. Prove this.

Similarly, if $\mathbf{n}'' = (n''_0, \ldots, n''_m)^*$ are the numbers of individuals in each age group that you can expect 2 years hence, then $\mathbf{n}'' = \mathbf{M}^2\mathbf{n}$.

Exercise 1b. Prove this.

In general, if $\mathbf{n}^{(r)} = (n_0^{(r)}, \ldots, n_m^{(r)})*$ is the expected population r years from now, then

$$\mathbf{n}^{(r)} = \mathbf{M}^r \mathbf{n}. \tag{1}$$

Exercise 1c. Prove this.

Our examples may help bring things down to earth.

Exercise 2. Given 1000 beetles$_1$ in each of the age groups of example 1, what distribution can you expect (a) in 1 year? (b) 2 years? (c) 3 years? (d) r years? (e) What if you had started with a distribution of 6000 of age 0–1 year, 3000 of age 1–2 years, and 1000 of age 2–3 years?

Exercise 3. Try exercise 2(a) through (e) for beetles$_2$. Don't expect such good luck in (d). However, this example has the interesting feature that no matter how you start out, you ultimately end up with a stable population distributed according to some scalar multiple of the vector $(6, 3, 1)*$.

Exercise 4. Try exercises 2(a) through (e) for beetles$_3$.

It is interesting to note that if $\mathbf{M}$ has an inverse $\mathbf{M}^{-1}$, then equation (1) holds for negative integers as well. For example, given $\mathbf{M}^{-1}$ and a distribution of $\mathbf{n} = (n_0, \ldots, n_m)*$ this year, it came (insofar as our assumptions are correct) from a distribution of $\bar{\mathbf{n}} = (\bar{n}_0, \ldots, \bar{n}_m)*$ last year, where $\bar{\mathbf{n}} = \mathbf{M}^{-1}\mathbf{n}$.

Exercise 5(a). Prove this. (b) Given the first distribution of beetles$_1$ in Exercise 2, what was last year's distribution? (c) Similarly for beetles$_2$ (this is discussed further in Exercise 6). (d) Find a population distribution of beetles$_3$ (with a positive number of members, a nonnegative number in each age group) for which there won't be any more beetles$_3$ next year. Why does this prove that $\mathbf{M}_3$ has no inverse?

Exercise 6. Since the inverse of a matrix is unique, Exercise 5c says that to get this year's equal distribution of beetles$_2$, one would have had to start out with a negative number of beetles$_2$ in the 3rd age group last year (i.e., our model was not applicable to the beetle population last year).

Find conditions on the relative number of beetles$_2$ that it would be possible to have in the three age classes last year, in order to get a valid distribution this year (i.e., find conditions on the entries of $\bar{\mathbf{n}}$ in order that both its entries and those of $\mathbf{n}$ are all nonnegative and still $\mathbf{M}_2\bar{\mathbf{n}} = \mathbf{n}$, that is, $\bar{\mathbf{n}} = \mathbf{M}_2^{-1}\mathbf{n}$).

For beetles$_3$ the postreproductive females contribute nothing (except perhaps themselves) to future populations. That is, $F_3 = F_4 = 0$. Thus, we can partition $\mathbf{M}_3$ as

$$\mathbf{M}_3 = \left(\begin{array}{ccc|cc} \multicolumn{3}{c|}{\mathbf{A}} & \multicolumn{2}{c}{\mathbf{0}} \\ 0 & 0 & 6 & 0 & 0 \\ \frac{1}{2} & 0 & 0 & 0 & 0 \\ 0 & \frac{1}{3} & 0 & 0 & 0 \\ \hline 0 & 0 & \frac{1}{3} & 0 & 0 \\ 0 & 0 & 0 & \frac{1}{10} & 0 \\ \multicolumn{3}{c|}{\mathbf{B}} & \multicolumn{2}{c}{\mathbf{C}} \end{array} \right),$$

where the matrix **A** (which should be familiar) contains the essential information concerning future population. In general, given **M** as above, suppose that the last age group containing fertile females is "h years to $(h + 1)$ years old", i.e., that

$$F_h \neq 0 \quad \text{but} \quad F_{h+1} = F_{h+2} = \cdots = 0.$$

Partition **M** so that an $(h + 1) \times (h + 1)$ square submatrix is in the upper left-hand corner,

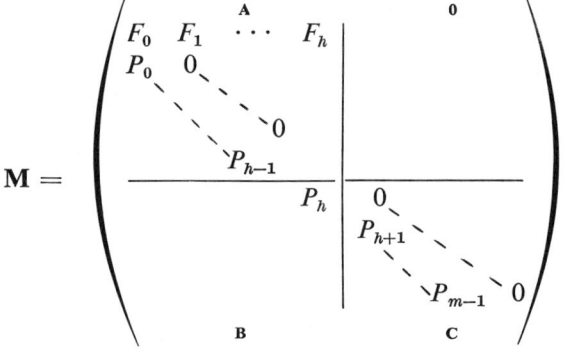

Exercise 7.
  a) Show that for this matrix

$$\mathbf{M} = \begin{pmatrix} \mathbf{A} & \mathbf{0} \\ \mathbf{B} & \mathbf{C} \end{pmatrix},$$

the matrix **A** completely determines the changes in the reproductive and pre-reproductive age groups of future populations.
  b) Prove that

$$\mathbf{M}^2 = \begin{pmatrix} \mathbf{A}^2 & \mathbf{0} \\ \blacksquare & \mathbf{C}^2 \end{pmatrix}.$$

Illustrate on $\mathbf{M}_3$.
  c) Prove that for any r,

$$\mathbf{M}^r = \begin{pmatrix} \mathbf{A}^r & \mathbf{0} \\ \blacksquare & \mathbf{C}^r \end{pmatrix}.$$

  d) Calculate $\mathbf{C}^2$; show that for some r, $\mathbf{C}^r = \mathbf{0}$ (this says that individuals in the postreproductive age contribute nothing to the population after their death).

REFERENCES

Leslie, P. H., "On the use of matrices in certain population mathematics", *Biometrika*, **33**, 1945, pp. 183–212.
———, "Some further notes on the use of matrices in population mathematics", *Biometrika*, **35**, 1948, pp. 213–245.

c. Game Theory; An example.

The game of morra seems to have existed in Italy throughout history (one Roman characterized another as being so honest that you could play morra with him in the dark). It is played by two persons as follows: Each player presents either 1 or 2 fingers and simultaneously calls out a number. If the number called by a player is the same as the total number of fingers which both have shown, then he wins an amount corresponding to this number. If both players guess correctly, neither collects.

There are some obvious strategies to avoid, such as calling 2 and playing 2. We summarize those strategies which offer any promise of success in the form of a "payoff matrix": if you adopt the strategy indicated next to the ith row and your opponent adopts that above the jth column, then you win an amount corresponding to the i,j-entry of the matrix (if the i,j-entry is negative, then you pay him).

Payoff matrix for morra:

$$\text{Opponent} \begin{cases} \text{Show} & 1 & 1 & 2 & 2 \\ \text{Call} & 2 & 3 & 3 & 4 \end{cases}$$

$$\text{You} \begin{cases} \text{Show 1} & \text{Call 2} \\ 1 & 3 \\ 2 & 3 \\ 2 & 4 \end{cases} \begin{pmatrix} 0 & 2 & -3 & 0 \\ -2 & 0 & 0 & 3 \\ 3 & 0 & 0 & -4 \\ 0 & -3 & 4 & 0 \end{pmatrix} = \mathbf{M}.$$

For example, if you adopt the last strategy, "show 2, call 4", and your opponent kindly adopts his 3rd strategy "show 2, call 3", then you win 4 somethings.

Exercise 1. (a) Show that the payoff matrix represents the only sensible strategies.
(b) Verify that the payoffs are as indicated.

Let $\mathbf{s}_1 = (1, 0, 0, 0)$, $\mathbf{s}_2 = (0, 1, 0, 0)$, $\mathbf{s}_3 = (0, 0, 1, 0)$ and $\mathbf{s}_4 = (0, 0, 0, 1)$, and let $\mathbf{s}_1^*, \ldots, \mathbf{s}_4^*$ denote the column vectors which are their transposes. We shall call $\mathbf{s}_i$ **your ith pure strategy** and $\mathbf{s}_j^*$ **the jth pure strategy of your opponent.** This is because, if you play the strategy indicated in row i and your opponent the strategy indicated in row j, then you receive an amount $\mathbf{s}_i \mathbf{M} \mathbf{s}_j^*$.

Exercise 2. Prove this fact.

We shall assume that your opponent is every bit as wily and subtle as you are, and that you have no special information as to his psychological makeup. Then it is clear that there is no best way to play any individual round of morra; one could only hope to develop long-term approaches.

What is the best you can expect to do in playing morra for an evening? The game of morra is completely symmetrical: whatever approach one player adopts, the other player can adopt with equal success.

Exercise 3. Express this as a fact about the matrix $\mathbf{M}$.

Ultimately, you can expect to win about half the time if both players are sufficiently astute. Thus, *the best you can hope to do is to break even.* Of course, you can always do worse than this; there are losing long-term strategies (e.g., always play the same $\mathbf{s}_i$).

How can you be assured of at least breaking even in the long run? You will certainly wish to play the pure strategies in some sort of random fashion, since your opponent might guess any nonrandom pattern. We shall thus try to decide an appropriate probability p_i for playing each pure strategy s_i. Let $s = (p_1, p_2, p_3, p_4)$ be a probability vector; that is, each $p_i \geq 0$ and $p_1 + \cdots + p_4 = 1$. We shall call s a (**mixed**) **strategy for you** and a probability column vector

$$t = (q_1, q_2, q_3, q_4)^*$$

a (**mixed**) **strategy for your opponent.** If you adopt such a mixed strategy s, we will take this to mean that you play s_1 (show 1, call 2) with probability $p_1, \ldots, s_4$ (show 2, call 4) with probability p_4. If you adopt a mixed strategy s and your opponent a mixed strategy t, then some elementary probability theory shows that you can expect to win on the average an amount equal to sMt.

For example, suppose your opponent adopts the strategy

$$t = (0, \tfrac{4}{7}, \tfrac{3}{7}, 0)^*.$$

He will "show 1, call 3" four-sevenths of the time and "show 2, call 3" three-sevenths of the time. Then if you take as your strategy $s = (p_1, p_2, p_3, p_4)$ your expected winnings will be

$$(p_1, p_2, p_3, p_4)\mathbf{M}(0, \tfrac{4}{7}, \tfrac{3}{7}, 0)^* = \tfrac{1}{7}p_1 + 0p_2 + 0p_3 + 0p_4$$
$$= \tfrac{1}{7}p_1.$$

Thus, if you adopt p_1 (show 1, call 2) with positive probability, you can expect to *lose* an amount $\tfrac{1}{7}p_1$. On the other hand, if you therefore decide to adopt s_1 with 0 probability, then your expected winnings will be 0. Thus, the strategy $(0, \tfrac{4}{7}, \tfrac{3}{7}, 0)^*$ is a best strategy for your opponent; you can never expect to win from him (over the long run), and should you occasionally adopt s_1, you have a positive probability of losing.

From what we have previously observed, you can obtain a best possible strategy by simply borrowing your opponent's and playing $s = (0, \tfrac{4}{7}, \tfrac{3}{7}, 0)$.

Exercise 4. Show that another best possible strategy for you is $s = (0, \tfrac{3}{5}, \tfrac{2}{5}, 0)$.

Exercise 5. Show that if p is any number satisfying $\tfrac{4}{7} \leq p \leq \tfrac{3}{5}$, then $s = (0, p, (1-p), 0)$ is a best possible strategy.

Exercise 6. Show that the only strategies with which you can hope to break even are those in Exercise 5.

Exercise 7. In the familiar game of "odds and evens", you and your opponent simultaneously show either 1 or 2 fingers. You win a penny if the total is an odd number; your opponent wins a penny if it is an even number. (a) Set up a payoff matrix for odds and evens. (b) Give a complete discussion of "best" strategies.

REFERENCES

Gale, D., *The Theory of Linear Economic Models.* New York: McGraw-Hill, 1960.
Shubik, M. (ed.), *Game Theory and Related Approaches to Social Behavior.* New York: Wiley, 1964.

d. Markov Chains

We shall have to content ourselves with a rough study of this topic, since a full appreciation involves more probability theory than we have time to develop. Intuitively speaking, a **finite Markov Chain** deals with a system which can be in any one of n different states (n fixed). At regular time intervals, the system moves from

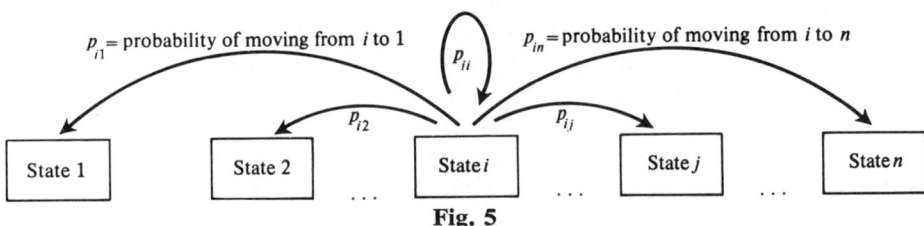

Fig. 5

one of the states to another (or stays put) with fixed probabilities attached to each option. Let p_{ij} denote the probability that if the system is in state i, then it will move to state j for the next time interval, $i,j = 1, \ldots, n$ (Fig. 5). The matrix

$$\mathbf{P} = \begin{pmatrix} p_{11} & \cdots & p_{1j} & \cdots & p_{1n} \\ \cdot & & \cdot & & \cdot \\ \cdot & & \cdot & & \cdot \\ p_{i1} & \cdots & p_{ij} & \cdots & p_{in} \\ \cdot & & \cdot & & \cdot \\ \cdot & & \cdot & & \cdot \\ p_{n1} & \cdots & p_{nj} & \cdots & p_{nn} \end{pmatrix}$$

is called the **transition matrix** of the Markov Chain.

What is the probability of moving from state i to state j in two steps? From elementary probability theory, this is

(Probability of moving from state i to state 1, and then from state 1 to state j)

+ (Probability of moving from state i to state 2 and then from state 2 to state j)

+ $\cdots$ + (Probability of moving from state i to state n, and then from state n to state j)

$$= p_{i1}p_{1j} + p_{i2}p_{2j} + \cdots + p_{in}p_{nj}.$$

Thus, the probability of moving from state i to state j in two steps is just the i,j-entry of $\mathbf{P}^2$.

Similarly, the probability of starting in state i and landing in state j in m steps is the i,j-entry of $\mathbf{P}^m$.

Since the system can only move within the given n states, the probability is 1 that if it is in state i, then it will move to either state 1, or to state 2, or to ..., or to state n,

$$p_{i1} + \cdots + p_{in} = 1 \qquad \text{for each } i = 1, \ldots, n.$$

Thus, the sum of the elements of each *row* of $\mathbf{P}$ is 1. A square matrix with this property and the additional property that all entries are nonnegative is called a **stochastic matrix** (see the exercises which follow). It is clear that any stochastic matrix can serve as a transition matrix.

Let p_i be the probability that the system be in state i. (Intuitively speaking, if you observe the system at any time, there is a probability of p_i that you will find it in state i.) The list of probabilities for each state, or "probability distribution" for the n states, is given by the probability vector $\mathbf{p} = (p_1, \ldots, p_n)$. (This is a vector with nonzero components whose sum is 1.) The probability distribution for the system at the next step is then given by the vector $\mathbf{p}' = \mathbf{p}\mathbf{P}$.

Proof. The probability that the system is in state j at the next step is

Prob. of starting in state 1 and ending in state j
    + Prob. of starting in state 2 and ending in state j + $\cdots$
    + Prob. of starting in state n and ending in state j

$$= p_1 p_{1j} + p_2 p_{2j} + \cdots + p_n p_{nj}.$$

This is the component p'_j of the product $\mathbf{p}' = \mathbf{p}\mathbf{P}$. ∎

In an exercise below we ask you to show that if $\mathbf{p}$ is a probability vector and $\mathbf{P}$ is a stochastic matrix, then $\mathbf{p}\mathbf{P}$ is also a probability vector. In particular, note that $\mathbf{p} = \mathbf{e}_1 = (1, 0, \ldots, 0)$ means that the system starts in state 1 and hence

$$\mathbf{p}' = \mathbf{e}_1 \mathbf{p} = (p_{11}, p_{12}, \ldots, p_{1n})$$

does indeed give the probability distribution for the system at the first step. Similarly for the other basis vectors.

The probability distribution of the system at the end of the second step is now seen to be $(\mathbf{p}\mathbf{P})\mathbf{P} = \mathbf{p}\mathbf{P}^2$. By induction we prove that $\mathbf{p}\mathbf{P}^m$ is the probability distribution at the mth step. In particular we find again that $\mathbf{P}^m$ is the transition matrix for m steps.

Markov Chains have been widely applied (e.g., shortly after their invention, A. A. Markov used them in a study of vowel-consonant frequency in *Eugene Onegin*). Among more important uses, they occur in the Ehrenfest model of diffusion in physics, studies of in-breeding in genetics, mathematical learning theory in psychology, the Leontief model of the economy, and social mobility theory.

We shall take a more careful look at an application of Markov Chains to social mobility. Suppose that we divide society up into various classes and assume that the probability p_{ij} that a father in class i has a son in class j does not change with time (this is a very suspect assumption, but it nonetheless leads to reasonably good results). By treating the possible classes as states through which we are following a single family from generation to (single-sonned) generation, we obtain a Markov Chain. Here is an example of the resulting matrix of transition probabilities for a division of British society into seven classes (Prais, page 57; its derivation is discussed therein).

	1	2	3	4	5	6	7	Class
	.388	.146	.202	.062	.140	.047	.015	1. professional and high administrative
	.107	.267	.227	.120	.206	.053	.020	2. managerial and executive
	.035	.101	.188	.191	.357	.067	.061	3. higher grade supervisory and nonmanual
$\mathbf{P} =$	.021	.039	.112	.212	.430	.124	.062	4. lower grade supervisory and nonmanual
	.009	.024	.075	.123	.473	.171	.125	5. skilled manual and routine nonmanual
	.000	.013	.041	.088	.391	.312	.155	6. semiskilled manual
	.000	.008	.036	.083	.364	.235	.274	7. unskilled manual.

The social transition matrix in England, 1949 (the i,j-entry indicates the probability that if a father is in class i, then his son will be in class j).

For $\mathbf{e}_1 = (1, 0, \ldots, 0), \ldots, \mathbf{e}_7 = (0, \ldots, 0, 1)$ we see that $\mathbf{e}_i\mathbf{P} = (p_{i1}, \ldots, p_{i7})$, so, given a father in class i, $\mathbf{e}_i\mathbf{P}$ lists the probability that his son will be in class 1, class 2, ..., class 7. Similarly, $\mathbf{e}_i\mathbf{P}^2$ will list the probability that a father in class i will have grandsons in class 1, ..., class 7.

For another interpretation of the probability distribution, suppose that $p_1\%$ of society is currently in class 1, ..., $p_7\%$ of society is in class 7, so that the probability vector $\mathbf{P} = (p_1, \ldots, p_7)$ gives the current distribution of society among the various classes. Then $\mathbf{pP}$ will give the probability distribution a generation hence, $\mathbf{pP}^2$ the probability distribution two generations hence, etc. It is worthwhile asking if society is heading toward any specific population distribution (based on the given transition matrix $\mathbf{P}$). The theory of Markov Chains shows that there is (in this case) such a distribution $\bar{\mathbf{p}} = (\bar{p}_1, \ldots, \bar{p}_7)$, which $\mathbf{pP}^n$ approaches as n grows large; and, in fact, $\bar{\mathbf{p}}\mathbf{P} = \bar{\mathbf{p}}$. (In Chapter 6 we shall consider the question as to when there is a $\bar{\mathbf{p}}$ for which $\bar{\mathbf{p}}\mathbf{P} = \bar{\mathbf{p}}$, and if so, how to find it.)

For the above example,†

Current distribution is **p** = $(\underset{1}{.037}, \underset{2}{.043}, \underset{3}{.098}, \underset{4}{.148}, \underset{5}{.432}, \underset{6}{.131}, \underset{7}{.111})$,

Distribution in the next
generation is **pP** = $(.029, .046, .094, .131, .409, .170, .121)$,

Distribution approached
is $\bar{\mathbf{p}}$ = $(.023, .042, .088, .127, .409, .182, .129)$.

Markov Chain theory also provides such important statistical measures as the average number of generations which a family would spend in a class before "breaking out". For our example, this would be‡

$$(\underset{1}{1.63}, \underset{2}{1.36}, \underset{3}{1.23}, \underset{4}{1.27}, \underset{5}{1.90}, \underset{6}{1.45}, \underset{7}{1.38})$$

The theory gives much other valuable information, together with statistical measures of its accuracy (see the Kemeny-Snell reference, for example).

Markov Chain Exercises

1. Recall that a vector is called a **probability vector** if all its components are nonnegative and the sum of its components is 1. A square matrix is called a **stochastic matrix** if each of its rows is a probability vector.

Examples. $(\frac{1}{3}, 0, \frac{2}{3})$ is a probability vector;

$$\begin{pmatrix} \frac{1}{3} & \frac{2}{3} \\ \frac{1}{2} & \frac{1}{2} \end{pmatrix}$$

is stochastic. In the following problems all row vectors are of the same size ($= n$) and all square matrices are $n \times n$. Furthermore, **u** is the column vector consisting of n 1's:

$$\mathbf{u} = (1, 1, \ldots, 1)^*.$$

a) Let **p** be a vector with nonnegative entries. Show that **p** is a probability vector iff **pu** = 1 and **P** is a stochastic matrix iff **Pu** = **u**.
b) If **p** is a probability vector and **P** is a stochastic matrix, then **pP** is also a probability vector.
c) If **P** and **Q** are stochastic matrices, then **PQ** is also stochastic. In particular **P**², **P**³, ... are all stochastic.

2. Suppose that society is divided into three classes: U, M, and L, with social transition matrix§

$$\mathbf{P} = \begin{matrix} U \\ M \\ L \end{matrix} \begin{pmatrix} \overset{U}{\frac{2}{5}} & \overset{M}{\frac{1}{2}} & \overset{L}{\frac{1}{10}} \\ \frac{1}{10} & \frac{7}{10} & \frac{1}{5} \\ \frac{1}{10} & \frac{1}{2} & \frac{2}{5} \end{pmatrix}$$

---

† Prais, p. 58.
‡ Prais, p. 60.
§ This is a rounded-off version of Kemeny-Snell, page 191.

a) Show that **P** is a stochastic matrix.
b) Given an initial distribution of $\frac{1}{5}U$, $\frac{7}{10}M$, and $\frac{1}{10}L$, what will the population distribution be a year from now? two years? three years?
c) Find the distribution toward which the population is tending.

3. According to the rules for writing Samoan, a consonant never follows a consonant and a consonant has about a 50% chance of following a vowel. Let us regard Samoan writing as being a Markov Chain with two states, "vowelhood" and "consonantness". (This is from the Newman reference below.)
   a) Find the transition matrix **P**.
   b) Suppose for the moment that one always starts out with a consonant, so the next stage is a vowel. Express this in terms of **P** and $\mathbf{e}_1 = (1, 0)$, $\mathbf{e}_2 = (0, 1)$.
   c) Calculate the probabilities that the third letter of a word will be a consonant or a vowel.
   d) Suppose it is known that the first letter has a 50-50 chance of being a vowel. Calculate the probability that the third letter is a vowel.
   e) Calculate $\mathbf{e}_1\mathbf{P}$, $\mathbf{e}_1\mathbf{P}^2$, $\mathbf{e}_1\mathbf{P}^3$, $\mathbf{e}_1\mathbf{P}^4$ and see if these seem to approach a "steady state".
   f) Write out $\bar{\mathbf{p}}\mathbf{P} = \bar{\mathbf{p}}$ for $\bar{\mathbf{p}} = (p_1, p_2)$. Treating p_1 and p_2 as unknowns, solve the system of equations. Find the probability vector $\bar{\mathbf{p}}$ corresponding to the steady state.
   g) Discuss the meaning of the steady state $\bar{\mathbf{p}}$ in terms of the Samoan language.
   h) Show that $\mathbf{e}_1\mathbf{P}^m$ approaches $\bar{\mathbf{p}}$; that $\mathbf{e}_2\mathbf{P}^m$ approaches $\bar{\mathbf{p}}$; that $\mathbf{p}\mathbf{P}^m$ approaches $\bar{\mathbf{p}}$ for any probability vector **p**.

4. A state in a Markov Chain is said to be **absorbing** if, once entered, it cannot be left.
   a) Given a chain with states $s_1, \ldots, s_n$, suppose that s_i is absorbing. How is this reflected in the transition matrix **P**?
   b) Suppose that states $s_1, \ldots, s_m$ are absorbing while states $s_{m+1}, \ldots, s_n$ are not. Discuss the natural partitioning of the transition matrix **P** which results. What does it mean in terms of the matrix if it is impossible to get from the nonabsorbing states to the absorbing states? If it *is* possible? Write $\mathbf{P}^2$ in terms of the partitioning of **P**.
   c) Write $\mathbf{P}^k$.

REFERENCES

Kemeny, J. G. and J. L. Snell, *Finite Markov Chains*. Princeton, N.J.: Van Nostrand, 1970.
Newman, E. B., The Pattern of Vowels and Consonants in Various Languages, *Am. J. Psych.*, **64,** 1951, pp. 369–79.
Prais, S. J., "Measuring Social Mobility," *J. Roy. Stat. Soc.*, **118,** 1955, pp. 56–66. [*Note:* Prais' matrices are the transpose of ours; we are using right-hand notation for a change, which conforms to the usage of Kemeny-Snell.]

e. Permutation Matrices and Kinship Systems

The kinship system material follows after some relevant notions concerning permutation matrices.

A **permutation** is a one-to-one rearrangement of a finite set. A **permutation matrix** is a square matrix which has exactly one 1 in each row and column, and 0's elsewhere.

Examples. The $n \times n$ identity matrix

$$\mathbf{I} = \begin{pmatrix} 1 & & \\ & \ddots & \\ & & 1 \end{pmatrix};$$

the $n \times n$ row-interchange matrix

$$\mathbf{N}_{ij} = \begin{pmatrix} 1 & & & & & & \\ & \ddots & & & & & \\ & & 0 & \cdots & 1 & & \\ & & \vdots & \ddots & \vdots & & \\ & & 1 & \cdots & 0 & & \\ & & & & & \ddots & \\ & & & & & & 1 \end{pmatrix};$$

the following 4×4 matrices

$$\mathbf{N}_{12} = \begin{pmatrix} 0 & 1 & 0 & 0 \\ 1 & 0 & 0 & 0 \\ 0 & 0 & 1 & 0 \\ 0 & 0 & 0 & 1 \end{pmatrix}, \quad \mathbf{W} = \begin{pmatrix} 0 & 1 & 0 & 0 \\ 1 & 0 & 0 & 0 \\ 0 & 0 & 0 & 1 \\ 0 & 0 & 1 & 0 \end{pmatrix},$$

$$\mathbf{C} = \begin{pmatrix} 0 & 0 & 1 & 0 \\ 0 & 0 & 0 & 1 \\ 1 & 0 & 0 & 0 \\ 0 & 1 & 0 & 0 \end{pmatrix}, \quad \mathbf{D} = \begin{pmatrix} 0 & 0 & 0 & 1 \\ 0 & 0 & 1 & 0 \\ 0 & 1 & 0 & 0 \\ 1 & 0 & 0 & 0 \end{pmatrix}, \quad \mathbf{E} = \begin{pmatrix} 0 & 1 & 0 & 0 \\ 0 & 0 & 1 & 0 \\ 0 & 0 & 0 & 1 \\ 1 & 0 & 0 & 0 \end{pmatrix}.$$

We shall now develop a few relationships (if not a kinship) between permutations and permutation matrices. (There is a rather simple one between the rows of such a matrix and the rows of the identity matrix of the same size. Can you find it?) Recall that premultiplication of a matrix by $\mathbf{N}_{ij}$ interchanges the ith and jth rows of the matrix. In particular, for a column vector $\mathbf{V}$ in $\mathbb{R}_{n\times 1}$, $\mathbf{N}_{ij}\mathbf{V}$ differs from $\mathbf{V}$ only in that its ith and jth entries have been switched: if

$$\mathbf{V} = (\cdots v_i \cdots v_j \cdots)^*,$$

then

$$\mathbf{N}_{ij}\mathbf{V} = (\cdots v_j \cdots v_i \cdots)^*.$$

Consequently, for the basis for $\mathbb{R}_{n\times 1}$, with $\mathbf{e}_1 = (1, 0, \ldots, 0)^*, \ldots, \mathbf{e}_n = (0, \ldots, 0, 1)^*$, we have

$$\mathbf{N}_{ij}\mathbf{e}_i = \mathbf{e}_j, \quad \mathbf{N}_{ij}\mathbf{e}_j = \mathbf{e}_i, \quad \mathbf{N}_{ij}\mathbf{e}_k = \mathbf{e}_k \quad \text{if } k \neq i, j.$$

Hence the mapping $e_l \to N_{ij}e_l$ is a permutation of the (ordered) set $\{e_1, \ldots, e_n\}$. For example, with the 4×4 matrix N_{12} above, we have

$$N_{12}e_1 = e_2, \quad N_{12}e_2 = e_1, \quad N_{12}e_3 = e_3, \quad N_{12}e_4 = e_4,$$

a permutation of $\{e_1, e_2, e_3, e_4\}$.

This is a general fact:

Given an $n \times n$ permutation matrix P, the ordered set $Pe_1, \ldots, Pe_n$ is a permutation of the ordered set $\{e_1, \ldots, e_n\}$.

Exercise 1. Illustrate this on the 4×4 permutation matrices **W, C, D,** and **E** listed above. That is, calculate $Pe_1, \ldots, Pe_4$, with **P** in turn each of these matrices, and note that in each case you obtain a permutation of $\{e_1, \ldots, e_4\}$.

Exercise 2. Prove this fact for 3×3 permutation matrices. You must show that given any 3×3 permutation matrix **P**, (a) Pe_i is some e_j (b) each e_j is some Pe_i and (c) if $Pe_i = Pe_j$ then $i = j$ (one-to-oneness). (If you are unable to complete an abstract proof, you could actually list all 3×3 permutation matrices and show they work properly.)

Exercise 3. Prove this fact for $n \times n$ permutation matrices.

There is a converse to the above: **any permutation $e_1, \ldots, e_n \to e_{1'}, \ldots, e_{n'}$ of the vectors e_i in $\mathbb{R}_{n \times 1}$ can be achieved via a permutation matrix**, i.e., given such a permutation there is a permutation matrix **P** for which $Pe_1 = e_{1'}, \ldots, Pe_n = e_{n'}$.

Exercise 4. Prove this for $n = 3$ (again, if you're desperate you can list all permutations, but it's really not too hard to describe a matrix for which $Pe_i = e_{j'}$ and then to show it is a permutation matrix).

Exercise 5. Prove this for arbitrary n.

Thus, one reason that permutation matrices are so called is that, under matrix multiplication, they perform a permutation of the standard set of basis vectors. Another reason (you probably guessed above) is that the rows (or columns) of a permutation matrix are a permutation of the rows (or columns) of the identity matrix. Even though there is an intimate relationship between permutations and permutation matrices, it is often most convenient to work with permutations in a nonmatrix form.

Exercise 6. Although the permutation matrices interchange the basis vectors $e_1, \ldots, e_n$, there are nonzero vectors which are left fixed under multiplication by all permutation matrices. (a) Find a $V \neq 0$ for which $PV = V$ for all $n \times n$ permutation matrices **P**. (b) Find all such **V**. (c) Prove that you have found all such **V**.

Exercise 7. How many $n \times n$ permutation matrices are there for (a) $n = 3$? (b) $n = 4$? (c) arbitrary n?

Exercise 8. a) For the matrices listed in the above example, calculate W^2, E^2, **WC**, **WD**.
b) Prove that the product of any two 3×3 permutation matrices is a permutation matrix.
c) Prove that the product of any two $n \times n$ permutation matrices is a permutation matrix.

Exercise 9. (a) Prove that the transpose of a permutation matrix is a permutation matrix. (b) Calculate **DD*** and **EE***.

Exercise 10. (a) Prove that any permutation matrix has an inverse and this inverse is a permutation matrix. (b) When is the sum of two permutation matrices a permutation matrix?

Exercise 11. (*Uses groups*) (a) Prove that the set of all $n \times n$ permutation matrices forms a group under matrix multiplication. (b) For what n is the group commutative? (Prove.)

Exercise 12. We now outline a proof that for every permutation matrix there is a positive integer N for which $\mathbf{P}^N = \mathbf{I}$. (a) Show that there are positive integers r and s for which $\mathbf{P}^r = \mathbf{P}^s$. [*Suggestion:* Previous exercises show that $\mathbf{P}^k$ is a permutation matrix for any integer k and that there is only a finite number of permutation matrices.] (b) Assuming $r > s$, show that $\mathbf{P}^s(\mathbf{P}^{r-s} - \mathbf{I}) = \mathbf{0}$. (c) Justifying your steps, obtain that $\mathbf{P}^{r-s} - \mathbf{I} = \mathbf{0}$ and show that $N = r - s$ works as desired. (d) Find the minimal N for which $\mathbf{P}^N = \mathbf{I}$ for matrices **W**, **C**, **D**, **E**, and $\mathbf{N}_{12}$ above.

Kinship Systems

Some primitive societies are divided into clans in such a fashion that a man in clan i can only marry women in clan j, while his children will belong to a (possibly) different clan k. The basic rules followed in some of these societies can be construed as an axiom system (White, page 34).

Suppose that there are n clans in a society. Let W be the matrix which has a 1 as its i,j-entry if a man in clan i must marry a woman in clan j, and has 0's otherwise:

Wife matrix

$$\mathbf{W} = \text{Husband's clan } i \begin{pmatrix} & & \text{Wife's clan} \\ & & j \\ & & | \\ ---&1&--- \\ & & | \end{pmatrix}.$$

Similarly, let **C** be the matrix which has a 1 as its i,k-entry if a man in clan i has children in clan k, and 0's otherwise

Children matrix

$$\mathbf{C} = \text{Father's clan } i \begin{pmatrix} & & \text{Children's clan} \\ & & k \\ & & | \\ ---&1&--- \\ & & | \end{pmatrix}.$$

The rules in these societies imply that **W** and **C** are permutation matrices (White, page 35).

Example (R. R. Bush; White, page 159f).

In the Kariera system there are four clans. The matrices are

$$\mathbf{W}\colon \text{Clan of husband} \quad \begin{array}{c} \text{Clan of wife} \\ \begin{array}{cccc} 1 & 2 & 3 & 4 \end{array} \\ \begin{array}{c} 1 \\ 2 \\ 3 \\ 4 \end{array}\!\!\begin{pmatrix} 0 & 1 & 0 & 0 \\ 1 & 0 & 0 & 0 \\ 0 & 0 & 0 & 1 \\ 0 & 0 & 1 & 0 \end{pmatrix} \end{array},$$

$$\mathbf{C}\colon \text{Clan of father} \quad \begin{array}{c} \text{Clan of children} \\ \begin{array}{cccc} 1 & 2 & 3 & 4 \end{array} \\ \begin{array}{c} 1 \\ 2 \\ 3 \\ 4 \end{array}\!\!\begin{pmatrix} 0 & 0 & 1 & 0 \\ 0 & 0 & 0 & 1 \\ 1 & 0 & 0 & 0 \\ 0 & 1 & 0 & 0 \end{pmatrix} \end{array}.$$

These indicate, for example, that a man in clan 1 must marry a woman in clan 2, while his children are in clan 3.

Exercise 13. "For the Tarau there are also four clans, Pochana, Tlangsha, Thimasha, and Khuipu. For ease in pronounciation, we call these A, B, C, and D, respectively." The rules of marriage and descent are summarized as follows:

An A man marries a B woman and has A children
      B C B
      C D C
      D A D

(Bush in White, page 166). Set up the **W** and **C** matrices for the Tarau.

Various products of the **W** and **C** matrices (both together and separately) yield important explicit information.

Let us examine **WC**:

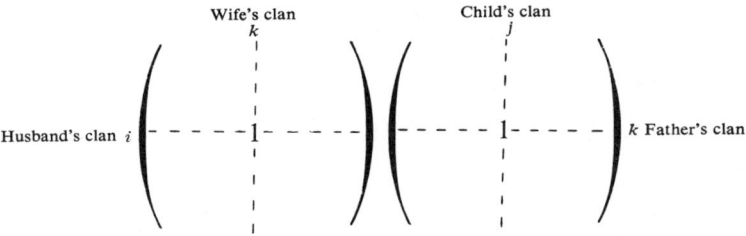

If the i,j-entry of **WC** is $\neq 0$, then

        a man of clan i marries a woman of clan k,

and

        a man of clan k has children in clan j.

Thus, **WC** lists the clan of the brother-in-law's children:

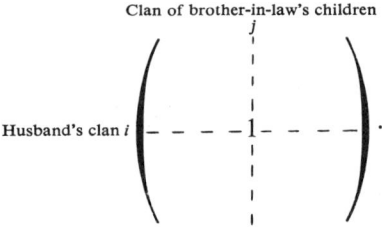

As another example, since **W** lists the clan of the wife according to the clan of the husband, **W*** will list the clan of the husband according to that of the wife:

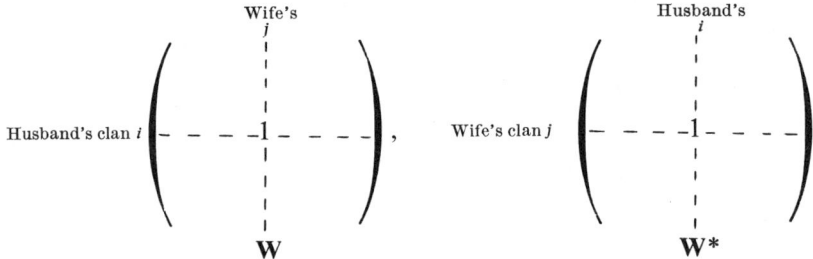

Since this is essentially an inverse operation, this lends credence to Exercise 10 which shows that for permutation matrices, $\mathbf{W}^* = \mathbf{W}^{-1}$.

Exercise 14. Find an interpretation for the matrices

a) **CW**   b) $\mathbf{W}^2$ c) $\mathbf{C}^2$

d) Show that the associativity of matrix multiplication implies that the clan of a man's son's grandson is the same as that of his grandson's son (thank goodness!).

Various relations among the products of **C**'s and **W**'s have important meanings. Using the above mentioned axioms one can show, for example, that

a man's son's daughter can marry his daughter's son (known as **matrilateral cross cousins**) iff **W** and **C** commute, **WC = CW**;

a man's son's son can marry his daughter's daughter (**patrilateral cross cousins**) iff **CW = W*C**   (White, pp. 39–42).

Exercise 15. Interpret (a) $\mathbf{W}^2 = \mathbf{I}$ (b) $\mathbf{C}^N = \mathbf{I}$.

Exercise 16. See if the Kariera allow
  a) matrilateral cross-cousin marriage,
  b) patrilateral cross-cousin marriage;
  c) do different clans simply switch wives? (If not, describe the method of exchange.)
  d) After how many generations will a man's descendants belong to his clan again?
    In each case show the matrix calculations.

Exercise 17. Do Exercise 16 for the Tarau.

REFERENCE

White, H., *An Anatomy of Kinship*. Englewood Cliffs, N.J.: Prentice-Hall, 1963.

There are appendices by André Weil and R. R. Bush (we use White's matrices and notation with Bush's examples; his matrices are different).

All decent functions are practically linear.†

CHAPTER 3 ## Linear Transformations

By now you should be quite familiar with the concept of a vector space over $\mathbb{R}$ and such attendant notions as subspace, linear independence, basis, dimension, and the like. Recall also that we have seen that any two vector spaces over the same field and which have the same finite dimension are algebraically identical. Since studying these notions, we have had an interlude during which you have become familiar with matrices and their basic operations.

We could go little further in our study of a single vector space without imposing further structure on it. We shall do this in Chapter 8. We shall now, however, take a different tack and study the transformations from one vector space to another which preserve the vector space operations of vector addition and scalar multiplication. This entire chapter (and much of the rest of the book) is devoted to the study of these "linear transformations". We shall later see that such transformations, considered geometrically, do map lines into lines, giving one justification for their name. The reason you probably haven't encountered these functions much before is that they are particularly simple in the one-dimensional case (of the form $f(x) = ax$, in fact). However, they are of great importance in both mathematics and its applications.

---

† A. M. Gleason in his lectures on "The Geometric Content of Advanced Calculus", CUPM Geometry Conference, Summer, 1967. Most other mathematicians, as well as people who apply mathematics, make similar statements. We chose Gleason because we could nail him down with a reference.

1. WHAT THEY ARE

Let $\mathcal{V}$ and $\mathcal{W}$ be vector spaces over the reals $\mathbb{R}$. We shall call a function† **T** from $\mathcal{V}$ into $\mathcal{W}$ a **linear transformation** if

i) for all vectors **X** and **Y** in $\mathcal{V}$,

$$\mathbf{T(X + Y) = T(X) + T(Y)}. \qquad \text{(Additivity)}$$

ii) for all vectors **X** in $\mathcal{V}$ and all scalars a in $\mathbb{R}$,

$$\mathbf{T}(a\mathbf{X}) = a\mathbf{T(X)}. \qquad \text{(Homogeneity)}$$

Linear transformations are also called **linear mappings** and **linear functions**.

Recall that the contents of 5.1 of Chapter 2 was that for a (fixed) matrix A in $\mathbb{R}_{m \times n}$ and for V in $\mathbb{R}_n$ ($= \mathbb{R}_{n \times 1}$) the mapping $\mathbf{T}(V) = AV$ is a linear transformation from $\mathbb{R}_n$ into $\mathbb{R}_m$. This is because matrix multiplication is distributive and homogeneous. We shall see in Section 3 that this is essentially what linear transformations are always like on finite-dimensional vector spaces. We consider a few more examples.

Examples
1. An isomorphism between two vector spaces over $\mathbb{R}$ is a linear transformation which satisfies the additional requirements of being one-to-one and onto.
2. Let $\mathcal{V}$ be a vector space over $\mathbb{R}$. The **identity function I** from $\mathcal{V}$ to $\mathcal{V}$ is defined by $\mathbf{I(X) = X}$ for all **X** in $\mathcal{V}$. It is very easy to see that **I** is a linear transformation (even an isomorphism) of $\mathcal{V}$ over $\mathbb{R}$ onto $\mathcal{V}$ over $\mathbb{R}$.
3. Let $\mathcal{V}$ be any vector space over $\mathbb{R}$ and let $\mathcal{W}$ be the subspace of $\mathcal{V}$ consisting of just the zero vector of $\mathcal{V}$. Define a mapping **Z** from $\mathcal{V}$ to $\mathcal{W}$ (there's just one way you can) by

$$\mathbf{Z(X) = 0} \quad \text{for all} \quad \mathbf{X} \text{ in } \mathcal{V}.$$

   One can quickly check that this **zero function Z** is a linear transformation of $\mathcal{V}$ over $\mathbb{R}$ onto $\mathcal{W}$ over $\mathbb{R}$.
4. Consider the mapping **T** from $\mathbb{R}_2$ into $\mathbb{R}$ given by $\mathbf{T}(x, y) = ax + by$, where a and b are (fixed) elements of $\mathbb{R}$. Now **T** is certainly a function (see Chapter 0, Section 4, if in doubt) so we need only check additivity:

$$\mathbf{T}[(x, y) + (x', y')] = \mathbf{T}(x + x', y + y')$$
$$= a(x + x') + b(y + y') = ax + ax' + by + by'$$
$$= ax + by + ax' + by' = \mathbf{T}(x, y) + \mathbf{T}(x', y'),$$

   and homogeneity,

$$\mathbf{T}(c(x, y)) = \mathbf{T}(cx, cy) = acx + bcy = c(ax + by) = c\mathbf{T}(x, y).$$

   You might contemplate how this example relates to matrices (see Exercise 7).
5. Let C be the set of all continuous function defined for $0 \leq x \leq 1$, considered as a vector space over the reals $\mathbb{R}$ in the usual way. Let C' be the set of all functions

---

† See Section 4 of Chapter 0 for a discussion of functions.

defined for $0 \leq x \leq 1$ whose derivatives are continuous on this interval. C' is easily seen to be a vector space over $\mathbb{R}$, relative to the usual definitions of function addition and scalar multiplication. Define a mapping $\mathbf{D}$ from C' to C by $\mathbf{D}f = f'$ for all f in C' (that is, $\mathbf{D}f$ is the function whose value at x is $f'(x)$). Then $\mathbf{D}$ is a linear transformation from C' over $\mathbb{R}$ to C over $\mathbb{R}$.

Let us carefully check that the $\mathbf{D}$ in the last example is indeed a linear transformation. First of all, given any f in C', $\mathbf{D}f$ is defined since f is supposed to be a differentiable function. Moreover $\mathbf{D}f$ is in C, since elements of C' have continuous derivatives. Furthermore, $\mathbf{D}f$ is a unique well-defined element of C (namely, the function whose value at x if $f'(x)$.) Thus, $\mathbf{D}$ is itself a function from C' to C. Further $\mathbf{D}$ is additive:

$$\mathbf{D}(f + g) \stackrel{\text{def.}}{=} (f + g)' \stackrel{\text{ancient calculus}}{=} f' + g'.$$

Note that the last equality just says

$$\frac{d}{dx}(f(x) + g(x)) = \frac{df(x)}{dx} + \frac{dg(x)}{dx}.$$

$\mathbf{D}$ is homogeneous:

$$\mathbf{D}(af) \stackrel{\text{def.}}{=} (af)' \stackrel{\text{ancient calculus}}{=} af',$$

which is to say

$$\frac{d}{dx}(af(x)) = a\frac{df(x)}{dx}.$$

Thus $\mathbf{A}$ is a linear transformation.

We shall spend most of the rest of this section looking at linear transformations geometrically. Since our purpose is to broaden your intuition, we shall feel free to use geometrical terms such as "coordinate axis", "parallel", "perpendicular", and "area" which are not yet part of our official linear algebra vocabulary. Except for 1.1 and 1.2, we shall have an informal discussion rather than precise mathematical proofs. Although all of this could be made rigorous, it does not seem worthwhile to do so at this time.

By considering the situation in euclidean n-space (where we can draw pictures at least for $n \leq 3$) we can derive an intuitive geometrical interpretation of what it means to be a linear transformation. If $\mathbf{T}$ is one, then $\mathbf{T}(a\mathbf{V}) = a\mathbf{T}(\mathbf{V})$ says that $\mathbf{T}$ maps $a\mathbf{V}$ onto a vector $\mathbf{T}(a\mathbf{V})$ whose relationship to $\mathbf{T}(\mathbf{V})$, in terms of relative magnitude and direction, is the same as the relationship of $a\mathbf{V}$ to $\mathbf{V}$ (Fig. 1).

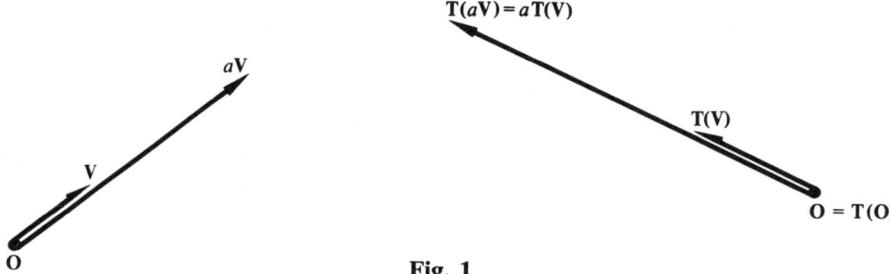

Fig. 1

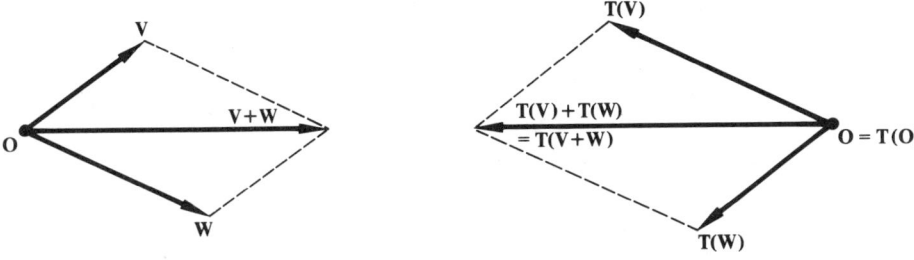

Fig. 2

Moreover, since **V** and a**V** are parallel vectors, one also has the weaker statement that **T** maps parallel vectors onto parallel vectors.

The equation $\mathbf{T(V + W) = T(V) + T(W)}$, with **V**, **W** in $\mathbb{R}$, says that **T** maps a parallelogram, together with its diagonals, onto a parallelogram together with its diagonals (Fig. 2).

Let us use this approach to give a geometrical proof that any reflection **R** in a line l through the origin of the euclidean plane is a linear transformation. We consider

$$\mathbf{R}(a\mathbf{V}) \quad \text{vs.} \quad a\mathbf{R}(\mathbf{V}) \qquad \text{and} \qquad \mathbf{R}(\mathbf{V + W}) \quad \text{vs.} \quad \mathbf{R(V) + R(W)}.$$

(See Fig. 3.) We have drawn the case when $a > 0$ (you consider $a < 0$), and it is evident that $a\mathbf{R(V)}$ is a vector of the same magnitude (namely, $|a| \times$ length of $\mathbf{R(V)}$) and direction as $a\mathbf{R(V)}$. Similarly, since the parallelogram with adjacent sides **V** and **W** is transformed intact by **R** onto the parallelogram with adjacent sides $\mathbf{R(V)}$ and $\mathbf{R(W)}$, we see that $\mathbf{R(V + W)}$ (which is the image of the diagonal of the first parallelogram) is actually

$$\mathbf{R(V + W) = R(V) + R(W)}.$$

In our study of analytic geometry, we shall see that any rigid motion of euclidean n-space which leaves the origin unchanged is a linear transformation.

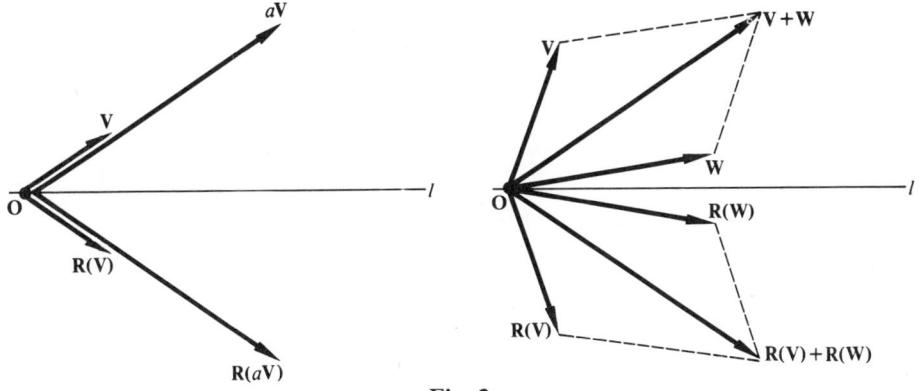

Fig. 3

This leaves out such worthy creatures as translations of the plane. However, it is easy to relate rigid motions which move the origin to those which do not. In fact, if **f** is a rigid motion, then we shall see that for all **X** in n-space,

$$\mathbf{f(X) = T(X) + f(0)},$$

where **T** is a linear transformation. Thus, any rigid motion is a linear transformation plus a constant value (**f(0)**, the image of the origin).

In general, a function **f** from $\mathbb{R}_n$ to $\mathbb{R}_m$ which is of the form

$$\mathbf{f(X) = T(X) + C}, \quad \text{for } \mathbf{X} \text{ in } \mathbb{R}_n,$$

where **T** is a linear transformation and **C** a fixed vector in $\mathbb{R}_m$, is called an **affine transformation**. Since affine transformations are so close to linear transformations, it suffices to study the latter, but you should be aware of the existence of the former since they are both common and useful.

The reason that the origin plays such an important role here is because any linear transformation maps the origin onto the origin:

1.1 For any vector spaces $\mathcal{V}$ and $\mathcal{W}$ over $\mathbb{R}$ and any linear transformation **T** from $\mathcal{V}$ into $\mathcal{W}$, if $\mathbf{0}_\mathcal{V}$ is the origin of $\mathcal{V}$ and $\mathbf{0}_\mathcal{W}$ the origin of $\mathcal{W}$, then $\mathbf{T(0_\mathcal{V}) = 0_\mathcal{W}}$.

We ask you to prove this in Exercise 2 (two ways).

The geometrical approach can also be important in understanding what a given linear transformation *does*:

Example 1. The linear transformation from $\mathbb{R}_3$ to $\mathbb{R}_3$ defined by $\mathbf{T(V) = AV}$, where

$$\mathbf{A} = \begin{pmatrix} 1 & 0 & 0 \\ 0 & 1 & 0 \\ 0 & 0 & 0 \end{pmatrix}$$

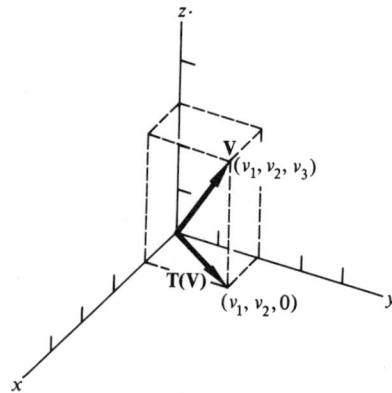

Fig. 4

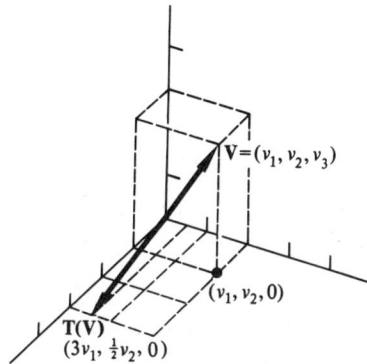

Fig. 5

maps $(v_1, v_2, v_3)^*$ to $(v_1, v_2, 0)^*$. Geometrically, it takes any vector **V** and projects it parallel to the z-axis onto a vector in the xy-plane (Fig. 4). Similarly, for

$$\mathbf{B} = \begin{pmatrix} 3 & 0 & 0 \\ 0 & \tfrac{1}{2} & 0 \\ 0 & 0 & 0 \end{pmatrix},$$

the linear transformation S defined by $\mathbf{S(V)} = \mathbf{BV}$ can be seen to take a vector **V** and project it (in a direction parallel to the z-axis) onto the xy-plane, and then stretch it by 3 in the x-direction while shrinking it by $\tfrac{1}{2}$ in the y-direction (Fig. 5).

We shall develop a number of techniques for obtaining geometrical information about linear transformations. For example, in Chapter 10 we shall see that every linear transformation from a finite-dimensional vector space over $\mathbb{R}$ into itself can be thought of as first performing a stretching in various directions, followed by a rotation about the origin.

We shall now examine a garland of linear transformations in the plane, each of which has an interesting geometrical interpretation. With l and l' as coordinate axes, the equation is $\mathbf{T}(x, y) = (x, -y)$.

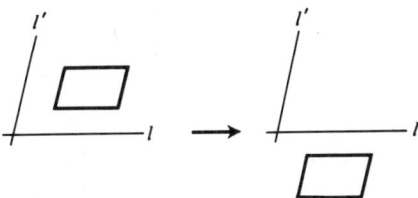

Fig. 6 Skew reflection in a line l.

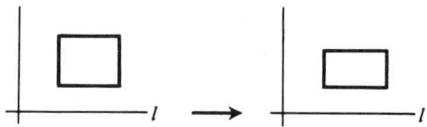

Fig. 7 Compression in a line *l*.

By taking *l* as one axis and a line perpendicular to *l* as another, the equation is $T(x, y) = (x, ky)$, with $k > 0$. (If $k > 1$ you might wish to call this a "stretching".)

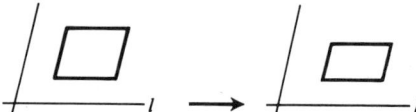

Fig. 8 Skew compression in a line *l*.

With appropriate choice of coordinate axes, this will have the same equation as a regular compression.

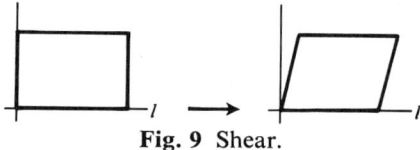

Fig. 9 Shear.

The equation is $T(x, y) = (x + ky, y)$. Note that there is an inverse mapping, a shear with coefficient $-k$. Moreover, a shear preserves area (check for parallelograms).

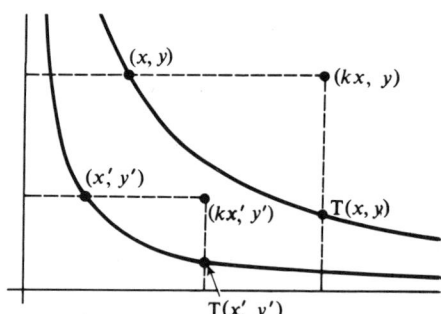

Fig. 10 Hyperbolic rotation.

Here $T(x, y) = (1/k\, x, ky)$. This particular linear transformation also happens to preserve area.

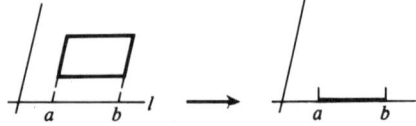

Fig. 11 Projection onto a line *l*.

As with compression, this can be either perpendicular or skew. As a matter of fact, this may be regarded as a rather extreme case of compression:

$$k = 0 \quad \text{and} \quad T(x, y) = (x, 0).$$

We shall discuss projections in a more general context in Supplement 1 to Section 4.

There are, of course, many other important geometrical linear transformations in the plane. For example, the following linear transformations are symmetries of the plane:

$$T(x, y) = (y, x); \quad T(x, y) = (-x, -y).$$

You probably used them in calculus to test if functions are odd or even. Do you recognize them? (See Exercise 1c.)

Finally, we note that the basic notion of linearity can be extended in the following useful manner:

1.2 T is a linear transformation on a vector space $\mathcal{V}$ iff for any $V_1, \ldots, V_n$ in $\mathcal{V}$ and $a_1, \ldots, a_n$ in $\mathbb{R}$,

$$T(a_1V_1 + \cdots + a_nV_n) = a_1T(V_1) + \cdots + a_nT(V_n).$$

Proof. 1. If the equality holds, one obtains additivity by taking $n = 2$ and $a_1 = a_2 = 1$. Homogeneity is the case $n = 1$.

2. The converse is a modest exercise in mathematical induction (Exercise 2c.) ∎

EXERCISES

1. In the following, if the given mapping is a linear transformation, then prove that it is. If it is not, then show that either it is not a function, or it is not additive, or it is not homogeneous. Which functions are affine but not linear?
   a) The **I** of example 2.
   b) The **Z** of example 3.
   c) Let T map $\mathbb{R}_2$ into $\mathbb{R}_2$ by $T(a, b) = (b, a)$. Describe geometrically. Same for $T(a, b) = (-a, -b)$.
   d) Let T map $\mathbb{R}_2$ into $\mathbb{R}_2$ by $T(x, y) = (x + a, y + b)$. Describe geometrically.
   e) Let T map $\mathbb{R}_2$ into $\mathbb{R}_2$ by $T(x, y) = (ax + by, cx + dy)$.
   f) Let T map $\mathbb{R}_3$ into $\mathbb{R}_3$ by $T(x, y, z) = (0, x, y)$. Describe geometrically.
   g) Let $\mathbb{C}$ be the vector space of complex numbers over $\mathbb{R}$. Let $T(a + bi) = a - bi$, a and b in $\mathbb{R}$. (That is, $T(Z) = \bar{Z}$, the complex conjugate of Z.)
   h) With C as in Example 5, and $\mathbb{R}$ considered as a vector space over $\mathbb{R}$, define T from C into $\mathbb{R}$ by $T(f) = f(0)$ for any f in C.
   i) The mapping T from C into C' defined by $T(f)(x) = \int_0^x f(t)\,dt$ (that is, $T(f)$ is the function whose value at any x is $\int_0^x f$).
   j) Define T from $\mathbb{R}_2$ into $\mathbb{R}$ by $T(a, b) = $ distance from the point (a, b) to the origin.
   k) Let T map $\mathbb{R}_2$ into $\mathbb{R}$ by $T(x, y) = \sqrt[3]{x^3 + y^3}$.

l) Let $\mathcal{V}$ be the subspace of $\mathbb{R}_2$ spanned by the vectors $V_1 = (1, 0)$ and $V_2 = (1, 1)$. Thus, any element of $\mathcal{V}$ is of the form $aV_1 + bV_2$, a, b in $\mathbb{R}$. Define T from $\mathcal{V}$ into $\mathbb{R}_2$ by

$$T(aV_1 + bV_2) = (a, b).$$

m) Let $\mathcal{W}$ be the subspace of $\mathbb{R}_2$ spanned by

$$W_1 = (1, 0) \quad \text{and} \quad W_2 = (-1, 0).$$

Define T from $\mathcal{W}$ into $\mathbb{R}_2$ by

$$T(aW_1 + bW_2) = (a, b).$$

2. Let $\mathcal{V}$ and $\mathcal{W}$ be vector spaces over $\mathbb{R}$, and T mapping $\mathcal{V}$ into $\mathcal{W}$ be a linear transformation. Prove that $T(0) = 0$, (a) using only homogeneity, (b) using only additivity, (c) prove 1.2.

3. Considering $\mathbb{R}$ as a vector space over itself, find (a) all linear transformations T from $\mathbb{R}$ into $\mathbb{R}$. [*Suggestion:* Consider $T(1)$ and homogeneity.] (b) Find all affine transformations.

4. a) Show geometrically that a rotation of the euclidean plane about the origin (through any given angle) is a linear transformation.
   b) Prove that translation of the euclidean plane is an affine transformation.

5. Describe the geometrical effect of the linear transformation T from $\mathbb{R}_2$ into $\mathbb{R}_2$ where $T(V) = AV$ and

a) $A = \begin{pmatrix} 1 & 0 \\ 0 & -1 \end{pmatrix}$ b) $A = \begin{pmatrix} 0 & -1 \\ 1 & 0 \end{pmatrix}$ c) $A = \begin{pmatrix} 0 & 1 \\ 1 & 0 \end{pmatrix}$

d) $A = \begin{pmatrix} 0 & -1 \\ 2 & 0 \end{pmatrix}$ e) $A = \begin{pmatrix} 2 & -1 \\ 1 & 2 \end{pmatrix}$ f) $A = \begin{pmatrix} 2 & 1 \\ 1 & 2 \end{pmatrix}$

6. Find a matrix A for which the linear transformation $T(V) = AV$ of $\mathbb{R}_2$ into $\mathbb{R}_2$ yields
   a) a reflection about the line $y = x$,
   b) a counterclockwise rotation about the origin through an angle of $\pi/2$.
   c) a reflection about the line $ax = by$,
   d) a counterclockwise rotation about the origin through an angle θ.

7. a) Show that the linear transformations in Exercise 6 may be obtained by matrix multiplication.
   b) Which of the linear transformations in Exercise 1 can be obtained by matrix multiplication? Specify the matrix in each case.
   c) Same for the examples.

Honors Projects.

1. Let $\mathcal{V}$ and $\mathcal{W}$ be two vector spaces over $\mathbb{C}$. Define **linear transformation of $\mathcal{V}$ into $\mathcal{W}$ over $\mathbb{C}$**.

Let T be the mapping of $\mathbb{C}$ into $\mathbb{C}$ considered in Exercise 1g, $T(z) = \bar{z}$ for z in $\mathbb{C}$. You were to show there that T is a linear transformation of $\mathbb{C}$ considered as a vector space over $\mathbb{R}$. Show that it is *not* a linear transformation of $\mathbb{C}$ considered as a vector space over $\mathbb{C}$.

2. Let $\mathcal{V}$ and $\mathcal{W}$ be two vector spaces over a field $\mathbb{F}$. Define **linear transformation of $\mathcal{V}$ into $\mathcal{W}$ over** $\mathbb{F}$.

As just illustrated, if $\mathbb{F}$ is a subfield of $\mathbb{F}_1$ one has no right to expect that a linear transformation over $\mathbb{F}$ will be a linear transformation over $\mathbb{F}_1$. If several fields are under consideration, it is necessary to write "linear transformation *over* $\mathbb{F}$" to keep things straight.

Note: All the material in this chapter carries over to spaces over the complex numbers (and even over an arbitrary field), except for the Dominant Diagonal Theorem of Section 4, Supplement 2 (which uses properties of ordering).

For the rest of this chapter we shall refrain from reminding you of these honors projects. Those interested should nonetheless work through the material for (1) vector spaces over $\mathbb{C}$, (2) vector spaces over an arbitrary field.

2. ARITHMETIC PROPERTIES

In this section we shall first investigate some strong similarities between linear transformations from one vector space to another and real-valued functions of a real variable (compare Chapter 0, Section 3). First of all, one can define scalar multiplication of linear transformations:

Given a linear transformation from the vector space $\mathcal{V}$ into the vector space $\mathcal{W}$, and given a scalar c in $\mathbb{R}$, let $c\mathbf{T}$ be that function from $\mathcal{V}$ into $\mathcal{W}$ defined by

$$(c\mathbf{T})(\mathbf{V}) = c(\mathbf{T}(\mathbf{V})) \qquad \text{for all } \mathbf{V} \text{ in } \mathcal{V}.$$

2.1 $c\mathbf{T}$ is a linear transformation from $\mathcal{V}$ into $\mathcal{W}$.

Proof. $c\mathbf{T}$ is surely a function from $\mathcal{V}$ into $\mathcal{W}$ since we have specified its value (namely, $c\mathbf{T}(\mathbf{V})$) for each $\mathbf{V}$ in $\mathcal{V}$. It is additive:

$$(c\mathbf{T})(\mathbf{V}_1 + \mathbf{V}_2) \stackrel{\text{defn.}}{=} c(\mathbf{T}(\mathbf{V}_1 + \mathbf{V}_2)) \stackrel{\text{T linear}}{=} c(\mathbf{T}(\mathbf{V}_1) + \mathbf{T}(\mathbf{V}_2))$$

$$\stackrel{\text{distr. of scalar mult.}}{=} c\mathbf{T}(\mathbf{V}_1) + c\mathbf{T}(\mathbf{V}_2) \stackrel{\text{defn.}}{=} (c\mathbf{T})(\mathbf{V}_1) + (c\mathbf{T})(\mathbf{V}_2).$$

It is homogeneous:

$$(c\mathbf{T})(a\mathbf{V}) \stackrel{\text{defn.}}{=} c(\mathbf{T}(a\mathbf{V})) \stackrel{\text{T linear}}{=} ca\mathbf{T}(\mathbf{V})$$

$$\stackrel{\text{comm. real mult.}}{=} ac\mathbf{T}(\mathbf{V}) \stackrel{\text{defn.}}{=} a((c\mathbf{T})(\mathbf{V})),$$

give or take a few vector space axioms. ∎

In like manner, for two linear transformations, $\mathbf{S}$ and $\mathbf{T}$ mapping $\mathcal{V}$ into $\mathcal{W}$ define their **sum S + T** by

$$(\mathbf{S} + \mathbf{T})(\mathbf{V}) = \mathbf{S}(\mathbf{V}) + \mathbf{T}(\mathbf{V}) \qquad \text{for all} \quad \mathbf{V} \text{ in } \mathcal{V}.$$

As you may have anticipated,

2.2 $S + T$ is a linear transformation from $\mathcal{V}$ into $\mathcal{W}$.

Proof. You provide the proof. It's just like 2.1. ∎

Example 1. Let T be the mapping from $\mathbb{R}_2$ into $\mathbb{R}_3$ defined by $T(x, y) = (x + y, x - y, 2x + y)$. T is easily seen to be linear (either directly or by relating it to a matrix multiplication). Then for any scalar c, cT is that linear transformation defined by

$$(cT)(x, y) = c(T(x, y)) = c(x + y, x - y, 2x + y)$$
$$= (cx + cy, cx - cy, 2(cx + cy)).$$

In like manner $S(x, y) = (x - y, x + y, -y)$ is easily seen to be a linear transformation from $\mathbb{R}_2$ into $\mathbb{R}_3$. Then $S + T$ is given by

$$(S + T)(x, y) = S(x, y) + T(x, y) = (x - y, x + y, -y) + (x + y, x - y, 2x + y)$$
$$= (2x, 2x, 2x).$$

Having endowed linear transformations with an addition and scalar multiplication, let us step back a moment and consider the collection of all such linear transformations from $\mathcal{V}$ into $\mathcal{W}$. Wonder-of-wonders (or possibly horror-of-horrors), we have:

2.3 The set of all linear transformations from a vector space $\mathcal{V}$ into a vector space $\mathcal{W}$ is itself a vector space under the operations of function addition and scalar multiplication.

Proof. In Exercise 1 we ask you to prove the more general result that the set of all functions (linear or not) from $\mathcal{V}$ into $\mathcal{W}$ forms a vector space. (The proof can be done as a slavish imitation of Chapter 0, 3.1, and 3.2.) Assuming you've proved this, we note that 2.1 and 2.2 tell us that the subset of all linear functions is closed under the vector space operations. Hence, this set is a subspace and, in particular, a vector space. ∎

Thus the collection of all linear transformations from one vector space to another forms a nice analogy to the vector space of real-valued functions defined on an interval. Common notations† are

$$\mathcal{L}(\mathcal{V}, \mathcal{W}) \quad \text{and} \quad \text{Hom}(\mathcal{V}, \mathcal{W}).$$

We shall not focus our attention on the set of all linear transformations as a vector space, but merely point out that 2.3 is a shorthand way of saying that linear transformations obey ten well-known axioms (such as commutativity of addition), and are subject to all the vector space theory developed in Chapter 1.

Let us now introduce another vector space $\mathcal{X}$ into the picture, and consider the **composition** of a linear transformation T from $\mathcal{V}$ into $\mathcal{W}$ with a linear transformation S from $\mathcal{W}$ into $\mathcal{X}$ (Fig. 12). Let ST be that function from $\mathcal{V}$ into $\mathcal{X}$

---

† "Hom" is for "homomorphism", another name for a linear transformation.

2 / Arithmetic properties   155

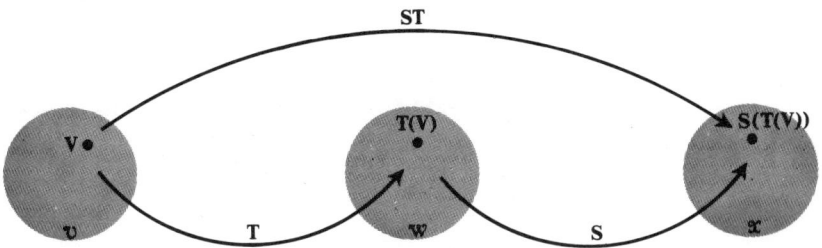

Fig. 12

defined by

$$(ST)(V) = S(T(V)) \quad \text{for all } V \text{ in } \mathcal{U}.$$

As ever,

2.4 ST is a linear transformation from $\mathcal{U}$ into $\mathcal{X}$.

Proof. $(ST)(V_1 + V_2) \stackrel{\text{defn.}}{=} S(T(V_1 + V_2)) \stackrel{\text{lin. of } T}{=} S(T(V_1) + T(V_2))$

$\stackrel{\text{lin. of } S}{=} S(T(V_1)) + S(T(V_2)) \stackrel{\text{defn.}}{=} (ST)(V_1) + (ST)(V_2).$

This establishes additivity. You attend to homogeneity. ∎

Example 2. With $T(x, y) = (x + y, x - y, 2x + y)$ the linear transformation from $\mathbb{R}_2$ into $\mathbb{R}_3$ in Example 1, let S be the linear transformation from $\mathbb{R}_3$ into $\mathbb{R}_1$ defined by $S(a, b, c) = a + b$. Then ST is the linear transformation from $\mathbb{R}_2$ into $\mathbb{R}_1$,

$$(ST)(x, y) = S(T(x, y)) = S(\overset{a}{\overline{x+y}}, \overset{b}{\overline{x-y}}, \overset{c}{\overline{2x+y}}) = 2x.$$

We now investigate some simple properties of this multiplication. First of all, it is homogeneous:

2.5 If T is a linear transformation from $\mathcal{U}$ into $\mathcal{W}$ and S a linear transformation from $\mathcal{W}$ into $\mathcal{X}$, and if c is any scalar, then

$$c(ST) = (cS)T = S(cT).$$

Proof. For any V in $\mathcal{U}$,

$$(c(ST))(V) = c((ST)(V)) = c(S(T(V)))$$
$$= (cS)(T(V)) = ((cS)(T))(V),$$

using the definition of composition and of scalar multiplication twice each. You prove the second equality. ∎

Secondly, our multiplication is distributive:

2.6 If T, T_1, and T_2 are linear transformations from $\mathcal{U}$ into $\mathcal{W}$ and S, S_1, and S_2 are linear transformations from $\mathcal{W}$ into $\mathcal{X}$, then

$$S(T_1 + T_2) = ST_1 + ST_2, \quad (S_1 + S_2)T = S_1T + S_2T.$$

Proof.

$$(S(T_1 + T_2))(V) \stackrel{\text{defn. } S(T_1 + T_2)}{=} S((T_1 + T_2)(V)) \stackrel{\text{defn. } T_1 + T_2}{=} S(T_1(V) + T_2(V))$$

$$\stackrel{\text{lin. of } S}{=} S(T_1(V)) + S(T_2(V)) \stackrel{\text{defn. } ST_1, ST_2}{=} (ST_1)(V) + (ST_2)(V)$$

$$\stackrel{\text{defn. } ST_1 + ST_2}{=} (ST_1 + ST_2)V.$$

Hence, $S(T_1 + T_2) = ST_1 + ST_2$. You do the other. ∎

Finally, our multiplication is associative.

2.7 If **T** is a linear transformation from $\mathcal{V}$ into $\mathcal{W}$, **S** is a linear transformation from $\mathcal{W}$ into $\mathcal{X}$, and **R** is a linear transformation from $\mathcal{X}$ into $\mathcal{Y}$, then

$$R(ST) = (RS)T.$$

Proof. Composition of functions is always associative (Chapter 0, 5.1). ∎

It is important to realize that the multiplication of linear transformations is not in general commutative. Given

$$\mathcal{V} \xrightarrow{T} \mathcal{W} \xrightarrow{S} \mathcal{X},$$

so that **ST** is defined, then **TS** will not even exist unless $\mathcal{V} = \mathcal{W} = \mathcal{X}$ (for example, to get **TS(V)** one must have **S(V)**, which necessitates **V** in $\mathcal{W}$). Even if **TS** and **ST** both exist, they are not likely to be the same linear transformation (cf. matrix multiplication).

Our last topic will be inverses, for which we must go back to considering single linear transformations. You recall that the two basic sets associated with any function are its domain and its range. For linear transformations, the latter is usually called the "image".

Given a linear transformation **T** from $\mathcal{V}$ into $\mathcal{W}$, the **image** $T(\mathcal{V})$ is the set of all images of vectors of $\mathcal{V}$ under **T**,

$$T(\mathcal{V}) = \{T(V) \mid V \text{ in } \mathcal{V}\}.$$

2.8 If **T** is a linear transformation from $\mathcal{V}$ into $\mathcal{W}$, then its image $T(\mathcal{V})$ is a subspace of $\mathcal{W}$.

Proof. $T(V_1) + T(V_2) = T(V_1 + V_2)$, since **T** is linear. Hence, $T(\mathcal{V})$ is closed under vector addition. In like manner, it is closed under scalar multiplication, and is hence a subspace. ∎

Example 3. For $T(x, y) = (x + y, x - y, 2x + y)$ as in the last few examples, the image

$$T(\mathbb{R}_2) = \{(x + y, x - y, 2x + y) \mid x, y \text{ in } \mathbb{R}\}.$$

Since

$$(x + y, x - y, 2x + y) = x(1, 1, 2) + y(1, -1, 1),$$

we see that we can also write

$$T(\mathbb{R}_2) = \text{span } \{(1, 1, 2), (1, -1, 1)\}.$$

Recall now that if **T** is any function from a set $\mathcal{V}$ into a set $\mathcal{W}$, then the **inverse function** T^{-1} is that function (if it exists) from the range $T(\mathcal{V})$ back to $\mathcal{V}$ defined by

$$T^{-1}(T(V)) = V \quad \text{for} \quad T(V) \text{ in } T(\mathcal{V}).$$

Moreover, the necessary and sufficient condition that T^{-1} exist is that **T** be one-to-one (for, **T** is one-to-one iff there is only one **V** which lands on T(V) for each T(V) in the range of **T**; hence T^{-1} will be a function, since for every element T(V) in *its* domain (that is, $T(\mathcal{V})$), there is only one corresponding element **V** in *its* range ($\mathcal{V}$); see Chapter 0, Section 5). In particular, if **T** is a linear transformation and $\mathcal{V}$ and $\mathcal{W}$ are vector spaces, then the inverse function T^{-1} will exist iff **T** is one-to-one. As you might expect in this best of all possible (linear) worlds,

2.9 If **T** is a one-to-one linear transformation from $\mathcal{V}$ into $\mathcal{W}$, then T^{-1} will be a (one-to-one) linear transformation from the vector space $T(\mathcal{V})$ back to $\mathcal{V}$.

Proof. As just observed, T^{-1} will be a function, so we need only check linearity. For additivity, this amounts to

$$T^{-1}(T(V_1) + T(V_2)) \stackrel{\text{add. of T}}{=} T^{-1}(T(V_1 + V_2)) \stackrel{\text{defn. } T^{-1}}{=} V_1 + V_2$$

$$\stackrel{\text{defn. } T^{-1}}{=} T^{-1}(T(V_1)) + T^{-1}(T(V_2)),$$

which says that $T^{-1}(W_1 + W_2) = T^{-1}(W_1) + T^{-1}(W_2)$ for any W_i in its domain. Homogeneity is yours. ∎

Example 4. Consider our trusty **T** in Example 3:

$$T(x, y) = T(x', y')$$

says that

$$(x + y, x - y, 2x + y) = (x' + y', x' - y', 2x' + y'),$$

so

$$x + y = x' + y', \quad x - y = x' - y', \quad 2x + y = 2x' + y',$$

whence it quickly follows that

$$x = x' \quad \text{and} \quad y = y'$$

(check). Hence **T** is one-to-one, so that T^{-1} exists from $T(\mathbb{R}_2)$ back to $\mathbb{R}_2$. We noticed in Example 3 that

$$T(x, y) = x(1, 1, 2) + y(1, -1, 1).$$

Hence, any element in $T(\mathbb{R}_2)$ is of this form. Hence, T^{-1} is that function defined on the subspace

$$T(\mathbb{R}_2) = \text{span } \{(1, 1, 2), (1, -1, 1)\}$$

of $\mathbb{R}_3$, for which $T^{-1}(x(1, 1, 2) + y(1, -1, 1)) = (x, y)$. Since $T(\mathbb{R}_2)$ is but a two-dimensional subspace of $\mathbb{R}_3$, T^{-1} does not exist in all of $\mathbb{R}_3$. Geometrically, Fig. 13 illustrates the situation.

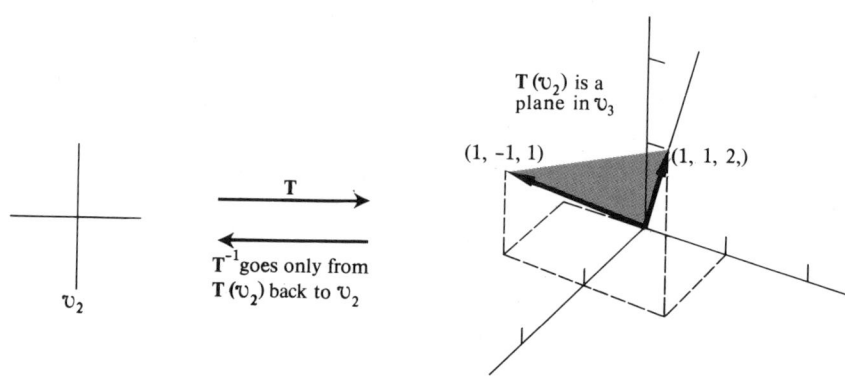

Fig. 13

Note that a one-to-one linear transformation **T** from $\mathcal{V}$ into $\mathcal{W}$ will be a one-to-one correspondence between $\mathcal{V}$ and $\mathbf{T}(\mathcal{V})$. In fact, it will be an isomorphism between these spaces (Chapter 1, Section 5). Hence $\mathcal{V}$ and $\mathbf{T}(\mathcal{V})$ will be essentially the same spaces (Fig. 14). (Compare Example 4.) Because of this one often concentrates on one-to-one linear transformations from a space onto itself.

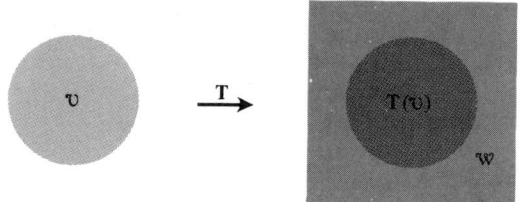

Fig. 14

EXERCISES

1. Prove that the set of all functions from one vector space into another forms a vector space under the usual definitions of function equality, addition, and scalar multiplication.

2. a) Prove the second equality in 2.5.
   b) Prove the other distributive law in 2.6.
   c) Prove that the image of a linear transformation is closed under scalar multiplication.
   d) Prove homogeneity in 2.9.
   Let **T** and **S** be the linear transformations from $\mathbb{R}_2$ to $\mathbb{R}_2$ given by

   $$\mathbf{T}(x, y) = (3x - 2y, x + y), \quad \mathbf{S}(x, y) = (-x + y, x + 3y).$$

For the following linear transformations, (a) determine the image of (x, y), (b) give a basis for the image space, (c) determine whether the linear transformation is invertible, and if so, describe its inverse.

3. $\mathbf{T} + 2\mathbf{S}$ 4. $\mathbf{TS}$ 5. $\mathbf{ST}$ 6. $\mathbf{T}^2 - 4\mathbf{T} + 5\mathbf{I}$

7. Let $\mathbf{S}, \mathbf{T}, \mathbf{U}$ be the following linear transformations:

$\mathbf{S}$ from $\mathbb{R}_2$ into $\mathbb{R}_2$ by $\mathbf{S}(x, y) = (-x + y, x + 3y)$,

$\mathbf{T}$ from $\mathbb{R}_2$ into $\mathbb{R}_3$ by $\mathbf{T}(x, y) = (x + y, 2x - y, x - 3y)$,

$\mathbf{U}$ from $\mathbb{R}_3$ into $\mathbb{R}_4$ by $\mathbf{U}(x, y, z) = (x + y, z, x - y, y + z)$.

Find the transformations

$$\mathbf{TS}, \quad \mathbf{U(TS)}, \quad \mathbf{UT}, \quad \mathbf{(UT)S},$$

and check the associativity.

8. Show that the linear transformation $\mathbf{T}(x, y) = (ax + by, cx + dy)$ has an inverse iff $ad - bc \neq 0$.

9. The following exercise shows how to factor a linear transformation from $\mathbb{R}_2$ to $\mathbb{R}_2$ into a product of shears and compressions. We shall work with $\mathbf{T}(x, y) = (ax + by, cx + dy)$ and assume that both $a \neq 0$ and $ad - bc \neq 0$. Define the following linear transformations:

$$\mathbf{T}_1(x, y) = \left(x + \frac{b}{a}y, y\right), \quad \mathbf{T}_2(x, y) = (ax, y),$$

$$\mathbf{T}_3(x, y) = \left(x, \frac{c}{ad - bc}x + y\right), \quad \mathbf{T}_4(x, y) = \left(x, \frac{ad - bc}{a}y\right).$$

a) Identify each of the $\mathbf{T}_i$'s geometrically and prove that

$$\mathbf{T} = \mathbf{T}_4\mathbf{T}_3\mathbf{T}_2\mathbf{T}_1.$$

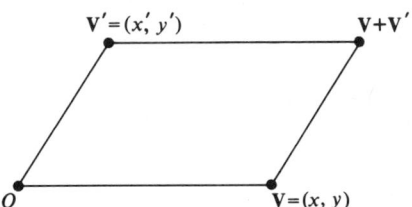

b) Consider the parallelogram with vertices

$$\mathbf{0}, \quad \mathbf{V}(= (x, y)), \quad \mathbf{V}'(= (x', y')), \quad \text{and} \quad \mathbf{V} + \mathbf{V}'(= (x + x', y + y')).$$

Since $\mathbf{T}$ is linear, the image of this will be the parallelogram with vertices

$$\mathbf{0}, \quad \mathbf{T}(\mathbf{V}), \quad \mathbf{T}(\mathbf{V}'), \quad \mathbf{T}(\mathbf{V}) + \mathbf{T}(\mathbf{V}').$$

Calculate the relationship of the area of the image parallelogram to that of the original by first solving this problem for each of the $\mathbf{T}_i$.

H²10. A simple excursion into the world of Generalized Abstract Nonsense: Let $\mathcal{U}$ be a fixed vector space. For any vector space $\mathcal{X}$ define

$$F(\mathcal{X}) = \text{Hom}(\mathcal{U}, \mathcal{X}).$$

(That is, all linear transformations from $\mathcal{U}$ into $\mathcal{X}$.)

a) Given any linear transformation **T** from $\mathcal{X}$ into $\mathcal{Y}$, show that the following $f(\mathbf{T})$ is a linear transformation from $F(\mathcal{X})$ into $F(\mathcal{Y})$: for S in $F(\mathcal{X})$, let $[f(\mathbf{T})](S) = \mathbf{T}S$.

Free picture :

b) Show that f is additive, i.e. $f(\mathbf{T}_1 + \mathbf{T}_2) = f(\mathbf{T}_1) + f(\mathbf{T}_2)$. This essentially proves that F is an additive covariant functor from the category of vector spaces and homomorphisms to itself (which in turn proves that we are capable of reaching the usual heights of obscurity expected in a mathematics text).

c) Show that if one defines $G(\mathcal{X}) = \text{Hom}(\mathcal{X}, \mathcal{U})$, then for any linear transformation **T** from $\mathcal{X}$ into $\mathcal{Y}$, the corresponding $g(\mathbf{T})$ maps $G(\mathcal{Y})$ into $G(\mathcal{X})$, thereby reversing directions (but one still has additivity). This means that G is an additive contravariant functor, whatever that means.

3. REPRESENTING LINEAR TRANSFORMATIONS

In this section we shall work exclusively with linear transformations on finite-dimensional vector spaces, where we are able to get a very concrete realization of these functions.

3.1 A linear transformation is completely determined by its effect on a basis. That is, given a basis $\{\mathbf{V}_1, \ldots, \mathbf{V}_n\}$ for the finite dimensional vector space $\mathcal{U}$, and given an arbitrary set of n elements $\{\mathbf{W}_1, \ldots, \mathbf{W}_n\}$ in the vector space $\mathcal{W}$, then there is one and only one linear transformation from $\mathcal{U}$ into $\mathcal{W}$ for which

$$\mathbf{T}(\mathbf{V}_1) = \mathbf{W}_1, \ldots, \mathbf{T}(\mathbf{V}_n) = \mathbf{W}_n.$$

Proof. We first manufacture one such linear transformation. For $\mathbf{V}$ in $\mathcal{U}$, $\mathbf{V}$ has a unique expression as

$$\mathbf{V} = a_1 \mathbf{V}_1 + \cdots + a_n \mathbf{V}_n$$

(Chapter 1, 5.1). We therefore define $\mathbf{T}(\mathbf{V}) = a_1 \mathbf{W}_1 + \cdots + a_n \mathbf{W}_n$. Then **T** is a function from $\mathcal{U}$ into $\mathcal{W}$, since there is a single element of $\mathcal{W}$ corresponding to each **V** in $\mathcal{U}$. To establish homogeneity, note that for **V** as above,

$$c\mathbf{V} = ca_1 \mathbf{V}_1 + \cdots + ca_n \mathbf{V}_n,$$

so that by definition,

$$\mathbf{T}(c\mathbf{V}) = ca_1 \mathbf{W}_1 + \cdots + ca_n \mathbf{W}_n,$$

which is the same as $c\mathbf{T}(\mathbf{V})$. You show that $\mathbf{T}$ is additive. Since
$$\mathbf{V}_i = 0\mathbf{V}_1 + \cdots + 1\mathbf{V}_i + \cdots + 0\mathbf{V}_n,$$
$\mathbf{T}(\mathbf{V}_i) = 0\mathbf{W}_1 + \cdots + 1\mathbf{W}_i + \cdots + 0\mathbf{W}_n = \mathbf{W}_i$, as it is supposed to.

To show that there is only one such function, suppose that $\mathbf{S}$ is another linear transformation for which
$$\mathbf{S}(\mathbf{V}_1) = \mathbf{W}_1, \ldots, \mathbf{S}(\mathbf{V}_n) = \mathbf{W}_n.$$
Then for $\mathbf{V} = a_1\mathbf{V}_1 + \cdots + a_n\mathbf{V}_n$,
$$\mathbf{S}(\mathbf{V}) = \mathbf{S}(a_1\mathbf{V}_1 + \cdots + a_n\mathbf{V}_n) = a_1\mathbf{S}(\mathbf{V}_1) + \cdots + a_n\mathbf{S}(\mathbf{V}_n)$$
$$= a_1\mathbf{W}_1 + \cdots + a_n\mathbf{W}_n = \mathbf{T}(\mathbf{V}).$$
Hence, $\mathbf{S}$ and $\mathbf{T}$ have the same effect on all $\mathbf{V}$. Hence, they are the same function. ∎

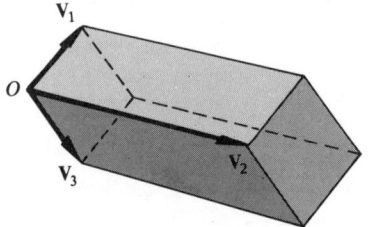

Fig. 15

There are several aspects of this important theorem which should be savored. For example,

if two linear transformations agree on a basis, then they agree everywhere.
Moreover,

under a linear transformation $\mathbf{T}$, the image of a basis spans the image $\mathbf{T}(\mathcal{V})$.
That is, $\mathbf{T}(\mathcal{V}) = \mathrm{span}\,\{\mathbf{T}(\mathbf{V}_1), \ldots, \mathbf{T}(\mathbf{V}_n)\}$.

To describe an arbitrary function, one must specify its value at each element of its domain. However, for a linear transformation, it is sufficient to specify its value only on the elements in a basis.

3.1 has a nice geometrical interpretation. For $n = 3$, the basis vectors $\mathbf{V}_1, \mathbf{V}_2, \mathbf{V}_3$ for $\mathcal{V}$ form adjacent edges of a parallelepiped (Fig. 15). 3.1 says that any linear transformation is completely determined by what happens to such a parallelepiped; all other images may be obtained by linearly extrapolating from this. The shape of the image of the parallelepiped completely determines the image of all of $\mathcal{V}$. In fact, you may choose any three vectors $\mathbf{W}_1, \mathbf{W}_2, \mathbf{W}_3$ in your image space $\mathcal{W}$ (they need not be linearly independent), and it will be possible to map $\mathbf{V}_1, \mathbf{V}_2, \mathbf{V}_3$ onto $\mathbf{W}_1, \mathbf{W}_2, \mathbf{W}_3$, respectively, and then extend to obtain a linear transformation in one and only one way.

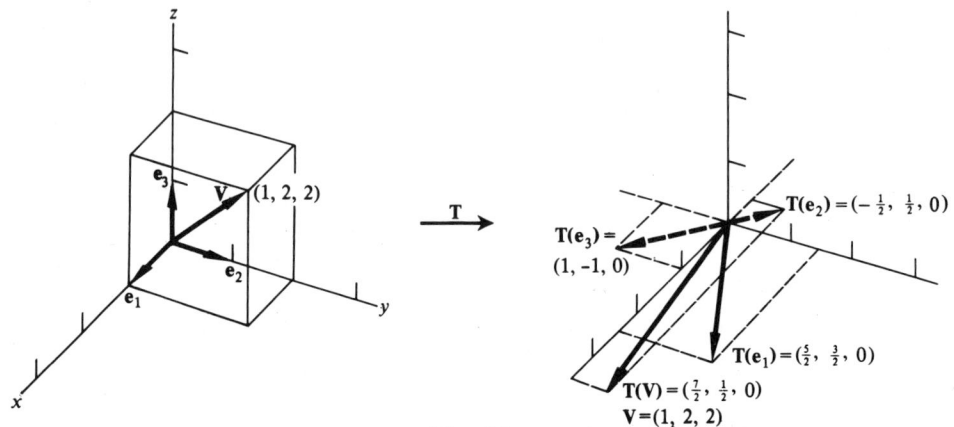

Fig. 16

Example 1. Let **T** be that linear transformation from $\mathbb{R}_3$ to $\mathbb{R}_3$, for which

$$T(e_1) = (\tfrac{5}{2}, \tfrac{3}{2}, 0), \quad T(e_2) = (-\tfrac{1}{2}, \tfrac{1}{2}, 0), \quad \text{and} \quad T(e_3) = (1, -1, 0).$$

Then for any other **V** in $\mathbb{R}_3$, $V = (x, y, z) = xe_1 + ye_2 + ze_3$, so

$$T(V) = xT(e_1) + yT(e_2) + zT(e_3)$$
$$= (\tfrac{5}{2}x - \tfrac{1}{2}y + z, \tfrac{3}{2}x + \tfrac{1}{2}y - z, 0).$$

For example, with $V = (1, 2, 2)$, $T(V) = (\tfrac{7}{2}, \tfrac{1}{2}, 0)$. Geometrically, **T** squashes $\mathbb{R}_3$ down into the xy-plane and then distorts things (Fig. 16). For an even better view of what happens, read on.

Example 2. Let **T** be the linear transformation from $\mathbb{R}_3$ into $\mathbb{R}_3$ for which

$$T(V_1) = 0, \quad T(V_2) = V_2, \quad \text{and} \quad T(V_3) = 2V_3,$$

where

$$V_1 = (0, 2, 1), \quad V_2 = (1, 3, 0), \quad \text{and} \quad V_3 = (1, 1, 0)$$

(see Fig. 17).

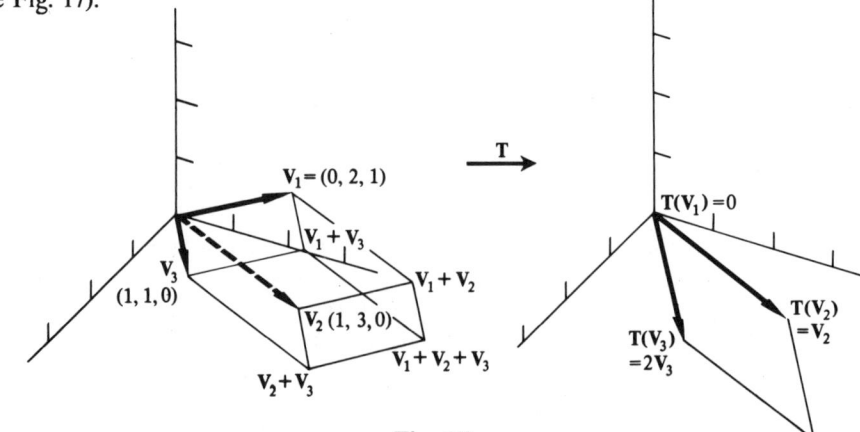

Fig. 17

3 / **Representing linear transformations** 163

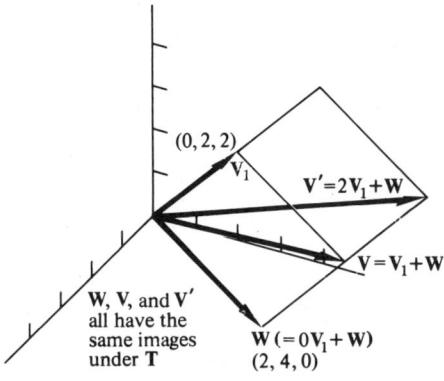

Fig. 18

Since V_1 is the only one of the V_i with nonzero third component, any vector V in $\mathbb{R}_3$ can be written in the form

$$V = cV_1 + W, \qquad \text{with } W \text{ in the } xy\text{-plane } (= \text{span } \{V_2, V_3\}).$$

Since T is linear,
$$T(V) = T(cV_1) + T(W) = cT(V_1) + T(W)$$
$$= 0 + T(W) = T(W).$$

Thus, T affects a vector V in $\mathbb{R}_3$ in the same fashion as it affects the vector in the xy-plane which is obtained from V by projecting in the direction of V_1 (Fig. 18).

What does T do to vectors in the xy-plane? It stretches by 2 in the direction of V_3 and leaves things unchanged in the direction of V_2.

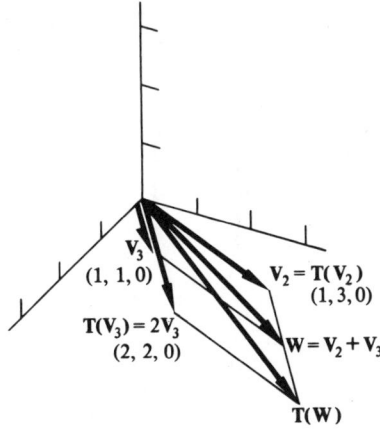

Thus, the total effect of T is first to project $\mathbb{R}_3$ onto the xy-plane in the direction of V_1, and then to stretch things by 2 in the direction of V_3.

One can check that if $V = (x, y, z)$, then

$$V = zV_1 + (-\tfrac{1}{2}x + \tfrac{1}{2}y - z)V_2 + (\tfrac{3}{2}x - \tfrac{1}{2}y + z)V_3,$$

so that
$$\mathbf{T(V)} = (-\tfrac{1}{2}x + \tfrac{1}{2}y - z)\mathbf{V}_2 + (3x - y + 2z)\mathbf{V}_3$$
$$= (\tfrac{5}{2}x - \tfrac{1}{2}y + z, \tfrac{3}{2}x + \tfrac{1}{2}y - z, 0).$$

Thus, this **T** is the same as the one in example 1! However, note that the parallelepiped formed from $\mathbf{V}_1, \mathbf{V}_2, \mathbf{V}_3$ is affected much more simply by **T** than the one formed from $\mathbf{e}_1, \mathbf{e}_2, \mathbf{e}_3$. We have seen that a fortuitous basis for $\mathcal{V}$ can lead to a simple geometric description of a linear transformation **T**. The method for finding such a basis (when possible) is the subject of Chapter 6.

We now specialize even further, and consider only linear transformations from one finite-dimensional vector space to another *in relation to specific bases for both spaces*. From Chapter 1, Section 5, we know that after bases have been chosen we might as well be working with $\mathbb{R}_n$ and $\mathbb{R}_m$, with respect to the standard bases. Consequently, we shall deal with $\mathbb{R}_n$ and $\mathbb{R}_m$ for much of the rest of this section. Since everything will be completely dependent on the specific bases chosen, we shall later have to take into account the effect of changing bases (Chapter 4, Section 4).

We are about to reintroduce matrices, and so we shall use column vectors. In particular, let us call our bases

$$\mathbf{e}_1 = \begin{pmatrix} 1 \\ 0 \\ \vdots \\ 0 \end{pmatrix}\}n, \quad \mathbf{e}_2 = \begin{pmatrix} 0 \\ 1 \\ 0 \\ \vdots \\ 0 \end{pmatrix}\}n, \ldots, \mathbf{e}_n = \begin{pmatrix} 0 \\ \vdots \\ 0 \\ 1 \end{pmatrix}\}n,$$

for $\mathbb{R}_n$, and

$$\mathbf{f}_1 = \begin{pmatrix} 1 \\ 0 \\ \vdots \\ 0 \end{pmatrix}\}m, \quad \mathbf{f}_2 = \begin{pmatrix} 0 \\ 1 \\ 0 \\ \vdots \\ 0 \end{pmatrix}\}m, \ldots, \mathbf{f}_m = \begin{pmatrix} 0 \\ \vdots \\ 0 \\ 1 \end{pmatrix}\}m,$$

for $\mathbb{R}_m$.

Recall that for any $m \times n$ matrix **A**, the mapping $\mathbf{T(V)} = \mathbf{AV}$ is a linear transformation from $\mathbb{R}_n$ into $\mathbb{R}_m$ (Chapter 2, Section 3). We next prove a converse to this.

3.2 Given any linear transformation **T** from $\mathbb{R}_n$ into $\mathbb{R}_m$, there is an $m \times n$ matrix **A** for which
$$\mathbf{T(V)} = \mathbf{AV}$$

for all **V** in $\mathbb{R}_n$. In fact, in terms of its columns,
$$\mathbf{A} = (\mathbf{T}(\mathbf{e}_1), \ldots, \mathbf{T}(\mathbf{e}_n)).$$

Proof. Given **T**, define **A** as indicated. For any **A**, $\mathbf{A}\mathbf{e}_i = i$th column of **A**. Hence,
$$\mathbf{A}\mathbf{e}_1 = \mathbf{T}(\mathbf{e}_1), \ldots, \mathbf{A}\mathbf{e}_n = \mathbf{T}(\mathbf{e}_n).$$
Thus, the linear transformation which sends **V** to **AV** has exactly the same effect on the basis $\{\mathbf{e}_1, \ldots, \mathbf{e}_n\}$ as does **T**. By 3.1 we know that
$$\mathbf{A}\mathbf{V} = \mathbf{T}(\mathbf{V}) \qquad \text{for all } \mathbf{V}. \quad \blacksquare$$

Notice in particular that the zero linear transformation corresponds to the zero matrix. Moreover, in considering linear transformations from $\mathbb{R}_n$ into itself, the identity mapping **I**, for which
$$\mathbf{I}(\mathbf{V}) = \mathbf{V} \qquad \text{for all } \mathbf{V},$$
corresponds to the identity matrix
$$\mathbf{I} = \begin{pmatrix} 1 & & 0 \\ & \ddots & \\ 0 & & 1 \end{pmatrix}.$$
We must be doing something right.

Although we are officially using column vectors in this context, it is sometimes worthwhile taking transposes to keep in touch with row vectors (recall that we're using $\mathbb{R}_n$ for both). We restate 3.2 in terms of components, using row vectors for a change:

any linear transformation from $\mathbb{R}_n$ into $\mathbb{R}_m$ is of the form
$$\mathbf{T}(x_1, \ldots, x_n) = (a_{11}x_1 + \cdots + a_{1n}x_n, \ldots, a_{m1}x_1 + \cdots + a_{mn}x_n). \quad (1)$$

Example 3. Let us consider the linear transformations of Example 1 for which (changing from row to column vectors)

$$\mathbf{T}(\mathbf{e}_1) = \begin{pmatrix} \frac{5}{2} \\ \frac{3}{2} \\ 0 \end{pmatrix}, \quad \mathbf{T}(\mathbf{e}_2) = \begin{pmatrix} -\frac{1}{2} \\ \frac{1}{2} \\ 0 \end{pmatrix}, \quad \mathbf{T}(\mathbf{e}_3) = \begin{pmatrix} 1 \\ -1 \\ 0 \end{pmatrix}.$$

Then the matrix **A** for which $\mathbf{T}(\mathbf{V}) = \mathbf{A}\mathbf{V}$ is
$$\mathbf{A} = \begin{pmatrix} \frac{5}{2} & -\frac{1}{2} & 1 \\ \frac{3}{2} & \frac{1}{2} & -1 \\ 0 & 0 & 0 \end{pmatrix},$$
as can be easily checked. The reason **A** happens to be square is because **T** is a mapping from $\mathbb{R}_3$ into itself.

The purpose of the next example is to show that in working with linear transformations from one finite-dimensional vector space to another, you should not be too hasty in choosing bases.

Example 4. The vectors from Example 2,
$$V_1 = (0, 1, 1), \quad V_2 = (1, 3, 0), \quad V_3 = (1, 1, 0),$$
are easily seen to constitute a basis for $\mathbb{R}_3$. Consequently, any V in $\mathbb{R}_3$ can be written in the form
$$V = a_1 V_1 + a_2 V_2 + a_3 V_3,$$
and we know that the mapping
$$V \to \begin{pmatrix} a_1 \\ a_2 \\ a_3 \end{pmatrix}$$
is an isomorphism; that is, we can treat a vector as if it were the triple of coordinates of its expression in terms of the V_i. In particular, V_1 corresponds to
$$\begin{pmatrix} 1 \\ 0 \\ 0 \end{pmatrix},$$
etc., so we can work with the V_i just as if they were the e_i in this case (and also the f_i, since the number of vectors is the same). Then, as we saw in Example 2,
$$T(V_1) = 0, \quad T(V_2) = V_2, \quad T(V_3) = 2V_3$$
would correspond to
$$T(e_1) = 0 = \begin{pmatrix} 0 \\ 0 \\ 0 \end{pmatrix}, \quad T(e_2) = f_2 = \begin{pmatrix} 0 \\ 1 \\ 0 \end{pmatrix}, \quad T(e_3) = 2f_3 = \begin{pmatrix} 0 \\ 0 \\ 2 \end{pmatrix}.$$
In making the correspondence, we would have for the matrix of T
$$\begin{pmatrix} 0 & 0 & 0 \\ 0 & 1 & 0 \\ 0 & 0 & 2 \end{pmatrix},$$
which is considerably simpler than the matrix of this linear transformation in Example 3.

Just as we can consider the elements of an abstract n-dimensional space as n-tuples (with respect to a specific basis), 3.2 says that we can consider an abstract linear transformation from one finite-dimensional space to another as just a matrix multiplication (again, after having chosen bases for the spaces).

For T mapping $\mathbb{R}_n$ into $\mathbb{R}_m$, we further inspect this correspondence between T and its matrix.

3.3 i) The correspondence
$$T \leftrightarrow A = (T(e_1), \ldots, T(e_n))$$
is a one-to-one correspondence between linear transformations (from $\mathbb{R}_n$ into $\mathbb{R}_m$) and $(m \times n)$ matrices.

ii) The correspondence is additive; i.e., the matrix of the sum of two linear transformations is the sum of the two matrices of the linear transformations. That is, if

$$T_1 \leftrightarrow A_1 \quad \text{and} \quad T_2 \leftrightarrow A_2,$$

then $T_1 + T_2 \leftrightarrow A_1 + A_2$.

iii) The correspondence is homogeneous; i.e. the matrix of a scalar multiple of a linear transformation is that scalar multiple of the matrix of the linear transformation. That is, if $T \leftrightarrow A$ then $cT \leftrightarrow cA$.

In short, the correspondence $T \leftrightarrow A$ is an isomorphism between the vector space of all linear transformations from $\mathbb{R}_n$ into $\mathbb{R}_m$ onto the vector space of all $m \times n$ matrices.

Proof. i) *One-to-one:* If also $S \leftrightarrow A = (S(e_1), \ldots, S(e_n))$, then, by definition of matrix equality,

$$S(e_1) = T(e_1), \ldots, S(e_n) = T(e_n).$$

By 3.1, $S = T$.

Onto: Given an $m \times n$ matrix A, define T by $T(V) = AV$. It is easy to see that $T \leftrightarrow A$ (Exercise 5b).

ii) $(T_1 + T_2)(e_1) = T_1(e_i) + T_2(e_i)$, for $i = 1, \ldots, n$. Hence,

$$T_1 + T_2 \leftrightarrow ((T_1 + T_2)(e_1), \ldots, (T_1 + T_2)(e_n))$$
$$= (T_1(e_1) + T_2(e_1), \ldots, T_1(e_n) + T_2(e_n))$$
$$\stackrel{\text{mx. add.}}{=} (T_1(e_1), \ldots, T_1(e_n)) + (T_2(e_2), \ldots, T_2(e_n))$$
$$= A_1 + A_2.$$

You finish up (Exercise 5b). ∎

3.4 The matrix of the product of two linear transformations is the product of the matrices: Let S be a linear transformation from $\mathbb{R}_m$ to $\mathbb{R}_l$ with matrix A and T a linear transformation from $\mathbb{R}_n$ to $\mathbb{R}_m$ with matrix B. Then ST has matrix AB (Fig. 19).

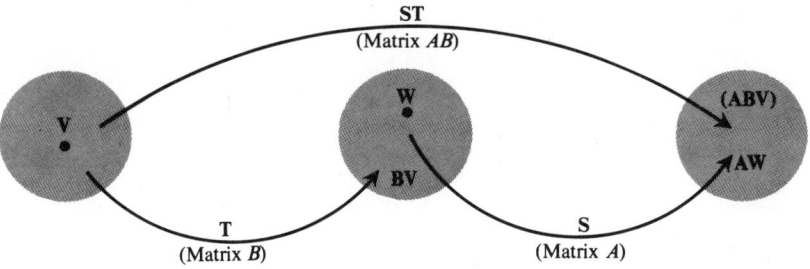

Fig. 19

Proof.
$$ST(V) = S(T(V)) = S(BV)$$
$$= A(BV) = (AB)V.$$
Thus, $ST \leftrightarrow AB$. ∎

Our final item will be to show that invertible matrices correspond to invertible linear transformations. Since the latter are isomorphisms between their domains and ranges, these two spaces are essentially the same. Thus, we may as well assume that our invertible linear transformations map $\mathbb{R}_n$ into $\mathbb{R}_n$. Since the matrix of any linear transformation from $\mathbb{R}_n$ into $\mathbb{R}_n$ is $n \times n$, this is in keeping with our considering invertibility only for square matrices.

Recall that to say that the $n \times n$ matrix A is invertible means that there is a matrix B for which
$$AB = BA = I,$$
the $n \times n$ identity matrix. To say that the linear transformation T is invertible means that there is a function T^{-1} for which
$$T^{-1}(T(V)) = V \quad \text{for all } V \text{ in } \mathbb{R}_n.$$

3.5 i) If T is an invertible linear transformation from $\mathbb{R}_n$ onto $\mathbb{R}_n$, then its matrix $A = (T(e_1), \ldots, T(e_n))$ is invertible.
ii) If A is an invertible $n \times n$ matrix, then the linear transformation $T(V) = AV$ is an invertible linear transformation from $\mathbb{R}_n$ onto $\mathbb{R}_n$.

Proof. i) Let B be the matrix of T^{-1}, so that $T^{-1}(V) = BV$. Then
$$V = T^{-1}(T(V)) = (BA)V,$$
so that $BA = I$ (why?). In like manner since $T^{-1}T = TT^{-1}$ (Chapter 0, 5.2),
$$V = T(T^{-1}(V)) = ABV,$$
so that $AB = I$. Hence A is invertible.
ii) For the matrix B satisfying $AB = BA = I$, define
$$S(V) = BV.$$
Then for all V in $\mathbb{R}_n$,
$$S(T(V)) = ABV = V$$
$$= BAV = T(S(V)).$$
Thus $S = T^{-1}$. ∎

EXERCISES

1. Let T be the linear transformation from $\mathbb{R}_3$ into $\mathbb{R}_3$ for which
$$T(e_1) = 2e_2, \quad T(e_2) = e_1, \quad T(e_3) = -e_3.$$

a) Write out $T(x, y, z)$.
b) Draw a picture illustrating the effect of T on the cube with adjacent edges e_1, e_2, e_3.
c) Describe the linear transformation geometrically.
d) Write out its matrix.

2. Let T be the linear transformation from $\mathbb{R}_2$ to $\mathbb{R}_3$ for which

$$T(x, y) = (x + y, 2x - y, x - 3y),$$

and S the linear transformation for which

$$S(x, y) = (y, x + 2y, y - x).$$

a) Find $(S + T)(x, y)$ and the matrices of S, T, and $S + T$.
b) Find $(\sqrt{3}T)(x, y)$ and the matrix of $\sqrt{3}T$.

3. Let U be the linear transformation from $\mathbb{R}_3$ to $\mathbb{R}_4$ for which

$$U(x, y, z) = (x + y, z, x - y, y + z).$$

a) Calculate $UT(x, y)$ for T as in Exercise 2.
b) Find the matrix of UT and verify that it is the product of the separate matrices.

4. Let S, T, U be the linear transformations in Exercise 7 of Section 2.
a) Find the matrices of these transformations.
b) Find the matrices of

$$TS, \quad U(TS), \quad UT, \quad (UT)S$$

by matrix multiplication. Compare with the results of Exercise 7 of Section 2.

5. a) Prove additivity in 3.1.
b) Finish the proof of 3.3.

6. Prove that if A is an $n \times n$ matrix and $AV = V$ for all V in $\mathbb{R}_n$, then $A = I$.

7. Let $\mathcal{L}(\mathcal{V}, \mathcal{W})$ denote the vector space of linear transformations from $\mathcal{V}$ into $\mathcal{W}$.
a) Find a specific isomorphism between

$$\mathcal{L}(\mathbb{R}_n, \mathbb{R}_m) \quad \text{and} \quad \mathbb{R}_{m \times n}.$$

b) Find a basis for $\mathcal{L}(\mathbb{R}_n, \mathbb{R}_m)$.

H 8. It is sometimes convenient to develop a relationship between matrices and linear transformations of infinite-dimensional vector spaces.
a) Emulate our discussion to obtain such a relationship for the linear transformations from $\mathbb{R}[x]$ into $\mathbb{R}[x]$. (Use the basis $\{1, x, x^2, \ldots\}$).
b) Obtain a matrix for the T for which $T(p(x)) = xp(x)$. Do the same for S defined by

$$S(a_0 + a_1 x + \cdots + a_n x^n) = a_1 + a_2 x + \cdots + a_n x^{n-1}.$$

c) Develop analogous definitions of matrix addition, scalar multiplication, and multiplication for the matrices obtained in (a).
d) Check your theory on the matrices of

$$S, \quad T, \quad ST, \quad \text{and} \quad TS.$$

9. Let C denote the vector space of convergent sequences (Chapter 1, Section 3, Exercise 10). For $\{a_n\}$ in C with $\lim_{n\to\infty} a_n = a$, define

$$T(\{a_n\}) = a.$$

(Thus, **T** assigns to each convergent sequence its limit.) Prove (a) that **T** is a linear transformation, (b) that **T** maps C onto $\mathbb{R}$.

10. Prove Eq. (1).

11. Do Exercises 3 through 6 of Section 2 by using matrices.

4. THE DIMENSION THEOREM; INVERTIBILITY

In this section we further explore the important question of invertibility of linear transformations (and matrices). For a short time we can consider linear transformations on arbitrary vector spaces, but we later concentrate again on the finite-dimensional case.

The main theoretical test for invertibility is based on the following notion:

If **T** is a linear transformation from $\mathcal{U}$ into $\mathcal{W}$, the **kernel** of **T** consists of all **V** in $\mathcal{U}$ which are sent to **0** by **T**:

$$\text{kernel } \mathbf{T} = \{\mathbf{V} \text{ in } \mathcal{U} \mid \mathbf{T}(\mathbf{V}) = \mathbf{0}\}.$$

Example 1. Let **T** be the linear transformation from $\mathbb{R}_3$ into $\mathbb{R}_2$,

$$\mathbf{T}(x, y, z) = (x - 2y + 2z, -x + 2y - 2z).$$

Then $\mathbf{T}(\mathbf{V}) = \mathbf{0}$ means that

$$x - 2y + 2z = 0 \quad \text{and} \quad -x + 2y - 2z = 0,$$

so that the kernel of **T** consists of all $\mathbf{V} = (x, y, z)$ for which $x - 2y + 2z = 0$,

$$\text{kernel of } \mathbf{T} = \{(x, y, z) \mid x - 2y + 2z = 0\}.$$

4.1 The kernel of a linear transformation is a subspace.

Proof. If **V** is in the kernel of **T** (so that $\mathbf{T}(\mathbf{V}) = \mathbf{0}$), then

$$\mathbf{T}(c\mathbf{V}) = c\mathbf{T}(\mathbf{V}) = c\mathbf{0} = \mathbf{0},$$

so that the kernel is closed under scalar multiplication. You take care of closure under addition (Exercise 6a). ∎

The reason for our present interest in the kernel is based on the following:

4.2 $\mathbf{T}(\mathbf{V}_1) = \mathbf{T}(\mathbf{V}_2)$ iff $\mathbf{V}_1 - \mathbf{V}_2$ is in the kernel; that is, $\mathbf{V}_1$ and $\mathbf{V}_2$ have the same image under **T** iff

$$\mathbf{V}_1 = \mathbf{V}_2 + \mathbf{K},$$

where $\mathbf{T}(\mathbf{K}) = \mathbf{0}$.

Proof. 1. If $T(V_1) = T(V_2)$, then

$$0 = T(V_1) - T(V_2) = T(V_1) + (-1)T(V_2)$$
$$= T(V_1 + (-1)V_2)) = T(V_1 - V_2),$$

so that $V_1 - V_2$ is in the kernel.

2. The converse is yours to prove (Exercise 6b). ∎

Example 1 (Continued). We have

$$T(x, y, z) = (x - 2y + 2z, -x + 2y - 2z) = (x - 2y + 2z)(1, -1),$$

so the image space is spanned by the vector $(1, -1)$. Let us determine all the vectors which land on $(2, -2)$; see Fig. 20. $V_1 = (0, 0, 1)$ is one such creature, as is $V_2 = (0, 1, 2)$.

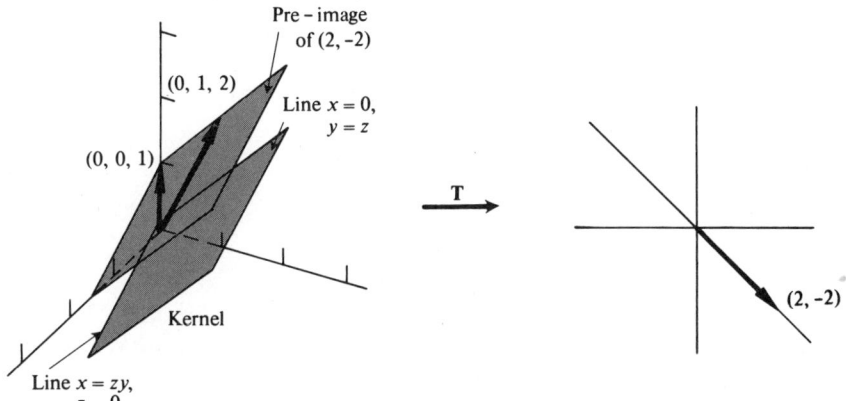

Fig. 20

Moreover,

$$V_1 - V_2 = (0, -1, -1),$$

which is in the kernel (where it belongs). A little experimentation will convince you that those vectors in $\mathbb{R}_3$ which map onto $(2, -2)$ in $\mathbb{R}_2$ have their heads on a plane parallel to the plane of the kernel. We shall explore these matters further in our discussion of analytic geometry.

The relationship between the kernel and invertibility is as follows:

4.3 Let T be a linear transformation from a vector space $\mathcal{V}$ into a vector space $\mathcal{W}$. Then there is an inverse linear transformation T^{-1} (from the image $T(\mathcal{V})$ onto $\mathcal{V}$) iff the kernel of T is $\{0\}$.

Proof. 1. Suppose the kernel of T is $\{0\}$. By 4.2,

$$T(V_1) = T(V_2)$$

iff $V_1 - V_2$ is in the kernel; so, in this case,

$$T(V_1) = T(V_2) \quad \text{iff } V_1, -V_2 = 0,$$

that is, iff $V_1 = V_2$. This says that **T** is one-to-one, so we know T^{-1} exists as a linear transformation from $T(\mathcal{V})$ onto $\mathcal{V}$.

2. You prove the converse (Exercise 6c). ∎

We have thus far associated two subspaces with a linear transformation **T**, namely, its image $T(\mathcal{V})$ and its kernel. The following fundamental and far-reaching theorem links the sizes of these subspaces in the finite-dimensional case:

4.4 *The Dimension Theorem.* Let **T** be a linear transformation from a vector space $\mathcal{V}$ into a vector space $\mathcal{W}$, and suppose that $\mathcal{V}$ is finite-dimensional. Then so are $T(\mathcal{V})$ and the kernel of **T**. Moreover,

$$\text{dimension of kernel} + \text{dimension of image} = \text{dimension of } \mathcal{V}.$$

Proof. Since the kernel is a subspace of $\mathcal{V}$ it has finite dimension, say k (Chapter 1, 4.9). Let $\{V_1, \ldots, V_k\}$ be a basis for the kernel and extend this to a basis $\{V_1, \ldots, V_k, V_{k+1}, \ldots, V_n\}$ for $\mathcal{V}$ (Chapter 1, 4.5). We must show that

[the dimension of the kernel $(= k)$] + [the dimension of $T(\mathcal{V})$] $= n$,

that is, that $T(\mathcal{V})$ has dimension $n - k$. We have artfully constructed our basis for $\mathcal{V}$; consider its image under **T**:

$$T(V_1)(= 0), \ldots, T(V_k)(= 0), \quad T(V_{k+1})(\neq 0), \ldots, T(V_n)(\neq 0),$$

since the first k V's are in the kernel and the last $(n - k)$ are not. To prove the theorem, we must find a basis for $T(\mathcal{V})$ containing $n - k$ elements. We recommend $\{T(V_{k+1}), \ldots, T(V_n)\}$. This set spans $T(\mathcal{V})$: given $T(V)$ in $T(\mathcal{V})$ we must have

$$V = a_1 V_1 + \cdots + a_n V_n,$$

so that

$$T(V) = T(a_1 V_1 + \cdots + a_n V_n)$$
$$= a_1 T_1(V_1) + \cdots + a_k T(V_k) + a_{k+1} T(V_{k+1}) + \cdots + a_n T(V_n)$$
$$= 0 + \cdots + 0 + a_{k+1} T(V_{k+1}) + \cdots + a_n T(V_n).$$

Moreover, the set is linearly independent, for, if

$$0 = a_{k+1} T(V_{k+1}) + \cdots + a_n T(V_n) = T(a_{k+1} V_{k+1} + \cdots + a_n V_n), \tag{1}$$

then $a_{k+1} V_{k+1} + \cdots + a_n V_n$ is in the kernel. Since $\{V_1, \ldots, V_k\}$ is a basis for the kernel, there are scalars b_i for which

$$a_{k+1} V_{k+1} + \cdots + a_n V_n = b_1 V_1 + \cdots + b_k V_k,$$

or

$$-b_1 V_1 - \cdots - b_k V_k + a_{k+1} V_{k+1} + \cdots + a_n V_n = 0.$$

Since $\{V_1, \ldots, V_n\}$ is a basis for $\mathcal{V}$, all coefficients are 0, especially the a's. Looking back to (1), we see that this shows that the $\{T(V_{k+1}), \ldots, T(V_n)\}$ are linearly independent, and hence they do form a basis of $T(\mathcal{V})$. ∎

For those of you who enjoy terminology, the dimension of the kernel of a linear transformation **T** is often called the **nullity** of **T**, while the dimension of its image $T(\mathcal{V})$ is called the **rank** of **T**. Thus, the Dimension Theorem can be stated in the form

$$\text{rank of } \mathbf{T} + \text{nullity of } \mathbf{T} = \text{dimension of domain of } \mathbf{T}.$$

In keeping with our usual austerity when it comes to terminology, we shall normally not use these words.

Example 1 (Further cont.). We saw that the kernel of

$$T(x, y, z) = (x - 2y + 2z, -x + 2y - 2z) = (x - 2y + 2z)(1, -1)$$

has a kernel consisting of all vectors (x, y, z) satisfying

$$x - 2y + 2z = 0.$$

One can check that this kernel has as basis the vectors

$$\mathbf{V}_1 = (-3, 0, 1), \qquad \mathbf{V}_2 = (0, 1, 1),$$

so its dimension is 2. Following the proof of 4.4, we can extend to a basis for $\mathbb{R}_3$ by tacking on $\mathbf{V}_3 = (0, 0, 1)$, for example. Then $\{\mathbf{V}_1, \mathbf{V}_2, \mathbf{V}_3\}$ is a basis for $\mathbb{R}_3$, with $\{\mathbf{V}_1, \mathbf{V}_2\}$ a basis for the kernel of **T**, and $\{T(\mathbf{V}_3) = (2, -2)\}$ a basis for the image of **T**. In this context, the Dimension Theorem tells us that $2 \, (= \text{dimension of kernel}) + 1 \, (= \text{dimension of image}) = 3 \, (= \text{dimension of domain})$.

As we noticed after 3.1, the image of a basis spans the image $T(\mathcal{V})$. Hence, if we know a basis $\mathbf{V}_1, \ldots, \mathbf{V}_n$ for $\mathcal{V}$ (and hence its dimension) and calculate the dimension of $\text{span}\{T(\mathbf{V}_1), \ldots, T(\mathbf{V}_n)\} = T(\mathcal{V})$ by the method of Chapter 2, Section 1, say, then we know the dimension of the kernel:

$$\text{dimension kernel } \mathbf{T} = \text{dimension domain} - \text{dimension image}.$$

In particular, since the zero space is the only vector space of dimension zero, from 4.3 we have:

4.5 **T** is an invertible linear transformation from $\mathcal{V}$ onto $T(\mathcal{V})$ iff the dimension of its image is equal to the dimension of its domain.

As an important special case of 4.5, we note that a linear transformation from $\mathbb{R}_n$ into $\mathbb{R}_m$ has an inverse iff the dimension of $T(\mathbb{R}_n)$, a subspace of $\mathbb{R}_m$, has dimension n. It follows that if **T** is invertible, then $m \geq n$. Moreover, $\mathbf{T}^{-1}$ will exist on all of $\mathbb{R}_m$ iff $T(\mathbb{R}_n) = \mathbb{R}_m$ and dimension $T(\mathbb{R}_n) = n$. That is, $\mathbf{T}^{-1}$ will exist on all of $\mathbb{R}_m$ iff $m = n$ and **T** is *onto* $\mathbb{R}_m$.

If we are willing to choose fixed bases for domain and image space (i.e., work from $\mathbb{R}_n$ into $\mathbb{R}_m$), we saw in the last section that any linear transformation **T** can be given by matrix multiplication:

$$\mathbf{T(V)} = \mathbf{AV}$$

where $\mathbf{A} = (\mathbf{T}(\mathbf{e}_1), \ldots, \mathbf{T}(\mathbf{e}_n))$ is the matrix comprised of the images of the basis for the domain. For such a linear transformation we observe that the kernel of $\mathbf{T}$ consists of all $\mathbf{V}$ in $\mathbb{R}_n$ for which $\mathbf{AV} = \mathbf{0}$. Moreover, the image space $\mathbf{T}(\mathbb{R}_n)$ is spanned by the columns $\mathbf{T}(\mathbf{e}_1), \ldots, \mathbf{T}(\mathbf{e}_n)$ of $\mathbf{A}$. Recall that the space spanned by the columns of $\mathbf{A}$ is the **column space** of $\mathbf{A}$. We have just observed that it is the same as the image space of the associated linear transformation. One often calls the dimension of the column space of $\mathbf{A}$ the **column rank** of $\mathbf{A}$. Thus, the column rank of $\mathbf{A}$ is the same as the dimension of the image of its associated linear transformation.

The Basis Algorithm of Chapter 2 gives us a method for calculating not only the column rank of $\mathbf{A}$, but also a basis for the column space. Note that this requires application of the Basis Algorithm to the *transposed* matrix

$$\mathbf{A}^* = \begin{pmatrix} \mathbf{T}(\mathbf{e}_1)^* \\ \cdot \\ \cdot \\ \cdot \\ \mathbf{T}(\mathbf{e}_n)^* \end{pmatrix};$$

the nonzero rows of the resulting matrix then form a basis for the column space (when stood back on end). We shall see in the next chapter that the row rank of $\mathbf{A}$ is the same as its column rank, so, if this is all that is desired, the Basis Algorithm can be applied directly to $\mathbf{A}$.

In the following we collect some relationships between invertibility of a linear transformation and properties of its matrix. These are simple applications of our previous results (and their proof is Exercise 7a).

4.6 Let $\mathbf{T}$ be a linear transformation from $\mathbb{R}_n$ to $\mathbb{R}_m$, say $\mathbf{T}(\mathbf{V}) = \mathbf{AV}$. Then

i) $\mathbf{T}$ has an inverse from $\mathbf{T}(\mathbb{R}_n)$ back to $\mathbb{R}_n$ iff the (column) rank of $\mathbf{A} = n$;

ii) $\mathbf{T}$ has an inverse from all of $\mathbb{R}_m$ back to $\mathbb{R}_n$ iff $m = n$ (that is, $\mathbf{A}$ is square) and the (column) rank of $\mathbf{A} = n$.

The Dimension Theorem has the following matrix interpretation:

For an $m \times n$ matrix $\mathbf{A}$,

$n =$ (the column rank of $\mathbf{A}$) + (the dimension of the subspace of $\mathbb{R}_n$ of $\mathbf{V}$ for which $\mathbf{AV} = \mathbf{0}$).

We are confining our discussion of matrix inverses to $n \times n$ matrices, and we spend the remainder of this section discussing square matrices. These are matrices belonging to linear transformations from $\mathbb{R}_n$ into $\mathbb{R}_n$. We saw in 3.5 that invertible linear transformations have invertible matrices, and conversely. From this we obtain an interesting matrix form of 4.3:

An $n \times n$ matrix $\mathbf{A}$ has an inverse iff $\mathbf{AV} = \mathbf{0}$ only for $\mathbf{V} = \mathbf{0}$.

Of more practical interest is the following:

4.7 Criterion for invertibility. An $n \times n$ matrix has an inverse iff its column rank (or row rank) is n.

Proof. 1. Suppose that **A** has an inverse. Then by 4.3 the kernel of $T(V) = AV$ is $\{0\}$. By 4.4 the dimension of the image of **T** ($=$ column rank **A**) is the dimension of the domain of **T** ($= n$).

Since **A** is invertible iff **A*** is (Chapter 2, 4.4) and since the row space of **A** is the column space of **A***, we see that the row rank of **A** is also n.

2. Conversely, if the column rank is n, this means that

$$T(V) = AV$$

maps $\mathbb{R}_n$ *onto* $\mathbb{R}_n$. Hence, by 4.5, T^{-1} exists from $\mathbb{R}_n$ onto $\mathbb{R}_n$. Hence by 3.5, **A** is invertible.

If we are given that the row rank of **A** is n, argue on **A*** instead. ∎

Example 2. If

$$A = \begin{pmatrix} a_1 & 0 & \cdots & 0 \\ 0 & \ddots & & \vdots \\ \vdots & & \ddots & 0 \\ 0 & \cdots & 0 & a_n \end{pmatrix}$$

is a diagonal matrix, clearly its column rank is n iff all a_i are nonzero. Thus, A^{-1} will exist iff all a_i are nonzero. It is easy to calculate A^{-1} in this case. In Supplement 2 we shall use 4.4 to derive a condition for **A** to be invertible if it is an "almost" diagonal matrix.

We next use 4.4 to prove a property of matrix arithmetic which is carried over from Chapter 2:

4.8 If an $n \times n$ matrix has a right inverse, then it has a left inverse and conversely. In fact, the two are the same; that is, if $AB = I$, then $BA = I$.

Proof. Let

$$S(V) = AV \quad \text{and} \quad T(V) = BV.$$

Then the kernel of **T** is $\{0\}$: If $T(V) = 0$, then

$$0 = S(T(V)) = ABV = IV = V.$$

Since **T** maps $\mathbb{R}_n$ into $\mathbb{R}_n$ and has zero kernel, the dimension of its image is n; that is, **T** maps $\mathbb{R}_n$ *onto* $\mathbb{R}_n$. Hence, **T** has an inverse T^{-1} from $\mathbb{R}_n$ onto $\mathbb{R}_n$. Now

$$T^{-1}(T(V)) = V \quad \text{and} \quad S(T(V)) = V.$$

Since **T** is onto, T^{-1} and **S** agree on all of $\mathbb{R}_n$; that is, $S = T^{-1}$. From Chapter 0, 5.2, we know that

$$T^{-1}T = TT^{-1},$$

so that $TS(V) = T(T^{-1}(V))$ for all **V**. But also

$$T(S(V)) = BAV.$$

Hence $BA = I$ (why?) and we are done. ∎

There are several other results which are closely akin to the fundamental notions we have proved in this section. Since these results are sometimes useful, we present them now. Since their proofs are easy to derive from our theorems, we leave the pleasure for you (Exercise 7).

4.9 A linear transformation from a vector space $\mathcal{V}$ into a vector space $\mathcal{W}$ has an inverse $\mathbf{T}^{-1}$ from $\mathcal{W}$ back to $\mathcal{V}$ iff the image of a basis of $\mathcal{V}$ is a basis for $\mathcal{W}$; that is, if $\{\mathbf{V}_1, \ldots, \mathbf{V}_n\}$ is a basis for $\mathcal{V}$, then $\{\mathbf{T}(\mathbf{V}_1), \ldots, \mathbf{T}(\mathbf{V}_n)\}$ is a basis for $\mathcal{W}$.

4.10 If $\mathbf{T}$ is a linear transformation from a finite-dimensional vector space $\mathcal{V}$ *into itself*, then $\mathbf{T}$ is invertible iff $\mathbf{T}$ maps $\mathcal{V}$ onto $\mathcal{V}$.

To stress the importance of the assumptions of finite dimensionality in 4.10, let us look at an infinite-dimensional case where it does not hold.

Example 3. Consider the transformation $\mathbf{T}$ on $\mathbb{R}[x]$ which increases the degree of each term of a polynomial by 1,

$$\mathbf{T}(a_0 + a_1 x + \cdots + a_n x^n) = a_0 x + a_1 x^2 + \cdots + a_n x^{n+1}.$$

It is easy to see that $\mathbf{T}$ is linear and one-to-one but not *onto* (Exercise 12).

EXERCISES

1. Find the kernel of each of the following linear transformations, and state whether the linear transformation is invertible.
   a) $T(x, y) = (x + y, 2x - y)$
   b) $T(x, y) = (x + y, 2x + 2y)$
   c) $T(x, y) = (x + y, 2x - y, x - 3y)$
   d) $T(x, y) = (x + y, 2x + 2y, 3x + 3y)$
   e) $T(x, y, z) = (x + 2y + z, x - y - 3z)$
   f) $T(x, y, z) = (x + 2y + 3z, x + 2y + 3z)$.
2. (a) through (f). Find the image space of each linear transformation in Exercise 1. Check your work using Exercise 1 and the Dimension Theorem. Draw a picture showing the kernel and a picture showing the image space.
3. Find the dimension of the kernel and of the image for the linear transformations,
   a) $T(x, y) = ax + by$. Be sure to consider all possible cases.
   b) $T(x_1, x_2, x_3, x_4, x_5) = (x_1 + ax_4 + ax_5, x_2 + bx_4 + b'x_5, x_3 + cx_4 + c'x_5)$.
   Find a basis for the kernel and for the image.
   c) $T(x, y, z) = (x, y, z, ax + by + cz, a'x + b'y + c'z)$.
4. Let C denote the vector space of functions continuous for $0 \leq x \leq 1$, and let C' denote the space of functions possessing continuous derivatives on that interval. Let $\mathbf{D}$ be the differentiation operator $\mathbf{D}(f(x)) = f'(x)$, which we saw in Example 5, Section 1, to be a linear transformation from C' into C.
   a) Determine the kernel of this linear transformation.
   b) Determine the dimension of the kernel.

5. Give a matrix interpretation of
   a) 4.1   b) 4.2   c) 4.9   d) 4.10.
6. Prove (a) closure under addition in 4.1, (b) the converse to 4.2, (c) the converse to 4.3.
7. (a) Prove 4.6. (b) Prove 4.9. (c) Prove 4.10.
8. a) Prove that any linear transformation from $\mathbb{R}_3$ into $\mathbb{R}_2$ has a nonzero kernel.
   b) Prove that any linear transformation from $\mathbb{R}_n$ into $\mathbb{R}_m$ has a nonzero kernel if $n > m$.
9. a) Prove that if $\mathbf{A}$ is any 2×3 matrix, then there is a nonzero column vector $\mathbf{V}$ for which $\mathbf{AV} = \mathbf{0}$.
   b) Prove that if $\mathbf{A}$ is any $m \times n$ matrix, with $m < n$, then for any positive integer r there is a nonzero $n \times r$ matrix $\mathbf{B}$ for which $\mathbf{AB} = \mathbf{0}$.
10. Let $\mathbf{T}$ be a linear transformation from $\mathbb{R}_n$ into $\mathbb{R}_m$ and suppose that $\{\mathbf{T}(\mathbf{e}_{k+1}), \ldots, \mathbf{T}(\mathbf{e}_n)\}$ forms a basis for the image $\mathbf{T}(\mathbb{R}_n)$. Why doesn't $\{\mathbf{e}_1, \ldots, \mathbf{e}_k\}$ necessarily form a basis for the kernel? (That is, compare with the proof of 4.4.) Give an example.
11. Why is the dimension of $\mathcal{W}$ of no importance in the Dimension Theorem? Explain and give examples.
12. Prove the statements in Example 4.

Supplement 1: Some Consequences of the Dimension Theorem

We now apply the Dimension Theorem (or, more correctly, its proof) to finding a particularly simple form for the matrix of a linear transformation. We then discuss the very pleasant class of linear transformations known as projections (again applying the proof; this time to split up the space upon which a projection acts). We finally use this proof to decompose any linear transformation into a projection followed by an isomorphism.

Our first theorem should strongly reinforce our dictum that whenever possible a premature choice of basis should be avoided.

4.11 Given a linear transformation $\mathbf{T}$ from an n-dimensional vector space $\mathcal{V}$ into an m-dimensional vector space $\mathcal{W}$, there are bases for $\mathcal{V}$ and $\mathcal{W}$ relative to which the matrix of $\mathbf{T}$ has the form

$$\left. \begin{array}{c} \begin{pmatrix} 1 & 0 & \cdots & 0 & \vline & 0 & \cdots & 0 \\ 0 & \ddots & & 0 & \vline & & & \\ \vdots & & \ddots & \vdots & \vline & & & \\ 0 & \cdots & 0 & 1 & \vline & 0 & \cdots & 0 \\ \hline 0 & \cdots & \cdots & 0 & \vline & 0 & \cdots & 0 \\ \vdots & & & \vdots & \vline & \vdots & & \vdots \\ 0 & \cdots & \cdots & 0 & \vline & 0 & \cdots & 0 \end{pmatrix} \end{array} \right\} r = \left(\begin{array}{c|c} \mathbf{I}_r & \mathbf{0} \\ \hline \mathbf{0} & \mathbf{0} \end{array} \right),$$

where r is the dimension of the image of $\mathbf{T}$.

Proof. As in the proof of the Dimension Theorem, if we take a basis $\{V_1, \ldots, V_k\}$ for the kernel of **T** and extend to a basis $\{V_1, \ldots, V_k, V_{k+1}, \ldots, V_n\}$ for $\mathcal{V}$, then $\{T(V_{k+1}), \ldots, T(V_n)\}$ forms a basis for the image of **T**. Extend this basis to a basis for $\mathcal{W}$, which we denote as $\{T(V_n), \ldots, T(V_{k+1}), W_1, \ldots, W_l\}$. Let the dimension of the image of **T** be r; thus $r = n - k$. Now identify

$$V_n, \ldots, V_{k+1}, V_k, \ldots, V_1 \quad \text{with } e_1, \ldots, e_n, \text{ respectively,}$$

and identify

$$T(V_n), \ldots, T(V_{k+1}), W_1, \ldots, W_l \quad \text{with } f_1, \ldots, f_m, \text{ respectively,}$$

where $f_i = (0, \ldots, 0, 1, 0, \ldots, 0)^*$. The effect of **T** relative to these bases is

$$T(e_1) = f_1, \ldots, T(e_r) = f_r, T(e_{r+1}) = 0, \ldots, T(e_n) = 0.$$

Hence, the matrix of **T**, which is $(T(e_1), \ldots, T(e_n))$ is as indicated in the statement of this theorem. ∎

Example 4. In the version of Example 1 which occurs right after 4.6, we can take the basis for the image $\{W = T(V_3)\}$ and extend it to a basis for $\mathbb{R}_2$, by tacking on $W' = (1, 0)$, for example. Then

$$T(V_3) = W, \quad T(V_2) = 0, \quad T(V_1) = 0;$$

when these vectors are appropriately identified with the **e**'s and **f**'s, the matrix of **T** is

$$A = \begin{pmatrix} 1 & 0 & 0 \\ 0 & 0 & 0 \end{pmatrix}.$$

We hasten to add that 4.11 is not an immensely practical theorem (at least at this time); one cannot usually find the requisite basis for the kernel or make the basis extensions without expending sufficient energy to solve the given problem in other ways. We shall return to this question in Chapter 4. You should compare 4.11 to 3.4 of Chapter 4, and see the discussion following 4.4 of Chapter 4.

Projections; A Decomposition Theorem

We are going to show how any linear transformation **T** on a finite-dimensional vector space can be written as a product of two linear transformations which are of a simple and appealing nature. Although the statements of the two theorems we present should be both novel and interesting, the proofs essentially involve no more than a recapitulation of the proof of the Dimension Theorem.

The geometrical notion of a projection takes the following algebraic form: a **projection** is a linear transformation **P** from a vector space into itself for which $P^2 = P$. Immediate examples include the zero linear transformation and the identity mapping of $\mathcal{V}$. More substantial examples follow.

Example 5. Let $\{V_1, V_2, V_3\}$ be any basis for $\mathbb{R}_3$ and let **P** be that linear transformation on $\mathbb{R}_3$ for which

$$P(V_1) = 0, \quad P(V_2) = V_2, \quad P(V_3) = V_3.$$

Then, for any $V = aV_1 + bV_2 + cV_3$,

$$P^2(V) = P(P(V)) = P(bV_2 + cV_3) = bV_2 + cV_3 = P(V)$$

(or just note that P^2 is a linear transformation and $P^2 = P$ on the basis). As a specific case (so that we can draw a picture), let

$$V_1 = (0, 2, 2), \quad V_2 = (1, 3, 0), \quad V_3 = (1, 1, 0).$$

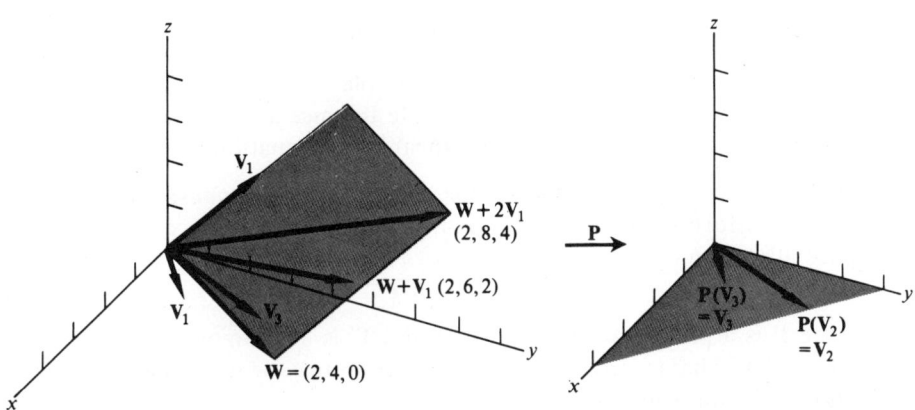

Fig. 21

It is easy to see that P projects any vector in $\mathbb{R}_3$ along a direction parallel to the vector V_1 and into the xy-plane, and behaves like the identity mapping on elements in the plane of V_2 and V_3.

Recall Chapter 1, Supplement to Section 3, where we found that a vector space $\mathcal{V}$ is the *direct sum* of subspaces $\mathcal{W}$ and $\mathcal{X}$, notation

$$\mathcal{V} = \mathcal{W} \oplus \mathcal{X},$$

iff every element V of $\mathcal{V}$ can be written in one and only one way as

$$V = W + X \quad \text{with} \quad W \text{ in } \mathcal{W} \text{ and } X \text{ in } \mathcal{X}.$$

4.12 Decomposition theorem for projection: If P is a projection defined on a finite-dimensional vector space $\mathcal{V}$, then $\mathcal{V}$ is the direct sum of the kernel and the image of P,

$$\mathcal{V} = \mathcal{K} \oplus P(\mathcal{V}), \quad \mathcal{K} \text{ the kernel of } P.$$

Moreover, when restricted to its kernel, P is the zero transformation, and when restricted to its image, it is the identity transformation:

$$P(V) = 0 \quad \text{if } V \text{ is in } \mathcal{V}; \quad P(V) = V \quad \text{if } V \text{ is in } P(\mathcal{V}).$$

Proof. Let $\{V_1, \ldots, V_k\}$ be a basis for the kernel of **P** and extend it to a basis $\{V_1, \ldots, V_k, V_{k+1}, \ldots, V_n\}$ for $\mathcal{V}$. As for the Dimension Theorem, we see that $\{V_{k+1}, \ldots, V_n\}$ forms a basis for $\mathbf{P}(\mathcal{V})$. Moreover, it easily follows that

$$\mathcal{V} = \mathcal{K} + \mathbf{P}(\mathcal{V}),$$

and that the sum is direct (Exercise 3). *Any* linear transformation behaves like the zero linear transformation on its kernel. Let **V** be in $\mathbf{P}(\mathcal{V})$, so that $\mathbf{V} = \mathbf{P}(\mathbf{W})$. Then

$$\mathbf{P}(\mathbf{V}) = \mathbf{P}(\mathbf{P}(\mathbf{W})) = \mathbf{P}^2(\mathbf{W}) = \mathbf{P}(\mathbf{W}) = \mathbf{V}. \quad \blacksquare$$

You can check this theorem on the above example.

Although projections may seem too simple and nice to be connected intimately with the ordinary run-of-the-mill ornery linear transformation, we now show:

4.13 *The decomposition theorem for linear transformations*: Let **T** be any linear transformation from a finite-dimensional vector space $\mathcal{V}$ to a vector space $\mathcal{W}$. Then

$$\mathbf{T} = \mathbf{T'P},$$

where **P** is a projection defined in $\mathcal{V}$ and **T'** is an isomorphism from $\mathbf{P}(\mathcal{V})$ onto $\mathbf{T}(\mathcal{V})$. That is, any linear transformation may be written as a projection followed by an isomorphism.

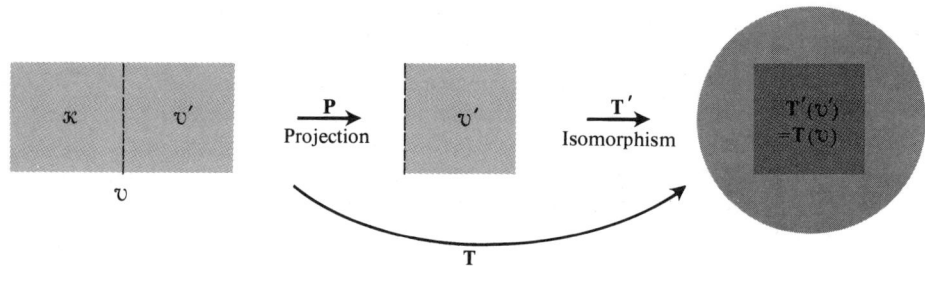

Fig. 22

Proof. Let $\mathcal{K}$ be the kernel of **T**. Exactly as in 4.12, extend a basis $\{V_1, \ldots, V_k\}$ for $\mathcal{K}$ to a basis $\{V_1, \ldots, V_k, V_{k+1}, \ldots, V_n\}$ for $\mathcal{V}$, let

$$\mathcal{V}' = \text{span } \{V_{k+1}, \ldots, V_n\},$$

and show that $\mathcal{V} = \mathcal{K} \oplus \mathcal{V}'$. Define **P** from $\mathcal{V}$ into $\mathcal{V}'$ by

$$\mathbf{P}(V_i) = \begin{cases} 0 & i = 1, \ldots, k, \\ V_i & i = k+1, \ldots, n, \end{cases} \quad \text{for the above basis.}$$

P is then a projection. Define a linear transformation **T'** from $\mathcal{V}'$ into $\mathbf{T}(\mathcal{V})$ by $\mathbf{T'}(V_i) = \mathbf{T}(V_i)$, for $i = k+1, \ldots, n$. Then **T'** is an isomorphism from $\mathcal{V}'$ onto $\mathbf{T}(\mathcal{V})$ and $\mathbf{T} = \mathbf{T'P}$. We leave the details for you. $\blacksquare$

Thus any linear transformation from $\mathcal{V}$ into $\mathcal{W}$ behaves in a very simple fashion: It first acts as a projection onto a subspace $\mathcal{V}'$ of $\mathcal{V}$, and then acts as an isomorphism between this $\mathcal{V}'$ and $T(\mathcal{V})$.

Example 6. Let T map $\mathbb{R}_3$ into $\mathbb{R}_3$ by

$$T(V_1) = 0, \quad T(V_2) = V_2, \quad T(V_3) = 2V_3,$$

with $V_1 = (0, 2, 2)$, $V_2 = (1, 3, 0)$, $V_3 = (1, 1, 0)$, as in Example 2. Then the kernel $\mathcal{K}$ of T is spanned by V_1; we can extend this to the basis $\{V_1, V_2, V_3\}$ for $\mathbb{R}_3$, and let

$$\mathcal{V}' = \text{span } \{V_2, V_3\}.$$

Then the projection P in the proof of the theorem is the projection P discussed in Example 2. The isomorphism T' from $\mathcal{V}'$ onto $T(\mathcal{V})$ is defined by

$$T'(V_2) = V_2, \quad T'(V_3) = 2V_3.$$

EXERCISES

1. Let P be a projection on $\mathcal{V}$ and suppose a basis $\{V_1, \ldots, V_k, V_{k+1}, \ldots, V_n\}$ is chosen for $\mathcal{V}$, as in 4.12, with $\{V_1, \ldots, V_k\}$ a basis for the kernel and $\{V_{k+1}, \ldots, V_n\}$ a basis for the image. With respect to this basis, $\mathcal{V}$ will be essentially the same as $\mathbb{R}_n$. What will the matrix of T be if we make the identification

$$V_i = e_i = f_i?$$

If A is this matrix, note that $A^2 = A$.

2. If P is a projection on the vector space $\mathcal{V}$, and I is the identity mapping, prove that $I - P$ is also a projection. Show that $(I - P)P = 0 = P(I - P)$. Projections P and P' for which $PP' = P'P = 0$ are called **orthogonal**. Prove that the sum of any two orthogonal projections is a projection. With bases chosen for $\mathcal{V}$ as in Exercise 1, what is the matrix of P vs. that of $I - P$? Note that the matrix units $E_{11}, \ldots, E_{nn}$ of Chapter 2, Section 3, Exercise 6, are mutually orthogonal projections. What is their sum?

3. (a) In 4.12, show that $\mathcal{V} = \mathcal{K} + P(\mathcal{V})$. (b) Show that the sum is direct.

4. Write out the details of the proof of 4.13.

Supplement 2: Almost Diagonal Matrices

We now use the results of this section to prove that if the diagonal entries of a matrix are sufficiently large in relation to the other entries in the rows (or columns), then the matrix is invertible. In addition to providing a cheap (sufficient) test for invertibility, our result is fundamental for some very pretty results in future supplements.

Our theorem can be stated thus: A matrix A is invertible if each diagonal entry is larger in absolute value than the sum of the absolute values of the other entries in its row.

4.14 The dominant diagonal theorem. Given a matrix

$$A = \begin{pmatrix} a_{11} & \cdots & & & a_{1n} \\ & \ddots & & & \\ a_{i1} & & a_{ii} & & a_{in} \\ & & & \ddots & \\ a_{n1} & & \cdots & & a_{nn} \end{pmatrix},$$

suppose that

$$|a_{ii}| > |a_{i1}| + \cdots + |a_{i,i-1}| + |a_{i,i+1}| + \cdots + |a_{in}| \tag{1}$$

for $i = 1, \ldots, n$. Then **A** is invertible. (Moreover, this is also true if rows are replaced by columns.)

Proof. Suppose the hypothesis is satisfied and the conclusion is false. Then by the matrix form of 4.3, there is some $\mathbf{V} \neq \mathbf{0}$ for which $\mathbf{AV} = \mathbf{0}$. For convenience, suppose that the first coordinate of $\mathbf{V} = (v_1, \ldots, v_n)^*$ is largest in absolute value, i.e.,

$$|v_1| \geq |v_i|, \quad i = 2, \ldots, n.$$

The first entry of $\mathbf{AV}$ is $0 = a_{11}v_1 + \cdots + a_{1n}v_n$. Then

$$-a_{11}v_1 = a_{12}v_2 + \cdots + a_{1n}v_n,$$

so that

$$|a_{11}| \, |v_1| = |-a_{11}v_1| = |a_{12}v_2 + \cdots + a_{1n}v_n|$$

$$\overset{\text{Why}_1?}{\leq} |a_{12}| \, |v_2| + \cdots + |a_{1n}| \, |v_n|$$

$$\overset{\text{Why}_2?}{\leq} |a_{12}| \, |v_1| + \cdots + |a_{1n}| \, |v_1|$$

$$= (|a_{12}| + \cdots + |a_{1n}|) \, |v_1|.$$

Since $v_1 \neq 0$, $|a_{11}| \leq |a_{12}| + \cdots + |a_{1n}|$. But this is contrary to the assumption (1) with $n = 1$. Hence, if the hypothesis holds, so does the conclusion. ∎

Exercise: Answer the questions in the above proof.

*5. DUAL SPACES

If $\mathcal{V}$ is an arbitrary vector space over $\mathbb{R}$, then the **dual space** $\mathcal{V}^*$ of $\mathcal{V}$ is defined to be the set of all linear transformations from $\mathcal{V}$ into $\mathbb{R}$. Elements of $\mathcal{V}^*$ are sometimes called **linear functionals**. Such a creature $\mathbf{V}^*$ is thus a function from $\mathcal{V}$ into $\mathbb{R}$ satisfying

$$\mathbf{V}^*(\mathbf{V}_1 + \mathbf{V}_2) = \mathbf{V}^*(\mathbf{V}_1) + \mathbf{V}^*(\mathbf{V}_2), \qquad \mathbf{V}^*(c\mathbf{V}) = c\mathbf{V}^*(\mathbf{V})$$

for **V**'s in $\mathcal{U}$ and c in $\mathbb{R}$. We know that $\mathcal{U}^*$ is a vector space, and if $\mathcal{U}$ is of finite dimension n, then by 2.3

dimension $\mathcal{U}^*$ = (dimension $\mathcal{U}$ over $\mathbb{R}$) (dimension $\mathbb{R}$ over $\mathbb{R}$)
$$= n \cdot 1 = n, \quad \text{by 3.3.}$$

If $\mathcal{U}$ is finite-dimensional over $\mathbb{R}$ with basis $\{\mathbf{V}_1, \ldots, \mathbf{V}_n\}$, and if $\mathbf{V}^*$ is in $\mathcal{U}^*$, suppose that
$$\mathbf{V}^*(\mathbf{V}_1) = b_1, \ldots, \mathbf{V}^*(\mathbf{V}_n) = b_n.$$
This specifies $\mathbf{V}^*$, since we have given its value on a basis for $\mathcal{U}$ over $\mathbb{R}$ in terms of a basis for $\mathbb{R}$ over $\mathbb{R}$ (namely, $\{1\}$). Note that if we identify $\mathbf{V}_1, \ldots, \mathbf{V}_n$ with $\mathbf{e}_1, \ldots, \mathbf{e}_n$, respectively, then the matrix of $\mathbf{V}^*$ is $(b_1, \ldots, b_n)$. Thus, if
$$\mathbf{W} = a_1 \mathbf{V}_1 + \cdots + a_n \mathbf{V}_n$$
is any vector in $\mathcal{U}$, the matrix form of the action of $\mathbf{V}^*$ on $\mathbf{W}$ is just

$$(b_1, \ldots, b_n) \begin{pmatrix} a_1 \\ \cdot \\ \cdot \\ \cdot \\ a_n \end{pmatrix} = a_1 b_1 + \cdots + a_n b_n.$$

How tame dual spaces are in matrix form!

If $\{\mathbf{V}_1, \ldots, \mathbf{V}_n\}$ is a basis for $\mathcal{U}$ over $\mathbb{R}$, define $\mathbf{V}_1^*, \ldots, \mathbf{V}_n^*$ to be the linear transformations for which
$$\mathbf{V}_i^*(\mathbf{V}_j^*) = \delta_{ij}$$
(i.e., $\mathbf{V}_i^*(\mathbf{V}_j) = 0$ if $i \neq j$ and $\mathbf{V}_i^*(\mathbf{V}_i) = 1$).

5.1 The dual basis. The set $\{\mathbf{V}_1^*, \ldots, \mathbf{V}_n^*\}$ forms a basis for $\mathcal{U}^*$ over $\mathbb{R}$.

Proof. It suffices to show that the set is linearly independent (why$_1$?). Suppose then that
$$c_1 \mathbf{V}_1^* + c_2 \mathbf{V}_2^* + \cdots + c_n \mathbf{V}_n^* = 0.$$
This is a function in $\mathcal{U}^*$ which, when applied to $\mathbf{V}_i$ in $\mathcal{U}$, shows that $c_i = 0$ (why$_2$?). Hence, the set is linearly independent. ∎

The basis $\{\mathbf{V}_1^*, \ldots, \mathbf{V}_n^*\}$ for $\mathcal{U}^*$ is called the **dual basis** to $\{\mathbf{V}_1, \ldots, \mathbf{V}_n\}$. Note that the matrix of $\mathbf{V}_1^*$ is $(1, 0, \ldots, 0), \ldots$, the matrix of $\mathbf{V}_n^*$ is $(0, \ldots, 0, 1)$. Thus, in matrix form $\mathbf{V}_i^*(\mathbf{V}_j)$ is just

$$(0, \ldots, 0, \overset{i}{1}, 0, \ldots, 0) \begin{pmatrix} 0 \\ \cdot \\ \cdot \\ \cdot \\ 1 \\ 0 \\ \cdot \\ \cdot \\ \cdot \\ 0 \end{pmatrix}^{\!\!j} = \delta_{ij},$$

as it is supposed to be.

Not content with considering merely $\mathcal{V}^*$, we could form its dual space:

$\mathcal{V}^{**}$ = all linear transformations from $\mathcal{V}^*$ to $\mathbb{R}$.

If $\mathcal{V}$ is of finite dimension n, then a continuation of our above argument shows that $\mathcal{V}^{**}$ is also of dimension n. Thus $\mathcal{V}$ and $\mathcal{V}^{**}$ will be isomorphic (and $\mathcal{V}^*$ too). However, there is a *natural isomorphism* between $\mathcal{V}$ and $\mathcal{V}^{**}$, that is, one that does not depend on a choice of bases. These is no such natural isomorphism between $\mathcal{V}$ and $\mathcal{V}^*$, but there is a "natural" way of identifying $\mathcal{V}$ and $\mathcal{V}^{**}$.

5.2 The natural isomorphism between $\mathcal{V}$ and $\mathcal{V}^{**}$. For any V in $\mathcal{V}$ let T_V be the mapping of $\mathcal{V}^*$ into $\mathbb{R}$ defined by

$$T_V(W^*) = W^*(V)$$

for W^* in $\mathcal{V}^*$. Then T_V is in $\mathcal{V}^{**}$ and the mapping $\mathbf{f}$ from $\mathcal{V}$ into $\mathcal{V}^{**}$, defined by $\mathbf{f}(V) = T_V$, is an isomorphism of $\mathcal{V}$ into $\mathcal{V}^{**}$.

If $\mathcal{V}$ is finite-dimensional, then $\mathbf{f}$ is an isomorphism of $\mathcal{V}$ *onto* $\mathcal{V}^{**}$.

Proof. A good, substantial exercise (with hints; see Exercise 3). ∎

We have already observed that the dual space to the vector space of column vectors is just the corresponding vector space of row vectors, and the operation of elements of $\mathcal{V}^*$ on $\mathcal{V}$ is just the matrix product.

The obvious isomorphism between $\mathbb{R}_n$ and $\mathbb{R}_n^*$ given by

$$\begin{pmatrix} a_1 \\ \cdot \\ \cdot \\ \cdot \\ a_n \end{pmatrix} \to (a_1, \ldots, a_n),$$

although very reasonable, is not "natural" in the sense defined above, since it depends very heavily on the form of

$$\begin{pmatrix} a_1 \\ \cdot \\ \cdot \\ \cdot \\ a_n \end{pmatrix}$$

with respect to a particular basis.

There is an interesting difference between the infinite- and finite-dimensional situations. If $\mathcal{V}$ is infinite-dimensional over $\mathbb{R}$, then $\mathcal{V}^*$ is infinite-dimensional over $\mathbb{R}$, of a *higher-order infinity:*

$$\text{dimension } \mathcal{V}^* \neq \text{dimension } \mathcal{V}.$$

It follows that the isomorphism of 5.2 is onto a *proper* subspace of $\mathcal{V}$ (see [1]).

We shall now discuss another natural isomorphism (heaven help us!), this time between the vector spaces $\mathcal{L}(\mathcal{V}, \mathcal{W})$, of linear transformations from $\mathcal{V}$ into another vector space $\mathcal{W}$, and $\mathcal{L}(\mathcal{W}^*, \mathcal{V}^*)$ of linear transformations from $\mathcal{W}^*$ into $\mathcal{V}^*$. (If the dimension of $\mathcal{V}$ is n and that of $\mathcal{W}$ is m, then $\mathcal{L}(\mathcal{V}, \mathcal{W})$ is isomorphic to $\mathbb{R}_{m \times n}$ and $\mathcal{L}(\mathcal{W}^*, \mathcal{V}^*)$ is isomorphic to $\mathbb{R}_{n \times m}$. Can you, therefore, guess the form this natural isomorphism will take if bases are chosen?)

Given $\mathbf{T}$ in $\mathcal{L}(\mathcal{V}, \mathcal{W})$, our isomorphism is to produce an element $\mathbf{T}^*$ in $\mathcal{L}(\mathcal{W}^*, \mathcal{V}^*)$. Thus, given a $\mathbf{W}^*$ in $\mathcal{W}^*$, $\mathbf{T}^*(\mathbf{W}^*)$ is to be in $\mathcal{V}^*$, a linear mapping from $\mathcal{V}$ into $\mathbb{R}$. Thus, $\mathbf{T}^*(\mathbf{W}^*)$ is to be a function evaluated at elements $\mathbf{V}$ of $\mathcal{V}$. What should $[\mathbf{T}^*(\mathbf{W}^*)](\mathbf{V})$ be? Given the collection of items $\mathbf{T}$, $\mathbf{W}^*$, and $\mathbf{V}$, $\mathbf{T}(\mathbf{V})$ is in $\mathcal{W}$, so $\mathbf{W}^*$ can act on it; we can at least form $\mathbf{W}^*(\mathbf{T}(\mathbf{V}))$. We do so:

Given $\mathbf{T}$ in $\mathcal{L}(\mathcal{V}, \mathcal{W})$ the **dual mapping** $\mathbf{T}^*$ is defined by

$$[\mathbf{T}^*(\mathbf{W}^*)](\mathbf{V}) = \mathbf{W}^*(\mathbf{T}(\mathbf{V})) \qquad \text{for all} \quad \mathbf{W}^* \text{ in } \mathcal{W}^*, \mathbf{V} \text{ in } \mathcal{V}.$$

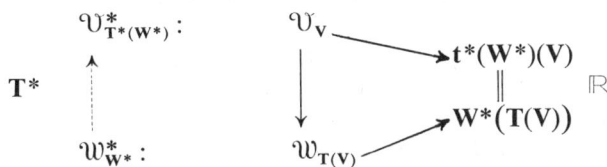

5.3 The mapping $\mathbf{T} \to \mathbf{T}^*$ is an isomorphism of $\mathcal{L}(\mathcal{V}, \mathcal{W})$ onto $\mathcal{L}(\mathcal{W}^*, \mathcal{V}^*)$. If $\mathcal{V}$ and $\mathcal{W}$ are finite-dimensional and if dual bases are chosen for $\mathcal{V}$ and $\mathcal{V}^*$ and for $\mathcal{W}$ and $\mathcal{W}^*$, then the matrix of $\mathbf{T}^*$ relative to the appropriate bases is the transpose of the matrix of $\mathbf{T}$ relative to its appropriate bases.

Proof. We shall abandon you to the first part of the proof, each step of which is quite straightforward; it is, however, somewhat tricky to decide exactly what the steps ought to be (see Exercise 4).

To prove the second part, let $\mathcal{V}$ have basis $\{\mathbf{V}_1, \ldots, \mathbf{V}_n\}$, which we identify with $\{\mathbf{e}_1, \ldots, \mathbf{e}_n\}$ so that $\mathcal{V}$ becomes $\mathbb{R}_n$, and let $\mathcal{W}$ have basis $\{\mathbf{W}_1, \ldots, \mathbf{W}_m\}$, which we identify with $\{f_1, \ldots, f_m\}$, and denote by $\{\mathbf{e}_1^*, \ldots, \mathbf{e}_n^*\}$ the dual basis for $\mathcal{V}^*$, and by $\{f_1^*, \ldots, f_m^*\}$ the dual basis for $\mathcal{W}^*$. We wish to show that, if $\mathbf{A}$ is the matrix of $\mathbf{T}$, then $\mathbf{A}^*$ is the matrix of $\mathbf{T}^*$. Suppose, then, that

$$\mathbf{T}(\mathbf{V}) = \mathbf{A}\mathbf{V}, \quad \mathbf{A} = (\mathbf{T}(\mathbf{e}_1), \ldots, \mathbf{T}(\mathbf{e}_n)),$$

and

$$\mathbf{T}^*(\mathbf{W}^*) = \mathbf{B}\mathbf{W}^*, \quad \mathbf{B} = (\mathbf{T}^*(\mathbf{f}_1^*), \ldots, \mathbf{T}^*(\mathbf{f}_m^*)).$$

Now, by definition, $\mathbf{T}^*(\mathbf{W}^*)$ is that element of $\mathcal{V}^*$ for which

$$[\mathbf{T}^*(\mathbf{W}^*)](\mathbf{V}) = \mathbf{W}^*(\mathbf{T}(\mathbf{V})),$$

so that, in particular,

$$[\mathbf{T}^*(\mathbf{f}_j^*)](\mathbf{e}_i) = \mathbf{f}_j^*(\mathbf{T}(\mathbf{e}_i)) = a_{ij},$$

the i, j-entry of $\mathbf{A}$. But also,
$$[\mathbf{T}^*(\mathbf{f}_j^*)](\mathbf{e}_i) = b_{ji},$$
the j, i-entry of $\mathbf{B}$. Thus $a_{ij} = b_{ji}$ for $j = 1, \ldots, m$ and $i = 1, \ldots, n$. Hence $\mathbf{B} = \mathbf{A}^*$. ∎

In summary, in matrix form with respect to dual bases, if the action of $\mathbf{T}$ on a typical element $\mathbf{V}$ of $\mathcal{U}$ is represented by $\mathbf{AV}$, $\mathbf{A}$ an $m \times n$ matrix, then the action of $\mathbf{T}^*$ on a typical element $\mathbf{W}^*$ of $\mathcal{W}^*$ can be represented by $\mathbf{A}^*\mathbf{W}^* = (\mathbf{WA})^*$.

EXERCISES

1. Answer the why's in the proof of 5.1.
2. Prove that given any $\mathbf{V} \neq \mathbf{0}$ in $\mathcal{U}$ there is a $\mathbf{W}^*$ in $\mathcal{U}^*$ for which $\mathbf{W}^*(\mathbf{V}) \neq 0$. [*Hint:* Any linearly independent set can be extended . . .]
3. In the proof of 5.2,
   a) Show that $\mathbf{T_V}$ is a function (i.e., if $\mathbf{W}_1^*$ and $\mathbf{W}_2^*$ are the same function, then $\mathbf{T_V}(\mathbf{W}_1^*) = \mathbf{T_V}(\mathbf{W}_2^*)$.
   b) Show that $\mathbf{T_V}$ is a linear mapping.
   c) Show that $\mathbf{f} \colon \mathbf{V} \to \mathbf{T_V}$ is a function.
   d) Show that $\mathbf{f}$ is linear.
   e) Show that $\mathbf{f}$ is one-to-one.
   f) Show that if $\mathcal{U}$ is finite-dimensional, then $\mathbf{f}$ is onto.
H4. Prove 5.3 (be sure you warm up on Exercise 3).
5. Prove that if $\mathbf{S}$ in $\mathcal{L}(\mathcal{U}, \mathcal{W})$ and $\mathbf{T}$ is in $\mathcal{L}(\mathcal{W}, \mathcal{X})$, then $(\mathbf{TS})^* = \mathbf{S}^*\mathbf{T}^*$.

REFERENCE

1. Jacobson, N., *Lectures in Abstract Algebra, Vol. II*, Chapter 9. Princeton, N.J.; Van Nostrand, 1953.

... find, if you are not literally cooefficient, how minney combinaisies and permutandies can be played on the international surd! pthwndxrclzp!, hids cubid rute being extructed, taking anan illitterettes, ififif at a tom. Answers, (for teasers only). Ten, twent, thirt, see, ex and three icky totchty ones. From solation to solution.†

CHAPTER 4 Systems of Linear Equations

In this chapter we study the ubiquitous and exceedingly important problem of solving systems of linear equations. In the first section, we adopt an elementary approach which relies only on the first two sections of Chapter 2. We bring our heavier machinery of matrices and linear transformations to bear in Section 2. We then reconsider the solution problem, using column operations, and we end the chapter by describing the effect of basis changes on the matrix of a linear transformation. Sections 3 and 4 are optimal, although Section 4 will be of some later use.

Our basic solution procedure is the cornerstone of a number of computer techniques, but as usual, the material cannot be directly applied to the computer with impunity. An analysis and treatment of the computational difficulties involves most of the linear algebra theory developed in the remainder of this book.

1. AN ELEMENTARY APPROACH

Consider m linear equations in n unknowns $x_1, x_2, \ldots, x_n$:

$$a_{11}x_1 + \cdots + a_{1n}x_n = y_1,$$

$\qquad\qquad a_{11}, \ldots, a_{mn}, \quad y_1, \ldots, y_m, \quad$ given (real numbers).‡ $\quad$ (1)

$$a_{m1}x_1 + \cdots + a_{mn}x_n = y_m,$$

Our problem is to find the set of all solutions $(x_1, \ldots, x_n)$, the **solution set**. Note that the solution set could be empty, i.e., there may be no solutions.

---

† James Joyce, *Finnegan's Wake*, Compass Books, 1959.
‡ The numbers actually could be complex, or even from an arbitrary field.

Examples.

1. $\left.\begin{array}{l}x + y = 1\\ x + y = 2\end{array}\right\}$ No solution
2. $0 \cdot x + 0 \cdot y = 1$ No solution
3. $0 \cdot x + 0 \cdot y = 0$ Here any pair (x, y) is a solution
4. $x - y = 0, x + y = 1$ Here the solution set is just $(\frac{1}{2}, \frac{1}{2})$
5. $x - z = 0, y - z = 1$ Here the solution set is all triples $(k, k + 1, k)$, where k is any real number.

We make the following definition:

Two systems of equations in n unknowns $(x_1, \ldots, x_n)$ are called **equivalent** if they have the same solution set.

Thus the systems in 1 and 2 of the above example are equivalent. Our procedure for handling this problem (usually attributed to Gauss) is based on the following remarks:

1.1 Permuting the given equations replaces the system by an equivalent system.

1.2 Multiplying any equation by a nonzero number replaces the system by an equivalent system.

1.3 Replacing any equation of the system by that equation plus any multiple of another equation of the system gives us a new system equivalent to the given one.

Compare these remarks to the corresponding ones in Section 1 of Chapter 2. If anything, they are even easier to prove here. For example,

Proof of 1.3. Suppose we change the system of equations (1) by adding $c \times$ (the second equation) to the first equation, giving us a system

$$(a_{11}x_1 + \cdots + a_{1n}x_n) + c(a_{21}x_1 + \cdots + a_{2n}x_n) = y_1 + cy_2,$$
$$a_{21}x_1 + \cdots + a_{2n}x_n = y_2,$$
$$\vdots$$
$$a_{m1}x_1 + \cdots + a_{mn}x_n = y_m;$$

or

$$(a_{11} + ca_{21})x_1 + \cdots + (a_{1n} + ca_{2n})x_n = y_1 + cy_2,$$
$$a_{21}x_1 + \cdots + a_{2n}x_n = y_2,$$
$$\vdots \tag{2}$$
$$a_{m1}x_1 + \cdots + a_{mn}x_n = y_m.$$

We must show that any solution to (1) is also a solution to (2), and conversely. Thus, let $(\bar{x}_1, \ldots, \bar{x}_n)$ be some specific solution to (1), so that
$$a_{i1}\bar{x}_1 + \cdots + a_{in}\bar{x}_n = y_i \quad \text{for } i = 1, \ldots, m.$$
The only difference between (2) and (1) is in the first equation, but by definition
$$(a_{11}\bar{x}_1 + \cdots + a_{1n}\bar{x}_n) + c(a_{21}\bar{x}_1 + \cdots + a_{2n}\bar{x}_n) = y_1 + cy_2,$$
so that $(\bar{x}_1, \ldots, \bar{x}_n)$ is a solution to (2) as well as to (1). We leave to you the equally difficult task of showing that any solution of (2) is a solution of (1). ∎

We shall show that, by using these facts, probably a number of times, we can replace the given system by an equivalent system for which the solution set is easily found. This is the entire basis of our solution procedure. In this section we shall hone up your intuition, and then codify the procedure in the form of an algorithm in the next section.

Example 1

$$\begin{aligned} x + 3y - 2z &= 5, \\ 2x + 4y + 3z &= -3 \\ -x + y - 12z &= 8. \end{aligned} \qquad \begin{matrix} -2 & 1 \\ \downarrow & \\ & \downarrow \end{matrix}$$

We apply remark 1.3 by adding -2 times the first equation to the second one, and at the same time adding the first equation to the third one. We thus get the equivalent system:

$$\begin{aligned} x + 3y - 2z &= 5, \\ -2y + 7z &= -13, \\ 4y - 14z &= +13. \end{aligned} \qquad \begin{matrix} \\ 2 \\ \downarrow \end{matrix}$$

Now we add 2 times the (new) second equation to the (new) third one and get another equivalent system

$$\begin{aligned} x + 3y - 2z &= 5, \\ -2y + 7z &= -13, \\ 0 &= -13. \end{aligned}$$

The last equation should, of course, be interpreted as
$$0 \cdot x + 0 \cdot y + 0 \cdot z = -13.$$
Since the solution set for this system is clearly empty we conclude that the same is true for the given system; i.e. the system has no solutions.

The operations just mentioned can be expeditiously performed by carrying out corresponding operations (already known to the reader from Chapter 2) on certain matrices. For this purpose we introduce the **coefficient matrix** of the system (1):

$$A = \begin{pmatrix} a_{11} & a_{12} & \cdots & a_{1n} \\ \cdot & & & \cdot \\ \cdot & & & \cdot \\ \cdot & & & \cdot \\ a_{m1} & a_{m2} & \cdots & a_{mn} \end{pmatrix} \qquad \text{(this is an } m \times n \text{ matrix)}$$

and the **augmented matrix** of the system (1):

$$\begin{pmatrix} a_{11} & a_{12} & \cdots & a_{1n} & | & y_1 \\ \vdots & & & \vdots & | & \vdots \\ a_{m1} & a_{m2} & \cdots & a_{mn} & | & y_m \end{pmatrix}$$ (This is an $(m \times (n+1))$ matrix.)

Note that the augmented matrix can be presented in the partitioned form

$(A \mid Y)$ where Y is the column vector $\begin{pmatrix} y_1 \\ \vdots \\ y_m \end{pmatrix}$.

In our example

$$A = \begin{pmatrix} 1 & 3 & -2 \\ 2 & 4 & 3 \\ -1 & 1 & -12 \end{pmatrix}$$

and the augmented matrix is

$$\begin{pmatrix} 1 & 3 & -2 & | & 5 \\ 2 & 4 & 3 & | & -3 \\ -1 & 1 & -12 & | & 8 \end{pmatrix}.$$

It should now be clear that the operations described in remarks 1.1, 1.2, 1.3 of this section correspond exactly to the elementary row transformations on the augmented matrix described in remarks 1.1, 1.2, 1.3 in Section 1 of Chapter 2. Thus we might have done our example by operating on the augmented matrix:

$$\begin{pmatrix} 1 & 3 & -2 & | & 5 \\ 2 & 4 & 3 & | & -3 \\ -1 & 1 & -12 & | & 8 \end{pmatrix} \to \begin{pmatrix} 1 & 3 & -2 & | & 5 \\ 0 & -2 & 7 & | & -13 \\ 0 & 4 & -14 & | & 13 \end{pmatrix} \to \begin{pmatrix} 1 & 3 & -2 & | & 5 \\ 0 & -2 & 7 & | & -13 \\ 0 & 0 & 0 & | & -13 \end{pmatrix}.$$

Notice in particular that the elementary row transformation, when performed on the augmented matrix, will automatically give you the corresponding row transformations of the coefficient matrix **A** and also of the column vector **Y**.

Recall that pivoting (the operation of Chapter 2, Section 1, which results in a column of zeros except for one 1) is accomplished by a succession of elementary row operations. Each row operation preserves the solution set of the corresponding system, by 1.1 through 1.3. Hence

1.4 If the augmented matrix of a system of linear equations is pivoted upon, then the resulting matrix is the augmented matrix of an equivalent system of linear equations.

Since the Basis Algorithm involves a succession of pivotings, it follows that, if one applies the Basis Algorithm to an augmented matrix, there results the augmented matrix of an equivalent system.

Our next example is one where the solution set is not empty:

Example 2
$$x + 3y - 2z = 5,$$
$$2x + 4y + 3z = -3,$$
$$-x + y - 12z = 21.$$

Here the augmented matrix is
$$\begin{pmatrix} 1 & 3 & -2 & | & 5 \\ 2 & 4 & 3 & | & -3 \\ -1 & 1 & -12 & | & 21 \end{pmatrix}.$$

(This is also an example considered in Chapter 2, Section 1.) The row operations yield

$$\begin{pmatrix} 1 & 3 & -2 & | & 5 \\ 0 & -2 & 7 & | & -13 \\ 0 & 0 & 0 & | & 0 \end{pmatrix}; \quad \text{that is,} \quad \begin{aligned} x + 3y - 2z &= 5, \\ -2y + 7z &= -13, \\ 0 &= 0. \end{aligned}$$

It is clear that this system has solutions. In fact, z can be taken equal to any real number; the second equation then gives
$$y = \tfrac{13}{2} + \tfrac{7}{2}z;$$

and finally the first equation gives
$$x = -\tfrac{29}{2} - \tfrac{17}{2}z.$$

The same result can be obtained by completing the pivoting. We get

$$\begin{pmatrix} 1 & 0 & \tfrac{17}{2} & | & -\tfrac{29}{2} \\ 0 & 1 & -\tfrac{7}{2} & | & \tfrac{13}{2} \\ 0 & 0 & 0 & | & 0 \end{pmatrix}; \quad \text{that is,} \quad \begin{aligned} x + \tfrac{17}{2}z &= -\tfrac{29}{2}, \\ y - \tfrac{7}{2}z &= \tfrac{13}{2}, \\ 0 &= 0, \end{aligned}$$

for which the complete solution can be at once written in the form

$$\begin{aligned} x &= -\tfrac{29}{2} - \tfrac{17}{2}z, \\ y &= \tfrac{13}{2} + \tfrac{7}{2}z, \\ z &= z, \end{aligned} \quad \text{or, in vector form,} \quad \begin{pmatrix} x \\ y \\ z \end{pmatrix} = \begin{pmatrix} -\tfrac{29}{2} \\ \tfrac{13}{2} \\ 0 \end{pmatrix} + z \begin{pmatrix} \tfrac{17}{2} \\ \tfrac{7}{2} \\ 1 \end{pmatrix},$$

where z can be any real number.

Example 3
$$x + 3y - 2z = 5,$$
$$2x + 4y + 3z = -3,$$
$$-x + y - 11z = 8.$$

The augmented matrix is
$$\begin{pmatrix} 1 & 3 & -2 & | & 5 \\ 2 & 4 & 3 & | & -3 \\ -1 & 1 & -11 & | & 8 \end{pmatrix},$$

and the usual steps yield
$$\begin{pmatrix} 1 & 0 & 0 & | & 96 \\ 0 & 1 & 0 & | & -39 \\ 0 & 0 & 1 & | & -13 \end{pmatrix}.$$
This at once gives the unique solution
$$x = 96, \quad y = -39, \quad z = -13.$$

It should now be clear how this works in general. If we apply the Basis Algorithm to the augmented matrix $(\mathbf{A} \mid \mathbf{Y})$ of a system (1), there results the augmented matrix $(\mathbf{A}' \mid \mathbf{Y}')$ of an equivalent system and one for which the solutions are readily apparent.

It is convenient to take the matrix $(\mathbf{A}' \mid \mathbf{Y}')$ after applying the Basis Algorithm, and interchange its rows (using 1.2), so that:

the first row contains the pivot entry farthest to the left, the second row contains the pivot entry next farthest to the left, and so on.

There results a matrix of the form

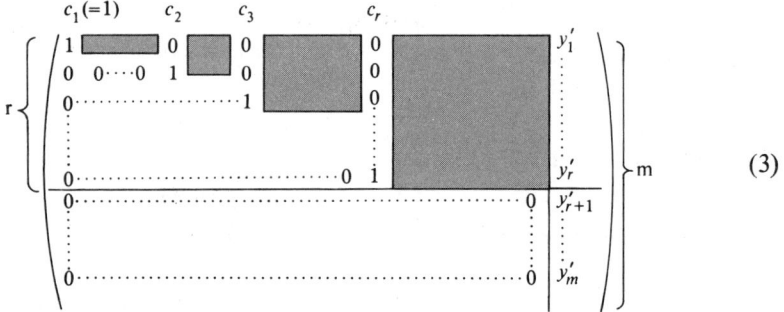

(3)

Here r is the row rank of $\mathbf{A}$, and $c_1, \ldots, c_r$ are the columns pivoted upon in achieving this form.†

The "A part" of the above matrix is said to be in **row-echelon** form. This is characterized by

1. the first nonzero entry in a row (if one occurs) is a 1;
2. above and below such first entry 1's are all 0's;
3. these first entry 1's are arranged in descending stair-step fashion from left to right (hence, "echelon").

The operations on the augmented matrix have replaced the vector $\mathbf{Y} = (y_1, \ldots, y_m)^*$ by the vector $\mathbf{Y}' = (y_1', \ldots, y_m')^*$.

---

† If column 1 is not pivoted upon (that is, $c_1 \neq 1$), then all coefficients of x_1 are 0 in the system of equations. Thus, we may as well assume that $c_1 = 1$.

Since the vector $\mathbf{Y}'$ is important in the discussion, it is well to note that, not merely is $\mathbf{Y}'$ readily determined if $\mathbf{Y}$ is given, but it is possible for any given $\mathbf{Y}'$ to find a vector $\mathbf{Y}$ that will yield it. This follows from the fact that each of the elementary row operations that were used in getting $\mathbf{Y}'$ from $\mathbf{Y}$ is reversible so that one can obtain a vector $\mathbf{Y}$ by applying the inverses in turn to a given vector $\mathbf{Y}'$. Note further that *the vector* $\mathbf{Y}' = \mathbf{0}$ *iff the vector* $\mathbf{Y} = \mathbf{0}$. This is because the elementary row operations cannot reduce a nonzero column to all zeros, nor create a nonzero column from a zero column.

We have actually arrived at the following:

1.5 The Solution Algorithm. Given a system of linear equations (1), take its augmented matrix and put it in row-echelon form.

i) If the resulting matrix (3) has any of $y'_{r+1}, \ldots, y'_m$ nonzero, then there is no solution.

ii) If $y'_{r+1} = \cdots = y'_m = 0$, then the original system has solutions. They may all be found as follows:

Write out the system of equations corresponding to the row-echelon matrix. Solve for those x_i whose columns have been pivoted upon, in terms of the remaining x_j and y'_i. Then every solution to the original system is obtained by assigning values to the remaining x_j. (This specifies the value of the x_i whose columns are pivoted upon.) Each assignment of values to these x_j yields a solution.

Proof. Let us assume that the x_i whose columns have been pivoted upon are $x_1, \ldots, x_r$. [There is no loss of generality here; if, instead, the x_i are $x_{c_1}, \ldots, x_{c_r}$ (as depicted in (3)), simply go back and interchange the positions of these x_i and their coefficients in the original equations (1) with $x_1, \ldots, x_r$ respectively. The effect of this is to interchange columns in the coefficient matrix so that the columns pivoted upon are the first r columns, and we have the desired situation.] Thus, we may assume that the row-echelon matrix (3) is

$$\left(\begin{array}{ccccccc|c} 1 & 0 & \cdots & 0 & b_{11} & \cdots & b_{1,n-r} & y'_1 \\ 0 & 1 & & & b_{21} & & b_{2,n-r} & y'_2 \\ \vdots & & \ddots & & \vdots & & \vdots & \vdots \\ 0 & \cdots & 0 & 1 & b_{r1} & \cdots & b_{r,n-r} & y'_r \\ \hline 0 & & & & & & 0 & y'_{r+1} \\ \vdots & & & & \vdots & & \vdots & \vdots \\ 0 & & & & & & 0 & y'_m \end{array}\right) \quad (3)'$$

(The b_{ij} correspond to the boxes appearing in (3); we've simply collected them all together.) The system corresponding to this matrix is

$$
\begin{aligned}
x_1 + b_{11}x_{r+1} + \cdots + b_{1,n-r}x_n &= y_1', \\
x_2 + b_{21}x_{r+1} + \cdots + b_{2,n-r}x_n &= y_2', \\
&\vdots \\
x_r + b_{r1}x_{r+1} + \cdots + b_{r,n-r}x_n &= y_r', \\
0 &= y_{r+1}', \\
&\vdots \\
0 &= y_m'.
\end{aligned}
\tag{4}
$$

If any of $y_{r+1}', \ldots, y_m'$ is nonzero, then there can be no solution, since (4) explicitly requires that each of these be zero.

Suppose from now on that $y_{r+1}' = 0, \ldots, y_m' = 0$. We follow the algorithm and solve for $x_1, \ldots, x_r$:

$$
\begin{aligned}
x_1 &= y_1' - b_{11}x_{r+1} - \cdots - b_{1,n-r}x_n, \\
x_2 &= y_2' - b_{21}x_{r+1} - \cdots - b_{2,n-r}x_n, \\
&\vdots \\
x_r &= y_r' - b_{r1}x_{r+1} - \cdots - b_{r,n-r}x_n.
\end{aligned}
\tag{5}
$$

It is to be emphasized that this is just another way of writing the system (4), and that this system is equivalent to (1). Thus, any n-tuple $(x_1, \ldots, x_n)$ which satisfies (1) will also satisfy (5), and conversely. Consequently, if we choose any values whatsoever for $x_{r+1}, \ldots, x_n$ and plug them into (5), there result values $x_1, \ldots, x_r$ for which $(x_1, \ldots, x_n)$ is a solution to (1). Thus, each assignment of values according to the Solution Algorithm yields a solution. Conversely, if $(x_1, \ldots, x_n)$ is a solution to (1), then it also is a solution to (4) and hence satisfies (5). Thus, any solution may be obtained in this fashion. ∎

Example 4. The system of equations

$$
\begin{aligned}
x + 2y + z + w &= 1, \\
x + 2y + 2z + 5w &= 0, \\
2x + 4y + 3z + 6w &= 1
\end{aligned}
$$

has the augmented matrix

$$\begin{pmatrix} 1 & 2 & 1 & 1 & | & 1 \\ 1 & 2 & 2 & 5 & | & 0 \\ 2 & 4 & 3 & 6 & | & 1 \end{pmatrix} \xrightarrow[\text{(No row interchanges necessary)}]{\text{Basis Algorithm}} \begin{pmatrix} 1 & 2 & 0 & -3 & | & 2 \\ 0 & 0 & 1 & 4 & | & -1 \\ 0 & 0 & 0 & 0 & | & 0 \end{pmatrix}$$

The system corresponding to the latter is

$$x + 2y - 3w = 2,$$
$$z + 4w = -1,$$
$$0 = 0.$$

The variables x and z have had their columns pivoted upon. Solving for them in terms of the remaining variables (y and w), we have

$$x = 2 - 2y + 3w,$$
$$z = -1 \quad\quad - 4w.$$

Any y and w will therefore yield values of x and z which satisfy the system corresponding to the row-echelon matrix, and hence to the original system. In vector form, we could write any solution $(x, y, z, w)^*$ as

$$\begin{pmatrix} x \\ y \\ z \\ w \end{pmatrix} = \begin{pmatrix} 2 \\ \\ -1 \\ \end{pmatrix} \begin{pmatrix} -2y & +3w \\ y & \\ & -4w \\ & w \end{pmatrix} = \begin{pmatrix} 2 \\ 0 \\ -1 \\ 0 \end{pmatrix} + y \begin{pmatrix} -2 \\ 1 \\ 0 \\ 0 \end{pmatrix} + w \begin{pmatrix} 3 \\ 0 \\ -4 \\ 1 \end{pmatrix}.$$

Vector Form of Solutions

Assume that the system (1) has solutions. As we have observed in various examples, it is possible (and sometimes exceedingly convenient) to write the set of solutions in vector form. Let us suppose that we have obtained the solutions by means of the Solution Algorithm. Just as in the proof of this algorithm, let us assume that the first r columns of the augmented matrix are those pivoted upon (we can either reorder the x's or rename them, so that this is the case). The Solution Algorithm then leads us to the expression (5) for those x_i whose columns have been pivoted upon, in terms of the y_i' and the remaining x_j; solutions to (1) are then obtained by giving values to these remaining x_j.

We can write any solution as a vector $(x_1, \ldots, x_r, x_{r+1}, \ldots, x_n)^*$, where we are assuming that $x_1, \ldots, x_r$ are determined by specifying the values of

$x_{r+1}, \ldots, x_n$ in (5). Substituting the expressions for $x_1, \ldots, x_r$ from (5) we have

$$\begin{pmatrix} x_1 \\ \cdot \\ \cdot \\ \cdot \\ x_r \\ x_{r+1} \\ \cdot \\ \cdot \\ \cdot \\ x_n \end{pmatrix} = \begin{pmatrix} y'_1 - b_{11}x_{r+1} - \cdots - b_{1,n-r}x_n \\ \cdot \\ \cdot \\ \cdot \\ y'_r - b_{r1}x_{r+1} - \cdots - b_{r,n-r}x_n \\ x_{r+1} \\ \cdot \\ \cdot \\ \cdot \\ x_n \end{pmatrix}$$

$$= \begin{pmatrix} y'_1 \\ \cdot \\ \cdot \\ y'_r \\ 0 \\ \cdot \\ \cdot \\ 0 \end{pmatrix} + x_{r+1}\begin{pmatrix} -b_{11} \\ \cdot \\ \cdot \\ -b_{r1} \\ 1 \\ 0 \\ \cdot \\ \cdot \\ 0 \end{pmatrix} + \cdots + x_n \begin{pmatrix} -b_{1,n-r} \\ \cdot \\ \cdot \\ -b_{r,n-r} \\ 0 \\ \cdot \\ \cdot \\ 0 \\ 1 \end{pmatrix}$$

$$= \bar{X} + x_{r+1}Z_1 + \cdots + x_n Z_{n-r}. \tag{6}$$

($\bar{X}$ and the Z_j are defined to be the vectors immediately above.)

In general, if the pivot entries are not in the first r columns, but are interspersed among the columns of the coefficient matrix, the 0's in (6) will be interspersed among the other entries in the vectors, and the pivoted-upon x's $(x_1, \ldots, x_r)$ and the non-pivoted-upon $(x_{r+1}, \ldots, x_n)$ will not be lumped together in so tidy a fashion. Example 4 illustrates this. In any event, one concludes that *all solutions to the system can be written as a particular vector $\bar{X}$ plus a linear combination of specific vectors $Z_1, \ldots, Z_{n-r}$.* One can always transform the situation into (6) by renumbering the x's. Assuming that this has been done, we shall continue working with (6).

Since (6) is a rewriting of (5), every solution to (1) can be written in the vector form (6) (as always, perhaps after renumbering the x's), and each vector (6) is a solution to (1). We now ascertain the specific role of $\bar{X}$ and the Z_i. If we take

$$x_{r+1} = 0, \ldots, x_n = 0$$

in (6), we see that *the vector $\bar{X}$ is a particular solution to* (1); this could be seen directly in (4). We next investigate the meaning of the Z_i.

Homogeneous Systems

We say that the system of linear equations (1) is **homogeneous** if each $y_i = 0$:

$$a_{11}x_1 + \cdots + a_{1n}x_n = 0$$
$$\vdots$$
$$a_{m1}x_1 + \cdots + a_{mn}x_n = 0.$$

A homogeneous system always has at least one solution, namely

$$(x_1, \ldots, x_n) = (0, \ldots, 0).$$

We shall see later that the solutions to a nonhomogeneous system are directly related to its homogeneous counterpart.

We have previously observed that if (1) is homogeneous ($\mathbf{Y} = \mathbf{0}$), then the transformed system (3) is homogeneous ($\mathbf{Y}' = \mathbf{0}$), and conversely. Thus, the vector $\bar{\mathbf{X}}$ in (6) will be $\bar{\mathbf{X}} = \mathbf{0}$ iff the original system is homogeneous. By making appropriate choices for $x_{r+1}, \ldots, x_n$, we see that the $\mathbf{Z}_i$ in (6) are solutions to the homogeneous counterpart to (4), and hence to the homogeneous counterpart to (1). We can say even more:

1.6 The set of all solutions to a homogeneous system of linear equations in n unknowns forms a subspace of $\mathbb{R}_n$.

Proof. Starting out with a homogeneous system (1), we see from the Solution Algorithm that all solutions appear in (6) (with $\bar{\mathbf{X}} = \mathbf{0}$),

$$\mathbf{X} = x_{r+1}\mathbf{Z}_1 + \cdots + x_n\mathbf{Z}_{n-r}, \tag{7}$$

and all such creatures are solutions. Thus, the solution set is span $\{\mathbf{Z}_1, \ldots, \mathbf{Z}_{n-r}\}$, which we know to be a subspace. ∎

This particular result is easy to prove directly; we ask you to do this in Exercise 21. The same is true of the following:

1.7 Any solution to a system of linear equations (1) is of the form $\bar{\mathbf{X}} + \mathbf{Z}$, where $\bar{\mathbf{X}}$ is some particular solution and $\mathbf{Z}$ is a solution to the corresponding homogeneous system. Moreover, anything of this form is a solution.

Proof. Just let $\mathbf{Z} = x_{r+1}\mathbf{Z}_1 + \cdots + x_n\mathbf{Z}_{n-r}$ in (6). ∎

The next is harder to prove directly:

1.8 The subspace of $\mathbb{R}_n$ of solutions to a homogeneous system in n unknowns is of dimension $(n - r)$, where r is the row rank of the coefficient matrix of the system.

Proof. Since elementary row operations are used, the row rank of the coefficient matrix is the same as that of the row-echelon matrix into which it is transformed.

We have been calling this r (= the number of its nonzero rows), and this furnishes us with $(n - r)$ Z_i's in (6) (perhaps after reordering the x's). We know that the Z_i's span the solution set (1.6). Since each Z_i contains a 1 where the others all have 0's, we see that the Z_i's are linearly independent. Hence, they form a basis and the dimension is $(n - r)$. ∎

The Rank and Solutions

It is sometimes naively felt that the existence and number of solutions to a system of linear equations is determined by the number of equations vs. the number of unknowns. A glance at our first simple examples will show that this is not the case. There is another crucial ingredient: the rank of the matrices involved. We now see how these things all relate. We shall return to many of these notions for a more sophisticated interpretation in Section 2. For now our main technique will be to stare at the transformed system (4) and the three numbers

$m = $ number of equations, $n = $ number of unknowns

$r = $ row rank of the coefficient matrix $\mathbf{A}$.

In terms of (3), the necessary and sufficient condition that solutions exist to the original system (1) is that

$$y'_{r+1} = 0, \quad y'_{r+2} = 0, \quad \ldots, \quad y'_m = 0.$$

We can transform this solution criterion on the y'_i to one about the original system itself:

1.9 Rank and solutions. The system of linear equations

$$a_{11}x_1 + \cdots + a_{1n}x_n = y_1$$
$$\vdots \tag{1}$$
$$a_{m1}x_1 + \cdots + a_{mn}x_n = y_m$$

has a solution iff the row rank of the augmented matrix

$$(\mathbf{A} \mid \mathbf{Y}) = \begin{pmatrix} a_{11} & \cdots & a_{1n} & y_1 \\ \vdots & & & \\ a_{m1} & \cdots & a_{mn} & y_m \end{pmatrix}$$

is the same as the row rank as the coefficient matrix

$$\mathbf{A} = \begin{pmatrix} a_{11} & \cdots & a_{1n} \\ \vdots & & \\ a_{m1} & \cdots & a_{mn} \end{pmatrix}.$$

Proof. We transform the augmented matrix $(\mathbf{A} \mid \mathbf{Y})$ into the matrix (3), which is of the form

$$\left(\begin{array}{c|c} & y'_1 \\ & \cdot \\ \mathbf{C} & \cdot \\ & \cdot \\ & y'_r \\ \hline & y'_{r+1} \\ & \cdot \\ \mathbf{0} & \cdot \\ & \cdot \\ & y'_n \end{array} \right),$$

where the rows of $\mathbf{C}$ are linearly independent, by means of elementary row operations. Hence, the two matrices have the same row rank. This will also be the row rank of $\mathbf{C}$ ($= r$) unless some one of $y'_{r+1}, \ldots, y'_n$ is nonzero; this will be the case iff (1) has no solution, by the above remark. But

$$\begin{pmatrix} \mathbf{C} \\ \mathbf{0} \end{pmatrix}$$

is obtained from $\mathbf{A}$ by elementary row operations, so they have the same row rank. Hence, $\mathbf{A}$ and $(\mathbf{A} \mid \mathbf{Y})$ will have the same rank unless (1) has no solutions. ∎

If $r = m$, it is clear from (4) that solutions will always exist no matter what the values of $y'_1, \ldots, y'_m$ are (hence, no matter what $y_1, \ldots, y_m$ are). Conversely, if we know that the system has solutions for all values of $y_1, \ldots, y_m$, then $r = m$. For, if $r > m$, we can find a vector

$$\mathbf{Y} = (y_1, \ldots, y_m)$$

for which the system has no solution by starting with a vector

$$\mathbf{Y}' = (y'_1, \ldots, y'_m)$$

that has, say $y'_{r+1} \neq 0$, and then finding the corresponding vector $\mathbf{Y}$ that has the desired property. In short,

1.10 The system of equations

$$a_{11}x_1 + \cdots + a_{1n}x_n = y_1$$
$$\cdot$$
$$\cdot \qquad\qquad (1)$$
$$\cdot$$
$$a_{m1}x_1 + \cdots + a_{mn}x_n = y_m$$

has a solution for any values of $y_1, \ldots, y_m$ iff the rank r of the coefficient matrix A is equal to the number of equations m.

A casual look at (4) will also reveal (Exercise 22) the following:

1.11 If the system of equations (1) has any solution, this solution is unique iff the rank r of the coefficient matrix $\mathbf{A}$ is equal to the number of unknowns n.

Pasting the last two items together, we see that the system has a unique solution for all values of $y_1, \ldots, y_n$ iff the number of unknowns, equations, and the rank of the coefficient matrix are all the same: $m = n = r$.

EXERCISES

In problems 1–8 find the complete solution set. Write your solutions in the vector form (6) so as to display a particular solution and a basis for the space of solutions to the homogeneous system.

1. $\begin{aligned} x + 2z &= 1 \\ y - 3z &= 2 \end{aligned}$

2. $\begin{aligned} x + y + z &= 0 \\ x + 2y + 3z &= 0 \end{aligned}$

3. $\begin{aligned} x \phantom{{}+y} - 2z &= 1 \\ 2x + y + 3z &= -1 \\ 4x + y - z &= 1 \end{aligned}$

4. $\begin{aligned} x \phantom{{}+y} - 2z &= 1 \\ 2x + y + 3z &= 2 \\ 4x + y - z &= 1 \end{aligned}$

5. $\begin{aligned} x_1 \phantom{{}+x_2} - 2x_3 &= 1 \\ 2x_1 + x_2 + 3x_3 &= -1 \\ 4x_1 + x_2 - x_3 &= 1 \\ -x_1 + 2x_2 + x_3 &= 1 \end{aligned}$

6. $\begin{aligned} x_1 + 2x_2 + 4x_3 - x_4 &= 1 \\ x_2 + x_3 + 2x_4 &= 1 \\ -2x_1 + 3x_2 - x_3 + x_4 &= 1 \end{aligned}$

7. $\begin{aligned} x_2 - 2x_3 + 3x_4 &= 0 \\ x_1 - 3x_2 + 2x_3 + x_4 &= 4 \\ -2x_1 + 4x_2 + x_3 + 2x_4 &= -2 \\ 3x_1 - 5x_2 + 4x_3 + 2x_4 &= 1 \end{aligned}$

8. $\begin{aligned} x_1 + 2x_2 + 4x_3 - x_4 + x_5 &= 0 \\ x_2 + x_3 + 2x_4 + x_5 &= 0 \\ -2x_1 + 3x_2 - x_3 + x_4 + x_5 &= 0 \\ x_1 - x_2 + x_3 + x_4 + x_5 &= 0. \end{aligned}$

9. Find a basis for the intersection of the spaces $\mathfrak{X}$ and $\mathfrak{Y}$ in problem 6 of Section 1 in Chapter 2 by using the methods of the present Section.

10. Find the complete solution of the system
$$\begin{aligned} x_1 + 3x_2 - 2x_3 + 5x_4 &= y_1, \\ 2x_1 + 4x_2 + 3x_3 - 3x_4 &= y_2, \\ -x_1 + x_2 - 12x_3 + 8x_4 &= y_3, \end{aligned}$$
for any y_1, y_2, y_3.

In problems 11, 12, and 13 find the conditions on y_1, y_2, y_3, y_4 required so that solutions exist, and find the solutions.

11. $\begin{aligned} x_1 + 2x_2 - x_3 &= y_1 \\ 3x_1 + 4x_2 + x_3 &= y_2 \\ -2x_1 + 3x_2 - 12x_3 &= y_3 \\ 5x_1 - 3x_2 + 8x_3 &= y_4 \end{aligned}$

12. $\begin{aligned} x_1 + 2x_2 + 4x_3 &= y_1 \\ x_2 + x_3 &= y_2 \\ -2x_1 + 3x_2 - x_3 &= y_3 \\ x_1 - x_2 + x_3 &= y_4 \end{aligned}$

13. $\begin{aligned} x_2 - 2x_3 + 3x_4 &= y_1 \\ x_1 - 3x_2 + 2x_3 + x_4 &= y_2 \\ x_1 - x_2 - 2x_3 + 7x_4 &= y_3 \\ x_1 \phantom{{}-x_2} - 4x_3 + 10x_4 &= y_4 \end{aligned}$

14. Let $a_1, a_2, \ldots, a_n$ be not all zero. Find a basis for the solution space of the equation
$$a_1 x_1 + a_2 x_2 + \cdots + a_n x_n = 0.$$

15. Let (a_1, a_2, a_3), (b_1, b_2, b_3) be linearly independent. Prove that the solution set of
$$a_1 x_1 + a_2 x_2 + a_3 x_3 = 0,$$
$$b_1 x_1 + b_2 x_2 + b_3 x_3 = 0,$$
is given by
$$x_1 = (a_2 b_3 - a_3 b_2)t,$$
$$x_2 = (a_3 b_1 - a_1 b_3)t,$$
$$x_3 = (a_1 b_2 - a_2 b_1)t,$$
where t is arbitrary.

16. Prove (a) 1.1, (b) 1.2, (c) the rest of 1.3.
17. Show that equivalence of systems of linear equations in n unknowns is an equivalence relation (see Chapter 0).
18. Prove 1.4.
19. Prove 1.7.
20. Given a system (1) of linear equations, with (a) more unknowns than equations, either prove the following or give a counterexample.
    i) There is always a solution for every $\mathbf{Y} = (y_1, \ldots, y_m)$.
    ii) There is never a solution for some $\mathbf{Y}$.
    iii) There is a unique solution for all $\mathbf{Y}$.
    iv) There is never a unique solution for some $\mathbf{Y}$.
    v) If a solution exists, then it is unique.
    b) Same as (a), with fewer unknowns than equations.
    c) Same as (a) with an equal number of unknowns and equations.
21. a) Prove 1.6 directly from (1).
    b) Prove 1.7 directly from (1).
22. Prove 1.11.

2. LINEAR EQUATIONS, MATRICES, AND LINEAR TRANSFORMATIONS

A system of linear equations
$$a_{11} x_1 + \cdots + a_{1n} x_n = y_1$$
$$\vdots \qquad\qquad\qquad \vdots \tag{1}$$
$$a_{m1} x_1 + \cdots + a_{mn} x_n = y_m$$
can be exhibited more compactly in matrix form. Let
$$\mathbf{A} = \begin{pmatrix} a_{11} & \cdots & a_{1n} \\ \vdots & & \vdots \\ a_{m1} & \cdots & a_{mn} \end{pmatrix}, \quad \mathbf{X} = \begin{pmatrix} x_1 \\ \vdots \\ x_n \end{pmatrix}, \quad \mathbf{Y} = \begin{pmatrix} y_1 \\ \vdots \\ y_m \end{pmatrix}.$$

Then it is a mere matter of matrix multiplication and vector equality to see that (1) is the same as $AX = Y$. Thus, the linear equations problem in matrix terms is

given an $m \times n$ matrix A and a vector Y in $\mathbb{R}_m$,
find all X in $\mathbb{R}_n$ for which $AX = Y$.

We now recall the solution procedure evolved in Section 1. We take the augmented matrix $(A \mid Y)$ and use elementary row operations to put the A part in row-echelon form. This gives a matrix

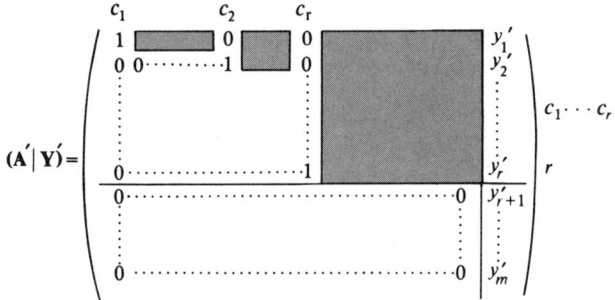

and the solutions of $A'X = Y'$ are exactly those of $AX = Y$. Moreover, there are solutions iff all of $y'_{r+1}, \ldots, y'_m$ are zero.

In matrix terms, our solution procedure transforms the matrix A and vector Y simultaneously into a matrix A' and vector Y' for which $AX = Y$ iff $A'X = Y'$, and for which the solutions (if any) to the latter are easily found. At the end of this section we shall demonstrate how all solutions (or lack thereof) can be immediately read off the matrix $(A' \mid Y')$.

We next turn our attention to the very important problem of inverting matrices, and show that our technique for solving systems of linear equations provides a way. If A is an $n \times n$ matrix with inverse A^{-1}, suppose that the columns of A^{-1} are $X_1, \ldots, X_n$, so that

$$A^{-1} = (X_1 \mid \cdots \mid X_n).$$

Then $AA^{-1} = I$ says that

$$A(X_1 \mid \cdots \mid X_n) = (AX_1 \mid \cdots \mid AX_n) = (e_1 \mid \cdots \mid e_n).$$

Thus,

$$AX_1 = e_1, \ldots, AX_n = e_n.$$

Conversely, if these equations have solutions $X_1, \ldots, X_n$, then it is immediate that the matrix $(X_1 \mid \cdots \mid X_n)$ is actually A^{-1}. Thus to invert A we must solve $AX_i = e_i$, $i = 1, \ldots, n$; and these equations have solutions iff A has an inverse. To solve such an equation, we would form the matrix $(A \mid e_i)$ and proceed to do row operations on the A part. Note, however, that we can do these equations *en masse*: just form

$$(A \mid e_1 \mid e_2 \mid \cdots \mid e_n) = (A \mid I)$$

and transform the **A** part (since only row operations are involved, the integrity of each column is not jeopardized). In summary,

2.1 Matrix Inversion Alogrithm. Given an $n \times n$ matrix **A**, form $(\mathbf{A} \mid \mathbf{I})$ and use row operations to put the **A** part into row-echelon form. Suppose this results in a matrix $(\mathbf{A}' \mid \mathbf{B})$. If **A**$'$ has a zero row, then **A** is not invertible. If **A**$'$ has no zero rows, then $\mathbf{B} = \mathbf{A}^{-1}$.

Proof. You supply the details. ∎

Note that the row-echelon form **A**$'$ of an invertible matrix **A** is $\mathbf{A}' = \mathbf{I}$, the identity matrix.

Example 1. Let

$$A = \begin{pmatrix} 1 & 3 & -2 \\ 2 & 4 & 3 \\ -1 & 1 & -11 \end{pmatrix}.$$

We operate on

$$(\mathbf{A} \mid \mathbf{I}) = \begin{pmatrix} 1 & 3 & -2 & | & 1 & 0 & 0 \\ 2 & 4 & 3 & | & 0 & 1 & 0 \\ -1 & 1 & -11 & | & 0 & 0 & 1 \end{pmatrix}$$

by pivoting as shown:

$$\begin{pmatrix} ① & 3 & -2 & | & 1 & 0 & 0 \\ 2 & 4 & 3 & | & 0 & 1 & 0 \\ -1 & 1 & -11 & | & 0 & 0 & 1 \end{pmatrix} \to \begin{pmatrix} 1 & 3 & -2 & | & 1 & 0 & 0 \\ 0 & ⊖2 & 7 & | & -2 & 1 & 0 \\ 0 & 4 & -13 & | & 1 & 0 & 1 \end{pmatrix} \to$$

$$\begin{pmatrix} 1 & 0 & \frac{17}{2} & | & -2 & \frac{3}{2} & 0 \\ 0 & 1 & -\frac{7}{2} & | & 1 & -\frac{1}{2} & 0 \\ 0 & 0 & 1 & | & -3 & 2 & 1 \end{pmatrix} \to \begin{pmatrix} 1 & 0 & 0 & | & \frac{47}{2} & -\frac{31}{2} & -\frac{17}{2} \\ 0 & 1 & 0 & | & -\frac{19}{2} & \frac{13}{2} & \frac{7}{2} \\ 0 & 0 & 1 & | & -3 & 2 & 1 \end{pmatrix}.$$

Hence **A** is invertible and

$$A^{-1} = \begin{pmatrix} \frac{47}{2} & -\frac{31}{2} & -\frac{17}{2} \\ -\frac{19}{2} & \frac{13}{2} & \frac{7}{2} \\ -3 & 2 & 1 \end{pmatrix}.$$

Example 2.

$$A = \begin{pmatrix} 1 & 3 & -2 \\ 2 & 4 & 3 \\ -1 & 1 & -12 \end{pmatrix}, \quad (\mathbf{A} \mid \mathbf{I}) = \begin{pmatrix} ① & 3 & -2 & | & 1 & 0 & 0 \\ 2 & 4 & 3 & | & 0 & 1 & 0 \\ -1 & 1 & -12 & | & 0 & 0 & 1 \end{pmatrix} \to$$

$$\begin{pmatrix} 1 & 3 & -2 & | & 1 & 0 & 0 \\ 0 & ⊖2 & 7 & | & -2 & 1 & 0 \\ 0 & 4 & -14 & | & 1 & 0 & 1 \end{pmatrix} \to \begin{pmatrix} 1 & 0 & \frac{17}{2} & | & -2 & \frac{3}{2} & 0 \\ 0 & 1 & -\frac{7}{2} & | & 1 & -\frac{1}{2} & 0 \\ 0 & 0 & 0 & | & -3 & 2 & 1 \end{pmatrix},$$

so that **A** is *not* invertible (in fact, $r = 2$). Compare with the examples in Section 1.

If **A** is an $n \times n$ invertible matrix, then the system of linear equations $\mathbf{AX} = \mathbf{Y}$ has one very obvious solution, $\mathbf{X} = \mathbf{A}^{-1}\mathbf{Y}$. Moreover, it is apparent that this will be the only solution, and that there is a solution for any **Y**. This is in keeping with our observation following 1.11, that if the number of equations, unknowns, and rank are all the same, then there is a unique solution for any **Y**.

We now recall that matrix multiplication of a vector is a linear transformation. Thus, we can reformulate the linear equations problem from the matrix terms:

given **A** and **Y**, find all **X** for which **AX** = **Y**,

to the linear transformation problem:
given **T** from $\mathbb{R}_n$ into $\mathbb{R}_m$, and given **Y** in $\mathbb{R}_m$, find all **X** in $\mathbb{R}_n$ for which **T(X)** = **Y**.

The solution procedure of Section 1 which transforms **AX** = **Y** into the equivalent system **A'X** = **Y'** clearly transforms the linear transformation problem of finding **X** for which **T(X)** = **Y** to finding all **X** for which **T'(X)** = **Y'** (where, of course, **T'(X)** = **A'X**). We thus have a different linear transformation **T'** but the same solutions **X**.

In particular, we note that homogeneous systems translate into the problem:

find all **X** for which **T(X)** = **0**; that is, find the kernel of **T**.

We have considerable machinery from Chapter 3, Section 4, which we can bring to bear on the problem. For example, the next result makes use of the Dimension Theorem from that section.

2.2 *Row-rank = column rank.* For any matrix **A**, its row rank is the same as its column rank.

Proof. Given an $m \times n$ matrix **A**, define a linear transformation **T** from $\mathbb{R}_n$ into $\mathbb{R}_m$ by **T(V)** = **AV**. Then the image of **T** is spanned by $T(e_1), \ldots, T(e_n)$, the columns of **A**. Hence, the dimension of the image space of **T** is the same as the column rank of **A**. By 1.8, if r is the row rank of **A**, then the dimension of the kernel of **T** is $n - r$. Hence,

column rank **A** = dimension of image of **T**

$$\stackrel{\text{Dim. Th.}}{=} n - (\text{dimension kernel of } T)$$
$$= n - (n - r) = r$$
$$= \text{row rank of } \mathbf{A}. \quad \blacksquare$$

We now apply the material from Chapter 3, Section 4, to look again at some of the elementary facts at the end of Section 1. For example, we just noticed that the set of solutions to a homogeneous system translates into the kernel of a linear transformation, and we knew that the kernel of a linear transformation is a subspace; the fact that the set of solutions to a homogeneous system forms a subspace was expressed in 1.6. Moreover, the property that

$T(V_1) = T(V_2)$ iff $V_1 = V_2 + K$,   **K** in the kernel of T

(Chapter 3, 4.2) translates into the matrix form

$$AX_1 = AX_2 \quad \text{iff} \quad X_1 = X_2 + Z, \text{ where } AZ = 0.$$

But this is the same as our 1.7:

Any solution to $AX = Y$ is of the form $X = \bar{X} + Z$ where $\bar{X}$ is a particular solution and Z is a solution to the homogeneous system.

The linear transformation problem of finding X for which $T(X) = Y$, Y given, will have a solution iff Y is in the image of T.

Example 3. Consider

$$2x - 6y = 2,$$
$$-x + 3y = -1,$$

which in matrix terms is

$$\begin{pmatrix} 2 & -6 \\ -1 & 3 \end{pmatrix} \begin{pmatrix} x \\ y \end{pmatrix} = \begin{pmatrix} 2 \\ -1 \end{pmatrix}.$$

The linear transformation $T(V) = AV$, A the matrix of the system, then has the indicated kernel and image:

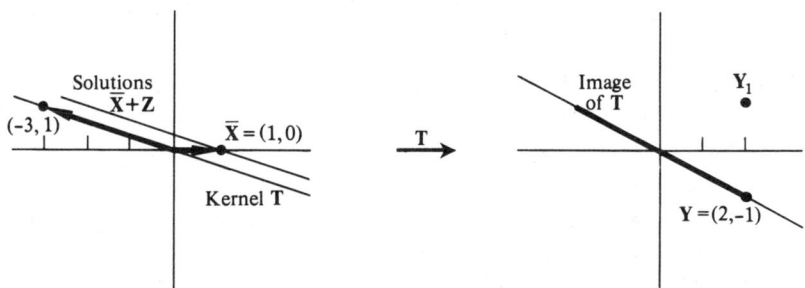

Fig. 1

Since we have taken the vector $Y = (2, -1)^*$, which is in the image of T, we know solutions exist. One solution to $AX = Y$ will be $\bar{X} = (1, 0)^*$. Any other solution will be of the form $\bar{X} + Z$, where Z is in the kernel.

If instead, we took $Y_1 = (2, 1)^*$, which is *not* in the image of T, then of course the system

$$2x - 6y = 2,$$
$$-x + 3y = +1,$$

can have no solution.

In matrix terms, to say that $T(\bar{X}) = Y$ has a solution means that Y is a linear combination of the columns of A (where $T(X) = AX$); that is, the span of the columns of A must contain Y. The dimension of the column spaces of A and $(A \mid Y)$ are the same; that is, the column rank of A and $(A \mid Y)$ are the same. By

2.2, this will be true iff the row ranks of $\mathbf{A}$ and $(\mathbf{A} \mid \mathbf{Y})$ are the same. This, of course, is 1.9.

If $\mathbf{A}$ is an $m \times n$ matrix then the system of equations $\mathbf{AX} = \mathbf{Y}$ will have a solution for all $\mathbf{Y}$ in $\mathbb{R}_m$ iff the linear transformation $\mathbf{T(X)} = \mathbf{AX}$ is onto. By Section 4 of Chapter 3, this will be the case iff the rank of $\mathbf{A}$ is m. This is 1.10.

Finally, the system of equations $\mathbf{AX} = \mathbf{Y}$ will have a unique solution (provided it has any solutions) iff the linear transformation $\mathbf{T(X)} = \mathbf{AX}$ is one-to-one. By Section 4 of Chapter 3, we know $\mathbf{T}$ will be one-to-one iff the dimension of its kernel is 0, or (by the Dimension Theorem) iff the dimension of the image of $\mathbf{T}$ is the same as the dimension of the domain. If $\mathbf{A}$ is $m \times n$, this is the same as saying that the (column) rank of $\mathbf{A}$ is n. This is 1.11.

The remainder of this section deals with more specialized topics which you may wish to omit at first reading. We discuss the following:

1. A means of reading off a basis for the column space of a matrix ($=$ image space of the corresponding linear transformation) directly from the row-echelon form (this is 2.3).

2. A means of reading off the solutions to a system of linear equations directly from the row-echelon matrix (this is 2.4), together with a timely Computer Warning.

3. Another necessary and sufficient condition that a system of linear equations have a solution (this is 2.5).

It is often desirable to calculate a basis for the column space of a matrix $\mathbf{A}$ (for example, only those $\mathbf{Y}$ which are linear combinations of such a basis can be solutions to $\mathbf{AX} = \mathbf{Y}$). We have one technique for finding such a basis: apply the Basis Algorithm to $\mathbf{A}^*$. Actually, a basis for the column space of A is a natural by-product of the Solution Algorithm (in which one also calculates a basis for the row space of $\mathbf{A}$). We shall use the fact that row rank $=$ column rank to prove this.

2.3 A Basis for the column space. Suppose that $\mathbf{A}$ is transformed into row-echelon form $\mathbf{A}'$ by pivoting on columns $c_1, \ldots, c_r$ (along with appropriate row interchanges):

$$\mathbf{A}' = \begin{pmatrix} c_1 & c_2 & \cdots & c_r & \\ 1 & 0 & & 0 & \\ & 1 & & \cdot & \\ & & & \cdot & \\ & & & \cdot & \\ & & & 1 & \\ \hline 0 & & & & 0 \\ \cdot & & & & \cdot \\ \cdot & & & & \cdot \\ \cdot & & & & \cdot \\ 0 & & & & 0 \end{pmatrix}$$

Then columns $c_1, \ldots, c_r$ *of* $\mathbf{A}$ are a basis for its column space. [*Warning:* The columns $c_1, \ldots, c_r$ *of* $\mathbf{A}'$ are most unlikely to form a basis for the column space of $\mathbf{A}$.]

Proof. Let $\bar{\mathbf{A}}$ be the matrix whose columns are columns $c_1, \ldots, c_r$ of $\mathbf{A}$. Placing $\bar{\mathbf{A}}$ in row-echelon form yields the matrix

$$\begin{pmatrix} \overset{1}{1} & \overset{2}{0} & \overset{3}{0} & \cdots & \overset{r}{0} \\ 0 & 1 & 0 & & \cdot \\ 0 & 0 & 1 & & \cdot \\ \cdot & & & & \cdot \\ \cdot & & & & 1 \\ \hline 0 & \cdots & & \cdots & 0 \\ \cdot & & & & \cdot \\ \cdot & & & & \cdot \\ 0 & \cdots & & \cdots & 0 \end{pmatrix}$$

(The row operations do the same job whether these columns are considered to belong to $\mathbf{A}$ or $\bar{\mathbf{A}}$.) By the Basis Algorithm, the row rank of $\bar{\mathbf{A}}$ is r. Hence, so is its column rank. Thus its columns are a basis for the column space of $\mathbf{A}$. ∎

We now show how to read off any and all solutions to $\mathbf{AX} = \mathbf{Y}$ from the transformed matrix $(\mathbf{A}' \mid \mathbf{Y}')$. Although this does save a step, if confusion results you should not hesitate to return to the method of Section 1 (which requires only that $(\mathbf{A}' \mid \mathbf{Y}')$ be translated back into a system of linear equations).

We know that a particular solution $\bar{\mathbf{X}}$ is obtained by spreading out the first r entries of

$$\mathbf{Y}' = (y_1', \ldots, y_r', 0, \ldots, 0)^*$$

into the entries $c_1, \ldots, c_r$ of a vector in $\mathbb{R}_n$. We show that all other solutions are obtained by similarly spreading out columns of $\mathbf{A}'$ not pivoted upon, subtracting a unit vector e_i in $\mathbb{R}_n$, and adding linear combinations of such creatures to the solution $\bar{\mathbf{X}}$. We give a precise recipe:

2.4 Given the system of linear equations $\mathbf{AX} = \mathbf{Y}$, with

$$\mathbf{A} = \begin{pmatrix} a_{11} & \cdots & a_{1n} \\ \cdot & & \cdot \\ \cdot & & \cdot \\ a_{m1} & \cdots & a_{mn} \end{pmatrix}, \quad \mathbf{X} = \begin{pmatrix} x_1 \\ \cdot \\ \cdot \\ x_n \end{pmatrix}, \quad \mathbf{Y} = \begin{pmatrix} y_1 \\ \cdot \\ \cdot \\ y_m \end{pmatrix},$$

to find all solutions (if any), take the augmented matrix $(A \mid Y)$ and put the A part into row-echelon form:

$$(A' \mid Y') = \begin{pmatrix} \overset{c_1}{1} & \cdots & \overset{c_2}{0} & \cdots & \overset{c_r}{0} & \cdots & y'_1 \\ 0 & & 1 & & 0 & & y'_2 \\ \cdot & & 0 & & \cdot & & \cdot \\ \cdot & & \cdot & & \cdot & & \cdot \\ 0 & & 0 & & 1 & & y'_r \\ 0 & \cdots & & & & 0 & y'_{r+1} \\ \cdot & & & & & & \cdot \\ \cdot & & & & & & \cdot \\ 0 & & & & & 0 & y'_m \end{pmatrix} \begin{smallmatrix} c_1,\ldots,c_r \text{ the columns} \\ \text{pivoted upon.} \end{smallmatrix}$$

Then solutions exist iff $y'_{r+1}, \ldots, y'_m = 0$. If there are solutions,

i) One solution is

$$\bar{X} = (\overset{c_1}{y'_1}, 0, \ldots, \overset{c_2}{y'_2}, 0, \ldots, \overset{c_r}{y'_r}, 0, \ldots, 0)^*.$$

ii) All other solutions are of the form $\bar{X} + Z$, where Z is a linear combination of certain Z_i, where each Z_i is obtained from a column of A' which has not been pivoted upon. In fact, if i is not one of $c_1, \ldots, c_r$ and

$$A' = \begin{pmatrix} 1 & 0 & a'_{1i} & & 0 \\ 0 & 1 & a'_{2i} & & \\ & 0 & & & \\ \cdot & \cdot & \cdot & & \cdot \\ \cdot & \cdot & \cdot & & \cdot \\ \cdot & \cdot & \cdot & & \cdot \\ 0 & 0 & a'_{ri} & & 1 \\ c_1 & c_2 & i & & c_r \end{pmatrix},$$

then

$$Z_i = (a'_{1i}, 0, \ldots, a'_{2i}, 0, \ldots, -1, 0, \ldots, a'_{ri}, 0, \ldots)^*.$$

(These Z_i are the negatives of those of Section 1.)

Proof. We saw in Section 1 that there are solutions iff $y'_{r+1}, \ldots, y'_m$ are zero, so suppose these are zero. By simply performing the matrix multiplications (Exercise 21), one can verify that $A'\bar{X} = Y'$. Thus, as we know, $A\bar{X} = Y$. In like manner, check that $A'Z_i = 0$, so that $AZ_i = 0$.

Now the Z_i are clearly linearly independent (why?) and there are $(n - r)$ of them (why?). By 1.8 the space of vectors satisfying $AZ = 0$ has dimension $(n - r)$.

Hence, our current Z_i form a basis. By 1.7 any solution to $AX = Y$ is of the form $X = \bar{X} + Z$, where Z is a linear combination of the Z_i. ∎

This method does have the advantage over the rather informal Solution Algorithm of Section 1, that it can be easily told to a computer (Exercise 21).

While a computer is a natural creature to enlist in solving systems of linear equations, we must at least tell one horror story for scientists.

Let us suppose that we wish to solve a system of linear equations (or to invert a matrix, perhaps). It turns out that there are modest-sized systems (say, of 5 or 6 equations in as many unknowns) which are simply beyond the capabilities of a computer working with 10-decimal-place arithmetic. The computer will give a "solution" but it may be so bad that even the signs of the entries are incorrect. Unfortunately, if the computer is asked to check its "solutions" by means of the appropriate matrix multiplication, the errors may well accumulate in reverse so that the "solution" seems to check! Even more unfortunately, it is quite possible to encounter such systems in the real world.

If you must use a computer, we suggest that, as a matter of policy, you see if there are other aspects of the problem which will give some indication as to the accuracy of a computed solution.

If it seems likely that you will encounter such problems in your work, hie yourself off to a numerical analysis course. The subject is too lengthy and sophisticated for us to treat further, but there are a number of topics we shall study in this book which are very important in numerical analysis. A good reference is G. E. Forsythe and C. B. Moler, *Computer Solutions of Linear Algebraic Systems* (Prentice-Hall, 1967).

Our Gaussian elimination approach of Section 1 has many computer virtues aside from its simplicity. It is the basis for several computer methods and is convenient for hand calculations as well.

As our final topic in this section, we develop a more subtle condition that a vector Y be such that the equation $AX = Y$ have solutions. Our study will be based on events happening to the left of the matrix A, for a change.

In particular, we define the **left kernel** of A to be the set of Z for which $ZA = 0$. If A is $m \times n$, then the dimension of this subspace of $\mathbb{R}_m$ is $m - r$, where r is still the rank of A (you supply the argument). A basis for the left kernel can, of course, be found by pivoting on the *columns*.

Example 4.

$$A = \begin{pmatrix} \boxed{1} & 3 & -2 & 5 \\ 2 & 4 & 3 & -3 \\ -1 & 1 & -12 & 21 \end{pmatrix} \xrightarrow{\text{Column pivoting}} \begin{pmatrix} 1 & 0 & 0 & 0 \\ 2 & \boxed{-2} & 7 & -13 \\ -1 & 4 & -14 & 26 \end{pmatrix}$$

$$\xrightarrow{\text{Column pivoting}} \begin{pmatrix} 1 & 0 & 0 & 0 \\ 0 & 1 & 0 & 0 \\ 3 & -2 & 0 & 0 \end{pmatrix}.$$

Hence a basis of the left kernel is $(-3, 2, 1)$.

Pivoting on the *columns* of a matrix gives us, of course, a basis for the column space. Since this column space is the image space of the transformation $T(X) = AX$, we can thus find a basis for the image space by using the results in Chapter 2. Thus in our example we find that

$$\begin{pmatrix} 1 \\ 0 \\ 3 \end{pmatrix} \quad \text{and} \quad \begin{pmatrix} 0 \\ 1 \\ -2 \end{pmatrix}$$

form a basis, and so do the vectors

$$\begin{pmatrix} 1 \\ 2 \\ -1 \end{pmatrix} \quad \text{and} \quad \begin{pmatrix} 3 \\ 4 \\ 1 \end{pmatrix},$$

which are columns of **A** itself.

We continue the study of the image space of the transformation $T(X) = AX$; we have just seen a way one can find a basis for it (and we also have 2.3). This space, whose dimension is the rank r of **A**, is exactly the set of vectors **Y** for which the system of equations $AX = Y$ has solutions. We therefore have obtained, in 1.9, a necessary and sufficient condition that **Y** belongs to the image space. Another form of this condition is given in the following theorem:

2.5 A vector **Y** in $\mathbb{R}_m$ is in the image space of **A** (i.e. $AX = Y$ has a solution) iff $ZY = 0$ for any **Z** in the left kernel of **A**.

Proof. If $Y = AX$ and $ZA = 0$, then

$$ZY = Z(AX) = (ZA)X = 0,$$

so the condition is certainly necessary.

To prove the converse, let $Z_1, Z_2, \ldots, Z_{m-r}$ be a basis of the left kernel of **A**. Then $ZY = 0$ for any **Z** in the left kernel iff

$$Z_1 Y = 0, \quad Z_2 Y = 0, \ldots, Z_{m-r} Y = 0.$$

We must show that any vector **Y** satisfying these equations is in the image space. To do this let $\mathcal{Y}$ be the set of vectors **Y** satisfying these equations: they clearly form a subspace of $\mathbb{R}_m$ and, as we have just seen, the image space is a subspace of $\mathcal{Y}$. In fact, $\mathcal{Y}$ is the kernel of the linear transformation

$$T(V) = \begin{pmatrix} Z_1 \\ \cdot \\ \cdot \\ \cdot \\ Z_{m-r} \end{pmatrix} V.$$

The rank of the matrix of Z_i's is $m - r$, since its rows are linearly independent. Hence the dimension of $\mathcal{Y}$ is

$$m - (m - r) = r.$$

Since the dimension of the image space is also r, we conclude that the image space is actually $\mathcal{Y}$ itself. ∎

For
$$A = \begin{pmatrix} 1 & 3 & -2 & 5 \\ 2 & 4 & 3 & -3 \\ -1 & 1 & -12 & 21 \end{pmatrix}$$
it is easy to see that
$$\begin{pmatrix} y_1 \\ y_2 \\ y_3 \end{pmatrix}$$
is an image iff $-3y_1 + 2y_1 + y_3 = 0$.

EXERCISES

For the matrices A in problems 1–11 find:
a) the rank,
b) a basis for the kernel of $T(V) = AV$,
c) a basis for the image space of the transformation given by the matrix (use 2.4), and
d) a basis for the left kernel.

1. $\begin{pmatrix} 1 & 0 & 2 \\ 0 & 1 & -3 \end{pmatrix}, \begin{pmatrix} 1 & 0 \\ 0 & 1 \\ 2 & -3 \end{pmatrix}$
2. $\begin{pmatrix} 1 & 1 & 1 \\ 1 & 2 & 3 \end{pmatrix}, \begin{pmatrix} 1 & 1 \\ 1 & 2 \\ 1 & 3 \end{pmatrix}$

3. $\begin{pmatrix} 1 & 0 & -2 \\ 2 & 1 & 3 \\ 4 & 1 & -1 \end{pmatrix}$
4. $\begin{pmatrix} 1 & 0 & -2 \\ 2 & 1 & 3 \\ 4 & 1 & -1 \\ -1 & 2 & 1 \end{pmatrix}$

5. $\begin{pmatrix} 1 & 2 & 4 & -1 \\ 0 & 1 & 1 & 2 \\ -2 & 3 & -1 & 1 \end{pmatrix}$
6. $\begin{pmatrix} 1 & 3 & -2 & 5 \\ 2 & 4 & 3 & -3 \\ -1 & 1 & -12 & 8 \end{pmatrix}$

7. $\begin{pmatrix} 1 & 2 & -1 \\ 3 & 4 & 1 \\ -2 & 3 & -12 \\ 5 & -3 & 8 \end{pmatrix}$
8. $\begin{pmatrix} 1 & 2 & 4 \\ 0 & 1 & 1 \\ -2 & 3 & -1 \\ 1 & -1 & 1 \end{pmatrix}$

9. $\begin{pmatrix} 1 & 2 & 1 & 3 \\ 2 & 4 & 2 & 6 \\ -1 & -2 & -1 & -3 \end{pmatrix}$
10. $\begin{pmatrix} 1 & 2 & -1 & 1 \\ 3 & 4 & 1 & 2 \\ -2 & 3 & -12 & 5 \\ 5 & -3 & 21 & -2 \end{pmatrix}$

11. $\begin{pmatrix} 0 & 1 & -2 & 3 \\ 1 & -3 & 2 & 1 \\ 1 & -1 & -2 & 7 \\ 1 & 0 & -4 & 10 \end{pmatrix}$

12. Let (a_1, a_2, a_3), (b_1, b_2, b_3) be linearly independent. Find the condition on c_1, c_2, c_3 that the equations
$$a_1 x + b_1 y = c_1,$$
$$a_2 x + b_2 y = c_2,$$
$$a_3 x + b_3 y = c_3,$$
have a common solution (which is necessarily unique). What is the geometric meaning of this? [*Hint:* Use the result of Exercise 15 in Section 1.]

13. Let $V_1, V_2, \ldots, V_{m_1}$ be a basis of a subspace $\mathfrak{X}$ of $\mathbb{R}_n$ and $W_1, W_2, \ldots, W_{m_2}$ be a basis of a subspace $\mathcal{Y}$ of $\mathbb{R}_n$. Show how to find a basis for the intersection $\mathfrak{X} \cap \mathcal{Y}$ by the methods of this chapter (Exercise 9 of Section 1 is an example). Prove that the dimension of $\mathfrak{X} \cap \mathcal{Y}$ is
$$m_1 + m_2 - \dim(\mathfrak{X} + \mathcal{Y});$$
compare this to Exercise 8 in Chapter 2, Section 2.

14. Another example to illustrate Exercise 13: In $\mathbb{R}_4$, let

$V_1 = (1, 0, 2, 1),$ $\quad W_1 = (1, 1, 0, 2);$
$V_2 = (0, 2, 4, -2)$ $\quad W_2 = (2, 1, -1, 1);$
$V_3 = (-2, 1, 3, -2),$ $\quad W_3 = (0, -2, 1, 3).$

First show that V_1, V_2, V_3 are linearly independent, so that V_1, V_2, V_3 form a basis of a subspace $\mathfrak{X}$, and that W_1, W_2, W_3 are linearly independent so that W_1, W_2, W_3 form a basis of a subspace $\mathcal{Y}$. Then find a basis for the intersection $\mathfrak{X} \cap \mathcal{Y}$.

15. Given the system of equations $\mathbf{AX} = \mathbf{Y}$, prove that one and only one of the following occur: (i) the system has solutions or (ii) $\mathbf{Z}$ exists, so that $\mathbf{ZA} = \mathbf{0}$ and $\mathbf{ZY} = \mathbf{1}$.

In problems 16 through 20 see whether the matrix is invertible and if it is, find its inverse.

16. $\mathbf{A} = \begin{pmatrix} 1 & 2 & 3 \\ 1 & 4 & 9 \\ 1 & 1 & 1 \end{pmatrix}$
17. $\mathbf{A} = \begin{pmatrix} 2 & 0 & 1 \\ 3 & 1 & 2 \\ 0 & 1 & 1 \end{pmatrix}$
18. $\mathbf{A} = \begin{pmatrix} 1 & 0 & -2 \\ 2 & 1 & 3 \\ 4 & 1 & -1 \end{pmatrix}$

19. $\mathbf{A} = \begin{pmatrix} 0 & 1 & -2 & 3 \\ 1 & -3 & 2 & 1 \\ -2 & 4 & 1 & 2 \\ 3 & -5 & 4 & 2 \end{pmatrix}$
20. $\mathbf{A} = \begin{pmatrix} 0 & 1 & -2 & 3 \\ 1 & -3 & 2 & 1 \\ 1 & -1 & -2 & 7 \\ 1 & 0 & -4 & 10 \end{pmatrix}$

21. a) Finish the proof of the Solution Algorithm (2.4).
    b) Make up a flow diagram for solving systems of linear equations.

22. If $\mathbf{A}'$ is obtained from $\mathbf{A}$ by elementary row operations, show that the linear transformations
$$T(X) = AX \quad \text{and} \quad T'(X) = A'X$$
have the same kernels.

23. Use 2.5 to obtain the conditions required in Exercises 11, 12, 13 of Section 1.

*3. MATRIX EQUIVALENCE; ANOTHER SOLUTION PROCEDURE

In this section we give an interpretation for our solution procedure in terms of matrix multiplication. In addition, we develop an alternative technique for solving systems of equations.

In the previous sections we have seen how one can change from the system of linear equations $\mathbf{AX} = \mathbf{Y}$ to an equivalent system $\mathbf{A'X} = \mathbf{Y'}$ from which solutions are readily apparent. In fact, we apply elementary row operations to $(\mathbf{A} \mid \mathbf{Y})$, obtaining the matrix $(\mathbf{A'} \mid \mathbf{Y'})$. Now recall that each elementary row operation can be accomplished by premultiplication by an elementary matrix (Chapter 2, 4.2). Thus, if $\mathbf{E}_1, \ldots, \mathbf{E}_k$ are the elementary matrices which correspond to the operations which change $(\mathbf{A} \mid \mathbf{Y})$ into $(\mathbf{A'} \mid \mathbf{Y'})$, we have

$$(\mathbf{A'} \mid \mathbf{Y'}) = \mathbf{E}_k \cdots \mathbf{E}_2 \mathbf{E}_1 (\mathbf{A} \mid \mathbf{Y}).$$

Letting $\mathbf{P} = \mathbf{E}_k \cdots \mathbf{E}_2 \mathbf{E}_1$, we have

$$(\mathbf{A'} \mid \mathbf{Y'}) = \mathbf{P}(\mathbf{A} \mid \mathbf{Y}) = (\mathbf{PA} \mid \mathbf{PY}),$$

so that $\mathbf{A'} = \mathbf{PA}$ and $\mathbf{Y'} = \mathbf{PY}$.

We also saw in 2.4 that if $\mathbf{A}$ was invertible, then we could apply elementary row operations to $(\mathbf{A} \mid \mathbf{I}_n)$, $\mathbf{I}_n$ the $n \times n$ identity matrix, to obtain $(\mathbf{I} \mid \mathbf{A}^{-1})$. With $\mathbf{P}$ again the product of the appropriate elementary matrices, we see that

$$(\mathbf{I} \mid \mathbf{A}^{-1}) = \mathbf{P}(\mathbf{A} \mid \mathbf{I}) = (\mathbf{PA} \mid \mathbf{PI}) = (\mathbf{PA} \mid \mathbf{P}),$$

so that in this case $\mathbf{P} = \mathbf{A}^{-1}$. Since $\mathbf{A}$ can be any invertible matrix so can $\mathbf{A}^{-1}$, and we have as a corollary

3.1 Any invertible matrix can be written as the product of elementary matrices.

Let us return to $\mathbf{AX} = \mathbf{Y}$, which we have changed to $\mathbf{A'X} = \mathbf{Y'}$, where

$$\mathbf{A'} = \mathbf{PA}, \quad \mathbf{Y'} = \mathbf{PY}.$$

The discussion of inverses indicates how we could even find $\mathbf{P}$: simply attach $\mathbf{I}_n$ to $(\mathbf{A} \mid \mathbf{Y})$ and perform row operations:

$$(\mathbf{A} \mid \mathbf{Y} \mid \mathbf{I}_n) \rightarrow (\mathbf{A'} \mid \mathbf{Y'} \mid \mathbf{P}).$$

Since $\mathbf{P}$ is the product of invertible matrices, $\mathbf{P}$ itself is invertible.

We say that $\mathbf{A'}$ is **row-equivalent** to $\mathbf{A}$ if there is an invertible matrix $\mathbf{P}$ for which $\mathbf{A'} = \mathbf{PA}$. This is easily seen to be an equivalence relation, as defined in Chapter 0 (see Exercise 21). We can thus interpret our solution procedure as one of replacing $\mathbf{A}$ by a matrix $\mathbf{A'}$ which is row-equivalent to $\mathbf{A}$ and for which systems of linear equations are easier to solve. (At the same time we replace $\mathbf{Y}$ by $\mathbf{Y'}$ so that the *systems* $\mathbf{AX} = \mathbf{Y}$ and $\mathbf{A'X} = \mathbf{Y'}$ are equivalent; note that the latter is obtained by premultiplying the former by $\mathbf{P}$.)

Of course there is the analogous notion of "column equivalent" matrices, arising from performing elementary column operations. It is interesting to inquire how this relates to the solution of linear equations. We shall consider this, but since the reader is by now used to performing row manipulations (at least the writers are), we shall couch things in terms of row operations on $\mathbf{A}^*$ rather than column operations on $\mathbf{A}$.

We shall first discuss the problem of finding the kernel of $\mathbf{T}(\mathbf{V}) = \mathbf{AV}$ and then, as an added attraction, show how the problem of finding solutions of systems $\mathbf{AX} = \mathbf{Y}$ can be reduced to the problem of finding kernels.

3.2 The kernel algorithm. Given an $m \times n$ matrix $\mathbf{A}$, to find the kernel of the linear transformation $\mathbf{T}(\mathbf{V}) = \mathbf{AV}$, put the $\mathbf{A}^*$ part of the matrix $(\mathbf{A}^* \mid \mathbf{I}_n)$ into row-echelon form:

$$(\mathbf{A}^* \mid \mathbf{I}_n) \rightarrow (\mathbf{B}^* \mid \mathbf{Q}^*) = \begin{pmatrix} \mathbf{B}_1^* & \mathbf{Q}_1^* \\ \cdot & \cdot \\ \cdot & \cdot \\ \cdot & \cdot \\ \mathbf{B}_r^* & \mathbf{Q}_r^* \\ 0 & \mathbf{Q}_{r+1}^* \\ \cdot & \cdot \\ \cdot & \cdot \\ 0 & \mathbf{Q}_n^* \end{pmatrix},$$

where r is the rank of $\mathbf{A}$. Then a basis for the kernel consists of

$$\{\mathbf{Q}_{r+1}, \ldots, \mathbf{Q}_n\},$$

the last $(n - r)$ columns of $\mathbf{Q}$.

Proof. A by now familiar argument shows that for some product of elementary matrices which we choose to call $\mathbf{Q}^*$, we have

$$\mathbf{Q}^*(\mathbf{A}^* \mid \mathbf{I}) = (\mathbf{B}^* \mid \mathbf{Q}^*) \text{ with } \mathbf{B}^* = \mathbf{Q}^*\mathbf{A}^*.$$

We can rewrite the latter as $\mathbf{AQ} = \mathbf{B}$ (why$_1$?), and note that the form of $\mathbf{B}$ is $(\mathbf{B}_1, \ldots, \mathbf{B}_r, 0, \ldots, 0)$. Hence, if $\mathbf{Q} = (\mathbf{Q}_1, \ldots, \mathbf{Q}_n)$, we must have

$$\mathbf{A}(\mathbf{Q}_1, \ldots, \mathbf{Q}_r, \ldots, \mathbf{Q}_n) = (\mathbf{AQ}_1, \ldots, \mathbf{AQ}_r, \ldots, \mathbf{AQ}_n)$$
$$= (\mathbf{B}_1, \ldots, \mathbf{B}_r, 0, \ldots, 0).$$

Thus $\mathbf{AQ}_{r+1} = 0, \ldots, \mathbf{AQ}_n = 0$, and $\mathbf{Q}_{r+1}, \ldots, \mathbf{Q}_n$ are in the kernel of $\mathbf{T}$. Since $\mathbf{Q}$ is invertible (why$_2$?), its columns are linearly independent (why$_3$?), and $\mathbf{Q}_{r+1}, \ldots, \mathbf{Q}_n$ are consequently a basis for the kernel of $\mathbf{T}$ (why$_4$?). ∎

Example 1. For

$$A = \begin{pmatrix} 1 & 3 & -2 & 5 \\ 2 & 6 & 3 & -3 \\ -1 & -3 & -12 & 21 \end{pmatrix}, \quad (A^* \mid I) = \begin{pmatrix} 1 & 2 & -1 & 1 & 0 & 0 & 0 \\ 3 & 6 & -3 & 0 & 1 & 0 & 0 \\ -2 & 3 & -12 & 0 & 0 & 1 & 0 \\ 5 & -3 & 21 & 0 & 0 & 0 & 1 \end{pmatrix}$$

$$\rightarrow \begin{pmatrix} 1 & 2 & -1 & 1 & 0 & 0 & 0 \\ 0 & 0 & 0 & -3 & 1 & 0 & 0 \\ 0 & 7 & -14 & 2 & 0 & 1 & 0 \\ 0 & -13 & 26 & -5 & 0 & 0 & 1 \end{pmatrix}$$

$$\rightarrow \begin{pmatrix} 1 & 0 & 3 & \frac{3}{7} & 0 & -\frac{2}{7} & 0 \\ 0 & 0 & 0 & -3 & 1 & 0 & 0 \\ 0 & 1 & -2 & \frac{2}{7} & 0 & \frac{1}{7} & 0 \\ 0 & 0 & 0 & -\frac{9}{7} & 0 & \frac{13}{7} & 1 \end{pmatrix}$$

$$\rightarrow \begin{pmatrix} 1 & 0 & 3 & \frac{3}{7} & 0 & -\frac{2}{7} & 0 \\ 0 & 1 & -2 & \frac{2}{7} & 0 & \frac{1}{7} & 0 \\ 0 & 0 & 0 & -3 & 1 & 0 & 0 \\ 0 & 0 & 0 & -\frac{9}{7} & 0 & \frac{13}{7} & 1 \end{pmatrix} = (B^* \mid Q^*).$$

Hence,

$$Q = \begin{pmatrix} \frac{3}{7} & \frac{2}{7} & -3 & -\frac{9}{7} \\ 0 & 0 & 1 & 0 \\ -\frac{2}{7} & \frac{1}{7} & 0 & \frac{13}{7} \\ 0 & 0 & 0 & 1 \end{pmatrix}$$

and a basis for the kernel of $T(X) = AX$ is

$$\begin{pmatrix} -3 \\ 1 \\ 0 \\ 0 \end{pmatrix}, \begin{pmatrix} -\frac{9}{7} \\ 0 \\ \frac{13}{7} \\ 1 \end{pmatrix}.$$

It is evident that 3.2 is more convenient than the Solution Algorithm for finding kernels when it comes to reading off the basis from the final matrix. On the other hand, 3.2 involves more writing, in that one must carry along the extra Q^* part of the matrix.

Now suppose we wish to solve $AX = Y$. Rewrite this as $AX - Y = 0$, or in terms of partitioned matrices as

$$(A \mid Y)\begin{pmatrix} X \\ -1 \end{pmatrix} = 0.$$

Thus, a solution X to $AX = Y$ is the first part of a vector $Z = (X \mid -1)^*$ with last entry -1 and for which Z is in the kernel of

$$T(Z) = (A \mid Y)Z = 0.$$

The steps are reversible: If a solution to $(A \mid Y)Z = 0$ has the form $Z = (X \mid -1)^*$, then $AX = Y$. Thus, the problem of finding solutions to systems of linear equations is essentially the same as that of finding kernels. The solution procedure can be put in an interesting matrix form:

3.3 Solution algorithm (transpose forms). To solve $AX = Y$, where A is $m \times n$, place the A^* part of

$$\left(\frac{A^*}{Y^*} \,\middle|\, I_{n+1}\right)$$

in row-echelon form:

$$\left(\frac{A^*}{Y^*} \,\middle|\, I_{n+1}\right) \to \left(\begin{array}{c|c|c} B^* & Q^* & 0 \\ & & \vdots \\ & & 0 \\ \hline V^* & W^* & 1 \end{array}\right).$$

Then

i) $AX = Y$ has a solution iff $V = 0$.
ii) If $V = 0$, then one solution is $\bar{X} = -W$.
iii) If B has r nonzero columns, then these are a basis for the image of $T(X) = AX$.
iv) Any other solution to $AX = Y$ is of the form $X = \bar{X} + Z$ where Z is a linear combination of the last $n - r$ columns of Q.

Proof. You check to make sure things work out as indicated. ∎

Example 2. Let us solve

$$\begin{aligned} x_1 + 3x_2 - 2x_3 &= 5, \\ 2x_1 + 6x_2 + 3x_3 &= -3, \\ -x_1 - 3x_2 - 12x_3 &= 21, \end{aligned}$$

using 3.3.

$$\left(\frac{A^*}{Y^*} \,\middle|\, I_{3+1}\right) = \left(\begin{array}{ccc|cccc} 1 & 2 & -1 & 1 & 0 & 0 & 0 \\ 3 & 6 & -3 & 0 & 1 & 0 & 0 \\ -2 & 3 & -12 & 0 & 0 & 1 & 0 \\ 5 & -3 & 21 & 0 & 0 & 0 & 1 \end{array}\right) \xrightarrow{\text{as in Example 1}}$$

$$\left(\begin{array}{ccc|cccc} 1 & 0 & 3 & \frac{3}{7} & 0 & -\frac{2}{7} & 0 \\ 0 & 1 & -2 & \frac{2}{7} & 0 & \frac{1}{7} & 0 \\ 0 & 0 & 0 & -3 & 1 & 0 & 0 \\ \hline 0 & 0 & 0 & -\frac{9}{7} & 0 & \frac{13}{7} & 1 \end{array}\right) = \left(\begin{array}{c|c|c} B^* & Q^* & 0 \\ & & 0 \\ \hline V^* & W^* & 1 \end{array}\right).$$

Since $V^* = 0$, solutions exist. One solution is

$$\bar{X}^* = -W^* = (\tfrac{9}{7}, 0, -\tfrac{13}{7}).$$

The kernel of the corresponding linear transformation may be taken as the last $3-2$ columns of **Q**, i.e., from the last row of **Q***:
$$(-3, 1, 0).$$
All of this is simple to check.

One could also allow both row and column operations, of course. This would result in a matrix $\mathbf{A}'' = \mathbf{PAQ}$, where **P** and **Q** are invertible. Such matrices **A** and **A''** are said to be just plain **equivalent** (another equivalence relation; see Exercise 21). In Section 1 we saw that by performing row operations and a permutation of columns, we could put any matrix in the form
$$\left(\begin{array}{c|c} \mathbf{I}_r & \mathbf{B} \\ \hline 0 & 0 \end{array}\right).$$

One can do better in fact:

3.4 Given a matrix **A** of rank r, there exist invertible matrices **P** and **Q** for which
$$\mathbf{PAQ} = \left(\begin{array}{c|c} \mathbf{I}_r & 0 \\ \hline 0 & 0 \end{array}\right).$$
That is, any matrix of rank r is equivalent to a matrix of this form.

Proof. For you to provide. ∎

The system of linear equations $\mathbf{A}''\mathbf{X}'' = \mathbf{Y}''$ with
$$\mathbf{A}'' = \left(\begin{array}{c|c} \mathbf{I}_r & 0 \\ \hline 0 & 0 \end{array}\right)$$
is monumentally easy to solve. Given $\mathbf{AX} = \mathbf{Y}$, if we have **P** and **Q** so that
$$\mathbf{A}'' = \mathbf{PAQ},$$
then taking $\mathbf{Y}'' = \mathbf{PY}$ and $\mathbf{X}'' = \mathbf{Q}^{-1}\mathbf{X}$, we have such a system.

You should compare 3.4 with 4.11 in Chapter 3, Section 4, Supplement 1.

EXERCISES

1 through 10. Find the kernels of the linear transformations $T(V) = \mathbf{AV}$, **A** as in Exercises 1 through 10 of Section 2, using 3.2.

11 through 18. Solve the systems of linear equations in Exercises 1 through 8 of Section 1.

19. Find all solutions to the following systems of linear equations using 3.3. Check your work.

a)
	(1)	(2)	(3)
$x_1 + 3x_2 =$	$0,$	$1,$	4
$-2x_1 + x_2 =$	$0,$	$2,$	-1
$4x_1 - x_2 =$	$0,$	$-1,$	3

b) $\begin{aligned} x_1 + 2x_2 - x_3 + 3x_4 &\overset{(1)}{=} 0, &\overset{(2)}{=} 2, &\overset{(3)}{=} 1 \\ 2x_1 + 3x_2 - 2x_3 + 4x_4 &= 0, &= -1, &= 0 \\ x_1 + x_2 - x_3 + 2x_4 &= 0, &= -3, &= 0 \end{aligned}$

c) $\begin{aligned} x_1 + 2x_2 + x_3 &\overset{(1)}{=} 0, &\overset{(2)}{=} 1, &\overset{(3)}{=} 1 \\ 2x_1 + 3x_2 + x_3 &= 0, &= 2, &= 2 \\ -x_1 - 2x_2 - x_3 &= 0, &= 1, &= -1 \\ 3x_1 + 4x_2 + 2x_3 &= 0, &= 2, &= 2 \end{aligned}$

d) $\begin{aligned} x_1 + 2x_2 + 4x_3 &\overset{(1)}{=} 0, &\overset{(2)}{=} 3, &\overset{(3)}{=} 4 \\ x_2 + x_3 &= 0, &= 1, &= 1 \\ -2x_1 + 3x_2 - x_3 &= 0, &= -2, &= -1 \\ x_1 - x_2 + x_3 &= 0, &= -1, &= 1 \end{aligned}$

e) $\begin{aligned} x_1 \quad - 2x_3 + x_4 &\overset{(1)}{=} 0, &\overset{(2)}{=} 3, &\overset{(3)}{=} 3 \\ 2x_1 + x_2 + 3x_3 - x_4 &= 0, &= 1, &= 1 \\ 4x_1 + x_2 - x_3 + x_4 &= 0, &= 7, &= -2 \end{aligned}$

20. Check those cases in Exercise 19 for which you obtained no solution by using 2.5.
21. (a) Prove that row equivalence is an equivalence relation. (b) Prove that matrix equivalence is.
22. Answer the why's in 3.2.
23. (a) Prove 3.3. (b) Discuss 3.3 in terms of the matrix multiplication

$$\left(\begin{array}{c|c} A & Y \\ \hline & I_{n+1} \end{array}\right) \left(\begin{array}{c|c} Q & -X \\ \hline V & 1 \end{array}\right)$$

24. Prove 3.4.
25. Find invertible matrices P, Q so that

$$PAQ = \left(\begin{array}{c|c} I_r & 0 \\ \hline 0 & 0 \end{array}\right).$$

for each of the following matrices; make use of results of previous problems whenever possible.

a) $A = \begin{pmatrix} 1 & 1 & 1 \\ 1 & 2 & 3 \end{pmatrix}$ b) $A = \begin{pmatrix} 1 & 3 & -2 & 5 \\ 2 & 4 & 3 & -3 \\ -1 & 1 & -12 & 8 \end{pmatrix}$ c) $A = \begin{pmatrix} 1 & 2 & -1 \\ 3 & 4 & 1 \\ -2 & 3 & -12 \\ 5 & -3 & 8 \end{pmatrix}$

d) $A = \begin{pmatrix} 1 & 2 & -1 \\ 3 & 4 & 1 \\ -2 & 3 & -12 \\ 5 & -3 & 21 \end{pmatrix}$ e) $A = \begin{pmatrix} 1 & 2 & 1 & 3 \\ 2 & 4 & 2 & 6 \\ -1 & -2 & -1 & -3 \end{pmatrix}$ f) $A = \begin{pmatrix} 0 & 1 & -2 & 3 \\ 1 & -3 & 3 & 1 \\ 1 & -1 & -2 & 7 \\ 1 & 0 & -4 & 10 \end{pmatrix}$

26. Given a square matrix A and (invertible) matrices P, Q so that $PAQ = I$, show how to express A^{-1} in terms of P and Q.
27. Let A be a matrix of rank r. Prove that A can be expressed as the sum of r matrices each of which is of rank 1.

28. Prove that a matrix is of rank 1 iff it can be exhibited in the form

$$\begin{pmatrix} p_m \\ \vdots \\ p_1 \end{pmatrix} (q_1, q_2, \ldots, q_n),$$

where neither of these vectors is zero. (Problem 9 in Section 2 gives an illustration.)

29. a) Prove that the last $n - r$ columns of the matrix $\mathbf{Q}$ in 3.4 are a basis of the kernel of $\mathbf{A}$; likewise the last $m - r$ rows of $\mathbf{P}$ are a basis of the left kernel.
    b) Use (a) to prove 2.5.

*4. CHANGE OF BASIS

We now turn to a more geometrical interpretation of the solution procedures. Recall that for a linear transformation $\mathbf{T}$ from $\mathbb{R}_n$ into $\mathbb{R}_m$, the matrix $\mathbf{A} = (\mathbf{T}(\mathbf{e}_1), \ldots, \mathbf{T}(\mathbf{e}_n))$ satisfies $\mathbf{T}(\mathbf{V}) = \mathbf{AV}$. We have called $\mathbf{A}$ "the matrix of $\mathbf{T}$ with respect to the standard bases for $\mathbb{R}_n$ and $\mathbb{R}_m$". If, however, we wish to utilize a different basis $\{\mathbf{V}_1, \ldots, \mathbf{V}_n\}$ for $\mathbb{R}_n$, under the correspondence

$$a_1 \mathbf{V}_1 + \cdots + a_n \mathbf{V}_n \to (a_1, \ldots, a_n),$$

we can consider $\mathbf{V}_1, \ldots, \mathbf{V}_n$ as if they were the standard basis $\mathbf{e}_1, \ldots, \mathbf{e}_n$. What effect would this have on the matrix of $\mathbf{T}$? What if we were to make a similar basis change in $\mathbb{R}_m$?

To answer these questions, suppose that

$$\alpha = \{\mathbf{V}_1, \ldots, \mathbf{V}_n\}$$

is a basis for $\mathbb{R}_n$ and that

$$\beta = \{\mathbf{W}_1, \ldots, \mathbf{W}_m\}$$

is a basis for $\mathbb{R}_m$. If

$$\mathbf{T}(\mathbf{V}_1) = a_{11}\mathbf{W}_1 + a_{21}\mathbf{W}_2 + \cdots + a_{m1}\mathbf{W}_m,$$
$$\mathbf{T}(\mathbf{V}_2) = a_{12}\mathbf{W}_1 + a_{22}\mathbf{W}_2 + \cdots + a_{m2}\mathbf{W}_m,$$
$$\vdots \qquad \qquad \vdots \tag{1}$$
$$\mathbf{T}(\mathbf{V}_n) = a_{1n}\mathbf{W} + a_{2n}\mathbf{W}_2 + \cdots + a_{mn}\mathbf{W}_m,$$

then we call the matrix

$$\mathbf{M}_{\beta\alpha}(\mathbf{T}) = \begin{pmatrix} a_{11} & a_{12} & \cdots & a_{1n} \\ a_{21} & a_{22} & \cdots & a_{2n} \\ \vdots & \vdots & & \vdots \\ a_{m1} & a_{m2} & \cdots & a_{mn} \end{pmatrix}$$

the matrix of T with respect to the bases α and β. Note that we have taken the *transpose* of the matrix occurring naturally in (1). This does, however, correspond to our previous work with the standard bases.

$$\mathbf{M}_{\beta\alpha}(\mathbf{T}) = \begin{pmatrix} \text{Coefficients} \\ \text{of} \\ \mathbf{T}(\mathbf{V}_1) \\ \text{in terms} \\ \text{of } \beta \end{pmatrix}, \begin{matrix} \text{Coefficients} \\ \text{of} \\ \mathbf{T}(\mathbf{V}_2) \\ \text{in terms} \\ \text{of } \beta \end{matrix}, \ldots, \begin{matrix} \text{Coefficients} \\ \text{of} \\ \mathbf{T}(\mathbf{V}_n) \\ \text{in terms} \\ \text{of } \beta \end{matrix} \end{pmatrix}$$

Since there is one and only one linear transformation satisfying (1), we see that the correspondence between linear transformations from $\mathbb{R}_n$ into $\mathbb{R}_m$ and $m \times n$ matrices given by

$$\mathbf{T} \to \mathbf{M}_{\beta\alpha}(\mathbf{T})$$

is a one-to-one correspondence.

Suppose now that we have in addition a linear transformation $\mathbf{S}$ from $\mathbb{R}_m$ into $\mathbb{R}_l$ which, in terms of the basis

$$\gamma = \{\mathbf{X}_1, \ldots, \mathbf{X}_l\}$$

for $\mathbb{R}_l$ is given by

$$\mathbf{S}(\mathbf{W}_1) = b_{11}\mathbf{X}_1 + \cdots + b_{l1}\mathbf{X}_l,$$
$$\vdots \qquad \vdots$$
$$\mathbf{S}(\mathbf{W}_m) = b_{1m}\mathbf{X}_1 + \cdots + b_{lm}\mathbf{X}_l,$$

(2)

so that

$$\mathbf{M}_{\gamma\beta}(\mathbf{S}) = \begin{pmatrix} b_{11} & \cdots & b_{1m} \\ \vdots & & \vdots \\ b_{l1} & \cdots & b_{lm} \end{pmatrix}.$$

We therefore have

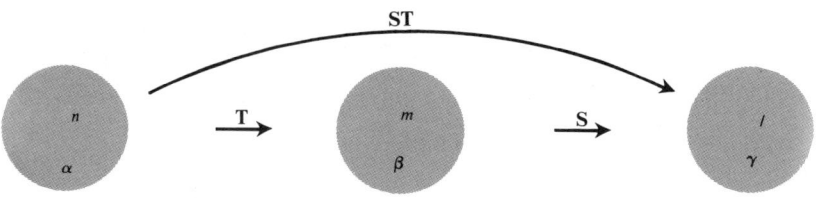

Fig. 2

We can inquire as to the relationship between

$\mathbf{M}_{\gamma\beta}(\mathbf{S})$ $\qquad$ $\mathbf{M}_{\beta\alpha}(\mathbf{T})$ $\qquad$ $\mathbf{M}_{\gamma\alpha}(\mathbf{ST})$
(which is $l \times m$), (which is $m \times n$), and (which is $l \times n$).

As you might hope, we have

4.1 $\mathbf{M}_{\gamma\beta}(\mathbf{S})\mathbf{M}_{\beta\alpha}(\mathbf{T}) = \mathbf{M}_{\gamma\alpha}(\mathbf{ST})$.

Proof. We have indicated the first two matrices following (1) and (2). To get the matrix of **ST** we calculate

$$(\mathbf{ST})(\mathbf{V}_j) = \mathbf{S}(\mathbf{T}(\mathbf{V}_j)) = \mathbf{S}(a_{1j}\mathbf{W}_1 + \cdots + a_{mj}\mathbf{W}_m)$$
$$= a_{1j}\mathbf{S}(\mathbf{W}_1) + \cdots + a_{mj}\mathbf{S}(\mathbf{W}_m)$$
$$= a_{1j}(b_{11}\mathbf{X}_1 + \cdots + b_{l1}\mathbf{X}_l) + \cdots + a_{mj}(b_{1m}\mathbf{X}_1 + \cdots + b_{lm}\mathbf{X}_l)$$
$$= (b_{11}a_{1j} + \cdots + b_{1m}a_{mj})\mathbf{X}_1 + \cdots + (b_{l1}a_{1j} + \cdots + b_{lm}a_{mj})\mathbf{X}_l.$$

Thus, the jth column of $\mathbf{M}_{\gamma\alpha}(\mathbf{ST})$ is the same as the jth column of $\mathbf{M}_{\gamma\beta}(\mathbf{S})\mathbf{M}_{\beta\alpha}(\mathbf{T})$. Since this is true for any j, the assertion is proved. ∎

Armed with the ability to multiply matrices of linear transformations with respect to different bases (and the fact that the correspondence between matrices and linear transformations is one-to-one), we can proceed to our main discussion.

Given bases

$$\alpha = \{\mathbf{V}_1, \ldots, \mathbf{V}_n\}, \qquad \alpha' = \{\mathbf{V}'_1, \ldots, \mathbf{V}'_n\}$$

for $\mathcal{U}$, we know we can write one basis in terms of the other

$$\mathbf{V}'_1 = q_{11}\mathbf{V}_1 + q_{21}\mathbf{V}_2 + \cdots + q_{n1}\mathbf{V}_n,$$
$$\vdots \qquad \vdots \qquad \qquad (1)$$
$$\mathbf{V}'_n = q_{1n}\mathbf{V}_1 + q_{2n}\mathbf{V}_2 + \cdots + q_{nn}\mathbf{V}_n.$$

The matrix

$$\mathbf{Q} = \begin{pmatrix} q_{11} & q_{12} & \cdots & q_{1n} \\ q_{21} & & & \vdots \\ \vdots & & & \vdots \\ q_{n1} & & \cdots & q_{nn} \end{pmatrix}$$

is called the **change of basis matrix from basis α to basis α'**. Thus

$$\mathbf{Q} = \begin{pmatrix} \text{Coefficients} \\ \text{of } \mathbf{V}'_1 \\ \text{in terms of } \alpha \end{pmatrix}, \ldots, \begin{pmatrix} \text{Coefficients} \\ \text{of } \mathbf{V}'_n \\ \text{in terms of } \alpha \end{pmatrix}.$$

Example 1. The vectors
$$\alpha: V_1 = (1, 1, 1), \quad V_2 = (1, 1, 0), \quad V_3 = (1, 0, 0)$$
are easily seen to be a basis for $\mathbb{R}_3$, as are
$$\alpha': V_1' = (3, 2, 1), \quad V_2' = (0, 1, 2), \quad V_3' = (0, 1, 0).$$
Moreover, one has that
$$V_1' = V_1 + V_2 + V_3, \quad V_2' = 2V_1 - V_2 - V_3, \quad V_3' = V_2 - V_3.$$
Hence, the change of basis matrix from α to α' is
$$\begin{pmatrix} 1 & 2 & 0 \\ 1 & -1 & 1 \\ 1 & -1 & -1 \end{pmatrix}.$$

4.2 Any change of basis matrix Q has an inverse. In fact, if $Q = M_{\alpha\alpha'}(I_\mathcal{V})$, then
$$Q^{-1} = M_{\alpha'\alpha}(I_\mathcal{V}).$$
Proof. $M_{\alpha\alpha'}(I_\mathcal{V}) M_{\alpha'\alpha}(I_\mathcal{V}) = M_{\alpha\alpha}(I_\mathcal{V}^2) = M_{\alpha\alpha}(I_\mathcal{V}) = I.$ ∎

Even more entertaining is the converse of this theorem:

4.3 Any invertible matrix is a change-of-basis matrix. In fact, if $Q = (q_{ij})$ is any invertible matrix in $\mathbb{R}_{n \times n}$, and if $\mathcal{V}$ is any n-dimensional vector space over $\mathbb{R}$ with basis
$$\alpha = \{V_1, \ldots, V_n\},$$
then $\alpha' = \{V_1', \ldots, V_n'\}$ is also a basis for $\mathcal{V}$, where
$$V_1' = q_{11}V_1 + \cdots + q_{n1}V_n, \quad \cdots, \quad V_n' = q_{1n}V_1 + \cdots + q_{nn}V_n.$$
Proof. Yours. You might try writing out
$$c_1 V_1' + \cdots + c_n V_n' = 0,$$
collecting coefficients of $V_1, \ldots, V_n$, and seeing what this implies about either the c's or the columns of Q. ∎

Example 2. With Q as in Example 1, let $\alpha = \{e_1, e_2, e_3\}$. Then the theorem guarantees that
$$\alpha' = \{Qe_1, Qe_2, Qe_3\}$$
forms a basis. In this case, α' is just the columns of Q.

Example 3. Take $P = \begin{pmatrix} 3 & -2 \\ 1 & 1 \end{pmatrix}$ in $\mathbb{R}_{2\times 2}$. It is clearly invertible (why?). Hence, its columns are a basis for $\mathbb{R}_2$ and by the theorem we may consider P a change-of-basis matrix from any basis for $\mathbb{R}_2$ to its image under P. For example,
$$\beta' = \left\{ \begin{pmatrix} \frac{1}{5} \\ -\frac{1}{5} \end{pmatrix}, \begin{pmatrix} \frac{2}{5} \\ \frac{3}{5} \end{pmatrix} \right\}$$

is a basis for $\mathbb{R}_2$, and its image under **P** is

$$\beta = \left\{ \begin{pmatrix} 1 \\ 0 \end{pmatrix}, \begin{pmatrix} 0 \\ 1 \end{pmatrix} \right\}$$

(check and see). Hence, $\mathbf{P}^{-1}$ should take basis β to β'. Now

$$\mathbf{P}^{-1} = \begin{pmatrix} \frac{1}{5} & \frac{2}{5} \\ -\frac{1}{5} & \frac{3}{5} \end{pmatrix},$$

which does the trick, to no one's surprise. You should verify that

$$\mathbf{P} = \mathbf{M}_{\beta\beta'}(\mathbf{I}) \quad \text{and} \quad \mathbf{P}^{-1} = \mathbf{M}_{\beta'\beta}(\mathbf{I}).$$

Returning now to the notion of matrix equivalence from the previous section, we have

4.4 Two $m \times n$ matrices **A** and **A**$'$ are equivalent iff they represent the same linear transformation relative to (possibly) different bases.

Proof. If $\mathbf{A} = \mathbf{M}_{\beta\alpha}(\mathbf{T})$ and $\mathbf{A}' = \mathbf{M}_{\beta'\alpha'}(\mathbf{T})$ for some **T** mapping $\mathcal{V}$ into $\mathcal{W}$, $\mathcal{W}$ with bases β and β', $\mathcal{V}$ with bases α and α', then define $\mathbf{P} = \mathbf{M}_{\beta'\beta}(\mathbf{I}_\mathcal{W})$ and $\mathbf{Q} = \mathbf{M}_{\alpha\alpha'}(\mathbf{I}_\mathcal{V})$; you finish up. Don't forget the converse. ∎

It is now easy to see that 3.4 is the same as 4.11 of Chapter 3: Choosing a basis

$$\alpha = \{\mathbf{V}_1, \ldots, \mathbf{V}_n\}$$

for $\mathcal{V}$ for which $\mathbf{V}_{r+1}, \ldots, \mathbf{V}_n$ are a basis for the kernel of **T**, and then extending $\mathbf{T}(\mathbf{V}_1), \ldots, \mathbf{T}(\mathbf{V}_r)$ to a basis β for $\mathcal{W}$, one has

$$\mathbf{M}_{\beta\alpha}(\mathbf{T}) = \left(\begin{array}{c|c} \mathbf{I}_r & \mathbf{0} \\ \hline \mathbf{0} & \mathbf{0} \end{array} \right).$$

There are of course "one-sided" analogs of 4.4. For example,

Two matrices **A** and **A**$'$ are row-equivalent iff they represent the same linear transformation relative to the same basis in the domain but with (possibly) different bases in the range space.

One part of the proof might run as follows: Given **A** and **A**$'$, row-equivalent $m \times n$ matrices with $\mathbf{A}' = \mathbf{PA}$, take the natural bases

$$\alpha = \{\mathbf{e}_1, \ldots, \mathbf{e}_n\} \text{ for } \mathbb{R}_n \quad \text{and} \quad \beta = \{\mathbf{f}_1, \ldots, \mathbf{f}_m\} \text{ for } \mathbb{R}_m$$

($\mathbf{f}_i = (0, \ldots, \overset{i}{1}, 0, \ldots, 0)^*$). Let $\beta' = \{\mathbf{Pf}_1, \ldots, \mathbf{Pf}_m\}$. Then for **T** mapping $\mathbb{R}_n$ into $\mathbb{R}_m$ by $\mathbf{T}(\mathbf{V}) = \mathbf{AV}$, one has

$$\mathbf{M}_{\beta\alpha}(\mathbf{T}) = \mathbf{A} \quad \text{and} \quad \mathbf{M}_{\beta'\beta}(\mathbf{I}_{\mathbb{R}_m}) = \mathbf{P},$$

so that

$$\mathbf{M}_{\beta'\alpha}(\mathbf{T}) = \mathbf{M}_{\beta'\beta}(\mathbf{IT}) = \mathbf{M}_{\beta'\beta}(\mathbf{I})\mathbf{M}_{\beta\alpha}(\mathbf{T}) = \mathbf{PA} = \mathbf{A}'.$$

Thus, **A** and **A**$'$ do represent the same linear transformation, as indicated above.

This discussion should offer a clear geometrical insight into our solution procedures. In that of Sections 1 and 2 we change $\mathbf{AX} = \mathbf{Y}$ to $\mathbf{A'X} = \mathbf{Y'}$, with $\mathbf{A'}$ row equivalent to $\mathbf{A}$; Let us consider the linear transformation $\mathbf{T}$ from $\mathbb{R}_n$ into $\mathbb{R}_m$ defined by $\mathbf{T(V)} = \mathbf{AV}$. 4.4 shows that we are changing the basis for the image space $\mathbb{R}_m$ to one which allows us to see solutions more easily. (Note that this change of basis does affect the image vector $\mathbf{Y}$; it is transformed into $\mathbf{Y'}$, while our desired $\mathbf{X}$ in the domain remains unchanged.)

In like manner we see that column manipulations of $\mathbf{A}$ in the solution procedure of Section 3 result in a change of basis in the domain $\mathbb{R}_n$ of $\mathbf{T(V)} = \mathbf{AV}$. The $\mathbf{Q}$ in 3.2 is the change-of-basis matrix, and since its last $(n - r)$ columns are a basis for the kernel of $\mathbf{T}$, we see that the column solution procedure works so as to bring in a basis for the kernel as part of a basis for the space. If we change bases in the domain, then we certainly transform the $\mathbf{X}$ in $\mathbf{AX} = \mathbf{Y}$. In fact, if $\mathbf{A'} = \mathbf{AQ}$, we change this system to $\mathbf{A'X'} = \mathbf{Y}$ where $\mathbf{X'} = \mathbf{Q^{-1}X}$ (or $\mathbf{X} = \mathbf{QX'}$, which is what we're really after). The algorithm 3.3 affords us a convenient method of calculating a particular solution $\mathbf{QX'}$ at the same time as we calculate $\mathbf{Q}$.

When it comes to solving systems of linear equations, it suffices to make basis changes in either the domain or the range of the corresponding linear transformation. Although changes in both spaces result in the extremely simple form of 3.4, it is unnecessary to go to the extra work.

We close this section with a description of how to effect the various row and column matrix manipulations through changes of bases in the underlying vector spaces.

4.5 Let $\mathcal{U}$ and $\mathcal{W}$ be vector spaces over $\mathbb{R}$ with bases

$$\alpha = \{\mathbf{V}_1, \ldots, \mathbf{V}_n\} \quad \text{and} \quad \beta = \{\mathbf{W}_1, \ldots, \mathbf{W}_m\},$$

respectively, and suppose that $\mathbf{T}$ is a linear transformation from $\mathcal{U}$ into $\mathcal{W}$ with matrix

$$\mathbf{M}_{\beta\alpha}(\mathbf{T}) = \begin{pmatrix} a_{11} & \cdots & a_{1n} \\ \vdots & & \vdots \\ a_{m1} & \cdots & a_{mn} \end{pmatrix}.$$

i) To interchange the ith and jth *rows* of $\mathbf{M}_{\beta\alpha}(\mathbf{T})$, replace β by the basis

$$\beta' = \{\mathbf{W}_1, \ldots, \overset{i}{\mathbf{W}_j}, \ldots, \overset{j}{\mathbf{W}_i}, \ldots, \mathbf{W}_m\}.$$

(That is, interchange the ith and jth elements of the basis β.)

ii) To interchange the ith and jth *columns* of $\mathbf{M}_{\beta\alpha}(\mathbf{T})$, interchange the ith and jth elements of the basis α.

iii) To pivot on the element a_{ij} ($\neq 0$), replace β by the basis β', which is the same as β except that its ith element is

$$\mathbf{W}'_i = a_{1j}\mathbf{W}_1 + \cdots + a_{mj}\mathbf{W}_m.$$

(That is, the ith element of the new basis consists of the sum of the old basis elements with coefficients from the jth column of $\mathbf{M}_{\beta\alpha}(\mathbf{T})$.)

iv) To *column* pivot on the element a_{ij} ($\neq 0$), replace α by the basis α', which is given by

$$\mathbf{V}'_k = \mathbf{V}_k - a_{ik}/a_{ij}\mathbf{V}_j, \quad \text{for } k \neq j,$$
$$\mathbf{V}'_j = 1/a_{ij}\mathbf{V}_j.$$

Proof. Exercise 6. Note that there are several possible ways of proceeding. One can write out $\mathbf{M}_{\beta'\alpha}(\mathbf{T})$ or $\mathbf{M}_{\beta\alpha'}(\mathbf{T})$ directly and verify that the coefficients are as they are supposed to be; or one can calculate the particular product of elementary matrices which causes the desired change and verify that this product matrix is the change-of-basis matrix for the appropriate basis change. ∎

Example. To appreciate the apparent lack of symmetry between the basis change necessary for a row-pivot in comparison with the column-pivot basis change, let us column-pivot on the element a_{23} (assumed $\neq 0$) in the matrix

$$\mathbf{A} = \begin{pmatrix} a_{11} & a_{12} & a_{13} \\ a_{21} & a_{22} & a_{23} \end{pmatrix}.$$

To calculate the matrix $\mathbf{B}$ for which $\mathbf{AB}$ is the pivoted matrix, form

$$\begin{pmatrix} \mathbf{A} \\ \hline \mathbf{I} \end{pmatrix}$$

and pivot on the 2.3 entry of this matrix to get

$$\begin{pmatrix} \mathbf{AB} \\ \hline \mathbf{B} \end{pmatrix}$$

(why?). In this case

$$\mathbf{B} = \begin{pmatrix} 1 & 0 & 0 \\ 0 & 1 & 0 \\ -a_{21}/a_{23} & -a_{22}/a_{23} & 1/a_{23} \end{pmatrix},$$

and this is indeed the change-of-basis matrix for

$$\mathbf{V}'_1 = \mathbf{V}_1 - a_{21}/a_{23}\mathbf{V}_3, \quad \mathbf{V}'_2 = \mathbf{V}_2 - a_{22}/a_{23}\mathbf{V}_3, \quad \mathbf{V}'_3 = 1/a_{23}\mathbf{V}_3.$$

To calculate the basis change for a corresponding row-pivot, let us consider

$$\mathbf{A}^* = \begin{pmatrix} a_{11} & a_{21} \\ a_{12} & a_{22} \\ a_{13} & a_{23} \end{pmatrix}.$$

Then $\mathbf{B^*A^*}$ results in pivoting on a_{23}. However, recall that for $\mathbf{A'} = \mathbf{PAQ}$, $\mathbf{Q}$ is the change of basis from α to α', while $\mathbf{P}$ is the change of basis matrix from β' to β. In particular then, $\mathbf{B^*}$ gives a basis change from β' to β, while we are after a basis change from β to β'; we want $(\mathbf{B^*})^{-1}$ (why?). This inverse is easy to calculate. One obtains

$$(\mathbf{B^*})^{-1} = \begin{pmatrix} 1 & 0 & a_{21} \\ 0 & 1 & a_{22} \\ 0 & 0 & a_{23} \end{pmatrix}.$$

Thus
$$\mathbf{W}_1' = \mathbf{W}_1, \qquad \mathbf{W}_2' = \mathbf{W}_2, \qquad \mathbf{W}_3' = a_{21}\mathbf{W}_1 + a_{22}\mathbf{W}_2 + a_{23}\mathbf{W}_3,$$

in agreement with the theory, if not to the taste of the basis changer.

EXERCISES

1. Let
$$\mathbf{V}_1 = \begin{pmatrix} 2 \\ 3 \\ 0 \end{pmatrix}, \quad \mathbf{V}_2 = \begin{pmatrix} 0 \\ 1 \\ 1 \end{pmatrix}, \quad \mathbf{V}_3 = \begin{pmatrix} 1 \\ 2 \\ 1 \end{pmatrix}$$
be vectors in $\mathbb{R}_3$.
a) Show that these can be used as a basis for $\mathbb{R}_3$.
b) Find the matrix $\mathbf{P}$ which gives the new coordinates $\mathbf{X}'$ in terms of the old ones,
$$\mathbf{X}' = \mathbf{PX}.$$
c) Find the new coordinates for the following vectors
$$\mathbf{X} = (1, -1, 2)^*, \qquad \mathbf{X} = (1, 0, 0)^*, \qquad \mathbf{X} = (0, 1, 0)^*, \qquad \mathbf{X} = \mathbf{e}_3.$$
d) Find $\mathbf{X}$ for $\mathbf{X}' = (2, 1, -3)^*$.

2. Let $\mathbf{W}_1 = \begin{pmatrix} 2 \\ -1 \end{pmatrix}$, $\mathbf{W}_2 = \begin{pmatrix} 3 \\ 1 \end{pmatrix}$ be vectors in $\mathbb{R}_2$. Show that these can be used as a basis in $\mathbb{R}_2$ and write down the matrix $\mathbf{Q}$ which gives $\mathbf{X} = \mathbf{QX}'$.

3. Let
$$\mathbf{A} = \begin{pmatrix} 2 & -1 \\ 1 & 1 \\ -2 & 3 \end{pmatrix}$$
give a linear transformation $\mathbf{T}:\mathbb{R}_2 \to \mathbb{R}_3$ by $\mathbf{T}(\mathbf{X}) = \mathbf{AX}$.
a) Make the change of basis in $\mathbb{R}_3$ described in Exercise 1 and calculate the matrix which now gives this transformation (do not change bases in $\mathbb{R}_2$.)
b) Make the change of basis in $\mathbb{R}_2$ described in Exercise 2 and calculate the matrix which now gives the transformation (do not change basis in $\mathbb{R}_3$.)
c) Make the changes of basis in both $\mathbb{R}_2$ and $\mathbb{R}_3$ described in Exercises 1 and 2 and calculate the matrix which now describes the transformation.

4. Prove 4.2.

5. (a) Finish proving 4.3. (b) Finish proving 4.4.
6. (a) through (d). Prove 4.5(i)–(iv).
7. Construct basis changes which will result in
   a) multiplying a row by a constant $c \neq 0$,
   b) multiplying a column by a constant $c \neq 0$,
   c) adding a multiple of one row to another row,
   d) adding a multiple of one column to another row.

> As this is quite an extraordinary discovery, I have not yet spoken to any others about it, but here is what it is.†

CHAPTER 5 **Determinants**

Determinants, alas, have fallen upon hard times. Whereas a few years ago they were overused, today the pendulum has swung the other way and especially pure mathematicians expend considerable energy avoiding the creatures.

The proper emphasis undoubtedly lies (rather unimaginatively) between these two extremes: determinants have a certain theoretical, if not computational, value. Moreover, they are part of the basic scientific and mathematical vocabulary.

For those who wish to de-emphasize determinants (whether through inclination or lack of time), we begin with a summary of the main results. As a sop to those who rather enjoy determinants (as we do ourselves), we have appended a chapter to the summary.

SUMMARY

The determinant is a certain real-valued function on the set of $n \times n$ matrices $\mathbb{R}_{n \times n}$; alternatively it is a real-valued function of n vector variables (i.e., of the rows, or columns, of an $(n \times n)$ matrix).

Notation:

$$\det \mathbf{A}, \quad \det \begin{pmatrix} a_{11} & \cdots & a_{1n} \\ \cdot & & \cdot \\ \cdot & & \cdot \\ \cdot & & \cdot \\ a_{n1} & \cdots & a_{nn} \end{pmatrix}, \quad \det(\mathbf{A}_1, \ldots, \mathbf{A}_n), \quad \det(\mathbf{A}_1, \ldots, \mathbf{A}_n)^*,$$

---

†Letter from Leibniz to l'Hôpital on determinants, 1693.

depending upon what we wish to emphasize. (Another fairly common notation is $|\mathbf{A}|$.) Geometrically, the determinant gives the n-dimensional volume of the parallelotope with adjacent edges the rows (or columns) of $\mathbf{A}$ (Section 1).

Defining Properties (Section 1)

We define det $\mathbf{A}$ as the unique function from $\mathbb{R}_{n \times n}$ to $\mathbb{R}$ satisfying the following three conditions on the rows of matrices (or columns—see 3.4):

i)
$$\det \mathbf{I} = \det \begin{pmatrix} \mathbf{e}_1 \\ \cdot \\ \cdot \\ \cdot \\ \mathbf{e}_n \end{pmatrix} = 1;$$

ii) interchanging rows (or columns) changes the sign:

$$\det \begin{pmatrix} \mathbf{A}_1 \\ \cdot \\ \cdot \\ \mathbf{A}_i \\ \cdot \\ \mathbf{A}_j \\ \cdot \\ \cdot \\ \mathbf{A}_n \end{pmatrix} = -\det \begin{pmatrix} \mathbf{A}_1 \\ \cdot \\ \cdot \\ \mathbf{A}_j \\ \cdot \\ \mathbf{A}_i \\ \cdot \\ \cdot \\ \mathbf{A}_n \end{pmatrix} \qquad (\textit{Alternating property});$$

iii) the determinant is linear in each row (or column):

$$\det \begin{pmatrix} \mathbf{A}_1 \\ \cdot \\ c\mathbf{A}_i \\ \cdot \\ \mathbf{A}_n \end{pmatrix} = c \det \begin{pmatrix} \mathbf{A}_1 \\ \cdot \\ \mathbf{A}_i \\ \cdot \\ \mathbf{A}_n \end{pmatrix}, \quad \det \begin{pmatrix} \mathbf{A}_1 \\ \cdot \\ \mathbf{A}_i + \mathbf{A}'_i \\ \cdot \\ \mathbf{A}_n \end{pmatrix} = \det \begin{pmatrix} \mathbf{A}_1 \\ \cdot \\ \mathbf{A}_i \\ \cdot \\ \mathbf{A}_n \end{pmatrix} + \det \begin{pmatrix} \mathbf{A}_1 \\ \cdot \\ \mathbf{A}'_i \\ \cdot \\ \mathbf{A}_n \end{pmatrix}$$

(*Multilinear* property)

Functional Properties of the Determinant

$$\text{Det } \mathbf{A} \neq 0 \quad \text{iff } \mathbf{A} \text{ is invertible} \quad \text{(this is 2.4)}.$$

Hence, $\det \mathbf{A} = 0$ iff one row of $\mathbf{A}$ is a linear combination of other rows (2.4); or

$$\det \mathbf{A} = 0 \quad \text{iff rank } A < n.$$
$$\text{Det } \mathbf{AB} = \det \mathbf{A} \det \mathbf{B} \quad (3.1);$$

but in general $\det (\mathbf{A} + \mathbf{B}) \neq \det \mathbf{A} + \det \mathbf{B}$. Hence,

$$\det \mathbf{A}^{-1} = (\det \mathbf{A})^{-1}, \tag{3.2}$$

$$\det \mathbf{A}^* = \det \mathbf{A} \quad (* = \text{transpose}) \tag{3.3}$$

Hence all row properties are column properties (3.4).

Expressions for the Determinant

$$\det \begin{pmatrix} a_{11} & \cdots & a_{1n} \\ \cdot & & \cdot \\ \cdot & & \cdot \\ \cdot & & \cdot \\ a_{n1} & \cdots & a_{nn} \end{pmatrix} = \sum \det(\mathbf{e}_{i_1}, \ldots, \mathbf{e}_{i_n}) a_{i_1 1} a_{i_2 2} \cdots a_{i_n n}$$
summed over all permutations $(i_1, \ldots, i_n)$ of $(1, \ldots, n)$,

and

$$\det(\mathbf{e}_{i_1}, \ldots, \mathbf{e}_{i_n}) = \pm 1;$$

$+1$ if it takes an even number of interchanges to get from $(\mathbf{e}_{i_1}, \ldots, \mathbf{e}_{i_n})$ to $(\mathbf{e}_1, \ldots, \mathbf{e}_n)$; -1 if it takes an odd number (Section 5).

It is also possible to write an $n \times n$ determinant as a sum of $(n-1) \times (n-1)$ determinants (and consequently these as a sum of $(n-2) \times (n-2)$ determinants, and so on down to 1×1 determinants, yielding the above expression).

The i, j-*cofactor* of $\mathbf{A}$,

$$A_{ij} = (-1)^{i+j} \det \begin{pmatrix} a_{11} & \cdots & a_{1n} \\ & \vdots & \\ \cdots & + & \cdots \\ & \vdots & \\ a_{n1} & \cdots & a_{nn} \end{pmatrix}_i,$$

the ith row and the jth column of $\mathbf{A}$ having been deleted. Then (by 4.2)

$$\begin{aligned} \det \mathbf{A} &= a_{11}A_{11} + \cdots + a_{1n}A_{1n} && \text{(1st row expansion)}, \\ \det \mathbf{A} &= a_{21}A_{21} + \cdots + a_{2n}A_{2n} && \text{(2nd row expansion)}, \\ &\vdots \\ \det \mathbf{A} &= a_{n1}A_{n1} + \cdots + a_{nn}A_{nn} && (n\text{th row expansion}). \end{aligned}$$

Corresponding expansions hold for columns as well; just interchange the order of subscripts.

Calculation of Determinants

If

$$\mathbf{A}' = \begin{pmatrix} a_{11} & & \\ & \ddots & \\ 0\text{'s} & & a_{nn} \end{pmatrix}$$

is *triangular*, then $\det \mathbf{A}' = a_{11} a_{22} \cdots a_{nn}$ (2.1).

The Determinant Algorithm (2.3): Given an arbitrary $n \times n$ matrix $\mathbf{A}$, transform $\mathbf{A}$ into a triangular matrix $\mathbf{A}'$, using the elementary row operations of row addition and row interchange. Then

$$\det \mathbf{A} = (-1)^i \det \mathbf{A}', \qquad i = \text{the number of row interchanges used.}$$

It is sometimes convenient to use this algorithm in conjunction with expansion by cofactors. If

$$\mathbf{A} = \begin{pmatrix} \mathbf{B} & \mathbf{X} \\ 0 & \mathbf{C} \end{pmatrix},$$

where $\mathbf{B}$ and $\mathbf{C}$ are square matrices, then $\det \mathbf{A} = \det \mathbf{B} \det \mathbf{C}$ (2.5).

Applications

The first two applications are of more theoretical than practical value, especially for large n.

Cramer's Rule (3.5): Given a system of n equations in n unknowns, in matrix terms $\mathbf{AX} = \mathbf{Y}$, where $\mathbf{A} = (\mathbf{A}_1, \ldots, \mathbf{A}_n)$, this system has a unique solution iff $\det \mathbf{A} \neq 0$. If this latter holds, then the unique solution is

$$x_1 = \frac{\det(\mathbf{Y}, \mathbf{A}_2, \ldots, \mathbf{A}_n)}{\det \mathbf{A}}, \qquad x_2 = \frac{\det(\mathbf{A}_1, \mathbf{Y}, \mathbf{A}_3, \ldots, \mathbf{A}_n)}{\det \mathbf{A}},$$

$$\ldots, \qquad x_n = \frac{\det(\mathbf{A}_1, \ldots, \mathbf{A}_{n-1}, \mathbf{Y})}{\det \mathbf{A}}.$$

The inverse of a matrix in terms of determinants is:

$$\mathbf{A}^{-1} = (\det \mathbf{A})^{-1} \begin{pmatrix} A_{11} & \cdots & A_{1n} \\ \vdots & & \vdots \\ A_{n1} & \cdots & A_{nn} \end{pmatrix}^*,$$

where A_{ij} is the i,j-cofactor of **A** (4.5). The matrix

$$\begin{pmatrix} A_{11} & \cdots & A_{1n} \\ \cdot & & \cdot \\ \cdot & & \cdot \\ \cdot & & \cdot \\ A_{n1} & \cdots & A_{nn} \end{pmatrix}^{*}$$

is called the *adjoint matrix* of **A**.

The *van der Monde determinant* (Section 4) is

$$\det \begin{pmatrix} 1 & 1 & \cdots & 1 \\ a_1 & a_2 & & a_n \\ a_1^2 & a_2^2 & & a_n^2 \\ \cdot & \cdot & & \cdot \\ \cdot & \cdot & & \cdot \\ \cdot & \cdot & & \cdot \\ a_1^{n-1} & a_2^{n-1} & & a_n^{n-1} \end{pmatrix}.$$

Its most important feature is that if the $a_1, \ldots, a_n$ are all different, then $\det \mathbf{A} \neq 0$ (actual value given in (1), Section 4).

If **A** is an $m \times n$ matrix, then **A** is of rank r iff there is an $r \times r$ submatrix $\mathbf{A}_r$ of **A** obtained by striking out $m - r$ rows and $n - r$ columns, for which $\det \mathbf{A}_r \neq 0$, and if $\det \mathbf{A}_s = 0$ for any $s \times s$ submatrix thus obtained where $s > r$ (by 4.6).

Everything mentioned above also holds for matrices with complex entries (except that you cannot draw pictures for the geometry). For **A** in $\mathbb{C}_{n \times n}$, $\det \mathbf{A}$ is a complex number, of course.

1. DEFINITIONS AND THE LIKE

Geometrically speaking, the determinant is a function which gives the oriented volume of the parallelotope† determined by n vectors $\mathbf{A}_1, \ldots, \mathbf{A}_n$ in $\mathbb{R}_n$. That is, the parallelotope with adjacent edges $\mathbf{A}_1, \ldots, \mathbf{A}_n$; we shall hereafter identify such a parallelotope with the matrix

$$\mathbf{A} = (\mathbf{A}_1, \ldots, \mathbf{A}_n)^{*}$$

whose rows are these adjacent sides (see Fig. 1).

Let us lay aside the word "oriented" for the moment and determine what simple properties one might reasonably demand merely from a volume function

---

† Parallelotope is the general term for the class of figures which consist of parallelograms in the plane and parallelepipeds in 3-space.

1 / Definitions and the like

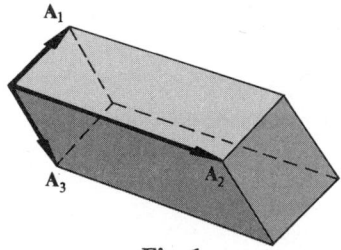

Fig. 1

defined on such parallelotopes. First of all, it seems sensible to require that the n-cube determined by $e_1, \ldots, e_n$ have volume $= 1$:

i) Volume of $(e_1, \ldots, e_n)^* = 1$ (Fig. 2).

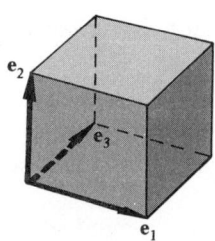

Volume = 1

Fig. 2

Next, if we stretch one edge of the parallelotope by an amount c (> 0), then we should increase the volume by a ratio of c:

ii) Volume $(A_1, \ldots, cA_i, \ldots, A_n) = c(\text{volume } (A_1, \ldots A_i, \ldots, A_n))$ (Fig. 3).

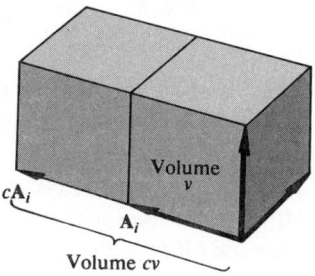

Fig. 3

Also, at least when edges A_i' and A_i'' have the same direction, we might ask that if two parallelotopes differ only in the ith edge, then the volume of the parallelotope with edges (A_i' and A_i'') be the sum of the volumes of those with edges A_i' and A_i'':

iii) volume $(A_1, \ldots, A_i' + A_i'', \ldots, A_n)$
 $=$ volume $(A_1, \ldots, A_i', \ldots, A_n) +$ volume $(A_1, \ldots, A_i'', \ldots, A_n)$ (Fig. 4).

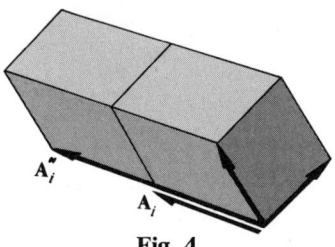

Fig. 4

However, if A_i' and A_i'' point in different directions, then the requirements of a volume function are not quite so clear. For example, if $A_i'' = -A_i'$, then the edge

$$A_i' + A_i'' = 0$$

and any parallelotope in n-space which has a zero edge should be counted as having zero n-dimensional volume.

By "oriented" volume, we have in mind regular volume with a plus or minus

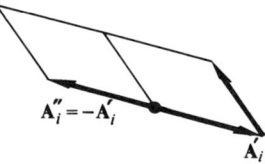

Fig. 5

sign. Although this may sound more complicated to deal with, the inherent algebraic manipulations are actually simpler, and the consequences are more far-reaching. Moreover, we can always recover nonoriented volumes by simply taking absolute values. A good model to keep in mind here is the definite integral $\int_a^b f$ of a nonnegative function f, which we sometimes loosely say represents the area under the curve f. Of course, this is true only if $b \geq a$; otherwise, we have the negative of this area: $\int_a^b f$ represents the *oriented* area (Fig. 6).

We shall require that our oriented volume (or determinant function) satisfy conditions which are slight extensions of (i) through (iii) above (it may not be entirely clear exactly what one should expect from oriented volume; we have the hindsight that this will bring us to everyone else's notion of determinant).

1 / Definitions and the like 235

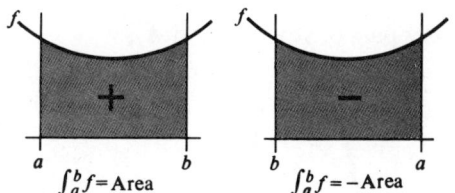

Fig. 6

Officially, the **determinant function** is that function on $n \times n$ matrices (or parallelotopes determined by n vectors in n-space, if you wish) which satisfies the following:

1. Determinant $(\mathbf{e}_1, \ldots, \mathbf{e}_n)^* = 1$; i.e. determinant $\mathbf{I} = 1$, where $\mathbf{I}$ is the $n \times n$ identity matrix.
2. *Multilinearity:* For each row $\mathbf{A}_i$,

$$\text{determinant} \begin{pmatrix} \mathbf{A}_1 \\ \vdots \\ c\mathbf{A}_i \\ \vdots \\ \mathbf{A}_n \end{pmatrix} = c \text{ determinant} \begin{pmatrix} \mathbf{A}_1 \\ \vdots \\ \mathbf{A}_i \\ \vdots \\ \mathbf{A}_n \end{pmatrix},$$

$$\text{determinant} \begin{pmatrix} \mathbf{A}_1 \\ \vdots \\ \mathbf{A}'_i + \mathbf{A}''_i \\ \vdots \\ \mathbf{A}_n \end{pmatrix} = \text{determinant} \begin{pmatrix} \mathbf{A}_1 \\ \vdots \\ \mathbf{A}'_i \\ \vdots \\ \mathbf{A}_n \end{pmatrix} + \text{determinant} \begin{pmatrix} \mathbf{A}_1 \\ \vdots \\ \mathbf{A}''_i \\ \vdots \\ \mathbf{A}_n \end{pmatrix}.$$

3. *Alternating:* For any pair of rows $\mathbf{A}_i$ and $\mathbf{A}_j$,

$$\text{determinant}\begin{pmatrix} \mathbf{A}_1 \\ \vdots \\ \mathbf{A}_i \\ \vdots \\ \mathbf{A}_j \\ \vdots \\ \mathbf{A}_n \end{pmatrix} = -\text{determinant}\begin{pmatrix} \mathbf{A}_1 \\ \vdots \\ \mathbf{A}_j \\ \vdots \\ \mathbf{A}_i \\ \vdots \\ \mathbf{A}_n \end{pmatrix}$$

for any $\mathbf{A}$'s in $\mathbb{R}_{n \times n}$ and scalar c (positive, negative, or zero). (The last condition is analogous to $\int_a^b f = -\int_b^a f$.)

Briefly, the determinant function on $n \times n$ matrices is an alternating, multilinear function of the rows, and has the value 1 on the $n \times n$ identity matrix. *Notation:* For

$$\mathbf{A} = \begin{pmatrix} \mathbf{A}_1 \\ \vdots \\ \mathbf{A}_n \end{pmatrix} = \begin{pmatrix} a_{11} & \cdots & a_{1n} \\ \vdots & & \vdots \\ a_{n1} & \cdots & a_{nn} \end{pmatrix}$$

we write

$$\det \mathbf{A} \quad \text{or} \quad \det \begin{pmatrix} \mathbf{A}_1 \\ \vdots \\ \mathbf{A}_n \end{pmatrix} \quad \text{or} \quad \det(\mathbf{A}_1, \ldots, \mathbf{A}_n)^*.$$

Another common notation is $|\mathbf{A}|$ or

$$\begin{vmatrix} a_{11} & \cdots & a_{1n} \\ \vdots & & \vdots \\ a_{n1} & \cdots & a_{nn} \end{vmatrix}$$

In regard to conditions (2) and (3), we shall later see that the determinant is an alternating, multilinear function of the *columns* of a matrix as well.

We were a bit presumptuous in defining the determinant to be "that function for which...". For, our conditions might be overly restrictive so that there are no

such functions (perhaps for some values of n). On the other hand, our assumptions may be so mild that several functions satisfy these conditions. Thus, we must show that for each n there is one and only one function having the above properties. For $n = 1$, there is only one row (which cannot be alternated) and the multilinearity (which becomes linearity) shows that
$$\det(a) = a \det(1) = a \cdot 1 = a.$$
Thus, the unique function on 1×1 matrices satisfying the determinant definition is $\det(a) = a$. As a further sample, we shall shortly show existence and uniqueness for $n = 2$. The following theorem contains useful consequences of the definition:

1.1 i) If two rows of a matrix $\mathbf{A}$ are the same then $\det \mathbf{A} = 0$.

ii) If one row of the matrix $\mathbf{A}$ consists of zeros, then $\det \mathbf{A} = 0$.

iii) If one row of the matrix $\mathbf{A}$ is a linear combination of the other rows, then $\det \mathbf{A} = 0$.

(This theorem is true for any alternating, multilinear function of the rows of a matrix and does not presuppose the existence or uniqueness of determinants.)

Proof. i) For the first part, interchange the rows which are the same and obtain $\det \mathbf{A} = -\det \mathbf{A}$. Hence $\det \mathbf{A} = 0$.

ii) If row $\mathbf{A}_i = (0, \ldots, 0)$, then $\mathbf{A}_i = 0\mathbf{A}_i$, so
$$\det(\mathbf{A}_1, \ldots, \mathbf{A}_i, \ldots, \mathbf{A}_n)^* = \det(\mathbf{A}_1, \ldots, 0\mathbf{A}_i, \ldots, \mathbf{A}_n)^*$$
$$= 0 \det \mathbf{A} = 0.$$

iii) For you. Replace the row by its linear combination; use multilinearity and (i). ∎

In the next section we shall encounter a less pessimistic companion theorem, which gives conditions under which the determinant is nonzero.

Geometrically, 1.1 states that if the parallelotope determined by $\mathbf{A}_1, \ldots, \mathbf{A}_n$ is collapsed (one edge a linear combination of the others) then the n-dimensional volume of this parallelotope is 0 (Fig. 7).

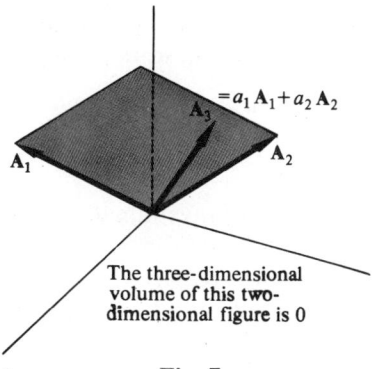

Fig. 7

We shall now settle the existence and uniqueness issues for determinants of 2×2 matrices. Let
$$\begin{pmatrix} a_{11} & a_{12} \\ a_{21} & a_{22} \end{pmatrix}$$
be such a matrix, with rows
$$\mathbf{A}_1 = (a_{11}, a_{12}), \qquad \mathbf{A}_2 = (a_{21}, a_{22}).$$
We can then write
$$\mathbf{A}_1 = a_{11}(1, 0) + a_{12}(0, 1) = a_{11}\mathbf{e}_1 + a_{12}\mathbf{e}_2. \tag{1}$$
$$\mathbf{A}_2 = a_{21}(1, 0) + a_{22}(0, 1) = a_{21}\mathbf{e}_1 + a_{22}\mathbf{e}_2. \tag{2}$$
Hence,
$$\det \begin{pmatrix} a_{11} & a_{12} \\ a_{21} & a_{22} \end{pmatrix}$$
$$= \det (\mathbf{A}_1, \mathbf{A}_2)^* \stackrel{(1)}{=} \det(a_{11}\mathbf{e}_1 + a_{12}\mathbf{e}_2, \mathbf{A}_2)^*$$
$$\stackrel{\text{linearity}}{=} a_{11} \det(\mathbf{e}_1, \mathbf{A}_2)^* + a_{12} \det(\mathbf{e}_2, \mathbf{A}_2)^*$$
$$\stackrel{(2)}{=} a_{11} \det(\mathbf{e}_1, a_{21}\mathbf{e}_1 + a_{22}\mathbf{e}_2)^* + a_{12} \det(\mathbf{e}_2, a_{21}\mathbf{e}_1 + a_{22}\mathbf{e}_2)^*$$
$$\stackrel{\text{linearity}}{=} a_{11}a_{21} \det(\mathbf{e}_1, \mathbf{e}_1)^* + a_{11}a_{22} \det(\mathbf{e}_1, \mathbf{e}_2)^* + a_{12}a_{21} \det(\mathbf{e}_2, \mathbf{e}_1)^*$$
$$+ a_{12}a_{22} \det(\mathbf{e}_2, \mathbf{e}_2)^*$$
$$\stackrel{1.1}{=} a_{11}a_{22} \det(\mathbf{e}_1, \mathbf{e}_2)^* + a_{12}a_{21} \det(\mathbf{e}_2, \mathbf{e}_1)^*$$
$$\stackrel{\text{alternating property}}{=} a_{11}a_{22} \det(\mathbf{e}_1, \mathbf{e}_2)^* - a_{12}a_{21} \det(\mathbf{e}_1, \mathbf{e}_2)^*$$
$$= (a_{11}a_{22} - a_{12}a_{21}) \det(\mathbf{e}_1, \mathbf{e}_2)^* = a_{11}a_{22} - a_{12}a_{21}.$$
Thus, if 2×2 determinants do indeed exist, we know what they must be:
$$\det \begin{pmatrix} a_{11} & a_{12} \\ a_{21} & a_{22} \end{pmatrix} = a_{11}a_{22} - a_{12}a_{21}.$$
We have thus settled the question of uniqueness before we know whether 2×2 determinants exist! Fortunately, all we need to do is show that the function
$$f((a_{11}, a_{12}), (a_{21}, a_{22})) = a_{11}a_{22} - a_{21}a_{12} \tag{3}$$
satisfies conditions (i), (ii), and (iii) of the definition, a task we leave for the student (Exercise 1).

Presumably, formula (3) is somehow intimately connected with the area of the parallelogram (Fig. 8).

We close this section with some remarks on functions of several variables in general, and on the alternating and multilinear conditions which appear in the definition of the determinant.

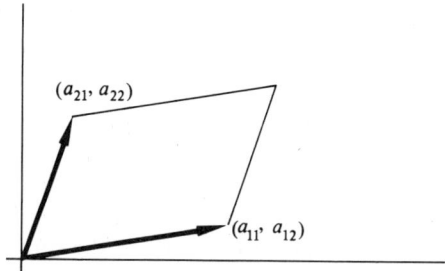

Fig. 8

The determinant may be considered an example of a function of n variables $f(\mathbf{V}_1, \ldots, \mathbf{V}_n)$, where $\mathbf{V}_1, \ldots, \mathbf{V}_n$ are taken from a vector space $\mathcal{V}$ over $\mathbb{R}$, and the value of the function, $\mathbf{W} = f(\mathbf{V}_1, \ldots, \mathbf{V}_n)$, lies in a vector space $\mathcal{W}$ over $\mathbb{R}$. Casually speaking, such a function is a rule which assigns to each set $\mathbf{V}_1, \ldots, \mathbf{V}_n$ of n vectors of $\mathcal{V}$ a unique vector $\mathbf{W}$ in $\mathcal{W}$. A picture:

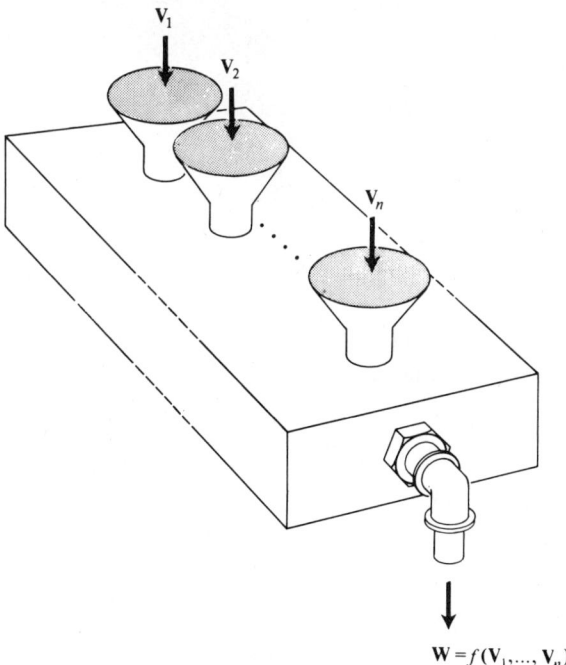

Fig. 9

Such a function is said to be **alternating** if whenever two of the arguments $V_1, \ldots, V_n$ are interchanged, the value of $W = f(V_1, \ldots, V_n)$ changes sign:

$$f(\cdots V_i \cdots V_j \cdots) = -f(\cdots V_j \cdots V_i \cdots) \quad \text{for all} \quad i \neq j.$$

For example, let $\mathcal{U} = \mathbb{R}$, $\mathcal{W} = \mathbb{R}$, and let

$$f(x, y) = \begin{cases} 1 & \text{if point } (x, y) \text{ is above the line } x = y \\ -1 & \text{if point } (x, y) \text{ is below the line } x = y \\ 0 & \text{if point } (x, y) \text{ is on the line } x = y. \end{cases}$$

Then f is an alternating function mapping pairs of vectors from $\mathcal{U}$ into $\mathcal{W}$, since interchanging x and y reflects the point (x, y) about the line $x = y$ (Fig. 10).

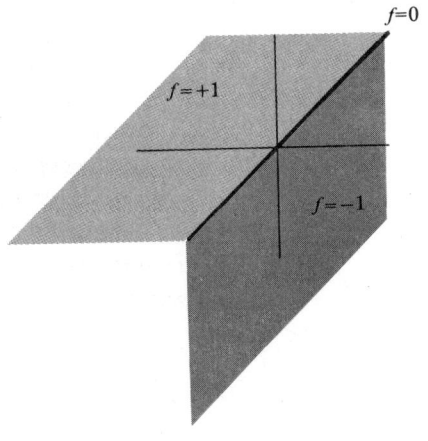

Fig. 10

A **multilinear** function $f(V_1, \ldots, V_n)$ is a function which is a linear transformation in each of its variables separately. That is,

$$f(V_1, \ldots, V_i + V_i', \ldots, V_n) = f(V_1, \ldots, V_i, \ldots, V_n) + f(V_1, \ldots, V_i', \ldots, V_n),$$

and

$$f(V_1, \ldots, aV_i, \ldots, V_n) = af(V_1, \ldots, V_i, \ldots, V_n).$$

That is, the function $T_i(X) = f(V_1, \ldots, V_{i-1}, X, V_{i+1}, \ldots, V_n)$ is a linear transformation for each $i = 1, \ldots, n$.

For example, it is easy to see that if $\mathcal{U} = \mathbb{R}$ and $\mathcal{W} = \mathbb{R}$, then

$$f(x, y, z) = xyz$$

is a multilinear function which maps triples of vectors from $\mathcal{U}$ into $\mathcal{W}$.

It is important to realize that if $f(V, W)$ is multilinear, then, for example,
$$f(V_1 + V_2, W) = f(V_1, W) + f(V_2, W)$$
and
$$f(V, W_1 + W_2) = f(V, W_1) + f(V, W_2);$$
but in general (see Exercise 7),
$$f(V_1 + V_2, W_1 + W_2) \neq f(V_1, W_1) + f(V_2, W_2). \tag{4}$$

EXERCISES

1. Prove that the function (3) satisfies the definition of determinant.
2. Relate (3) to the area of the parallelogram in Fig. 9.
3. Prove part (iii) of 1.1.
4. a) Show that a 2×2 matrix
$$A = \begin{pmatrix} a & b \\ c & d \end{pmatrix}$$
has an inverse iff $\det A \neq 0$.
   b) Write out A^{-1} in terms of the entries of A and $\det A$.
5. Prove that the only alternating multilinear function from triples of vectors from $\mathbb{R}_2$ into $\mathbb{R}$ is the function $f(A_1, A_2, A_3) = 0$. [*Hint:* Emulate our discussion showing the uniqueness of 2×2 determinants.]
6. Find some alternating multilinear functions from pairs of vectors from $\mathbb{R}_3$ into $\mathbb{R}$ (same hint).
7. Expand the left-hand side of (4) to see why it is different from the right-hand side.
8. Prove that $f(x, y, z) = xyz$ is multilinear.

Honors Projects.

1. *Complex Numbers.* Verify that everything here holds for scalars from $\mathbb{C}$ as well as from $\mathbb{R}$.
2. *Arbitrary Fields.* Verify that everything holds for scalars from an arbitrary field for which $x \neq -x$ for all nonzero elements (cf. the field with two elements in Chapter 0, Section 6, where $x = -x$ for all elements). If $x = -x$ does not imply $x = 0$, then our proof of 1.1 breaks down. Where? Most determinant results do hold in such fields, but proofs must be modified.
    These same honors projects apply to the other sections of this chapter as well, and will not be restated hereafter.

2. CALCULATIONS

It would be natural to settle the question of existence and uniqueness before proceeding further. However, the proofs are somewhat involved, so we have postponed the discussion until Section 5 (where you may immediately turn if you wish). We summarize the proof now since one expression, (2), will be immediately useful.

One can write a row $\mathbf{A}_i$ of an $n \times n$ matrix $\mathbf{A}$ as

$$\mathbf{A}_i = (a_{i1}, a_{i2}, \ldots, a_{in}) = a_{i1}\mathbf{e}_1 + a_{i2}\mathbf{e}_2 + \cdots + a_{in}\mathbf{e}_n.$$

Since the determinant function is linear in each row,

$$\det \mathbf{A} = \det (\mathbf{A}_1, \ldots, \mathbf{A}_i, \ldots, \mathbf{A}_n)^*$$
$$= \det (\mathbf{A}_1, \ldots, a_{i1}\mathbf{e}_1 + \cdots + a_{in}\mathbf{e}_n, \ldots, \mathbf{A}_n)^*$$
$$= a_{i1} \det (\mathbf{A}_1, \ldots, \mathbf{e}_1, \ldots, \mathbf{A}_n)^* + \cdots$$
$$+ a_{in} \det (\mathbf{A}_1, \ldots, \mathbf{e}_n, \ldots, \mathbf{A}_n)^*.$$

Writing the other rows in terms of $\mathbf{e}_1, \ldots, \mathbf{e}_n$ and expanding in this fashion, one obtains that

$$\det \mathbf{A} = \sum a_{i_1 1} a_{i_2 2} \cdots a_{i_n n} \det (\mathbf{e}_{i_1}, \mathbf{e}_{i_2}, \ldots, \mathbf{e}_{i_n})^*. \tag{1}$$

The latter determinants will be 0 if any of the $\mathbf{e}_{i_j}$ are the same, by 1.1. Thus, the only terms which contribute anything to the summation are those for which $(\mathbf{e}_{i_1}, \ldots, \mathbf{e}_{i_n})$ is a rearrangement (i.e., permutation) of $(\mathbf{e}_1, \ldots, \mathbf{e}_n)$. Moreover, in this case

$$\det (\mathbf{e}_{i_1}, \ldots, \mathbf{e}_{i_n})^* = \pm 1.$$

(If an even number of row interchanges is necessary to get from $\det (\mathbf{e}_{i_1}, \ldots, \mathbf{e}_{i_n})^*$ to $\det (\mathbf{e}_1, \ldots, \mathbf{e}_n) = 1$, the value will be $+1$; if an odd number are necessary, it will be -1.) Thus, if we delete extraneous terms from (1), we have

$$\det \begin{pmatrix} a_{11} & \cdots & a_{1n} \\ \cdot & & \cdot \\ \cdot & & \cdot \\ \cdot & & \cdot \\ a_{n1} & \cdots & a_{nn} \end{pmatrix} = \sum \pm a_{i_1 1} a_{i_2 2} \cdots a_{i_n n} \quad \begin{array}{l} \text{summed over all permutations} \\ (i_1, \ldots, i_n) \text{ of } (1, \ldots, n) \text{ with} \\ \text{the sign } \pm 1 = \det (\mathbf{e}_{i_1}, \ldots, \mathbf{e}_{i_n}). \end{array} \tag{2}$$

The details of this derivation are worked out more carefully in Section 5, where we shall base our proof of the existence and uniqueness of the determinant function on the expression just obtained. We shall assume for now that there is one and only one function satisfying our definition of determinant.

Equation (2) is an important expression for the determinant, but it is somewhat cumbersome to use. Our most economical method for calculating determinants will be based on the following:

2.1 The determinant of a triangular matrix. If

$$A = \begin{pmatrix} a_{11} & & & \\ 0 & a_{22} & & \\ 0 & \cdot & \cdot & \\ & & \cdot & \cdot \\ 0 & \cdots & 0 & a_{nn} \end{pmatrix},$$

then $\det A = a_{11}a_{22} \cdots a_{nn}$. That is, the determinant of a triangular matrix (one with 0's below the main diagonal) is the product of the elements on the main diagonal.

In like manner, for a matrix of the form

$$A = \begin{pmatrix} a_{11} & 0 & \cdots & 0 \\ & \cdot & \cdot & \cdot \\ & & \cdot & \cdot \\ & & & 0 \\ & & & a_{nn} \end{pmatrix},$$

$\det A = a_{11}a_{22} \cdots a_{nn}$.

Proof. We are assuming that $a_{pq} = 0$ if $p > q$. We shall work with the 3 × 3 case, which is typical. By (2),

$$\det \begin{pmatrix} a_{11} & a_{12} & a_{13} \\ a_{21} & a_{22} & a_{23} \\ a_{31} & a_{32} & a_{33} \end{pmatrix} = \sum \pm a_{i1}a_{j2}a_{k3}, \text{ summed over all permutations } (i, j, k) \text{ of } (1, 2, 3).$$

Let us consider the term $a_{i1}a_{j2}a_{k3}$. Now $a_{i1} = 0$ if $i > 1$, so we may suppose the first term is a_{11}. In like manner, $a_{j2} = 0$ if $j > 2$, so we need only consider $j = 1, 2$. Now 1 has been spoken for, since (i, j, k) is a permutation of $(1, 2, 3)$, so we can take $j = 2$. This leaves $k = 3$. Hence,

$$\det A = \pm a_{11}a_{22}a_{33}.$$

From (1), the sign comes from $\det (e_1, e_2, e_3) = 1$. Thus

$$\det A = a_{11}a_{22}a_{33}.$$

In the $n \times n$ case (Exercise 3) one would consider the terms $a_{i_1 1} a_{i_2 2} \cdots a_{i_n n}$ in (2), and show that we can assume that

$$i_1 = 1,\; i_2 = 2,\ldots, i_n = n.$$

Then determine the sign as we have done.

You show that if a matrix has 0's above the main diagonal, then the determinant is the product of the diagonal entries. ∎

Example.

$$\det \begin{pmatrix} 2 & - & - \\ 0 & 4 & - \\ 0 & 0 & -1 \end{pmatrix} = -8,$$

no matter how you fill in the blanks.

Geometrically, 2.1 is an extension of the theorem that the volume of a parallelepiped is the product of the length of one side times the width times the height. For the example, let

$$\mathbf{V}_1 = (2, -, -),\qquad \mathbf{V}_2 = (0, 4, -),\qquad \mathbf{V}_3 = (0, 0, -1)$$

(you fill in the blanks):

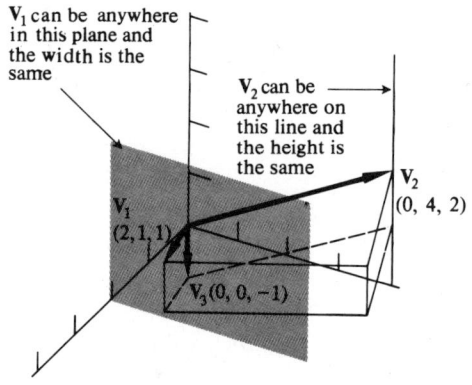

Fig. 11

It will now be useful to summarize the effect of the elementary row operations on the value of the determinant:

2.2 i) If $\mathbf{A}'$ is obtained from $\mathbf{A}$ by interchanging two rows, then $\det \mathbf{A}' = -\det \mathbf{A}$.

ii) If $\mathbf{A}'$ is obtained from $\mathbf{A}$ by multiplying a row by a scalar c, then $\det \mathbf{A}' = c \det \mathbf{A}$.

iii) If $\mathbf{A}'$ is obtained from $\mathbf{A}$ by adding a multiple of one row to another, then $\det \mathbf{A}' = \det \mathbf{A}$.

Proof. (i) is the alternating condition, (ii) comes from multilinearity. To prove (iii):

$$\det \begin{pmatrix} \mathbf{A}_1 \\ \vdots \\ \mathbf{A}_j \\ \vdots \\ c\mathbf{A}_j + \mathbf{A}_k \\ \vdots \\ \mathbf{A}_n \end{pmatrix} \stackrel{\text{multilinearity}}{=} c \det \begin{pmatrix} \mathbf{A}_1 \\ \vdots \\ \mathbf{A}_j \\ \vdots \\ \mathbf{A}_j \\ \vdots \\ \mathbf{A}_n \end{pmatrix} + \det \begin{pmatrix} \mathbf{A}_1 \\ \vdots \\ \mathbf{A}_j \\ \vdots \\ \mathbf{A}_k \\ \vdots \\ \mathbf{A}_n \end{pmatrix}$$

$$\stackrel{1.1}{=} \det \begin{pmatrix} \mathbf{A}_1 \\ \vdots \\ \mathbf{A}_j \\ \vdots \\ \mathbf{A}_k \\ \vdots \\ \mathbf{A}_n \end{pmatrix}. \quad \blacksquare$$

Now notice that we can transform any $n \times n$ matrix into triangular form by using the elementary row operations of row addition and row interchange: Given an $n \times n$ matrix $\mathbf{A}$

if there is a nonzero element in the first column, interchange its row with the first row and use it to clean out the remaining elements in the first column;

if there is a nonzero element in the second column (and not in the first row), interchange its row with the second row and use it to clean out the remaining elements in the first column;

.
.
.

if there is a nonzero element in the kth column (and not in the first $(k-1)$ rows), interchange its row with the kth row and use it to clean out the remaining elements in the kth column.

Since the determinant is an alternating function, each row interchange changes its sign. The "cleaning out" process involves adding a multiple of one row to another. Thus, as observed in 2.2, adding a multiple of one row to another does not affect the value of the determinant; we get the cleaning out process for free! We have only to keep track of the number of row interchanges in triangularizing the matrix. In summary:

2.3 *The determinant algorithm.* To calculate the value of det $\mathbf{A}$, for $\mathbf{A}$ an $n \times n$ matrix, use the elementary row operations of row addition and row interchange to transform $\mathbf{A}$ into a triangular matrix $\mathbf{A}'$. Then

$$\det \mathbf{A} = (-1)^i \det \mathbf{A}',$$

where i is the number of row interchanges used.

Example 1

$$\det \begin{pmatrix} 0 & 2 & 8 \\ 0 & 3 & 6 \\ 1 & 2 & -3 \end{pmatrix} = -\det \begin{pmatrix} 1 & 2 & -3 \\ 0 & 3 & 6 \\ 0 & 2 & 8 \end{pmatrix} = -\det \begin{pmatrix} 1 & 2 & -3 \\ 0 & 3 & 6 \\ 0 & 0 & 4 \end{pmatrix}$$
$$= -12.$$

Example 2

$$\det \begin{pmatrix} 2 & 8 & -4 & 4 \\ 1 & 4 & -1 & 3 \\ -1 & -4 & 3 & 0 \\ 0 & 0 & 8 & 5 \end{pmatrix} = \det \begin{pmatrix} 2 & 8 & -4 & 4 \\ 0 & 0 & 1 & 1 \\ 0 & 0 & 1 & 2 \\ 0 & 0 & 8 & 5 \end{pmatrix}.$$

We need go no further with the second example, since the zero in the second row–second column position will be retained throughout other row operations; the determinant is zero.

Our calculation procedure has some interesting theoretical consequences

2.4 i) det $\mathbf{A} \neq 0$ iff $\mathbf{A}$ has an inverse.

ii) det $\mathbf{A} \neq 0$ iff the rank of the $n \times n$ matrix $\mathbf{A}$ is n.

Proof. These statements say the same thing (Chapter 3, 4.6). We'll prove (ii). Transform $\mathbf{A}$ into a triangular matrix $\mathbf{A}'$ using elementary row operations. Then det $\mathbf{A} \neq 0$ iff det $\mathbf{A}' \neq 0$. Moreover, if

$$\mathbf{A}' = \begin{pmatrix} a'_{11} & & & \\ 0 & \cdot & & \\ \cdot & & \cdot & \\ \cdot & & & \cdot \\ 0 & \cdots & 0 & a'_{nn} \end{pmatrix},$$

then det $\mathbf{A}' = a'_{11} a'_{22} \cdots a'_{nn}$. By pivoting on the nonzero a'_{ii}, we see that

$$\text{rank } \mathbf{A} = n \quad \text{iff} \quad \text{all } a'_{ii} \neq 0 \text{ (why?)}.$$

Hence det $\mathbf{A}' \neq 0$ iff all $a'_{ii} \neq 0$, which is true iff rank $\mathbf{A}' = n$. Hence

$$\det \mathbf{A} \neq 0 \quad \text{iff} \quad \text{rank } \mathbf{A} = n. \quad \blacksquare$$

The following result is sometimes convenient:

2.5 If $\mathbf{A}$ is an $m \times m$ matrix and $\mathbf{B}$ an $n \times n$ matrix, then

$$\det \left(\begin{array}{c|c} \mathbf{A} & \mathbf{X} \\ \hline 0 & \mathbf{B} \end{array} \right) = \det \mathbf{A} \det \mathbf{B}.$$

Proof. In applying the Determinant Algorithm to the big matrix, one first applies it to $\mathbf{A}$ and then to $\mathbf{B}$ (you are asked for details in Exercise 6). $\blacksquare$

EXERCISES

1. Calculate the value of the following determinants:

a) $\det \begin{pmatrix} 1 & 0 & 2 & 0 \\ -1 & 1 & 0 & 1 \\ 3 & 0 & 1 & 2 \\ 4 & 1 & 2 & -1 \end{pmatrix}$
b) $\det \begin{pmatrix} 2 & 1 & 5 \\ 3 & -2 & 2 \\ 6 & -8 & -4 \end{pmatrix}$

c) $\det \begin{pmatrix} x-3 & 0 & -1 & 0 \\ 0 & x-3 & 0 & 0 \\ 0 & -1 & x-3 & 0 \\ 0 & 0 & 0 & x-3 \end{pmatrix}$

d) $\det \begin{pmatrix} 2 & -3 & 1 \\ 4 & 0 & 3 \\ 3 & 1 & 2 \end{pmatrix}$
e) $\det \begin{pmatrix} 1 & -1 & 3 & 4 \\ 0 & 1 & 0 & 1 \\ 2 & 0 & 1 & 2 \\ 0 & 1 & 2 & -1 \end{pmatrix}$

f) $\det \begin{pmatrix} 1 & 1 & 1 \\ 2 & 3 & 4 \\ 4 & 9 & 16 \end{pmatrix}$
g) $\det \begin{pmatrix} 1 & 1 & 1 & 1 \\ 1 & 2 & 3 & 4 \\ 1 & 4 & 9 & 16 \\ 1 & 8 & 27 & 64 \end{pmatrix}$

2. Calculate the value of the following 10 × 10 determinants:

a) $\det \begin{pmatrix} 0 & 0 & 0 & 0 & 0 & 1 & 0 & 0 & 0 & 0 \\ 0 & 0 & 0 & 0 & 0 & 0 & 2 & 0 & 0 & 0 \\ 0 & 0 & 0 & 0 & 0 & 0 & 0 & 3 & 0 & 0 \\ 0 & 0 & 0 & 0 & 0 & 0 & 0 & 0 & 4 & 0 \\ 5 & 0 & 0 & 0 & 0 & 0 & 0 & 0 & 0 & 5 \\ 0 & 6 & 0 & 0 & 0 & 0 & 0 & 0 & 0 & 0 \\ 0 & 0 & 7 & 0 & 0 & 0 & 0 & 0 & 0 & 0 \\ 0 & 0 & 0 & 8 & 0 & 0 & 0 & 0 & 0 & 0 \\ 0 & 0 & 0 & 0 & 9 & 0 & 0 & 0 & 0 & 0 \\ 0 & 0 & 0 & 0 & 0 & 10 & 0 & 0 & 0 & 0 \end{pmatrix}$

b) $\det \begin{pmatrix} 0 & 0 & 0 & 0 & 0 & 0 & 0 & 0 & 0 & 2 \\ 0 & 2 & 0 & 0 & 0 & 0 & 0 & 0 & 0 & 3 \\ 0 & 7 & \frac{1}{2} & 0 & 0 & 0 & 0 & 0 & 0 & 4 \\ 0 & 3 & 3 & 2 & 0 & 0 & 0 & 0 & 0 & 5 \\ 0 & 4 & 9 & 4 & \frac{1}{2} & 0 & 0 & 0 & 0 & 6 \\ 0 & 5 & 0 & 2 & 2 & 2 & 0 & 0 & 0 & 7 \\ 0 & 8 & 1 & 7 & 4 & 8 & \frac{1}{2} & 0 & 0 & 8 \\ 0 & e & 4 & 6 & 9 & \frac{3}{2} & 1 & 2 & 0 & 9 \\ 0 & 6 & \pi & 2 & 6 & 9 & 9 & 9 & \frac{1}{2} & 1 \\ 2 & 3 & 4 & 5 & 6 & 7 & 8 & 9 & 1 & 0 \end{pmatrix}$

c) $\det \begin{pmatrix} 0 & 2 & 0 & 0 & 0 & 0 & 0 & 0 & 0 & 0 \\ 0 & 0 & 2 & 0 & 0 & 0 & 0 & 0 & 0 & 0 \\ 0 & 0 & 0 & 2 & 0 & 0 & 0 & 0 & 0 & 0 \\ 0 & 0 & 0 & 0 & 2 & 0 & 0 & 0 & 0 & 0 \\ 0 & 0 & 0 & 0 & 0 & 2 & 0 & 0 & 0 & 0 \\ 0 & 0 & 0 & 0 & 0 & 0 & 2 & 0 & 0 & 0 \\ 0 & 0 & 0 & 0 & 0 & 0 & 0 & 2 & 0 & 0 \\ 0 & 0 & 0 & 0 & 0 & 0 & 0 & 0 & 2 & 0 \\ 0 & 0 & 0 & 0 & 0 & 0 & 0 & 0 & 0 & 2 \\ 1 & 0 & 0 & 0 & 0 & 0 & 0 & 0 & 0 & 0 \end{pmatrix}$

3. Prove 2.1 in the $n \times n$ case.
4. Make up a flow diagram for the Determinant Algorithm.
5. Answer the questions in the proof of 2.4.
6. Supply details for the proof of 2.5.

3. PRODUCTS AND TRANSPOSE

We shall first prove the following fundamental product formula:

3.1 For any $n \times n$ matrices **A** and **B**, det **AB** = det **A** det **B**.

3 / Products and transpose 249

Proof. CASE I. If $\det \mathbf{B} \neq 0$, then $\mathbf{B}^{-1}$ exists, by 2.4. For any $n \times n$ matrix $\mathbf{X} = (\mathbf{X}_1, \ldots, \mathbf{X}_n)^*$ with rows $\mathbf{X}_1, \ldots, \mathbf{X}_n$, define

$$f(\mathbf{X}) = f\begin{pmatrix} \mathbf{X}_1 \\ \vdots \\ \mathbf{X}_n \end{pmatrix} = \frac{1}{\det \mathbf{B}} \det \begin{pmatrix} \mathbf{X}_1 \mathbf{B} \\ \vdots \\ \mathbf{X}_n \mathbf{B} \end{pmatrix}.$$

Now f is alternating:

$$f\begin{pmatrix} \vdots \\ \mathbf{X}_i \\ \vdots \\ \mathbf{X}_j \\ \vdots \end{pmatrix} = \frac{1}{\det \mathbf{B}} \det \begin{pmatrix} \vdots \\ \mathbf{X}_i \mathbf{B} \\ \vdots \\ \mathbf{X}_j \mathbf{B} \\ \vdots \end{pmatrix}$$

$$= -\frac{1}{\det \mathbf{B}} \det \begin{pmatrix} \vdots \\ \mathbf{X}_j \mathbf{B} \\ \vdots \\ \mathbf{X}_i \mathbf{B} \\ \vdots \end{pmatrix} = -f\begin{pmatrix} \vdots \\ \mathbf{X}_j \\ \vdots \\ \mathbf{X}_i \\ \vdots \end{pmatrix}.$$

Further, f is multilinear (see Exercise 2), and

$$f(\mathbf{I}) = \frac{1}{\det \mathbf{B}} \det \begin{pmatrix} \mathbf{e}_1 \mathbf{B} \\ \vdots \\ \mathbf{e}_n \mathbf{B} \end{pmatrix} \stackrel{\text{(why?)}}{=} \frac{1}{\det \mathbf{B}} \det \mathbf{B} = 1.$$

Since the determinant is the only function which has these properties, we conclude that $f(\mathbf{X}) = \det \mathbf{X}$. In particular, $f(\mathbf{A}) = \det \mathbf{A}$, so

$$\det \mathbf{A} = f(\mathbf{A}) = \frac{1}{\det \mathbf{B}} \det \begin{pmatrix} \mathbf{A}_1\mathbf{B} \\ \vdots \\ \mathbf{A}_n\mathbf{B} \end{pmatrix} = \frac{1}{\det \mathbf{B}} \det \mathbf{AB}.$$

Hence, $\det \mathbf{A} \det \mathbf{B} = \det \mathbf{AB}$.

CASE II. $\det \mathbf{B} = 0$. Then by 2.4, the rank of $\mathbf{B}$ is less than n. Hence, the columns of $\mathbf{B}$ are not linearly independent. Hence, there are scalars $c_1, \ldots, c_n$ not all 0, for which

$$c_1 \mathbf{B}_1 + \cdots + c_n \mathbf{B}_n = 0,$$

where $\mathbf{B} = (\mathbf{B}_1, \ldots, \mathbf{B}_n)$. Now $\mathbf{AB} = (\mathbf{AB}_1, \ldots, \mathbf{AB}_n)$ and

$$c_1 \mathbf{AB}_1 + \cdots + c_n \mathbf{AB}_n = \mathbf{A}(c_1 \mathbf{B}_1 + \cdots + c_n \mathbf{B}_n) = \mathbf{A}0 = 0.$$

Hence, the columns of $\mathbf{AB}$ are not linearly independent; and the rank of $\mathbf{AB}$ is less than n, and $\det \mathbf{AB} = 0$. But also $\det \mathbf{A} \det \mathbf{B} = (\det \mathbf{A})0 = 0$, so we have finished. ∎

Thus, we have established the convenient and important fact that the determinant is multiplicative. On the other hand, the determinant is *not* additive in the usual sense: In general,

$$\det \mathbf{A} + \det \mathbf{B} \neq \det (\mathbf{A} + \mathbf{B}).$$

(See Exercise 7.) As a corollary to the above we obtain

3.2 The inverse of the determinant is the determinant of the inverse: If $\mathbf{A}$ has an inverse, then $\det \mathbf{A}^{-1} = 1/\det \mathbf{A}$.

Proof. Apply 3.1 to $\mathbf{AA}^{-1} = \mathbf{I}$. ∎

We next show that the determinant function behaves very nicely when it comes to transposes. We first practice on some of the elementary matrices: For the row interchange matrix

$$\mathbf{N}_{ij} = \begin{array}{c} \\ i \\ \\ j \\ \\ \end{array} \begin{pmatrix} 1 & & & & & & \\ & \ddots & \overset{i}{\vdots} & & \overset{j}{\vdots} & & \\ \cdots & & 0 & \cdots & 1 & \cdots & \\ & & \vdots & \ddots & \vdots & & \\ \cdots & & 1 & \cdots & 0 & \cdots & \\ & & \vdots & & \vdots & \ddots & \\ & & & & & & 1 \end{pmatrix},$$

we see that $\mathbf{N}_{ij}^* = \mathbf{N}_{ij}$ so that $\det \mathbf{N}_{ji} = \det \mathbf{N}_{ij}$. The row addition matrix

$$\mathbf{A}_{ij}(c) = \begin{pmatrix} 1 & & & & \\ & \ddots & & & \\ & & 1 & \cdots & -c \\ & & & \ddots & \\ & & & & 1 \\ & & & & & \ddots \\ & & & & & & 1 \end{pmatrix} \begin{matrix} \\ i \\ \\ j \\ \\ \\ \end{matrix}$$
$$\phantom{\mathbf{A}_{ij}(c) = xxxxxxxxxxxxx} i j$$

is triangular, as is $\mathbf{A}_{ij}(c)^*$, so their determinants are the same. In fact, in general

3.3 The determinant of the transpose is the same as the determinant of the original matrix: $\det \mathbf{A}^* = \det \mathbf{A}$.

Proof. By a succession of row interchanges and row additions, we can reduce any matrix to triangular form. Now these row operations may be accomplished by premultiplication by elementary matrices of the above types:

$$\mathbf{A}' = \begin{pmatrix} a'_{11} & & & & \\ 0 & & & & \\ \vdots & & & & \\ \vdots & & & & \\ 0 & \cdots & 0 & a'_{nn} \end{pmatrix} = \mathbf{E}_1 \mathbf{E}_2 \cdots \mathbf{E}_k \mathbf{A},$$

where the $\mathbf{E}$'s indicate row addition or row interchange matrices. Hence,

$$\mathbf{A} = \mathbf{E}_k^{-1} \mathbf{E}_{k-1}^{-1} \cdots \mathbf{E}_2^{-1} \mathbf{E}_1^{-1} \mathbf{A}' \quad (\text{why}_1?). \tag{1}$$

Hence,

$$\mathbf{A}^* = \mathbf{A}'^* (\mathbf{E}_1^{-1})^* (\mathbf{E}_2^{-1})^* \cdots (\mathbf{E}_k^{-1})^* \quad (\text{why}_2?). \tag{2}$$

Now the inverse of a row addition (or interchange) matrix is a row addition (or interchange) matrix (why$_3$?), and we saw that for such matrices

$$\det \mathbf{E}_j^{-1*} = \det \mathbf{E}_j^{-1}.$$

Moreover,

$$\det \mathbf{A}'^* = \det \begin{pmatrix} a'_{11} & 0 & \cdots & 0 \\ & \ddots & & \vdots \\ & & \ddots & 0 \\ & & & a'_{nn} \end{pmatrix} = \det \mathbf{A}' \quad (\text{why}_4?).$$

Applying 3.1 to (1) and (2) we therefore have that $\det \mathbf{A} = \det \mathbf{A}^*$ (why$_5$?) ∎

Since the transpose operator takes columns into rows, we obtain (you supply the details, in Exercise 5):

3.4 The determinant is an alternating, multilinear function of the *columns* of a matrix.

It follows therefore that all our results which deal with the rows of a matrix and the determinant can also be expressed as results about columns (Exercise 5). For example,

if two columns of a matrix are the same then the determinant is 0.

We close this section with a famous theorem on solutions to systems of linear equations with the same number of unknowns as equations:

3.5 Cramer's Rule. The system of n linear equations in n unknowns

$$a_{11}x_1 + \cdots + a_{1n}x_n = y_1$$
$$\vdots$$
$$a_{n1}x_1 + \cdots + a_{nn}x_n = y_n$$

has a unique solution iff the coefficient matrix

$$\mathbf{A} = \begin{pmatrix} a_{11} & \cdots & a_{1n} \\ \vdots & & \vdots \\ a_{n1} & \cdots & a_{nn} \end{pmatrix}$$

has nonzero determinant. Moreover, if $\det \mathbf{A} \neq 0$, this unique solution is

$$x_1 = \frac{\det \begin{pmatrix} y_1 & a_{12} & \cdots & a_{1n} \\ \vdots & \vdots & & \vdots \\ y_n & a_{n2} & \cdots & a_{nn} \end{pmatrix}}{\det \mathbf{A}}, \quad x_2 = \frac{\det \begin{pmatrix} a_{11} & y_1 & a_{13} & \cdots & a_{1n} \\ \vdots & \vdots & \vdots & & \vdots \\ a_{n1} & y_n & a_{n3} & \cdots & a_{nn} \end{pmatrix}}{\det \mathbf{A}},$$

$$\ldots, \quad x_n = \frac{\det \begin{pmatrix} a_{11} & \cdots & a_{1,n-1} & y_1 \\ \vdots & & \vdots & \vdots \\ a_{n1} & \cdots & a_{n,n-1} & y_n \end{pmatrix}}{\det \mathbf{A}}.$$

If $\det \mathbf{A} = 0$, the system will have either no solutions or an infinite number of solutions.

Proof. We know that the system has a unique solution iff $\mathbf{A}$ is invertible (why$_1$?), so the first part of Cramer's Rule follows from 2.4. Suppose then that $\det \mathbf{A} \neq 0$ and that $\mathbf{X} = (x_1, \ldots, x_n)^*$ is this unique solution. We must show that the x_i's may be expressed as indicated. Consider the scalar x_1, for example. Using column properties we have

$x_1 \det(\mathbf{A}_1, \ldots, \mathbf{A}_n)$

$\overset{(\text{why}_2?)}{=} x_1 \det(\mathbf{A}_1, \ldots, \mathbf{A}_n) + x_2 \det(\mathbf{A}_2, \mathbf{A}_2, \ldots, \mathbf{A}_n)$
$\qquad\qquad\qquad\qquad\qquad\qquad\qquad + \cdots + x_n \det(\mathbf{A}_n, \mathbf{A}_2, \ldots, \mathbf{A}_n)$

$\overset{(\text{why}_3?)}{=} \det(x_1\mathbf{A}_1 + x_2\mathbf{A}_2 + \cdots + x_n\mathbf{A}_n, \mathbf{A}_2, \ldots, \mathbf{A}_n)$

$= \det \begin{pmatrix} x_1 a_{11} + \cdots + x_n a_{1n} & a_{12} & \cdots & a_{1n} \\ \vdots & \vdots & & \vdots \\ x_1 a_{n1} + \cdots + x_n a_{nn} & a_{n2} & \cdots & a_{nn} \end{pmatrix}$

$\overset{(\text{why}_4?)}{=} \det \begin{pmatrix} y_1 & a_{12} & \cdots & a_{1n} \\ \vdots & \vdots & & \vdots \\ y_n & a_{n2} & \cdots & a_{nn} \end{pmatrix},$

where $\mathbf{A}_1, \ldots, \mathbf{A}_n$ are the columns of $\mathbf{A}$. If we let $\mathbf{Y} = (y_1, \ldots, y_n)^*$, then these equalities state that

$$x_1 \det \mathbf{A} = \det(\mathbf{Y}, \mathbf{A}_2, \ldots, \mathbf{A}_n),$$

so that

$$x_1 = \det(\mathbf{Y}, \mathbf{A}_2, \ldots, \mathbf{A}_n)/\det \mathbf{A},$$

which is the desired form. In like manner, all other x_i's are as indicated (Exercise 6). ∎

Cramer's Rule is of more theoretical than computational interest, at least for decent-sized values of n. Our computational objections are indicated in the next section. It is worth noting the 2×2 case of Cramer's Rule:

The equations $ax + by = c$, $a'x + b'y = c'$ have a unique solution iff the appropriate determinant is nonzero, and the solution is

$$x = \det \begin{pmatrix} c & b \\ c' & b' \end{pmatrix} \Big/ \det \begin{pmatrix} a & b \\ a' & b' \end{pmatrix}, \qquad y = \det \begin{pmatrix} a & c \\ a' & c' \end{pmatrix} \Big/ \det \begin{pmatrix} a & b \\ a' & b' \end{pmatrix}.$$

EXERCISES

1. Use Cramer's Rule (if possible) to solve the following systems of linear equations in the exercises to Chapter 4, Section 1:
   a) Exercise 1;   b) Exercise 3;   c) Exercise 4;
   d) Exercise 7;   e) Exercise 13

2. Show that the f in 3.1 is multilinear.

3. (a) Prove 3.2. (b) Check 3.2 on the matrices in those exercises you were assigned from among Exercises 16 through 20 of Chapter 4, Section 2.
4. Answer the questions in the proof of 3.3.
5. (a) Prove 3.4. (b) Go through the material in Sections 1 through 3 and restate any results concerning the rows of a matrix in terms of the columns.
6. Answer the questions in the proof of 3.5.
7. Prove by an example that, in general,
$$\det (\mathbf{A} + \mathbf{B}) \neq \det \mathbf{A} + \det \mathbf{B}.$$
8. a) Prove that the set of all elements of $\mathbb{R}_{n \times n}$ which have nonzero determinant forms a group under multiplication (definition in Supplement 2 of Chapter 0, Section 1).
   b) Show that the set of all elements of $\mathbb{R}_{n \times n}$ whose determinant is ± 1 forms a group under multiplication.
   c) Show that the group in (b) is noncommutative if $n = 2$.
   d) Show that it is noncommutative if $n > 1$.
   e) Find another set of elements in $\mathbb{R}_{n \times n}$ which forms a group under multiplication.

4. COFACTORS

In this section we shall develop some useful theoretical tools, which are sometimes mistaken for important computational techniques.

First of all, given an $n \times n$ matrix

$$\mathbf{A} = \begin{pmatrix} a_{11} & \cdots & a_{1n} \\ \vdots & & \vdots \\ a_{n1} & \cdots & a_{nn} \end{pmatrix},$$

we define the *i, j*-cofactor A_{ij} to be the determinant of the $(n-1) \times (n-1)$ matrix obtained from $\mathbf{A}$ by deleting its *i*th row and *j*th column and taken with the sign $(-1)^{i+j}$:

$$A_{ij} = (-1)^{i+j} \det \begin{pmatrix} a_{11} & \cdots & a_{1,j-1} & a_{1,j+1} & \cdots & a_{1n} \\ \vdots & & \vdots & \vdots & & \vdots \\ a_{i-1,1} & & & & & a_{i-1,n} \\ a_{i+1,1} & & & & & a_{i+1,n} \\ \vdots & & \vdots & \vdots & & \vdots \\ a_{n1} & \cdots & a_{n,j-1} & a_{n,j+1} & \cdots & a_{nn} \end{pmatrix}.$$

Thus, the cofactor is a scalar which is $\pm$ (a certain $(n-1) \times (n-1)$ determinant). We shall now prove a lemma which describes cofactors in terms of $n \times n$ determinants.

4.1 The i,j-cofactor is given by

$$A_{ij} = \det \begin{pmatrix} a_{11} & \cdots & a_{1,j-1} & 0 & a_{1,j+1} & \cdots & a_{1n} \\ \vdots & & \vdots & \vdots & \vdots & & \vdots \\ a_{i-1,1} & \cdots & a_{i-1,j-1} & 0 & a_{i-1,j+1} & \cdots & a_{i-1,n} \\ 0 & \cdots & 0 & 1 & 0 & \cdots & 0 \\ a_{i+1,1} & \cdots & a_{i+1,j-1} & 0 & a_{i+1,j+1} & \cdots & a_{i+1,n} \\ \vdots & & \vdots & \vdots & \vdots & & \vdots \\ a_{n1} & \cdots & a_{n,j-1} & 0 & a_{n,j+1} & \cdots & a_{nn} \end{pmatrix}.$$

That is, A_{ij} is the determinant of an $n \times n$ matrix obtained from $\mathbf{A}$ by replacing a_{ij} by 1, and all other entries in the ith row and jth column by 0's.

Proof. Using row and column interchanges, we have

$$\det \begin{pmatrix} a_{11} & \cdots & \overset{j}{0} & \cdots & a_{1n} \\ \vdots & & \vdots & & \vdots \\ 0 & \cdots & 1 & \cdots & 0 \\ \vdots & & \vdots & & \vdots \\ a_{n1} & \cdots & 0 & \cdots & a_{nn} \end{pmatrix} \; i$$

$$= (-1)^{i-1} \det \begin{pmatrix} 0 & \cdots & \overset{j}{1} & \cdots & 0 \\ a_{11} & \cdots & 0 & \cdots & a_{1n} \\ \vdots & & & & \vdots \\ a_{n1} & \cdots & 0 & \cdots & a_{nn} \end{pmatrix} \quad (1)$$

$$= (-1)^{i-1}(-1)^{j-1} \det \begin{pmatrix} 1 & 0 & \cdots & 0 \\ 0 & a_{11} & \cdots & a_{1n} \\ & \text{ith row and} & & \\ & \text{jth column} & & \\ & \text{missing} & & \\ 0 & a_{n1} & \cdots & a_{nn} \end{pmatrix}$$

$$\overset{2.5}{=} (-1)^{i+j} \det 1 \det \begin{pmatrix} a_{11} & \cdots & a_{1n} \\ & \text{ith row and} & \\ & \text{jth column} & \\ & \text{missing} & \\ a_{n1} & \cdots & a_{nn} \end{pmatrix} = A_{ij}. \quad \blacksquare$$

We can now prove a theorem on the expansion of a determinant by cofactors.†

4.2 If

$$A = \begin{pmatrix} a_{11} & \cdots & a_{1n} \\ \vdots & & \vdots \\ a_{n1} & \cdots & a_{nn} \end{pmatrix},$$

then $\det A = a_{1j}A_{1j} + a_{2j}A_{2j} + \cdots + a_{nj}A_{nj}$, for any $j = 1, \ldots, n$.

Proof. By the linearity of the jth column,

$$\det A = a_{1j} \det \begin{pmatrix} a_{11} & \cdots & 1 & \cdots & a_{1n} \\ a_{21} & \cdots & 0 & \cdots & a_{2n} \\ \vdots & & \vdots & & \vdots \\ a_{n1} & \cdots & 0 & \cdots & a_{nn} \end{pmatrix} + a_{2j} \det \begin{pmatrix} a_{11} & \cdots & 0 & \cdots & a_{1n} \\ a_{21} & \cdots & 1 & \cdots & a_{2n} \\ \vdots & & 0 & & \vdots \\ a_{n1} & \cdots & 0 & \cdots & a_{nn} \end{pmatrix}$$

$$+ \cdots + a_{nj} \det \begin{pmatrix} a_{11} & \cdots & 0 & \cdots & a_{1n} \\ \vdots & & \vdots & & \vdots \\ a_{n1} & \cdots & 1 & \cdots & a_{nn} \end{pmatrix}.$$

Now add appropriate multiples of the columns containing one 1 and the rest zeros to the other columns, so as to clean out the rows containing the 1's. By the column form of 2.2(iii) and by 4.1, the ith term in this sum is $a_{ij}A_{ij}$. ∎

Example. Expanding by the first column,

$$\det \begin{pmatrix} a_{11} & a_{12} & a_{13} \\ a_{21} & a_{22} & a_{23} \\ a_{31} & a_{32} & a_{33} \end{pmatrix} = a_{11}A_{11} + a_{21}A_{21} + a_{31}A_{31}$$

$$= a_{11}(-1)^{1+1} \det \begin{pmatrix} a_{22} & a_{23} \\ a_{32} & a_{33} \end{pmatrix}$$

$$+ a_{21}(-1)^{2+1} \det \begin{pmatrix} a_{12} & a_{13} \\ a_{32} & a_{33} \end{pmatrix}$$

$$+ a_{31}(-1)^{3+1} \det \begin{pmatrix} a_{12} & a_{13} \\ a_{22} & a_{23} \end{pmatrix}.$$

---

† The term **minor** for a cofactor without the sign $(-1)^{i+j}$ is a common one. The following theorem is often stated in this terminology as "expansion by minors".

From this one can easily obtain an expression of the form $\sum \pm a_{i1}a_{j2}a_{k3}$. Those of you who have previously seen the "method of diagonals" for calculating 3×3 determinants should check this formula (incidentally, there are no convenient and simple comparable "methods of diagonals" for 4×4 or larger determinants).

4.3 $\det A = a_{i1}A_{i1} + a_{i2}A_{i2} + \cdots + a_{in}A_{in}$, for any $i = 1, \ldots, n$.

Proof. For you (Exercise 2a). ∎

If we mismatch cofactors and elements, all comes to naught:

4.4 If $i \neq j$, then
$$a_{i1}A_{j1} + a_{i2}A_{j2} + \cdots + a_{in}A_{jn} = 0,$$
$$a_{1i}A_{1j} + a_{2i}A_{2j} + \cdots + a_{ni}A_{nj} = 0.$$

Proof. Yours again (Exercise 2b). We suggest for the first that you expand by cofactors a matrix whose ith and jth rows are both $(a_{i1}, a_{i2}, \ldots, a_{in})$. ∎

The **adjoint** matrix adj A of a matrix $A = (a_{ij})$ is the matrix whose i, j-entry is the j, i-cofactor of A. That is,

$$\operatorname{adj} A = \begin{pmatrix} A_{11} & A_{21} & \cdots & A_{n1} \\ A_{12} & & & \\ \vdots & & \ddots & \vdots \\ A_{1n} & & \cdots & A_{nn} \end{pmatrix} = \begin{pmatrix} A_{11} & A_{12} & \cdots & A_{1n} \\ A_{21} & & & \\ \vdots & & \ddots & \vdots \\ A_{n1} & & \cdots & A_{nn} \end{pmatrix}^{*}.$$

4.5
$$A \cdot \operatorname{adj} A = \begin{pmatrix} \det A & & 0 \\ & \ddots & \\ 0 & & \det A \end{pmatrix} = \det A \cdot I = \operatorname{adj} A \cdot A.$$

Hence, if A is invertible, then
$$A^{-1} = \frac{1}{\det A} \operatorname{adj} A.$$

Proof. Just multiply (Exercise 2c). ∎

Example.
$$A = \begin{pmatrix} a & b \\ c & d \end{pmatrix}$$
is invertible iff $\det A = ad - bc \neq 0$. In this case,
$$A^{-1} = \frac{1}{\det A} \begin{pmatrix} d & -b \\ -c & a \end{pmatrix},$$

as can be readily verified. On the other hand,

$$A_{11} = (-1)^2 \det \begin{pmatrix} \cancel{a} & \cancel{b} \\ \cancel{c} & d \end{pmatrix} = d, \qquad A_{12} = (-1)^3 \det \begin{pmatrix} \cancel{a} & \cancel{b} \\ c & \cancel{d} \end{pmatrix} = -c,$$

$$A_{21} = (-1)^3 \det \begin{pmatrix} \cancel{a} & b \\ \cancel{c} & \cancel{d} \end{pmatrix} = -b, \qquad A_{22} = (-1)^4 \det \begin{pmatrix} a & b \\ \cancel{c} & \cancel{d} \end{pmatrix} = a.$$

Hence,

$$\text{adj } \mathbf{A} = \begin{pmatrix} A_{11} & A_{21} \\ A_{12} & A_{22} \end{pmatrix} = \begin{pmatrix} d & -b \\ -c & a \end{pmatrix}.$$

Why do we not consider expansion by cofactors (4.2) a particularly useful method for calculating the determinant of a matrix? Unless you have a particularly choice matrix, for which the cofactors are all especially easy to calculate, the expansion merely replaces the calculation of one $n \times n$ determinant by the calculation of n $[(n-1) \times (n-1)]$ determinants, which, in turn, must be expanded in terms of $[(n-2) \times (n-2)]$ determinants, and so on. This process ultimately leads to our expression for determinant,

$$\det \mathbf{A} = \sum \pm a_{i_1 1} a_{i_2 2} \cdots a_{i_n n}, \qquad \text{over all permutations}$$

$(i_1, \ldots, i_n)$ of $(1, \ldots, n)$.

The direct evaluation of this expression involves $n!$ multiplications alone, a rather large number, in general, when compared to our method of the previous section which demands fewer than n^3 multiplications (and which can be improved upon). To get some notion of the orders of magnitude involved, suppose we were to calculate a 100×100 determinant on a computer which could do a multiplication in 10^{-8} seconds. Then (ignoring the time necessary to perform other operations) using our best method would take the computer

$$(100)^3 \text{ multiplications} \times 10^{-8} \text{ sec/mult.} = 0.01 \text{ sec.}$$

On the other hand, the direct method would take $100!$ multiplications $\times$ 10^{-8} sec/mult. Now by Stirling's Formula $100! \approx (100/e)^{100}\sqrt{2\pi 100}$, which is roughly 9×10^{157}. Hence,

$$100! \text{ multiplications} \times 10^{-8} \text{ sec/mult.} \approx 9 \times 10^{149} \text{ secs.}$$

There are approximately 3×10^7 seconds per year, so the total time necessary to expand a 100×100 determinant using the summation formula would be approximately 3×10^{142} years.

Why don't we suggest the calculation of $\mathbf{A}^{-1}$ by 4.5? To use 4.5 involves not only calculating det $\mathbf{A}$, but also adj $\mathbf{A}$, which is comprised of n^2 $[(n-1) \times (n-1)]$ determinants, about $n^3 + n^2(n-1)^3$ multiplications by our best method. Our Chapter 4 technique of calculating $\mathbf{A}^{-1}$ was based on the same pivoting procedure as used for calculating det $\mathbf{A}$, so again about n^4 multiplications are called for.

Why don't we recommend Cramer's Rule (3.5) for solving systems of linear equations? First of all, it's too specialized; it involves the same number of equations as unknowns, and the matrix $\mathbf{A}$ must be invertible. Secondly, it requires computing

$n + 1$ determinants, while our Chapter 4 method of Gaussian elimination essentially involves the same amount of work as the calculation of one determinant.

One should not think that to minimize the number of arithmetic operations involved in a method is the sole desideratum. Other important considerations include the amount of error a computer would introduce in using the method and special forms of systems (such as having a very large number of equations, or lots of zero coefficients, or both very large and very small coefficients, and the like). We also re-emphasize that in special cases, and in particular for $n = 2$ and 3, the theorems of Sections 3 and 4 may be very helpful computational tools. In fact, we shall discover this in the next chapter.

In computing a determinant the expansion by cofactors is often used in conjunction with our methods.

Example. Let us redo Example 1 of Section 2, expanding by cofactors on the first column:

$$\det \begin{pmatrix} 0 & 2 & 8 \\ 0 & 3 & 6 \\ 1 & 2 & -3 \end{pmatrix} = 1(-1)^{3+1} \det \begin{pmatrix} 2 & 8 \\ 3 & 6 \end{pmatrix} = (12 - 24) = -12.$$

The **van der Monde determinant** is

$$\det \begin{pmatrix} 1 & 1 & \cdots & 1 \\ a_1 & a_2 & \cdots & a_n \\ a_1^2 & a_2^2 & \cdots & a_n^2 \\ \vdots & \vdots & & \vdots \\ a_1^{n-1} & a_2^{n-1} & \cdots & a_n^{n-1} \end{pmatrix}.$$

Although rather special in appearance, it pops up in enough important situations to merit our attention. Its value can be seen to be

$$(a_2 - a_1)(a_3 - a_1)(a_4 - a_1) \cdots (a_n - a_1)$$
$$(a_3 - a_2)(a_4 - a_2) \cdots (a_n - a_2)$$
$$\vdots \tag{1}$$
$$(a_{n-1} - a_{n-2})(a_n - a_{n-2})(a_n - a_{n-1}) = \prod_{i>j}(a_i - a_j) \quad (\prod = \text{product}).$$

(See Exercise 3.)

Note that *if $a_1, \ldots, a_n$ are all different, then the value of the van der Monde determinant is nonzero.* We shall consider the 4×4 case:

$$\det \begin{pmatrix} 1 & 1 & 1 & 1 \\ a_1 & a_2 & a_3 & a_4 \\ a_1^2 & a_2^2 & a_3^2 & a_4^2 \\ a_1^3 & a_2^3 & a_3^3 & a_4^3 \end{pmatrix} \overset{\text{column pivot}}{=} \det \begin{pmatrix} 1 & 0 & 0 & 0 \\ a_1 & a_2 - a_1 & a_3 - a_1 & a_4 - a_1 \\ a_1^2 & a_2^2 - a_1^2 & a_3^2 - a_1^2 & a_4^2 - a_1^2 \\ a_1^3 & a_2^3 - a_1^3 & a_3^3 - a_1^3 & a_4^3 - a_1^3 \end{pmatrix}$$

$$= \det \begin{pmatrix} 1 & 0 & 0 & 0 \\ a_1 & (a_2 - a_1) \cdot 1 & (a_3 - a_1) \cdot 1 & (a_4 - a_1) \cdot 1 \\ a_1^2 & (a_2 - a_1)(a_2 + a_1) & (a_3 - a_1)(a_3 + a_1) & (a_4 - a_1)(a_4 + a_1) \\ a_1^3 & (a_2 - a_1)(a_2^2 + a_2 a_1 + a_1^2) & (a_3 - a_1)(a_3^2 + a_3 a_1 + a_1^2) & (a_4 - a_1)(a_4^2 + a_4 a_1 + a_1^2) \end{pmatrix}$$

2.4 multilin.
$$= (a_2 - a_1)(a_3 - a_1)(a_4 - a_1) \det \begin{pmatrix} 1 & 1 & 1 \\ a_2 + a_1 & a_3 + a_1 & a_4 + a_1 \\ a_2^2 + a_2 a_1 + a_1^2 & a_3^2 + a_3 a_1 + a_1^2 & a_4^2 + a_4 a_1 + a_1^2 \end{pmatrix}$$

add $-a_1$ row 2 to row 3
$$= (a_2 - a_1)(a_3 - a_1)(a_4 - a_1) \det \begin{pmatrix} 1 & 1 & 1 \\ a_2 + a_1 & a_3 + a_1 & a_4 + a_1 \\ a_2^2 & a_3^2 & a_4^2 \end{pmatrix}$$

add $-a_1$ row 1 to row 2
$$= (a_2 - a_1)(a_3 - a_1)(a_4 - a_1) \det \begin{pmatrix} 1 & 1 & 1 \\ a_2 & a_3 & a_4 \\ a_2^2 & a_3^2 & a_4^2 \end{pmatrix}.$$

Since the last term is a 3×3 van der Monde determinant, one can assume by induction (after checking the $n = 1$ case) that its value is $(a_2 - a_3)(a_2 - a_4) \times (a_3 - a_4)$, which yields the appropriate value for the 4×4 one.

We close with a more specialized result, which says that the rank of a matrix (not necessarily square) is the size of the largest nonzero subdeterminant. We shall make no official use of this theorem (with the exception of Exercises 13 and 14).

4.6 Let $\mathbf{A}$ be an $m \times n$ matrix of rank r. Then

i) There is an $r \times r$ submatrix $\mathbf{A}_r$ obtained from $\mathbf{A}$ by deleting $m - r$ rows and $n - r$ columns, for which $\det \mathbf{A}_r \neq 0$.

ii) If $\mathbf{A}_s$ is any $s \times s$ submatrix obtained from $\mathbf{A}$ by deleting $m - s$ rows and $n - s$ columns, and if $s > r$, then $\det \mathbf{A}_s = 0$.

Proof. Let $\mathbf{A}_s$ be the submatrix obtained from $\mathbf{A}$ by keeping the first s rows and columns. Thus, the columns of $\mathbf{A}_s$ are $\bar{\mathbf{A}}_1, \bar{\mathbf{A}}_2, \ldots, \bar{\mathbf{A}}_s$ and they look like the first s columns of $\mathbf{A}$ (namely, $\mathbf{A}_1, \ldots, \mathbf{A}_s$) with the last $m - s$ components chopped off. Now suppose $\det \mathbf{A}_s \neq 0$. Then $\bar{\mathbf{A}}_1, \ldots, \bar{\mathbf{A}}_s$ are linearly independent (why$_1$?). Hence, $\mathbf{A}_1, \ldots, \mathbf{A}_s$ are linearly independent (why$_2$?). [*Hint:* Just suppose some linear combination is 0 and reduce it to the $\bar{\mathbf{A}}_i$ situation.] Hence, if $\det \mathbf{A}_s \neq 0$, then $s \leq r$ (why$_3$?). The only difference between this and an arbitrary $s \times s$ submatrix $\mathbf{A}_s$ is one of notation (why$_4$?), so we have proved (ii).

To prove (i), suppose that $\mathbf{A}$ is of rank r and, in fact, that the first r columns $\mathbf{A}_1, \ldots, \mathbf{A}_r$ are linearly independent. Then $(\mathbf{A}_1, \ldots, \mathbf{A}_r)$ is an $m \times r$ matrix of rank r. Since its row rank must also be r, suppose that the first r rows are linearly independent. Then the matrix $\mathbf{A}_r$ consisting of these first r rows is of rank r, so that $\det \mathbf{A}_r \neq 0$, and $\mathbf{A}_r$ is obtained from $\mathbf{A}$ by deleting $m - r$ rows and $n - r$ columns. ∎

Another proof of this theorem is outlined in Exercise 7.

EXERCISES

1. Verify that the signs in (1) are correct.
2. Prove (a) 4.3; (b) 4.4; (c) 4.5.
3. Prove that the value of the van der Monde determinant is as stated (a) for $n = 3$, (b) for arbitrary n. [*Hint:* Begin by pivoting on columns on the first element of the first row, and then take out factors from the remaining columns. Expand by cofactors and reduce the resulting $(n - 1) \times (n - 1)$ determinant to a van der Monde determinant by row operations.]
4. Prove that to calculate
$$\sum \text{sign } (i_1, \ldots, i_n) a_{i_11} a_{i_22} a_{i_3 3} \cdots {}_n n$$
requires $n!$ multiplications.
5. Prove that our method of calculating determinants involves at most
$$\tfrac{1}{3} n(n^2 - 1) + 1$$
multiplications.
6. Prove the theorem on determinants of triangular matrices using expansion by cofactors.
7. We define the **determinant rank** of an $m \times n$ matrix A to be the dimension k of the largest submatrix A_k of A for which $\det A_k \neq 0$ (A_k obtained by deleting entire rows and columns, of course).
   a) Prove that equivalent matrices have the same determinant rank.
   b) Use this and the fact that the determinant rank of a matrix of the form
$$\begin{pmatrix} I_r & 0 \\ \hline 0 & 0 \end{pmatrix},$$
   where I_r is the $r \times r$ identity matrix, is the same as the rank of the matrix to give another proof of 4.6.
8. Calculate the adjoints of the following matrices A. Check your results by showing that $A \text{ adj } A$ is a diagonal matrix.

   a) $A = \begin{pmatrix} 1 & 2 & 0 \\ 1 & -4 & 1 \\ 2 & 1 & -3 \end{pmatrix}$ b) $A = \begin{pmatrix} 2 & 1 & 5 \\ 3 & -2 & 2 \\ 6 & -8 & -4 \end{pmatrix}$.

9. (a) through (g). Calculate the determinants in Exercise 1 of Section 2, using the most convenient means at your disposal.
10. (a) through (c). Same for Exercise 2 of Section 2.
11. Calculate the following determinants. Save your work for future use (although this may seem rather unlikely at the moment).

   a) $\det \begin{pmatrix} 3-x & 0 & 1 & 0 \\ 0 & 3-x & 0 & 0 \\ 0 & 1 & 3-x & 0 \\ 0 & 0 & 0 & 3-x \end{pmatrix}$ b) $\det \begin{pmatrix} 2-x & 1 & 0 \\ 0 & 2-x & 0 \\ 0 & 0 & 3-x \end{pmatrix}$

   c) $\det \begin{pmatrix} 3-x & -2 & 1 \\ 2 & 1-x & -1 \\ 2 & -6 & 4-x \end{pmatrix}$ d) $\det \begin{pmatrix} 3-x & 2 & 2 \\ -2 & 1-x & -6 \\ 1 & -1 & 4-x \end{pmatrix}$

12. More examples of the van der Monde type. Prove:

a) $\det \begin{pmatrix} 1 & 1 & 1 \\ a & b & c \\ a^3 & b^3 & c^3 \end{pmatrix} = (b-a)(c-a)(c-b)(a+b+c)$

b) $\det \begin{pmatrix} 1 & 1 & 1 \\ a^2 & b^2 & c^2 \\ a^3 & b^3 & c^3 \end{pmatrix} = (b-a)(c-a)(c-b)(bc+ab+ac)$

c) $\det \begin{pmatrix} 1 & 1 & 1 \\ a & b & c \\ a^4 & b^4 & c^4 \end{pmatrix} = (b-a)(c-a)(c-b)(a^2+b^2+c^2+ab+ac+bc)$

d) $\det \begin{pmatrix} 1 & 1 & 1 & 1 \\ a & b & c & d \\ a^2 & b^2 & c^2 & d^2 \\ a^4 & b^4 & c^4 & d^4 \end{pmatrix} = (b-a)(c-a)(d-a)(c-b)(d-b)(d-c)(a+b+c+d).$

13. If the square $n \times n$ matrix **A** has rank $n - 1$, show that any nonzero column of adj **A** gives a basis of the kernel of $T(V) = AV$.

14. Prove the following: (a) If **A** has rank n, then adj **A** has rank n. (b) If **A** has rank $n - 1$, then adj **A** has rank 1. (c) If **A** has rank $< n - 1$, then adj **A** has rank 0.

15. Prove that det (adj **A**) $= (\det A)^{n-1}$. [*Hint:* Use 4.5 but notice that there are two cases.]

16. Let **A** be an $(n-1) \times n$ matrix of rank $n - 1$ and let **A**′ be the $n \times n$ matrix whose first row is $(a_1, \ldots, a_n)$ where a_i are arbitrary and whose last $(n-1)$ rows are the rows of **A**. Prove that the kernel of $T(V) = AV$ has as a basis the cofactors of the first row of **A**′.

17. a) Given that **A** is an $n \times n$ matrix with integer entries and det **A** $= \pm 1$, show that A^{-1} has integer entries also.
    b) Given that **A** and A^{-1} both have integer entries, prove that det **A** $= \pm 1$.

5. PROOF OF EXISTENCE AND UNIQUENESS

For our discussion of existence and uniqueness, we consider first the situation for determinants of 3×3 matrices, which contains the flavor of the general case much more satisfactorily than does the 2×2 case of Section 1. In fact, we shall set things up so that a good understanding of 3×3 determinants should enable you to do the $n \times n$ case with only minor notational difficulties.

We shall first suppose that 3×3 determinants exist, and from the defining properties develop an explicit formula for such determinants. This will show that if they do exist, then they are necessarily unique. We then show that the function defined by the explicit formula does satisfy the conditions of the definition of determinant, and thus 3×3 determinants do exist.

5 / Proof of existence and uniqueness

So let $A = (A_1 A_2 A_3)^*$ be a 3×3 matrix with rows

$$A_1 = (a_{11}, a_{12}, a_{13}) = a_{11}e_1 + a_{12}e_2 + a_{13}e_3,$$
$$A_2 = (a_{21}, a_{22}, a_{23}) = a_{21}e_1 + a_{22}e_2 + a_{23}e_3,$$
$$A_3 = (a_{31}, a_{32}, a_{33}) = a_{31}e_1 + a_{32}e_2 + a_{33}e_3.$$

Using linearity on A_1 in $\det A = \det (A_1, A_2, A_3)^*$, we can express $\det A$ as a sum of terms of the form

$$a_{1i} \det (e_i, A_2, A_3)^*.$$

Using linearity on A_2, we can express each such term as a sum of terms of the form

$$a_{1i}a_{2j} \det (e_i, e_j, A_3)^*.$$

Finally, using linearity on A_3, we can express these as sums of terms of the form

$$a_{1i}a_{2j}a_{3k} \det (e_i, e_j, e_k)^*.$$

Now many of the 3^3 terms of this form will be zero—whenever at least two of i, j, k are the same (by 1.1). Dispensing with these unnecessary terms, we see that we can write

$$\det A = \sum a_{1i}a_{2j}a_{3k} \det (e_i, e_j, e_k)^*, \tag{1}$$

where the sum is taken over all possible ways of choosing i, j, k as distinct numbers from the set $\{1, 2, 3\}$, a total of $3!$ terms. That is,

$$\det A = a_{11}a_{22}a_{33} \det (e_1, e_2, e_3)^*$$
$$+ a_{11}a_{23}a_{32} \det (e_1, e_3, e_2)^* + a_{12}a_{21}a_{33} \det (e_2, e_1, e_3)^*$$
$$+ a_{12}a_{23}a_{31} \det (e_2, e_3, e_1)^*$$
$$+ a_{13}a_{21}a_{32} \det (e_3, e_1, e_2)^* + a_{13}a_{22}a_{31} \det (e_3, e_2, e_1)^*. \tag{2}$$

Recall that the 3×3 matrices $(e_i e_j e_k)$ with i, j, and k distinct, are called *permutation matrices*. The rows of a permutation matrix are a permutation of the rows of the identity matrix, so $\det (e_i, e_j, e_k)^* = \pm 1$. We shall denote by "sign (i, j, k)" the sign of this determinant (and we shall assume a result from permutation theory which says that sign (i, j, k) is uniquely defined; a proof is indicated below). For example,

$$\det (e_3, e_1, e_2)^* = \det \begin{pmatrix} 0 & 0 & 1 \\ 1 & 0 & 0 \\ 0 & 1 & 0 \end{pmatrix} = -\det \begin{pmatrix} 0 & 1 & 0 \\ 1 & 0 & 0 \\ 0 & 0 & 1 \end{pmatrix} = +\det \begin{pmatrix} 1 & 0 & 0 \\ 0 & 1 & 0 \\ 0 & 0 & 1 \end{pmatrix} = +1.$$

Thus, sign $(3, 1, 2) = +1$. We can therefore rewrite (1) as

$$\det A = \sum \text{sign}\,(i, j, k) a_{1i}a_{2j}a_{3k}, \tag{3}$$

summed over all permutations (i, j, k) of $(1, 2, 3)$.

Note that, for example, sign $(i, j, k) = -\text{sign}\,(j, i, k)$. It is clear that, in general, sign $(i, j, k) = (-1)^N$, where N is the number of interchanges necessary to change (e_i, e_j, e_k) to the identity matrix (e_1, e_2, e_3).

The permutation-theory result mentioned above essentially says that if you use a different series of interchanges to get from (i, j, k) to $(1, 2, 3)$, then the total number N' of interchanges is of the form $N' = N + 2M$, for some integer M, so that
$$(-1)^N = (-1)^{N'}.$$
For example,

$(3, 1, 2) \to (1, 3, 2) \to (1, 2, 3), \quad N = 2,$

$(3, 1, 2) \to (3, 2, 1) \to (2, 3, 1) \to (2, 1, 3) \to (1, 2, 3), \quad N' = 4.$

Uniqueness theorem for 3 × 3 determinants. There is at most one function satisfying the definition for the determinant of 3×3 matrices.

Proof. We have just shown that any such function must have form (3), which completely specifies its value on all 3×3 matrices. ∎

Existence theorem for 3 × 3 determinants. There exists a function satisfying the definition for the determinant of 3×3 matrices.

Proof. Our candidate, naturally enough, is the function

$$f(\mathbf{A}_1, \mathbf{A}_2, \mathbf{A}_3)^* = f\begin{pmatrix} a_{11} & a_{12} & a_{13} \\ a_{21} & a_{22} & a_{23} \\ a_{31} & a_{32} & a_{33} \end{pmatrix} = \sum \text{sign}\,(i, j, k) a_{i1} a_{j2} a_{k3}.$$

i) To show that f is alternating, let us, for example, interchange $\mathbf{A}_1$ and $\mathbf{A}_2$. Now the term

$$\text{sign}\,(i, j, k) \underset{\text{(from row 1)}}{a_{1i}} \underset{\text{(from row 2)}}{a_{2j}} \underset{\text{(from row 3)}}{a_{3k}},$$

occurring in $f(\mathbf{A}_1, \mathbf{A}_2, \mathbf{A}_3)^*$, will correspond to the term

$$\text{sign}\,(j, i, k) a_{2j} a_{1i} a_{3k}$$

from $f(\mathbf{A}_2, \mathbf{A}_1, \mathbf{A}_3)^*$. Since sign $(j, i, k) = -\text{sign}\,(i, j, k)$, we see that the displayed terms are negatives of each other. Since this gives a one-to-one correspondence between terms of $f(\mathbf{A}_1, \mathbf{A}_2, \mathbf{A}_3)^*$ and their negatives in $f(\mathbf{A}_2, \mathbf{A}_1, \mathbf{A}_3)^*$, we see that

$$f(\mathbf{A}_1, \mathbf{A}_2, \mathbf{A}_3)^* = -f(\mathbf{A}_2, \mathbf{A}_1, \mathbf{A}_3)^*.$$

A similar argument could be given for any other interchange of rows.

ii) To see that f is multilinear, let $\mathbf{B}_1 = (b_{11}, b_{12}, b_{13})$ and calculate

$$f(\mathbf{A}_1 + \mathbf{B}_1, \mathbf{A}_2, \mathbf{A}_3)^* = \sum \text{sign}\,(i, j, k)(a_{1i} + b_{1i}) a_{2j} a_{3k}$$
$$= \sum [\text{sign}\,(i, j, k) a_{1i} a_{2j} a_{3k} + \text{sign}\,(i, j, k) b_{1i} a_{2j} a_{3k}]$$
$$= f(\mathbf{A}_1, \mathbf{A}_2, \mathbf{A}_3)^* + f(\mathbf{B}_1, \mathbf{A}_2, \mathbf{A}_3)^*.$$

Moreover, since $b\mathbf{A}_1 = (ba_{11}, ba_{12}, ba_{13})$,

$$f(b\mathbf{A}_1, \mathbf{A}_2, \mathbf{A}_3)^* = \sum \text{sign } (i, j, k)(ba_{1i})a_{2j}a_{3k}$$
$$= b \sum \text{sign } (i, j, k)a_{1i}a_{2j}a_{3k}$$
$$= bf(\mathbf{A}_1, \mathbf{A}_2, \mathbf{A}_3)^*.$$

Thus, f is a linear function of the first row of $\mathbf{A}$. The same argument holds for other rows, or you can use the alternating property (i) if you wish.

iii) For

$$\mathbf{A} = \mathbf{I} = \begin{pmatrix} 1 & 0 & 0 \\ 0 & 1 & 0 \\ 0 & 0 & 1 \end{pmatrix},$$

$a_{rs} = 0$ if $r \neq s$ and $a_{rr} = 1$, so $f(\mathbf{e}_1, \mathbf{e}_2, \mathbf{e}_3) = \text{sign } (1, 2, 3)a_{11}a_{22}a_{33} = 1$, all other terms having some a_{rs} with $r \neq s$. ∎

For the general case of determinants of $n \times n$ matrices,

$$\mathbf{A} = \begin{pmatrix} a_{11} & \cdots & a_{1n} \\ \cdot & & \cdot \\ \cdot & & \cdot \\ \cdot & & \cdot \\ a_{n1} & \cdots & a_{nn} \end{pmatrix},$$

one can show in exactly the same manner as above that

$\det \mathbf{A} = \sum \text{sign } (i_1, i_2, \ldots, i_n)a_{1i_1}, a_{2i_2} \cdots a_{ni_n}$, summed over all permutations
$(i_1, \ldots, i_n)$ of $(1, 2, \ldots, n)$, (4)

where "sign $(i_1, i_2, \ldots, i_n)$" is now the determinant of the $n \times n$ permutation matrix $(\mathbf{e}_{i_1}\mathbf{e}_{i_2} \cdots \mathbf{e}_{i_n})^*$. Note especially that each summand in (4) consists of one element from each row and one element from each column of $\mathbf{A}$; and each such element occurs as a summand. From (4) there follows

Uniqueness theorem. There exists at most one function defined on $n \times n$ matrices and having values in $\mathbb{R}$, which satisfies the properties of the determinant function.

Existence theorem. There exists a function satisfying the definition of determinant for $n \times n$ matrices, namely, the function (4).
(The proof of this is also exactly analogous to that in the 3×3 case.)

PERMUTATIONS

Any permutation of $(1, 2, 3, \ldots, n)$ can be accomplished by a succession of transpositions (i.e., interchanges).

Example:
$(5, 3, 2, 1, 4) \to (5, 3, 1, 2, 4) \to (5, 1, 3, 2, 4) \to (1, 5, 3, 2, 4)$
$\to (1, 3, 5, 2, 4) \to (1, 3, 2, 5, 4) \to (1, 2, 3, 5, 4)$
$\to (1, 2, 3, 4, 5)$ (7 interchanges)

or

$(5, 3, 2, 1, 4) \to (5, 2, 3, 1, 4) \to (1, 2, 3, 5, 4) \to (1, 2, 3, 4, 5)$ (3 interchanges).

We must prove that, for a given permutation, the number of transpositions needed is either always *even* or always *odd*. The permutation is then called **even** or **odd** accordingly. To do this we consider the function

$$f(x_1, \ldots, x_n) = \prod_{i>j}(x_i - x_j) \quad (\prod = \text{product}),$$

and prove that interchanging any two x's changes the sign of f:

$f(x_1, \ldots, x_k, \ldots, x_l, \ldots, x_n) = -f(x_1, \ldots, x_l, \ldots, x_k, \ldots, x_n), \quad (k > l)$.

Example. For $n = 5$, $f(x_1, \ldots, x_5) = (x_2 - x_1)(x_3 - x_1)(x_4 - x_1)(x_5 - x_1)$
$(x_3 - x_2)(x_4 - x_2)(x_5 - x_2)$
$(x_4 - x_3)(x_5 - x_3)$
$(x_5 - x_4)$.

Clearly, $f(x_2, x_1, x_3, x_4, x_5) = -f(x_1, x_2, x_3, x_4, x_5)$.

The general proof goes the same way. We write

$$(x_1, \ldots, x_n) = \pm(x_l - x_k)\prod_{i \ne k,l}(x_k - x_i)\prod_{i \ne k,l}(x_l - x_i)\prod_{\substack{i \ne k,l \\ j \ne k,l \\ i>j}}(x_i - x_j)$$

and see at once that interchanging x_k and x_l changes the sign of f.

The effect on f of making a permutation of $x_1, \ldots, x_n$ is therefore to leave f unchanged if the number of transpositions needed is even, or to change f into $-f$ if the number needed is **odd**. Thus our theorem is proved since, for a given permutation, we must have either one or the other.

EXERCISES

1. Take $n = 4$ and list all 24 permutations of $(1, 2, 3, 4)$. Determine which are even and which are odd.
2. Prove that of the $n!$ permutations of $1, 2, \ldots, n$ one-half are even and one-half are odd. (Check this fact for $n = 2, 3, 4$).
3. a) Calculate sign (i, j, k) for all possible permutations of $(1, 2, 3)$.
   b) Use these values to write an expression for the determinant of a 3×3 matrix similar to equation (2) but with $+$ or $-$ in place of det (e_i, e_j, e_k).
4. a) Derive equation (4). (b) Prove the general uniqueness theorem.
5. Prove the general existence theorem.
6. a) How many terms should equation (3) have for 4×4 determinants? Why?
   b) Write out an expression for the determinant of a 4×4 matrix similar to the 3×3 expression derived in Exercise 3(b).

A large-scale use of this procedure [for finding eigenvalues] was recently undertaken . . . for studying different policies to be directed against the rabbit pest on New Zealand farmlands.†

CHAPTER 6 # The Eigenvalue Problem; Similar Matrices

We now turn to linear transformations from a vector space into itself (or square matrices, when translated into matrix terms). The basic problem we consider is that of finding those vectors (if any) upon which the linear transformation acts by scalar multiplication, and also to find the scalars involved (the "eigenvalues").

This quest leads us to methods for finding a particularly simple matrix for a linear transformation (although this can be quite difficult to carry out for certain matrices; see Section 5 on Jordan Form).

The following material is of great importance in many areas of science and mathematics.

1. EIGENVALUES

In this chapter, we shall deal exclusively with linear transformations from a vector space into itself. Given such a linear transformation **T** from a vector space $\mathcal{V}$ into $\mathcal{V}$, we shall first search for those vectors which **T** treats in an especially simple fashion. For example, those vectors fixed by **T** (that is, **T(V)** = **V**), or annihilated by **T** (that is, **T(V)** = **0**), or reflected about the origin by **T** (that is, **T(V)** = −**V**). In general, we might wish to search for those vectors which **T** stretches by a scalar amount:

$$\mathbf{T(V)} = \lambda \mathbf{V}, \quad \lambda \text{ (for the present) in } \mathbb{R};$$

λ, the Greek "lambda", seems the traditional symbol used here.

---

† Searle, S. R., *Matrix Algebra for the Biological Sciences*. New York: Wiley, 1966; page 185.

Given a linear transformation **T** from $\mathcal{V}$ into $\mathcal{V}$, a nonzero vector **V** in $\mathcal{V}$ is an **eigenvector** of **T** if

$$\mathbf{T(V)} = \lambda \mathbf{V}, \quad \lambda \text{ in } \mathbb{R}, \quad \mathbf{V} \neq \mathbf{0}.$$

The scalar λ is called the **eigenvalue** of T corresponding to **V**. We require **V** to be nonzero since $\mathbf{T(0)} = \lambda \mathbf{0}$ for any λ; each of our eigenvalues must have a *nonzero* eigenvector.

If you don't like mongrel words, here are some other common terms:

for eigenvalue: **characteristic value, proper value, latent root** (hence the λ);

for eigenvector: **characteristic vector, proper vector, latent vector.**

("Eigenvalue" didn't quite make it over from the German "Eigenwert".)

Example 1. Consider the linear transformation from $\mathbb{R}_2$ to itself

$$\mathbf{T}\begin{pmatrix} x \\ y \end{pmatrix} = \begin{pmatrix} 0 & 1 \\ 1 & 0 \end{pmatrix} \begin{pmatrix} x \\ y \end{pmatrix}.$$

Geometrically, we know that this represents a reflection about the line $x = y$. This should give us a clue as to where to look for its eigenvectors: anything on the line of reflection should be invariant (that is, have eigenvalue 1), while anything on the line $x = -y$ should be reflected about the origin (i.e., have eigenvalue -1). It is instantaneous to check that

$$\mathbf{T}\begin{pmatrix} x \\ x \end{pmatrix} = 1 \begin{pmatrix} x \\ x \end{pmatrix} \quad \text{for any } x \quad \text{and} \quad \mathbf{T}\begin{pmatrix} x \\ -x \end{pmatrix} = -1 \begin{pmatrix} x \\ -x \end{pmatrix}.$$

We shall see that these represent all the eigenvectors of **T**.

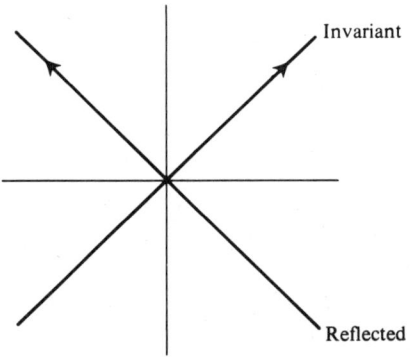

Fig. 1

The **eigenvalue problem** is: given **T** from $\mathcal{V}$ into itself, find all eigenvalues of **T** and their corresponding eigenvectors; it is called the eigenvalue problem rather than the eigenvector problem, since it is usually easier, at least in theory, to find the values first, after which the finding the vector is quite simple (in fact, in the finite-dimensional case it is the old problem of solving systems of linear equations). The practical difficulties which arise, together with the great importance of the problem, have resulted in a voluminous literature on the subject (see [1]).

To reduce the problem to something we can (theoretically) handle, we shall now confine ourselves to the finite-dimensional case, i.e., to linear transformations **T** from $\mathbb{R}_n$ into $\mathbb{R}_n$, for which, as we know, $\mathbf{T(V)} = \mathbf{AV}$ for some $n \times n$ matrix **A**. We say that **V** ($\neq \mathbf{0}$) is an **eigenvector of A with eigenvalue** λ if $\mathbf{AV} = \lambda \mathbf{V}$, in an obvious transferral of the linear transformation problem.

Our approach to the eigenvalue problem will be based upon the following *deus ex machina*: If **A** is an $n \times n$ matrix, then $\det(\mathbf{A} - x\mathbf{I})$ is a function (a polynomial in x which we shall study further), the roots of which are the eigenvalues of **A**.

1.1 λ is an eigenvalue of **A** iff λ is a root of the equation $\det(\mathbf{A} - x\mathbf{I}) = 0$.

Proof. 1. If λ is an eigenvalue, then $\mathbf{AV} = \lambda \mathbf{V}$ for some $\mathbf{V} \neq \mathbf{0}$. Hence,

$$\mathbf{AV} = (\lambda \mathbf{I})\mathbf{V} \quad \text{or} \quad (\mathbf{A} - \lambda \mathbf{I})\mathbf{V} = \mathbf{AV} - \lambda \mathbf{IV} = \mathbf{0}.$$

Since the matrix $\mathbf{A} - \lambda \mathbf{I}$ annihilates some nonzero vector, it is not invertible, so $\det(\mathbf{A} - \lambda \mathbf{I}) = 0$ and λ is a root.

2. Conversely, if λ satisfies $\det(\mathbf{A} - \lambda \mathbf{I}) = 0$, then $\mathbf{A} - \lambda \mathbf{I}$ is not invertible, so there is a $\mathbf{V} \neq \mathbf{0}$ for which

$$(\mathbf{A} - \lambda \mathbf{I})\mathbf{V} = \mathbf{0} \quad \text{or} \quad \mathbf{AV} - \lambda \mathbf{IV} = \mathbf{0}.$$

Thus, λ is an eigenvalue. ∎

Example 1 (*Cont.*). For

$$\mathbf{A} = \begin{pmatrix} 0 & 1 \\ 1 & 0 \end{pmatrix},$$

we have

$$\mathbf{A} - x\mathbf{I} = \begin{pmatrix} 0 & 1 \\ 1 & 0 \end{pmatrix} - \begin{pmatrix} x & 0 \\ 0 & x \end{pmatrix} = \begin{pmatrix} -x & 1 \\ 1 & -x \end{pmatrix},$$

so $\det(\mathbf{A} - x\mathbf{I}) = x^2 - 1 = (x-1)(x+1)$. This has roots $\lambda = \pm 1$, in keeping with our previous observations.

We are, therefore, led to attempt to solve the equation $\det(\mathbf{A} - x\mathbf{I}) = 0$ for x. Let us write out what this means: for

$$\mathbf{A} = \begin{pmatrix} a_{11} & a_{12} & \cdots & a_{1n} \\ a_{21} & a_{22} & \cdots & a_{2n} \\ \vdots & & & \vdots \\ a_{n1} & & \cdots & a_{nn} \end{pmatrix},$$

$$\det(\mathbf{A} - x\mathbf{I}) = \det \begin{pmatrix} a_{11} - x & a_{12} & \cdots & a_{1n} \\ a_{21} & a_{22} - x & \cdots & a_{2n} \\ \vdots & & & \vdots \\ a_{n1} & \cdots & \cdots & a_{nn} - x \end{pmatrix}.$$

If we write out the determinant as a sum of products of elements of the matrix, each summand contains exactly one element from each row and column; that is, we are applying to the matrix $\mathbf{A} - x\mathbf{I}$ the formula which, for the matrix $\mathbf{A}$, says:

$$\det \mathbf{A} = \sum \pm a_{i_1 1} a_{i_2 2} \cdots a_{i_n n}.$$

It is then apparent that $\det(\mathbf{A} - x\mathbf{I})$ is a polynomial in x.

The polynomial $\det(\mathbf{A} - x\mathbf{I})$ is called the **characteristic polynomial** of $\mathbf{A}$. We have seen that

λ is a (real) characteristic root (i.e., eigenvalue) of $\mathbf{A}$ iff λ is a (real) root of the characteristic polynomial.

The characteristic polynomial is of degree n: For, each summand in the determinant expansion is a product of n terms, each term is of degree at most 1 in x, and the only way you can get all terms of degree 1 is in the summand consisting of the product of the diagonal entries, which gives you $(-1)^n x^n$ plus other things. Thus, since a polynomial of degree n has at most n distinct roots:

1.2 An $n \times n$ matrix $\mathbf{A}$ has at most n eigenvalues. (Thus, a linear transformation $\mathbf{T}$ from $\mathbb{R}_n$ into itself has at most n eigenvalues.)

Of course, a matrix (or linear transformation) is likely to have an infinite number of eigenvectors. In fact,

1.3 If $\mathbf{V}$ is an eigenvector corresponding to a particular eigenvalue then so is $c\mathbf{V}$ for any $c \neq 0$.

Proof. To use the linear transformation formulation,

$$\mathbf{T}(c\mathbf{V}) = c\mathbf{T}(\mathbf{V}) = c\lambda\mathbf{V} = \lambda(c\mathbf{V}). \quad \blacksquare$$

1.4 If $\mathbf{V}$ and $\mathbf{W}$ are eigenvectors corresponding to the same eigenvalue, then so is $\mathbf{V} + \mathbf{W}$.

Proof. See Exercise 3a. $\blacksquare$

In sum,

the set of all eigenvectors corresponding to the same eigenvalue, together with $\mathbf{0}$, forms a subspace.

It is therefore sufficient to specify a basis for this subspace to determine all such eigenvectors. In Example 1, a basis for the subspace corresponding to eigenvalue $+1$ is

$$\begin{pmatrix} 1 \\ 1 \end{pmatrix};$$

to eigenvalue -1 is

$$\begin{pmatrix} 1 \\ -1 \end{pmatrix}.$$

1.5 For a matrix **A**:

i) The subspace corresponding to the eigenvalue 0 is the kernel of $T(V) = AV$.

ii) The subspace corresponding to the eigenvalue λ is the kernel of $T'(V) = (A - \lambda I)V$.

Proof. See Exercise 3b. ∎

We now outline a procedure for finding the eigenvalues and eigenvectors of a matrix **A**:

I. Write out det $(A - xI)$ as a polynomial in **A**.
II. find all roots $\lambda_1, \ldots, \lambda_n$ of the equation det $(A - xI) = 0$ (not all roots need be distinct).
III. for each root λ solve the system of equations

$$AV = \lambda V \quad \text{for } V = (v_1, \ldots, v_n)^*.$$

This is a system of homogeneous linear equations in the v_i, as may be seen by rewriting the system in the form $(A - \lambda I)V = 0$ (it saves a step to start out trying to solve the system in this form). The space of solutions to this homogeneous system will then be the space of eigenvectors corresponding to λ (together with **0**). As usual, one can simply specify a basis for this space.

Example 2. Let

$$A = \begin{pmatrix} \frac{5}{2} & -\frac{1}{2} & 1 \\ \frac{3}{2} & \frac{1}{2} & -1 \\ 0 & 0 & 0 \end{pmatrix}.$$

For step I we are to evaluate

$$\det (A - xI) = \det \begin{pmatrix} \frac{5}{2} - x & -\frac{1}{2} & 1 \\ \frac{3}{2} & \frac{1}{2} - x & -1 \\ 0 & 0 & -x \end{pmatrix}.$$

Since such calculations require a bit of effort, we shall indicate two methods.

det $(A - xI)$ via the Determinant Algorithm: We treat entries such as $\frac{5}{2} - x$ as if they were ordinary numbers and proceed accordingly. When possible one should avoid pivoting on entries involving x, since this introduces rational functions (quotients of polynomials), an added complication.†

$$\det \begin{pmatrix} \frac{5}{2}-x & -\frac{1}{2} & 1 \\ \frac{3}{2} & \frac{1}{2}-x & -1 \\ 0 & 0 & -x \end{pmatrix} = \det \begin{pmatrix} 0 & -\frac{2}{3}x^2+2x-\frac{4}{3} & -\frac{2}{3}x+\frac{8}{3} \\ \frac{3}{2} & \frac{1}{2}-x & -1 \\ 0 & 0 & -x \end{pmatrix}$$

(Multiply 2nd row by $-\frac{2}{3}(\frac{5}{2}-x)$ and add to 1st.)   (Interchange 1st and 2nd rows.)

$$= -\det \begin{pmatrix} \frac{3}{2} & \frac{1}{2}-x & -1 \\ 0 & -\frac{2}{3}x^2+2x-\frac{4}{3} & -\frac{2}{3}x+\frac{8}{3} \\ 0 & 0 & -x \end{pmatrix}$$

$$= (\tfrac{3}{2})(x)(-\tfrac{2}{3}x^2+2x-\tfrac{4}{3})$$

$$= x(-x^2+3x-2).$$

det $(A - xI)$ via expansion by cofactors: Let us expand by the last column, which looks promising:

$$\det \begin{pmatrix} \frac{5}{2}-x & -\frac{1}{2} & 1 \\ \frac{3}{2} & \frac{1}{2}-x & -1 \\ 0 & 0 & -x \end{pmatrix} = 1(-1)^4 A_{13} + (-1)(-1)^5 A_{23} + (-x)(-1)^6 A_{33}$$

$$= \det \begin{pmatrix} \frac{3}{2} & \frac{1}{2}-x \\ 0 & 0 \end{pmatrix} + \det \begin{pmatrix} \frac{5}{2}-x & -\frac{1}{2} \\ 0 & 0 \end{pmatrix} - x \det \begin{pmatrix} \frac{5}{2}-x & -\frac{1}{2} \\ \frac{3}{2} & \frac{1}{2}-x \end{pmatrix}$$

$$= -x[(\tfrac{5}{2}-x)(\tfrac{1}{2}-x) - \tfrac{3}{2}(-\tfrac{1}{2})] = -x(x^2-3x+2).$$

The last row would be even better, but less illustrative. In this particular case, step II is trivial:

$$-x(x^2-3x+2) = -x(x-1)(x-2),$$

so the eigenvalues are 0, 1, and 2.

Step III: To find the eigenvectors corresponding to 2, say, we solve $AV = 2V$ for V. If $V = (v_1, v_2, v_3)^*$ this says

$$\begin{pmatrix} \frac{5}{2} & -\frac{1}{2} & 1 \\ \frac{3}{2} & \frac{1}{2} & -1 \\ 0 & 0 & 0 \end{pmatrix} \begin{pmatrix} v_1 \\ v_2 \\ v_3 \end{pmatrix} = 2 \begin{pmatrix} v_1 \\ v_2 \\ v_3 \end{pmatrix},$$

or

$$\tfrac{5}{2}v_1 - \tfrac{1}{2}v_2 + v_3 = 2v_1,$$
$$\tfrac{3}{2}v_1 + \tfrac{1}{2}v_2 - v_3 = 2v_2,$$
$$0v_1 + 0v_2 + 0v_3 = 2v_3.$$

---

† The fact that $\frac{5}{2} - x$ is zero when $x = \frac{5}{2}$ causes no difficulty, since one can carry out the calculations for x which are different from the diagonal entries. The resulting characteristic polynomial will be the value of det $(A - xI)$ for $x \neq$ diagonal entries. By continuity this polynomial will then be det $(A - xI)$ for all x. Although this argument is valid when A has real or complex entries, it would have to be changed for some other fields.

or
$$\tfrac{1}{2}v_1 - \tfrac{1}{2}v_2 + v_3 = 0,$$
$$\tfrac{3}{2}v_1 - \tfrac{3}{2}v_2 - v_3 = 0,$$
$$0v_1 + 0v_2 - 2v_3 = 0,$$

a homogeneous system which has as solution $v_1 = v_2, v_3 = 0$. Thus, one eigenvector corresponding to the eigenvalue 2 is $(1, 1, 0)^*$, and all other such eigenvectors are multiples of this one.

By changing $\mathbf{AV} = 2\mathbf{V}$ into $(\mathbf{A} - 2\mathbf{I})\mathbf{V} = \mathbf{0}$ and then transforming into a matrix equation, you not only save a step; you also arrive at the matrix form upon which we based our solution techniques for systems of linear equations:

$$(\mathbf{A} - 2\mathbf{I})\mathbf{V} = \begin{pmatrix} \tfrac{1}{2} & -\tfrac{1}{2} & 1 \\ \tfrac{3}{2} & -\tfrac{3}{2} & -1 \\ 0 & 0 & -2 \end{pmatrix} \begin{pmatrix} v_1 \\ v_2 \\ v_3 \end{pmatrix} = \mathbf{0},$$

all ready for the solution procedure of Chapter 4. Similarly, one finds that an eigenvector corresponding to $\lambda = 1$ is $(1, 3, 0)^*$; to $\lambda = 0$, $(0, 2, 1)^*$.

Our method for finding eigenvalues and eigenvectors may appear deceptively simple. In higher-dimensional cases there are serious difficulties with steps I and II. Step I poses problems because characteristic polynomials do not lend themselves well to computer calculation. When it comes to step II, for polynomials of degree <5, there are special formulas for writing out the roots in terms of the coefficients. However, as Evariste Galois brilliantly proved (before being killed in a duel at the age of 21), there can exist no such formulas for polynomials of degree $n \geq 5$. The problem of finding the roots of polynomials is so difficult that one sometimes applies the advanced iterative techniques for finding eigenvalues to this problem, and not the other way around (cf. supplement 2 and the supplement to Section 2).

Another area of possible concern is that some of the roots of a real polynomial may be complex. However, the nature of a given problem normally dictates one of two responses: Either the complex eigenvalues are meaningless to the problem and may be safely ignored (although care must be taken here; sometimes it is at least necessary to check the size of the complex roots), or the complex eigenvalues are very much a part of the problem. In the latter case, you simply treat the matrix $\mathbf{A}$ as an $n \times n$ matrix with complex entries (since $\mathbb{C}$ contains $\mathbb{R}$, $\mathbb{C}_{n \times n}$ contains $\mathbb{R}_{n \times n}$) and search for complex eigenvectors in $\mathbb{C}_n$ corresponding to the complex eigenvalues. As we have pointed out in various asides, matrices with complex entries obey the same arithmetic laws as do real matrices, and $n \times n$ complex matrices have determinants (which are complex numbers) satisfying the same properties as do real determinants.

Example 3. Consider
$$\mathbf{A} = \begin{pmatrix} 0 & -1 \\ 1 & 0 \end{pmatrix}.$$

Here

$$\det(\mathbf{A} - x\mathbf{I}) = \det\begin{pmatrix} -x & -1 \\ 1 & -x \end{pmatrix} = x^2 + 1,$$

so the eigenvalues of **A** are i and $-i$. It is a cheerful exercise in multiplication by i to show, for instance, that an eigenvector corresponding to i is $(1, i)^*$; see Exercise 8. Recall that $T(V) = AV$ corresponds to rotation through 90°, so geometrically we would not expect it to have any (real) eigenvectors. However, when considered as a linear transformation on $\mathbb{C}_2$, our mapping stretches $(1, i)$ an amount i, demonstrating that it can be exceedingly difficult to visualize events in complex vector spaces.

Lest we have made the situation seem unduly bleak, let us end with a case where one can immediately read off the eigenvalues:

1.6 The eigenvalues of a triangular matrix

$$\begin{pmatrix} a_{11} & & & \\ 0 & \ddots & & \\ \vdots & & \ddots & \\ 0 & \cdots & 0 & a_{nn} \end{pmatrix}$$

are its diagonal entries $a_{11}, \ldots, a_{nn}$.

Proof. How do you calculate determinants? (See Exercise 1). ∎

EXERCISES

1. Prove that the eigenvalues of a triangular matrix are its diagonal entries.
2. Prove that **A** is not invertible iff it has 0 as an eigenvalue.
3. Prove (a) 1.4; (b) 1.5
4. Find the characteristic polynomial of

a) $\begin{pmatrix} a & b \\ c & d \end{pmatrix}$ b) $\begin{pmatrix} 3 & 0 & 1 & 0 \\ 0 & 3 & 0 & 0 \\ 0 & 1 & 3 & 0 \\ 0 & 0 & 0 & 3 \end{pmatrix}$ c) $\begin{pmatrix} 2 & 1 & 0 \\ 0 & 2 & 0 \\ 0 & 0 & 3 \end{pmatrix}$

d) $\begin{pmatrix} 3 & -2 & 1 \\ 2 & 1 & -1 \\ 2 & -6 & 4 \end{pmatrix}$ e) $\begin{pmatrix} 3 & 2 & 2 \\ -2 & 1 & -6 \\ 1 & -1 & 4 \end{pmatrix}$ f) $\begin{pmatrix} 5 & -1 & -1 \\ -1 & 3 & 1 \\ -1 & 1 & 3 \end{pmatrix}$

Save your work for future use.

5. (a) through (f) Find the eigenvalues of the matrices in Exercise 4.

6. (a) through (f) Find the eigenvectors belonging to each eigenvalue found in Exercise 5. Check your work.
7. By looking at the characteristic polynomial, how can you tell whether or not a matrix is invertible?
8. Find all the complex eigenvectors of the matrix in Example 3.

Supplement 1. Gershgorin's Theorem

In applications it is sometimes unnecessary to know an eigenvalue exactly; a rough estimate may do very nicely. We present here a particularly convenient and pretty method for obtaining some indication of the location of eigenvalues. Since even

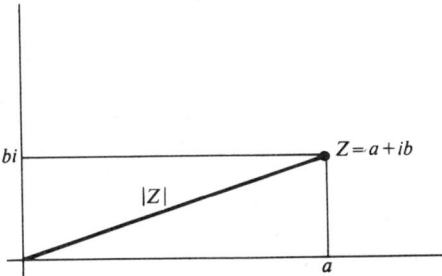

Fig. 2

the eigenvalues of real matrices are, in general, inhabitants of the complex plane rather than the real line, it is in this plane that we shall work. We therefore recall that for $x = a + ib$, the *modulus of z*, $|z| = \sqrt{a^2 + b^2}$, represents the distance from z to the origin; consequently, the set

$$\mathfrak{D} = \{\text{all complex } z \mid |z - a| \leq r, \quad r \text{ real}\}$$

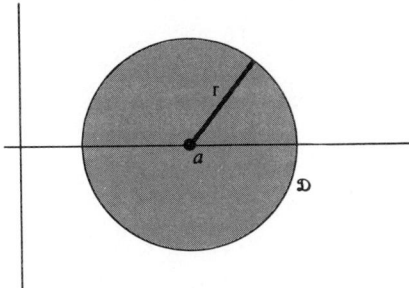

Fig. 3

will be a disk of radius r with center a. Let

$$A = \begin{pmatrix} a_{11} & \cdots & & & \cdots & a_{1n} \\ \vdots & & & & & \vdots \\ a_{i1} & \cdots & a_{ii} & \cdots & a_{in} \\ \vdots & & & & & \vdots \\ a_{n1} & \cdots & & & \cdots & a_{nn} \end{pmatrix}$$

be any $n \times n$ matrix with real (or complex) entries. Let

$$r_i = |a_{i1}| + \cdots + |a_{i,i-1}| + |a_{i,\,i+1}| + \cdots + |a_{in}|$$

denote the sum of the absolute values (or moduli in the complex case) of the elements of the ith row, except for the ii-element a_{ii}. Let

$$\mathfrak{D}_i = \{\text{complex } z \mid |z - a_{ii}| \leq r_i\}$$

be the disk around a_{ii} with radius r_i.

1.7 Gershgorin's Theorem. *The eigenvalues (real and complex) of A are to be found in the union of the discs $\mathfrak{D}_i$, $i = 1, \ldots, n$.*

Proof. Apply the Dominant Diagonal Theorem (Chapter 3, 4.14) to the matrix $A - xI$ to see that it will have an inverse whenever x is not in any of the disks. That is, when x is not in the union of the $\mathfrak{D}_i$,

$$\det(A - xI) \neq 0,$$

so x cannot be an eigenvalue. ∎

Example. For

$$A = \begin{pmatrix} \frac{5}{2} & -\frac{1}{2} & 1 \\ \frac{3}{2} & \frac{1}{2} & -1 \\ 0 & 0 & 0 \end{pmatrix},$$

we have

$$\mathfrak{D}_1 = \{z \mid |z - \tfrac{5}{2}| \leq \tfrac{3}{2}\},$$
$$\mathfrak{D}_2 = \{z \mid |z - \tfrac{1}{2}| \leq \tfrac{5}{2}\},$$
$$\mathfrak{D}_3 = \{z \mid |z - 0| \leq 0\}.$$

The actual roots (Example 2) are 0, 1, and 2.

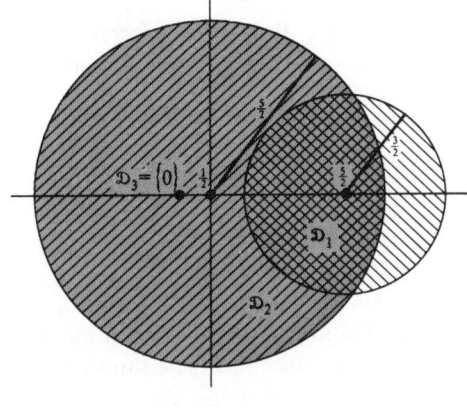

Fig. 4

EXERCISES (Cont.)

9. Show that Gershgorin's Theorem holds with the role of row replaced by column.
10. (a) through (f) Find the Gershgorin disks $\mathfrak{D}_i$ for each of the matrices in Exercise 4. Draw a picture in each case and locate the actual eigenvalues.

Supplement 2. The Companion Matrix; Cauchy's Polynomial Root Theorem

In this supplement we shall see one way in which eigenvalues can be applied to the study of roots of polynomials. We first show that given any polynomial $p(x)$, there is a matrix **P** whose characteristic polynomial is $p(x)$. We then apply Gershgorin's Theorem (Supplement 1) to obtain a famous bound for the roots of a polynomial. For convenience, we shall assume that the highest coefficient of $p(x)$ is 1 (if this is not the case for your polynomial, divide by the highest coefficient to make it so; this has no effect on the roots of the polynomial). We can therefore write

$$p(x) = a_0 + a_1 x + \cdots + a_{n-1} x^{n-1} + x^n.$$

We define the **companion matrix P** of $p(x)$ to be the $(n \times n)$ matrix

$$\mathbf{P} = \begin{pmatrix} 0 & 1 & 0 & \cdots & \cdots & 0 \\ 0 & 0 & 1 & & & \vdots \\ \vdots & & & \ddots & & \vdots \\ \vdots & & & & \ddots & 0 \\ 0 & \cdots & \cdots & \cdots & 0 & 1 \\ -a_0 & -a_1 & \cdots & & -a_{n-2} & -a_{n-1} \end{pmatrix}$$

1.8 The companion matrix **P** has characteristic polynomial $\pm p(x)$. In particular, the eigenvalues of **P** are the same as the roots of $p(x)$.

Outline of proof for $n = 4$. In this case we have

$$\mathbf{P} = \begin{pmatrix} 0 & 1 & 0 & 0 \\ 0 & 0 & 1 & 0 \\ 0 & 0 & 0 & 1 \\ -a_0 & -a_1 & -a_2 & -a_3 \end{pmatrix}$$

corresponding to $p(x) = x^4 + a_3 x^3 + a_2 x^2 + a_1 x + a_0$. Consider det $(\mathbf{P} - x\mathbf{I})$. If you add x(Column 2), x^2(Column 3), and x^3(Column 4) to Column 1, the value of the determinant is the same, but the first column becomes $(0, 0, 0, -p(x))^*$; try it and don't despair too soon. Expand the determinant by cofactors in this column to obtain

$$\det (\mathbf{P} - x\mathbf{I}) = -(-1)^5 p(x). \blacksquare$$

Exercise 11(a). Exhibit the details of the proof for the $n = 4$ case. (b) Prove the theorem for arbitrary n.

Exercise 12(a). Show that if $\lambda_1, \lambda_2, \lambda_3, \lambda_4$ are the roots of

$$x^4 + a_3 x^3 + a_2 x^2 + a_1 x + a_0 = 0,$$

then the eigenvectors of the companion matrix can be obtained from

$$\mathbf{V}_i = (1, \lambda_i, \lambda_i^2, \lambda_i^3)^*.$$

(b) Generalize to arbitrary n.

1.9 Cauchy's polynomial root theorem. Let

$$p(x) = x^n + a_{n-1} x^{n-1} + \cdots + a_1 x + a_0$$

be any polynomial with real (or complex) coefficients. Then all the roots (real and complex) of $p(x)$ lie within the circle about the origin in the complex plane of radius

$$r = \max \{|a_0|, 1 + |a_1|, 1 + |a_2|, \ldots, 1 + |a_{n-1}|\}.$$

Proof. Just apply Gershgorin (with column sums) to the companion matrix **P**. $\blacksquare$

For example, the polynomial $p(x) = 6 + 2x^2 - 6x^4 + 4x^5 - 3x^6 + x^8$ has all roots within a circle of radius

$$\max \{6, 1, 3, 7, 5, 4, 1\} = 7.$$

Honors Projects

1. Assuming you've kept up with such things as determinants of complex $n \times n$ matrices, show that all the material in this section carries over for such matrices. Show in addition that any matrix in $\mathbb{C}_{n \times n}$ has at least one eigenvector in $\mathbb{C}_n$.

Supplement 1, as pointed out, also applies to complex matrices, as does Supplement 2.

2. Some care is needed throughout this chapter in working over other fields. One may encounter a situation similar to that with a matrix with real entries: The eigenvalues of a matrix in $\mathbb{F}_{n \times n}$ may not belong to the "ground field" $\mathbb{F}$. Fortunately, just as we can consider the reals as being contained in the field of complex numbers, given a field $\mathbb{F}$ there is an "extension field" $\mathbb{F}_1$ containing $\mathbb{F}$ in which each polynomial with coefficients from $\mathbb{F}$ splits into linear factors (i.e., has all its roots). It is to $\mathbb{F}_1$ that one must go in search of all the eigenvalues of a matrix in $\mathbb{F}_{n \times n}$. Although the techniques of proving all the material in this section are the same for any field, the discussion of extension fields is beyond our scope.

Gershgorin's Theorem (Supplement 1) makes use of absolute values and inequalities. Field freaks will immediately recognize that this cannot be expected to hold in general. The same goes for Cauchy's Polynomial Root Test, although companion matrices work in all fields.

2. DISTINCT EIGENVALUES

Given a finite-dimensional vector space $\mathcal{V}$, we know that for any basis $\{\mathbf{V}_1, \ldots, \mathbf{V}_n\}$ we can treat $\mathcal{V}$ just as if it were $\mathbb{R}_n$ (and the $\mathbf{V}_i$'s were $\mathbf{e}_i$'s) by assigning to each vector $\mathbf{V}$ the n-tuple of coefficients in its expression in terms of the basis:

$$\mathbf{V} = a_1 \mathbf{V}_1 + \cdots + a_n \mathbf{V}_n \leftrightarrow (a_1, \ldots, a_n)^*.$$

Moreover, relative to this basis (now identified with the usual $\mathbf{e}_i$'s), any linear transformation $\mathbf{T}$ from $\mathcal{V}$ into itself has a matrix $\mathbf{A} = (\mathbf{T}(\mathbf{e}_1), \ldots, \mathbf{T}(\mathbf{e}_n))$.

Of course, different choices of bases will in general result in different matrices $\mathbf{A}$ for the linear transformation (see Chapter 4, Section 4 for details). As you are by now painfully aware, some matrices are a great deal more pleasant to deal with than others. A good portion of the rest of this chapter will be concerned with how to choose a basis so that the resulting matrix of a given linear transformation is as simple as possible. Eigenvalues and eigenvectors will play a central role throughout. One of the main reasons for this is the following:

2.1 Let $\lambda_1, \ldots, \lambda_m$ be *distinct* eigenvalues of the linear transformation $\mathbf{T}$. If $\mathbf{V}_1, \ldots, \mathbf{V}_m$ are eigenvectors corresponding to the respective eigenvalues, then they are linearly independent.

Proof. Suppose, on the contrary, that $\mathbf{V}_1, \ldots, \mathbf{V}_m$ are *not* linearly independent. Among the dependence relations

$$c_1 \mathbf{V}_1 + \cdots + c_m \mathbf{V}_m = \mathbf{0} \quad \text{with some } c_j \neq 0,$$

which hold for these $\mathbf{V}_j$'s, some may involve fewer nonzero c_j's than others. Choose one that involves the fewest possible nonzero coefficients, and renumber if necessary so that the relation is

$$c_1 \mathbf{V}_1 + \cdots + c_k \mathbf{V}_k = \mathbf{0}, \quad \text{all } c_j \neq 0. \tag{1}$$

We shall produce a contradiction by showing how to find a dependence relation which involves fewer nonzero coefficients than (1). Take (1) and apply **T** to obtain

$$\mathbf{T}(c_1\mathbf{V}_1) + \cdots + \mathbf{T}(c_k\mathbf{V}_k) = c_1\mathbf{T}(\mathbf{V}_1) + \cdots + c_k\mathbf{T}(\mathbf{V}_k)$$
$$= c_1\lambda_1\mathbf{V}_1 + \cdots + c_k\lambda_k\mathbf{V}_k = \mathbf{0}. \tag{2}$$

If one of the λ's is 0, this is already our shorter expression. If none of the λ's is 0, multiply (1) by λ_1 and subtract from (2) to obtain

$$(c_1\lambda_1\mathbf{V}_1 + \cdots + c_k\lambda_k\mathbf{V}_k) - \lambda_1(c_1\mathbf{V}_1 + \cdots + c_k\mathbf{V}_k)$$
$$= c_2(\lambda_2 - \lambda_1)\mathbf{V}_1 + \cdots + c_k(\lambda_k - \lambda_1)\mathbf{V}_k$$
$$= \mathbf{0} - \mathbf{0} = \mathbf{0}.$$

Since $\lambda_i - \lambda_1 \neq 0$ for $i = 2, \ldots, n$, this, then, is our shorter expression. Hence, there is no shortest expression of the form (1). Hence, there is no expression of the form (1), and the **V**'s are linearly independent. ∎

2.2 Let $\mathcal{V}$ be an n-dimensional vector space over $\mathbb{R}$. If a linear transformation **T** from $\mathcal{V}$ into $\mathcal{V}$ has n *distinct* eigenvalues $\lambda_1, \ldots, \lambda_n$ in $\mathbb{R}$, then the corresponding eigenvectors form a basis for $\mathcal{V}$ over $\mathbb{R}$.

Moreover, the matrix for the linear transformation **T** relative to this basis is the diagonal matrix

$$\begin{pmatrix} \lambda_1 & 0 & \cdots & 0 \\ 0 & \lambda_2 & & \\ \vdots & & \ddots & \vdots \\ 0 & \cdots & 0 & \lambda_n \end{pmatrix}.$$

Proof. You prove the first part; we shall work on the "moreover". Let $\mathbf{V}_i$ be an eigenvector corresponding to λ_i, so that

$$\mathbf{T}(\mathbf{V}_1) = \lambda_1\mathbf{V}_1, \ldots, \mathbf{T}(\mathbf{V}_n) = \lambda_n\mathbf{V}_n.$$

If we identify the $\mathbf{V}_i$ with $\mathbf{e}_i$, then $\mathbf{T}(\mathbf{e}_i) = \lambda_i\mathbf{e}_i$ so that the matrix of **T** is

$$(\mathbf{T}(\mathbf{e}_1), \ldots, \mathbf{T}(\mathbf{e}_n)) = (\lambda_1\mathbf{e}_1, \ldots, \lambda_n\mathbf{e}_n) = \begin{pmatrix} \lambda_1 & 0 & \cdots & 0 \\ 0 & \lambda_2 & & \vdots \\ \vdots & & \ddots & \vdots \\ & & & 0 \\ 0 & \cdots & 0 & \lambda_n \end{pmatrix}. \quad \blacksquare$$

Example 1. Consider $T(V) = AV$, where

$$A = \begin{pmatrix} \frac{5}{2} & -\frac{1}{2} & 1 \\ \frac{3}{2} & \frac{1}{2} & -1 \\ 0 & 0 & 0 \end{pmatrix}$$

is as in Example 2, Section 1. At that time we saw that A has eigenvalues 0, 1, and 2. Corresponding eigenvectors are

$$V_1 = (0, 2, 1)^*, \quad V_2 = (1, 3, 0)^*, \quad \text{and} \quad V_3 = (2, 2, 0)^*.$$

As in 2.2, if we replace the e_i's by the V_i's, the matrix of T becomes

$$\begin{pmatrix} 0 & 0 & 0 \\ 0 & 1 & 0 \\ 0 & 0 & 2 \end{pmatrix}.$$

Compare Examples 1 through 4 of Chapter 3, Section 3, where this same transformation is discussed and pictures are drawn.

Note carefully that 2.1 and 2.2 have absolutely nothing to say about eigenvectors corresponding to the same eigenvalue: they may be linearly independent, or they may not. A linear transformation which does not have n distinct eigenvalues may have a basis of eigenvectors or it may not.

Example 2. Let us contrast the behavior of the following linear transformations from $\mathbb{R}_2$ to $\mathbb{R}_2$:

$$T_1(V) = \begin{pmatrix} 2 & 0 \\ 0 & 2 \end{pmatrix} V, \quad T_2(V) = \begin{pmatrix} 2 & 1 \\ 0 & 2 \end{pmatrix} V.$$

T_1 has characteristic polynomial $(x - 2)^2$ as does T_2, so both have 2 as their only eigenvalues. However, *any* vector ($\neq 0$) in $\mathbb{R}_2$ is an eigenvector for T_1, so that T_1 in particular has a set of eigenvectors which form a basis for $\mathbb{R}_2$. On the other hand, if you attempt to solve $T_2(V) = 2V$, the only solutions are multiples of $(1, 0)^*$. Hence, there is no basis for $\mathbb{R}_2$ consisting of eigenvectors of T_2.

Besides showing that eigenvectors corresponding to the same eigenvalue may or may not be linearly independent, the above example also illustrates that eigenvalues do not tell us everything; two linear transformations may have the same eigenvalues but behave quite differently.

It is also worth noting that to apply 2.2 successfully to obtain a basis for $\mathbb{R}_n$, it is necessary that the eigenvalues of T all belong to $\mathbb{R}$. Our friend

$$T(V) = \begin{pmatrix} 0 & -1 \\ 1 & 0 \end{pmatrix} V$$

from Example 3, Section 1, has two distinct eigenvalues (i and $-i$) but absolutely no real eigenvectors (of course, if you wish to consider T as a linear transformation on $\mathbb{C}_2$ rather than $\mathbb{R}_2$, one can find a basis for $\mathbb{C}_2$ consisting of linearly independent eigenvectors of T).

We now turn to discussing the eigenvalues and eigenvectors of some matrices related to a given matrix **A**.

2.3 If **A** is invertible, then **A** and $\mathbf{A}^{-1}$ have the same eigenvectors.
λ is an eigenvalue of **A** iff λ^{-1} is an eigenvalue of $\mathbf{A}^{-1}$.

Proof. Suppose λ is an eigenvalue of **A** with eigenvector **V**. Thus, $\mathbf{AV} = \lambda \mathbf{V}$. Then
$$\mathbf{V} = \mathbf{A}^{-1}(\mathbf{AV}) = \mathbf{A}^{-1}(\lambda \mathbf{V}) = \lambda \mathbf{A}^{-1}\mathbf{V}.$$
Since $\lambda \neq 0$ (why?), $\mathbf{A}^{-1}\mathbf{V} = \lambda^{-1}\mathbf{V}$. ∎

Example 3. We can read off the eigenvalues of
$$\mathbf{A} = \begin{pmatrix} 1 & 2 \\ 0 & 3 \end{pmatrix}$$
and easily find that we have eigenvector

$\begin{pmatrix} 1 \\ 0 \end{pmatrix}$ corresponding to eigenvalue 1,

and

$\begin{pmatrix} 1 \\ 1 \end{pmatrix}$ corresponding to eigenvalue 3.

Our matrix **A** has inverse
$$\mathbf{A}^{-1} = \begin{pmatrix} 1 & -\frac{2}{3} \\ 0 & \frac{1}{3} \end{pmatrix}.$$
You can immediately see that it has eigenvector

$\begin{pmatrix} 1 \\ 0 \end{pmatrix}$ corresponding to 1

and

$\begin{pmatrix} 1 \\ 1 \end{pmatrix}$ corresponding to $\frac{1}{3}$.

2.4 The eigenvalues of **A***, the transpose of **A**, are the same as those of **A**. (The eigenvectors, however, are usually different.)

Proof.
$$\det(\mathbf{A} - x\mathbf{I}) = \det[(\mathbf{A} - x\mathbf{I})^*]$$
$$= \det(\mathbf{A}^* - (x\mathbf{I})^*) = \det(\mathbf{A}^* - x\mathbf{I}),$$
so the characteristic polynomials are the same. Hence, the roots of this polynomial are the same. ∎

Example 4. For our matrix **A** of Example 2,
$$\mathbf{A}^* = \begin{pmatrix} 1 & 0 \\ 2 & 3 \end{pmatrix}$$

and has eigenvector

$$\begin{pmatrix} 1 \\ -1 \end{pmatrix} \quad \text{corresponding to 1,}$$

and

$$\begin{pmatrix} 0 \\ 1 \end{pmatrix} \quad \text{corresponding to 3.}$$

Do you see any relation at all between eigenvectors of **A** and those of **A***?

The eigenvectors of **A** and **A*** are related in the following way, which is a bit more subtle than the previous relations (but which will be found useful, e.g., in the supplement to this section):

2.5 If $\mathbf{AV} = \lambda \mathbf{V}$ and $\mathbf{WA} = \mu \mathbf{W}$ (so that $\mathbf{A}^*\mathbf{W}^* = \mu \mathbf{W}^*$), then $\mathbf{WV} = 0$ if $\lambda \neq \mu$. That is, if **V** is an eigenvector of **A** corresponding to λ and if **W** is an eigenvector of **A*** corresponding to $\mu \neq \lambda$, then $\mathbf{WV} = 0$†.

Proof. $\mathbf{WAV} = \mathbf{W}(\mathbf{AV}) = \mathbf{W}(\lambda \mathbf{V}) = \lambda \mathbf{WV}$ and also $\mathbf{WAV} = (\mathbf{WA})\mathbf{V} = (\mu \mathbf{W})\mathbf{V} = \mu \mathbf{WV}$. Therefore,

$$0 = \lambda \mathbf{WV} - \mu \mathbf{WV} = (\lambda - \mu)\mathbf{WV}.$$

Hence, if $\lambda \neq \mu$ then the scalar $\mathbf{WV} = 0$. ∎

Example 5. The matrix

$$\mathbf{A} = \begin{pmatrix} \frac{2}{5} & \frac{1}{2} & \frac{1}{10} \\ \frac{1}{10} & \frac{7}{10} & \frac{1}{5} \\ \frac{1}{10} & \frac{1}{2} & \frac{2}{5} \end{pmatrix}$$

has eigenvalues $1, \frac{1}{5}, \frac{3}{10}$ with corresponding eigenvectors

$$\mathbf{V}_1 = \begin{pmatrix} 1 \\ 1 \\ 1 \end{pmatrix}, \quad \mathbf{V}_2 = \begin{pmatrix} -5 \\ 3 \\ -5 \end{pmatrix}, \quad \mathbf{V}_3 = \begin{pmatrix} 6 \\ -1 \\ -1 \end{pmatrix},$$

as can be readily verified. Appropriate "left" eigenvectors of **A** (or eigenvectors of **A***, if you wish) corresponding to these eigenvalues are

$$\mathbf{W}_1 = (8, 35, 13), \quad \mathbf{W}_2 = (0, 1, -1), \quad \mathbf{W}_3 = (1, 0, -1).$$

As promised by 2.5, $\mathbf{W}_j \mathbf{V}_i = 0$ for $i \neq j$, as you should check.

EXERCISES

1. Find bases relative to which the matrices of the following linear transformations are diagonal:
   a) $T(\mathbf{e}_1) = -3\mathbf{e}_1 + \mathbf{e}_2$
      $T(\mathbf{e}_2) = 2\mathbf{e}_1 - 2\mathbf{e}_2$
   b) $T(x, y, z) = (3x - 2y + z, 2x + y - z, 2x - 6y + 4z)$.

---

† In Chapter 8 we shall introduce the usual notion of perpendicularity, in terms of which we could interpret 2.5 as saying: The eigenvectors of **A** and **A*** corresponding to *different* eigenvectors are perpendicular.

c) $T(V_1) = 3V_1 - 2V_2 + V_3$,
   $T(V_2) = 2V_1 + V_2 - V_3$, V_1, V_2, V_3 some basis for $\mathbb{R}_3$.
   $T(V_3) = 2V_1 - 6V_2 + 4V_3$,

2. (a) through (e) Verify that 2.3 holds for those matrices in Exercise 4 of Section 1 which are invertible. You may use Exercise 2 to test for invertibility. In (a), assume that the inverse exists.

3. a) Verify that 2.4 holds for the matrices in (d) and (e) of Exercise 4 of Section 1.
   b) Verify that 2.5 holds for these same matrices.

4. a) Find a class of matrices for which the eigenvectors of A and A^* are the same.
   b) Verify 2.5 for Exercise 4(f) of Section 1.

5. Stochastic matrices were defined in Chapter 2, Section 5(d). Redefine this notion in terms of eigenvectors and eigenvalues.

Supplement to Section 2: The Power Method

The following approximative method searches for the largest eigenvalue and a corresponding eigenvector simultaneously. It is straightforward and often quite useful. Moreover, one sometimes needs to know only the largest eigenvalue (e.g., in the rabbit-hunting problem quoted at the beginning of this chapter). We then go on to describe the process of "deflation", whereby one is able to search for successively smaller eigenvalues. This uses some of the later material of the section in an entertaining fashion.

Suppose that A is an $n \times n$ matrix with n distinct† eigenvalues $\lambda_1, \ldots, \lambda_n$, including the complex ones. Suppose that λ_1 is the largest in modulus:

$$|\lambda_1| > |\lambda_i|, \quad i = 2, \ldots, n.$$

We wish to find λ_1 and an eigenvector V_1 associated with it. Let us make an initial guess that the vector V_1 is near V (using physical aspects of the problem, a ouija board, or whatever). By 2.2,

$$V = c_1 V_1 + c_2 V_2 + \cdots + c_n V_n \qquad \text{for some scalars } c_1, \ldots, c_n. \qquad (1)$$

Moreover,

$$\begin{aligned} AV &= c_1 \lambda_1 V_1 + c_2 \lambda_2 V_2 + \cdots + c_n \lambda_n V_n, \\ A^2 V &= c_1 \lambda_1^2 V_1 + c_2 \lambda_2^2 V_2 + \cdots + c_n \lambda_n^2 V_n, \\ &\quad \vdots \\ A^k V &= c_1 \lambda_1^k V_1 + c_2 \lambda_2^k V_2 + \cdots + c_n \lambda_n^k V_n. \end{aligned} \qquad (2)$$

---

† In Section 5 we shall see that it is unnecessary to require that the λ_i's be distinct; it is only necessary that $|\lambda_1|$ be strictly larger than the moduli of the others.

Now since $|\lambda_1| > |\lambda_i|$, we must have
$$|\lambda_i|/|\lambda_1| < 1,$$
where $i \geq 2$, so that $|\lambda_i|^k/|\lambda_1|^k$ can be made arbitrarily close to zero by choosing a large enough k (dig back into your calculus to recall why). Thus, for a sufficiently large k,
$$\mathbf{A}^k\mathbf{V} = \lambda_1^k(c_1\mathbf{V}_1 + c_2(\lambda_2/\lambda_1)^k\mathbf{V}_2 + \cdots + c_n(\lambda_n/\lambda_1)^k\mathbf{V}_n)$$
$$\approx \lambda_1^k c_1 \mathbf{V}_1.$$
Thus, all other terms being negligible, $\mathbf{A}^k\mathbf{V}$ is approximately $c_1\lambda_1^k\mathbf{V}_1$. (It sounds as if one might have to take a tremendously large k, but this is often not the case.) Further sophistry indicates that
$$\mathbf{A}^{k+1}\mathbf{V} \approx c_1\lambda_1^{k+1}\mathbf{V}_1.$$
Hence, the ratios of the coefficients in $\mathbf{A}^{k+1}\mathbf{V}$ to those in $\mathbf{A}^k\mathbf{V}$ should be approximately λ_1:

if $\quad \mathbf{A}^k\mathbf{V} = (v_1^{(k)}, \ldots, v_n^{(k)})^* \quad$ and $\quad \mathbf{A}^{k+1}\mathbf{V} = (v_1^{(k+1)}, \ldots, v_n^{(k+1)})^*$

then $\quad \lambda_1 \approx v_1^{(k+1)}/v_1^{(k)}, \ldots, \lambda_1 \approx v_n^{(k+1)}/v_n^{(k)}.$

In summary the power method involves calculating the effect of powers of $\mathbf{A}$ on the initial guess $\mathbf{V}$, and checking the ratios of the nonzero coefficients of successive terms $\mathbf{A}^{k+1}\mathbf{V}$ and $\mathbf{A}^k\mathbf{V}$. When these ratios
$$r_j = v_j^{(k+1)}/v_j^{(k)} \quad \text{(taken only for } v_j^{(k)} \neq 0\text{)}$$
are all fairly near one another ($|r_i - r_j|$ less than some tolerance level ϵ, say), then they should all be fairly near λ_1. You could take λ_1 to be their average value, and know you were within ϵ of *some* eigenvalue λ_1. For a corresponding eigenvector, one might choose $\mathbf{A}^k\mathbf{V}$ itself, since you know that
$$\mathbf{A}(\mathbf{A}^k\mathbf{V}) = \mathbf{A}^{k+1}\mathbf{V} \approx \lambda_1\mathbf{A}^k\mathbf{V}$$
(to within ϵ).

Example.
$$\mathbf{A} = \begin{pmatrix} 2 & 1 \\ 1 & 3 \end{pmatrix},$$
initial guess
$$\mathbf{V} = \begin{pmatrix} 1 \\ 1 \end{pmatrix}.$$
The eigenvalues are actually $(5 \pm \sqrt{5})/2$; the largest is $\lambda_1 \approx 3.618$.
$$\mathbf{AV} = \begin{pmatrix} 3 \\ 4 \end{pmatrix}, \quad \mathbf{AV}^2 = \begin{pmatrix} 10 \\ 15 \end{pmatrix}, \quad \mathbf{AV}^3 = \begin{pmatrix} 35 \\ 55 \end{pmatrix},$$

Since the entries are getting large, we divide by 5 and continue with

$$\mathbf{V}' = \begin{pmatrix} 7 \\ 11 \end{pmatrix} = \tfrac{1}{5}\mathbf{A}\mathbf{V}^3:$$

$$\mathbf{A}\mathbf{V}' = \begin{pmatrix} 25 \\ 40 \end{pmatrix}, \quad \mathbf{A}\mathbf{V}'^2 = \begin{pmatrix} 90 \\ 145 \end{pmatrix}; \quad \tfrac{90}{25} = 3.6; \quad \tfrac{145}{40} = 3.625.$$

Try again, factoring out another 5, and letting $\mathbf{V}'' = \tfrac{1}{5}\mathbf{A}\mathbf{V}'^2$

$$\mathbf{V}'' = \begin{pmatrix} 18 \\ 29 \end{pmatrix}, \quad \mathbf{A}\mathbf{V}'' = \begin{pmatrix} 65 \\ 105 \end{pmatrix}, \quad \tfrac{65}{18} = 3.61+; \quad \tfrac{105}{29} = 3.62+.$$

Try again, letting $\mathbf{V}''' = \tfrac{1}{5}\mathbf{A}\mathbf{V}''$

$$\mathbf{V}''' = \begin{pmatrix} 13 \\ 21 \end{pmatrix}, \quad \mathbf{A}\mathbf{V}''' = \begin{pmatrix} 47 \\ 76 \end{pmatrix}, \quad \tfrac{47}{13} = 3.615+; \quad \tfrac{76}{21} = 3.619+.$$

There are a number of difficulties with the power method (some of which can be taken care of; see the accompanying exercises). For example, if the coefficient c_1 of $\mathbf{V}_1$ in your original guess (1) was 0, you might end up finding some eigenvalue other than the largest one. (A computer might be of unexpected help here. Even if you start out with $\mathbf{V}_1$ coefficient 0, after a few rounds a computer would likely introduce a bit of "noise" so that the coefficient would no longer be 0, and it could proceed to blow up as it's supposed to do.) Nonetheless, the utter simplicity and usually good speed make this method often worth a try, even when there is no reason to believe the eigenvalues are distinct.

EXERCISES

1. Let

$$\mathbf{M}_1 = \begin{pmatrix} 0 & 0 & 6 \\ \tfrac{1}{2} & 0 & 0 \\ 0 & \tfrac{1}{3} & 0 \end{pmatrix} \quad \text{and} \quad \mathbf{M}_2 = \begin{pmatrix} 0 & 1 & 3 \\ \tfrac{1}{2} & 0 & 0 \\ 0 & \tfrac{1}{3} & 0 \end{pmatrix}.$$

These are the "population matrices" of Chapter 2, Section 5(b); given a population distribution $n = (n_1, n_2, n_3)^*$ of beetles$_i$, $\mathbf{M}_i \mathbf{n}$ is the population distribution a year hence.
  a) Start out with a particular population distribution $(n_1, n_2, n_3)^*$, your choice of n_i not all 0. What happens if you apply the power method to $\mathbf{M}_1$? Interpret in terms of stable populations of beetles$_1$.
  b) Find the eigenvalues of $\mathbf{M}_1$ and explain.
  c) Now do (a) for $\mathbf{M}_2$; see if you get close to a stable population fairly rapidly.
  d) Find the eigenvalues of $\mathbf{M}_2$ and explain.

2. (See also Section 5, Exercise 13.) Let

$$\mathbf{P} = \begin{pmatrix} \tfrac{2}{5} & \tfrac{1}{2} & \tfrac{1}{10} \\ \tfrac{1}{10} & \tfrac{7}{10} & \tfrac{1}{5} \\ \tfrac{1}{10} & \tfrac{1}{2} & \tfrac{2}{5} \end{pmatrix}$$

be as in Chapter 2, Section 5d; this matrix gives a model of the social mobility of a society. It turns out that society is heading toward a fixed distribution of social classes.

a) Use the power method to find this distribution (i.e., make a guess $V = (v_1, v_2, v_3)^*$ and calculate $P^k V$ for sufficiently large k to find this distribution.

b) What does "fixed distribution" mean in terms of eigenvalues? Calculate those of P, to verify why the power method works.

3. Suppose that the maximum eigenvalue λ_1 is not distinct, say $\lambda_1 = \lambda_2$, but the eigenvectors still span the space. Show that if either of the coefficients c_1 and c_2 are non-zero in (1), the initial guess V, then $A^k V$ will tend toward an eigenvector for λ_1 ($= \lambda_2$). Show how λ_1 can be found.

4. Suppose that $\lambda_1 = -\lambda_2$, but $|\lambda_1| > |\lambda_i|$ for $i = 3, \ldots, n$. Show how to amend the method so you can find λ_1, λ_2 and their corresponding eigenvectors.

H5. Consider one of our favorite matrices

$$A = \begin{pmatrix} 0 & -1 \\ 1 & 0 \end{pmatrix}.$$

If you start with a guess V with real entries, you will always have that $A^k V$ is real and consequently never get close to an eigenvector of A. In fact, the behavior of $A^k V$ is periodic, as one might guess from the geometrical meaning of A.

How could you amend the power method to find complex eigenvalues? Note that in general, if a real matrix has complex eigenvalues, they come in conjugate pairs.

6. (Bernoulli's method for finding a real root of a polynomial) Let x_1 be a (real) root, of maximum absolute value, of the polynomial

$$p(x) = a_0 + a_1 x + \cdots + a_{n-1} x^{n-1} + x^n$$

with real coefficients; i.e., for any other root x_i (real or complex),

$$|x_i| < |x_1|.$$

Construct the sequence u_m by taking $u_0, \ldots, u_{n-1}$ arbitrarily and then finding $u_n, u_{n+1}, \ldots$ from the recursion formula

$$u_m = -(a_{n-1} u_{m-1} + a_{n-2} u_{m-2} + \cdots + a_0 u_{m-n}), \quad \text{where } m = n, n+1, \ldots$$

Prove that

$$\lim_{m \to \infty} \frac{u_{m+1}}{u_m} = x_1.$$

[*Hint:* Apply the power method to the companion matrix of $p(x)$. You may assume that the roots of $p(x)$ are distinct.]

7. Apply Bernoulli's method to find the real root of $x^3 - x - 1 = 0$, to a reasonably good approximation.

8. Consider the Fibonacci sequence u_m defined by
$$u_0 = 1, \quad u_1 = 1, \quad u_m = u_{m-1} + u_{m-2} \ (m \geq 2).$$
Find $\lim_{m \to \infty} \frac{u_{m+1}}{u_m}$.

Deflation

Suppose we are again in the ideal situation with distinct eigenvalues, and that we have found $\mathbf{A}$'s largest eigenvalue λ_1 and a corresponding eigenvector $\mathbf{V}_1$. We now show how to replace $\mathbf{A}$ by a matrix $\mathbf{B}$ whose eigenvalues are those of $\mathbf{A}$, except that it has a 0 in place of λ_1. We could then use the power method to search for $\mathbf{B}$'s largest eigenvalue (which is $\mathbf{A}$'s next-to-largest), and so on.

Having found the largest real eigenvalue λ_1 of $\mathbf{A}$ (and a corresponding vector $\mathbf{V}_1$), we know that λ_1 is the largest for $\mathbf{A}^*$ as well (by 2.4). By our standard technique of solving a system of equations, it is easy to find an eigenvector $\mathbf{W}^*$ of $\mathbf{A}^*$ corresponding to λ_1, so suppose that
$$\mathbf{A}^*\mathbf{W}^* = \lambda_1 \mathbf{W}^*;$$
or $\mathbf{W}\mathbf{A} = \lambda_1 \mathbf{W}$ (for $\mathbf{W}$ a row vector). Since the eigenvectors of $\mathbf{A}$ are supposed to span $\mathbb{R}_n$, we can write
$$\mathbf{W}^* = c_1 \mathbf{V}_1 + \cdots + c_n \mathbf{V}_n.$$
Then
$$\mathbf{W}\mathbf{W}^* = \mathbf{W}(c_1 \mathbf{V}_1 + \cdots + c_n \mathbf{V}_n)$$
$$= c_1 \mathbf{W}\mathbf{V}_1 + \cdots + c_n \mathbf{W}\mathbf{V}_n \overset{2.5}{=} c_1 \mathbf{W}\mathbf{V}_1.$$

Since $\mathbf{W}\mathbf{W}^*$ is a positive scalar (why$_1$?), this shows that $\mathbf{W}\mathbf{V}_1 \neq 0$. Now since $\mathbf{W}\mathbf{V}_1$ is a nonzero scalar, we can replace $\mathbf{W}$ by $\mathbf{W}_1 = (\mathbf{W}\mathbf{V}_1)^{-1}\mathbf{W}$ and obtain another eigenvector for $\mathbf{A}^*$, and one for which $\mathbf{W}_1\mathbf{V}_1 = 1$ (why$_2$?).

Since $\mathbf{W}_1$ is $1 \times n$ and $\mathbf{V}_1$ is $n \times 1$, $\mathbf{V}_1\mathbf{W}_1$ is an $n \times n$ matrix (what are its entries?). Let $\mathbf{B} = \mathbf{A} - \lambda_1 \mathbf{V}_1 \mathbf{W}_1$. Then
$$\mathbf{B}\mathbf{V}_1 = \mathbf{A}\mathbf{V}_1 - \lambda_1 \mathbf{V}_1(\mathbf{W}_1 \mathbf{V}_1) = \lambda_1 \mathbf{V}_1 - \lambda_1 \mathbf{V}_1 = \mathbf{0},$$
while for $i \neq 1$,
$$\mathbf{B}\mathbf{V}_i = \mathbf{A}\mathbf{V}_i - \lambda_1 \mathbf{V}_1(\mathbf{W}_1 \mathbf{V}_i) \overset{2.5}{=} \lambda_i \mathbf{V}_i,$$
so that $\mathbf{B}$ is as desired.

Exercise 9. Answer the above questions.

Honors Projects. Show that the material in this section holds for (1) vector spaces over $\mathbb{C}$ (including some portions of the supplement; which and what?) (2) Same for vector spaces over an arbitrary $\mathbb{F}$, with no hope for the supplement, in general.

3. THE CHARACTERISTIC POLYNOMIAL

We now consider further the characteristic polynomial $p(x)$ of a matrix $\mathbf{A}$:

$$p(x) = \det (\mathbf{A} - x\mathbf{I}) = \begin{pmatrix} a_{11} - x & a_{12} & \cdots & a_{1n} \\ a_{21} & a_{22} - x & \cdots & a_{2n} \\ \vdots & & & \\ a_{n1} & \cdots & \cdots & a_{nn} - x \end{pmatrix}. \quad (1)$$

Let us also write $p(x)$ in the form of a polynomial in x,

$$p(x) = (-1)^n x^n + (-1)^{n-1} b_{n-1} x^{n-1} + (-1)^{n-2} b_{n-2} x^{n-2} + \cdots + (-1)^1 b_1 x + b_0.$$

(The $(-1)^k$'s are put in to make things work out nicely.) We shall derive some interesting properties of the coefficients b_0 and b_{n-1}. Now

$$b_0 = p(0) = \det (\mathbf{A} - 0\mathbf{I}) = \det \mathbf{A}.$$

Thus,

the constant coefficient of the characteristic polynomial is $\det \mathbf{A}$.

This is delightful in itself, but it also leads to

3.1 The determinant is the product of the eigenvalues: if the eigenvalues of $\mathbf{A}$ are $\lambda_1, \ldots, \lambda_n$ (each listed the number of times it occurs as a root of the characteristic polynomial), then

$$\det \mathbf{A} = \lambda_1 \lambda_2 \cdots \lambda_n.$$

Proof. Since the characteristic polynomial $p(x)$ has roots $\lambda_1, \ldots, \lambda_n$, it is a well-known result about polynomials that

$$p(x) = (\lambda_1 - x)(\lambda_2 - x) \cdots (\lambda_n - x).$$

Hence, $p(0) = \lambda_1 \lambda_2 \cdots \lambda_n$. We just noticed that $p(0) = \det \mathbf{A}$, as well. ∎

Even though some of the λ's are possibly complex, their product, $\det \mathbf{A}$, is real. For example, our friend

$$\begin{pmatrix} 0 & -1 \\ 1 & 0 \end{pmatrix}$$

has eigenvalues i, $-i$ and $i(-i) = 1 = \det \mathbf{A}$.

How does one obtain the coefficient b_{n-1} of $(-1)^{n-1} x^{n-1}$ in $p(x)$? It comes from terms in the determinant expansion which involve products of $(n-1)$ of the diagonal elements of (1). Now since each summand in the determinant expansion has one element from each row and each column, if $(n-1)$ of the elements are diagonal elements, then so must be the nth element. Thus, $(-1)^{n-1} b_{n-1} x^{n-1}$

comes from gathering up the terms of degree $(n - 1)$ in
$$(a_{11} - x)(a_{22} - x) \cdots (a_{nn} - x). \tag{2}$$
A moment's reflection (Exercise 1) shows that the coefficient of $(-1)^{n-1}x^{n-1}$ is $a_{11} + a_{22} + \cdots + a_{nn}$.

The **trace** of a matrix
$$A = \begin{pmatrix} a_{11} & \cdots & a_{1n} \\ & a_{22} & & \\ & & \ddots & \\ a_{n1} & \cdots & a_{nn} \end{pmatrix}$$
is the sum of the entries on its main diagonal:
$$\text{trace } A = a_{11} + a_{22} + \cdots + a_{nn}.$$
Thus, the characteristic polynomial of A is
$$(-x)^n + \text{trace } A(-x)^{n-1} + \cdots + \det A.$$
Comparing the trace of A with the coefficient of $(-x)^{n-1}$ is a very simple partial check that the characteristic polynomial has been correctly calculated. The trace is also intimately connected with the eigenvalues:

3.2 The trace is the sum of the eigenvalues: If A has eigenvalues $\lambda_1, \ldots, \lambda_n$, each counted according to its multiplicity as a root of the characteristic polynomial, then
$$\text{trace } A = \lambda_1 + \lambda_2 + \cdots + \lambda_n.$$

Proof. $p(x) = (\lambda_1 - x)(\lambda_2 - x) \cdots (\lambda_n - x)$. Just as for (2), the coefficient of $(-x)^{n-1}$ is $\lambda_1 + \lambda_2 + \cdots + \lambda_n$. By definition this is b_{n-1}, and we have seen that this is also $a_{11} + \cdots + a_{nn}$. ∎

The same remark that follows 3.1 applies here, and the same example should serve to illuminate it:
$$\begin{pmatrix} 0 & -1 \\ 1 & 0 \end{pmatrix}$$
has trace $0 + 0 = 0$ and the sum of its eigenvalues is also $0 = i + (-i)$.

Some important properties of the trace function:

3.3 i) The trace is a linear transformation from $\mathbb{R}_{n \times n}$ into $\mathbb{R}$. That is
$$\text{trace } (A + B) = \text{trace } A + \text{trace } B,$$
$$\text{trace } cA = c \text{ trace } A.$$
ii) trace $(AB) =$ trace (BA).

Proof. i) is easy. ii) involves writing out the diagonal elements of **AB** and **BA**. Try it (Exercise 2). For example,

$$\text{trace}\left[\begin{pmatrix} a_{11} & a_{12} \\ a_{21} & a_{22} \end{pmatrix}\begin{pmatrix} b_{11} & b_{12} \\ b_{21} & b_{22} \end{pmatrix}\right] = a_{11}b_{11} + a_{12}b_{21} + a_{21}b_{12} + a_{22}b_{22}$$

$$= b_{11}a_{11} + b_{12}a_{21} + b_{21}a_{12} + b_{22}a_{22}$$

$$= \text{trace}\left[\begin{pmatrix} b_{11} & b_{12} \\ b_{21} & b_{22} \end{pmatrix}\begin{pmatrix} a_{11} & a_{12} \\ a_{21} & a_{22} \end{pmatrix}\right]. \quad \blacksquare$$

We next give the full statement and partial proof of a famous theorem on the characteristic polynomial. The full proof is to be found in Section 6. There is a marvelously simple (but fallacious) proof given in Exercise 3.

3.4 The Cayley-Hamilton theorem. Let **A** be an $n \times n$ matrix with characteristic polynomial

$$p(x) = (-1)^n x^n + (-1)^{n-1} b_{n-1} x^{n-1} + \cdots (-1) b_1 x + b_0.$$

Then

$$p(\mathbf{A}) = (-1)^n \mathbf{A}^n + (-1)^{n-1} b_{n-1} \mathbf{A}^{n-1} + \cdots (-b_1)\mathbf{A} + b_0 \mathbf{I} = \mathbf{0}. \quad (3)$$

(Thus, any matrix is a "root" of its characteristic polynomial.)

Partial proof. We shall prove this for the case $n = 2$. Let

$$\mathbf{A} = \begin{pmatrix} a & b \\ c & d \end{pmatrix},$$

so that

$$p(x) = \det\begin{pmatrix} a - x & b \\ c & d - x \end{pmatrix}$$

$$= (a - x)(d - x) - bc = x^2 - (a + d)x + (ad - bc).$$

Then

$$p(\mathbf{A}) = \begin{pmatrix} a & b \\ c & d \end{pmatrix}\begin{pmatrix} a & b \\ c & d \end{pmatrix} - (a + d)\begin{pmatrix} a & b \\ c & d \end{pmatrix} + (ad - bc)\mathbf{I}$$

$$= \begin{pmatrix} a^2 + bc & ab + bd \\ ac + cd & bc + d^2 \end{pmatrix} - \begin{pmatrix} a^2 + ad & ab + bd \\ ac + cd & ad + d^2 \end{pmatrix} + \begin{pmatrix} ad - bc & 0 \\ 0 & ad - bc \end{pmatrix}$$

$$= \begin{pmatrix} 0 & 0 \\ 0 & 0 \end{pmatrix}. \quad \blacksquare$$

As a first application of this theorem, we shall show how to express the inverse of an invertible matrix **A** as a polynomial in **A**. **A** is invertible iff $\det \mathbf{A} \neq 0$ and $\det \mathbf{A} = b_0$ in (3). Thus, if **A** is invertible we can solve (3) for **I** and obtain

$$\mathbf{I} = -b_0^{-1}(-1)^n \mathbf{A}^n - b_0^{-1}(-1)^{n-1} b_{n-1} \mathbf{A}^{n-1} \cdots - b_0^{-1}(-1)^1 b_1 \mathbf{A}.$$

Factoring out **A** we have

$$\mathbf{I} = \mathbf{A}[(-1)^{n-1}b_0^{-1}\mathbf{A}^{n-1} + (-1)^{n-2}b_0^{-1}b_{n-1}\mathbf{A}^{n-2} + \cdots + b_0^{-1}b_1\mathbf{I}].$$

Thus,

If **A** is invertible then its inverse is a polynomial in **A**. In fact, if the characteristic polynomial of **A** is (1), then

$$\mathbf{A}^{-1} = b_0^{-1}((-1)^{n-1}\mathbf{A}^{n-1} + (-1)^{n-2}b_{n-1}\mathbf{A}^{n-2} + \cdots + b_1\mathbf{I}).$$

Example. For

$$\mathbf{A} = \begin{pmatrix} a & b \\ c & d \end{pmatrix},$$

the material in this section allows us to read off immediately the coefficients in the characteristic polynomial; it is $x^2 - (a+d)x + (ad-bc)$. If **A** has an inverse, it is therefore

$$\frac{1}{ad-bc}((-1)\mathbf{A} + (a+d)\mathbf{I}) = \frac{1}{ad-bc}\left[\begin{pmatrix} -a & -b \\ -c & -d \end{pmatrix} + \begin{pmatrix} a+d & 0 \\ 0 & a+d \end{pmatrix}\right]$$

$$= \frac{1}{ad-bc}\begin{pmatrix} d & -b \\ -c & a \end{pmatrix}.$$

In addition to calculating the inverse of a matrix as a polynomial in the matrix, the Cayley-Hamilton Theorem provides a check on one's calculation of the characteristic polynomial: just plug in the matrix and see if the zero matrix is obtained. Moreover, the theorem allows one to calculate $\mathbf{A}^{n+1}$ in terms of $\mathbf{A}, \mathbf{A}^2, \ldots, \mathbf{A}^n$. This might be convenient on a computer, where multiplication is more expensive than addition (provided you were given the characteristic polynomial, of course).

EXERCISES

1. Prove that the coefficient of x^{n-1} in (2) is as stated.
2. Prove (a) 3.3(i); (b) 3.3(ii).
3. Demolish the following "proof" of the Cayley-Hamilton Theorem: The characteristic polynomial of **A** is $p(x) = \det(\mathbf{A} - x\mathbf{I})$. Thus

$$p(\mathbf{A}) = \det(\mathbf{A} - \mathbf{A}\mathbf{I}) = \det(\mathbf{A} - \mathbf{A}) = \det(\mathbf{0}) = 0. \quad \text{Q.E.D.}$$

4. (a) through (f) Verify 3.1 on the matrices in Exercise 4 in Section 1.
5. (a) through (f) Verify 3.2 on these same matrices.
6. (a) through (f) Verify the Cayley-Hamilton Theorem on these matrices.
7. Prove that in general, trace $\mathbf{AB} \neq (\text{trace } \mathbf{A})(\text{trace } \mathbf{B})$.
8. Let **A** be in $\mathbb{R}_{n \times n}$, and consider **A** as belonging to $\mathbb{C}_{n \times n}$, so that it has all its eigenvalues. Use 3.1 to prove that if $\lambda = a + ib$ is an eigenvalue (a, b real), then so is its conjugate $\bar{\lambda} = a - ib$ (if you're rusty on this sort of thing, start by calculating $\lambda\bar{\lambda}$).

9. a) Find the eigenvalues of the $(2n \times 2n)$ matrix

$$A = \begin{pmatrix} 1 & 0 & 1 & 0 & \cdots & 1 & 0 \\ 0 & 1 & 0 & 1 & \cdots & 0 & 1 \\ 1 & 0 & 1 & 0 & \cdots & 1 & 0 \\ \vdots & & & & & & \vdots \\ 0 & 1 & 0 & 1 & & 0 & 1 \end{pmatrix}.$$

[*Suggestions*: $(1, 1, \ldots, 1)^*$ is an eigenvector; use the trace for another eigenvalue. Look at the rank for the others.]

b) Same problem for

$$B = \begin{pmatrix} 0 & 1 & \cdots & 1 \\ 1 & 0 & \cdots & 0 \\ \vdots & & & \vdots \\ 1 & 0 & \cdots & 0 \end{pmatrix}.$$

10. If A is invertible and $p(x)$ is the characteristic polynomial of A, prove that the characteristic polynomial of A^{-1} is

$$\frac{1}{\det A}(-x)^n p(1/x).$$

11. a) Write out the characteristic polynomial in full for the case $n = 3$. In particular, prove that b_1 is the sum of the cofactors of the diagonal elements of the matrix.

H*b) Prove (for any n) that the coefficient b_1 of the characteristic polynomial is the sum of the cofactors of the diagonal elements of the matrix. (If A is invertible you can use Exercise 10.)

Honors Projects. Show that the material in this section holds (1) for vector spaces over $\mathbb{C}$, (2) for vector spaces over an arbitrary field. However, you may need to go outside the field to an "extension field" to get a complete set of eigenvalues for 3.1 and 3.2.

4. SIMILARITY

We say that two $n \times n$ matrices A and B are **similar** if there is an invertible P for which $B = P^{-1}AP$. This is an equivalence relation on $\mathbb{R}_{n \times n}$. That is,

i) any matrix is similar to itself;
ii) if A is similar to B, then B is similar to A;
iii) if A is similar to B and B is similar to C, then A is similar to C.

[See Chapter 0, Section 1, Supplement 1; we ask you to prove all these properties in Exercise 2.] The notion of similarity should be compared to that of matrix equivalence (see Chapter 4, Section 4).

It is worth noting that for fixed invertible $\mathbf{P}$, the mapping $\mathbf{A} \to \mathbf{P}^{-1}\mathbf{A}\mathbf{P}$

i) is an isomorphism of the vector space of matrices $\mathbb{R}_{n \times n}$ onto itself, (1)

ii) preserves matrix multiplication,

$$\mathbf{A}_1 \mathbf{A}_2 \to \mathbf{P}^{-1}\mathbf{A}_1\mathbf{A}_2\mathbf{P} = (\mathbf{P}^{-1}\mathbf{A}_1\mathbf{P})(\mathbf{P}^{-1}\mathbf{A}_2\mathbf{P}). \tag{2}$$

(See Exercise 7.)
It is because of these properties that we have

4.1 Similar matrices have the same

        rank
        characteristic polynomial
        determinant
        trace
and   eigenvalues.

Proof. One may show that the ranks are the same either directly or by appealing to material in Chapter 4, Section 4. We leave the choice to you (Exercise 8).

To show that the characteristic polynomials are the same, suppose that $\mathbf{B} = \mathbf{P}^{-1}\mathbf{A}\mathbf{P}$. Then

$$\det (\mathbf{B} - x\mathbf{I}) \stackrel{\text{why?}}{=} \det \mathbf{P}^{-1}(\mathbf{B} - x\mathbf{I})\mathbf{P}$$
$$= \det (\mathbf{P}^{-1}\mathbf{B}\mathbf{P} - \mathbf{P}^{-1}(x\mathbf{I})\mathbf{P}) = \det (\mathbf{A} - x\mathbf{I}),$$

the last step holding because for any value of x,

$$\mathbf{P}^{-1}(x\mathbf{I})\mathbf{P} = x\mathbf{P}^{-1}\mathbf{I}\mathbf{P}.$$

Hence the characteristic polynomials are the same. Hence the roots of these polynomials (the eigenvalues) are the same, and the coefficients of degree zero (determinant) and coefficients of degree $(n-1)$ (i.e., trace) are all the same. ∎

In general the eigenvectors of similar matrices are *not* the same. In fact,

$\mathbf{V}$ is an eigenvector of $\mathbf{A}$ corresponding to the eigenvalue λ iff $\mathbf{P}^{-1}\mathbf{V}$ is an eigenvector of $\mathbf{B} = \mathbf{P}^{-1}\mathbf{A}\mathbf{P}$ corresponding to λ. (3)

(See Exercise 5.)

We now make a very important link between similar matrices and linear transformations. Recall (for the $(n+1)$st time) that if $\mathbf{T}$ is a linear transformation from a finite-dimensional vector space $\mathcal{V}$ into itself, and if we choose a basis for $\mathcal{V}$ (which we identify with $\mathbf{e}_1, \ldots, \mathbf{e}_n$), then $\mathbf{T}(\mathbf{V}) = \mathbf{A}\mathbf{V}$, where $\mathbf{A} = (\mathbf{T}(\mathbf{e}_1), \ldots, \mathbf{T}(\mathbf{e}_n))$. Of course, if we choose different bases for $\mathcal{V}$, we will obtain a different matrix for $\mathbf{T}$. This is where similarity enters in.

4.2 Two matrices are similar iff they are matrices of the same linear transformation from $\mathbb{R}_n$ into $\mathbb{R}_n$ relative to different bases:
  i) If $\mathbf{A}$ is similar to $\mathbf{B}$, and $\mathbf{T(V)} = \mathbf{AV}$ (i.e., with respect to the usual basis $\{\mathbf{e}_1, \ldots, \mathbf{e}_n\}$, $\mathbf{T}$ is matrix multiplication by $\mathbf{A}$), then there is a basis $\{\mathbf{e}'_1, \ldots, \mathbf{e}'_n\}$ relative to which $\mathbf{T(V)} = \mathbf{BV}$.
  ii) If $\mathbf{A}$ and $\mathbf{B}$ are the matrices of a linear transformation $\mathbf{T}$ relative to different bases, then $\mathbf{A}$ and $\mathbf{B}$ are similar.

Proof. (For this proof we shall make use of the notation and notions from the first part of Chapter 4, Section 4). (i) If $\mathbf{B} = \mathbf{P}^{-1}\mathbf{AP}$, let $\mathbf{e}'_1 = $ 1st column of $\mathbf{P}, \ldots, \mathbf{e}'_n = n$th column of $\mathbf{P}$; this is a basis for $\mathbb{R}_n$. Let

$$\alpha = \{\mathbf{e}_1, \ldots, \mathbf{e}_n\} \quad \text{and} \quad \alpha' = \{\mathbf{e}'_1, \ldots, \mathbf{e}'_n\}.$$

In the notation of Chapter 4, Section 4,

$$\mathbf{A} = \mathbf{M}_{\alpha\alpha}(\mathbf{T}) \quad \text{and} \quad \mathbf{P} = \mathbf{M}_{\alpha\alpha'}(\mathbf{I}),$$

so that $\mathbf{P}^{-1} = \mathbf{M}_{\alpha'\alpha}(\mathbf{I})$ (by Chapter 4, 4.1). Hence,

$$\mathbf{B} = \mathbf{P}^{-1}\mathbf{AP} = \mathbf{M}_{\alpha'\alpha}(\mathbf{I})\mathbf{M}_{\alpha\alpha}(\mathbf{T})\mathbf{M}_{\alpha\alpha'}(\mathbf{I}) = \mathbf{M}_{\alpha'\alpha'}(\mathbf{ITI}) = \mathbf{M}_{\alpha'\alpha'}(\mathbf{T}), \quad (4)$$

so that $\mathbf{B}$ is the matrix of $\mathbf{T}$ relative to the basis α'. (ii) If $\mathbf{A}$ is the matrix of $\mathbf{T}$ relative to α and $\mathbf{B}$ the matrix of $\mathbf{T}$ relative to α', let $\mathbf{P} = \mathbf{M}_{\alpha\alpha'}(\mathbf{I})$. Traversing (4) in the opposite direction, we see that $\mathbf{B} = \mathbf{P}^{-1}\mathbf{AP}$, so that $\mathbf{A}$ and $\mathbf{B}$ are similar. ∎

Example. Let $\mathbf{T}$ be the linear transformation from $\mathbb{R}_2$ into $\mathbb{R}_2$ for which

$$\mathbf{T}(\mathbf{e}_1) = 5\mathbf{e}_1 + 2\mathbf{e}_2, \quad \mathbf{T}(\mathbf{e}_2) = 3\mathbf{e}_1 + 4\mathbf{e}_2,$$

and let

$$\mathbf{e}'_1 = 2\mathbf{e}_1 + \mathbf{e}_2, \qquad \mathbf{e}_1 = -\mathbf{e}'_1 + \mathbf{e}'_2$$
$$\mathbf{e}'_2 = 3\mathbf{e}_1 + \mathbf{e}_2, \quad \text{so that} \quad \mathbf{e}_2 = 3\mathbf{e}'_1 - 2\mathbf{e}'_2.$$

Then

$$\mathbf{T}(\mathbf{e}'_1) = 13\mathbf{e}_1 + 8\mathbf{e}_2 = 11\mathbf{e}'_1 - 3\mathbf{e}'_1,$$
$$\mathbf{T}(\mathbf{e}'_2) = 18\mathbf{e}_1 + 10\mathbf{e}_2 = 12\mathbf{e}'_2 - 2\mathbf{e}'_2.$$

In matrix terms, the matrix of $\mathbf{T}$ with respect to $\alpha = \{\mathbf{e}_1, \mathbf{e}_2\}$ is

$$\mathbf{M}_{\alpha\alpha}(\mathbf{T}) = \mathbf{A} = \begin{pmatrix} 5 & 3 \\ 2 & 4 \end{pmatrix},$$

while with respect to the basis $\alpha' = \{\mathbf{e}'_1, \mathbf{e}'_2\}$ it is

$$\mathbf{M}_{\alpha'\alpha'}(\mathbf{T}) = \mathbf{A}' = \begin{pmatrix} 11 & 12 \\ -3 & -2 \end{pmatrix}.$$

The matrix for expressing the $\mathbf{e}'_i$ in terms of the $\mathbf{e}_j$ is

$$\mathbf{M}_{\alpha\alpha'}(\mathbf{I}) = \mathbf{P} = \begin{pmatrix} 2 & 3 \\ 1 & 1 \end{pmatrix},$$

and that for the e_i in terms of the e'_j is

$$M_{\alpha'\alpha}(I) = P^{-1} = \begin{pmatrix} -1 & 3 \\ 1 & -2 \end{pmatrix}.$$

It is easy to check that $A' = P^{-1}AP$; that is,

$$\begin{pmatrix} 11 & 12 \\ -3 & -2 \end{pmatrix} = \begin{pmatrix} -1 & 3 \\ 1 & -2 \end{pmatrix} \begin{pmatrix} 5 & 3 \\ 2 & 4 \end{pmatrix} \begin{pmatrix} 2 & 3 \\ 1 & 1 \end{pmatrix},$$

or

$$M_{\alpha'\alpha'}(T) = M_{\alpha'\alpha}(I) M_{\alpha\alpha}(T) M_{\alpha\alpha'}(I).$$

Linking up 4.1 and 4.2, we see that any matrix of a linear transformation T will, in particular, have the same characteristic polynomial, determinant, and trace. These are notions we have not previously attributed to linear transformations, but we are now free to do so:

> Given a linear transformation T from the finite-dimensional vector space $\mathcal{U}$ into itself, let A be any matrix of T (so that $T(V) = AV$ relative to some basis). Then we define the
>
> **characteristic polynomial** of T to be the characteristic polynomial of A,
> **determinant** of T to be the determinant of A,
> **trace** of T to be the trace of A.

Our remarks show that these definitions are consistent and unambiguous.

If you see this material again from a very sophisticated point of view, you will find that it is possible to introduce these notions for linear transformations directly, without recourse to matrices. This approach is not likely to be appreciated the first time around, however.

We next use some of the notions embodied in the proof of 4.2 to obtain a slight extension of 2.2.

4.3 If A is an $n \times n$ matrix with n linearly independent eigenvectors $V_1, \ldots, V_n$, then A is similar to a diagonal matrix. In fact if we define $P = (V_1, \ldots, V_n)$, then

$$P^{-1}AP = \begin{pmatrix} \lambda_1 & & 0 \\ & \cdot & \\ & & \cdot \\ 0 & & \lambda_n \end{pmatrix},$$

where λ_i is the eigenvalue of V_i. (In particular, this will hold if A has n distinct eigenvalues.) Conversely, if A is similar to a diagonal matrix, then it has n linearly independent eigenvectors.

Proof. i)

$$AP = (AV_1, \ldots, AV_n) = (\lambda_1 V_1, \ldots, \lambda_n V_n) = P \begin{pmatrix} \lambda_1 & & 0 \\ & \cdot & \\ & & \cdot \\ 0 & & \lambda_n \end{pmatrix},$$

and **P** is invertible since it is of rank n.

ii) To prove the converse, if

$$P^{-1}AP = \begin{pmatrix} \lambda_1 & & 0 \\ & \cdot & \\ & & \cdot \\ 0 & & \lambda_n \end{pmatrix},$$

this diagonal matrix has n linearly independent eigenvectors (namely, $e_1, \ldots, e_n$). Use (3) to obtain n eigenvectors for **A**; these are easily shown to be linearly independent (Exercise 9). ∎

For the computations involved in this connection, we have the following extension of 2.1.

4.4 Let **A** be an $n \times n$ matrix with distinct eigenvalues $\lambda_1, \lambda_2, \ldots, \lambda_r$. For each i suppose that the kernel of $T_i(V) = (A - \lambda_i I)V$ has dimension n_i. Choose bases for these subspaces. The $n_1 + n_2 + \cdots + n_r$ eigenvectors thus obtained are linearly independent. Moreover, this is the maximal number of linearly independent eigenvectors.

Proof. The proof of the first part is like that of 2.1 and is left to the reader (Exercise 9). For the "moreover" note that any eigenvector belongs to the kernel of one of the T_i. You finish up. ∎

If $n_1 + n_2 + \cdots + n_r$ happens to equal n, we can then use 4.3.

Example.

$$A = \begin{pmatrix} 8 & -6 & 6 \\ 9 & -7 & 9 \\ 0 & 0 & 2 \end{pmatrix}$$

has the characteristic polynomial $-(x + 1)(x - 2)^2$. Now $\lambda_1 = -1$ has eigenvector $(2, 3, 0)^*$ and no other linearly independent ones ($n_1 = 1$). But $\lambda_2 = 2$ has linearly independent eigenvectors

$$(-1, 0, 1)^*, \quad (1, 1, 0)^* \quad (\text{so } n_2 = 2).$$

Hence, we can take

$$P = \begin{pmatrix} 2 & -1 & 1 \\ 3 & 0 & 1 \\ 0 & 1 & 0 \end{pmatrix},$$

and find

$$AP = \begin{pmatrix} -2 & -2 & 2 \\ -3 & 0 & 2 \\ 0 & 2 & 0 \end{pmatrix} = P\begin{pmatrix} -1 & 0 & 0 \\ 0 & 2 & 0 \\ 0 & 0 & 2 \end{pmatrix},$$

so that

$$P^{-1}AP = \begin{pmatrix} -1 & 0 & 0 \\ 0 & 2 & 0 \\ 0 & 0 & 2 \end{pmatrix}.$$

Thus if A has n linearly independent eigenvectors, then it is, in fact, similar to a diagonal matrix. However, not every matrix has this many independent eigenvectors.

Example. Let

$$A = \begin{pmatrix} 2 & 1 \\ 0 & 2 \end{pmatrix}.$$

If A were similar to a diagonal matrix, it would have to be

$$B = \begin{pmatrix} 2 & 0 \\ 0 & 2 \end{pmatrix},$$

since both the eigenvalues of A are 2's. But if $B = P^{-1}AP$, then $A = PBP^{-1}$, and since $B = 2I$, it commutes with P and we have

$$A = BPP^{-1} = BI = B.$$

From this you can conclude that $0 = 1$. From this you can conclude anything.
Note the difference between the above A and

$$A' = \begin{pmatrix} 2 & 1 \\ 0 & 3 \end{pmatrix}.$$

Since A' has distinct eigenvalues, it is in fact similar to

$$\begin{pmatrix} 2 & 0 \\ 0 & 3 \end{pmatrix}$$

(see Exercise 10).

Another possible difficulty is the usual one that A may have complex roots for which it is necessary to consider complex eigenvectors. Even assuming that A has all real roots (or, treating A as belonging to $\mathbb{C}_{n \times n}$, so that complex roots cause no concern), the above example shows that A need not be similar to a diagonal matrix. In the next section we consider what one can do about matrices which suffer this affliction (which 4.3 shows we can attribute to a paucity of eigenvectors).

EXERCISES

1. a) Prove that similar matrices have similar transposes.
   b) Prove that similar invertible matrices have similar inverses.
2. Prove that similarity is an equivalence relation.
3. (b) through (f) For those matrices **A** in Exercise 4 of Section 1 which have distinct characteristic roots, find a **P** for which $\mathbf{P}^{-1}\mathbf{AP} = \mathbf{D}$ is a diagonal matrix. Check your work.
4. (a) through (c) What are the determinants, traces, and characteristic polynomials of the linear transformations in Exercise 1 of Section 2?
5. Prove (3).
6. Prove that

$$\mathbf{A} = \begin{pmatrix} 5 & -4 & 4 \\ 6 & -5 & 6 \\ 0 & 0 & 1 \end{pmatrix}$$

   is similar to a diagonal matrix **D** and find a change-of-basis matrix **P** for which $\mathbf{P}^{-1}\mathbf{AP} = \mathbf{D}$. Check your work.

7. Prove (1) and (2).
8. Finish the proof of 4.1.
9. a) Finish proving the converse of 4.3.
   b) Prove 4.4.
10. Find **P** for which

$$\mathbf{P}^{-1} \begin{pmatrix} 2 & 1 \\ 0 & 3 \end{pmatrix} \mathbf{P} = \begin{pmatrix} 2 & 0 \\ 0 & 3 \end{pmatrix}.$$

11. Prove that the following matrices are similar to diagonal matrices in $\mathbb{C}_{3\times3}$ and $\mathbb{C}_{4\times4}$, respectively; find the appropriate change-of-basis matrix in each case.

   a) $\begin{pmatrix} 0 & 1 & 0 \\ 0 & 0 & 1 \\ 1 & 0 & 0 \end{pmatrix}$
   b) $\begin{pmatrix} 0 & 1 & 0 & 0 \\ 0 & 0 & 1 & 0 \\ 0 & 0 & 0 & 1 \\ 1 & 0 & 0 & 0 \end{pmatrix}$

   c) $\begin{pmatrix} 0 & 1 & -1 \\ -1 & 0 & 1 \\ 1 & -1 & 0 \end{pmatrix}$

5*. AN INTRODUCTION TO JORDAN FORM

Diagonal matrices are just about as nice as matrices can be, so it is worthwhile inquiring as to when a given matrix is similar to a diagonal matrix. In 2.2 we saw that if a matrix has n distinct eigenvalues, then it is similar to a diagonal matrix. Of course, this is not a *necessary* condition; many diagonal matrices have some repeats among their diagonal entries. As 4.4 showed, the crucial point is the existence of enough linearly independent eigenvectors.

Your appreciation of diagonalizable matrices will be greatly enhanced by the study of this and the following section, in which we consider matrices which may be impossible to diagonalize. One such creature, you will recall, is the matrix

$$\begin{pmatrix} 2 & 1 \\ 0 & 2 \end{pmatrix},$$

which has only a single eigenvector, when considered as belonging to either $\mathbb{R}_{2\times 2}$ or $\mathbb{C}_{2\times 2}$. Our basic question will be:

what is the best you can do with a matrix which cannot be diagonalized?

We shall assume that our $n \times n$ matrix has a complete collection of n eigenvalues (not necessarily distinct), since otherwise further complications arise. If necessary, we can regard a matrix with real entries as belonging to $\mathbb{C}_{n\times n}$, where it will have all its eigenvalues, so this requirement of eigenvalues is not much of a restriction, at least from this point of view.

An answer to the above question which is satisfactory for many purposes can be stated quite simply; it is the Jordan Form in the following theorem. The actual proof that you can do this to a given matrix requires a bit of effort (expended in the next section). Actually getting a matrix there can require even more work. Fortunately for those primarily interested in applications, it is usually not necessary to know in detail the actual Jordan Form of a given matrix, although one may need to know the possible Jordan Forms to see what one may be up against (see the applications at the end of this section).

Suppose that the characteristic polynomial of an $n \times n$ matrix **A** factors completely (or regard **A** as being in $\mathbb{C}_{n\times n}$, so this is the case). Let it have as its *distinct* eigenvalues $\lambda_1, \ldots, \lambda_r$, so that we can write the characteristic polynomial† as

$$p(x) = (x - \lambda_1)^{m_1}(x - \lambda_2)^{m_2} \cdots (x - \lambda_r)^{m_r}, \quad \text{with } \lambda_1, \ldots, \lambda_r \text{ distinct.} \quad (1)$$

We shall call m_i the **multiplicity** of the root λ_i.

5.1 *The Jordan Form of a Matrix.* Let **A** be an $n \times n$ matrix with characteristic polynomial (1), so that it has eigenvalues

$$\lambda_1 \text{ (with multiplicity } m_1\text{)}, \ldots, \lambda_r \text{ (with multiplicity } m_r\text{)}.$$

---

† As written, we have $(-1)^n \times$ (the characteristic polynomial), a creature which does have the same roots, which is all that matters. We shall hereafter be quite relaxed about the distinction between the characteristic polynomial and its negative.

Then **A** is similar to a matrix

$$\mathbf{B} = \begin{pmatrix} \mathbf{J}_1 & & & 0 \\ & \mathbf{J}_2 & & \\ & & \ddots & \\ 0 & & & \mathbf{J}_r \end{pmatrix}, \quad \text{where each } \mathbf{J}_i = \begin{pmatrix} \lambda_i & \epsilon & 0 & \cdots & 0 \\ 0 & \lambda_i & \epsilon & & \vdots \\ \vdots & & \ddots & \ddots & 0 \\ & & & \ddots & \epsilon \\ 0 & \cdots & & 0 & \lambda_i \end{pmatrix},$$

and is an $m_i \times m_i$ matrix whose entries ϵ above the main diagonal are either 0's or 1's (usually a mixture of both).

A finer breakdown of the $\mathbf{J}_i$ will be given in the next section. For now, take as paradigms the matrices

$$\begin{pmatrix} 2 & 1 & 0 \\ 0 & 2 & 1 \\ 0 & 0 & 2 \end{pmatrix},$$

which has eigenvalue 2 with multiplicity 3 and essentially only one eigenvector (multiples of $\mathbf{e}_1$);

$$\begin{pmatrix} 2 & 0 & 0 \\ 0 & 2 & 1 \\ 0 & 0 & 2 \end{pmatrix},$$

which also has eigenvalue 2 with multiplicity 3, but two linearly independent eigenvectors ($\mathbf{e}_1$ and $\mathbf{e}_2$);

$$\begin{pmatrix} 2 & 0 & 0 \\ 0 & 2 & 0 \\ 0 & 0 & 2 \end{pmatrix},$$

which also has eigenvalue 2 with multiplicity 3, but has 3 linearly independent eigenvectors $\{\mathbf{e}_1, \mathbf{e}_2, \mathbf{e}_3\}$.

You should construe the above theorem as meaning

I. If **A** is real and its characteristic polynomial factors completely, then there is a real matrix **P** for which $\mathbf{B} = \mathbf{P}^{-1}\mathbf{A}\mathbf{P}$ is as indicated.

II. If **A** is complex (or real but with complex eigenvalues), then there is a complex **P** for which $\mathbf{B} = \mathbf{P}^{-1}\mathbf{A}\mathbf{P}$ is as indicated.

Example 1:

$$\mathbf{A} = \begin{pmatrix} 3 & 0 & 1 & 0 \\ 0 & 3 & 0 & 0 \\ 0 & 1 & 3 & 0 \\ 0 & 0 & 0 & 3 \end{pmatrix}$$

has det $(\mathbf{A} - x\mathbf{I}) = (x - 3)^4$.

A basis can be found (by you?) relative to which one obtains the similar matrix

$$B = \begin{pmatrix} 3 & 1 & 0 & 0 \\ 0 & 3 & 1 & 0 \\ 0 & 0 & 3 & 0 \\ 0 & 0 & 0 & 3 \end{pmatrix}.$$

As far as proving things go, until the next section we shall be content to give you the following modest (but illuminating) sample:

Proof of 5.1 in the 2×2 *case.* We must show that if A is a 2×2 matrix with characteristic polynomial $(x - \lambda_1)(x - \lambda_2)$, then it is similar to a matrix

$$B = \begin{pmatrix} \lambda_1 & \epsilon \\ 0 & \lambda_2 \end{pmatrix},$$

where $\epsilon = 0$ if $\lambda_1 \ne \lambda_2$ and ϵ is either 0 or 1 if $\lambda_1 = \lambda_2$. If $\lambda_1 \ne \lambda_2$, then we know, by Section 2, that A is similar to

$$B = \begin{pmatrix} \lambda_1 & 0 \\ 0 & \lambda_2 \end{pmatrix}.$$

Suppose then that $\lambda_1 = \lambda_2$. Call them both λ. Then the characteristic polynomial of A is $(x - \lambda)^2$. Moreover, from the Cayley-Hamilton Theorem (which you believe, at least for $n = 2$),

$$(A - \lambda I)^2 = 0.$$

If $A - \lambda I = 0$, then $A = \lambda I$ and A is similar to the matrix

$$\begin{pmatrix} \lambda & 0 \\ 0 & \lambda \end{pmatrix}.$$

Finally, if $A - \lambda I \ne 0$, then there is some vector V such that

$$(A - \lambda I)V = W \ne 0.$$

Since $(A - \lambda I)^2 = 0$,

$$(A - \lambda I)W = 0,$$

so $AW = \lambda W$ and $AV = W + \lambda V$. Then $\{W, V\}$ is easily seen to be a basis, and relative to this basis, $T(V) = AV$ has the matrix

$$\begin{pmatrix} \lambda & 1 \\ 0 & \lambda \end{pmatrix} \quad \blacksquare$$

The Jordan Form Theorem just quoted can be looked upon as an *existence theorem* in the sense that it may be quite difficult to carry out the details whereby a given matrix is put in the Jordan form. However, the following item of information is useful and is relatively easy to obtain. Each Jordan block J_i in 5.1 breaks up into "sub-blocks" of the form

$$\begin{pmatrix} \lambda_i & 1 & 0 & \cdots & 0 \\ 0 & \lambda_i & 1 & & \vdots \\ & & \ddots & \ddots & \\ & & & \ddots & 1 \\ 0 & \cdots & & 0 & \lambda_i \end{pmatrix},$$

with all 1's above the main diagonal. It is then easy to prove:

5.2 The number of sub-blocks for an eigenvalue λ_i of multiplicity m_i is equal to the dimension of the kernel of $\mathbf{A} - \lambda_i \mathbf{I}$.

We leave the proof to the reader (Exercise 10). For example, if the dimension of the kernel is m_i, then the sub-blocks must all be 1×1; and if the dimension is 1, then $\mathbf{J}_i$ consists of just one sub-block. See also Exercises 11 and 12.

A matrix or mapping for which $\mathbf{N}^k = \mathbf{0}$ for some k is called **nilpotent**. One important conclusion we obtain from the Jordan Form Theorem is the following:

5.3 If the characteristic polynomial of $\mathbf{A}$ factors completely as in (1), then $\mathbf{A}$ is similar to a matrix $\mathbf{B}$ for which $\mathbf{B} = \mathbf{D} + \mathbf{N}$, $\mathbf{D}$ and $\mathbf{N}$ matrices for which

$$\mathbf{D} = \begin{pmatrix} d_{11} & & 0 \\ & \cdot & \\ & & \cdot \\ 0 & & d_{nn} \end{pmatrix}$$

is diagonal; $\mathbf{N}$ is nilpotent (in fact $\mathbf{N}^{n-1} = \mathbf{0}$) and $\mathbf{DN} = \mathbf{ND}$.

(Thus, this theorem says that $\mathbf{A}$ is similar to the sum of a diagonal and a nilpotent matrix and that these matrices commute.)

Proof. For the 2×2 case [if you believe in the Jordan Form, you may try the $n \times n$ case yourself (Exercise 5)]. Suppose that

$$\mathbf{B} = \begin{pmatrix} \lambda_1 & \epsilon \\ 0 & \lambda_2 \end{pmatrix}$$

is the Jordan Form of $\mathbf{B}$. If $\lambda_1 \neq \lambda_2$, then $\epsilon = 0$, and we can take $\mathbf{N} = \mathbf{0}$. If $\lambda_1 = \lambda_2$ and $\epsilon = 0$, likewise. If $\lambda_1 = \lambda_2 = \lambda$ and $\epsilon = 1$, then

$$\mathbf{B} = \begin{pmatrix} \lambda & 0 \\ 0 & \lambda \end{pmatrix} + \begin{pmatrix} 0 & 1 \\ 0 & 0 \end{pmatrix}.$$

The first is diagonal,

$$\begin{pmatrix} 0 & 1 \\ 0 & 0 \end{pmatrix}^2 = \mathbf{0},$$

and they commute. ∎

As an application of 5.3, let us see how one can use it to calculate high powers of a matrix. Suppose that **A** has all its characteristic roots (if **A** is real, consider it as being complex so this is the case). Then by 5.3,

$$\mathbf{A} = \mathbf{P}^{-1}(\mathbf{D} + \mathbf{N})\mathbf{P}, \quad \mathbf{D} \text{ diagonal}, \quad \mathbf{N} \text{ nilpotent}, \quad \mathbf{DN} = \mathbf{ND}, \quad (2)$$

for some matrix **P**. We first note that it is easy to calculate powers of **D** and relatively easy for powers of **N** (Exercise 4). Moreover, since **D** and **N** commute,

$$(\mathbf{D} + \mathbf{N})^2 = \mathbf{D}(\mathbf{D} + \mathbf{N}) + \mathbf{N}(\mathbf{D} + \mathbf{N})$$
$$= \mathbf{D}^2 + \mathbf{DN} + \mathbf{ND} + \mathbf{N}^2 = \mathbf{D}^2 + 2\mathbf{DN} + \mathbf{N}^2.$$

As this might lead you to believe, you can likewise expand $(\mathbf{D} + \mathbf{N})^k$ by the Binomial Theorem: Since **D** and **N** commute,

$$(\mathbf{D} + \mathbf{N})^k = \mathbf{D}^k + \binom{k}{1}\mathbf{D}^{k-1}\mathbf{N} + \cdots + \binom{k}{1}\mathbf{DN}^{k-1} + \mathbf{N}^k. \quad (3)$$

(See Exercise 2.) Moreover, since **N** is nilpotent, for sufficiently large values of k, the terms on the right in (3) begin to vanish. Thus, it is not difficult to calculate $(\mathbf{D} + \mathbf{N})^k$.

To return to **A**, we find that something very pleasant happens if we substitute **A** from (2) in $\mathbf{A}^k$:

$$\mathbf{A}^k = \overbrace{\mathbf{A} \cdots \mathbf{A}}^{k} = \mathbf{P}^{-1}(\mathbf{D} + \mathbf{N})\mathbf{P}\mathbf{P}^{-1}(\mathbf{D} + \mathbf{N})\mathbf{P} \cdots \mathbf{P}^{-1}(\mathbf{D} + \mathbf{N})\mathbf{P}$$
$$= \mathbf{P}^{-1}(\mathbf{D} + \mathbf{N})^k \mathbf{P}. \quad (4)$$

Thus, if we know the **P** for which $\mathbf{PAP}^{-1}$ is in Jordan Form, then we can use (4) to calculate $\mathbf{A}^k$ with some convenience.

Example. Let $\mathbf{A} = \begin{pmatrix} 1 & 1 \\ -1 & 3 \end{pmatrix}$, whose characteristic polynomial is $x^2 - 4x + 4$, so that both eigenvalues are 2. There is just one linearly independent eigenvector which can be taken to be $(1, 1)^*$, so we know that the Jordan Form must be $\begin{pmatrix} 2 & 1 \\ 0 & 2 \end{pmatrix}$. We must now find **P** so that

$$\mathbf{A} = \mathbf{P}^{-1} \begin{pmatrix} 2 & 1 \\ 0 & 2 \end{pmatrix} \mathbf{P}$$

or

$$\mathbf{AP}^{-1} = \mathbf{P}^{-1} \begin{pmatrix} 2 & 1 \\ 0 & 2 \end{pmatrix}.$$

Therefore we shall take $(1, 1)^*$ for the first column of $\mathbf{P}^{-1}$. If **V** is the second column of $\mathbf{P}^{-1}$ we must have

$$\mathbf{AV} = \begin{pmatrix} 1 \\ 1 \end{pmatrix} + 2\mathbf{V}$$

or

$$(\mathbf{A} - 2\mathbf{I})\mathbf{V} = \begin{pmatrix} 1 \\ 1 \end{pmatrix},$$

and it turns out that we can take $\mathbf{V} = (1, 2)$.* Hence we have
$$\mathbf{P}^{-1} = \begin{pmatrix} 1 & 1 \\ 1 & 2 \end{pmatrix}, \quad \mathbf{P} = \begin{pmatrix} 2 & -1 \\ -1 & 1 \end{pmatrix}.$$
Finally
$$\mathbf{A}^k = \mathbf{P}^{-1} \begin{pmatrix} 2^k & k \cdot 2^{k-1} \\ 0 & 2^k \end{pmatrix} \mathbf{P}.$$

Assuming the Jordan Form Theorem (for arbitrary n), we can quite easily supply the missing proof to 3.4:

Proof of the Cayley-Hamilton Theorem. Given an $n \times n$ matrix with characteristic polynomial $p(x) = \det(\mathbf{A} - x\mathbf{I})$, we must show that $p(\mathbf{A}) = 0$. That is, if
$$p(x) = a_n x^n + a_{n-1} x^{n-1} + \cdots + a_1 x_1 + a_0,$$
we must show that
$$a_n \mathbf{A}^n + a_{n-1} \mathbf{A}^{n-1} + \cdots + a_1 \mathbf{A} + a_0 \mathbf{I} = 0.$$

If $\mathbf{A}$ is similar to $\mathbf{B}$, then $\mathbf{A} = \mathbf{PBP}^{-1}$ for some invertible $\mathbf{P}$. Now $\mathbf{A}^k = \mathbf{PB}^k\mathbf{P}^{-1}$ (why$_1$?), so
$$p(\mathbf{A}) = p(\mathbf{PBP}^{-1}) = a_n(\mathbf{PBP}^{-1})^n + \cdots + a_1 \mathbf{PBP}^{-1} + a_0 \mathbf{I}$$
$$= a_n \mathbf{PB}^n\mathbf{P}^{-1} + \cdots + a_1 \mathbf{PBP}^{-1} + a_0 \mathbf{PIP}^{-1}$$
$$= \mathbf{P}(a_n \mathbf{B}^n + \cdots + a_1 \mathbf{B} + a_0 \mathbf{I})\mathbf{P}^{-1} = \mathbf{P}p(\mathbf{B})\mathbf{P}^{-1}.$$
Hence, $p(\mathbf{A}) = 0$ iff $p(\mathbf{B}) = 0$ for any $\mathbf{B}$ similar to $\mathbf{A}$.

Suppose that the characteristic polynomial of $\mathbf{A}$ factors completely:
$$p(x) = (x - \lambda_1)^{m_1} \cdots (x - \lambda_r)^{m_r}, \quad \text{with } \lambda_1, \ldots, \lambda_r \text{ distinct.} \tag{5}$$
(If some of the roots λ_i are complex, change over to $\mathbb{C}_{n \times n}$ as usual).

Take $\mathbf{B}$ to be the Jordan Form of $\mathbf{A}$. From what we have observed, it suffices to show that $p(\mathbf{B}) = 0$. Now
$$\mathbf{B} = \begin{pmatrix} \mathbf{J}_1 & & \\ & \cdot & \\ & & \mathbf{J}_r \end{pmatrix} \quad \text{and} \quad \mathbf{B}^k = \begin{pmatrix} \mathbf{J}_1^k & & \\ & \cdot & \\ & & \mathbf{J}_r^k \end{pmatrix};$$
in fact,
$$p(\mathbf{B}) = \begin{pmatrix} p(\mathbf{J}_1) & & \\ & \cdot & \\ & & p(\mathbf{J}_r) \end{pmatrix}$$
(why$_2$?). Hence, it suffices to show that each Jordan block $\mathbf{J}_i$ satisfies $p(\mathbf{J}_i) = 0$. Now, since $p(x)$ is of the form (5), so will $p(\mathbf{J}_i)$ be:
$$p(\mathbf{J}_i) = (\mathbf{J}_i - \lambda_1 \mathbf{I})^{m_1} \cdots (\mathbf{J}_i - \lambda_i \mathbf{I})^{m_i} \cdots (\mathbf{J}_i - \lambda_r \mathbf{I})^{m_i}. \tag{6}$$

(See Exercise 8.) But

$$J_i - \lambda_i I = \begin{pmatrix} \lambda_i & \epsilon & & \\ & \ddots & \ddots & \\ & & \ddots & \epsilon \\ & & & \lambda_i \end{pmatrix} - \begin{pmatrix} \lambda_i & & & \\ & \ddots & & \\ & & \ddots & \\ & & & \lambda_i \end{pmatrix} = \begin{pmatrix} 0 & \epsilon & & \\ & \ddots & \ddots & \\ & & \ddots & \epsilon \\ & & & 0 \end{pmatrix} \Bigg\} m_i,$$

and thus $(J_i - \lambda_i I)^{m_i} = 0$ (why$_3$?). Hence, $p(J_i) = 0$ for each i. Thus,

$$p(B) = 0, \quad \text{and} \quad p(A) = 0. \quad \blacksquare$$

As another application of Jordan Form, it is easy to see that the Power Method (Supplement to Section 2) will carry over if there is a single eigenvalue of maximum absolute value, whether or not the other eigenvalues are the same (Exercise 9).

EXERCISES

1. Write out $E^{-1}AE$ for the three kinds of elementary matrices, when

$$A = \begin{pmatrix} a_{11} & \cdots & a_{1n} \\ \vdots & & \vdots \\ a_{n1} & \cdots & a_{nn} \end{pmatrix}$$

2. Prove (3).
3. Prove that if A is in $\mathbb{R}_{2\times 2}$ and has both eigenvalues in $\mathbb{R}$, then there is a *unique* matrix B similar to A which is in Jordan Form (up to the order of the eigenvalues).
4. a) Prove that if

$$N = \begin{pmatrix} 0 & & & \\ & \ddots & & \\ & & \ddots & \\ 0 & \cdots & \cdots & 0 \end{pmatrix},$$

then

$$N^2 = \begin{pmatrix} 0 & 0 & & & \\ & \ddots & \ddots & & \\ & & \ddots & & \\ & & & & 0 \\ 0 & \cdots & \cdots & \cdots & 0 \end{pmatrix}$$

b) Prove that N is nilpotent.

5. Assuming the existence of Jordan Form, prove 5.3 for arbitrary n.
6. If a and b are real, find the possible Jordan Forms for matrices
$$\begin{pmatrix} a & -b \\ b & a \end{pmatrix},$$
7. a) Answer the questions in the proof of the Cayley-Hamilton Theorem.
8. a) Prove that if for polynomials $p_i(x)$ one has $p_1(x)p_2(x) = p_3(x)$, then for any $n \times n$ matrix one has $p_1(A)p_2(A) = p_3(A)$.
   b) Prove (6).
9. Prove the last statement in this section concerning the Power Method.
10. Prove 5.2.
11. Prove that if $m_i = 1, 2,$ or 3 in 5.2, the Jordan block corresponding to λ_i is completely known when the dimension of the kernel of $A - \lambda_i I$ has been found. What can you say if $m_i = 4$?

The next two exercises use Section 1, Supplement 2.

12. a) Find the Jordan Form for the companion matrix of a polynomial.
    b) When is a matrix similar to the companion matrix of a polynomial?
13. (See also the supplement to Section 2, Exercise 2.) Let
$$P = \begin{pmatrix} \frac{2}{5} & \frac{1}{2} & \frac{1}{10} \\ \frac{1}{10} & \frac{7}{10} & \frac{1}{5} \\ \frac{1}{10} & \frac{1}{2} & \frac{2}{5} \end{pmatrix}$$
be as in Chapter 2, Section 5d; this matrix gives a model of the social mobility of a society. It turns out that according to this model, society is heading toward a fixed distribution of social classes; that is, P^k approaches some specific limit. Use the method of this section to calculate P^k and find the limit it is approaching.

14. Prove that a permutation matrix is always similar to a diagonal matrix, and that its eigenvalues are roots of unity ($\lambda^m = 1$).
15. Find M_2^k where M_2 is the matrix in Example 2, p. 128

6*. JORDAN FORM PROOFS; NILPOTENT MATRICES

The main purpose of this section is to prove the Jordan Form Theorem (5.1). It will be convenient to consider the problem in linear transformation terms:

Given a linear transformation T from the n-dimensional space $\mathcal{V}$ into $\mathcal{V}$†, suppose that T has eigenvalues $\lambda_1, \ldots, \lambda_r$ with multiplicities $m_1, \ldots, m_r$ respectively (with $m_1 + m_2 + \cdots + m_r = n$). Then there is a basis for $\mathcal{V}$ relative to

---

† If the eigenvalues do not all lie in $\mathbb{R}$, it is necessary to consider T as a linear transformation from $\mathbb{C}_n$ into $\mathbb{C}_n$; one then obtains the appropriate basis for $\mathbb{C}_n$. When necessary this substitution should be made throughout this section. If T has all real eigenvalues, we can have $\mathcal{V} = \mathbb{R}_n$.

which the matrix of **T** is of the form

$$B = \begin{pmatrix} J_1 & & 0 \\ & \ddots & \\ 0 & & J_r \end{pmatrix} \quad \text{where } J_i = \begin{pmatrix} \lambda_i & \epsilon & & \\ & \ddots & \ddots & \\ & & \ddots & \epsilon \\ & & & \lambda_i \end{pmatrix}, \quad (1)$$

where the ϵ's are 1's and 0's.

We call the J_i **Jordan blocks**.

We shall first break up $\mathcal{U}$ into a sum of subspaces $\mathcal{U}_1, \ldots, \mathcal{U}_r$, with each $\mathcal{U}_i$ corresponding to a Jordan block J_i. We then improve the choice of basis on each $\mathcal{U}_i$ so that the matrix of **T** relative to this refined basis will be as in (1). Recall from the Supplement Chapter 1, Section 3, that a space $\mathcal{U}$ is the *direct sum* of two subspaces $\mathcal{X}$ and $\mathcal{Y}$, written $\mathcal{U} = \mathcal{X} \oplus \mathcal{Y}$, if each **V** in $\mathcal{U}$ can be written uniquely as $\mathbf{V} = \mathbf{X} + \mathbf{Y}$, **X** in $\mathcal{X}$ and **Y** in $\mathcal{Y}$. Moreover, recall that the uniqueness part is equivalent to saying

$$\mathcal{X} \cap \mathcal{Y} = \{\mathbf{0}\}.$$

The following shows how to break up $\mathcal{U}$ into a direct sum of two subspaces in a rather surprising fashion:

6.1 Let **T** be a linear transformation from $\mathcal{U}$ into itself ($\mathcal{U}$ n-dimensional). Then $\mathbb{R}_n$ is the direct sum of subspaces $\mathcal{X}$ and $\mathcal{Y}$, each subspace being mapped into itself by **T**, and:

As a linear transformation from $\mathcal{X}$ into $\mathcal{X}$, **T** is nilpotent (i.e., there is an N for which $\mathbf{T}(\mathbf{X})^N = \mathbf{0}$ for all **X** in $\mathcal{X}$);

As a linear transformation from $\mathcal{Y}$ into $\mathcal{Y}$, **T** is invertible.

Proof. We first construct $\mathcal{X}$. Let $\mathcal{K}_1$ be the kernel of **T**, $\mathcal{K}_2$ the kernel of $\mathbf{T}^2$, $\mathcal{K}_3$ the kernel of $\mathbf{T}^3$, etc. Then

$$\mathcal{K}_1 \subset \mathcal{K}_2 \subset \mathcal{K}_3 \subset \cdots \subset \mathcal{U}$$

(why$_1$?). Since $\mathcal{U}$ is finite-dimensional, it follows that there is some k for which

$$\mathcal{K}_k = \mathcal{K}_{k+1} \quad (\text{why}_2?).$$

Suppose k is the first such integer. Then

$$\mathcal{K}_k = \mathcal{K}_{k+1} = \mathcal{K}_{k+2} = \cdots \quad \text{forever afterward} \quad (\text{why}_3?).$$

We let $\mathcal{X} = \mathcal{K}_k$ be the kernel of $\mathbf{T}^k$, and let $\mathcal{Y}$ be the image of $\mathbf{T}^k$. We shall show that these spaces satisfy the indicated properties. Now, certainly **T** maps $\mathcal{X}$ into $\mathcal{X}$ and $\mathcal{Y}$ into $\mathcal{Y}$ (why$_4$?). Moreover, for **X** in $\mathcal{X}$, $\mathbf{T}^k(\mathbf{X}) = \mathbf{0}$, so **T** is nilpotent on **X** and we are off to a flying start.

To see that **T** is invertible on $\mathcal{Y}$, it suffices to show that it has zero kernel there (why$_5$?). For **Y** in $\mathcal{Y} = T^k(\mathcal{U})$, suppose that $T(Y) = 0$. Then **Y** is in $\mathcal{K}_1$ ($\subset \mathcal{K}_k = \mathcal{X}$) and in $\mathcal{Y}$. Thus, it suffices to show that $\mathcal{X} \cap \mathcal{Y} = \{0\}$, which we have to do anyway for the direct-sum part. Suppose, then, that **Z** is in $\mathcal{X} \cap \mathcal{Y}$, so that $T^k(Z) = 0$ and $Z = T^k(V)$, simultaneously. Then

$$T^k(Z) = 0 = T^k(T^k(V)) = T^{2k}(V),$$

so that **V** is in $\mathcal{K}_{2k}$. But then **V** is also in $\mathcal{K}_k$ (why$_6$?), so that $T^k(V) = 0$. Since also $Z = T^k(V)$, we have shown that if **Z** is in $\mathcal{X} \cap \mathcal{Y}$, then $Z = 0$.

It remains only to show that each **V** in $\mathcal{U}$ can be written as $V = X + Y$, the locations of **X** and **Y** being the obvious ones. If $\mathcal{X}$ is m-dimensional, then $\mathcal{Y}$ is $(n - m)$-dimensional (why$_7$?). Thus, we have subspaces $\mathcal{X}$ and $\mathcal{Y}$ for which $\mathcal{X} \cap \mathcal{Y} = \{0\}$, and the sum of whose dimensions is n. It follows that

$$\mathcal{X} + \mathcal{Y} = \mathcal{U} \quad (\text{why}_8?),$$

so that in fact

$$\mathcal{U} = \mathcal{X} \oplus \mathcal{Y}. \quad \blacksquare$$

This is rather remarkable. By the repeated application of **T**, we are able to break up $\mathcal{U}$ into two pieces, one where **T**'s behavior is as "bad" as possible (i.e., **T** is nilpotent) and one where it is as "good" as possible (i.e., **T** is invertible).

We now apply 6.1 to splitting up $\mathcal{U}$ into subspaces corresponding to Jordan blocks.

6.2 *Decomposition of a space into components corresponding to the eigenvalues of a linear transformation.* Let **T** be a linear transformation of $\mathcal{U}$ into $\mathcal{U}$, and suppose that the characteristic polynomial of **T** factors as

$$p(x) = (x - \lambda_1)^{m_1}(x - \lambda_2)^{m_2} \cdots (x - \lambda_r)^{m_r},$$

where $\lambda_1, \ldots, \lambda_r$ are all distinct. Then there is a basis relative to which the matrix of **T** is

$$A = \begin{pmatrix} A_1 & & & \\ & A_2 & & \\ & & \ddots & \\ & & & A_r \end{pmatrix},$$

where A_i is an $m_i \times m_i$ matrix and $A_i - \lambda_i I$ is a nilpotent matrix.

Proof. Apply 6.1 to the linear transformation $T - \lambda_i I$ to obtain subspaces $\mathcal{U}_1$ and $\mathcal{Y}_1$ of $\mathcal{U}$, for which

$$\mathcal{U} = \mathcal{U}_1 \oplus \mathcal{Y}_1,$$

with $T - \lambda_1 I$ mapping the respective subspaces into themselves, nilpotent on $\mathcal{U}_1$ and invertible on $\mathcal{Y}_1$. Then **T** maps $\mathcal{U}_1$ into $\mathcal{U}_1$ and $\mathcal{Y}_1$ into $\mathcal{Y}_1$ (why$_1$?). Choose a basis for $\mathcal{U}_1$ and one for $\mathcal{Y}_1$, so that together they constitute a basis for $\mathcal{U}$

(why$_2$?). It follows (why$_3$?) that the matrix of **T** relative to this basis for $\mathcal{U}$ is

$$\left(\begin{array}{c|c} \text{Matrix of } \mathbf{T} \text{ as a linear} \\ \text{transformation on } \mathcal{U}_1 \text{ (call it } \mathbf{A}_1\text{)} & \mathbf{0} \\ \hline \mathbf{0} & \begin{array}{c} \text{Matrix of } \mathbf{T} \text{ as a linear} \\ \text{transformation on } \mathcal{Y}_1 \text{ (call it } \mathbf{B}_1\text{)} \end{array} \end{array} \right).$$

Moreover, $\mathbf{A}_1 - \lambda_1 \mathbf{I}$ is nilpotent (why$_4$?) and $\mathbf{B}_1 - \lambda_1 \mathbf{I}$ is invertible (why$_5$?).

We next show that $\mathbf{A}_1$ is an $m_1 \times m_1$ matrix. From Chapter 5, we see that the characteristic polynomial of **T**

$$p(x) = (x - \lambda_1)^{m_1} \cdots (x - \lambda_r)^{m_r} = \text{(characteristic polynomial of } \mathbf{A}_1\text{)} \quad (2)$$
$$\times \text{(characteristic polynomial of } \mathbf{B}_1\text{)}.$$

Since $\mathbf{B}_1 - \lambda_1 \mathbf{I}$ is invertible, the characteristic polynomial of $\mathbf{B}_1$ does not have $x - \lambda_1$ as a factor (why$_6$?). Hence, from (2), the characteristic polynomial of $\mathbf{A}_1 = (x - \lambda_1)^{m_1} \times$ (something else, perhaps). If there is in fact, something else, it has an $(x - \lambda_i)$ factor, $i \neq 1$. Then $\mathbf{A}_1$ would have another eigenvalue λ_i besides λ_1. Since $\mathbf{A}_1 - \lambda_1 \mathbf{I}$ is nilpotent, this is patently impossible (why$_7$?). [*Suggestion:* take an eigenvector **V** corresponding to λ_i, take s so that

$$\mathbf{W} = (\mathbf{A}_1 - \lambda_1 \mathbf{I})^s \mathbf{V} \neq \mathbf{0} \quad \text{but} \quad (\mathbf{A}_1 - \lambda_1 \mathbf{I})\mathbf{W} = \mathbf{0},$$

and show that **W** is an eigenvector for both λ_1 and λ_i.] Hence, the characteristic polynomial of $\mathbf{A}_1$ is $(x - \lambda_1)^{m_1}$. Hence, $\mathbf{A}_1$ is an $m_1 \times m_1$ matrix.

Now consider **T** acting on $\mathcal{Y}_1$. Its characteristic polynomial is $(x - \lambda_2)^{m_2} \cdots (x - \lambda_r)^{m_r}$ from (2) and what we've just observed. If we repeat the argument in $\mathbf{T} - \lambda_2 \mathbf{I}$, we can obtain a basis for $\mathcal{Y}_1$ relative to which the matrix of **T** acting in $\mathcal{Y}_1$ is of the form

$$\left(\begin{array}{c|c} \mathbf{A}_2 & \mathbf{0} \\ \hline \mathbf{0} & \mathbf{B}_2 \end{array} \right).$$

Moreover, if we replace our basis for $\mathcal{U}$ obtained above by the basis for $\mathcal{U}_1$ together with this basis for $\mathcal{Y}_1$, we obtain a matrix for **T** of the form

$$\begin{pmatrix} \mathbf{A}_1 & & \\ & \mathbf{A}_2 & \\ & & \mathbf{B} \end{pmatrix}$$

(why$_8$?). Continuing (or inducting), we obtain the desired basis for $\mathcal{U}$. ∎

For our final Jordan Form ingredient (as suggested by the above theorem), we shall see how to construct a basis relative to which a given nilpotent transformation assumes a particularly docile form. Given a vector space $\mathcal{U}$ with basis $\{\mathbf{V}_1, \ldots, \mathbf{V}_n\}$, the linear transformation **N** on $\mathcal{U}$ defined by

$$\mathbf{N}(\mathbf{V}_1) = \mathbf{0}, \, \mathbf{N}(\mathbf{V}_2) = \mathbf{V}_1, \, \mathbf{N}(\mathbf{V}_3) = \mathbf{V}_2, \ldots, \mathbf{N}(\mathbf{V}_n) = \mathbf{V}_{n-1} \quad (3)$$

will certainly be nilpotent (each application of **N** chops off a $\mathbf{V}_i$ and moves the remaining **V**'s one unit to the left; hence $\mathbf{N}^n$ will do in all the **V**'s and consequently anything in $\mathcal{V}$). If **N** is a nilpotent linear transformation on $\mathcal{V}$ relative to which there exists such a basis upon which **N** acts as in (3), we call **N** a **cyclic nilpotent** mapping and say that the **V**'s are a **cyclic basis**. Note that if **N** is cyclic nilpotent, then its matrix relative to this basis is

$$\begin{pmatrix} 0 & 1 & & & \\ & \ddots & \ddots & & \\ & & \ddots & \ddots & \\ & & & \ddots & 1 \\ & & & & 0 \end{pmatrix}$$

(which is of course called a **cyclic** nilpotent matrix).

6.3 The decomposition of a space relative to cyclic nilpotent components. If **N** is any nilpotent linear transformation on $\mathcal{V}$, then there are subspaces $\mathcal{V}_1, \ldots, \mathcal{V}_k$ of $\mathcal{V}$ upon each of which **N** is a cyclic nilpotent mapping. Moreover, $\mathcal{V}$ is the direct sum of these subspaces:

$$\mathcal{V} = \mathcal{V}_1 \oplus \cdots \oplus \mathcal{V}_k.$$

Consequently, if we choose cyclic bases for each $\mathcal{V}_i$ and string them together to obtain a basis for $\mathbb{R}_n$, the matrix of **N** relative to this basis is

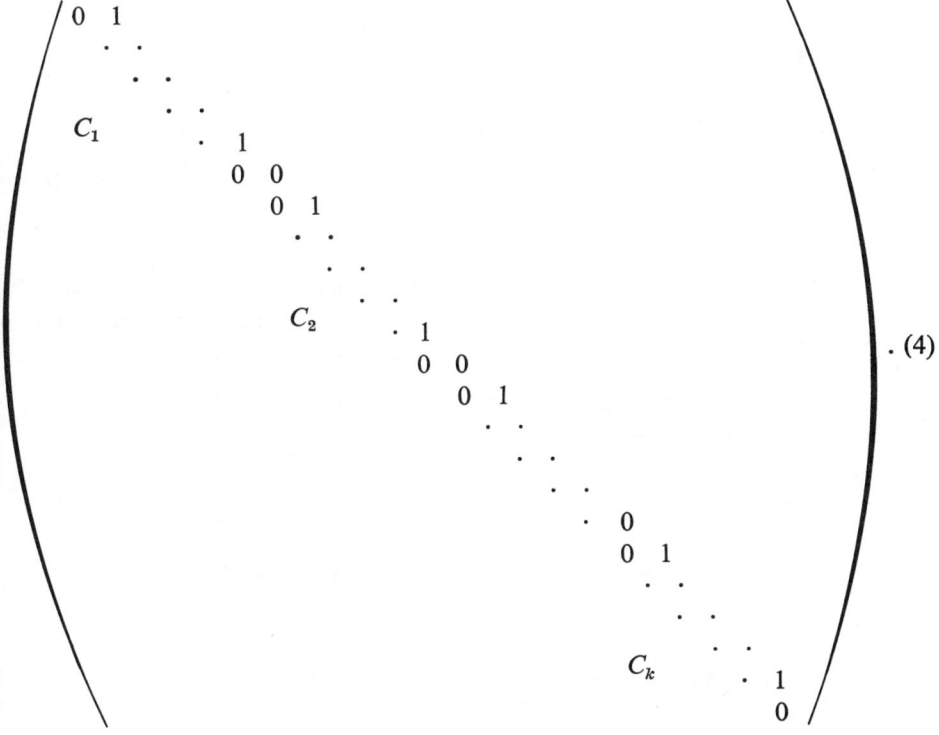

. (4)

Proof (for a special case). The notational complications in this proof are ghastly. Fortunately, all the essential ideas are contained in the following case: We shall suppose that $N^3 = 0$, $N^2 \neq 0$ (this says that all the cyclic nilpotent matrices in (4) are at most 3×3). Then, as in the proof of 6.1,

$$\mathcal{K}_1 = \text{kernel } N \subset \mathcal{K}_2 = \text{kernel } N^2 \subset \mathcal{U} = \text{kernel } N^3.$$

For further simplification let us consider the (extra) special case when the dimension of $\mathcal{K}_1 = 5$, the dimension of $\mathcal{K}_2 = 9$, and $n = 11$. Construct a basis for the 11-dimensional space $\mathcal{U}$ by taking a basis for $\mathcal{K}_1$, extending to a basis for $\mathcal{K}_2$, and extending again to a basis for $\mathcal{U}$:

$$\{\underbrace{\underbrace{V_1, V_2, V_3, V_4, V_5,}_{\text{basis for } \mathcal{K}_1} V_6, V_7, V_8, V_9,}_{\text{basis for } \mathcal{K}_2} V_{10}, V_{11}\} \quad \text{basis for } \mathcal{U}. \tag{5}$$

Since V_{10} and V_{11} are not in $\mathcal{K}_2$, they are the longest-lived of all of the basis vectors under N, and we shall use them for the last vectors in cyclic bases of maximal length 3. We shall, however, have to rearrange the rest of the basis in (5) to obtain these cyclic bases, and we must also see if there are any other cyclic bases of length <3 which must be picked up.

Step 1. $N(V_{10})$ and $N(V_{11})$ are in $\mathcal{K}_2$ but not $\mathcal{K}_1$, for

$$N^2(N(V_{10})) = N^3(V_{10}) = 0,$$

but if $N^2(V_{10}) = 0$, then V_{10} is in $\mathcal{K}_2$, contrary to (5). Similarly for V_{11}.

Step 2. $\{V_1, \ldots, V_5, N(V_{10}), N(V_{11})\}$ is a linearly independent subset of $\mathcal{K}_2$: If

$$a_1 V_1 + \cdots + a_5 V_5 + b N(V_{10}) + c N(V_{11}) = 0,$$

apply N to obtain that $b V_{10} + c V_{11}$ is in $\mathcal{K}_2$, and hence is a linear combination of $V_1, \ldots, V_9$ which is impossible unless $b = c = 0$. Hence the a's must be 0, too. Extend this linearly independent set to a basis for $\mathcal{K}_2$, say:

$$\{\underbrace{V_1, V_2, V_3, V_4, V_5,}_{\text{basis for } \mathcal{K}_1} N(V_{10}), N(V_{11}), W_1, W_2\} \quad \text{basis for } \mathcal{K}_2. \tag{6}$$

Step 3. We next replace the basis $V_1, \ldots, V_5$ for $\mathcal{K}_1$. First of all,

$$\{N^2(V_{10}), N^2(V_{11}), N(W_1), N(W_2)\} \tag{7}$$

is linearly independent: if

$$a_1 N^2(V_{10}) + a_2 N^2(V_{11}) + b_1 N(W_1) + b_2 N(W_2) = 0,$$

then we can pull off an **N** and say that

$$a_1 N(V_{10}) + a_2 N(V_{11}) + b_1 W_1 + b_2 W_2$$

is in $\mathcal{K}_1$; hence it is a linear combination of $V_1, \ldots, V_5$. But this contradicts (6). Extend (7) to a basis for $\mathcal{K}_1$:

$$\{N^2(V_{10}), N^2(V_{11}), N(W_1), N(W_2), X_1\} \quad \text{basis for } \mathcal{K}_1. \tag{8}$$

Step 4. The set

$$\{\underbrace{N^2(V_{10}), N^2(V_{11}), N(W_1), N(W_2), X_1}_{\text{basis for } \mathcal{K}_1}, N(V_{10}), N(V_{11}), W_1, W_2, V_{10}, V_{11}\} \tag{9}$$

$$\text{basis for } \mathcal{K}_2$$

is a basis for $\mathcal{V}$. For, if the appropriate linear combination is **0**, strike twice with **N** to get the coefficients of V_{10} and V_{11} zero, then once to get those of $N(V_{10})$, $N(V_{11})$, W_1, W_2 all zero (by (8)), leaving only terms which belong to a basis (8). Hence (9) is linearly independent and hence a basis.

Step 5 (and last). Rearrange the basis to obtain

$$\begin{pmatrix} 0 & 1 & 0 & & & & & & & \\ 0 & 0 & 1 & & & & & & & \\ 0 & 0 & 0 & & & & & & & \\ & & & 0 & 1 & 0 & & & & \\ & & & 0 & 0 & 1 & & & & \\ & & & 0 & 0 & 0 & & & & \\ & & & & & & 0 & 1 & & \\ & & & & & & 0 & 0 & & \\ & & & & & & & & 0 & 1 \\ & & & & & & & & 0 & 0 \\ & & & & & & & & & & 0 \end{pmatrix} \cdot \blacksquare$$

Proof of the Jordan Form Theorem (which is stated at the beginning of this section). Given **T** with eigenvalues $\lambda_1, \ldots, \lambda_r$ of multiplicities $m_1, \ldots, m_r$, use 6.2 to obtain a basis relative to which the matrix of **T** is

$$\begin{pmatrix} A_1 & & \\ & \ddots & \\ & & A_r \end{pmatrix},$$

with $A_i - \lambda_i I$ an $m_i \times m_i$ nilpotent matrix. Next, for $i = 1, \ldots, r$ apply 6.3 to the linear transformation $T - \lambda_i I$ restricted to the subspace $\mathcal{V}_i$ corresponding to the A_i. This yields a basis for $\mathcal{V}_i$ relative to which the matrix of $T - \lambda_i I$

(restricted to λ_i) is

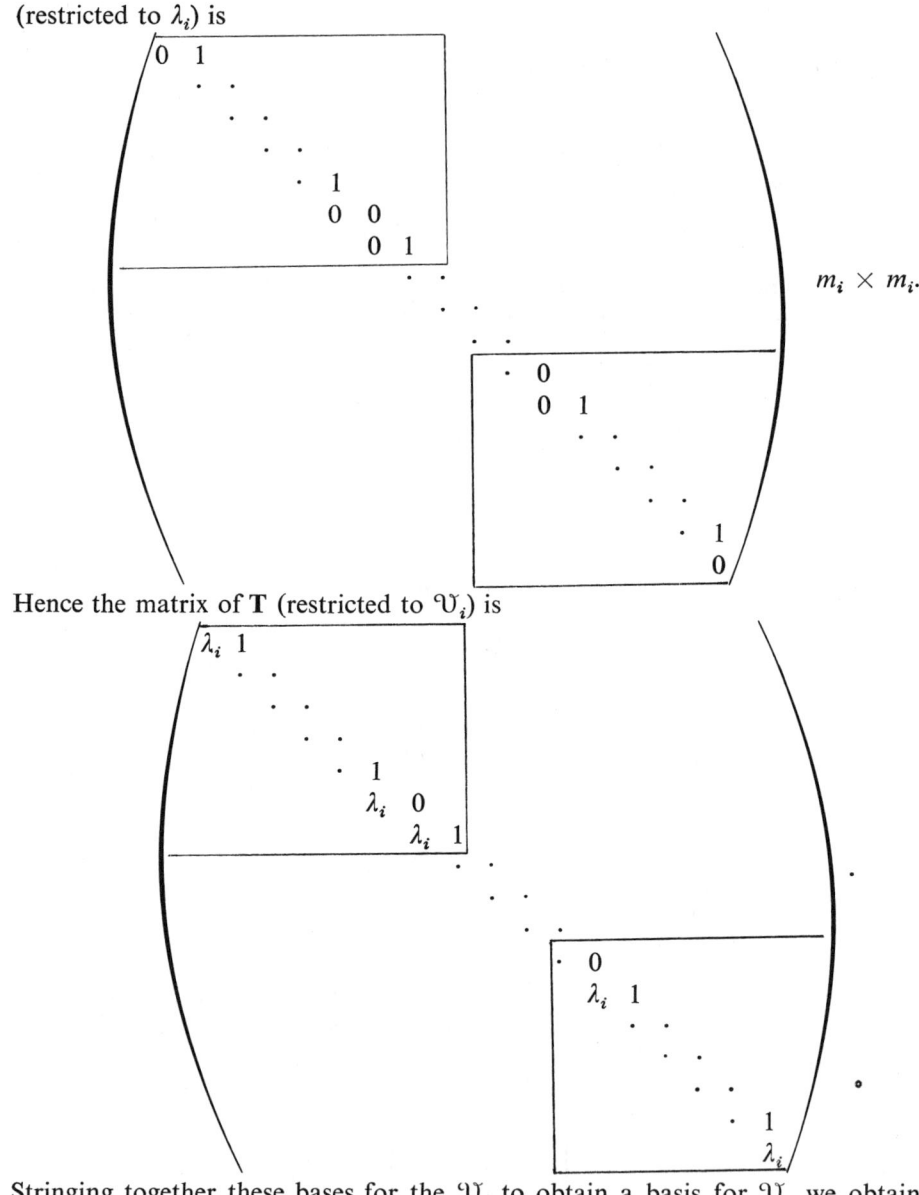

Hence the matrix of **T** (restricted to $\mathcal{U}_i$) is

Stringing together these bases for the $\mathcal{U}_i$ to obtain a basis for $\mathcal{U}$, we obtain a matrix for **T** in Jordan Form. ∎

EXERCISES

1. Answer the questions in the proof of 6.1.
2. Answer the questions in the proof of 6.2.
H3. Give a general proof of 6.3, if you dare.

"Why take chances," replied the line without much conviction. "I'm dependable. I know where I'm going. I've got dignity!"†

CHAPTER 7 Lines, Planes and Other Flats

SUMMARY

In this chapter we shall be concerned with lines, planes, and their higher-dimensional analogs. Naturally, every notion which is not a part of our basic vector space axiom system must be defined, and we shall attempt to give intuitive justification for the definitions we choose. These definitions are a genuine matter of choice; there are a number of equivalent possibilities which we prove as theorems, but which could be taken as definitions.

In our linear algebra context, it seems reasonable to link up the "one-dimensional" aspect of lines with one-dimensional subspaces. These latter are fine candidates for lines through the origin, and we shall consider an arbitrary line to be one which has been pushed aside by some vector. Thus we shall define a "line through P" to be all points of the form $P + V$, where V is in a one-dimensional subspace $\mathcal{U}$ of $\mathbb{R}_n$. $\mathcal{U}$ will be called the "direction space".

A plane should be connected with two-dimensional subspaces in a similar fashion. We shall in fact consider a "plane through P" to be all points of the form $P + V$, V in $\mathcal{U}$, a two-dimensional subspace of $\mathbb{R}_n$ (again called $\mathcal{U}$ the "direction space").

The general higher-dimensional analog of lines and planes we shall call a "flat". Special attention will be paid to those consisting of points of the form $P + V$, V from an $(n-1)$-dimensional subspace $\mathcal{U}$ of $\mathbb{R}_n$. These latter are called "hyperplanes"; some of the important properties of planes in $\mathbb{R}_3$ hold because they are hyperplanes (that is, translates of $3-1$-dimensional subspaces) and only incidentally because they are planes (i.e., translates of 2-dimensional subspaces).

---

† N. Juster, *The Dot and the Line*. New York: Random House, 1963.

316   Lines, planes, and other flats

We now present a summary of the main approaches to lines, planes, flats, and hyperplanes, especially in $\mathbb{R}_3$. This is both to give an overview for those working through the chapter, and also to indicate the most important features for those who will not be able (or do not choose) to consider the chapter in detail.

Lines in $\mathbb{R}_3$

  Geometrical approach (and definition).   All points of the form $\mathbf{P} + \mathbf{V}$, $\mathbf{V}$ in $\mathcal{V}$, $\mathcal{V}$ one dimensional (Fig. 1).

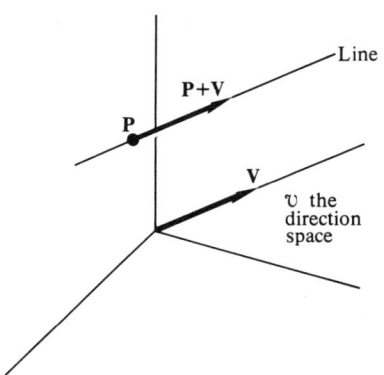

Fig. 1

Parametric equations. All points $\mathbf{P} + t\mathbf{V}_1$, $\mathbf{V}_1$ fixed ($\neq 0$), t in $\mathbb{R}$ (t is called the *parameter*). In terms of coordinates, all points (x, y, z) with

$$x = p_1 + tv_1,$$
$$y = p_2 + tv_2,$$
$$z = p_3 + tv_3.$$

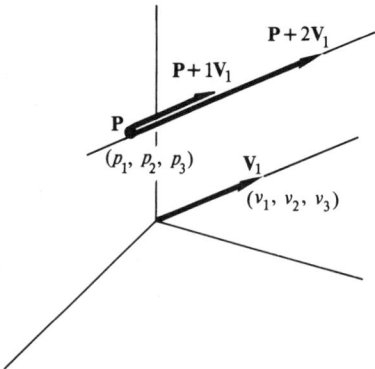

Fig. 2

Here V_1 is a basis for the one dimensional subspace $\mathcal{U}$ in the geometrical approach (Fig. 2).

Affine equations. All points $(1 - \lambda)\mathbf{P} + \lambda\mathbf{Q}$, λ real. The basis V_1 for the subspace $\mathcal{U}$ in the geometrical approach is $V_1 = \mathbf{Q} - \mathbf{P}$ (Fig. 3). In coordinates, all points (x, y, z) with

$$x = (1 - \lambda)p_1 + \lambda q_1,$$
$$y = (1 - \lambda)p_2 + \lambda q_2,$$
$$z = (1 - \lambda)p_3 + \lambda q_3.$$

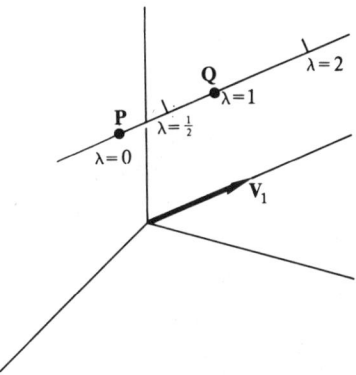

Fig. 3

Solutions to system of linear equations. $\mathbf{AX} = \mathbf{Y}$, $\mathbf{X} = (x, y, z)^*$, $\mathbf{A}$ of rank 2. Typically, all solutions to

$$a_1 x + a_2 y + a_3 z = c_1,$$
$$b_1 x + b_2 y + b_3 z = c_2.$$

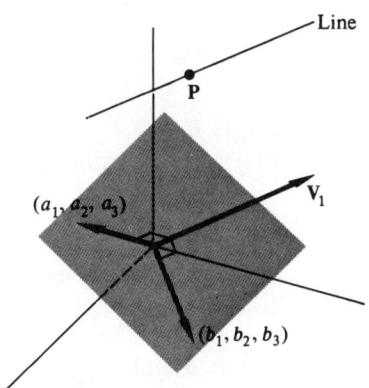

Fig. 4

318   Lines, planes, and other flats

Here the basis V_1 for the one-dimensional subspace in the geometrical approach consists of a nonzero solution to the homogeneous system $AX = 0$ (the space of solutions is one-dimensional). The point P is any particular solution, so $AP = Y$ (Fig. 4). Given a line in geometrical form, $P + V$, V in $\mathcal{U}$, take a basis

$$V_1 = (v_1, v_2, v_3)^*$$

for $\mathcal{U}$. Then the rows (a_1, a_2, a_3) and (b_1, b_2, b_3) of A are any independent solutions to

$$av_1 + bv_2 + cv_3 = 0,$$

and $Y = AP$.

Planes in $\mathbb{R}_3$

Geometrical approach (and definition). All points of the form $P + V$, V in a two dimensional subspace $\mathcal{U}$ (Fig. 5).

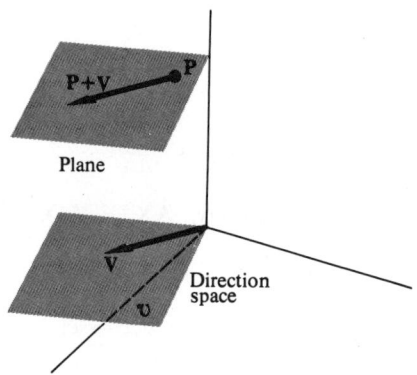

Fig. 5

Parametric equations. All points $P + t_1V_1 + t_2V_2$, V_1, V_2 linearly independent, t_1, t_2 in $\mathbb{R}$ (Fig. 6). In terms of coordinates, all points (x, y, z) with

$$x = p_1 + t_1v_1 + t_2v_1',$$
$$y = p_2 + t_1v_2 + t_2v_2',$$
$$z = p_3 + t_1v_3 + t_2v_3'.$$

Here $\{V_1, V_2\}$ is a basis for the two dimensional subspace in the geometrical approach.

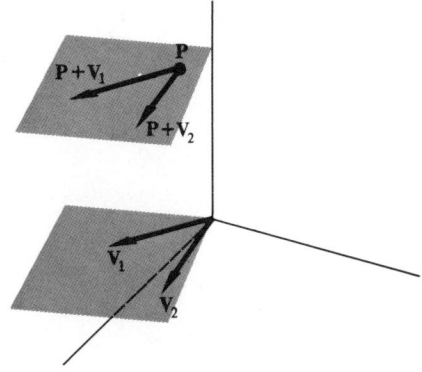

Fig. 6

Affine equations. All points $\lambda_1 \mathbf{P} + \lambda_2 \mathbf{Q} + \lambda_3 \mathbf{R}$ where $\lambda_1 + \lambda_2 + \lambda_3 = 1$ (Fig. 7). Substituting into the affine equation

$$\mathbf{V}_1 = \mathbf{Q} - \mathbf{P}, \quad \mathbf{V}_2 = \mathbf{R} - \mathbf{P},$$

with

$$t_1 = \lambda_2, \quad t_2 = \lambda_3,$$

yields the parametric equations. In coordinate form, the affine equations are

$$x = \lambda_1 p_1 + \lambda_2 q_1 + \lambda_3 v_1,$$
$$y = \lambda_1 p_2 + \lambda_2 q_2 + \lambda_3 r_2,$$
$$z = \lambda_1 p_3 + \lambda_2 q_3 + \lambda_3 r_3,$$

where $\lambda_1 + \lambda_2 + \lambda_3 = 1$.

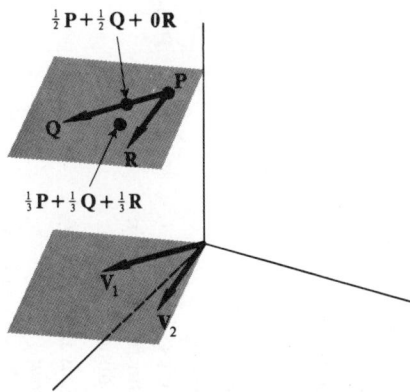

Fig. 7

Summary for $\mathbb{R}_n$

	Point	Line	Plane	k-flat	Hyperplane	$\mathbb{R}_n$
Geometrical definition all points $\mathbf{P} + \mathbf{V}$, $\mathbf{V}$ in a subspace $\mathcal{V}$ of dimension	0	1	2	k	$n-1$	n
Parametric form All points with $\mathbf{V}_i$ linearly independent, t_i in $\mathbb{R}$	$\mathbf{P}$	$\mathbf{P} + t_1\mathbf{V}_1$	$\mathbf{P} + t_1\mathbf{V}_1 + t_2\mathbf{V}_2$	$\mathbf{P} + t_1\mathbf{V}_1 + \cdots + t_k\mathbf{V}_k$	$\mathbf{P} + t_1\mathbf{V}_1 + \cdots + t_{n-1}\mathbf{V}_{n-1}$	$\mathbf{P} + t_1\mathbf{V}_1 + \cdots + t_n\mathbf{V}_n$
Affine form All points λ_i's sum to 1 points in n_o lower dimensional flat	$\lambda\mathbf{P}$	$\lambda_1\mathbf{P} + \lambda_2\mathbf{Q}$	$\lambda_1\mathbf{P}_1 + \lambda_2\mathbf{P}_2 + \lambda_3\mathbf{P}_3$	$\lambda_1\mathbf{P}_1 + \cdots + \lambda_k\mathbf{P}_k$	$\lambda_1\mathbf{P}_1 + \cdots + \lambda_{n-1}\mathbf{P}_{n-1}$	$\lambda_1\mathbf{P}_1 + \cdots + \lambda_n\mathbf{P}_n$
Solutions to system of equations $\mathbf{AX} = \mathbf{Y}$ with rank $\mathbf{A} =$ (and $\mathbf{Y} = \mathbf{AP}$ with $\mathbf{P}$ as above).	n	$n-1$	$n-2$	$n-k$	1	0

Solutions to a system of linear equations. All solutions to $\mathbf{AX} = \mathbf{Y}$, $\mathbf{X} = (x, y, z)^*$, $\mathbf{A}$ of rank 1. Typically, all solutions to $ax + by + cz = d$ (Fig. 8). $\mathbf{A} = (a, b, c)$ is a basis for the solution space of

$$av_1 + bv_2 + cv_3 = 0,$$
$$av'_1 + bv'_2 + cv'_3 = 0,$$

where $\mathbf{V}_1 = (v_1, v_2, v_3)$ and $\mathbf{V}_2 = (v'_1, v'_2, v'_3)$ are, as above, a basis for the direction space, and $d = \mathbf{AP}$.

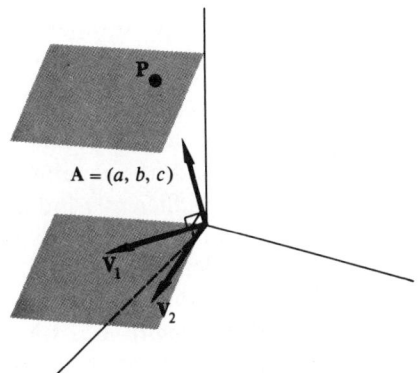

Fig. 8

1. FLATS

Recall that real n-space $\mathbb{R}_n$ is the vector space of n-tuples $(a_1, \ldots, a_n)$ of real numbers under componentwise addition and scalar multiplication. At least for $n = 3$ we can visualize elements of $\mathbb{R}_n$ as the cartesian coordinates of points in our usual geometric space. Thus, to each n-tuple $(a_1, \ldots, a_n)$, a_i in $\mathbb{R}$, there corresponds a point $\mathbf{P}$ with cartesian coordinates $(a_1, \ldots, a_n)$, and conversely.

From a slightly different point of view, we recall that to each point $\mathbf{P}$ with coordinates $(a_1, \ldots, a_n)$ we can assign a vector $\mathbf{OP}$ whose tail is at the origin $\mathbf{O}$ and whose head is at $\mathbf{P}$. This one-to-one correspondence has the added property that the addition and scalar multiplication of n-tuples is the same as that of vectors:

If $\mathbf{OP} \leftrightarrow (a_1, \ldots, a_n)$ and $\mathbf{OQ} \leftrightarrow (b_1, \ldots, b_n)$,

then $\mathbf{OP} + \mathbf{OQ} \leftrightarrow (a_1 + b_1, \ldots, a_n + b_n)$, $\quad c(\mathbf{OP}) \leftrightarrow (ca_1, \ldots, ca_n)$,

for any c in $\mathbb{R}$.

Let us summarize some of the attributes of these three ways of looking at the world:

	Virtues	Vices
Point	Convenient geometric interpretation for $n \leq 3$	No intrinsic interpretation of addition or scalar multiplication
Vector	Convenient geometric interpretation for $n \leq 3$; addition and scalar multiplication possible	Calculations often not convenient
n-tuple	Convenient for calculations; easy to generalize	Not as easy to visualize geometrically; a poor choice of coordinate systems can make a simple problem horrible

Most readers are probably familiar with the rudiments of analytic geometry in the plane, and know precisely what is meant by points, lines, and vectors in that context. Some of you may also have a nodding acquaintance with the geometry of 3-space. We trust that everyone is now sufficiently indoctrinated with the use of the axiomatic method so as to know what to expect: it is not enough to work with half-precise intuitive notions—we must define our terms. This is true even when working with such strong concepts as point, line, and plane. Even though you may feel you know what these creatures *really* are, when nailed to the wall we would not expect most readers to be able, at this moment, to supply appropriate definitions even for 3-space.

Our plan of action will be as follows: We shall utilize $\mathbb{R}_n$, $n \leq 3$ (where we have the geometric visualization afforded us by points and vectors), to make intuitively plausible definitions. These will be expressible either in terms of linear algebra or analytically (that is, in terms of n-tuples), and can in these ways be carried over to precise definitions in $\mathbb{R}_n$ for arbitrary n.

As a simple example of our more axiomatic approach, based on our experience with the one-to-one correspondence between points and n-tuples for $n \leq 3$, we shall *define* a **point P** in $\mathbb{R}_n$ to be an n-tuple $(a_1, \ldots, a_n)$. To preserve our notion of points as passive creatures merely sitting in n-space, as opposed to vectors which can be added, scalar-multiplied and otherwise abused, we must make a separate definition for vectors. Using the only available materials, the **vector PQ from the point P** $= (a_1, \ldots, a_n)$ **to the point Q** $= (b_1, \ldots, b_n)$ (or *with tail at* **P** *and head at* **Q**) is defined to be the n-tuple

$$\mathbf{PQ} = (b_1 - a_1, \ldots, b_n - a_n).$$

Thus, the difference between points, vectors, and n-tuples in $\mathbb{R}_n$ is merely one of attitude. Note especially that $\mathbf{PQ} = \mathbf{Q} - \mathbf{P}$ (subtraction being allowed for **P** and **Q** as n-tuples). (Fig. 9.)

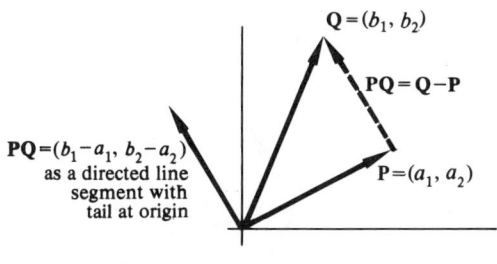

Fig. 9

In particular, the vector **OP** from **O** to **P** has coordinates the same as the point **P**. We shall be quite relaxed about the distinction between the vector **OP** and the point **P**, and about identifying vectors, points, and n-tuples in general. Any distinctions we do make are only for intuitive motivation. We are being axiomatic, but the aim is to assist clear thinking, not to smother intuition.

Having exhausted the subject of points, we turn to lines. Recall that in elementary analytic geometry a line was defined to be the set of all points (x, y) which satisfied an equation $ax + by = c$, with a, b not both equal to zero. As you will see in retrospect, the most immediate generalization does not work here (you may know that a line in $\mathbb{R}_3$ is *not* the set of solutions to an equation $ax + by + cz = d$).

How should we regard lines? We previously noted that a line through the origin was just a one-dimensional vector $\mathcal{V}$ of $\mathbb{R}_2$. Suppose, then, that we are confronted by a line $\mathcal{L}$ which does not pass through the origin; let $\mathcal{L}$ be all solutions to $ax + by = c$. It seems reasonable to regard $\mathcal{L}$ as a line which would have liked to go through the origin but which has been shoved out of the way (**translated** is the technical term). In fact, if we choose any point $\mathbf{P} = (x_0, y_0)$ on $\mathcal{L}$, it is easy to see that we may translate back by the vector **OP**, and obtain a one-dimensional subspace $\mathcal{V}$. That is, if we subtract **OP** from vector **OQ**, with **Q** on $\mathcal{L}$, the set

$$\{\mathbf{OQ} - \mathbf{OP} \mid \mathbf{Q} \text{ on } \mathcal{L}\} = \{\mathbf{PQ} = \mathbf{Q} - \mathbf{P} \mid \mathbf{Q} \text{ on } \mathcal{L}\}$$

is a one-dimensional subspace (see Exercise 2).

Example 1. Let $\mathcal{L}$ be the set of solutions to $2x + y = 4$, and $\mathbf{P} = (1, 2)$ the chosen point on $\mathcal{L}$ (Fig. 10). For any $\mathbf{Q} = (x, y)$ on $\mathcal{L}$, consider $\mathbf{PQ} = (x - 1, y - 2)$. Since the equation for $\mathcal{L}$ can also be written as

$$2(x - 1) + (y - 2) = 0,$$

we see that

$$\mathcal{V} = \{\mathbf{PQ} \mid \mathbf{P} \text{ fixed on } \mathcal{L}, \mathbf{Q} \text{ arbitrary on } \mathcal{L}\}$$

is a one-dimensional subspace of $\mathbb{R}_2$ ($\mathcal{V}$ consists of the set of solutions of a nontrivial homogeneous linear equation; $\mathcal{V}$ is the kernel of a linear transformation of rank 1.)

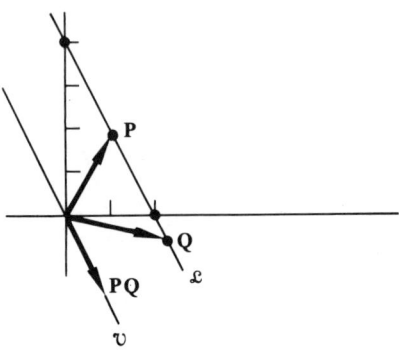

Fig. 10

Conversely, it is equally easy to see that starting with a one-dimensional subspace $\mathcal{U}$ of $\mathbb{R}_2$, and any point **P**, the set of pairs of numbers of the form **P** + **V**, **V** in $\mathcal{U}$ forms a line in $\mathbb{R}_2$.

Example 2. Let $\mathcal{U}$ = span $\{(2, 3)\}$ and **P** $= (-1, 2)$. Then let

$$\mathcal{L} = \{\mathbf{P} + \mathbf{V} \mid \mathbf{V} \text{ in } \mathcal{U}\} = \{(-1, 2) + k(2, 3) \mid k \text{ in } \mathbb{R}\}$$
$$= \{(-1 + 2k, \quad 2 + 3k) \mid k \text{ in } \mathbb{R}\}$$

(Fig. 3). To see that $\mathcal{L}$ is a line we observe that $\mathcal{U}$ is the set of solutions to the homogeneous equation $3x + 2y = 0$. Letting $c = 3(-1) - 2(2) = -7$ (for reasons that will become clear in Section 3), it is easy to verify that $\mathcal{L}$ is the set of solutions to $3x + 2y = -7$.

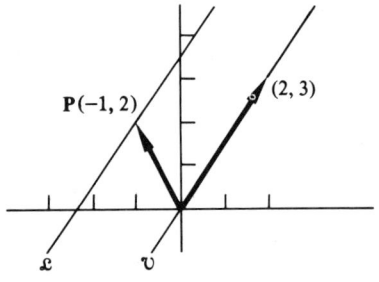

Fig. 11

You might, therefore, be tempted (correctly this time) to make the following definition.

A **line** $\mathcal{L}$ in $\mathbb{R}_n$ is the collection of all n-tuples of the form **P** + **V**, where **P** is a point in $\mathbb{R}_n$ and **V** is taken from some one-dimensional vector subspace $\mathcal{U}$ of

$\mathbb{R}_n$. Symbolically, we write

$$\mathcal{L} = \mathbf{P} + \mathcal{V}.$$

The subspace $\mathcal{V}$ is called the **direction space** of $\mathcal{L}$.

Note that we cannot consider the vector $\mathbf{P} + \mathbf{V}$ as lying *along* $\mathcal{L}$ (unless $\mathcal{L}$ passes through the origin); the *head* of the vector $\mathbf{P} + \mathbf{V}$ is on $\mathcal{L}$ (when its tail is at the origin). It is probably thus less confusing to consider a line as a collection of points, not a collection of vectors.

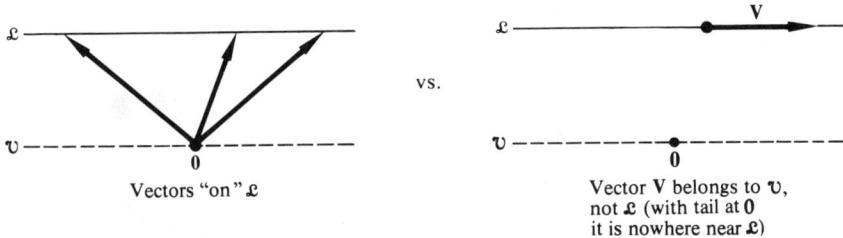

Fig. 12

1.1 The direction space $\mathcal{V}$ of a line $\mathcal{L}$ in $\mathbb{R}_n$ is unique. That is, if

$$\mathcal{L} = \mathbf{P} + \mathcal{V} \quad \text{and also} \quad \mathcal{L} = \mathbf{Q} + \mathcal{W},$$

then $\mathcal{V} = \mathcal{W}$.

Proof. If $\mathbf{P} = \mathbf{Q}$ the sets $\mathcal{V}$ and $\mathcal{W}$ are surely the same. If $\mathbf{P} \neq \mathbf{Q}$, then

$$\mathbf{P} + \mathbf{V} = \mathbf{Q} \quad \text{for some} \quad \mathbf{V} \neq \mathbf{Q} \text{ in } \mathcal{V}$$

and

$$\mathbf{P} = \mathbf{Q} + \mathbf{W} \quad \text{for some} \quad \mathbf{W} \neq \mathbf{0} \text{ in } \mathcal{W}.$$

Hence $\mathbf{P} - \mathbf{Q} = -\mathbf{V} = \mathbf{W} \neq \mathbf{0}$ is in $\mathcal{V}$ and in $\mathcal{W}$. Since $\mathcal{V}$ and $\mathcal{W}$ are one-dimensional, $\mathcal{V} = \mathcal{W}$. ∎

1.2 If $\mathcal{L} = \mathbf{P} + \mathcal{V}$ is a line in $\mathbb{R}_n$ and if $\mathbf{Q}$ is a point on $\mathcal{L}$, then also $\mathcal{L} = \mathbf{Q} + \mathcal{V}$.

Proof. $\mathbf{Q} = \mathbf{P} + \mathbf{V}$ for some $\mathbf{V}$ in $\mathcal{V}$. For any other $\mathbf{R} = \mathbf{P} + \mathbf{W}$, $\mathbf{W}$ in $\mathcal{V}$, one then has

$$\mathbf{R} = \mathbf{P} + \mathbf{W} = \mathbf{Q} - \mathbf{V} + \mathbf{W} = \mathbf{Q} + (\mathbf{W} - \mathbf{V}),$$

so $\mathbf{R}$ is in $\mathbf{Q} + \mathcal{V}$. Conversely, it is easy to show that if $\mathbf{R}$ is in $\mathbf{Q} + \mathcal{V}$, then $\mathbf{R}$ is on $\mathcal{L}$. ∎

Having seen your way through lines, it is easy to find an appropriate definition for planes. Let us work first in $\mathbb{R}_3$ where we can draw pictures for comfort. Clearly, a plane through the origin in $\mathbb{R}_3$ ought to be nothing more than a two-dimensional subspace of $\mathbb{R}_3$. Moreover, a plane not through the origin ought to be a plane through the origin pushed aside by some vector. Finally, no matter what you happen to think should be true about planes in $\mathbb{R}_n$, n arbitrary (if you grant that they should exist there), they should somehow be translated two-dimensional subspaces. Thus,

A **plane** $\mathscr{P}$ in $\mathbb{R}_n$ is the set of all n-tuples of the form $\mathbf{P} + \mathbf{V}$ where $\mathbf{P}$ is some fixed point in $\mathbb{R}_n$ and $\mathbf{V}$ is from some two-dimensional subspace $\mathcal{V}$ of $\mathbb{R}_n$. We call $\mathcal{V}$ the **direction space** of $\mathscr{P}$ and write

$$\mathscr{P} = \mathbf{P} + \mathcal{V}.$$

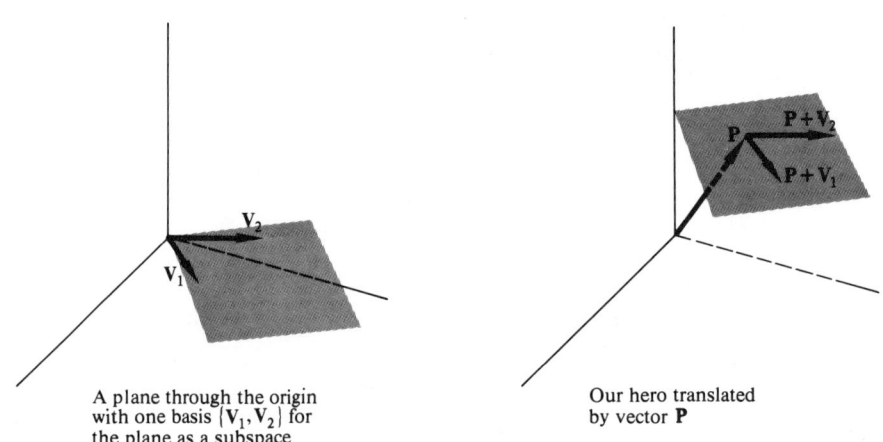

A plane through the origin with one basis $\{\mathbf{V}_1, \mathbf{V}_2\}$ for the plane as a subspace

Our hero translated by vector $\mathbf{P}$

Fig. 13

Example 3. The plane perpendicular to the y-axis in $\mathbb{R}_3$ and passing through the point $(0, 2, 0)$ has as direction space the xz-plane. One translation vector is $\mathbf{P} = (0, 2, 0)$ Another is $\mathbf{Q} = (1, 2, 1)$. Each vector in the direction space is of the form

$$h(1, 0, 0) + k(0, 0, 1) = (h, 0, k), \qquad h, k \text{ in } \mathbb{R},$$

so each vector in our plane is of the form

$$\mathbf{Q} + (h, 0, k) = (h + 1, 2, k + 1).$$

Since this plane can also be described as all points (x, y, z) with $y = 2$, or just by the equation $y = 2$, we have apparently not said the last word yet. (See Fig. 14, p. 327.)

1.3 The direction space of a plane is unique: If $\mathscr{P} = \mathbf{P} + \mathcal{V}$ and also $\mathscr{P} = \mathbf{Q} + \mathcal{W}$, then $\mathcal{V} = \mathcal{W}$.

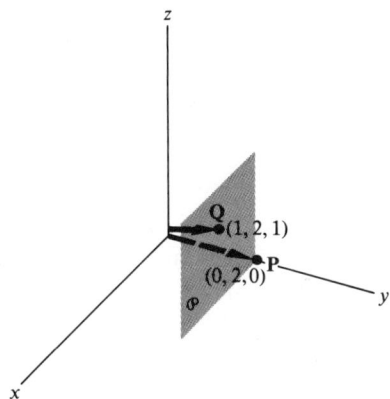

Fig. 14

Proof. As in 1.1, this is immediate if $\mathbf{P} = \mathbf{Q}$; and if $\mathbf{P} \neq \mathbf{Q}$, then $\mathbf{P} - \mathbf{Q}$ is in both $\mathcal{V}$ and $\mathcal{W}$. Thus, $\mathbf{P} - \mathbf{Q} = \mathbf{V}_1$ in $\mathcal{V}$, so that $\mathbf{P} = \mathbf{Q} + \mathbf{V}_1$. Hence any point of the form $\mathbf{P} + \mathbf{V}$ can be written as

$$\mathbf{P} + \mathbf{V} = \mathbf{Q} + \mathbf{V}_1 + \mathbf{V} = \mathbf{Q} + \mathbf{V}', \qquad \mathbf{V}' \text{ in } \mathcal{V}.$$

Conversely, any point $\mathbf{Q} + \mathbf{V}'$ is also of the form $\mathbf{P} + \mathbf{V}$. Thus

$$\mathbf{P} + \mathcal{V} = \{\mathbf{P} + \mathbf{V}\} = \{\mathbf{Q} + \mathbf{V}'\} = \mathbf{Q} + \mathcal{V}.$$

Since this is also $\mathbf{Q} + \mathcal{W}$, it follows quickly that $\mathcal{V} = \mathcal{W}$. ∎

1.4 A plane can be represented by any of its points. If $\mathcal{F} = \mathbf{P} + \mathcal{V}$ is a plane and $\mathbf{Q}$ is on $\mathcal{F}$, then $\mathcal{F} = \mathbf{Q} + \mathcal{V}$.

Proof. Follow 1.2. ∎

You will note that in $\mathbb{R}_n$ for $n > 3$ there lurk higher-dimensional analogs of lines and planes. We shall call these creatures "flats" (the terminology is not yet standardized; some people call them *linear varities* or *affine subspaces*. Our name is shorter). Formally a **k-flat** $\mathcal{F}$ in $\mathbb{R}_n$ is the set of all n-tuples of the form $\mathbf{P} + \mathbf{V}$, where $\mathbf{P}$ is some point of $\mathbb{R}_n$ and $\mathbf{V}$ is from a k-dimensional subspace $\mathcal{V}$ of $\mathbb{R}_n$. $\mathcal{V}$ is called the **direction space** of $\mathcal{F}$, and we write $\mathcal{F} = \mathbf{P} + \mathcal{V}$.

Besides points, lines, and planes, one other type of k-flat deserves special mention. In $\mathbb{R}_n$, an $(n-1)$-flat is called a **hyperplane**. Thus, in $\mathbb{R}_3$ a hyperplane is just a plane, in $\mathbb{R}_2$ a hyperplane is a line, while in $\mathbb{R}_4$ a three-dimensional subspace is a particular example of a hyperplane. We shall see on several later occasions why we single out hyperplanes for special consideration.

Our cast of characters:

point	= 0-flat
line	= 1-flat
plane	= 2-flat
⋮	⋮
translated k-dimensional subspace	= k-flat
⋮	⋮
hyperplane	= $(n-1)$-flat
$\mathbb{R}_n$	= n-flat.

Just as for lines and planes, we have

1.5 *The direction space of a flat is unique.* That is, if
$$\mathcal{F} = \mathbf{P} + \mathcal{V} \quad \text{and also} \quad \mathcal{F} = \mathbf{Q} + \mathcal{W},$$
then $\mathcal{V} = \mathcal{W}$.

Proof. Exercise 1b. ∎

1.6 *A flat may be represented by any of its points.* If $\mathbf{P}$ is any point in a flat $\mathcal{F}$, then $\mathcal{F} = \mathbf{P} + \mathcal{V}$, where $\mathcal{V}$ is the direction space of $\mathcal{F}$.

Proof. Exercise 1c. ∎

EXERCISES

1. a) Prove that for vectors $\mathbf{PQ} + \mathbf{QR} = \mathbf{PR}$. b) Prove 1.5. c) Prove 1.6.
2. Finish the details of the proof that if $\mathcal{L}$ is a line in $\mathbb{R}_2$ and $\mathbf{P}$ is any point of $\mathcal{L}$, then
$$\mathcal{V} = \{\mathbf{OQ} - \mathbf{OP} \mid \mathbf{Q} \text{ on } \mathcal{L}\}$$
is a one-dimensional subspace.
3. Prove that if $\mathcal{V}$ is a one-dimensional subspace of $\mathbb{R}_2$ and if $\mathbf{P}$ is any point of $\mathbb{R}_2$, then the set
$$\mathcal{L} = \{\mathbf{P} + \mathbf{V} \mid \mathbf{V} \text{ in } \mathcal{V}\}$$
is a line.
4. a) Find a basis for $\mathcal{V}$ in Example 1 and verify that $\mathcal{L} = \mathbf{P} + \mathcal{V}$.
   b) Verify that the $\mathcal{L}$ in Example 2 is the set of solutions to $3x - 2y = -7$.
5. Write the following lines in the form $\mathcal{L} = \mathbf{P} + \mathcal{V}$.
   a) The line through $(1, 2)$ in $\mathbb{R}_2$ and parallel to the x-axis.
   b) The line through $(1, 2, 3)$ in $\mathbb{R}_3$ and parallel to the x-axis.
   c) The line through $(1, 2)$ in $\mathbb{R}_2$ and making an angle of $45°$ with the x-axis.
   d) The line through $(1, 2)$ and $(2, 3)$.

6. Draw pictures of the following lines:
   a) $\mathcal{L} = (2, 3) + \text{span}\{(2, -1)\}$.
   b) $\mathcal{L} = (1, 2, 3) + \text{span}\{(1, 1, 1)\}$.
   c) $\mathcal{L} = (1, 2) + \text{kernel}(T(x, y) = y - x)$.
7. Show that $2x + 3y - z = 10$ is the equation of a plane in $\mathbb{R}_3$. Find a basis for its direction space and show how to write it in the form $\mathbf{P} + \mathcal{V}$.

Honors Projects. Throughout this chapter, the only special feature of the field of real numbers $\mathbb{R}$ which we shall use is our ability to draw pictures in $\mathbb{R}_2$ and $\mathbb{R}_3$. As standing honors projects we therefore invite you to check that the definitions make sense and the theorems are true for vector spaces

1) over the complex numbers $\mathbb{C}$ and
2) for vector spaces over any field $\mathbb{F}$.

2. PARAMETRIC EQUATIONS AND AFFINE EQUATIONS

There are several convenient methods of describing lines, planes, and other flats in terms of linear equations.

Let us take first a line $\mathcal{L} = \mathbf{P} + \mathcal{V}$ in $\mathbb{R}_n$. Since $\mathcal{V}$ is one-dimensional, $\mathcal{V}$ has a single basis vector $\mathbf{V}_1$ and

$$\mathcal{V} = \{t\mathbf{V}_1 \mid t \text{ in } \mathbb{R}\}.$$

Hence, any point $\mathbf{X}$ in $\mathcal{L}$ is of the form

$$\mathbf{X} = \mathbf{P} + t\mathbf{V}_1, \qquad t \text{ in } \mathbb{R}. \tag{1}$$

In terms of coordinates, if

$$\mathbf{P} \text{ is } (p_1, \ldots, p_n), \qquad \mathbf{V}_1 \text{ is } (v_1, \ldots, v_n), \qquad \text{and} \qquad \mathbf{X} \text{ is } (x_1, \ldots, x_n),$$

then (1) becomes

$$\begin{aligned} x_1 &= p_1 + tv_1, \\ & \vdots \\ x_n &= p_n + tv_n, \end{aligned} \qquad t \text{ in } \mathbb{R}. \tag{1'}$$

These equations ((1) or (1)′) are called the **parametric equations** of a line. Note that they exhibit a point P on the line and a basis vector $\mathbf{V}$ for the direction space. Here t is called the **parameter** and may be intuitively thought of as showing you where you are at time t. (See Fig. 15, p. 330.)

For our favorite n, we have $\mathbf{X} = (x, y, z)$ and we can rewrite (1)′ as

$$\begin{aligned} x &= p_1 + tv_1, \\ y &= p_2 + tv_2, \\ z &= p_3 + tv_3, \end{aligned} \qquad \text{Parametric equations for a line in } \mathbb{R}_3.$$

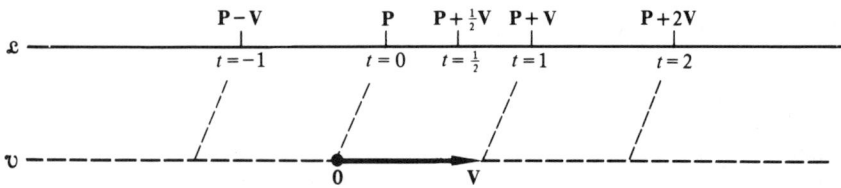

Fig. 15

Example 1. Let us take $\mathbf{P} = (3, 0, 2)$ and $\mathbf{V}_1 = (1, 2, 2)$. Then the line $\mathcal{L} = \mathbf{P} + \mathcal{V}$ has parametric equations

$$x = 3 + t, \quad y = 2t, \quad z = 2 + 2t.$$

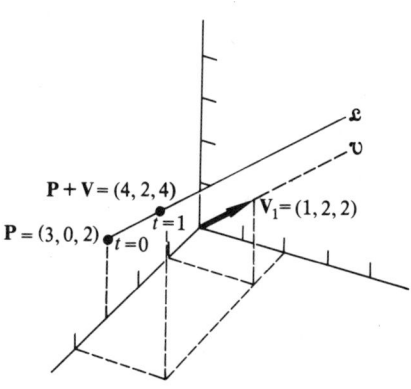

Fig. 16

Since a two-dimensional subspace $\mathcal{V}$ of $\mathbb{R}_n$ has as basis two vectors $\mathbf{V}_1$, $\mathbf{V}_2$, it is of the form

$$\{t_1 \mathbf{V}_1 + t_2 \mathbf{V}_2 \mid t_1, t_2 \text{ in } \mathbb{R}\}.$$

Hence, any plane $\mathcal{P} = \mathbf{P} + \mathcal{V}$ in $\mathbb{R}_n$ will consist of points of the form

$$\mathbf{X} = \mathbf{P} + t_1 \mathbf{V}_1 + t_2 \mathbf{V}_2, \quad t_1, t_2 \text{ in } \mathbb{R}. \tag{2}$$

In terms of obvious coordinates,

$$x_1 = p_1 + t_1 v_1 + t_2 v_1',$$
$$\vdots \qquad \vdots$$
$$\vdots \qquad \vdots \tag{2}'$$
$$\vdots \qquad \vdots$$
$$x_n = p_n + t_1 v_n + t_2 v_n'.$$

2 / Parametric equations and affine equations

These are the **parametric equations** of a plane in $\mathbb{R}_n$. Here t_1 and t_2 are the **parameters**. Note that the parametric equations display a point **P** on the plane and also give basis vectors $\mathbf{V}_1$, $\mathbf{V}_2$ for the direction space. In particular for $n = 3$,

$$\begin{aligned} x &= p_1 + t_1v_1 + t_2v'_1, \\ y &= p_2 + t_1v_2 + t_2v'_2, \\ z &= p_3 + t_1v_3 + t_2v'_3, \end{aligned} \quad \text{Parametric equations for a plane in } \mathbb{R}_3.$$

Example 2. Let

$$\mathbf{P} = (1, 2, \tfrac{1}{2}), \quad \mathbf{V}_1 = (2, 0, 1), \quad \mathbf{V}_2 = (0, 0, 1).$$

The parametric equations for the plane

$$\mathcal{P} = \{\mathbf{P} + t_1\mathbf{V}_1 + t_2\mathbf{V}_2 \mid t_i \text{ in } \mathbb{R}\}$$

are

$$x = 1 + 2t_1, \quad y = 2, \quad z = \tfrac{1}{2} + t_1 + t_2.$$

(Find some points on $\mathcal{P}$ for various values of t_1 and t_2.)

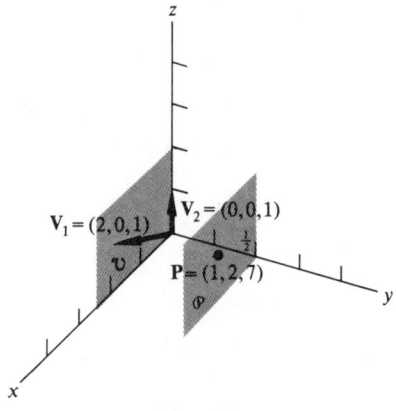

Fig. 17

As in the example in the previous section, we can also describe this plane more simply as the set of points for which $y = 2$.

By this time the reader can surely concoct his own discussion of parametric equations for k-flats and we abandon him to that project.

Affine Equations

We next investigate an alternative way of describing lines, planes, and other flats by means of "affine equations". The λ's in the following are not meant to suggest eigenvalues; they are simply the traditional symbol used in both contexts.

2.1 The two-point formula for a line. Given two distinct points **P** and **Q** in $\mathbb{R}_n$, there is a unique line through **P** and **Q**. It consists of all points with coordinates of the form

$$\lambda P + (1 - \lambda)Q, \quad \lambda \text{ real.} \tag{3}$$

Proof. Let $V = Q - P$. Then

$$\lambda P + (1 - \lambda)Q = P + [-P + \lambda P + (1 - \lambda)Q]$$
$$= P + [(1 - \lambda)Q + (\lambda - 1)P] = P + (1 - \lambda)(Q - P)$$
$$= P + (1 - \lambda)V, \quad (V = Q - P).$$

Thus, (3) is a point on the line $\mathcal{L} = P + \text{span } \{V\}$. Conversely (alias Exercise 4a), any point on $\mathcal{L}$ is of the form (3). We have therefore obtained a line $\mathcal{L}$ which passes through P ($\lambda = 1$) and Q ($\lambda = 0$). Using 1.1, it is easy to show (Exercise 11(b)) that there is only one line through P and Q. ∎

Example 3. Consider the line with equation $2x + y = 4$, as in Section 1, Eq. 1.

$$P = (1, 2) \quad \text{and} \quad Q = (3, -2)$$

are two points on the line and

$$\lambda P + (1 - \lambda)Q = (\lambda + 3(1 - \lambda), 2\lambda - 2(1 - \lambda))$$
$$= (3 - 2\lambda, 4\lambda - 2).$$

This is an alternative way of describing the line since, in fact,

$$2(3 - 2\lambda) + (4\lambda - 2) = 4.$$

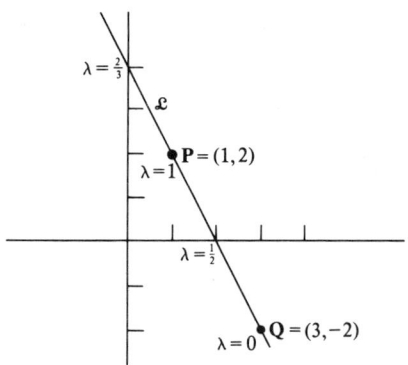

Fig. 18

In order to formulate a corresponding theorem for planes, it is necessary first to consider collinear points. We say that points P, Q, and R are **collinear** if they are on the same line in $\mathbb{R}_n$. The basic result is:

2.2 P, Q, and R are collinear iff the vectors $Q - P$ and $R - P$ are *not* linearly independent.

2 / Parametric equations and affine equations

Proof. 1. If **P**, **Q**, and **R** are collinear, then they are on a line $\mathbf{P} + \mathcal{V}$, so that

$$\mathbf{Q} = \mathbf{P} + \mathbf{V}_1 \quad \text{and} \quad \mathbf{R} = \mathbf{P} + \mathbf{V}_2,$$

where $\mathbf{V}_i$ are from the same subspace $\mathcal{V}$. Since $\mathcal{V}$ is one-dimensional,

$$\mathbf{V}_1 = \mathbf{Q} - \mathbf{P} \quad \text{and} \quad \mathbf{V}_2 = \mathbf{R} - \mathbf{P}$$

are not linearly independent.

2. Conversely, if $\mathbf{Q} - \mathbf{P}$ and $\mathbf{R} - \mathbf{P}$ are not linearly independent, suppose that

$$\mathbf{Q} - \mathbf{P} = \lambda(\mathbf{R} - \mathbf{P}) \quad \text{for some scalar } \lambda.$$

Then $\mathbf{Q} = \mathbf{P} + \lambda(\mathbf{R} - \mathbf{P}) = \lambda\mathbf{R} + (1 - \lambda)\mathbf{P}$, so that $\mathbf{Q}$ is on the line through **P** and **R**. ∎

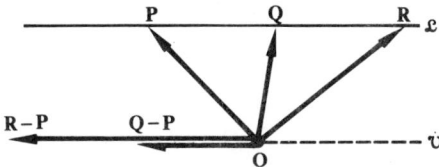

Fig. 19

Corresponding to 2.1, we now have

2.3 Three-point formula for a plane. Given three noncollinear points **P**, **Q**, and **R** in $\mathbb{R}_n$, there is a unique plane passing through these points. It consists of all points whose coordinates are of the form

$$\lambda_1 \mathbf{P} + \lambda_2 \mathbf{Q} + \lambda_3 \mathbf{R}, \quad \text{where } \lambda_1 + \lambda_2 + \lambda_3 = 1. \tag{4}$$

Proof. Let $\mathbf{V} = \mathbf{Q} - \mathbf{P}$ and $\mathbf{W} = \mathbf{R} - \mathbf{P}$. Then

$$\lambda_1 \mathbf{P} + \lambda_2 \mathbf{Q} + \lambda_3 \mathbf{R} \stackrel{(\text{why}_1\,?)}{=} \mathbf{P} + \lambda_2 \mathbf{V} + \lambda_3 \mathbf{W}.$$

Hence, all points (4) form a plane (why$_2$?). Uniqueness follows as in 2.1 (why$_3$?). ∎

Example 4. The points

$$\mathbf{P} = (0, 2, 0), \quad \mathbf{Q} = (1, 2, 1) \quad \text{and} \quad \mathbf{R} = (2, 2, 1)$$

are not collinear since $\mathbf{Q} - \mathbf{P} = (1, 0, 1)$ and $\mathbf{R} - \mathbf{P} = (2, 0, 1)$ are linearly independent. The plane through these points consists of all points

$$\lambda_1 \mathbf{P} + \lambda_2 \mathbf{Q} + \lambda_3 \mathbf{R}, = (\lambda_2 + 2\lambda_3, 2, \lambda_2 + \lambda_3),$$

$$\lambda_1 + \lambda_2 + \lambda_3 = 1$$

that is, the plane $y = 2$.

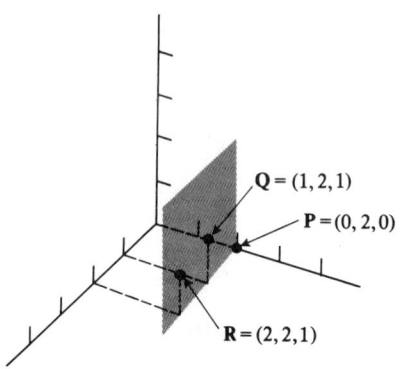

Fig. 20

In general, we have

2.4 i) Given points $P_0, P_1, \ldots, P_k$ in $\mathbb{R}_n$, the set of all points with coordinates of the form

$$\lambda_0 P_0 + \lambda_1 P_1 + \cdots + \lambda_k P_k, \quad \text{with } \lambda_0 + \lambda_1 + \cdots \lambda_k = 1$$

is a flat $\mathcal{F}$.
ii) If $P_0, P_1, \ldots, P_k$ are not contained in a $(k-1)$-flat, $\mathcal{F}$ will be a k-flat.
iii) These points will not lie in a $(k-1)$-flat iff the vectors $\{P_1 - P_0, P_2 - P_0, \ldots, P_k - P_0\}$ are linearly independent.

Proof. Exercise 11a. ∎

We close this section with a characterization of flats by means of lines, which shows what flats really are.

2.5 A nonempty set S in $\mathbb{R}_n$ is a flat iff for every pair of distinct points P and Q in S, S contains the line joining P and Q.

Proof. 1. Suppose S is a flat. Then for distinct P and Q in S, we have $S = P + \mathcal{V}$ and $Q = P + V_1$ for some V_1 in $\mathcal{V}$, $V_1 \neq 0$. Since tV_1 will be in $\mathcal{V}$ for any t we see that the set of points $P + tV_1$ will all lie in S. But this is a line containing both P and Q.

2. Conversely, suppose that S contains the line through each two of its distinct points. We show that S is of the form $P + \mathcal{V}$, $\mathcal{V}$ a subspace of $\mathbb{R}_n$. Choose any point P in S and let

$$\mathcal{V} = \{Q - P \mid Q \text{ in } S\}.$$

It suffices to show that $\mathcal{V}$ is a subspace. To show that $\mathcal{V}$ is closed under addition, take any

$$V = Q - P \quad \text{and} \quad W = R - P \quad \text{in } \mathcal{V}.$$

Then

$$V + W = Q - P + R - P = [(\tfrac{1}{2}Q + \tfrac{1}{2}R) + (\tfrac{1}{2}Q + \tfrac{1}{2}R) - P] - P$$
$$= [2S - P] - P, \quad \text{where } S = \tfrac{1}{2}Q + \tfrac{1}{2}R.$$

By hypothesis and 2.1, S is in $\mathcal{S}$, and so is $T = 2S - P$ (take $\lambda = -1$). Hence,

$$V + W = T - P \quad \text{is in } \mathcal{U},$$

and $\mathcal{U}$ is closed under addition.

Similarly, if $V = Q - P$ is in $\mathcal{U}$, you use the expression

$$aV = a(Q - P) = [(1 - a)P + aQ] - P$$

to show that $\mathcal{U}$ is closed under scalar multiplication (Exercise 11c). Hence $\mathcal{U}$ is a subspace and $\mathcal{S} = P + \mathcal{U}$ is a flat. ∎

EXERCISES

1. Discuss and define parametric equations for k-flats in $\mathbb{R}_n$.
2. Find parametric equations for the plane through

    $$P = (1, -1, 2), \quad Q = (2, 1, 4), \quad \text{and} \quad R = (3, 0, 1) \quad \text{in } \mathbb{R}_3.$$

3. (a) through (d) Write the lines of Section 1, Exercise 5, in parametric form.
4. a) Connect up the parametric form of a line with the affine form $\lambda P + (1 - \lambda)Q$. That is, show how to pass from each to the other.
    b) Same for planes.
5. a) Let $\mathcal{L}$ be the line through

    $$P = (1, 2) \quad \text{and} \quad Q = (2, 3).$$

    Write the point where $\mathcal{L}$ crosses the x-axis in the form $\lambda P + (1 - \lambda)Q$. Same for the y-axis. Draw a picture.
    b) Let $\mathcal{L}$ be the line through

    $$P = (3, 1, 1) \quad \text{and} \quad Q = (1, 2, 3).$$

    Write the points where $\mathcal{L}$ intersects the planes

    $$x = 0, \quad y = 0, \quad \text{and} \quad z = 0$$

    in the form $\lambda P + (1 - \lambda)Q$. Draw a picture.
6. What can you say about all points of the form $\lambda P + (1 - \lambda)Q$ when:

    a) $0 < \lambda < 1$?,   b) $\lambda > 1$?,   c) $\lambda < 0$? Illustrate.

    Parts (a)–(c) make use of some specific properties of $\mathbb{R}$. Discuss.

7. Write the plane through $(1, -1, 2)$, $(2, 1, 4)$, and $(3, 0, 1)$ in affine form. Draw a picture.
8. a) Show that $3x - 2y + 5z = 15$ is the equation of a plane in $\mathbb{R}_3$.
   b) Find a basis for its direction space, and write a general point in the form $\mathbf{P} + \mathbf{V}$.
   c) Write it in the form $\lambda_1 \mathbf{P} + \lambda_2 \mathbf{Q} + \lambda_3 \mathbf{R}$.
9. Find the plane determined by the points
$$\mathbf{P}_0 = (1, 0, -2, 3), \quad \mathbf{P}_1 = (7, 8, 9, 5), \quad \text{and} \quad \mathbf{P}_2 = (7, -3, -2, 5).$$
   Give both parametric and affine equations. Draw a picture.
10. Find the 3-flat in $\mathbb{R}_4$ determined by the points $\mathbf{P}_0, \mathbf{P}_1, \mathbf{P}_2$ in Exercise 9 and $\mathbf{P}_3 = (3, 4, -3, 6)$. Give both parametric and affine equations.
11. a) Prove 2.4.
    b) Finish proving 2.1.
    c) Finish proving 2.5.

3. SOLUTIONS TO SYSTEMS OF LINEAR EQUATIONS

We shall now proceed to tie together our work in analytic geometry and in systems of linear equations. Let us consider a system of equations:

$$a_{11}x_1 + \cdots + a_{1n}x_n = b_1,$$
$$a_{21}x_1 + \cdots + a_{2n}x_n = b_2,$$
$$\vdots$$
$$a_{m1}x_1 + \cdots + a_{mn}x_n = b_m.$$

In matrix form this is $\mathbf{AX} = \mathbf{B}$, where

$$\mathbf{A} = \begin{pmatrix} a_{11} & \cdots & a_{1n} \\ \vdots & & \vdots \\ a_{m1} & \cdots & a_{mn} \end{pmatrix}, \quad \mathbf{X} = \begin{pmatrix} x_1 \\ \vdots \\ x_n \end{pmatrix}, \quad \mathbf{B} = \begin{pmatrix} b_1 \\ \vdots \\ b_m \end{pmatrix}.$$

3.1 Solutions and flats. The system of linear equations $\mathbf{AX} = \mathbf{B}$ either (i) has no solution or (ii) has solutions. In case (ii) the set of all solutions constitutes a k-flat in $\mathbb{R}_n$, where $k = n - r$, r the rank of $\mathbf{A}$. The direction space is the kernel of $T(\mathbf{X}) = \mathbf{AX}$.

Proof. This has essentially been done in Chapter 4. If there are any solutions, let $\mathbf{X}_0$ be one. Then every other solution is of the form $\mathbf{X} = \mathbf{X}_0 + \mathbf{Z}$, $\mathbf{Z}$ in the kernel $\mathcal{K}$ of $T(\mathbf{X}) = \mathbf{AX}$ (Chapter 4). The set of all solutions is thus $\mathbf{X}_0 + \mathcal{K}$, and the dimension of $\mathcal{K}$ is $n - r$ (Chapter 4). Hence, the set of all solutions is a k-flat with direction space $\mathcal{K}$, the kernel of $T(\mathbf{W}) = \mathbf{AX}$. ∎

There are many special cases of this theorem which you should savor: If there are any solutions at all, then

the set of all solutions to one nontrivial equation† in n unknowns is a hyperplane in $\mathbb{R}_n$;
the set of all solutions to a system of equations in n unknowns whose matrix is of rank $n - 2$, is a plane;
the set of all solutions to a system of equations in n unknowns whose matrix is of rank $n - 1$, is a line;
the set of all solutions to a system of equations in n unknowns whose matrix is of rank n, is a point

(but we knew this already). As *special* special cases, let us note that *in general*, the set of all (x_1, x_2, x_3) for which

$$a_1 x_1 + a_2 x_2 + a_3 x_3 = y \text{ is a plane in } \mathbb{R}_3;$$

all (x_1, x_2, x_3) for which

$$\left. \begin{array}{l} a_1 x_1 + a_2 x_2 + a_3 x_3 = y_1 \\ b_1 x_1 + b_2 x_2 + b_3 x_3 = y_2 \end{array} \right\} \text{ is a line in } \mathbb{R}_3;$$

all (x_1, x_2, x_3) for which

$$\left. \begin{array}{l} a_1 x_1 + a_2 x_2 + a_3 x_3 = y_1 \\ b_1 x_1 + b_2 x_2 + b_3 x_3 = y_2 \\ c_1 x_1 + c_2 x_2 + c_3 x_3 = y_3 \end{array} \right\} \text{ is a point in } \mathbb{R}_3$$

(but watch that "in general").

Example 1. The solutions to

$$x_1 - 2x_2 + 4x_3 = 1,$$
$$3x_1 + x_2 - x_3 = 0,$$

are of the form $(\frac{1}{7}, -\frac{3}{7}, 0) + c(-\frac{2}{7}, \frac{13}{7}, 1)$ which we now recognize is a line through $(\frac{1}{7}, -\frac{3}{7}, 0)$ with direction space spanned by $(-\frac{2}{7}, \frac{13}{7}, 1)$.

Given points $\mathbf{P}_0, \mathbf{P}_1, \ldots, \mathbf{P}_m$ in $\mathbb{R}_n$, *now written as column vectors*, we shall now see how to express the minimal flat containing them as the solution set for a system of linear equations $\mathbf{AX} = \mathbf{Y}$. By now you realize that we would like the direction space of this flat to be spanned by

$$\mathbf{V}_1 = \mathbf{P}_1 - \mathbf{P}_0, \quad \mathbf{V}_2 = \mathbf{P}_2 - \mathbf{P}_0, \ldots, \mathbf{V}_m = \mathbf{P}_m - \mathbf{P}_0.$$

Moreover, we would like these direction space vectors to satisfy the associated homogeneous system $\mathbf{AX} = \mathbf{0}$. Furthermore, we would like to take $\mathbf{A}$ of maximal

---

† By a **nontrivial equation**, we mean one with not all zero coefficients: $a_1 x_1 + \cdots + a_n x_n = y$, not all $a_i = 0$.

rank (so that the flat will be minimal). Now let
$$\mathbf{V}_1 = (v_{11}, \ldots, v_{1n})^*, \ldots, \mathbf{V}_m = (v_{m1}, \ldots, v_{mn})^*$$
and consider the system of homogeneous equations

$$\begin{aligned} z_1 v_{11} + \cdots + z_n v_{1n} &= 0, \\ &\vdots \\ z_1 v_{m1} + \cdots + z_n v_{mn} &= 0. \end{aligned} \tag{1}$$

The solutions to this system form a subspace of $\mathbb{R}_n$; let
$$\mathbf{A}_1 = (a_{11}, \ldots, a_{1n}), \ldots, \mathbf{A}_r = (a_{r1}, \ldots, a_{rn})$$
be a basis for this subspace. Let

$$\mathbf{A} = \begin{pmatrix} \mathbf{A}_1 \\ \vdots \\ \mathbf{A}_r \end{pmatrix} \quad (\text{rank } \mathbf{A} = r).$$

Then

$$\mathbf{A} \mathbf{V}_i = \begin{pmatrix} \mathbf{A}_1 \mathbf{V}_i \\ \vdots \\ \mathbf{A}_r \mathbf{V}_i \end{pmatrix} = \begin{pmatrix} 0 \\ \vdots \\ 0 \end{pmatrix} = \mathbf{0},$$

the rows of $\mathbf{A}$ are linearly independent, and any solution to (1) can be written in terms of the rows of $\mathbf{A}$.

Consider the solutions to $\mathbf{AX} = \mathbf{Y}$, where $\mathbf{Y} = \mathbf{AP}_0$. This system has solutions (for example, $\mathbf{X} = \mathbf{P}_0$), so that by 3.1, the set of all solutions forms a k-flat $\mathcal{F}$, with $k = n - r$. As the proof of 3.1 indicates, this flat is of the form $\mathbf{P}_0 + \mathcal{K}$, where $\mathcal{K}$ is the kernel of $T(\mathbf{X}) = \mathbf{AX}$. Now the $\mathbf{V}_i$ are in $\mathcal{K}$ since $\mathbf{AV}_i = \mathbf{0}$; in fact, the $\mathbf{V}_i$ span $\mathcal{K}$ (Exercise 10). Hence, each
$$\mathbf{P}_i = \mathbf{P}_0 + (\mathbf{P}_i - \mathbf{P}_0) = \mathbf{P}_0 + \mathbf{V}_i$$
is in $\mathcal{F}$. No h-flat with $h < k$ can contain the $\mathbf{P}_i$ since the direction space must contain the $\mathbf{V}_i$ since the $\mathbf{V}_i$ span $\mathcal{K}$.

To recapitulate, to find the smallest k-flat containing $\mathbf{P}_0, \mathbf{P}_1, \ldots, \mathbf{P}_m$: let
$$\mathbf{V}_i = \mathbf{P}_i - \mathbf{P}_0 = (v_{i1}, \ldots, v_{in})^*,$$

let $A_1, \ldots, A_r$ be a basis for the solution space of (1), let

$$A = \begin{pmatrix} A_1 \\ \cdot \\ \cdot \\ \cdot \\ A_r \end{pmatrix},$$

and let $Y = AP_0$. Then the desired flat consists of all solutions X to $AX = Y$. Another treatment of this problem is given in 3.4 of the next chapter.

Example 2. Let $\mathcal{F}$ be all points $P = P_0 + sV + tW$, where

$V = (2, -1, 0)^*$, $W = (2, 0, -1)^*$ and $P_0 = (1, 1, 2)^*$.

Thus, $P_1 = V + P_0 = (3, 0, 2)^*$, $P_2 = W + P_0 = (3, 1, 1)^*$, $V_1 = V$, $V_2 = W$.

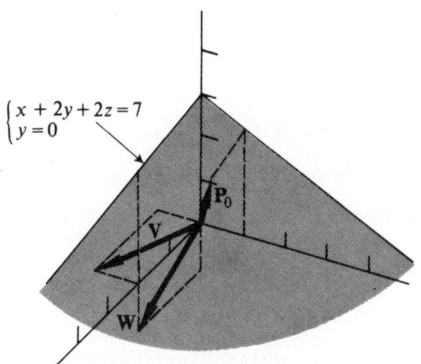

Fig. 21

The system of linear equations 1) is

$$\begin{cases} 2z_1 - z_2 = 0 \\ 2z_1 - z_3 = 0 \end{cases} \tag{1}$$

Its solution set is spanned by $A = (A_1) = (1, 2, 2)$. Since $AP = 7$ the equation describing $\mathcal{F}$ is

$$x + 2y + 2z = 7.$$

Can you discover the geometrical meaning of A in the picture? (You will in the next chapter.)

We have seen that flats and solutions to systems of linear equations are one and the same. We now capitalize on this.

3.2 i) The intersection of two flats in $\mathbb{R}_n$ is either empty or another flat.
  ii) Any flat is the intersection of hyperplanes.

Proof. i) If $\mathcal{F}_1$ is the set of solutions to $A_1 X = Y_1$ and $\mathcal{F}_2$ is all solutions to $A_2 X = Y_2$, then their intersection $\mathcal{F}_1 \cap \mathcal{F}_2$ is the set of all solutions to *both* systems of equations, and (i) follows from 3.1 (note that $\mathcal{F}_1 \cap \mathcal{F}_2$ is all solutions to

$$\begin{pmatrix} A_1 \\ A_2 \end{pmatrix} X = \begin{pmatrix} Y_1 \\ Y_2 \end{pmatrix}.$$

ii) Given the flat $\mathcal{F}$ consisting of solutions to the system

$$a_{11} x_1 + \cdots + a_{1n} x_n = y_1,$$
$$\vdots$$
$$a_{m1} x_1 + \cdots + a_{mn} x_n = y_m,$$

let $\mathcal{H}_1$ be all solutions to

$$a_{11} x_1 + \cdots + a_{1n} x_n = y_1,$$

$\ldots$, $\mathcal{H}_m$ be all solutions to

$$a_{m1} x_1 + \cdots + a_{mn} x_n = y_m.$$

Clearly $\mathcal{F} = \mathcal{H}_1 \cap \cdots \cap \mathcal{H}_m$. ∎

Example 3. The plane $\mathcal{S}$ in the previous example and the plane $y = 0$ intersect at all points satisfying

$$x + 2y + 2z = 7,$$
$$y = 0.$$

Since the equations are independent and there are two of them, this is a line. Considered in the plane $y = 0$, it is the line

$$x + 2z = 7.$$

(What is $x + 2z = 7$ in 3-space?)

Example 4. Consider the planes in 4-space given by

$$\mathcal{S}_1 = \begin{cases} x_1 + x_3 = 0, \\ x_1 + x_4 = 0, \end{cases} \quad \text{and} \quad \mathcal{S}_2 = \begin{cases} x_2 + x_1 = 0, \\ x_2 + x_4 = 0. \end{cases}$$

It is easy to see that these planes intersect in a single point (Exercise 11). Thus, you have to be careful not to expect too much from your intuition in spaces of dimension $n > 3$.

Hyperplanes are particularly pleasant to express as solutions to systems of linear equations, since they require only one equation. We use this in the following theorem:

3.3 The determinantal expression for a hyperplane.
i) Given two distinct points

$$P = (p_1, p_2) \quad \text{and} \quad Q = (q_1, q_2) \quad \text{in } \mathbb{R}_2,$$

the equation for the line through **P** and **Q** is

$$\det \begin{pmatrix} p_1 & q_1 & x \\ p_2 & q_2 & y \\ 1 & 1 & 1 \end{pmatrix} = 0.$$

ii) Given three noncollinear points

$$\mathbf{P} = (p_1, p_2, p_3), \qquad \mathbf{Q} = (q_1, q_2, q_3), \qquad \text{and} \qquad \mathbf{R} = (r_1, r_2, r_3) \quad \text{in } \mathbb{R}_3,$$

the equation for the plane through these points is

$$\det \begin{pmatrix} p_1 & q_1 & r_1 & x \\ p_2 & q_2 & r_2 & y \\ p_3 & q_3 & r_3 & z \\ 1 & 1 & 1 & 1 \end{pmatrix} = 0.$$

iii) In general, given $n + 1$ points $\mathbf{P}_1, \ldots, \mathbf{P}_{n+1}$ in $\mathbb{R}_n$ which do not lie in an $(n - 1)$-flat, the equation for the hyperplane through these points is

$$\det \begin{pmatrix} \mathbf{P}_1 & \mathbf{P}_2 & \cdots & \mathbf{P}_{n+1} & \mathbf{X} \\ 1 & 1 & & 1 & 1 \end{pmatrix} = 0,$$

where $\mathbf{P}_i = (p_1^{(i)}, \ldots, p_n^{(i)})^*$ and $\mathbf{X} = (x_1, \ldots, x_n)^*$.

Proof. (ii) Contains the essential ideas. Suppose for the moment that

$$\det \begin{pmatrix} \mathbf{P} & \mathbf{Q} & \mathbf{R} & \mathbf{X} \\ 1 & 1 & 1 & 1 \end{pmatrix} = 0 \tag{2}$$

is in fact the equation of a (hyper)plane in $\mathbb{R}_3$. If we substitute **P** for **X** we obtain

$$\det \begin{pmatrix} \mathbf{P} & \mathbf{Q} & \mathbf{R} & \mathbf{P} \\ 1 & 1 & 1 & 1 \end{pmatrix},$$

which is 0 since two of its columns are the same. Thus **P** satisfies the equation of the plane. Similarly, **Q** and **R** are on the plane, so if (2) *is* the equation of a plane, then it is the right one.

Expanding the determinant in (2) by cofactors using the last column, we see that we do have one equation in three unknowns, so that it is a plane in $\mathbb{R}_3$ if it is nontrivial (i.e., not $0x + 0y + 0z = 0$) and consistent (i.e., not $0x + 0y + 0z = d$, $d \neq 0$). Thus, if we can show that at least one of the coefficients of x, y, or z in (2) is nonzero, we know we have the equation of a plane. To see this note that

$$\operatorname{rank} \begin{pmatrix} \mathbf{P} & \mathbf{Q} & \mathbf{R} \\ 1 & 1 & 1 \end{pmatrix} = \operatorname{rank} \begin{pmatrix} \mathbf{P} & \mathbf{Q} - \mathbf{P} & \mathbf{R} - \mathbf{P} \\ 1 & 0 & 0 \end{pmatrix}$$

(why$_1$?), and this rank is 3 (why$_2$?). Hence, there is a vector $(\mathbf{S}, 0)^*$ which, when added to this set, preserves linear independence (why$_3$?). Hence,

$$\det \begin{pmatrix} \mathbf{P} & \mathbf{Q} & \mathbf{R} & \mathbf{S} \\ 1 & 1 & 1 & 0 \end{pmatrix} \neq 0$$

(why$_4$?). Expanding the determinant by the last column shows that one of the coefficients of x, y, or z in (2) is nonzero (why$_5$?). ∎

Finally, we call your attention to our discussion of parallel flats in Exercises 16–20.

EXERCISES

1. Let $\mathcal{F}_1$ be the flat
$$x_1 = 3 + t,$$
$$x_2 = 1 + t,$$
$$x_3 = 2 - t,$$
$$x_4 = 5 - t,$$
and $\mathcal{F}_2$ the flat
$$x_1 = 1 + 6t_1 + 6t_2,$$
$$x_2 = 8t_1 - 3t_2,$$
$$x_3 = -2 + 11t_1,$$
$$x_4 = 3 + 2t_1 + 2t_2.$$

a) Write the equations in t, t_1, t_2 which give the points of intersection of $\mathcal{F}_1$ and $\mathcal{F}_2$.
b) What are the possibilities for the intersection $\mathcal{F}_1 \cap \mathcal{F}_2$?
c) Find which of the possibilities in (b) actually obtains.

2. Find an equation for the plane containing the points $(1, -2, 2)$, $(2, 1, 4)$, and $(3, 0, 1)$.

3. Find the parametric equations of the plane $a_1x_1 + a_2x_2 + a_3x_3 = C$ in $\mathbb{R}_3$.

4. Let $\mathbf{U} = (2, 0, 2, -1)$, $\mathbf{V} = (3, 1, 0, -3)$, $\mathbf{W} = (0, 0, 1, 1)$.
   a) Find the flat through $\mathbf{P}_0 = (5, -2, 6, -3)$ with direction space spanned by $\mathbf{U}$, $\mathbf{V}$, and $\mathbf{W}$. Find the linear equation giving this flat.
   b) Same for the flat through $\mathbf{P}_0$ with direction space spanned by $\mathbf{U}$ and $\mathbf{V}$.

5. Let $\mathcal{V}$ be spanned by
$$\begin{pmatrix} 6 \\ 8 \\ 11 \\ 2 \end{pmatrix} \text{ and } \begin{pmatrix} 6 \\ -3 \\ 0 \\ 2 \end{pmatrix}$$
and let
$$\mathbf{P} = \begin{pmatrix} 1 \\ 0 \\ -2 \\ 3 \end{pmatrix}.$$

a) Find a system of equations which describe the flat $\mathbf{P} + \mathcal{V}$.
b) Use this result to do Exercise 1.

6. Find the intersections of the planes in Exercises 7 and 8 of Section 2. Draw a picture.
7. Find the intersection of the flat in Exercise 5 (above) with the flat in (a) Exercise 4(a), (b) Exercise 4(b).
8. Write the line through $\mathbf{P} = (3, 1, 1)$ and $\mathbf{Q} = (1, 2, 3)$ as an intersection of (hyper)planes. Draw a picture.
9. (a) through (e) Find the flats which are the solution sets to the systems of equations in Chapter 4, Exercise 19, Section 3. Draw pictures when possible.
10. Let $\mathbf{V}_1, \ldots, \mathbf{V}_m$ span a subspace of dimension k of $\mathbb{R}_n$. Let

$$\mathbf{B} = \begin{pmatrix} \mathbf{V}_1 \\ \cdot \\ \cdot \\ \cdot \\ \mathbf{V}_m \end{pmatrix}.$$

a) Why does $T_1(\mathbf{X}) = \mathbf{BX}$ have kernel of dimension $r = n - k$? Let $\mathbf{A}_1, \ldots, \mathbf{A}_r$ be a basis for the kernel of T_1, and let

$$\mathbf{A} = \begin{pmatrix} \mathbf{A}_1 \\ \cdot \\ \cdot \\ \cdot \\ \mathbf{A}_r \end{pmatrix}.$$

b) Why does $T_2(\mathbf{X}) = \mathbf{AX}$ have kernel of dimension $k = n - r$?
c) Why do $\mathbf{V}_1^*, \ldots, \mathbf{V}_m^*$ span the kernel of T_2?

11. Show that the planes in Example 4 intersect in a single point.
12. a) Answer the question in the proof of 3.3.
    b) Prove 3.3 for arbitrary n.
13. Verify 3.3 for (a) Exercise 5d, Section 1, (b) Exercise 2, Section 2, (c) Exercise 10, Section 2.
14. Prove the following theorem: Let $\mathcal{F} = \mathbf{X}_0 + \mathcal{V}$ be a flat in $\mathbb{R}_n$ and suppose that $\mathcal{V}$ is spanned by $\mathbf{V}_1, \ldots, \mathbf{V}_t$. Let

$$\mathbf{B} = (\mathbf{V}_1, \ldots, \mathbf{V}_t)$$

and suppose that $\mathbf{W}_1, \ldots, \mathbf{W}_m$ span the left kernel of $\mathbf{B}$. Let

$$\mathbf{A} = \begin{pmatrix} \mathbf{W}_1 \\ \cdot \\ \cdot \\ \cdot \\ \mathbf{W}_m \end{pmatrix}.$$

Then $\mathcal{F}$ is the set of all solutions to $\mathbf{AX} = \mathbf{Y}$, where $\mathbf{Y} = \mathbf{AX}_0$.

Parallel Flats

We make the following natural definitions.

Two lines $\mathbf{P} + \mathcal{V}$ and $\mathbf{Q} + \mathcal{W}$ are **parallel** if $\mathbf{Q}$ is not in $\mathbf{P} + \mathcal{V}$ and $\mathcal{W} = \mathcal{V}$.

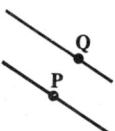

Two k-flats $\mathbf{P} + \mathcal{V}$ and $\mathbf{Q} + \mathcal{W}$ are **parallel** if $\mathbf{Q}$ is not in $\mathbf{P} + \mathcal{V}$ and $\mathcal{W} = \mathcal{V}$.

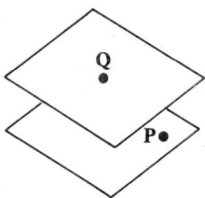

Let

$$\mathbf{P} + \mathcal{V} \text{ be a } k\text{-flat},$$

$$\mathbf{Q} + \mathcal{W} \text{ be an } l\text{-flat},$$

where $l < k$. Then $\mathbf{Q} + \mathcal{W}$ is **parallel** to $\mathbf{P} + \mathcal{V}$ if $\mathbf{Q}$ is not in $\mathbf{P} + \mathcal{V}$ and $\mathcal{W}$ is a subspace of $\mathcal{V}$.

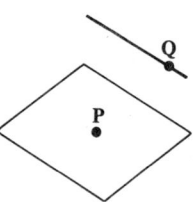

Prove.

15. Two parallel flats have no point in common.
16. Two hyperplanes are parallel iff they have no point in common.
17. a) Two nonparallel planes in $\mathbb{R}_3$ intersect in a line.
    b) Two nonparallel hyperplanes in $\mathbb{R}_n$ ($n \geq 2$) intersect in a $(n - 2)$-flat.
18. A line is parallel to a hyperplane iff it has no point in common with the hyperplane.
19. Two nonparallel lines intersect iff they lie in the same plane.
20. Two parallel lines determine a plane in which they both lie.

Length, Breadth and Thickness take up the whole of Space. Nor can Fansie imagine how there should be a Fourth Local Dimension beyond these Three.†

CHAPTER 8 **Euclidean Spaces**

We become even more geometrical in this chapter, and discuss angles, distance, and perpendicularity in n-space. This is tied in with our work in the previous chapter in Section 3. We then proceed to consider transformations of n-space which preserve these new basic geometrical properties.

From here on in, certain aspects of the real numbers become rather crucial (usually having to do with the ordering properties). We shall consequently dispense with honors projects which ask you to extend the work to arbitrary fields. It is, however, often possible to modify the material so that it can be extended to vector spaces over the complex numbers (usually by replacing absolute value by modulus and x^2 by $z\bar{z}$). We discuss this in several supplements.

1. DISTANCE, ANGLES, AND DOT PRODUCT

We shall now proceed to equip $\mathbb{R}_n$ with notions of distance, angle, and other pleasant accoutrements. Actually, all this entails is taking advantage of the dot product. To advertise our essential use of the dot product, we shall even change names:

By **Euclidean n-space** we shall mean the vector space $\mathbb{R}_n$, $\mathbb{R}$ the reals, together with the usual **dot product** of vectors in $\mathbb{R}_n$: for

$$\mathbf{V} = (v_1, \ldots, v_n), \quad \mathbf{W} = (w_1, \ldots, w_n) \quad \text{in } \mathbb{R}_n,$$

---

† John Wallis (1685).

their dot product is the scalar

$$\mathbf{V} \cdot \mathbf{W} = v_1 w_1 + \cdots + v_n w_n.$$

The dot product is also called the **inner product** and the **scalar product** of vectors, and notations such as $(\mathbf{V}, \mathbf{W})$ and $\langle \mathbf{V}, \mathbf{W} \rangle$ are often encountered.† While the mere recognition of something already there may not seem to warrant a name change, this does make sense from a slightly more abstract viewpoint, which is touched upon in Supplement 1 to Section 2.

There are two theorems standing between us and a realization of the geometrical entities *length* and *angle*. The first theorem appears to be a modest compendium of information about the dot product. It actually has a deeper significance, as shown in Supplement 1 to Section 2. In fact, when proving results about dot products, it is only necessary to use the three properties mentioned in 1.1. Except for this theorem the fact that $\mathbf{V} \cdot \mathbf{W}$ is actually the scalar

$$v_1 w_1 + \cdots + v_n w_n$$

is only useful for computational purposes.

1.1 *Basic properties of the dot product.* For any vectors $\mathbf{V}, \mathbf{W}, \mathbf{X}$ in $\mathbb{R}_n$, and any scalars a, b in $\mathbb{R}$,

i) $\mathbf{V} \cdot \mathbf{W} = \mathbf{W} \cdot \mathbf{V}$ (the dot product is commutative).

ii) $\mathbf{V} \cdot \mathbf{V} \geq 0$, and $\mathbf{V} \cdot \mathbf{V} = 0$ iff $\mathbf{V} = \mathbf{0}$.

iii) $(a\mathbf{V} + b\mathbf{W}) \cdot \mathbf{X} = a(\mathbf{V} \cdot \mathbf{X}) + b(\mathbf{W} \cdot \mathbf{X})$,

$\mathbf{V} \cdot (a\mathbf{W} + b\mathbf{X}) = a(\mathbf{V} \cdot \mathbf{W}) + b(\mathbf{V} \cdot \mathbf{X})$ (the dot product is multilinear).

Proof. Just use the basic definition (Exercise 1). ∎

We define the **norm** $|\mathbf{V}|$ of a vector $\mathbf{V}$ in $\mathbb{R}_n$ to be

$$|\mathbf{V}| = \sqrt{\mathbf{V} \cdot \mathbf{V}}.$$

Thus, for $\mathbf{V} = (v_1, \ldots, v_n)$,

$$|\mathbf{V}| = \sqrt{v_1^2 + v_2^2 + \cdots + v_n^2}.$$

In particular, for $\mathbf{V} = (v_1, v_2)$ in $\mathbb{R}_2$, $|\mathbf{V}|$ is nothing but the length of $\mathbf{V}$, so we're hot on the trail of a notion of distance. Another common notation for the norm of $\mathbf{V}$ is $\|\mathbf{V}\|$.

Although its importance is not apparent at first glance, the next theorem is absolutely essential in our introduction of geometrical notions into $\mathbb{R}_n$.

---

† We use the term "Dot product" in honor of our faithful secretary, Mrs. Dorothy Blythe.

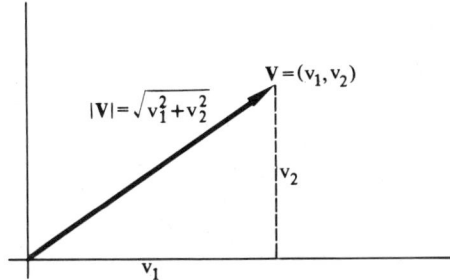

Fig. 1

1.2 The Schwarz inequality

$$|\mathbf{V} \cdot \mathbf{W}| \leq |\mathbf{V}|\,|\mathbf{W}| \qquad \text{for any } \mathbf{V}, \mathbf{W} \text{ in } \mathbb{R}_n.$$

In terms of components,

$$|v_1 w_1 + \cdots + v_n w_n| \leq \sqrt{v_1^2 + \cdots + v_n^2}\,\sqrt{w_1^2 + \cdots + w_n^2};$$

or, using summation notation,

$$\left|\sum_1^n v_i w_i\right| \leq \sqrt{\sum_1^n v_i^2}\,\sqrt{\sum_1^n w_i^2}.$$

(Note which of the symbols are "absolute value" and which are "norm".)

Proof. Notice this is trivial if $\mathbf{W} = \mathbf{O}$. If $\mathbf{W} \neq \mathbf{O}$, we can prove the inequality nicely by means of a trick. For any real x, the quantity $|\mathbf{V} + x\mathbf{W}|^2$ is a nonnegative real number (why$_1$?). This is true in particular when

$$x = -(\mathbf{V} \cdot \mathbf{W})/(\mathbf{W} \cdot \mathbf{W}).$$

Plugging in this value of x, using the definition of norm, and expanding, we are led to

$$(\mathbf{V} \cdot \mathbf{V})(\mathbf{W} \cdot \mathbf{W}) - (\mathbf{V} \cdot \mathbf{W})^2 \geq 0$$

(why$_2$?), which is what we're after (why$_3$?). ∎

The Schwarz Inequality is sometimes called the Cauchy-Schwarz Inequality (or the Cauchy-Bunyakovsky Inequality, by the Russians).

We now have the technical tools necessary to discuss distance and angles. In $\mathbb{R}_2$, at least, the distance from the point† $\mathbf{V}$ to the point $\mathbf{W}$ is just $|\mathbf{V} - \mathbf{W}|$. While

---

† Recall that in terms of our discussion in the previous chapter, both points and vectors in $\mathbb{R}_n$ are defined to be n-tuples of real numbers; the difference is in attitude. It makes more intuitive sense to talk of the length of a *vector* and the angle between two *vectors*, but of the distance between two *points*.

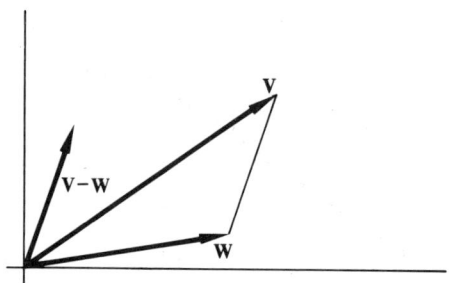

Fig. 2

there is no *a priori* notion of distance in $\mathbb{R}_n$, $n > 3$, it might seem reasonable to define length and distance as follows:

the **length** of a vector **V** in $\mathbb{R}_n$ is $|\mathbf{V}|$.
the **distance** between points **V** and **W** in $\mathbb{R}_n$ is $|\mathbf{V} - \mathbf{W}|$.

This reasonableness is substantiated by the following theorem, which says that these concepts behave sensibly.

1.3 Basic properties of the norm. For any **V**, **W** in $\mathbb{R}_n$ and a in $\mathbb{R}$,
   i) $|\mathbf{V}|$ is a real number, $|\mathbf{V}| \geq 0$, and $|\mathbf{V}| = 0$ iff $\mathbf{V} = \mathbf{0}$.
   ii) $|a\mathbf{V}| = |a|\,|\mathbf{V}|$ (again note the differences between norm and absolute value).

   iii) $|\mathbf{V} - \mathbf{W}| = |\mathbf{W} - \mathbf{V}|$.

   iv) $|\mathbf{V} + \mathbf{W}| \leq |\mathbf{V}| + |\mathbf{W}|$ (the *triangle inequality*).

Proof. You may do (i), (ii), and (iii); starting with the definition it's hard to go wrong. For (iv),

$$|\mathbf{V} + \mathbf{W}|^2 = (\mathbf{V} + \mathbf{W}) \cdot (\mathbf{V} + \mathbf{W}) = (\mathbf{V} + \mathbf{W}) \cdot \mathbf{V} + (\mathbf{V} + \mathbf{W}) \cdot \mathbf{W}$$
$$= \mathbf{V} \cdot \mathbf{V} + \mathbf{W} \cdot \mathbf{V} + \mathbf{V} \cdot \mathbf{W} + \mathbf{W} \cdot \mathbf{W}$$
$$= |\mathbf{V}|^2 + 2\mathbf{V} \cdot \mathbf{W} + |\mathbf{W}|^2$$

(Schwarz inequality)
$$\leq |\mathbf{V}|^2 + 2|\mathbf{V}|\,|\mathbf{W}| + |\mathbf{W}|^2 = (|\mathbf{V}| + |\mathbf{W}|)^2.$$

Take square roots. ∎

The properties of the norm listed in 1.3 are actually the key to calculus. It is now possible to talk about limits and continuity in a meaningful way (see Exercise 14).

1 / Distance, angles, and dot product

A **unit vector** **U** is any vector of length 1, $|\mathbf{U}| = 1$. Given any vector $\mathbf{V} \neq \mathbf{0}$, it is easy to see that $\mathbf{U} = (1/|\mathbf{V}|)\mathbf{V}$ is a unit vector (since **U** is a positive multiple of **V**, it is to be thought of as in the same direction as **V**).

Given vectors **V**, **W** in $\mathbb{R}_n$, we now introduce the concept of the angle between **V** and **W**. Actually, it is easier to get the cosine of the angle, and this is usually what is needed rather than the angle itself. We shall use the Law of Cosines,

$$c^2 = a^2 + b^2 - 2ab\cos\theta,$$

in $\mathbb{R}_2$ to see what should be done. Take a triangle in $\mathbb{R}_2$ comprised of vectors **V**, **W** and $\mathbf{V} - \mathbf{W}$, and suppose that

$$|\mathbf{V}| = a, \qquad |\mathbf{W}| = b, \quad \text{and} \quad |\mathbf{V} - \mathbf{W}| = c.$$

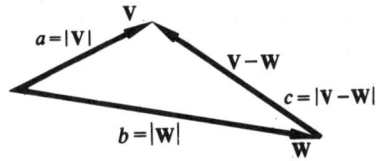

Fig. 3

If we put this into the Law of Cosines, use the definition of norm, and solve for $\cos\theta$, we find when the smoke clears away that

$$\cos\theta = \frac{\mathbf{V}\cdot\mathbf{W}}{|\mathbf{V}||\mathbf{W}|} \tag{1}$$

(see Exercise 6). This is a choice candidate for a definition in $\mathbb{R}_n$, all the more so since the Schwarz Inequality shows us that the rather necessary condition

$$-1 \leq \frac{\mathbf{V}\cdot\mathbf{W}}{|\mathbf{V}||\mathbf{W}|} \leq 1, \quad \text{if } \mathbf{V}, \mathbf{W} \neq \mathbf{0}, \tag{2}$$

holds. Thus, there is exactly one θ, with $0 \leq \theta \leq \pi$, for which

$$\cos\theta = \frac{\mathbf{V}\cdot\mathbf{W}}{|\mathbf{V}||\mathbf{W}|}.$$

We are all set:

Given nonzero vectors **V**, **W** in $\mathbb{R}_n$, the **angle between V and W** is that angle θ for which

$$\cos\theta = \frac{\mathbf{V}\cdot\mathbf{W}}{|\mathbf{V}||\mathbf{W}|}, \qquad \text{where } 0 \leq \theta \leq \pi. \tag{1}$$

In (1), $\cos \theta$ is unchanged if we replace $\mathbf{V}$ by $a\mathbf{V}$, $a > 0$ (Exercise 6b), and similarly for $\mathbf{W}$. Hence, we can replace $\mathbf{V}$ by the unit vector

$$\mathbf{U} = \frac{1}{|\mathbf{V}|} \mathbf{V}$$

without affecting the angle between this vector and another vector $\mathbf{W}$.

Suppose that $\mathbf{U} = (u_1, \ldots, u_n)$ is the unit vector in the direction of $\mathbf{V}$. We can now ascribe some meaning to the coordinates u_i. For the cosine of the angle θ_i between $\mathbf{U}$ and $\mathbf{e}_i$ is

$$\cos \theta_i = \frac{\mathbf{U} \cdot \mathbf{e}_i}{|\mathbf{U}| |\mathbf{e}_i|} = u_i.$$

Thus

the coordinates of a *unit* vector $\mathbf{U} = (u_1, \ldots, u_n)$ are the cosines of the angles which $\mathbf{U}$ makes with the vectors $\mathbf{e}_1, \ldots, \mathbf{e}_n$:

$$\mathbf{U} = \cos \theta_1 \mathbf{e}_1 + \cdots + \cos \theta_n \mathbf{e}_n, \qquad 0 \leq \theta_i \leq \pi.$$

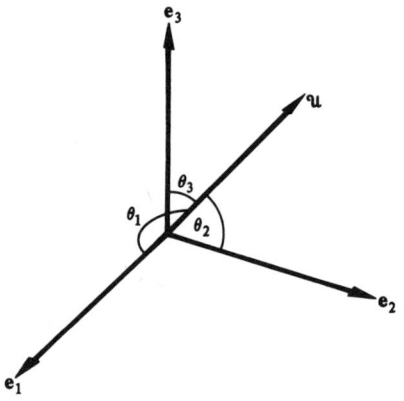

Fig. 4

The u_i are accordingly called the **direction cosines** of the vector $\mathbf{U}$ (and the θ_i the **direction angles**, although these are rarely needed). If you reverse the direction of $\mathbf{U}$ (that is, replace it by $-\mathbf{U}$), note that you change the signs of the direction cosines (this replaces the angle θ_i by $(\pi - \theta_i)$).

To return to (4), let us now take

$$\mathbf{V}_1 = (\cos \alpha_1, \ldots, \cos \alpha_n)$$

to be a unit vector in the direction of $\mathbf{V}$ and

$$\mathbf{W}_1 = (\cos \beta_1, \ldots, \cos \beta_n)$$

to be a unit vector in the direction of **W** (as noted, this does not affect (1)). We then have

the cosine of the angle between vectors **V** and **W** is

$$\cos\theta = \cos\alpha_1 \cos\beta_1 + \cdots + \cos\alpha_n \cos\beta_n,$$

where the $\cos\alpha_i$ are the direction cosines of a unit vector in the direction of **V** and the $\cos\beta_i$ those of a unit vector in the direction of **W**.

EXERCISES

1. Prove 1.1.
2. Prove 1.3, parts (i), (ii), and (iii).
3. Find the length of the vector **PQ** where **P** is $(1, 0, -2, 3)$, and **Q** is $(7, 8, 9, 5)$. Find the cosine of the angle between **P** and **Q**.
4. Let $\mathbf{A} = (0, 0, 0)$, $\mathbf{B} = (1, 2, 2)$, $\mathbf{C} = (2, 1, -1)$.
    a) Find the lengths of **AB**, **BC**, **AC**.
    b) Find the cosines of the angles A, B, C of the triangle **ABC**.
    c) Find the sines of the angles A, B, C and verify the law of sines.
    d) Verify that $A + B + C = 180°$. [*Hint:* Do not compute the angles.]
5. a) Prove the law of cosines for a triangle **ABC** in $\mathbb{R}_n$.
    b) Prove the law of sines for a triangle **ABC** in $\mathbb{R}_n$.
6. a) Derive Equation (1) in the context of Figure (3).
    b) Show that (1) is unchanged if **V** is replaced by $a\mathbf{V}$.
7. Prove Equation (2).
8. Prove that if $\mathbf{V} \neq 0$, then

    $$\frac{1}{|\mathbf{V}|}\mathbf{V}$$

    is a unit vector.
9. a) If $\mathbf{V} = (v_1, v_2)$ and $\mathbf{W} = (w_1, w_2)$ are vectors in $\mathbb{R}_2$, define

    $$\mathbf{V} \circ \mathbf{W} = v_1 w_1 + v_1 w_2 + v_2 w_1 + 2 v_2 w_2.$$

    Show that this "dot product" also satisfies 1.1(i) through (iii).
    b) Define "length" with respect to this dot product, and calculate the "length" of the vector $(0, 1)$.
    c) Define the "angle" between two vectors and calculate the "angle" between $(0, 1)$ and $(1, 0)$.
10. a) Let $\mathcal{V}$ be the set of all continuous real-valued functions defined for $0 \leq x \leq 1$, and considered as a vector space over $\mathbb{R}$ in the usual way. Define a "dot product" on $\mathcal{V}$ by

    $$f \circ g = \int_0^1 f(x)g(x)\, d(x),$$

    for f, g in $\mathcal{V}$. Verify that this "dot product" satisfies the conditions of 1.1.

b) Define the "distance" between two functions in $\mathcal{V}$. Illustrate with the functions x and x^2.

c) Define the "angle" between two functions in $\mathcal{V}$. Illustrate with x and x^2.

11. For **V**, **W** in $\mathbb{C}_n$, if you defined
$$\mathbf{V} \cdot \mathbf{W} = v_1 w_1 + \cdots + v_n w_n,$$
which conditions in 1.1 would fail to hold?

12. a) Answer the why's in the proof of the Schwarz Inequality (1.2).

b) Prove that $|\mathbf{V} + x\mathbf{W}|$ is a nonnegative quadratic function which attains its absolute minimum at.
$$x = \frac{-\mathbf{V} \cdot \mathbf{W}}{\mathbf{W} \cdot \mathbf{W}}.$$

13. a) Prove that the Schwarz Inequality becomes an equality if **V** and **W** are not linearly independent.

b) Prove that the Schwarz Inequality is an equality only if **V** and **W** are not linearly independent.

H14. Let $\mathcal{V}$ be any vector space over $\mathbb{R}$ with a norm $|\ |$ satisfying the conditions of 1.3. Let **f** be a function mapping $\mathcal{V}$ into $\mathbb{R}$.

a) Define
$$\text{"}\lim_{\mathbf{V} \to \mathbf{V}_0} \mathbf{f}(\mathbf{V})\text{"}.$$

[*Suggestion:* If you get stuck anywhere in this exercise, emulate your elementary calculus book.]

b) Prove that for a in $\mathbb{R}$,
$$\lim_{\mathbf{V} \to \mathbf{V}_0} a\mathbf{f}(\mathbf{V}) = a \lim_{\mathbf{V} \to \mathbf{V}_0} \mathbf{f}(\mathbf{V}).$$

c) Prove that if **g** is also a function from $\mathcal{V}$ into $\mathbb{R}$, then
$$\lim_{\mathbf{V} \to \mathbf{V}_0} [\mathbf{f}(\mathbf{V}) + \mathbf{g}(\mathbf{V})] = \lim_{\mathbf{V} \to \mathbf{V}_0} \mathbf{f}(\mathbf{V}) + \lim_{\mathbf{V} \to \mathbf{V}_0} \mathbf{g}(\mathbf{V}).$$

d) Define "**f** is continuous at $\mathbf{V}_0$".

e) Prove that if **f** is continuous at $\mathbf{V}_0$, so is $a\mathbf{f}$.

f) Prove that if **f** and **g** are continuous at $\mathbf{V}_0$, so is $\mathbf{f} + \mathbf{g}$.

g) Prove that any linear transformation is everywhere continuous.

Actually, one can very simply define limits and continuity for functions **f** from any vector space $\mathcal{V}$ over $\mathbb{R}$ with norm $|\ |_v$ into a vector space $\mathcal{W}$ over $\mathbb{R}$ with norm $|\ |_w$ (this is Exercise 14(h)). Moreover, the arithmetic corresponding to (b), (c), (e), (f) still holds (this is Exercise 14(i)). Also, (g) still holds.

2. ORTHOGONALITY

We have just introduced angles into $\mathbb{R}_n$. Given any nonzero vectors **V** and **W** in $\mathbb{R}_n$, the angle between them is that unique θ, $0 \leq \theta \leq \pi$, for which
$$\cos \theta = \frac{\mathbf{V} \cdot \mathbf{W}}{|\mathbf{V}| |\mathbf{W}|}.$$

Of all the angles which you will encounter in n-space, by far the handiest is $\pi/2$. You can immediately see that the angle between **V** and **W** is $\pi/2$ iff $\mathbf{V} \cdot \mathbf{W} = 0$. Accordingly,

two nonzero vectors **V** and **W** are **perpendicular** if $\mathbf{V} \cdot \mathbf{W} = 0$. Such vectors are also called **orthogonal**. We write $\mathbf{V} \perp \mathbf{W}$.

Given any vector $\mathbf{V} \neq \mathbf{0}$ in $\mathbb{R}_n$, it is easy to find a $\mathbf{W} \neq \mathbf{0}$ which is orthogonal to **V** (how?). For example, $(4, 2)$ is orthogonal to $(1, -2)$ in $\mathbb{R}_2$. You might

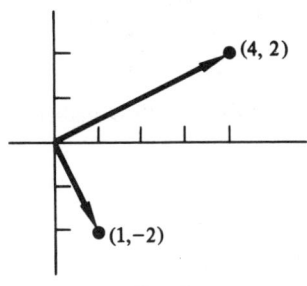

Fig. 5

speculate as to the set of vectors orthogonal to a given vector. We shall consider this in the next section.

Given two nonzero vectors **V** and **W** in $\mathbb{R}_n$, we can write **V** as the sum of a vector $a\mathbf{W}$ parallel to **W** and a vector $\mathbf{V} - a\mathbf{W}$ perpendicular to **W**. In fact, the appropriate scalar is just

$$a = \frac{\mathbf{V} \cdot \mathbf{W}}{|\mathbf{W}|^2}$$

(Exercise 1). The vector $a\mathbf{W}$ is called the **projection of V on W**.

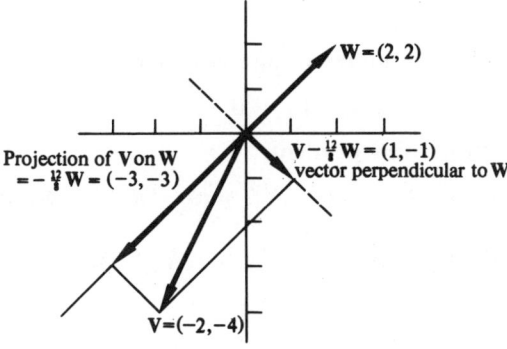

Fig. 6

Intuitively, the projection $a\mathbf{W}$ is the shadow of $\mathbf{V}$ on $\mathbf{W}$ obtained by shining a light which is an infinite distance away in a direction perpendicular to $\mathbf{W}$. Again, given nonzero $\mathbf{V}$ and $\mathbf{W}$ we can write

$$\mathbf{V} = \left(\frac{\mathbf{V}\cdot\mathbf{W}}{|\mathbf{W}|^2}\right)\mathbf{W} + \left(\mathbf{V} - \left(\frac{\mathbf{V}\cdot\mathbf{W}}{|\mathbf{W}|^2}\right)\mathbf{W}\right)$$
$$= (\text{projection of } \mathbf{V} \text{ on } \mathbf{W}) + (\mathbf{V} - \text{projection of } \mathbf{V} \text{ on } \mathbf{W})$$
$$= (\text{a vector parallel to } \mathbf{W}) + (\text{a vector perpendicular to } \mathbf{W});$$

and this displays $\mathbf{V}$ as a sum of orthogonal vectors. The coefficient

$$\frac{\mathbf{V}\cdot\mathbf{W}}{|\mathbf{W}|^2} = \frac{\mathbf{V}\cdot\mathbf{W}}{\mathbf{W}\cdot\mathbf{W}}$$

will often be encountered in our study of orthogonality.

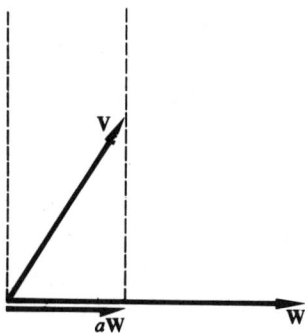

Fig. 7

Sets of mutually orthogonal vectors are particularly convenient to deal with, as the following two theorems show. The vectors

$$\mathbf{e}_1 = (1, 0, \ldots, 0), \quad \mathbf{e}_2 = (0, 1, 0, \ldots, 0), \quad \ldots, \quad \mathbf{e}_n = (0, \ldots, 0, 1)$$

form such a set.

2.1 If $\mathbf{V}_1, \ldots, \mathbf{V}_m$ are nonzero mutually orthogonal vectors in $\mathbb{R}_n$, that is,

$$\mathbf{V}_i \cdot \mathbf{V}_j = 0 \quad \text{if } i \neq j,$$

then they are linearly independent.

Proof. Suppose that $c_1\mathbf{V}_1 + \cdots + c_m\mathbf{V}_m = \mathbf{0}$. Take the dot product of this with $\mathbf{V}_1$ to obtain

$$0 = \mathbf{0}\cdot\mathbf{V}_1 = (c_1\mathbf{V}_1 + \cdots + c_m\mathbf{V}_m)\cdot\mathbf{V}_1$$
$$= c_1(\mathbf{V}_1\cdot\mathbf{V}_1) + c_2(\mathbf{V}_2\cdot\mathbf{V}_1) + \cdots + c_m(\mathbf{V}_m\cdot\mathbf{V}_1)$$
$$= c_1(\mathbf{V}_1\cdot\mathbf{V}_1) + c_2 0 + \cdots + c_m 0$$
$$= c_1(\mathbf{V}_1\cdot\mathbf{V}_1).$$

Since $\mathbf{V}_1 \cdot \mathbf{V}_1 \neq 0$, $c_1 = 0$. By an exceedingly subtle and complicated argument, one then shows that $c_2, \ldots, c_m$ are also 0. ∎

2.2 Corollary. If $\{\mathbf{V}_1, \ldots, \mathbf{V}_n\}$ are nonzero mutually orthogonal vectors in $\mathbb{R}_n$, then they are a basis for $\mathbb{R}_n$.

We shall call a basis $\{\mathbf{V}_1, \ldots, \mathbf{V}_n\}$ for $\mathbb{R}_n$ which consists of mutually orthogonal vectors, an **orthogonal basis** for $\mathbb{R}_n$. An example is $\{\mathbf{e}_1, \mathbf{e}_2, \ldots, \mathbf{e}_n\}$. If each of the vectors in an orthogonal basis has length 1, we celebrate this fact by calling it an **orthonormal basis**. Our preceding example, for instance, is one. Thus, $\{\mathbf{V}_1, \ldots, \mathbf{V}_n\}$ is an orthonormal basis iff $\mathbf{V}_i \cdot \mathbf{V}_j = \delta_{ij}$, where as you recall the *Kronecker delta* is defined by

$$\delta_{ij} = \begin{cases} 1 & \text{if } i = j, \\ 0 & \text{if } i \neq j. \end{cases}$$

Note that if $\{\mathbf{V}_1, \ldots, \mathbf{V}_n\}$ is an *orthogonal* basis, then

$$\left\{ \frac{1}{|\mathbf{V}_1|} \mathbf{V}_1, \ldots, \frac{1}{|\mathbf{V}_n|} \mathbf{V}_n \right\}$$

is an *orthonormal* basis (Exercise 2). Thus, given an orthogonal basis, an orthonormal basis is easy to come by.

Given just any old basis $\{\mathbf{V}_1, \ldots, \mathbf{V}_n\}$ for $\mathbb{R}_n$ and a vector $\mathbf{V}$, we know that it is possible to write

$$\mathbf{V} = c_1 \mathbf{V}_1 + \cdots + c_n \mathbf{V}_n.$$

In general it may take a good deal of Gaussian elimination-type work to obtain the coefficient c_i. Not so with orthonormal bases:

2.3 Expression of a vector in terms of an orthonormal basis. If $\{\mathbf{V}_1, \ldots, \mathbf{V}_n\}$ is an orthonormal basis for $\mathbb{R}_n$, and if $\mathbf{V}$ is any vector in $\mathbb{R}_n$, then

$$\mathbf{V} = (\mathbf{V} \cdot \mathbf{V}_1)\mathbf{V}_1 + (\mathbf{V} \cdot \mathbf{V}_2)\mathbf{V}_2 + \cdots + (\mathbf{V} \cdot \mathbf{V}_n)\mathbf{V}_n.$$

$\left(\text{If the basis is merely orthogonal, the coefficients are } \dfrac{\mathbf{V} \cdot \mathbf{V}_i}{\mathbf{V}_i \cdot \mathbf{V}_i} \right)$.

Proof. We know $\mathbf{V} = c_1 \mathbf{V}_1 + \cdots + c_n \mathbf{V}_n$ for some choice of c's. Take the dot product with $\mathbf{V}_i$ to obtain

$$\mathbf{V} \cdot \mathbf{V}_i = c_i (\mathbf{V}_i \cdot \mathbf{V}_i),$$

which is what we're after, if $\mathbf{V}_i \cdot \mathbf{V}_i = 1$. The statement in parentheses follows similarly. ∎

2.4 Let $\{\mathbf{V}_1, \ldots, \mathbf{V}_n\}$ be an orthonormal basis for $\mathbb{R}_n$. If

$$\mathbf{X} = x_1 \mathbf{V}_1 + \cdots + x_n \mathbf{V}_n \quad \text{and} \quad \mathbf{Y} = y_1 \mathbf{V}_1 + \cdots + y_n \mathbf{V}_n,$$

then $\mathbf{X} \cdot \mathbf{Y} = x_1 y_1 + \cdots + x_n y_n$.

That is, with respect to any orthonormal basis, the dot product has the same expression as with respect to the standard orthonormal basis $\{e_1, \ldots, e_n\}$.

Proof. $(x_i V_i) \cdot (y_j V_j) = x_i y_j (V_i \cdot V_j) = x_i y_j \delta_{ij}$. Now just use the fact that the dot product is multilinear (Exercise 3). ∎

Having demonstrated some of the virtues of orthogonal and orthonormal bases, what do you do if you are presented with an ordinary old-fashioned basis? The process indicated in the theorem below, although bearing the weight of two names, is so natural that we can barely resist leaving not only the proof, but also the statement of the theorem for you.

2.5 *The Gram-Schmidt process.* Given an arbitrary basis $\{V_1, \ldots, V_n\}$ of $\mathbb{R}_n$, the following vectors form an orthogonal basis:

$$W_1 = V_1$$

$$W_2 = V_2 - \frac{V_2 \cdot W_1}{W_1 \cdot W_1} W_1$$

$$W_3 = V_3 - \frac{V_3 \cdot W_2}{W_2 \cdot W_2} W_2 - \frac{V_3 \cdot W_1}{W_1 \cdot W_1} W_1$$

.
.
.

$$W_n = V_n - \frac{V_n \cdot W_{n-1}}{W_{n-1} \cdot W_{n-1}} W_{n-1} - \cdots - \frac{V_n \cdot W_1}{W_1 \cdot W_1} W_1.$$

Proof. For $n = 2$ we observed before that $W = V_1$ is perpendicular to

$$W_2 = V_2 - (W_1 \cdot V_2 / W_2 \cdot W_2) W_1.$$

Before abandoning you to an inductive proof (Exercise 4), we shall attend to the situation for $n = 3$.

Suppose then that we have transformed the basis $\{V_1, V_2, V_3\}$ into $\{W_1, W_2, V_3\}$ with $W_1 \perp W_2$. We are hoping to show that the appropriate constants c_1 and c_2 will make

$$W_3 = V_3 - c_1 W_1 - c_2 W_2$$

perpendicular to both W_1 and W_2. If this is the case,

$$0 = W_3 \cdot W_1 = (V_3 - c_1 W_1 - c_2 W_2) \cdot W_1$$
$$= V_3 \cdot W_1 - c_1 W_1 \cdot W_1.$$

Hence, $c_1 = (V_3 \cdot W_1)/(W_1 \cdot W_1)$. Similarly, $c_2 = (V_3 \cdot W_2)/(W_2 \cdot W_2)$. Proceeding backward, it is easy to see that this choice of c's actually works. ∎

Note that at any stage, W_k is a linear combination of V_k and $W_1, \ldots, W_{k-1}$. The only way that W_k could be 0 would be if V_k were a linear combination of $W_1, \ldots, W_{k-1}$. Since the V's are a basis this would be impossible (why?).

As the proof of 2.5 shows, all you need remember is that each of the orthogonal W_i's is expressed in terms of V_i and the preceding W_j's; the actual coefficients can be quickly derived (they are our usual "orthogonal coefficients", in fact).

Example. Let
$$V_1 = (1, 0, 1), \quad V_2 = (3, 1, -1), \quad V_3 = (1, 2, 3).$$
There is only one central idea involved in changing this set into an orthogonal set: you subtract from each vector a scalar multiple of each of the preceding, so that the resulting vectors are perpendicular. Thus, our set will become
$$W_1 = V_1, \quad W_2 = V_2 - aW_1, \quad W_3 = V_3 - bW_1 - cW_2,$$
and we must determine a, b, c so that these vectors are orthogonal. Rather than looking at the formula in the theorem, let us proceed directly:
$$0 = W_1 \cdot W_2 = V_1 \cdot V_2 - aV_1 \cdot V_1$$
$$= (3 - 1) - a(1 + 1) = 2 - 2a.$$
Thus $a = 1$ and $W_2 = V_2 - 1V_1 = (2, 1, -2)$. Next
$$0 = W_1 \cdot W_3 = V_1 \cdot V_3 - bV_1 \cdot V_1 - cW_1 \cdot W_2$$
$$= 1 + 3 - b(1 + 1) + 0,$$
so $b = 2$, and we need only determine c:
$$0 = W_2 \cdot W_3 = W_2 \cdot V_3 - bW_2 \cdot W_1 - cW_2 \cdot W_2$$
$$= (2 + 2 - 6) - b \cdot 0 - c(4 + 1 + 4),$$
and $c = -\frac{2}{9}$. Hence, we have determined
$$W_3 = V_3 - 2W_1 + (\tfrac{2}{9})W_2 = (-\tfrac{5}{9}, \tfrac{20}{9}, \tfrac{5}{9}).$$
One can (and should) check that $W_1 \cdot W_2 = W_2 \cdot W_3 = W_1 \cdot W_3 = 0$.

2.6 Corollary. There exists an orthogonal basis for $\mathbb{R}_n$ containing any given nonzero vector V.

Proof. Extend the linearly independent set $\{V\}$ to a basis and apply the Gram-Schmidt Process. ∎

Given any basis for $\mathbb{R}_n$ one can transform it into an orthonormal basis, of course, simply by applying the Gram-Schmidt Process and then normalizing each vector in the orthogonal basis. You might try this on the above example.

The Gram-Schmidt Process can be applied to sets that are not linearly independent. As you might expect,

2.7 When applied to an arbitrary set of vectors $\{V_1, \ldots, V_m\}$ in $\mathbb{R}_n$, the Gram-Schmidt Process produces an orthogonal basis for the subspace spanned by these vectors.

Proof. If one of the W_j's encountered in orthogonalizing turns out to be **0**, this says that the corresponding V_i is a linear combination of the preceding W_j's. Thus, one can delete V_i. In this fashion, one obtains an orthogonal set $W_1, \ldots, W_k$ and each V_j is a linear combination of the W's. ∎

EXERCISES

1. Prove that if $W \neq 0$, the vector $V - aW$ is perpendicular to W for any choice of V, where
$$a = (V \cdot W)/(W \cdot W).$$

2. Prove that if $\{V_1, \ldots, V_n\}$ is an orthogonal basis, then
$$\left\{\frac{1}{|V_1|} V_1, \ldots, \frac{1}{|V_n|} V_n\right\}$$
is an orthonormal basis.

3. Finish the proof of 2.4.
4. Finish the proof of the Gram-Schmidt Process.
5. a) Find an orthogonal basis for the space spanned by
$$U = (\tfrac{1}{3}, \tfrac{2}{3}, \tfrac{2}{3}, 0), \quad V = (7, -1, 2, 2), \quad W = (1, 1, 3, 2).$$
   b) Find an orthonormal basis for the space.
   c) Extend this orthonormal basis to an orthonormal basis for $\mathbb{R}_4$.

6. a) If V and W are not linearly independent, what ought to be the cosine of the angle between them?
   b) Prove your conjecture.
   c) Is it a necessary and sufficient condition? If so, prove.

7. Make up a flow diagram for the Gram-Schmidt Process.

Supplement 1. Inner Product Spaces

From a slightly more sophisticated point of view, we could have defined a **euclidean space** to be any vector space $\mathcal{U}$ over $\mathbb{R}$ which is equipped with an **inner product**, that is, a mapping $\langle\,,\,\rangle$ which takes pairs of vectors from $\mathcal{U}$ into scalars in $\mathbb{R}$ and satisfies

i) $\langle V, W \rangle = \langle W, V \rangle$, for any V, W in $\mathcal{U}$;

ii) $\langle V, V \rangle \geq 0$ and $\langle V, V \rangle = 0$ iff $V = 0$;

iii) for any a, b in $\mathbb{R}$,
$$\langle aV + bW, X \rangle = a\langle V, X \rangle + b\langle W, X \rangle,$$

and
$$\langle \mathbf{V}, a\mathbf{W} + b\mathbf{X}\rangle = a\langle \mathbf{V}, \mathbf{W}\rangle + b\langle \mathbf{V}, \mathbf{X}\rangle.$$

(Any resemblance between this and 1.1 is purely intentional.) The chief advantages of this approach to euclidean space are:

1. It is basis-free; you are not bogged down with having to relate everything to $(1, 0, \ldots, 0), \ldots, (0, \ldots, 1)$.
2. It applies equally well to infinite-dimensional euclidean spaces, a subject of considerable importance (see Supplement 2).
3. Both of the previous advantages hold when you discuss *complex* inner-product spaces (see Supplement 3).

It would be very beneficial for you to go back and check that all of the theorems in Sections 1 and 2 hold for our more general inner product. Note that Exercises 9 and 10 of Section 1 give inner products which are not our ordinary "dot product". (However, read on.)

In particular, notice that orthogonal bases exist, and the Gram-Schmidt Process will still hold. Now suppose that $\mathcal{U}$ is finite-dimensional over $\mathbb{R}$. Given a basis $\{\mathbf{V}_1, \ldots, \mathbf{V}_n\}$, use the Gram-Schmidt Process to transform it into an orthogonal basis $\{\mathbf{V}'_1, \ldots, \mathbf{V}'_n\}$, and then use the standard trick of changing this to an orthonormal basis $\{\mathbf{W}_1, \ldots, \mathbf{W}_n\}$. Thus,

$$\langle \mathbf{W}_i, \mathbf{W}_j\rangle = 0 \quad \text{if } i \neq j \quad \text{and} \quad \langle \mathbf{W}_i, \mathbf{W}_i\rangle = 1.$$

Now suppose that

$$\mathbf{U} = a_1\mathbf{W}_1 + \cdots + a_n\mathbf{W}_n \quad \text{and} \quad \mathbf{V} = b_1\mathbf{W}_1 + \cdots + b_n\mathbf{W}_n$$

are any two vectors in $\mathcal{U}$. Then, just as in 2.4,

$$\langle \mathbf{U}, \mathbf{V}\rangle = a_1 b_1 + \cdots + a_n b_n.$$

Thus, given a finite-dimensional inner-product space $\mathcal{U}$ over $\mathbb{R}$, there exists a basis $\mathbf{W}_1, \ldots, \mathbf{W}_n$, relative to which the inner product behaves exactly as if it were the dot product and the vectors $\mathbf{W}_i$ were the vectors

$$\mathbf{e}_i = (0 \cdots \overset{i}{1} \cdots 0).$$

Using the isomorphism which sends $\mathbf{W}_i$ into $\mathbf{e}_i$, $i = 1, \ldots, n$, we obtain

2.8 If $\mathcal{U}$ is an n-dimensional inner-product space, then $\mathcal{U}$ is isomorphic to euclidean n-space under an isomorphism which preserves inner products; i.e., if $\mathbf{T}$ is the isomorphism, then

$$\langle \mathbf{V}, \mathbf{W}\rangle = \mathbf{T}(\mathbf{V}) \cdot \mathbf{T}(\mathbf{W}) \quad \text{for all } \mathbf{V} \text{ and } \mathbf{W} \text{ in } \mathcal{U};$$

such an isomorphism is called an **isometric isomorphism** since it preserves distances.

The inner product in Exercise 9 of Section 1 was concocted by using the basis
$$\{V_1 = (1, 0), V_2 = (1, 1)\}$$
for $\mathbb{R}_2$, which should explain the peculiar results of (b) and (c) of that exercise.

Supplement 2. Legendre Polynomials

It is possible to equip the space $\mathcal{C}$ of all continuous functions on an interval $a \leq x \leq b$ with an inner product that satisfies the requirements of Supplement 1. The usual way is to take

$$\langle f, g \rangle = \int_a^b f(x)g(x)\, dx \qquad \text{for } f \text{ and } g \text{ in } \mathcal{C}.$$

It is easy to verify that (i), (ii), and (iii) in Supplement 1 do indeed hold.

We shall now consider the subspace of $\mathcal{C}$ consisting of all (real) polynomial functions on the interval $-1 \leq x \leq 1$; i.e., we are discussing $\mathbb{R}[x]$ on this interval; we leave it to the reader to show that any interval $a \leq x \leq b$ can be reduced to this case by a simple change of variable. It is then possible to construct an orthogonal basis for this subspace by applying the Gram-Schmidt Process to the known basis $\{1, x, x^2, \ldots, x^n, \ldots\}$. The process applies without modification but, of course, does not come to an end. The resulting polynomials when properly normalized are called **Legendre polynomials**; the standard notation for the nth one is $P_n(x)$. The conventional normalization is to make $P_n(1) = 1$ rather than making the basis orthonormal. It turns out that this is possible since $P_n(1) \neq 0$ (see 4 in the list below). A few properties of these important polynomials are given here; you may want to try proving some of them. One encounters them again in the study of differential equations, for example.

In summary, the Legendre polynomials are the orthogonal basis for the space of polynomials $\mathbb{R}[x]$ (with the above inner product) obtained by applying the Gram-Schmidt process to the basis

$$\{1, x, x^2, \ldots, x_n, \ldots\}$$

and requiring that the resulting polynomials

$$\{P_0(x), P_1(x), \ldots, P_n(x), \ldots\}$$

satisfy $P_n(1) = 1$, for $n = 0, 1, 2, \ldots$. You should calculate a few polynomials directly (the first few are listed in 8 below) and then feel free to use the recursion formula in 7.

Some properties (alias exercises) of Legendre polynomials $P_n(x)$ follow:

1. $P_n(x)$ is a polynomial of degree n.
2. $P_n(x)$ is an *even* polynomial if n is even; $P_n(x)$ is an *odd* polynomial if n is odd.
3. $\int_{-1}^{+1} P_n(x) P_m(x)\, dx = 0 \quad$ if $m \neq n$.

4. $P_n(x)$ has exactly n real roots in the interval $-1 < x < 1$ for $n > 0$.
5. $\int_{-1}^{+1} P_n^2(x)\,dx = \dfrac{2}{2n+1}$.
6. The coefficient of x^n in $P_n(x)$ is $(2n)!/(2^n n!\, n!)$.
7. $(n+1)P_{n+1}(x) = (2n+1)x \cdot P_n(x) - nP_{n-1}(x)$ (recursion formula).
8. $P_0(x) = 1$, $P_1(x) = x$, $P_2(x) = (\tfrac{1}{2})(3x^2 - 1)$, $P_3(x) = (\tfrac{1}{2})(5x^3 - 3x)$,
   $P_4(x) = (\tfrac{1}{8})(35x^4 - 30x^2 + 3)$, $P_5(x) = (\tfrac{1}{8})(63x^5 - 70x^3 + 15x)$.
9. $(1-x^2)P_n''(x) - 2xP_n'(x) + n(n+1)P_n(x) = 0$ (Legendre's differential equation).

Supplement 3. Complex Inner Product Spaces

As you may have discovered in Exercise 11 of Section 1, the straightforward definition of dot product for vectors from $\mathbb{C}_n$ has its problems. For example, in $\mathbb{C}_2$ if we let

$$(z_1, z_2) \cdot (w_1, w_2) = z_1 w_1 + z_2 w_2 \quad \text{for } z_i, w_i \text{ in } \mathbb{C},$$

we would have $(1, i) \cdot (1, i) = 0$. Thus, in addition to violating inner-product condition (ii), $(1, i)$ would be perpendicular to itself! The mind boggles. A way around this is as follows:

For $\mathbf{V} = (v_1, \ldots, v_n)$, we first define

$$\bar{\mathbf{V}} = (\bar{v}_1, \ldots, \bar{v}_n),$$

where $\bar{v}_j$ is the complex conjugate of v_j. We then define **complex euclidean n-space** to be $\mathbb{C}_n$ together with the **complex dot product** $\langle \mathbf{V}, \mathbf{W} \rangle = \bar{\mathbf{V}} \cdot \mathbf{W}$, where $\cdot$ indicates the usual dot product:

$$\langle \mathbf{V}, \mathbf{W} \rangle = \overline{(v_1, \ldots, v_n)} \cdot (w_1, \ldots, w_n) = \bar{v}_1 w_1 + \cdots + \bar{v}_n w_n,$$

v_i, w_i in $\mathbb{C}$. The following properties can easily be proved for $\mathbf{V}$, $\mathbf{W}$ in $\mathbb{C}_n$, a, b in $\mathbb{C}$:

i) $\langle \mathbf{V}, \mathbf{W} \rangle = \overline{\langle \mathbf{W}, \mathbf{V} \rangle}$ (we've had to give up commutativity).

ii) $\langle \mathbf{V}, \mathbf{V} \rangle \geq 0$ and $\langle \mathbf{V}, \mathbf{V} \rangle = 0$ iff $\mathbf{V} = \mathbf{0}$.

iii) $\langle (a\mathbf{V} + b\mathbf{W}), \mathbf{X} \rangle = \bar{a}\langle \mathbf{V}, \mathbf{X} \rangle + \bar{b}\langle \mathbf{W}, \mathbf{X} \rangle$ (which is not quite distributivity);

$\langle \mathbf{V}, a\mathbf{W} + b\mathbf{X} \rangle = a\langle \mathbf{V}, \mathbf{W} \rangle + b\langle \mathbf{V}, \mathbf{X} \rangle$.

In order to obtain a norm which is a nonnegative real number, we again define

$$|\mathbf{V}| = \langle \mathbf{V}, \mathbf{V} \rangle^{1/2} = \sqrt{\bar{v}_1 v_1 + \cdots + \bar{v}_n v_n}.$$

With this norm, the Schwarz Inequality holds with | | meaning "modulus of complex number" in appropriate spots (but the proof has a slight twist, which you should straighten out). 1.3 then holds, with the same minor changes necessary in its proof.

Although the notion of distance can thus be very reasonably introduced into complex euclidean space, angles are something else again, since the expression for the cosine of the angle between the vectors **V** and **W** would involve the complex number $\langle \mathbf{V}, \mathbf{W} \rangle$. However, one special angle still remains loyal, and we can define **V** and **W** to be **perpendicular** if $\langle \mathbf{V}, \mathbf{W} \rangle = 0$. Moreover, using this notion of orthogonality, Section 2 and its attendant notions go through as before.

From the slightly more abstract approach of inner products, we could define a **complex inner product space** (often called **unitary space**) to be a vector space $\mathcal{U}$ over $\mathbb{C}$ with an inner product satisfying the following conditions:

i) $\langle \mathbf{V}, \mathbf{W} \rangle = \overline{\langle \mathbf{W}, \mathbf{V} \rangle}$

ii) $\langle \mathbf{V}, \mathbf{V} \rangle \geq 0$ and equality holds iff $\mathbf{V} = \mathbf{0}$.

iii) $\langle a\mathbf{V} + b\mathbf{W}, \mathbf{X} \rangle = \bar{a}\langle \mathbf{V}, \mathbf{X} \rangle + \bar{b}\langle \mathbf{W}, \mathbf{X} \rangle$.

$\langle \mathbf{V}, a\mathbf{W} + b\mathbf{X} \rangle = a\langle \mathbf{V}, \mathbf{W} \rangle + b\langle \mathbf{V}, \mathbf{X} \rangle$.

The discussions in Supplements 1 and 2 then go through.

3.* ORTHOGONAL COMPLEMENTS AND FLATS AGAIN

If $\mathcal{U}$ is any subspace of $\mathbb{R}_n$, then the **orthogonal complement** $\mathcal{U}^\perp$ of $\mathcal{U}$ is the set of all vectors in $\mathbb{R}_n$ which are perpendicular to every vector in $\mathcal{U}$:

$$\mathcal{U}^\perp = \{\mathbf{W} \text{ in } \mathbb{R}_n \mid \mathbf{W} \cdot \mathbf{V} = 0 \quad \text{for all } \mathbf{V} \text{ in } \mathcal{U}\}.$$

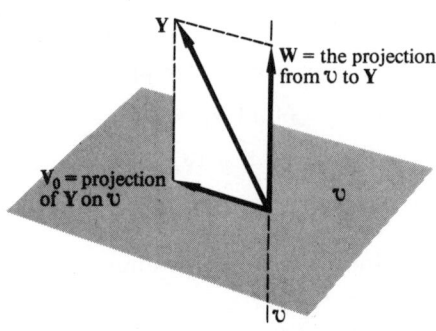

Fig. 8

3.1 Properties of orthogonal complements

i) The orthogonal complement $\mathcal{U}^\perp$ of any subspace of $\mathbb{R}_n$ is also a subspace of $\mathbb{R}_n$.
ii) If an orthogonal basis $\{V_1, \ldots, V_k\}$ for $\mathcal{U}$ is extended to an orthogonal basis
$$\{V_1, \ldots, V_k, V_{k+1}, \ldots, V_n\} \quad \text{for } \mathbb{R}_n,$$
then the vectors $\{V_{k+1}, \ldots, V_n\}$ form an orthogonal basis for $\mathcal{U}^\perp$.

Proof. (i) If W_1 and W_2 are in $\mathcal{U}^\perp$ and a_1 and a_2 are in $\mathbb{R}$, then
$$(a_1 W_1 + a_2 W_2) \cdot V = a_1(W_1 \cdot V) + a_2(W_2 \cdot V) = a_1 0 + a_2 0 = 0,$$
for any V in $\mathcal{U}$ so $a_1 W_1 + a_2 W_2$ is in $\mathcal{U}^\perp$. Hence $\mathcal{U}^\perp$ is a subspace.

ii) Given the orthogonal basis $\{V_1, \ldots, V_k\}$ for $\mathcal{U}$ which has been extended to an orthogonal basis
$$\{V_1, \ldots, V_k, V_{k+1}, \ldots, V_n\} \quad \text{for } \mathbb{R}_n.$$
Then $V_{k+1}, \ldots, V_n$ are in $\mathcal{U}^\perp$ since they are each orthogonal to a basis for $\mathcal{U}$. Moreover, if W is in $\mathcal{U}^\perp$, then
$$W = c_1 V_1 + \cdots + c_n V_n.$$
It follows from 2.3 that $c_1, \ldots, c_k$ are all 0, so actually
$$W = c_{k+1} V_{k+1} + \cdots + c_n V_n.$$
Thus, anything in $\mathcal{U}^\perp$ is a linear combination of $V_{k+1}, \ldots, V_n$, and clearly every linear combination of $V_{k+1}, \ldots, V_n$ is in $\mathcal{U}^\perp$. Hence
$$\mathcal{U}^\perp = \text{span } \{V_{k+1}, \ldots, V_n\}. \quad \blacksquare$$

We shall use the second part of this theorem as a basic theoretical tool in investigating orthogonal complements. It is also worthwhile noting how one might actually go about extending an orthogonal basis $\{V_1, \ldots, V_k\}$ for $\mathcal{U}$ to one for $\mathbb{R}_n$. Simply take
$$V_1, \ldots, V_k, \quad e_1, \ldots, e_n,$$
and apply the Gram-Schmidt process to it, throwing out any zero vectors encountered. Of course, $V_1, \ldots, V_k$ will be unchanged by this and by 2.7 the resulting set will be an orthogonal basis
$$\{V_1, \ldots, V_k, V_{k+1}, \ldots, V_n\} \quad \text{for } \mathbb{R}_n.$$
Although this does yield a reasonably practical method for attacking many of the problems we shall encounter, the Gram-Schmidt process is more lengthy than Gaussian elimination, so that other techniques are often more useful. We reexamine this approach in the following section.

Several pleasantries follow directly from 3.1. For example, if $\mathcal{U}$ is any subspace of $\mathbb{R}_n$, then from the symmetrical form of (ii) we have
$$\mathcal{U}^{\perp\perp} = \mathcal{U} \tag{1}$$

(Exercise 8). It is also apparent from (ii) that

$$(\text{dimension of } \mathcal{U}) + (\text{dimension of } \mathcal{U}^\perp) = n.$$

Of more immediate use is the following:

3.2 If $\mathcal{U}$ is any subspace of $\mathbb{R}_n$, then

i) any vector $\mathbf{Y}$ in $\mathbb{R}_n$ can be written in the form

$$\mathbf{Y} = \mathbf{V} + \mathbf{W}, \quad \mathbf{V} \text{ in } \mathcal{U}, \quad \mathbf{W} \text{ in } \mathcal{U}^\perp;$$

ii) this expression is unique; if also $\mathbf{Y} = \mathbf{V}' + \mathbf{W}'$ with $\mathbf{V}$ in $\mathcal{U}$ and $\mathbf{W}$ in $\mathcal{U}^\perp$, then $\mathbf{V} = \mathbf{V}'$ and $\mathbf{W} = \mathbf{W}'$. (Equivalently, the only vector common to both $\mathcal{U}$ and $\mathcal{U}^\perp$ is the zero vector.)

(In the language of the supplement to Chapter 1, Section 3,

i) says that $\mathbb{R}_n$ is the *sum* of $\mathcal{U}$ and $\mathcal{U}^\perp$,

$$\mathbb{R}_n = \mathcal{U} + \mathcal{U}^\perp;$$

ii) says that $\mathbb{R}_n$ is the *direct* sum,

$$\mathbb{R}_n = \mathcal{U} \oplus \mathcal{U}^\perp.)$$

Proof. Take an orthogonal basis for $\mathcal{U}$, $\{\mathbf{V}_1, \ldots, \mathbf{V}_k\}$, and extend it to an orthogonal basis $\{\mathbf{V}_1, \ldots, \mathbf{V}_n\}$ for $\mathbb{R}_n$ so we can use 3.1. Thus, any $\mathbf{Y}$ in $\mathbb{R}_n$ can be written

$$\mathbf{Y} = c_1 \mathbf{V}_1 + \cdots + c_k \mathbf{V}_k + c_{k+1} \mathbf{V}_{k+1} + \cdots + c_n \mathbf{V}_n.$$

Let

$$\mathbf{V} = c_1 \mathbf{V}_1 + \cdots + c_k \mathbf{V}_k \quad \text{and} \quad \mathbf{W} = c_{k+1} \mathbf{V}_{k+1} + \cdots + c_n \mathbf{V}_n.$$

This takes care of (i). For (ii), if also $\mathbf{Y} = \mathbf{V}' + \mathbf{W}'$, $\mathbf{V}'$ in $\mathcal{U}$ and $\mathbf{W}'$ in $\mathcal{U}^\perp$, write each in terms of the basis for its subspace and add together to get another expression for $\mathbf{Y}$ in terms of this basis. Comparing to the displayed expression, we see that $\mathbf{V} = \mathbf{V}'$ and $\mathbf{W} = \mathbf{W}'$.

You show that this is equivalent to saying that the intersection of $\mathcal{U}$ and $\mathcal{U}^\perp$ is zero (Exercise 9). ∎

As a slight generalization of our previous usage, if $\mathbf{Y} = \mathbf{V}_0 + \mathbf{W}$, $\mathbf{V}_0$ in $\mathcal{U}$, $\mathbf{W}$ in $\mathcal{U}^\perp$, $\mathbf{V}_0$ is called the **projection**† **of Y on** $\mathcal{U}$ (and $\mathbf{W}$ the **perpendicular** from $\mathcal{U}$ to $\mathbf{Y}$).

---

† In the supplement to Chapter 3, Section 4, we defined a "projection" to be any linear transformation satisfying $\mathbf{T}^2 = \mathbf{T}$. Given a subspace $\mathcal{U}$ of $\mathbb{R}_n$, we now define the **orthogonal projection of** $\mathbb{R}_n$ **on** $\mathcal{U}$ to be the mapping for which $\mathbf{T}(\mathbf{Y}) = \mathbf{V}$, where $\mathbf{Y} = \mathbf{V} + \mathbf{W}$ is the expression of $\mathbf{Y}$ in terms of an element from $\mathcal{U}$ and one from $\mathcal{U}^\perp$. It is easy to see that $\mathbf{T}$ is a projection in the sense of Chapter 3 (Exercise 10). However, not every projection from Chapter 3 is an orthogonal projection. The necessary and sufficient condition that a projection be an orthogonal projection is that its domain be perpendicular to its kernel (using 4.12 of Chapter 3 this is also easy; see Exercise 10). Draw yourself a picture to see that this makes geometrical sense.

3 / **Orthogonal complements and flats again** 365

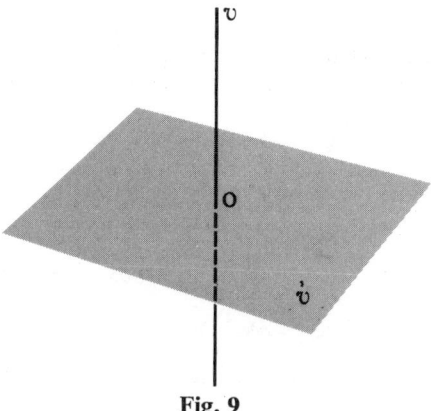

Fig. 9

By 3.2 every vector **Y** in $\mathbb{R}_n$ can be written as the sum of its projection on the subspace $\mathcal{V}$ and the perpendicular from $\mathcal{V}$ to **Y**.

3.3 Given a subspace $\mathcal{V}$ and a vector **Y** in $\mathbb{R}_n$, the projection V_0 of **Y** on $\mathcal{V}$ is the vector of $\mathcal{V}$ which is closest to **Y** in that the distance from **Y** to vectors of $\mathcal{V}$ is at a minimum at V_0.

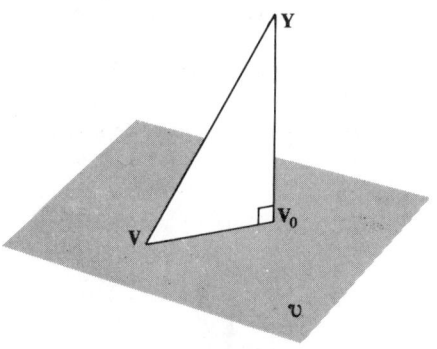

Fig. 10

Proof. Consider the number $|Y - V|$ for **V** in $\mathcal{V}$. Since $V_0 - V$ is in $\mathcal{V}$ while $Y - V_0$ is in $\mathcal{V}^\perp$, it is easy to show that

$$|V - Y|^2 = |V - V_0|^2 + |V_0 - Y|^2. \qquad (2)$$

This, in fact, is the Pythagorean Theorem; just write $V - Y = (V - V_0) + (V_0 - Y)$ and expand $(V - Y) \cdot (V - Y)$, using the fact that the "cross-product" term disappears (Exercise 11). From (2), since $|V - V_0|^2 \geq 0$, we see that $|V - Y| \geq |V_0 - Y|$ for any **V** in $\mathcal{V}$, thus proving the theorem. ∎

Example. Let us determine the nearest point in $\mathcal{V} = \text{span}\{V_1, V_2\}$ to **Y** where

$$V_1 = (2, 1, 0), \qquad V_2 = (-1, 2, 0),$$
$$Y = (1, 2, 3).$$

We first take an orthogonal basis for $\mathcal{V}$ (let's try with the one given rather than the more obvious $\{e_1, e_2\}$), and then extend to an orthogonal basis for $\mathbb{R}_3$ by adding on, say, $W = (0, 0, 1)$, which must form a basis for $\mathcal{V}^\perp$. We then know that

$$Y = aV_1 + bV_2 + cW,$$

and the coefficients are easy to determine since the basis is orthogonal. Taking dot products in turn,

$$Y \cdot V_1 = 4 = aV_1 \cdot V_1 = 5a; \qquad a = \tfrac{4}{5},$$
$$Y \cdot V_2 = 3 = bV_2 \cdot V_2 = 5b; \qquad b = \tfrac{3}{5},$$
$$Y \cdot W = 3 = cW \cdot W = c; \qquad c = 3.$$

Thus

$$Y = (\tfrac{4}{5})V_1 + (\tfrac{3}{5})V_2 + 3W,$$

and the projection of Y on $\mathcal{V}$ is

$$V_0 = (\tfrac{4}{5})V_1 + (\tfrac{3}{5})V_2 = (1, 2, 0),$$

while the perpendicular from Y to $\mathcal{V}$ is $3W = (0, 0, 3)$. Thus, the distance from Y to $\mathcal{V}$ is

$$|Y - V_0| = |3W| = 3,$$

in solemn agreement with simpler ways of viewing the situation.

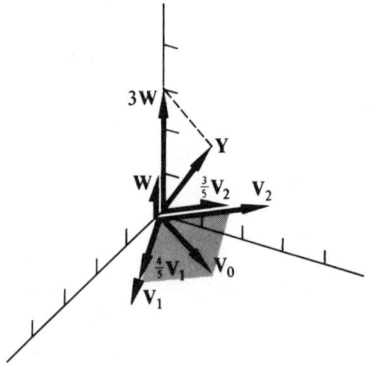

Fig. 11

For the rest of this section, we use the euclidean concepts we have developed to shed further light on the nature of flats.

3 / Orthogonal complements and flats again

3.4 Given a flat $\mathcal{F} = \mathbf{P} + \mathcal{V}$ with direction space $\mathcal{V}$, let $\{\mathbf{W}_1, \ldots, \mathbf{W}_m\}$ span $\mathcal{V}^\perp$, and let

$$\mathbf{A} = \begin{pmatrix} \mathbf{W}_1^* \\ \cdot \\ \cdot \\ \cdot \\ \mathbf{W}_m^* \end{pmatrix}.$$

Then $\mathcal{F}$ consists of all solutions $\mathbf{X}$ to $\mathbf{AX} = \mathbf{AP}$.

Proof. Since $\mathbf{AX} = \mathbf{AP}$ is a system of equations which has a solution (namely, $\mathbf{P}$), we know from the previous chapter that the solutions form a flat, in fact, the flat $\mathbf{P} + \mathcal{V}'$, where $\mathcal{V}'$ is the kernel of the linear transformation $T(V) = \mathbf{A}V$. Now the rows of $\mathbf{A}$ span $\mathcal{V}^\perp$, so this kernel consists precisely of $\mathcal{V}^{\perp\perp}$ (Exercise 12). Since by (2) $\mathcal{V}^{\perp\perp} = \mathcal{V}$, the flat of solutions is $\mathbf{P} + \mathcal{V}$. ∎

There is one case of particular interest here. If $\mathcal{H} = \mathbf{P} + \mathcal{V}$ is a hyperplane, then $\mathcal{V}$ is of dimension $(n - 1)$, so $\mathcal{V}^\perp$ is spanned by the single vector $\mathbf{W}$. Our theorem says that

$$\mathcal{H} = \{\mathbf{X} \mid \mathbf{WX} = \mathbf{WP}\}.$$

Thus,

the hyperplane $\mathcal{H} = \mathbf{P} + \mathcal{V}$ is the set of all solutions to

$$w_1 x_1 + \cdots + w_n x_n = y,$$

where $\mathbf{W} = (w_1, \ldots, w_n)$ is any nonzero vector in $\mathcal{V}^\perp$ and

$$y = \mathbf{W} \cdot \mathbf{P}.$$

The vector $\mathbf{W}$ in $\mathcal{V}^\perp$ is often said to be **normal** to $\mathcal{H}$. The nice thing about a hyperplane, of course, is that there is only one vector normal to its direction space (up to scalar multiples).

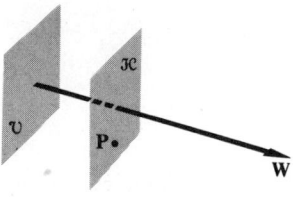

Fig. 12

In particular, you can specify a plane in $\mathbb{R}_3$ by means of a point $\mathbf{P}$ on it and a vector $\mathbf{W}$ perpendicular to its direction space: the plane is all solutions to

$$ax + by + cz = (ap_1 + bp_2 + cp_3) \quad \text{or}$$
$$a(x - p_1) + b(y - p_2) + c(z - p_3) = 0,$$

where $\mathbf{W} = (a, b, c)$ is a vector perpendicular to the direction space of the plane and $\mathbf{P} = (p_1, p_2, p_3)$ is any point on the plane. (If $\mathbf{W}$ is a unit vector, $|\mathbf{W}| = 1$, this is sometimes called the **normal form** of the equation of a plane in $\mathbb{R}_3$.)

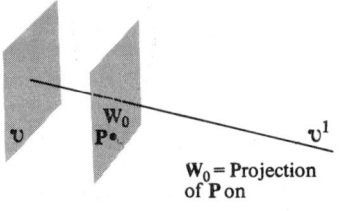

Fig. 13

3.5 A flat $\mathcal{F} = \mathbf{P} + \mathcal{V}$ consists of all $\mathbf{X}$ in $\mathbb{R}_n$ for which
$$\mathbf{W} \cdot \mathbf{X} = \mathbf{W} \cdot \mathbf{P} \qquad \text{for all } \mathbf{W} \text{ in } \mathcal{V}^\perp.$$
That is, $\mathcal{F}$ consists of all points which make the same projection on $\mathcal{V}^\perp$ as does $\mathbf{P}$.

Proof. This is just a reformulation of 3.4 (Exercise 13a) together with the observation that $\mathbf{P}$ and $\mathbf{X}$ have the same projections $\mathbf{P}$ on $\mathcal{V}^\perp$; $\mathbf{P} - \mathbf{X}$ is in $\mathcal{V}$, which is true iff $(\mathbf{P} - \mathbf{X}) \cdot \mathbf{W} = 0$ for all $\mathbf{W}$ in $\mathcal{V}^\perp$, which is true iff $\mathbf{P} \cdot \mathbf{W} = \mathbf{X} \cdot \mathbf{W}$ (Exercise 13b.) ∎

Another picture for 3.5

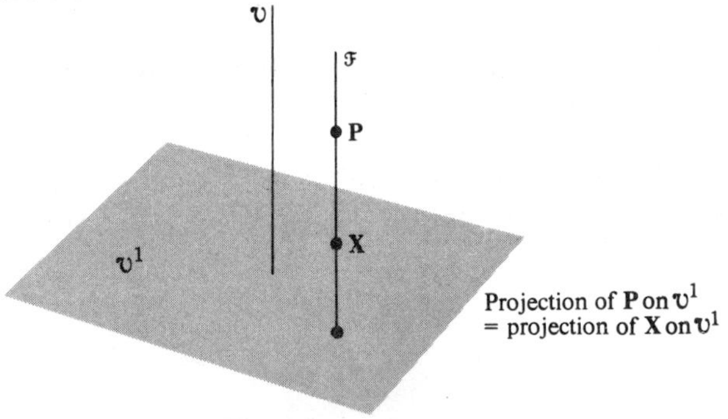

Fig. 14

Example. Let $\mathbf{P} = (1, 2, 1)$ and $\mathcal{V} = \text{span}\{\mathbf{e}_1, \mathbf{e}_3\}$ be the xz-plane in $\mathbb{R}_3$. Then
$$\mathcal{V}^\perp = y\text{-axis} = \text{span of } \{\mathbf{e}_2\},$$
and 3.5 says that
$$\mathbf{P} + \mathcal{V} = \{\mathbf{X} \mid \mathbf{e}_2 \cdot \mathbf{X} = \mathbf{e}_2 \cdot \mathbf{P}\} = \{\mathbf{X} = (x, y, z) \mid y = 2\}.$$
(See Fig. 15.)

As we have observed, a hyperplane $\mathcal{H}$ is the set of all $\mathbf{X}$ in $\mathbb{R}_n$ for which $\mathbf{X} \cdot \mathbf{W} = c$ for some fixed $\mathbf{W}$ in $\mathbb{R}_n$ and scalar c in $\mathbb{R}$. We noticed the meaning of

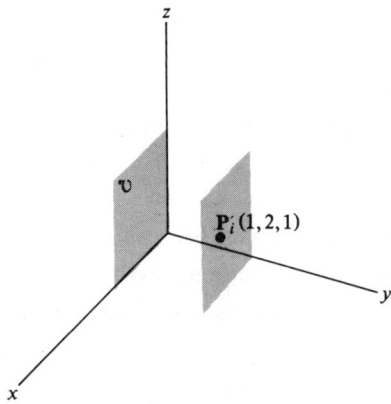

Fig. 15

W above; if we normalize W (so |W| = 1), then c gives the distance the hyperplane lies above or below its direction space (Exercise 14).

EXERCISES

1. Given any arbitrary set S in $\mathbb{R}_n$, define the orthogonal complement $S^\perp$ of S to be all W in $\mathbb{R}_n$ for which $S \cdot W = 0$ for all S in S.
   a) Prove that $S^\perp$ is a subspace of $\mathbb{R}_n$; (b) Prove that $S^\perp = (\text{span } S)^\perp$.

2. Let
$$\mathcal{V} = \{(x, y) \text{ in } \mathbb{R}_2 \mid x = y\}$$
and let $X = (1, 2)$. (a) Write $X = V + W$, W in $\mathcal{V}^\perp$. (b) Show that V actually is the closest vector in $\mathcal{V}$ to X by calculating the distance from a point in $\mathcal{V}$.

3. Find an equation of the plane in $\mathbb{R}_3$ which contains the point $(1, 1, 2)$ and is perpendicular to the vector $(2, 1, 3)$.

4. a) Prove that $\mathcal{V}^{\perp\perp} = \mathcal{V}$ for any subspace of $\mathbb{R}_n$.
   b) If $\mathcal{V}$ and $\mathcal{W}$ are subspaces of $\mathbb{R}_n$ and if $\mathcal{V}$ is contained in $\mathcal{W}$, prove that $\mathcal{W}^\perp$ is contained in $\mathcal{V}^\perp$.

5. Find the distance from the plane $2x + 3y - z = 10$ to the point $(1, 1, 2)$.

6. Find $\mathcal{V}^\perp$ if $\mathcal{V} = \text{span } \{(2, -1, 0), (2, 0, -1)\}$.

7. Write out a formula for the distance
   a) from the point (x_0, y_0) to the line
   $$ax + by + c = 0;$$
   b) for the distance from the point (x_0, y_0, z_0) to the plane
   $$ax + by + cz + d = 0.$$

8. Prove (1).
9. a) Prove that the sum of the dimensions of $\mathcal{V}$ and $\mathcal{V}^\perp$ is n.
   b) Finish the proof of 3.2.
10. Prove the results stated in the footnote on page 364.
11. Prove (2).
12. Complete the proof of 3.4.
13. (a) and (b) Complete the indicated parts of the proof of 3.5.
14. Prove the last statement before the exercises.
15. a) Prove that if a line in $\mathbb{R}_2$ intersects the x- and y-axes at a and b, respectively, then
$$a = \frac{d}{\cos \alpha} \quad \text{and} \quad b = \frac{d}{\sin \alpha},$$
where d is the directed distance from the line to the origin and α is the angle between the x-axis and the normal to the line.
   b) Similarly, show that if a plane intersects an axis in $\mathbb{R}_3$, then the intercept is $\frac{d}{\cos \alpha}$, where d is the distance from the origin to the plane and α the angle between the normal and the axis.
   c) Generalize to higher dimensions. [K. O. May, *American Math. Monthly*, **55**, 1948, p. 156.]
16. Derive an expression for the distance between two lines
$$\mathcal{L}_1 = \mathbf{P}_1 + \text{span} \{\mathbf{V}_1\} \quad \text{and} \quad \mathcal{L}_2 = \mathbf{P}_2 + \text{span} \{\mathbf{V}_2\}.$$
17. a) Show that in $\mathbb{R}_3$ the equation of the plane through points $\mathbf{P}$ and $\mathbf{Q}$ and perpendicular to the plane with normal $\mathbf{N} = (a, b, c)$ is
$$\det \begin{pmatrix} \mathbf{X} & \mathbf{P} & \mathbf{Q} & \mathbf{N} \\ 1 & 1 & 1 & 0 \end{pmatrix} = 0,$$
where $\mathbf{X} = (x, y, z)$ (cf. Chapter 7, 3.3).
   b) Similarly, the plane through $\mathbf{P}$ and perpendicular to the line of intersection of the planes with normals $\mathbf{N}_1$ and $\mathbf{N}_2$ is
$$\det \begin{pmatrix} \mathbf{X} & \mathbf{P} & \mathbf{N}_1 & \mathbf{N}_2 \\ 1 & 1 & 0 & 0 \end{pmatrix} = 0.$$
   c) Interpret the equation
$$\det \begin{pmatrix} \mathbf{X} & \mathbf{N}_1 & \mathbf{N}_2 & \mathbf{N}_3 \\ 1 & 0 & 0 & 0 \end{pmatrix} = 0.$$
[R. R. Stoll, *American Math. Monthly*, **61**, 1954, p. 255.]

4.* DISTANCE FROM POINTS TO SUBSPACES; LEAST SQUARES

In 3.3 we saw that the point in a subspace $\mathcal{V}$ closest to a given point $\mathbf{Y}$ is the projection $\mathbf{V}_0$ of $\mathbf{Y}$ on $\mathcal{V}$, and the distance from $\mathbf{Y}$ to the subspace is then $|\mathbf{V}_0 - \mathbf{Y}|$.

4 / Distance from points to subspaces; least squares

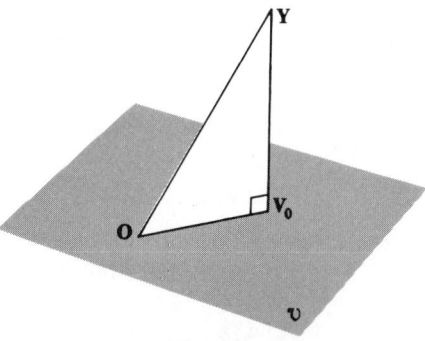

Fig. 16

While the problem of finding this projection and the attendant distance can be sometimes handled by orthogonalizing, we proceed to derive more useful expressions for the distance and the projection. To get hold of $\mathbf{V}_0$, suppose we know a basis $\{\mathbf{V}_1, \ldots, \mathbf{V}_k\}$ for $\mathcal{U}$ (not necessarily an orthogonal basis). Then by 3.2

$$\mathbf{Y} = \mathbf{V}_0 + \mathbf{W} = x_1\mathbf{V}_1 + \cdots + x_k\mathbf{V}_k + \mathbf{W}, \qquad (1)$$

for some $\mathbf{W}$ in $\mathcal{U}^\perp$. Since $\mathbf{W}$ is perpendicular to $\mathcal{U}$, we can transform this into a system of linear equations by just taking the dot product of (1) with each of $\mathbf{V}_1, \ldots, \mathbf{V}_k$ successively:

$$\begin{aligned}
\mathbf{Y} \cdot \mathbf{V}_1 &= x_1\mathbf{V}_1 \cdot \mathbf{V}_1 + x_2\mathbf{V}_2 \cdot \mathbf{V}_1 + \cdots + x_k\mathbf{V}_k \cdot \mathbf{V}_1, \\
\mathbf{Y} \cdot \mathbf{V}_2 &= x_1\mathbf{V}_1 \cdot \mathbf{V}_2 + x_2\mathbf{V}_2 \cdot \mathbf{V}_2 + \cdots + x_k\mathbf{V}_k \cdot \mathbf{V}_2, \\
&\quad \vdots \\
\mathbf{Y} \cdot \mathbf{V}_k &= x_1\mathbf{V}_1 \cdot \mathbf{V}_k + x_2\mathbf{V}_2 \cdot \mathbf{V}_k + \cdots + x_k\mathbf{V}_k \cdot \mathbf{V}_k.
\end{aligned} \qquad (2)$$

Given a specific $\mathbf{Y}$ and a basis $\{\mathbf{V}_1, \ldots, \mathbf{V}_k\}$ for $\mathcal{U}$, the above system of equations yields a unique solution $(x_1, \ldots, x_k)$ (just as it should; see 4.2 below). The vector $x_1\mathbf{V}_1 + \cdots + x_k\mathbf{V}_k$ must then also satisfy (1). (Since the $\mathbf{V}_i$'s are a basis, there is a unique set of x_i's with which one can write the vector $\mathbf{V}_0$; these x_i's will also satisfy (2), which has a unique solution as we shall see in 4.2.) Thus, for the solution $(x_1, \ldots, x_n)$ to (2) we have

$$\mathbf{V}_0 = x_1\mathbf{V}_1 + \cdots + x_k\mathbf{V}_k.$$

You might note that the system (2) takes on a particularly attractive form if the $\mathbf{V}_i$'s are orthogonal; but with an eye to actual use we shall not assume orthogonality here.

Having found the point $\mathbf{V}_0$ of $\mathcal{U}$ nearest $\mathbf{Y}$, the actual distance $|\mathbf{Y} - \mathbf{V}_0|$ from $\mathbf{Y}$ to $\mathcal{U}$ is sometimes a bit easier to get from the following:

$$|\mathbf{Y} - \mathbf{V}_0|^2 = |\mathbf{Y}|^2 - |\mathbf{V}_0|^2. \qquad (3)$$

This follows from (2) of Section 3 by putting $\mathbf{V} = \mathbf{0}$. We now cast this in a matrix form

4.1 Distance from a point to a subspace. Given a point $\mathbf{Y}$ and a subspace $\mathcal{U}$ of $\mathbb{R}_n$ with basis $\{\mathbf{V}_1, \ldots, \mathbf{V}_k\}$, let

$$\mathbf{A} = (\mathbf{V}_1, \ldots, \mathbf{V}_k)$$

be the $n \times k$ matrix whose columns are this basis. Then the point of $\mathcal{U}$ nearest $\mathbf{Y}$ (that is, the projection of $\mathbf{Y}$ on $\mathcal{U}$) has coordinates $\mathbf{V}_0 = \mathbf{AX}$, where $\mathbf{X}$ is the (unique) solution to

$$(\mathbf{A}^*\mathbf{A})\mathbf{X} = \mathbf{A}^*\mathbf{Y}. \tag{4}$$

The actual distance from $\mathbf{Y}$ to $\mathcal{U}$ is

$$|\mathbf{Y} - \mathbf{V}_0|^2 = \mathbf{Y} \cdot \mathbf{Y} - \mathbf{AX} \cdot \mathbf{AX} = |\mathbf{Y}|^2 - \mathbf{X}^*(\mathbf{A}^*\mathbf{A})\mathbf{X}. \tag{5}$$

Proof. Just show that (2) is (4) and (3) is (5) (Exercise 10). ∎

As you might expect, it is almost equally easy to obtain an expression for the distance from a point $\mathbf{Y}$ to a flat

$$\mathcal{F} = \mathbf{P} + \mathcal{U};$$

just translate everything back by $\mathbf{P}$ and calculate the distance from $\mathbf{Y} - \mathbf{P}$ to $\mathcal{U}$ (Exercise 13).

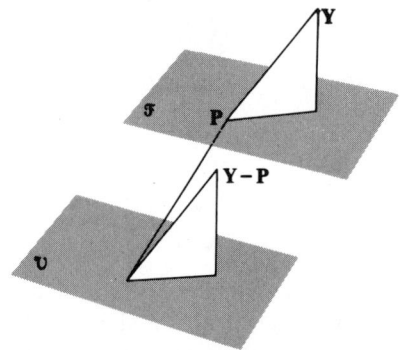

Fig. 17

The fact that the system (2) has a unique solution can be seen from the following result:

4.2 If $\mathbf{V}_1, \ldots, \mathbf{V}_k$ are linearly independent vectors in $\mathbb{R}_n$, then the matrix

$$\begin{pmatrix} \mathbf{V}_1 \cdot \mathbf{V}_1 & \mathbf{V}_1 \cdot \mathbf{V}_2 & \cdots & \mathbf{V}_1 \cdot \mathbf{V}_k \\ \vdots & & & \vdots \\ \mathbf{V}_k \cdot \mathbf{V}_1 & \mathbf{V}_k \cdot \mathbf{V}_2 & \cdots & \mathbf{V}_k \cdot \mathbf{V}_k \end{pmatrix}$$

has an inverse.

Proof. It suffices to consider the case where $k = 3$, since it contains all the necessary ideas. Given $\mathbf{V}_1, \mathbf{V}_2, \mathbf{V}_3$ linearly independent, Gram-Schmidt them to obtain

$$\mathbf{V}'_1 = \mathbf{V}_1, \qquad \mathbf{V}'_2 = \mathbf{V}_2 - a\mathbf{V}'_1, \qquad \mathbf{V}'_3 = \mathbf{V}_3 - b'\mathbf{V}'_1 - c'\mathbf{V}'_2 = \mathbf{V}_3 - b\mathbf{V}_1 - c\mathbf{V}_2.$$

which are orthogonal. Then

$$\det \begin{pmatrix} \mathbf{V}_1 \cdot \mathbf{V}_1 & \mathbf{V}_1 \cdot \mathbf{V}_2 & \mathbf{V}_1 \cdot \mathbf{V}_3 \\ \mathbf{V}_2 \cdot \mathbf{V}_1 & \mathbf{V}_2 \cdot \mathbf{V}_2 & \mathbf{V}_2 \cdot \mathbf{V}_3 \\ \mathbf{V}_3 \cdot \mathbf{V}_1 & \mathbf{V}_3 \cdot \mathbf{V}_2 & \mathbf{V}_3 \cdot \mathbf{V}_3 \end{pmatrix}$$

(why$_1$?)
$$= \det \begin{pmatrix} \mathbf{V}_1 \cdot \mathbf{V}_1 & \mathbf{V}_1 \cdot \mathbf{V}_2 - a\mathbf{V}_1 \cdot \mathbf{V}_1 & \mathbf{V}_1 \cdot \mathbf{V}_3 - b\mathbf{V}_1 \cdot \mathbf{V}_1 - c\mathbf{V}_1 \cdot \mathbf{V}_2 \\ \mathbf{V}_2 \cdot \mathbf{V}_1 & \mathbf{V}_2 \cdot \mathbf{V}_2 - a\mathbf{V}_2 \cdot \mathbf{V}_1 & \mathbf{V}_2 \cdot \mathbf{V}_3 - b\mathbf{V}_2 \cdot \mathbf{V}_1 - c\mathbf{V}_2 \cdot \mathbf{V}_2 \\ \mathbf{V}_3 \cdot \mathbf{V}_1 & \mathbf{V}_3 \cdot \mathbf{V}_2 - a\mathbf{V}_3 \cdot \mathbf{V}_1 & \mathbf{V}_3 \cdot \mathbf{V}_3 - b\mathbf{V}_3 \cdot \mathbf{V}_1 - c\mathbf{V}_3 \cdot \mathbf{V}_2 \end{pmatrix}$$

(why$_2$?)
$$= \det \begin{pmatrix} \mathbf{V}_1 \cdot \mathbf{V}'_1 & \mathbf{V}_1 \cdot \mathbf{V}'_2 & \mathbf{V}_1 \cdot \mathbf{V}'_3 \\ \mathbf{V}_2 \cdot \mathbf{V}'_1 & \mathbf{V}_2 \cdot \mathbf{V}'_2 & \mathbf{V}_2 \cdot \mathbf{V}'_3 \\ \mathbf{V}_3 \cdot \mathbf{V}'_1 & \mathbf{V}_3 \cdot \mathbf{V}'_2 & \mathbf{V}_3 \cdot \mathbf{V}'_3 \end{pmatrix}$$

(why$_3$?)
$$= \det \begin{pmatrix} \mathbf{V}'_1 \cdot \mathbf{V}'_1 & \mathbf{V}'_1 \cdot \mathbf{V}'_2 & \mathbf{V}'_1 \cdot \mathbf{V}'_3 \\ \mathbf{V}'_2 \cdot \mathbf{V}'_1 & \mathbf{V}'_2 \cdot \mathbf{V}'_2 & \mathbf{V}'_2 \cdot \mathbf{V}'_3 \\ \mathbf{V}'_3 \cdot \mathbf{V}'_1 & \mathbf{V}'_3 \cdot \mathbf{V}'_2 & \mathbf{V}'_3 \cdot \mathbf{V}'_3 \end{pmatrix} \overset{\text{(why}_4\text{?)}}{\neq} 0.$$

Hence the matrix has an inverse. ∎

An Application: Least Squares

Suppose that a certain physical quantity x is to be determined and n measurements are taken to obtain it. If the results are $y_1, y_2, \ldots, y_n$, what is the best estimate we can make for x? There is no problem, of course, if all the y_i are equal, but this is hardly likely. It is natural to try to find a number x_0 so that the vector whose components are $x - y_i$ is "small" in some sense. If we interpret this to mean that the euclidean length of this vector is to be as small as possible, we take

$$u = \sqrt{(x - y_1)^2 + \cdots + (x - y_n)^2},$$

and find the value x_0 that minimizes† u. It is a simple exercise (Exercise 16) in elementary calculus to find the value of x that does this; in fact, we readily get

$$x_0 = \frac{1}{n}(y_1 + \cdots + y_n),$$

and the resulting minimum value of u is

$$u_0 = \sqrt{\sum y_i^2 - nx_0^2}.$$

This is the simplest example of the method of least squares.

---

† Note that u has its minimum (if it exists) where u^2 has its minimum. It is often more convenient to use u^2, so as to avoid taking square roots. However, we want the norm properties that the square root yields.

It is an interesting fact that x_0 is actually the average of the n measurements. The size of u_0 determines the precision of these measurements.

We could have found this result immediately by noting that the problem consists in minimizing the distance in $\mathbb{R}_n$ from the point
$$\mathbf{Y} = (y_1, \ldots, y_n)^*$$
to the line through the origin with direction $\mathbf{V} = (1, 1, \ldots, 1)^*$. For any point in the subspace spanned by $\mathbf{V}$ is of the form
$$x\mathbf{V} = (x, \ldots, x)^*,$$
and $|\mathbf{Y} - x\mathbf{V}|$ is just the above expression which we are trying to minimize.

Example. Let
$$y_1 = 2.3, \quad y_2 = 3.1, \quad y_3 = 1.9, \quad y_4 = 2.5, \quad y_5 = 2.2.$$
We find $x_0 = 2.4$ and $u_0 = \sqrt{0.8}$ (a pretty good fit).

The median is sometimes used to get a quick answer to this problem. It turns out that the median minimizes the sum of the absolute values of the components $x - y_i$, that is, $\sum |x - y_i|$ (the "taxi-cab norm"; see Chapter 10). Thus, in our example the median is 2.3, and the sum of the absolute values of the deviations of the measurements from this is 1.5.

As a more significant problem, let us consider fitting a linear function to n pairs of data $(x_1, y_1), \ldots, (x_n, y_n)$. This means that we must find a and b so that $a + bx$ yields, for each x_i, the best approximation to y_i. For example, we could be measuring n position coordinates $y_1, \ldots, y_n$ of a moving object corresponding to n specified values $x_1, \ldots, x_n$ of the time. If it is assumed that the motion takes place with constant velocity, we are asked to find the best equation of the motion, in particular, the velocity b.

Here we are to bring the vector whose components are $a + bx_i$ as near as we can to the vector
$$\mathbf{Y} = (y_1, \ldots, y_n)^*$$
by properly choosing a and b. The method of least squares does this by minimizing the euclidean length of the vector whose components are $a + bx_i - y_i$. Thus, if we let
$$\mathbf{V}_1 = (1, \ldots, 1)^* \quad \text{and} \quad \mathbf{V}_2 = (x_1, \ldots, x_n)^*,$$
then
$$a\mathbf{V}_1 + b\mathbf{V}_2 - \mathbf{Y} = (a + bx_1 - y_1, a + bx_2 - y_2, \ldots, a + bx_n - y_n)^*;$$
making this vector as short as possible in euclidean length for all possible choices of a and b is the same as finding the vector in
$$\mathcal{V} = \text{span } \{\mathbf{V}_1, \mathbf{V}_2\} = \{a\mathbf{V}_1 + b\mathbf{V}_2 \mid a, b \text{ in } \mathbb{R}\},$$
which is as close as possible to the vector $\mathbf{Y}$.

4 / Distance from points to subspaces; least squares

To align this with the computational scheme of (4), we let $\mathbf{A} = (\mathbf{V}_1, \mathbf{V}_2)$ so that

$$\mathbf{A}^*\mathbf{A} = \begin{pmatrix} 1 & \cdots & 1 \\ x_1 & \cdots & x_n \end{pmatrix} \begin{pmatrix} 1 & x_1 \\ \cdot & \cdot \\ \cdot & \cdot \\ \cdot & \cdot \\ 1 & x_n \end{pmatrix} = \begin{pmatrix} n & \sum x_i \\ \sum x_i & \sum x_i^2 \end{pmatrix},$$

where $\sum$ means $\sum_{i=1}^n$. Thus, we must solve (4) which becomes

$$\mathbf{A}^*\mathbf{A}\mathbf{X} = \begin{pmatrix} n & \sum x_i \\ \sum x_i & \sum x_i^2 \end{pmatrix} \begin{pmatrix} a \\ b \end{pmatrix} = \begin{pmatrix} \sum y_i \\ \sum x_i y_i \end{pmatrix} = \mathbf{A}^*\mathbf{Y}, \quad \text{where } \mathbf{X} = \begin{pmatrix} a \\ b \end{pmatrix}.$$

Hence the desired values of a and b are obtained from the equations

$$na + b\sum x_i = \sum y_i, \qquad (6)$$
$$a\sum x_i + b\sum x_i^2 = \sum x_i y_i.$$

If $\mathbf{V}_1$ and $\mathbf{V}_2$ are linearly independent, then 4.2 assures us that there is a unique solution $(a, b)^*$, so that the closest vector is

$$\mathbf{V}_0 = a\mathbf{V}_1 + b\mathbf{V}_2.$$

By (5) of 4.2, the distance from $\mathbf{V}_0$ to $\mathbf{Y}$ is then

$$|\mathbf{Y} - \mathbf{V}_0| = \sqrt{\mathbf{Y}\cdot\mathbf{Y} - \mathbf{X}^*\mathbf{A}^*\mathbf{Y}} = \sqrt{\sum y_i^2 - a\sum y_i - b\sum x_i y_i}.$$

Note that $\mathbf{V}_1$ and $\mathbf{V}_2$ will be linearly independent if any two of the measured x_i's are different. If they are all the same (which is hardly likely in an experimental situation), then we are back in the previous problem of finding the best estimate for a single measured quantity.

In this problem of fitting a linear function, it is convenient to simplify the situation by first moving the origin to the point

$$\bar{x} = \frac{1}{n}\sum x_i, \qquad \bar{y} = \frac{1}{n}\sum y_i \qquad \text{(the center of gravity)}.$$

This means that we put

$$x_i' = x_i - \bar{x}, \qquad y_i' = y_i' = y_i - \bar{y}$$

and then start all over again. Since $\sum x_i' = 0$ and $\sum y_i' = 0$, we get much simpler equations than (1) to solve. The solutions are

$$a = 0 \qquad b = \frac{\sum x_i' y_i'}{\sum x_i'^2},$$

and $y' = bx'$ gives the desired linear function. In terms of the original coordinates this is

$$y - \bar{y} = b(x - \bar{x}).$$

Numerical Example. Let x be the time and y the position coordinate measurements given in the first two columns of the following table. We have calculated the entries in the remaining columns.

$n=5$	x	y	x'	y'	$x'y'$	x'^2	(for later use) x'^3	x'^4	$x'^2 y'$
	0	1.2	−2	−1.8	3.6	4	−8	16	−7.2
	1	2.1	−1	−0.9	0.9	1	−1	1	−0.9
	2	3.8	0	0.8	0	0	0	0	0
	3	2.9	1	−0.1	−0.1	1	1	1	−0.1
	4	5.0	2	2.0	4.0	4	8	16	8.0
Sum	10	15.0	0	0	8.4	10	0	34	−0.2

We find

$$\bar{x} = 2, \quad \bar{y} = 3.0,$$
$$x' = x - 2, \quad y' = y - 3,$$
$$b = 0.84 \quad \text{(the estimated velocity)};$$
$$y' = 0.84x', \quad y - 3 = 0.84(x - 2).$$

It is, at least in principle, just as easy to fit a polynomial of any degree to a set of data. We shall illustrate this by fitting a quadratic polynomial $a + bx + cx^2$ to the data (x_i, y_i), $i = 1, 2, \ldots, n$ ($n > 3$). Here $k = 3$,

$$\mathbf{A} = \begin{pmatrix} 1 & x_1 & x_1^2 \\ \cdot & \cdot & \cdot \\ \cdot & \cdot & \cdot \\ \cdot & \cdot & \cdot \\ 1 & x_n & x_n^2 \end{pmatrix}, \quad \mathbf{Y} = \begin{pmatrix} y_1 \\ \cdot \\ \cdot \\ y_n \end{pmatrix}, \quad \mathbf{X} = \begin{pmatrix} a \\ b \\ c \end{pmatrix},$$

$$\mathbf{A}^*\mathbf{A} = \begin{pmatrix} n & \sum x_i & \sum x_i^2 \\ \sum x_i & \sum x_i^2 & \sum x_i^3 \\ \sum x_i^2 & \sum x_i^3 & \sum x_i^4 \end{pmatrix}, \quad \mathbf{A}^*\mathbf{Y} = \begin{pmatrix} \sum y_i \\ \sum x_i y_i \\ \sum x_i^2 y_i \end{pmatrix}.$$

We will have a unique solution if $\mathbf{A}$ is of rank 3, which will be the case if at least three of the x_i are different (Exercise 7). (As noted, one would expect this in an experimental situation.) It is again convenient to move the origin to $(\bar{x}, \bar{y})$, as in the case of the linear approximation.

As a numerical illustration, we shall fit a quadratic polynomial to the data given before. The additional calculations are given there in the last three columns. Hence

$$\mathbf{A}^*\mathbf{A} = \begin{pmatrix} 5 & 0 & 0 \\ 0 & 10 & 0 \\ 10 & 0 & 34 \end{pmatrix}, \quad \mathbf{A}^*\mathbf{Y} = \begin{pmatrix} 0 \\ 8.4 \\ -0.2 \end{pmatrix},$$

and we get

$$a = \tfrac{1}{35}, \qquad b = 0.84, \qquad c = -\tfrac{1}{70}.$$

If we assume that the motion takes place with constant acceleration, we could use this to estimate the acceleration.

The general least-squares problem is:

> Given a matrix **A** and a vector **Y**, find the vector **X** (if it exists) for which **AX** is as close as possible to **Y**; that is, |**AX** − **Y**| is as small as possible.

This is exactly the problem we solved in 4.1, if the columns of **A** are linearly independent. We saw that the solution **X** to the least-squares problem was the solution to the system of linear equations **A*AX** = **A*Y**. The general least-squares problem is usually thought of as finding a "best" approximate solution to a system of linear equations (i.e., one makes **AX** as close to **Y** as possible).

An important practical modification of the least-squares process consists in using a **weighted norm**. This means that we would use for the norm of a vector $\mathbf{X} = (x_1, \ldots, x_n)$ in $\mathbb{R}_n$,

$$|\mathbf{X}|_w = \sqrt{w_1 x_1^2 + \cdots + w_n x_n^2}.$$

Here $w_1, \ldots, w_n$ are given positive "weights" which are assigned to measurements in accordance to the known (or suspected) accuracy, or the reverse, of given pieces of equipment. It is easy to go through the whole process using this norm, or it is also possible to reduce the weighted-norm case to the case we have just discussed. We shall merely quote the main result, which is that $|\mathbf{AX} - \mathbf{Y}|_w$ is minimized by the vector $\mathbf{X}_0$ obtained by solving

$$(\mathbf{A*WA})\mathbf{X}_0 = \mathbf{A*WY};$$

here **W** is the diagonal matrix

$$\begin{pmatrix} w_1 & 0 & \cdots & 0 \\ 0 & w_2 & & \cdot \\ \cdot & & \cdot & \cdot \\ \cdot & & & \cdot \\ \cdot & & & \cdot \\ 0 & \cdots & \cdots & w_n \end{pmatrix}.$$

One may consider this "weighted norm" as another way of measuring distance in $\mathbb{R}_n$. Such possibilities are further discussed in Chapter 10.

As an example we shall fit a linear function $a + bx$ to data (x_i, y_i), using the weighted norm with weights $w_1, \ldots, w_n$. Here **A** and **Y** are the same as before, and we get

$$\mathbf{A*WA} = \begin{pmatrix} w & \sum w_i x_i \\ \sum w_i x_i & \sum w_i x_i^2 \end{pmatrix}, \qquad \mathbf{A*WY} = \begin{pmatrix} \sum w_i y_i \\ \sum w_i x_i y_i \end{pmatrix},$$

where we have put $w = \sum w_i$ (the total weight). In this case we move the origin to the center of mass

$$\bar{x} = \frac{1}{w} \sum w_i x_i, \qquad \bar{y} = \frac{1}{w} \sum w_i y_i,$$

by putting $x_i' = x_i - \bar{x}$, $y_i' = y_i - \bar{y}$. The result (in terms of the new coordinates) is

$$a = 0, \qquad b = \frac{\sum w_i x_i' y_i'}{\sum w_i x_i'^2}.$$

The reader might try this out on the data given before with weights

$$w_1 = 2, \qquad w_2 = 2, \qquad w_3 = 3, \qquad w_4 = 2, \qquad w_5 = 1; \qquad w = 10.$$

Fitting of quadratic and higher-degree polynomials can obviously be carried out with the weighted norm much as it was before.

Finally, there is nothing special about polynomials here; one can use least-squares techniques for finding the coefficients of any linear combination of functions of the experimental data. For example, if the data (x_i, y_i, z_i) seems to warrant it, one could approximate $y_i \log z_i$ by

$$a \sin x_i + b e^{y_i} + c x_i y_i z_i + d$$

(see Exercise 4).

EXERCISES

1. Let data (x_i, y_i, z_i), $i = 1, 2, \ldots, n$, be given. Fit a function of the form $a + bx + cy$ to these data so as to yield the best value of z (in the sense of least squares) for each pair (x_i, y_i). That is, find expressions 4) and 5) in this case.
2. Do Exercise 1 using a weighted norm.
3. Fit a third-degree polynomial $a + bx + cx^2 + dx^3$ to data (x_i, y_i).
4. Show how to make the least-squares approximation indicated just before the exercises.
5. Carry out the numerical example with the weighted norm suggested in the text.
6. Prove the result stated for least squares with weighted norm.
7. Prove that the matrix A that occurs in the fitting of a quadratic polynomial is of rank 3 if at least three of the x_i are distinct.
8. Prove that the square of the minimum value of the norm, in the case of fitting a linear function to a set of data, can be written in the form $(1 - \rho^2) \sum y_i'^2$, where

$$\rho = \frac{\sum x_i' y_i'}{\sqrt{\sum x_i'^2 \sum y_i'^2}}$$

(ρ is called the **coefficient of correlation**). Find ρ for the numerical example in the text. ρ is a measure of the "goodness of fit": if $\rho = \pm 1$ the fit is perfect; if ρ is near 0 the fit is terrible. Note that ρ is the cosine of the angle between two vectors.
9. Prove (3).
10. Prove 4.1.
11. Answer the questions in the proof of 4.2.
12. Use 4.2 to prove that if $\mathbf{A}$ is an $n \times k$ matrix of rank k ($< n$), then $\mathbf{A}^*\mathbf{A}$ is of rank k.
13. Show that the point in the flat $\mathscr{F} = \mathbf{P} + \mathscr{V}$ which is closest to a given point $\mathbf{Y}$ is

$$\mathbf{P}_0 = \mathbf{AX} + \mathbf{P},$$

where $\mathbf{X}$ is the solution to

$$(\mathbf{A}^*\mathbf{A})\mathbf{X} = \mathbf{A}^*(\mathbf{Y} - \mathbf{P}),$$

and $\mathbf{A} = (\mathbf{V}_1, \ldots, \mathbf{V}_k)$, $\mathbf{V}$'s a basis for $\mathscr{V}$. Thus, the minimum distance from $\mathbf{Y}$ to $\mathscr{F}$ is

$$|\mathbf{Y} - \mathbf{P}_0| = \sqrt{|\mathbf{Y}|^2 - |\mathbf{P}_0|^2}.$$

14. Find an expression for a point on the line $\mathcal{L}$ (through the origin) closest to a given point $\mathbf{Y}$. Find a formula for the distance.
15. Given an orthogonal basis for its direction space, find an expression for a point on the plane $\mathscr{P}$ (through the origin) closest to a given point $\mathbf{Y}$. Find a formula for the distance.
16. In the first least-squares problem, use elementary calculus to prove that u_0 is the desired minimum.

5. ORTHOGONAL LINEAR TRANSFORMATIONS

A linear transformation $\mathbf{T}$ from $\mathbb{R}_n$ to $\mathbb{R}_n$ is **orthogonal** if for any $\mathbf{V}$, $\mathbf{W}$ in $\mathbb{R}_n$

$$\mathbf{T}(\mathbf{V}) \cdot \mathbf{T}(\mathbf{W}) = \mathbf{V} \cdot \mathbf{W}.$$

Thus, orthogonal mappings are those which preserve the dot product of vectors. In particular,

$$\mathbf{T}(\mathbf{V}) \cdot \mathbf{T}(\mathbf{V}) = \mathbf{V} \cdot \mathbf{V},$$

so $\mathbf{T}(\mathbf{V}) = \mathbf{0}$ iff $\mathbf{V} = \mathbf{0}$, by the "positivity" of $\mathbf{V} \cdot \mathbf{V}$. Since $\mathbb{R}_n$ is finite-dimensional, this means that an orthogonal linear transformation is an isomorphism of $\mathbb{R}_n$ onto $\mathbb{R}_n$. (See Section 4, Chapter 3.)

Since distances and angles are defined in terms of the dot product, it follows that

orthogonal mappings preserve distance: $|\mathbf{T}(\mathbf{V}) - \mathbf{T}(\mathbf{W})| = |\mathbf{V} - \mathbf{W}|$;

orthogonal mappings preserve angles: the angle between $\mathbf{V}$ and $\mathbf{W}$ is the same as the angle between $\mathbf{T}(\mathbf{V})$ and $\mathbf{T}(\mathbf{W})$;

orthogonal mappings preserve orthogonality.

Thus, orthogonal mappings preserve all the basic properties of euclidean space; they are quite important.

It is interesting to note that a linear transformation is orthogonal iff it merely preserves distance (Exercise 1). (In fact, the assumption of the preservation of distance is so strong that any mapping which preserves distance and maps **0** to **0** must be linear as well; this is 6.2.) Mappings which preserve distance are said to be **isometric**; another name for an orthogonal linear transformation is an **isometric isomorphism**.

On the other hand, there are angle-preserving mappings which are not orthogonal. For example, the similarity mapping $T(V) = 2V$, for all V in $\mathbb{R}_n$, which stretches vectors by a factor of 2, is such an angle-preserving transformation (Exercise 2).

It would be better if orthogonal linear transformations were called "orthonormal", as the following theorem shows:

5.1 A linear transformation T from $\mathbb{R}_n$ into $\mathbb{R}_n$ is orthogonal iff it maps orthonormal bases onto orthonormal bases. It is sufficient to check that the image of one orthonormal basis is orthonormal.

Proof. If T is orthogonal and $\{V_1, \ldots, V_n\}$ is an orthonormal basis, then

$$T(V_i) \cdot T(V_j) = V_i \cdot V_j = \delta_{ij}.$$

Thus, $\{T(V_1), \ldots, T(V_n)\}$ is an orthonormal basis for $\mathbb{R}_n$.

Conversely, suppose that $\{V_1, \ldots, V_n\}$ and $\{T(V_1), \ldots, T(V_n)\}$ are orthonormal bases for $\mathbb{R}_n$. If

$$X = x_1 V_1 + \cdots + x_n V_n \quad \text{and} \quad Y = y_1 V_1 + \cdots + y_n V_n,$$

then

$$X \cdot Y = x_1 y_1 + \cdots + x_n y_n,$$

by 2.4. Similarly, since

$$T(X) = x_1 T(V_1) + \cdots + x_n T(V_n)$$

and

$$T(Y) = y_1 T(V_1) + \cdots + y_n T(V_n),$$

we have that $T(X) \cdot T(Y) = x_1 y_1 + \cdots + x_n y_n = X \cdot Y$ and T is orthogonal. ∎

Example 1. In $\mathbb{R}_2$ consider the linear transformation $T(V) = BV$, where

$$B = \begin{pmatrix} 1 & 0 \\ 0 & -1 \end{pmatrix}.$$

This maps the basis

$$\left\{ e_1 = \begin{pmatrix} 1 \\ 0 \end{pmatrix}, \quad e_2 = \begin{pmatrix} 0 \\ 1 \end{pmatrix} \right\}$$

onto the orthonormal basis

$$\left\{ \begin{pmatrix} 1 \\ 0 \end{pmatrix}, \begin{pmatrix} 0 \\ -1 \end{pmatrix} \right\}$$

and is thus an orthogonal mapping. What does it mean geometrically? Since a linear transformation preserves scalar multiplication and parallelograms, it is sufficient to check its geometrical effect on a basis:

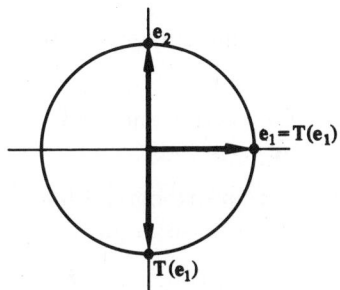

Fig. 18

Thus, **T** is just reflection about the x-axis, a process which ought to take orthonormal bases onto orthonormal bases. In a similar fashion, one can see that

$$\mathbf{T(V)} = \begin{pmatrix} 0 & 1 \\ 1 & 0 \end{pmatrix} \mathbf{V}$$

represents a reflection about the line $x = y$.

Example 2. Still in $\mathbb{R}_2$, let $\mathbf{T(V)} = \mathbf{BV}$ where

$$\mathbf{B} = \begin{pmatrix} \cos\theta & \sin\theta \\ \sin\theta & -\cos\theta \end{pmatrix}.$$

Since

$$\mathbf{T(e_1)} = \begin{pmatrix} \cos\theta \\ \sin\theta \end{pmatrix}, \quad \mathbf{T(e_2)} = \begin{pmatrix} \sin\theta \\ -\cos\theta \end{pmatrix}$$

is an orthonormal basis, **T** is an orthogonal mapping. What does it do? Again using the fact that linear transformations preserve scalar multiples and parallelograms, it is enough to check the effect of **T** on a basis. In this case it is convenient to choose the basis $\{\mathbf{e_1}, \mathbf{V}\}$,

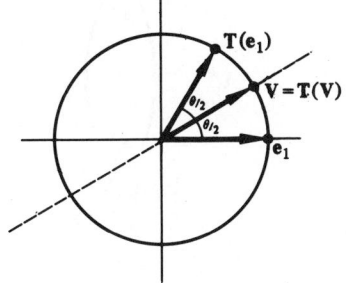

Fig. 19

where $V = (\cos \theta/2, \sin \theta/2)^*$. It is a matter of trivial trigonometry to check that $T(V) = V$. Moreover, $T(e_1)$ is e_1 reflected about V. Thus, T is simply reflection about a line through the origin making an angle of $\theta/2$ with the x-axis. Note that our previous examples are special cases.

We shall now derive some information about matrices of orthogonal transformations.

5.2 A is the matrix of an orthogonal transformation relative to an orthonormal basis iff $A^* = A^{-1}$.

Proof. 1. If A is the matrix of an orthogonal linear transformation, suppose for convenience that the orthonormal basis is $\{e_1, \ldots, e_n\}$, so

$$A = (T(e_1), \ldots, T(e_n)).$$

Hence,

$$A^* = \begin{pmatrix} T(e_1)^* \\ \vdots \\ T(e_n)^* \end{pmatrix}.$$

The ij-entry of A^*A is $A_i^*A_j$, where A_k is the kth column of A. But this is

$$T(e_i)^*T(e_j) = T(e_i) \cdot T(e_j) = e_i \cdot e_j = \delta_{ij}.$$

Thus,

$$A^*A = \begin{pmatrix} 1 & & 0 \\ & \ddots & \\ 0 & & 1 \end{pmatrix}$$

and $A^* = A^{-1}$.

2. Conversely, if $A^* = A^{-1}$ and if we write $A = (A_1, \ldots, A_n)$ in terms of its columns, then

$$A^* = \begin{pmatrix} A_1^* \\ \vdots \\ A_n^* \end{pmatrix},$$

so that $A_i \cdot A_j = \delta_{ij}$. Define T on $\mathbb{R}_n$ by $T(V) = AV$, V written as an n-tuple in terms of the usual basis $\{e_1, \ldots, e_n\}$ for $\mathbb{R}_n$. Then the matrix of T is A and

$$\{T(e_1) = A_1, \ldots, T(e_n) = A_n\}$$

is an orthonormal basis. Hence T is orthogonal. ∎

We restate for emphasis that $A^* = A^{-1}$ means that the columns of A are an orthonormal basis for $\mathbb{R}_n$. You should check this and the above theorem on the matrices in the previous examples.

We shall call a matrix for which $A^* = A^{-1}$, naturally enough, an **orthogonal matrix**. Note that this is an easy condition to test; just multiply A^*A. Furthermore, since $A^*A = I$ iff $AA^* = I$, it follows that if A is orthogonal then so is A^* ($= A^{-1}$).

Example 3. Let $T(V) = AV$ where

$$A = \begin{pmatrix} \cos\theta & -\sin\theta \\ \sin\theta & \cos\theta \end{pmatrix}.$$

Then

$$A^*A = \begin{pmatrix} \cos\theta & \sin\theta \\ -\sin\theta & \cos\theta \end{pmatrix} \begin{pmatrix} \cos\theta & -\sin\theta \\ \sin\theta & \cos\theta \end{pmatrix} = \begin{pmatrix} 1 & 0 \\ 0 & 1 \end{pmatrix}.$$

Hence, A is orthogonal. To check the geometrical meaning of $T(V) = AV$, note that $T(e_1) = (\cos\theta, \sin\theta)^*$ and $T(e_2) = (\cos(\theta + \pi/2), \sin(\theta + \pi/2))^*$. Thus, T is a counterclockwise rotation through an angle θ.

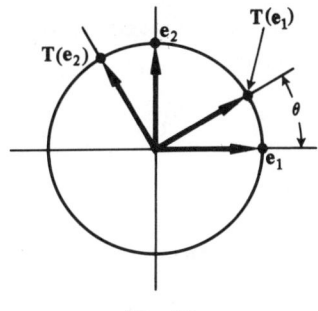

Fig. 20

5.3 If A is an orthogonal matrix, then

(i) $\det A = \pm 1$, (ii) the only possible real eigenvalues of A are ± 1.

Proof. (i) $\det A = \det A^* = \det A^{-1} = (\det A)^{-1}$. Thus

$$(\det A)(\det A^{-1}) = (\det A)^2 = 1.$$

Thus $\det A = \pm 1$.

ii) If $T(V) = AV$, we show the equivalent fact that the real eigenvalues of T are ± 1. For, if $T(V) = \lambda V$, then

$$|T(V)| = |\lambda V| = |\lambda|\,|V|.$$

But since T is orthogonal,

$$|T(V)| = |V|.$$

Hence $|\lambda| |V| = |V|$. Thus, if $V \neq 0$ (which must be true for an eigenvector),
$$|\lambda| = 1.$$
Thus, $\lambda = \pm 1$. ∎

The proof of (ii) may be amended to show that the complex eigenvalues of an orthogonal matrix are of modulus 1, $|\lambda| = 1$. Thus, they lie on the unit circle $|z| = 1$ in the complex plane. See the supplement to this section.

You should check this theorem on the matrices of our previous examples. When will **A** in Example 3 (= rotation through angle θ) have real eigenvalues? We shall now determine the form of all 2×2 orthogonal matrices:

5.4 Any orthogonal matrix in $\mathbb{R}_{2\times 2}$ has one of the following two forms:

$$\begin{pmatrix} \cos\theta & -\sin\theta \\ \sin\theta & \cos\theta \end{pmatrix} \quad \text{or} \quad \begin{pmatrix} \cos\theta & \sin\theta \\ \sin\theta & -\cos\theta \end{pmatrix}$$

counterclockwise rotation through an angle θ reflection in a line making an angle $\theta/2$ with the x-axis.

Proof. Let
$$\mathbf{A} = \begin{pmatrix} a & b \\ c & d \end{pmatrix}$$
be an orthogonal matrix. As you have proved many times before, the general form of the inverse of any 2×2 matrix is
$$\frac{1}{ad-bc}\begin{pmatrix} d & -b \\ -c & a \end{pmatrix} = \frac{1}{\det A}\begin{pmatrix} d & -b \\ -c & a \end{pmatrix}.$$
Since $\mathbf{A}^{-1} = \mathbf{A}^* = \begin{pmatrix} a & c \\ b & d \end{pmatrix}$ we have:

i) If $\det \mathbf{A} = 1$, $a = d$, $b = -c$, then
$$\mathbf{A} = \begin{pmatrix} a & -c \\ c & a \end{pmatrix}.$$
Since $\det \mathbf{A}$ also is $a^2 + c^2$, we get $a^2 + c^2 = 1$, and we can therefore find θ so that $a = \cos\theta$, $c = \sin\theta$. Thus we have
$$\mathbf{A} = \begin{pmatrix} \cos\theta & -\sin\theta \\ \sin\theta & \cos\theta \end{pmatrix}.$$

ii) If $\det \mathbf{A} = -1$, you show that we get the second form. ∎

Geometrically, this theorem should come as no surprise. For, if $\mathbf{T}(\mathbf{V}) = \mathbf{A}\mathbf{V}$, **A** a 2×2 orthogonal matrix, then, since $|\mathbf{T}(\mathbf{e}_1)| = 1$, $\mathbf{T}(\mathbf{e}_1)$ must be on the unit circle. Since $\mathbf{T}(\mathbf{e}_1) \cdot \mathbf{T}(\mathbf{e}_2) = 0$, $\mathbf{T}(\mathbf{e}_2)$ must be perpendicular to $\mathbf{T}(\mathbf{e}_1)$. Thus, having affixed $\mathbf{T}(\mathbf{e}_1)$, there are only two alternatives for $\mathbf{T}(\mathbf{e}_2)$. Either it is 90° clockwise from $\mathbf{T}(\mathbf{e}_1)$, and we have a rotation; or it is 90° counterclockwise and we have a reflection.

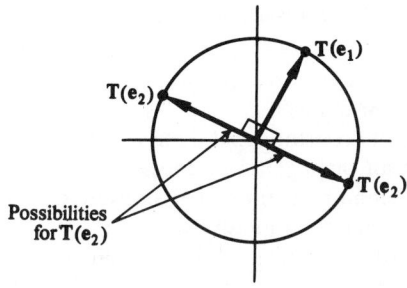

Fig. 21

In Section 7 we shall discover what can be said about the form of $n \times n$ orthogonal matrices. We discuss the 3×3 case in the next section.

EXERCISES

1. Show that if **T** is a linear transformation in $\mathbb{R}_n$, and $|T(V)| = |V|$ for all **V** in $\mathbb{R}_n$, then **T** is orthogonal. [*Hint:* Write out $|V - W| = |T(V) - T(W)|$ and compare to $|V|$ and $|W|$ in terms of dot products.]

2. Show that the mapping $T(V) = 2V$ preserves angles but not distances.

3. Verify that the following matrices are orthogonal. Describe the linear transformation $T(V) = AV$ geometrically whenever possible.

   a) $\begin{pmatrix} \frac{3}{5} & -\frac{4}{5} \\ \frac{4}{5} & \frac{3}{5} \end{pmatrix}$
   b) $\begin{pmatrix} \frac{3}{5} & \frac{4}{5} \\ \frac{4}{5} & -\frac{3}{5} \end{pmatrix}$

   c) $\begin{pmatrix} -1 & 0 & 0 \\ 0 & \frac{3}{5} & -\frac{4}{5} \\ 0 & \frac{4}{5} & \frac{3}{5} \end{pmatrix}$
   d) $\begin{pmatrix} 1 & 0 & 0 \\ 0 & \frac{3}{5} & \frac{4}{5} \\ 0 & \frac{4}{5} & -\frac{3}{5} \end{pmatrix}$

   e) $\begin{pmatrix} 1/\sqrt{2} & \frac{2}{3} & 1/3\sqrt{2} \\ 0 & \frac{1}{3} & -4/3\sqrt{2} \\ 1/\sqrt{2} & -\frac{2}{3} & -1/3\sqrt{2} \end{pmatrix}$
   f) $\begin{pmatrix} \frac{29}{45} & -\frac{4}{9} & \frac{28}{45} \\ \frac{28}{45} & \frac{7}{9} & -\frac{4}{45} \\ -\frac{4}{9} & \frac{4}{9} & \frac{7}{9} \end{pmatrix}$

4. Show that the following matrices are of the form indicated in 5.4:

   $\begin{pmatrix} 1 & 0 \\ 0 & 1 \end{pmatrix}, \quad \begin{pmatrix} 1 & 0 \\ 0 & -1 \end{pmatrix}, \quad \begin{pmatrix} -1 & 0 \\ 0 & -1 \end{pmatrix}.$

   What is the geometrical meaning of the corresponding transformations?

5. Give a geometrical proof that reflection about any line $ax + by = 0$ in $\mathbb{R}_2$ is an orthogonal mapping.

6. Show that clockwise rotation about the origin is an orthogonal mapping and find its matrix.

7. Prove that the second matrix in 5.4 is actually reflection about the x-axis followed by rotation through an angle θ.
8. a) Prove that the product of any two orthogonal matrices is an orthogonal matrix.
   b) Prove that (a) is false for sums.
   c) Prove that the inverse of an orthogonal matrix is orthogonal.
   d) Prove that the set of all orthogonal matrices (for any fixed n) forms a group under multiplication (definition at the end of the exercises of Section 1, Chapter 0).
   e) Prove that the group is not commutative if $n > 1$.
   f) Prove that all orthogonal matrices with determinant $+1$ form a group (the **special orthogonal group**).
9. Let R_θ be counterclockwise rotation about the origin through an angle θ in $\mathbb{R}_2$.
   a) Prove by matrix multiplication that
   $$R_\theta R_\varphi = R_{\theta+\varphi}.$$
   b) What is the matrix of $(R_\theta)^{-1}$?
   c) When will R_θ have any (real) eigenvectors? First think this out geometrically; then prove what you discover.
10. Find the matrix which performs rotation through an angle θ about the z-axis in $\mathbb{R}_3$. Verify that it is orthogonal.

Supplement 1 to Section 5. Unitary Mappings

The complex analog of orthogonal mapping is also of very great importance. To fully appreciate this section, you should first read Supplement 3 to Section 2 on complex inner-product spaces.

A linear transformation **T** on $\mathbb{C}_n$ is called **unitary** if it preserves the complex dot product:

$$\langle \mathbf{V}, \mathbf{W} \rangle = \bar{\mathbf{V}} \cdot \mathbf{W} = \overline{\mathbf{T}(\mathbf{V})} \cdot \mathbf{T}(\mathbf{W}) = \langle \mathbf{T}(\mathbf{V}), \mathbf{T}(\mathbf{W}) \rangle$$

for **V**, **W** in $\mathbb{C}_n$, where $\bar{\mathbf{V}} \cdot \mathbf{W} = \bar{v}_1 w_1 + \cdots + \bar{v}_n w_n$ if

$$\mathbf{V} = (v_1, \ldots, v_n) \quad \text{and} \quad \mathbf{W} = (w_1, \ldots, w_n).$$

5.1 goes through for unitary transformations exactly as before. 5.2 is true if transpose is replaced by conjugate transpose: for unitary matrices $\bar{\mathbf{A}}^* = \mathbf{A}^{-1}$. The valid conclusions for 5.3 are: (i) $|\det A| = 1$ and (ii) $|\lambda| = 1$ for any eigenvalue, where $|\ |$ denotes the modulus of a complex number. 5.4 comes out even stronger: given any unitary **T** there exists an orthonormal basis for $\mathbb{C}_n$ relative to which the matrix of **T** is diagonal.

It would be very healthy for the reader to supply proofs for these theorems. The only one which might cause any difficulty is the last, in which case you should check the Jordan Form section of Chapter 6 and test the possible Jordan Form matrices that a unitary transformation could have (cf. Section 7 of this chapter).

Supplement 2 to Section 5. Orthogonal Matrices Expressed in Terms of Parameters

We show how most orthogonal matrices can be expressed in terms of *skew-symmetric* matrices, i.e., ($B^* = -B$) and then use these matrices to introduce parameters. We shall give skeletons of proofs; you should provide the flesh. First, two preliminary results:

Lemma 1. If B is skew-symmetric, then $I + B$ is invertible.

Proof. Suppose that there is a $V \neq 0$ for which $BV = -V$. Then $B^*BV = -B^2V = -V$. Hence

$$|V|^2 = |BV|^2 = BV \cdot BV = (BV)^*BV = V^*B^*BV = -|V|^2,$$

and we have a positive number equal to a negative. Hence B has no eigenvalue -1, and $I + B$ is invertible. ∎

Lemma 2. If C is any matrix for which $I + C$ is invertible, then $I - C$ commutes with $(I + C)^{-1}$,

$$(I - C)(I + C)^{-1} = (I + C)^{-1}(I - C).$$

Proof. By the Cayley-Hamilton Theorem, $(I + C)^{-1}$ is a polynomial in $I + C$, and hence a polynomial in C; but any two polynomials in C commute. ∎

We can now prove our main result relating skew symmetric and orthogonal matrices:

5.5 Let A be an orthogonal matrix which does not have -1 as an eigenvalue. Then there is a skew-symmetric matrix B for which

$$A = (I - B)(I + B)^{-1}.$$

Conversely, if A is of this form where B is skew-symmetric, then A is orthogonal (and does not have -1 as an eigenvalue).

Proof. 1. If A does not have -1 as an eigenvalue, then $I + A$ is invertible. Define $B = (I - A)(I + A)^{-1}$. Since $A^* = A^{-1}$, and

$$(I + A^{-1})^{-1} = (A^{-1}(A + I))^{-1} = (A + I)^{-1}A,$$

it is easy to see that $B^* = -B$. Moreover, one can solve back from the definition of B to get $A = (I - B)(I + B)^{-1}$, using both lemmas.

2. Suppose that B is skew symmetric. Define $A = (I - B)(I + B)^{-1}$. Calculate A^*A to see that A is orthogonal. Write $A = -I + 2(I + B)^{-1}$ to see that $I + A$ is invertible (i.e., -1 is not an eigenvalue of A). ∎

Since $B = (I + A)^{-1}(I - A)$ it is straightforward to show that

$$AV = V \quad \text{iff} \quad BV = 0.$$

Now let us use our results to describe a large class of orthogonal matrices (namely, those without eigenvalue -1) in terms of parameters. These parameters will simply be the entries of the corresponding skew-symmetric matrix **B**. For convenience we shall consider only the 3×3 case. As an exercise you should first try $n = 2$ to see that our results agree with that of the section, and then look at $n = 4$ to flex your muscles.

For $n = 3$ we can write any skew-symmetric matrix as

$$\mathbf{B} = \begin{pmatrix} 0 & c & -b \\ -c & 0 & a \\ b & -a & 0 \end{pmatrix}.$$

Then $\mathbf{V} = (a, b, c)^*$ will be annihilated by **B** so it will be an invariant vector for **A**. (This, together with the fact that det **A** is $+1$, will show in the following section that **A** must be a rotation about **V**.) It is easy to write the matrix **A** in terms of the parameters a, b, c:

$$\mathbf{I} + \mathbf{B} = \begin{pmatrix} 1 & c & -b \\ -c & 1 & a \\ b & -a & 1 \end{pmatrix},$$

$$\det (\mathbf{I} + \mathbf{B}) = 1 + a^2 + b^2 + c^2,$$

(reaffirming that $\mathbf{I} + \mathbf{B}$ is always invertible);

$$(\mathbf{I} + \mathbf{B})^{-1} = \frac{1}{1 + a^2 + b^2 + c^2} \begin{pmatrix} 1 + a^2 & ab - c & ac + b \\ ab + c & 1 + b^2 & bc - a \\ ac - b & bc + a & 1 + c^2 \end{pmatrix}.$$

Hence

$$\mathbf{A} = \frac{1}{1 + a^2 + b^2 + c^2} \times \begin{pmatrix} 1 + a^2 - b^2 - c^2 & 2(ab - c) & 2(ac + b) \\ 2(ab + c) & 1 + b^2 - a^2 - c^2 & 2(bc - a) \\ 2(ac - b) & 2(bc + a) & 1 + c^2 - a^2 - b^2 \end{pmatrix}.$$

From the material at the end of the next section, we see that the cosine of the angle of rotation of **A** can be read off from the trace and is

$$\cos \theta = \frac{1 - a^2 - b^2 - c^2}{1 + a^2 + b^2 + c^2}.$$

6.* RIGID MOTIONS, ESPECIALLY IN 2- AND 3-SPACE

In this section we shall study these functions **F** (linear or not) from $\mathbb{R}_n$ onto $\mathbb{R}_n$ which preserve distance: For any **P**, **Q** in $\mathbb{R}_n$,

$$|\mathbf{F}(\mathbf{P}) - \mathbf{F}(\mathbf{Q})| = |\mathbf{P} - \mathbf{Q}|.$$

Such distance-preserving mappings are also called **rigid motions** (and also **isometries**). For example, we noted in the last section that an orthogonal linear transformation (i.e., a linear transformation which preserves dot products) also preserves distances. An example of a rigid motion which is not linear is the **translation F** in $\mathbb{R}_n$ defined by

$$F(P) = P + P_0, \quad P_0 \text{ a fixed point in } \mathbb{R}_n.$$

This will be linear only if $P_0 = 0$ (Exercise 2), but it is always a rigid motion:

$$|F(P) - F(Q)| = |(P + P_0) - (Q + P_0)| = |P - Q|.$$

It turns out that between orthogonal linear transformations and translations, we now have all the basic ingredients of rigid motions (see 6.2).

If a function fixes the origin $(F(0) = 0)$, then the assumption that it preserves distances is such a strong demand that it will automatically force the function to preserve dot products (and hence angles) as well:

6.1 If **F** is a rigid motion of $\mathbb{R}_n$ and if, in addition, $F(0) = 0$, then **F** also preserves dot products. That is, if for all **P** and **Q** in $\mathbb{R}_n$,

$$|F(P) - F(Q)| = |P - Q|,$$

and if $F(0) = 0$, then $F(P) \cdot F(Q) = P \cdot Q$.

Proof. For any points **P** and **Q**

$$|P - Q^2| = (P - Q) \cdot (P - Q) = P \cdot P - 2P \cdot Q + Q \cdot Q \qquad (1)$$

and also

$$|F(P) - F(Q)|^2 = F(P) \cdot F(P) - 2F(P) \cdot F(Q) + F(Q) \cdot F(Q). \qquad (2)$$

If **F** fixes **0** then

$$F(P) \cdot F(P) = (F(P) - F(0)) \cdot (F(P) - F(0))$$
$$= |F(P) - F(0)|^2 = |P - 0|^2 = P \cdot P.$$

Similarly, $F(Q) \cdot F(Q) = Q \cdot Q$. If $(1) = (2)$, we obtain $-2F(P) \cdot F(Q) = -2P \cdot Q$, whence the theorem may be wrested. ∎

It is an immediate corollary that any linear rigid motion is an orthogonal linear transformation.

However, we have even bigger things in store:

6.2 Any rigid motion of $\mathbb{R}_n$ is an orthogonal mapping followed by a translation: If **F** is a rigid motion of $\mathbb{R}_n$, then for any **P** in $\mathbb{R}_n$,

$$F(P) = P_0 + T(P),$$

where **T** is an orthogonal linear transformation.

The translating vector is $P_0 = F(0)$, so in particular a rigid motion is a linear transformation iff it maps 0 onto 0.

Proof. Define a mapping T on $\mathbb{R}_n$ by

$$T(P) = F(P) - F(0).$$

We shall show first that T is a linear transformation. Note that $T(0) = 0$, so we're off to a good start. Moreover, it is easy to see that T is a rigid motion (Exercise 3). We next show that if $\{V_1, \ldots, V_n\}$ is a basis for $\mathbb{R}_n$, then so is $\{T(V_1), \ldots, T(V_n)\}$. For, suppose that

$$cT(V_1) + \cdots + c_n T(V_n) = 0 \qquad \text{for scalars } c_1, \ldots, c_n \text{ in } \mathbb{R}.$$

Then, for any X in $\mathbb{R}_n$,

$$0 = T(X) \cdot 0 = T(X) \cdot (c_1 T(V_1) + \cdots + c_n T(V_n))$$
$$= c_1 (T(X) \cdot T(V_1)) + \cdots + c_n (T(X) \cdot T(V_n))$$
$$\stackrel{6.1}{=} c_1 (X \cdot V_1) + \cdots + c_n (X \cdot V_n) = X \cdot (c_1 V_1 + \cdots + c_n V_n).$$

Hence the vector $c_1 V_1 + \cdots + c_n V_n$ is orthogonal to any X in $\mathbb{R}_n$. Hence it is 0 (Exercise 3). Since the V's are a basis, the c's are 0. Since the $T(V)$'s are linearly independent and there are n of them, they are a basis for $\mathbb{R}_n$.

We can now show that T is additive. For any X and Y in $\mathbb{R}_n$ and any $i = 1, \ldots, n$,

$$T(V_i) \cdot [T(X + Y) - T(X) - T(Y)]$$
$$= T(V_i) \cdot T(X + Y) - T(V_i) \cdot T(X) - T(V_i) \cdot T(Y)$$
$$= V_i \cdot (X + Y) - V_i \cdot X - V_i \cdot Y = 0.$$

Thus, $T(X + Y) - T(X) - T(Y)$ is a vector which is perpendicular to every element of a basis for $\mathbb{R}_n$. Hence,

$$T(X + Y) - T(X) - T(Y) = 0$$

(Exercise 3b), which is additivity. Similarly, one can show that T is homogeneous (Exercise 3c).

Thus, T is a linear transformation and also a rigid motion. By 6.1, it is thus an orthogonal mapping. Moreover, F is T followed by the translation

$$P \rightarrow P + F(0),$$

so the theorem is proved. ∎

We now take a closer look at the rigid motions of the euclidean plane. We have observed that any *linear* rigid motion is an orthogonal mapping. By 5.4 we know that relative to an orthonormal basis the matrix of any orthogonal mapping of $\mathbb{R}_2$ has one of two forms:

$$A = \begin{pmatrix} \cos\theta & -\sin\theta \\ \sin\theta & \cos\theta \end{pmatrix} \quad \text{or} \quad B = \begin{pmatrix} \cos\theta & \sin\theta \\ \sin\theta & -\cos\theta \end{pmatrix}. \tag{3}$$

A: Rotation about the origin through an angle θ

B: Reflection about a line through the origin and making an angle of $\theta/2$ with the x-axis.

Using these matrices, from 6.2 we immediately obtain that any rigid motion is either a reflection or a rotation followed by a translation. It turns out, however, that the net effect can be very nicely described.

We shall use our linear knowledge to discuss such notions as rotations about points other than the origin, and reflections about lines other than those through the origin. It makes geometrical sense that one can accomplish rotation about a point Q by first translating from Q to 0 (i.e., adding $-Q$), rotating about 0, and then translating back from 0 to Q (i.e., adding $+Q$).† We accordingly define (counterclockwise) **rotation about Q through an angle** θ to be the mapping

$$F(P) = A(P - Q) + Q, \quad A \text{ as in (3)}.$$

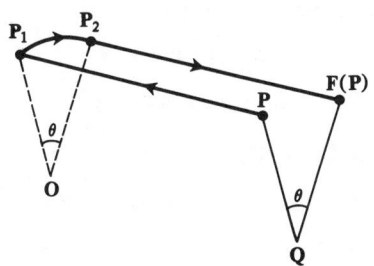

Fig. 22

It is easy to see that this mapping agrees with our previous discussion when $Q = 0$, leaves Q fixed $(F(Q) = Q)$ and only Q, and satisfies any other rotational desiderata you might wish to set up (Exercise 4). Note also that we can write this rotation as

$$F(P) = AP + P_0,$$

where $P_0 = Q - AQ$.

In a similar fashion, we could accomplish reflection about a line

$$\mathcal{L} = Q + \mathcal{V},$$

by translating by $-Q$, reflecting about $\mathcal{V}$, and translating back by $+Q$: We therefore define **reflection about** $\mathcal{L} = Q + \mathcal{V}$ to be the transformation

$$F(P) = B(P - Q) + Q,$$

---

† Note that we are simply moving the origin from 0 to Q.

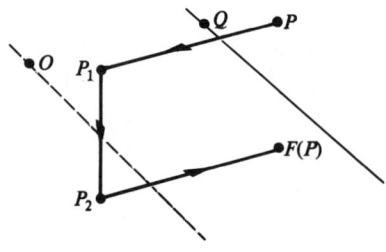

Fig. 23

where **B** (as in (3)) is reflection about the line $\mathcal{U}$ through **0** and making an angle of $\theta/2$ with the positive x-axis. You should verify such items as:
F fixes the points of $\mathcal{L}$,
$$\mathbf{F(P) = P} \qquad \text{for } \mathbf{P} \text{ on } \mathcal{L};$$
when applied twice, **F** gets you back home ($\mathbf{F^2(P) = P}$);
F agrees with our previous use of reflection if $\mathcal{L}$ passes through **0**, etc. (Exercise 4). Note that we can write **F** as
$$\mathbf{F(P) = BP + P_0},$$
where $\mathbf{P_0 = Q - BQ}$. Moreover, since $\mathbf{B^2 = I}$,
$$\mathbf{BP_0 = B(Q - BQ) = BQ - B^2Q = BQ - Q}$$
$$= -\mathbf{(Q - BQ) = -P_0}.$$

It is necessary to consider one more creature. We define **glide reflection along the line** $\mathcal{L} = \mathbf{Q} + \mathcal{U}$ **in an amount** $\mathbf{V_0}$ to be reflection in $\mathcal{L}$ followed by translation by $\mathbf{V_0}$, where $\mathbf{V_0}$ is in $\mathcal{U}$; that is,
$$\mathbf{G(P) = F(P) + V_0}, \ \mathbf{F} \text{ the above reflection, } \mathbf{V_0} \text{ in } \mathcal{U}.$$
Thus,
$$\mathbf{G(P) = BP + (P_0 + V_0)},$$
where $\mathbf{BP_0 = -P_0}$ and $\mathbf{BV_0 = V_0}$ (since $\mathbf{V_0}$ is in $\mathcal{U}$).

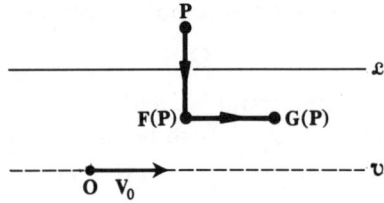

Fig. 24

We can now state our results in $\mathbb{R}_2$:

6.3 Classification of rigid motions in the plane. Any rigid motion F in $\mathbb{R}_2$ is one of the following four types (with A and B as in (3)):
  i) $F(P) = P + P_0$ (translation by a vector $P_0 = F(0)$);
  ii) $F(P) = AP + P_0$, A having $\theta \neq 2k\pi$
  (rotation through an angle θ about a point Q, where $P_0 = Q - AQ$);
  iii) $F(P) = BP + P_0$, with $BP_0 = -P_0$ (reflection about the line $\mathcal{L} = \frac{1}{2}P_0 + \mathcal{V}$; if $P_0 \neq 0$, then $\mathcal{V}$ is the set of all vectors perpendicular to P_0);
  iv) $F(P) = BP + P_0$, with $BP_0 \neq -P_0$ (glide reflection along the line $\mathcal{L} = \frac{1}{4}(P_0 - BP_0) + \mathcal{V}$, $\mathcal{V} = \text{span}\{P_0 + BP_0\}$, in an amount $\frac{1}{2}(P_0 + BP_0)$).

Proof. We know that any rigid motion is either of the form

$$F(P) = AP + P_0 \quad \text{or} \quad F(P) = BP + P_0,$$

with A, B as in (3). In terms of our definition, we must show the following (the actual proofs we leave for you in Exercise 4).

I. If $F(P) = AP + P_0$ and $A \neq I$, then this is rotation about a point Q. That is, there is a unique Q satisfying

$$P_0 = Q - AQ,$$

so that $F(P) = A(P - Q) + Q$, our definition of rotation about Q.

II. If $F(P) = BP + P_0$, and if $BP_0 = -P_0$, then this is reflection about $\mathcal{L} = Q + \mathcal{V}$, where

$$P_0 = Q - BQ.$$

That is, this last equation has a solution Q (it has a whole line of them in fact) so that the transformation is of the form

$$F(P) = B(P - Q) + Q,$$

which is a reflection in a line through Q. Moreover, it must be shown that this line of reflection is $\frac{1}{2}P_0 + \mathcal{V}$. $\mathcal{V} = P_0^\perp$, if $P_0 \neq 0$.

III. Finally, if $F(P) = BP + P_0$ with $BP_0 \neq -P_0$, then note that we can write

$$P_0 = \underbrace{\tfrac{1}{2}(P_0 - BP_0)}_{P_1} + \underbrace{\tfrac{1}{2}(P_0 + BP_0)}_{V_1} = P_1 + V_1.$$

Show that $BP_1 = -P_1$ and $PV_1 = V_1$, so that

$$F_1(P) = BP + P_1$$

is a reflection and $F(P) = F_1(P) + V_1$ is a glide reflection. Finally, show it is the glide reflection promised in the statement of the theorem. ∎

Since rigid motions preserve distances, they ought to preserve areas as well. You will recall that the area of the parallelogram

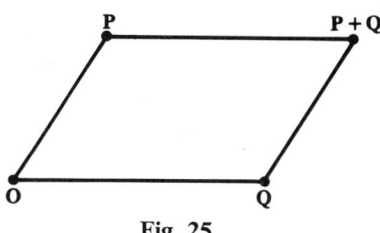

Fig. 25

is $|\det(P, Q)|$. (More generally, any parallelogram should be a translation of a parallelogram with a vertex at the origin; we leave it for you to show that the area is preserved under the translation $X \to P_0 + X$.)

In general, if we apply the rigid motion F to the above parallelogram, we obtain another parallelogram.† From 6.3,

$$F(X) = CX + F(0),$$

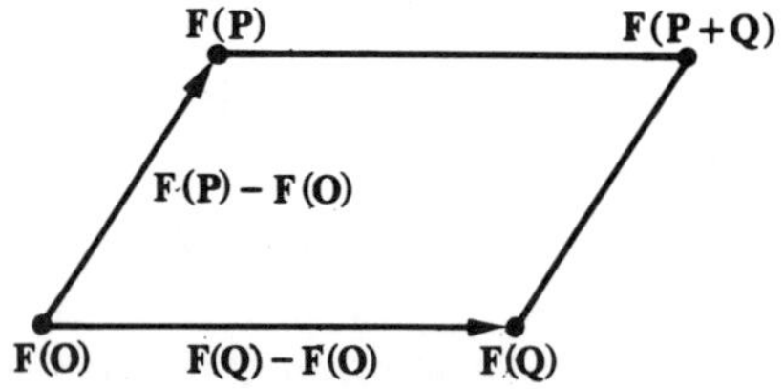

Fig. 26

where C is an orthogonal matrix. The area of this image parallelogram is therefore

$$|\det(F(P) - F(0), F(Q) - F(0))| = |\det(CP, CQ)|$$
$$= |\det C|\,|\det(P, Q)| = |\det(P, Q)|,$$

---

† In our rather informal treatment of parallelotopes in general, we have heretofore only considered those with one vertex at the origin. An appropriate definition of the general species would be a translate of this particular creature.

since det $C = \pm 1$. Thus, **F** preserves the area of parallelograms†.

It is worthwhile distinguishing between those rigid motions $F(X) = CP + F(0)$ for which det $C = +1$ (i.e., the linear part of **F** is the rotation with matrix **A**) and those for which det $C = -1$ (i.e., the linear part of **F** is the reflection with matrix **B**). We call the former **orientation preserving** and the latter **orientation reversing**. Intuitively speaking, a rigid motion is orientation preserving if the order of the vertices of the image of a parallelogram is the same as the order in the original parallelogram, when traversed counterclockwise; a rigid motion is orientation reversing if this order is reversed‡:

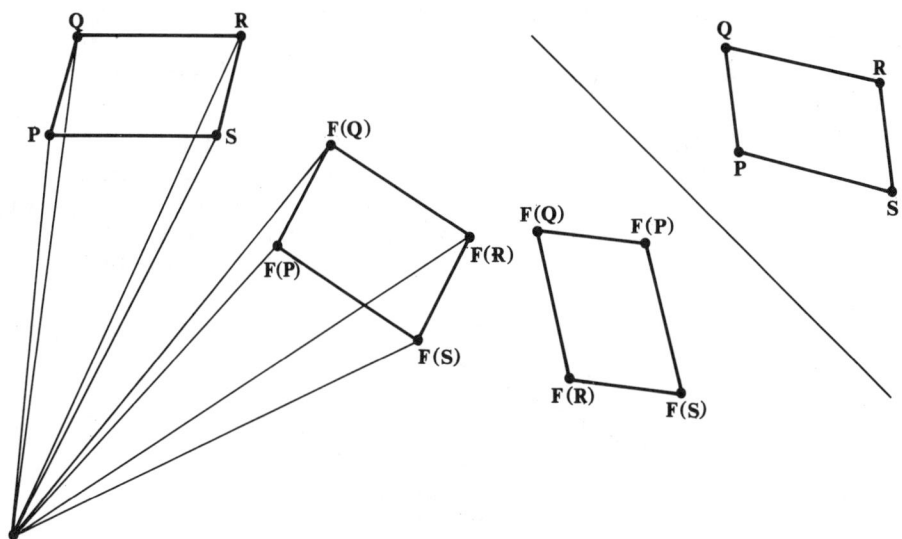

Fig. 27

We summarize all this information, along with useful facts about the fixed points of the rigid motion, in the accompanying table (p. 396).

For our final foray into the plane, let us note (informally) that translation by a vector P_0 can be realized by two reflections in lines perpendicular to P_0, and

---

† Recall that an *affine transformation* is one of the form $F(P) = CP + P_0$, where **C** is any $n \times n$ matrix. Our calculation actually shows that
  the area of the image of a parallelogram under this affine transformation is $|(\det C)|$ times the area of the original parallelogram.
‡ To be precise it would be necessary to define pieces of this intuitive definition such as "counterclockwise" in terms of the determinant of an appropriate parallelogram. This is another example of a very visible concept which takes some mild sophistication to make precise.

Rigid Motion in $\mathbb{R}_2$

	Transformation	Fixed Points	Matrix Form (matrices as in (3))	Remarks
Orientation preserved (determinant of linear part = 1)	Translation	None	$F(P) = P + P_0$.	
	Rotation	One point	$F(P) = AP + P_0$, $A \neq I$.	To find point of rotation, solve $F(Q) = Q$ for Q. To find angle of rotation, check A.
Orientation reversed (determinant of linear part = −1)	Reflection	A line	$F(P) = BP + P_0$, $BP_0 = -P_0$	Line is $\frac{1}{2}P_0 + \mathcal{V}$; $\mathcal{V} = \{V \mid P_0 \cdot V = 0\}$ if $P_0 \neq 0$.
	Glide reflection	None	$F(P) = BP + P_0$, $BP_0 \neq -P_0$.	Line is $\frac{1}{4}(P_0 - BP_0) + \mathcal{V}$; $\mathcal{V} = \text{span}\{P_0 + BP_0\}$; amount of glide is $\frac{1}{2}(P_0 + BP_0)$.

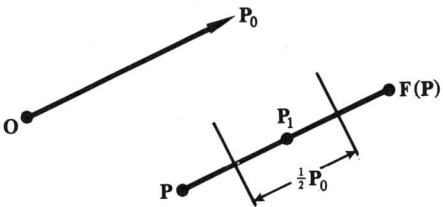

Fig. 28

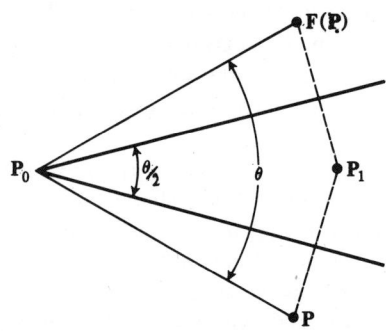

Fig. 29

spaced a distance of $\frac{1}{2}|\mathbf{P}_0|$ apart (Fig. 28). Moreover, any rotation through an angle θ at a point $\mathbf{P}_0$ can be expressed by two reflections in lines through $\mathbf{P}_0$ making an angle of $\theta/2$ with each other. Consequently, any glide reflection can be realized as the product of three reflections.

Rigid Motions in $\mathbb{R}_3$

We now turn to rigid motions in 3-space. We still have that

$$F(P) = T(P) + F(0),$$

where **T** is an orthogonal linear transformation, so we again look to classifying the linear rigid motions **T**. This we do in terms of the dimension of the subspace $\mathcal{V}$ of vectors fixed by **T**,

$$\mathcal{V} = \{\mathbf{V} \text{ in } \mathbb{R}_3 \,|\, T(\mathbf{V}) = \mathbf{V}\}.$$

CASE I. Dimension $\mathcal{V} = 3$. Then **T** is the identity linear transformation.

CASE II. Dimension $\mathcal{V} = 2$. Let $\{\mathbf{V}_1, \mathbf{V}_2\}$ be an orthonormal basis for $\mathcal{V}$ and extend it to an orthonormal basis $\{\mathbf{V}_1, \mathbf{V}_2, \mathbf{W}\}$ for $\mathbb{R}_3$ (thus, $\mathcal{V}^\perp$ is spanned by **W**). Then, since **W** is perpendicular to $\mathbf{V}_1$ and $\mathbf{V}_2$, $T(\mathbf{W})$ will be perpendicular to $T(\mathbf{V}_1) = \mathbf{V}_1$ and to $T(\mathbf{V}_2) = \mathbf{V}_2$. Hence $T(\mathbf{W})$ is in $\mathcal{V}^\perp$. Hence $T(\mathbf{W}) = k\mathbf{W}$. Since **T** preserves length, $k = \pm 1$. If $k = 1$, we would be back in Case I, so $k = -1$ and $T(\mathbf{W}) = -\mathbf{W}$. Thus, **T** is reflection in a plane. The matrix of **T** relative to the orthonormal basis is

$$\mathbf{B} = \begin{pmatrix} 1 & 0 & 0 \\ 0 & 1 & 0 \\ 0 & 0 & -1 \end{pmatrix}.$$

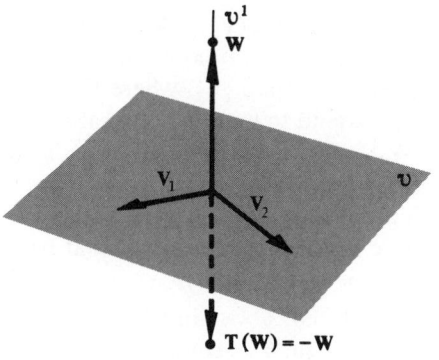

Fig. 30

CASE III. Dimension $\mathcal{U} = 1$. Let **V** span $\mathcal{U}$ with $|\mathbf{V}|$ and extend to an orthonormal basis $\{\mathbf{V}, \mathbf{W}_1, \mathbf{W}_2\}$ for $\mathbb{R}_3$, where $\mathcal{U}^\perp = \text{span } \{\mathbf{W}_1, \mathbf{W}_2\}$. Then **T** maps $\mathcal{U}^\perp$ onto $\mathcal{U}^\perp$ (why?) and, restricted to $\mathcal{U}^\perp$, **T** is an orthogonal mapping of this plane. From our work in $\mathbb{R}_2$ we know the possibilities here: it is either a rotation about **0** or a reflection in a line through **0**. However, if it were a reflection in such a line $\mathcal{L}$ (spanned by $\mathbf{V}'$, say), then $\mathcal{U}$ would contain $\mathbf{V}'$ (which is perpendicular to **V**), so the dimension of $\mathcal{U}$ would be greater than one and we would be back in a previous case. Hence, **T** acts as a rotation in the plane $\mathcal{U}^\perp$. Relative to the above orthonormal basis, the matrix of **T** has the form

$$\mathbf{A} = \begin{pmatrix} 1 & 0 & 0 \\ 0 & \cos\theta & -\sin\theta \\ 0 & \sin\theta & \cos\theta \end{pmatrix}, \quad \cos\theta \neq 1.$$

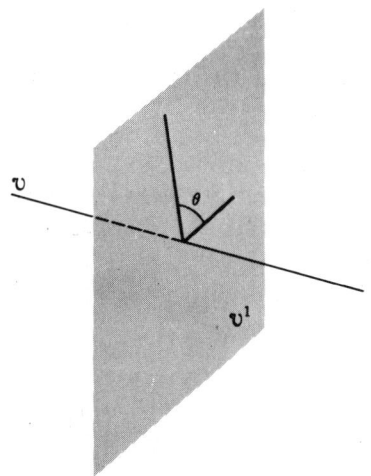

Fig. 31

CASE IV. Dimension $\mathcal{U} = 0$. Here $\mathbf{T}(\mathbf{V}) \neq \mathbf{V}$ for all $\mathbf{V} \neq \mathbf{0}$. Since the characteristic polynomial of **T** is of degree 3, we know it has a real eigenvalue $\lambda = \pm 1$ and since $\lambda \neq 1$, $\lambda = -1$. Let $\mathcal{W} = \text{span } \{\mathbf{V}\}$, where **V** is an eigenvector corresponding to this eigenvalue, and extend to an orthonormal basis $\{\mathbf{V}, \mathbf{W}_1, \mathbf{W}_2\}$ for $\mathbb{R}_3$, so that $\mathcal{W}^\perp = \text{span } \{\mathbf{W}_1, \mathbf{W}_2\}$. As in the previous case, **T** cannot be a reflection on $\mathcal{W}^\perp$ since we would then have dimension $\mathcal{U} > 0$. Hence, **T** is a rotation in the plane $\mathcal{W}^\perp$ together with a reflection in this plane (such a creature is often called a **rotary reflection**). Relative to the stated orthonormal basis the matrix of **T** is

$$\mathbf{C} = \begin{pmatrix} -1 & 0 & 0 \\ 0 & \cos\theta & -\sin\theta \\ 0 & \sin\theta & \cos\theta \end{pmatrix}, \quad \cos\theta \neq 1.$$

(Note that if $\cos\theta = 1$, this would be similar to the matrix **B** in Case II.)

6 / Rigid motions, especially in 2- and 3-space

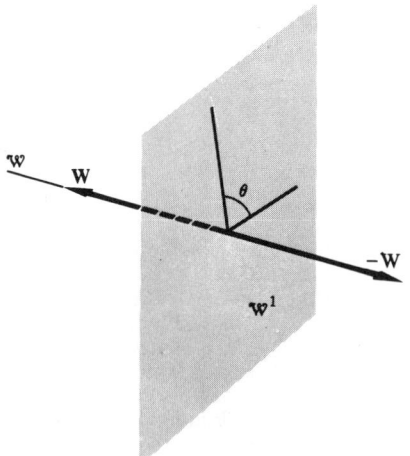

Fig. 32

As for $\mathbb{R}_2$, we can and shall define a rigid motion **F** in $\mathbb{R}_3$ to be **orientation preserving** if the determinant of its linear part is $+1$ and **orientation reversing** if the determinant of its linear part is -1. Intuitively speaking, **F** is orientation preserving provided that when you curl your fingers in the direction from e_1 to e_2, and your thumb points toward e_3, then it is impossible to change hands after applying **F**.

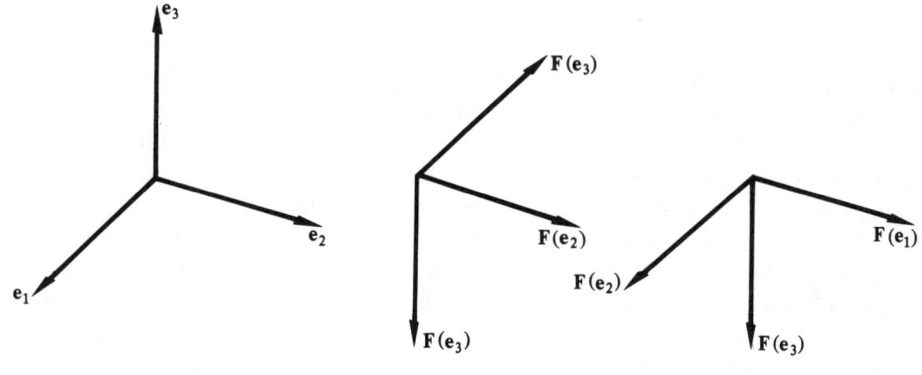

Fig. 33

We know that a rigid motion is of the form $F(P) = MP + P_0$, **M** similar to a matrix of type **A**, **B**, or **C** as above. By analyzing the various possibilities (Exercise 5), one obtains the classification in the accompanying table p. 400.

Rigid Motion in $\mathbb{R}_3$

	Transformation	Fixed points	Description	Matrix form (matrices only similar *to above*)	Dimension of space of vectors fixed by the linear part
Orientation preserved	Translation	None		$F(P) = P + P_0$	3
	Plane rotation	Line	Rotation about axis $\mathcal{L}$ = {solutions to $F(Q) = Q$}	$F(P) = AP + P_0$	1
	Screw displacement	None	Rotation about axis followed by translation along the axis	$F(P) = AP + P_0$	1
Orientation reversed	Reflection in a plane	Plane	Reflection in the plane of solutions to $F(Q) = Q$.	$F(P) = BP + P_0$ $BP_0 = -P_0$.	2
	Glide reflection	None	Reflection in a plane followed by translation along a vector in the plane	$F(P) = BP + P_0$ $BP_0 \neq -P_0$	2
	Rotary reflection	One point	Reflection in a plane followed by rotation about line perpendicular to the plane	$F(P) = CP + P_0$.	0

In order to classify a given rigid motion $F(P) = MP + P_0$ in $\mathbb{R}_3$, it is only necessary to check det M to see if F is orientation preserving or reversing, and then check the fixed points of F (i.e., see if $F(P) = P$ has no solutions, a unique solution, a line of solutions, or a plane of solutions). The above table then immediately specifies F (since it is hard to confuse screw displacements and translations). It is then possible to make a finer description of a rigid motion in terms of: angle and axis of rotation for a rotation; axis of rotation and amount of displacement for a screw displacement; plane of reflection and amount of translation for a glide reflection; and plane of reflection and angle of rotation for a rotary reflection.

As opposed to the matrices of the linear rigid motions in $\mathbb{R}_2$ which are exactly as stated in (3), a specific basis must be chosen in order to get the matrices of a linear rigid motion in $\mathbb{R}_3$ in the form **A**, **B**, or **C**. That is, the given matrix is merely *similar* to one of these matrices. It is possible, however, to read off the cosine of the angle θ displayed in **A** or **C** right from the original matrix, provided you know whether the determinant is $+1$ or -1. One can sometimes tell whether the given matrix is orientation preserving or reversing without calculating the determinant:

6.4 Let **M** be a 3×3 orthogonal matrix;
  i) If det $M = +1$, then **M** is similar to a matrix of type **A**, and trace $M = 1 + 2\cos\theta$.

6 / Rigid motions, especially in 2- and 3-space

ii) If det $\mathbf{M} = -1$, then $\mathbf{M}$ is similar to a matrix of type $\mathbf{B}$ (and trace $\mathbf{M} = 1$) or of type $\mathbf{C}$ (and trace $\mathbf{M} = -1 + 2\cos\theta$).

iii) If trace $\mathbf{M} > 1$, then det $\mathbf{M} = +1$ and $\mathbf{M}$ is similar to a matrix of type $\mathbf{A}$ with $\cos\theta = \frac{1}{2}$ (trace $\mathbf{M} - 1$).

iv) If trace $\mathbf{M} < -1$, then det $\mathbf{M} = -1$ and $\mathbf{M}$ is similar to a matrix of type $\mathbf{C}$ with $\cos\theta = \frac{1}{2}(1 + \text{trace } \mathbf{M})$.

Proof. Exercise 6. ∎

Note that det $\mathbf{M}$ can be easily calculated by checking any nonzero element of the adjoint matrix, for

$$\mathbf{M}^{-1} = \frac{1}{\det \mathbf{M}} \text{adjoint } \mathbf{M} = \pm \text{adjoint } \mathbf{M};$$

since also $\mathbf{M}^{-1} = \mathbf{M}^*$, we see that the ij-entry of $\mathbf{M}$ is (det $\mathbf{M}$) × (the ij-cofactor); thus, just compare any nonzero entry with its cofactor.

EXERCISES

1. In the following, describe the rigid motion $F(\mathbf{P}) = \mathbf{MP} + \mathbf{P}_0$. (The matrices are orthogonal; see Section 5, Exercise 3.) In (d) through (i) use 6.4 whenever helpful.

a) $\mathbf{M} = \begin{pmatrix} \frac{3}{5} & -\frac{4}{5} \\ \frac{4}{5} & \frac{3}{5} \end{pmatrix}$, $\mathbf{P}_0 = \begin{pmatrix} 6 \\ 2 \end{pmatrix}$

b) $\mathbf{M} = \begin{pmatrix} \frac{3}{5} & \frac{4}{5} \\ \frac{4}{5} & -\frac{3}{5} \end{pmatrix}$, $\mathbf{P}_0 = \begin{pmatrix} 3 \\ -2 \end{pmatrix}$

c) same matrix as (b), $\mathbf{P}_0 = \begin{pmatrix} 2 \\ -4 \end{pmatrix}$

d) $\mathbf{M} = \begin{pmatrix} -1 & 0 & 0 \\ 0 & \frac{3}{5} & -\frac{4}{5} \\ 0 & \frac{4}{5} & \frac{3}{5} \end{pmatrix}$, $\mathbf{P}_0 = \begin{pmatrix} 3 \\ 6 \\ 2 \end{pmatrix}$

e) $\mathbf{M} = \begin{pmatrix} 1 & 0 & 0 \\ 0 & \frac{3}{5} & \frac{4}{5} \\ 0 & \frac{4}{5} & -\frac{3}{5} \end{pmatrix}$, $\mathbf{P}_0 = \begin{pmatrix} 1 \\ 3 \\ -2 \end{pmatrix}$,

f) $\mathbf{M}$ same as (e) $\mathbf{P}_0 = \begin{pmatrix} 0 \\ 2 \\ -4 \end{pmatrix}$

g) $\mathbf{M} = \begin{pmatrix} 1/\sqrt{2} & \frac{2}{3} & 1/3\sqrt{2} \\ 0 & \frac{1}{3} & -4/3\sqrt{2} \\ 1/\sqrt{2} & -\frac{2}{3} & -1/3\sqrt{2} \end{pmatrix}$, $\mathbf{P}_0 = \begin{pmatrix} 1 \\ -2 \\ 3 \end{pmatrix}$

h) $\mathbf{M} = \begin{pmatrix} \frac{29}{45} & -\frac{4}{9} & \frac{28}{45} \\ \frac{28}{45} & \frac{7}{9} & -\frac{4}{45} \\ -\frac{4}{9} & \frac{4}{9} & \frac{7}{9} \end{pmatrix}$, $\mathbf{P}_0 = \begin{pmatrix} 1 \\ -2 \\ 3 \end{pmatrix}$

i) $\mathbf{M}$ same as (h), $\mathbf{P}_0 = \begin{pmatrix} 2 \\ 1 \\ -2 \end{pmatrix}$.

2. Show that a translation is linear iff the translating vector $\mathbf{P}_0 = \mathbf{0}$
3. a) Prove that if $\mathbf{F}$ is a rigid motion, then so is $\mathbf{T(P)} = \mathbf{F(P)} - \mathbf{F(0)}$.
   b) Prove that the only vector orthogonal to every vector in a basis is the zero vector.
   c) Show that the $\mathbf{T}$ in 6.2 is homogeneous.
4. a) Prove the stated properties of plane rotations,
   b) of plane reflections;
   c) prove 6.3.
5. Fill in the details showing that the rigid motions of $\mathbb{R}_3$ are as in the table. Draw pictures for each case.
6. Prove 6.4.
7. In $\mathbb{R}_3$ first rotate about the z-axis through an angle θ and then rotate about the y-axis through an angle φ. This combination is also a rotation.
   a) Find the matrix of all three rotations.
   b) Find the axis and angle of rotation for the resulting rotation. Try the following supplement if you get stuck.
8. a) Prove that if $\mathbf{F}_1$ and $\mathbf{F}_2$ are rigid motions, then so is $\mathbf{F}_1\mathbf{F}_2$.
   b) Show that the set of all rigid motions in $\mathbb{R}_n$ forms a group under multiplication.
9. Let

$$\mathbf{M} = \begin{pmatrix} -\frac{3}{5} & 0 & \frac{4}{5} \\ \frac{4}{5} & 0 & \frac{3}{5} \\ 0 & 1 & 0 \end{pmatrix}, \quad \mathbf{P}_0 = \begin{pmatrix} x_0 \\ y_0 \\ z_0 \end{pmatrix}.$$

   a) Show that $\mathbf{M}$ is orthogonal and that $\mathbf{F(P)} = \mathbf{MP} + \mathbf{P}_0$ is either a rotation about an axis or a screw displacement.
   b) Find the condition on $\mathbf{P}_0$ that $\mathbf{F(P)}$ be a rotation and find the axis of rotation.

Supplement to Section 6. Rotation in $\mathbb{R}_3$

Rotations and the Cross Product

We wish to write down the transformation that gives a rotation about an axis through the origin in $\mathbb{R}_3$. To do this neatly we shall need the **cross product** of two vectors $\mathbf{V}, \mathbf{W}$, which is defined as follows: Let

$$\mathbf{i} = (1, 0, 0), \quad \mathbf{j} = (0, 1, 0), \quad \mathbf{k} = (0, 0, 1)$$

be the standard basis for $\mathbb{R}_3$; then if $\mathbf{V} = v_1\mathbf{i} + v_2\mathbf{j} + v_3\mathbf{k}$, $\mathbf{W} = w_1\mathbf{i} + w_2\mathbf{j} + w_3\mathbf{k}$, the cross product of $\mathbf{V}$ and $\mathbf{W}$ is defined as follows:

$$\mathbf{V} \times \mathbf{W} = \det \begin{pmatrix} v_2 & v_3 \\ w_2 & w_3 \end{pmatrix} \mathbf{i} + \det \begin{pmatrix} v_3 & v_1 \\ w_3 & w_1 \end{pmatrix} \mathbf{j} + \det \begin{pmatrix} v_1 & v_2 \\ w_1 & w_2 \end{pmatrix} \mathbf{k}.$$

(This is sometimes called the **vector product**; it is a product restricted to $\mathbb{R}_3$.) We shall need some simple properties of this cross product. These are easily proved and their proof is left to the reader.

i) $c\mathbf{V} \times \mathbf{W} = c(\mathbf{V} \times \mathbf{W})$, (Homogeneity)
ii) $\mathbf{V} \times \mathbf{V} = 0, \mathbf{W} \times \mathbf{V} = -\mathbf{V} \times \mathbf{W}$ (Anti-commutativity)
iii) $(\mathbf{V} + \mathbf{W}) \times \mathbf{X} = \mathbf{V} \times \mathbf{X} + \mathbf{W} \times \mathbf{X}$, (Distributivity)
  $\mathbf{X} \times (\mathbf{V} + \mathbf{W}) = \mathbf{X} \times \mathbf{V} + \mathbf{X} \times \mathbf{W}$,
iv) $\mathbf{V} \cdot (\mathbf{V} \times \mathbf{W}) = 0, \mathbf{W} \cdot (\mathbf{V} \times \mathbf{W}) = 0$. (The cross product is perpendicular to the plane of the two vectors).
v) $\mathbf{V} \times \mathbf{W} = 0$ iff $\mathbf{V}, \mathbf{W}$ are not linearly independent.
vi) $|\mathbf{V} \times \mathbf{W}|^2 = (\mathbf{V} \cdot \mathbf{V})(\mathbf{W} \cdot \mathbf{W}) - (\mathbf{V} \cdot \mathbf{W})^2$.
vii) If $\mathbf{V} \neq 0$ and $\mathbf{W} \neq 0$, $|\mathbf{V} \times \mathbf{W}| = |\mathbf{V}| |\mathbf{W}| \sin \theta$, where θ is the angle between $\mathbf{V}$ and $\mathbf{W}$. (The norm of the cross product is the area of the parallelogram determined by the two vectors).
viii) If $\mathbf{V} \cdot \mathbf{W} = 0$, $|\mathbf{V} \times \mathbf{W}| = |\mathbf{V}| |\mathbf{W}|$.
ix) $(\mathbf{V} \times \mathbf{W}) \cdot \mathbf{X} = \mathbf{V} \cdot (\mathbf{W} \times \mathbf{X}) = \det (\mathbf{V}, \mathbf{W}, \mathbf{X})$.

We now take a unit vector $\mathbf{V}$, $|\mathbf{V}| = 1$, and determine how to rotate a given vector $\mathbf{X}$ about $\mathbf{V}$ through the angle θ to get a vector $\mathbf{Y}$. To do this we first take the projection of $\mathbf{X}$ on $\mathbf{V}$ which is $(\mathbf{X} \cdot \mathbf{V})\mathbf{V}$ and set

$$\mathbf{U} = \mathbf{X} - (\mathbf{X} \cdot \mathbf{V})\mathbf{V},$$

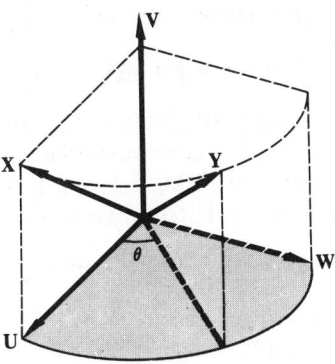

Fig. 34

so that $U \cdot V = 0$ and, of course,
$$X = U + (X \cdot V)V.$$
Then put $W = V \times U$ so that
$$W \cdot U = 0, \quad W \cdot V = 0, \quad \text{and} \quad |W| = |U|.$$
To rotate X about V we rotate U, the projection of X in the UW plane, into $(\cos \theta)U + (\sin \theta)W$. Hence the resulting vector is Y, where
$$Y = (X \cdot V)V + (\cos \theta)U + (\sin \theta)W.$$

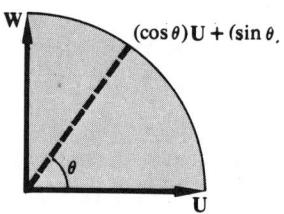

Fig. 35

Since $W = V \times U = V \times X - (X \cdot V)V \times V = V \times X$, this can be written
$$Y = (X \cdot V)(1 - \cos \theta)V + (\cos \theta)X + (\sin \theta)V \times X. \tag{1}$$

It is easy to write down the matrix of this linear transformation, though possibly not very useful. In particular the trace turns out to be $1 + 2 \cos \theta$ (as it should).

Rotations and Quaternions

It is possible to express rotations in terms of quaternion multiplication. To do this, we remind the reader that a **quaternion** is an object of the form
$$\mathbf{q} = a_0 + a_1\mathbf{i} + a_2\mathbf{j} + a_3\mathbf{k}, \quad a_i \text{ in } \mathbb{R}.$$

(See also the supplement to Chapter 10, Section 4.) Equality, addition, and scalar multiplication of quaternions are done componentwise, so that the set of all quaternions forms a vector space over $\mathbb{R}$ with basis $1, \mathbf{i}, \mathbf{j}, \mathbf{k}$; note that $\mathbb{R}_3$ is a subspace. Multiplication is defined so as to be distributive and satisfy

$$\mathbf{ij} = -\mathbf{ji} = \mathbf{k}, \quad \mathbf{jk} = -\mathbf{kj} = \mathbf{i}, \quad \mathbf{ki} = -\mathbf{ik} = \mathbf{j}, \quad \mathbf{i}^2 = \mathbf{j}^2 = \mathbf{k}^2 = -1.$$

The resulting multiplication is associative (why do you need only check for various combinations of $\mathbf{i}, \mathbf{j}$, and $\mathbf{k}$?). We can write
$$\mathbf{q} = a_0 + V \quad \text{where} \quad V = a_1\mathbf{i} + a_2\mathbf{j} + a_3\mathbf{k}$$

is a "pure vector" (that is, in $\mathbb{R}_3$). It is easy to check that the quaternion product of two pure vectors **V** and **U** is

$$\mathbf{VU} = -(\mathbf{V} \cdot \mathbf{U}) + (\mathbf{V} \times \mathbf{U});$$

and this fact makes it easy to multiply two general quaternions.

We also define $\bar{\mathbf{q}} = a_0 - \mathbf{V}$ (the **conjugate** of **q**) and easily prove

$$\overline{\mathbf{q}_1 \mathbf{q}_2} = \bar{\mathbf{q}}_2 \bar{\mathbf{q}}_1,$$

as well as $\mathbf{q}\bar{\mathbf{q}} = a_0^2 + (\mathbf{V} \cdot \mathbf{V}) = a_0^2 + a_1^2 + a_2^2 + a_3^2 = \bar{\mathbf{q}}\mathbf{q}$. We define the quaternion **norm** of **q** to be $|\mathbf{q}| = \sqrt{\mathbf{q}\bar{\mathbf{q}}}$. Since $|\mathbf{q}|^2$ is the sum of the squares of the coefficients, we see that **q** has a multiplicative inverse

$$\mathbf{q}^{-1} = \frac{1}{|\mathbf{q}^2|} \bar{\mathbf{q}} \quad \text{iff} \quad \mathbf{q} \neq \mathbf{0}.$$

We also find $|\mathbf{q}_1 \mathbf{q}_2|^2 = |\mathbf{q}_1|^2 |\mathbf{q}_2|^2$ (when written out in detail, this gives *Euler's identity*). Note also that **q** is a pure vector (i.e., in $\mathbb{R}_3$) iff $\bar{\mathbf{q}} = -\mathbf{q}$.

A quaternion **q** with $|\mathbf{q}| = 1$ is called a **unit quaternion**. It is of the form

$$\mathbf{q} = a + b\mathbf{V}, \quad \mathbf{V} \text{ a unit vector in } \mathbb{R}_3, \quad \text{with } a^2 + b^2 = 1.$$

Now let **V** be a unit vector in $\mathbb{R}_3$ and $\mathbf{q} = a + b\mathbf{V}$ a unit quaternion ($a^2 + b^2 = 1$).

If **X** is a vector in $\mathbb{R}_3$, then we consider the mapping

$$\mathbf{X} \to \mathbf{Y}, \quad \text{where} \quad \mathbf{Y} = \mathbf{q}\mathbf{X}\bar{\mathbf{q}} = (a + b\mathbf{V})\mathbf{X}(a - b\mathbf{V}). \tag{2}$$

Since $\bar{\mathbf{Y}} = \mathbf{q}(-\mathbf{X})\bar{\mathbf{q}} = -\mathbf{Y}$, we see that **Y** is also a vector in $\mathbb{R}_3$. Hence (2) is a linear transformation from $\mathbb{R}_3$ into $\mathbb{R}_3$, which is clearly *onto* (why clearly?).

Furthermore, $\mathbf{Y}\bar{\mathbf{Y}} = \mathbf{q}\mathbf{X}\bar{\mathbf{q}}\mathbf{q}\bar{\mathbf{X}}\bar{\mathbf{q}} = \mathbf{X}\bar{\mathbf{X}}$ so that $|\mathbf{Y}| = |\mathbf{X}|$ (norms in $\mathbb{R}_3$). Therefore (2) is orthogonal. Also $\mathbf{X} = \mathbf{V}$ gives

$$\mathbf{Y} = (a^2 + b^2)\mathbf{V} = \mathbf{V},$$

so that **V** is invariant. Hence (2) is a rotation about **V**.

All that remains is to figure out the angle of rotation. To do this we put

$$a = \cos \varphi, \quad b = \sin \varphi \quad (\text{since } a^2 + b^2 = 1).$$

To simplify the calculations we choose (orthonormal) axes so that $\mathbf{V} = \mathbf{k}$. After lengthy but simple calculations, (2) becomes

$$\mathbf{Y} = \mathbf{X} \cos 2\varphi + (\mathbf{X} \cdot \mathbf{V})(1 - \cos 2\varphi)\mathbf{V} + (\sin 2\varphi)(\mathbf{V} \times \mathbf{X}).$$

When we compare this to (1), we see that any rotation can be expressed in this way and the angle of rotation θ is $\theta = 2\varphi$. Thus,

any rotation **R** *in* $\mathbb{R}_3$ *is of the form* $\mathbf{R}(\mathbf{X}) = \mathbf{q}\mathbf{X}\bar{\mathbf{q}}$ *where* **q** *is a unit quaternion.*

As an application we shall work out the product of two rotations in $\mathbb{R}_3$. Let
$$\mathbf{q}_1 = \cos \theta/2 + (\sin \theta/2)\mathbf{V}, \quad |\mathbf{V}| = 1,$$
$$\mathbf{q}_2 = \cos \varphi/2 + (\sin \varphi/2)\mathbf{U}, \quad |\mathbf{U}| = 1,$$
be the quaternions that give these rotations. Then
$$\mathbf{Y} = \mathbf{q}_2(\mathbf{q}_1 \mathbf{X} \bar{\mathbf{q}}_1)\bar{\mathbf{q}}_2 = (\mathbf{q}_1 \mathbf{q}_2)\mathbf{X}(\bar{\mathbf{q}}_2 \bar{\mathbf{q}}_1)$$
gives the product of the two rotations. The quaternion for the product is therefore $\mathbf{q}_2\mathbf{q}_1$, and this works out to be
$$\mathbf{q}_2\mathbf{q}_1 = \big(\cos \theta/2 \cos \varphi/2 - (\sin \theta/2 \sin \varphi/2)\mathbf{U} \cdot \mathbf{V}\big)$$
$$+ (\sin \theta/2 \cos \varphi/2)\mathbf{V} + (\cos \theta/2 \sin \varphi/2)\mathbf{U}$$
$$+ (\sin \theta/2 \sin \varphi/2)\mathbf{U} \times \mathbf{V}.$$
If we put $\mathbf{q}_2\mathbf{q}_1 = \cos \psi/2 + (\sin \psi/2)\mathbf{W}, |\mathbf{W}| = 1$, we get
$$\cos \psi/2 = \cos \theta/2 \cos \varphi/2 - \sin \theta/2 \sin \varphi/2 \cos \epsilon,$$
$$\sin \psi/2\, \mathbf{W} = (\sin \theta/2 \cos \varphi/2)\mathbf{V} + (\cos \theta/2 \sin \varphi/2)\mathbf{U} + (\sin \theta/2 \sin \varphi/2)\mathbf{U} \times \mathbf{V}.$$
Here $\cos \epsilon = \mathbf{U} \cdot \mathbf{V}$, where ϵ is the angle between $\mathbf{V}$ and $\mathbf{U}$, the two axes of rotation.

7.* RIGID MOTIONS IN $\mathbb{R}_n$

In this section we first show how a linear rigid motion decomposes $\mathbb{R}_n$ into a sum of planes (restricted to which the motion is a rotation), together with a small subspace upon which it may not be a rotation. We then investigate reflections in $\mathbb{R}_n$, and we finally show that any rigid motion is a product of reflections.

As in $\mathbb{R}_3$, we define a rigid motion $\mathbf{F}(\mathbf{P}) = \mathbf{AP} + \mathbf{P}_0$, $\mathbf{A}$ an orthogonal matrix, to be

orientation preserving if det $\mathbf{A} = +1$,

and

orientation reversing if det $\mathbf{A} = -1$.

Since the product of two rigid motions which preserve orientation is a rigid motion which preserves orientation (Exercise 3), it follows that we cannot hope to generate all rigid motions as products of rotations (in planes, say). However, the following theorem shows that it is not too fanciful to consider rigid motions as being generalized rotations.

7.1 Given any orthogonal linear transformation $\mathbf{T}$ in $\mathbb{R}_n$, one can decompose $\mathbb{R}_n$ as a sum of mutually orthogonal planes $\mathfrak{I}_1, \ldots, \mathfrak{I}_k$ passing through $\mathbf{0}$

and a subspace $\mathcal{V}$ of dimension ≤ 2, in such a fashion that $\mathbf{T}$ is a rotation in each plane $\mathfrak{F}_j$ and $\mathbf{T}$ is either a reflection or the identity on $\mathcal{V}$:

$$\mathbb{R}_n = \mathfrak{F}_1 + \cdots + \mathfrak{F}_k + \mathcal{V},$$

the dimension of each $\mathfrak{F}_j = 2$, and dimension $\mathcal{V} = 0, 1,$ or 2, and $\mathbf{T}$ restricted to each $\mathfrak{F}_j$ is a rotation, and $\mathbf{T}$ is a reflection or the identity on $\mathcal{V}$.†

In matrix terms there is an orthonormal basis for $\mathbb{R}_n$ relative to which the matrix of $\mathbf{T}$ is

$$\begin{pmatrix} A_1 & 0 & \cdots & & 0 \\ 0 & \ddots & & & \vdots \\ \vdots & & \ddots & & \vdots \\ \vdots & & & A_k & 0 \\ 0 & & \cdots & & B \end{pmatrix},$$

where

$$\mathbf{A}_j = \begin{pmatrix} \cos\theta_j & -\sin\theta_j \\ \sin\theta_j & \cos\theta_j \end{pmatrix} \quad \text{and} \quad \mathbf{B} = \begin{pmatrix} 1 & 0 \\ 0 & -1 \end{pmatrix},$$

or $\quad \mathbf{B} = (1), \quad$ or $\quad \mathbf{B} = (-1), \quad$ or $\quad \mathbf{B}$ isn't there.

Proof. (In this proof we use some information from Section 5, Supplement 1, and consider linear transformations as acting on complex vector spaces.) Suppose $\mathbf{T}(\mathbf{V}) = \mathbf{A}\mathbf{V}$. Regard $\mathbf{A}$ as a matrix in $\mathbb{C}_{n \times n}$, where it will be a unitary matrix. Hence, by the supplement to Section 5, its eigenvalues will satisfy $|\lambda| = 1$. Hence, each eigenvalue λ is of the form

$$\lambda = \cos\theta + i\sin\theta$$

for some real θ. For such an eigenvalue, let $\mathbf{V}$ be an eigenvector. We can separate real and pure imaginary components in the n-tuple $\mathbf{V}$ and write $\mathbf{V} = \mathbf{X} + i\mathbf{Y}$ where $\mathbf{X}$ and $\mathbf{Y}$ have real entries. Then

$$\mathbf{AV} = \mathbf{A}(\mathbf{X} + i\mathbf{Y}) = \mathbf{AX} + i\mathbf{AY},$$

while

$$\lambda\mathbf{V} = (\cos\theta + i\sin\theta)(\mathbf{X} + i\mathbf{Y})$$
$$= (\cos\theta\,\mathbf{X} - \sin\theta\,\mathbf{Y}) + i(\sin\theta\,\mathbf{X} + \cos\theta\,\mathbf{Y}).$$

Equating these expressions for $\mathbf{AV}$ and $\lambda\mathbf{V}$, and separating into real and pure imaginary parts, we have

$$\mathbf{AX} = \cos\theta\,\mathbf{X} - \sin\theta\,\mathbf{Y}, \qquad \mathbf{AY} = \sin\theta\,\mathbf{X} + \cos\theta\,\mathbf{Y}.$$

---

† This decomposition is in fact a *direct* sum, as those who have studied this notion will recognize. This means that any vector $\mathbf{W}$ in $\mathbb{R}_n$ can be written *uniquely* in the form

$$\mathbf{W} = \mathbf{P}_1 + \cdots + \mathbf{P}_k + \mathbf{V}, \qquad \mathbf{P}_i \text{ in } \mathfrak{F}_i, \qquad \mathbf{V} \text{ in } \mathcal{V}.$$

This looks very promising, and span $\{X, Y\}$ will be an appropriate plane for our decomposition if X and Y are linearly independent. If $\cos\theta = \pm 1$ (that is, $\lambda = \pm 1$), then these equations could be satisfied by X and Y which are not linearly independent. We now outline how to show that under any other circumstances, X and Y are linearly independent (see Exercise 4). First recall that if λ is an eigenvalue of A corresponding to V, then λ^{-1} is an eigenvalue of A^{-1} corresponding to V. If $\lambda = \lambda^{-1}$ then $\lambda = \pm 1$ and we are in the nonindependent situation mentioned above. Otherwise, if $\lambda \neq \lambda^{-1}$ then

$$\lambda V^*V = V^*\lambda V = V^*AV = (A^*V)^*V = (A^{-1}V)^*V = (\lambda^{-1}V)^*V = \lambda^{-1} V^*V.$$

Thus if, $\lambda \neq \lambda^{-1}$

$$V^*V = V \cdot V = 0 \ (\textit{real} \text{ dot product}),$$

so for $V = X + iY$,

$$0 = V \cdot V = (X + iY) \cdot (X + iY) = X \cdot X - Y \cdot Y + 2i(X \cdot Y).$$

Taking the pure imaginary part of this equation, we obtain $2i(X \cdot Y) = 0$, so $X \cdot Y = 0$ and X is orthogonal to Y. Hence, they are linearly independent, if not 0. Since $X + iY$ is an eigenvector, not both are 0, and the real part of the equation yields $X \cdot X = Y \cdot Y$, so that *neither* is 0. By replacing X by $X/|X|$ and Y by $Y/|Y|$ we obtain an orthonormal basis for a plane in $\mathbb{R}_n$ relative to which the matrix of T looks like one of the A_j.

In general, $\mathcal{V} = \text{span } \{X, Y\}$ will be a subspace of $\mathbb{R}_n$ which is mapped into itself by T. We can write

$$\mathbb{R}_n = \mathcal{V} + \mathcal{V}^\perp$$

and obtain that $\mathcal{V}^\perp$ is also mapped into itself by T (Exercise 4). Replace $\mathbb{R}_n$ by $\mathcal{V}^\perp$ and argue again on $\mathcal{V}^\perp$ to tear off a piece of it (or use induction). One obtains that

$$\mathbb{R}_n = \mathcal{V}_1 + \mathcal{V}_2 + \cdots + \mathcal{V}_l,$$

where the $\mathcal{V}_j$'s are subspaces mapped into themselves by T, and either $\mathcal{V}_j$ is a plane (and the matrix of T relative to an appropriate basis looks like an A_j), or $\mathcal{V}_j$ is one-dimensional and has matrix (1) or (-1). Pair up one-dimensional subspaces upon which T has eigenvalue $+1$ to obtain two-dimensional spaces upon which T has matrix

$$\begin{pmatrix} 1 & 0 \\ 0 & 1 \end{pmatrix} = \begin{pmatrix} \cos 0 & -\sin 0 \\ \sin 0 & \cos 0 \end{pmatrix},$$

and similarly for subspaces upon which T has eigenvalue -1. You may be left with a one- or two-dimensional space upon which the matrix of T has the form B given in the statement of the theorem, so the theorem is proved. ∎

Compare the statement of this theorem with the results obtained for $\mathbb{R}_3$ in the previous section.

Reflections

In $\mathbb{R}_2$ we could have discussed reflection in a point as well as reflection in a line. However, it is clear that this is the same as rotation about the point through an angle of π. Similarly, in $\mathbb{R}_3$ we could have considered reflections in points and in lines as well as in planes (Exercise 5; what do these other types of reflections amount

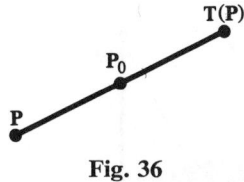

Fig. 36

to in our classification scheme?). The property common to all these types of reflections is that if you apply one twice, you get back where you started:

$$\mathbf{R}^2(\mathbf{P}) = \mathbf{R}(\mathbf{R}(\mathbf{P})) = \mathbf{P} \quad \text{for any } \mathbf{P};$$

i.e., $\mathbf{R}^2 = \mathbf{I}, \mathbf{R} \neq \mathbf{I}$. We can take this as our definition: A rigid motion $\mathbf{R}$ in $\mathbb{R}_n$ is a **reflection** if $\mathbf{R}^2 = \mathbf{I}$ and $\mathbf{R} \neq \mathbf{I}$.

Let us now work toward seeing that this is a sensible definition. We know that since a reflection is a rigid motion, we can write

$$\mathbf{R}(\mathbf{P}) = \mathbf{T}(\mathbf{P}) + \mathbf{P}_0,$$

where $\mathbf{T}$ is an orthogonal linear transformation. Hence,

$$\mathbf{P} = \mathbf{R}^2(\mathbf{P}) = \mathbf{R}(\mathbf{T}(\mathbf{P}) + \mathbf{P}_0) = \mathbf{T}(\mathbf{T}(\mathbf{P}) + \mathbf{P}_0) + \mathbf{P}_0$$
$$= \mathbf{T}^2(\mathbf{P}) + (\mathbf{T}(\mathbf{P}_0) + \mathbf{P}_0).$$

Since $\mathbf{T}^2$ is linear, $\mathbf{T}^2(\mathbf{0}) = \mathbf{0}$, so that $\mathbf{T}(\mathbf{P}_0) + \mathbf{P}_0 = \mathbf{0}$ and so $\mathbf{T}^2(\mathbf{P}) = \mathbf{P}$. Thus, for the reflection $\mathbf{R}(\mathbf{P}) = \mathbf{T}(\mathbf{P}) + \mathbf{P}_0$,

$$\mathbf{T}^2 = \mathbf{I} \quad \text{and} \quad \mathbf{T}(\mathbf{P}_0) = -\mathbf{P}_0.$$

Thus, the linear part of a reflection is also a reflection (why can't $\mathbf{T} = \mathbf{I}$?). We next analyze linear reflections:

7.2 Let $\mathbf{T}$ be a linear reflection (that is, an orthogonal transformation $\neq \mathbf{I}$ for which $\mathbf{T}^2 = \mathbf{I}$). Then there is an orthonormal basis for $\mathbb{R}_n$ relative to which $\mathbf{T}$ has as matrix

$$A = \begin{pmatrix} 1 & 0 & \cdots & & & 0 \\ 0 & \ddots & & & & \\ \vdots & & 1 & & & \vdots \\ & & & -1 & & \\ & & & & \ddots & 0 \\ 0 & \cdots & & & 0 & -1 \end{pmatrix}.$$

The fixed points of **T** form a flat, in fact a subspace $\mathcal{V}$. For **W** in the orthogonal complement $\mathcal{V}^\perp$ of this subspace, $\mathbf{T(W)} = -\mathbf{W}$. Any vector **P** in $\mathbb{R}_n$ can be (uniquely) written as

$$\mathbf{P} = \mathbf{V} + \mathbf{W},$$

where **T** fixes **V** and $\mathbf{T(W)} = -\mathbf{W}$.

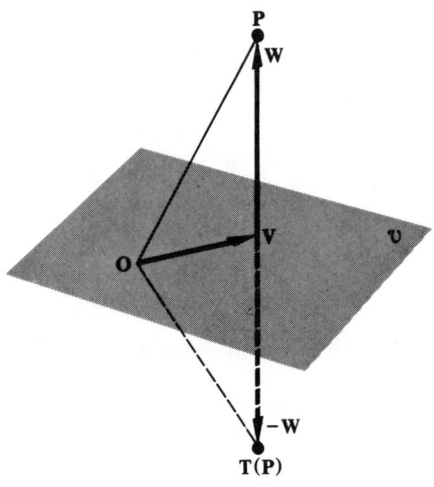

Fig. 37

Proof. Choose an orthonormal basis so that the matrix of **T** is as in 7.1:

$$\begin{pmatrix} \mathbf{A}_1 & & & & 0 \\ & \cdot & & & \\ & & \cdot & & \\ & & & \cdot & \\ 0 & & & \mathbf{A}_k & \\ & & & & \mathbf{B} \end{pmatrix}.$$

Then $\mathbf{T}^2 = \mathbf{I}$ implies that $\mathbf{A}_i^2 = \mathbf{I}$ (and $\mathbf{B}^2 = \mathbf{I}$ too). It then follows from the form of the $\mathbf{A}_i$ that each $\mathbf{A}_i = \pm \mathbf{I}$ (why?). Then the basis can be easily rearranged so that the matrix is of the form stated in the theorem (how?).

If we let $\mathcal{V}$ be the set of vectors fixed by **T**, then $\mathcal{V}$ is spanned by those vectors in the (rearranged) basis with eigenvalue 1, and the rest of the basis vectors span the orthogonal complement $\mathcal{V}^\perp$ (why?). Since these latter vectors all have eigenvalue -1, we see that for **W** in $\mathcal{V}^\perp$,

$$\mathbf{T(W)} = -\mathbf{W}.$$

You finish up. ∎

It is now a simple matter to analyze an arbitrary reflection.

7.3 The fixed points of a reflection $\mathbf{R}(\mathbf{P}) = \mathbf{T}(\mathbf{P}) + \mathbf{P}_0$ form a flat
$$\mathcal{F} = \tfrac{1}{2}\mathbf{P}_0 + \mathcal{V},$$
where $\mathcal{V}$ is the space of fixed points of $\mathbf{T}$, and $\mathbf{P}_0$ is in $\mathcal{V}^\perp$.

Proof. If $\mathbf{P} = \mathbf{V} + \mathbf{W}$, $\mathbf{V}$ in $\mathcal{V}$, $\mathbf{W}$ in $\mathcal{V}^\perp$, and if $\mathbf{P}$ is fixed by $\mathbf{R}$ then
$$\mathbf{V} + \mathbf{W} = \mathbf{P} = \mathbf{R}(\mathbf{P}) = \mathbf{T}(\mathbf{P}) + \mathbf{P}_0 = \mathbf{T}(\mathbf{V} + \mathbf{W}) + \mathbf{P}_0 = \mathbf{V} - \mathbf{W} + \mathbf{P}_0,$$
so $2\mathbf{W} = \mathbf{P}_0$, which says that $\mathbf{P} = \tfrac{1}{2}\mathbf{P}_0 + \mathbf{V}$. By applying $\mathbf{R}$, we see that all such points are fixed. Hence, the fixed points of $\mathbf{R}$ are the flat $\tfrac{1}{2}\mathbf{P}_0 + \mathcal{V}$. We noticed before that $\mathbf{T}(\mathbf{P}_0) = -\mathbf{P}_0$, and 7.2 says that $\mathbf{P}_0$ is thus in $\mathcal{V}^\perp$. ∎

If the reflection $\mathbf{R}$ has the flat $\mathcal{F}$ as its fixed points, we say that $\mathbf{R}$ is **reflection in the flat** $\mathcal{F}$. The following picture shows that this is intuitively reasonable.

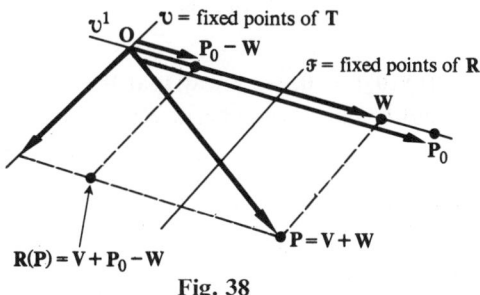

Fig. 38

Note that reflection in a k-flat is not necessarily an orientation-reversing rigid motion (see Exercise 8).

Reflections in hyperplanes are of special importance.

7.4 Any rigid motion in $\mathbb{R}_n$ is the product of at most $(n + 1)$ reflections in hyperplanes.

If the rigid motion fixes the origin then at most n reflections are needed.

Proof. Given a rigid motion $\mathbf{F}$ in $\mathbb{R}_n$, if $\mathbf{F}(0) \neq 0$, let $\mathbf{R}$ be the reflection in the hyperplane orthogonal to $\mathbf{F}(0)$ and passing through the point $\tfrac{1}{2}\mathbf{F}(0)$ (Fig. 39). Then $\mathbf{RF}$ is a rigid motion (why?). Hence, if we can show that
$$\mathbf{RF} = \mathbf{R}_1\mathbf{R}_2 \cdots \mathbf{R}_n$$
is the product of at most n reflections in hyperplanes, then since $\mathbf{R}^{-1} = \mathbf{R}$,
$$\mathbf{F} = \mathbf{R}\mathbf{R}_1\mathbf{R}_2 \cdots \mathbf{R}_n$$
is the product of $(n + 1)$ reflections (or less).

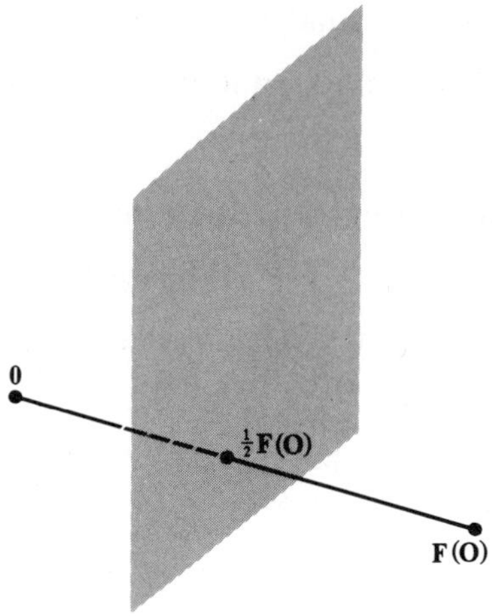

Fig. 39

Since $\mathbf{RF}(\mathbf{0}) = \mathbf{R}(\mathbf{F}(\mathbf{0})) = \mathbf{0}$, we know that $\mathbf{RF} = \mathbf{T}$ is a linear rigid motion. Choose the orthonormal basis guaranteed in 7.1 relative to which

$$\mathbb{R}_n = \mathfrak{T}_1 + \cdots + \mathfrak{T}_k + \mathfrak{V},$$

where the $\mathfrak{T}_i$ are planes, and the matrix of $\mathbf{T}$ is

$$\mathbf{A} = \begin{pmatrix} \mathbf{A}_1 & & & & 0 \\ & \cdot & & & \\ & & \cdot & & \\ & & & \cdot & \\ 0 & & & \mathbf{A}_k & \\ & & & & \mathbf{B} \end{pmatrix},$$

where

$$\mathbf{A}_i = \begin{pmatrix} \cos \theta_i & -\sin \theta_i \\ \sin \theta_i & \cos \theta_i \end{pmatrix}.$$

Now in any plane $\mathfrak{T}$, it is easy to see that a rotation about the origin through an angle θ can be obtained by reflecting in the x-axis and then reflecting again in a line making an angle of $\theta/2$ with that axis (Fig. 40). In our official matrix terms we have

reflection about the x-axis has matrix

$$\begin{pmatrix} 1 & 0 \\ 0 & -1 \end{pmatrix};$$

reflection about the line making an angle $\theta/2$ with the x-axis has matrix

$$\begin{pmatrix} \cos \theta & \sin \theta \\ \sin \theta & -\cos \theta \end{pmatrix}.$$

7 / Rigid motions in $\mathbb{R}_n$

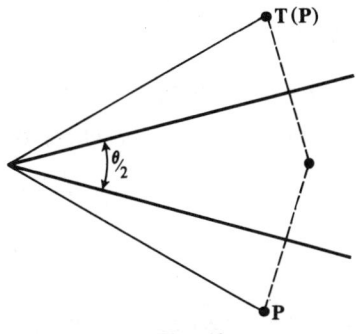

Fig. 40

It is easy to check that these matrices do give reflections and that their product is

$$\begin{pmatrix} \cos\theta & -\sin\theta \\ \sin\theta & \cos\theta \end{pmatrix},$$

which is rotation through an angle θ.

Returning to $\mathbb{R}_n$, we let

$$\mathbf{R}_i = \begin{pmatrix} 1 & & & & & & & \\ & \ddots & & & & & & \\ & & 1 & & & & & \\ & & & \cos\theta_i & \sin\theta_i & & & \\ & & & \sin\theta_i & -\cos\theta_i & & & \\ & & & & & 1 & & \\ & & & & & & \ddots & \\ & & & & & & & 1 \end{pmatrix},$$

$$\mathbf{S}_i = \begin{pmatrix} 1 & & & & & & & \\ & \ddots & & & & & & \\ & & 1 & & & & & \\ & & & 1 & 0 & & & \\ & & & 0 & -1 & & & \\ & & & & & 1 & & \\ & & & & & & \ddots & \\ & & & & & & & 1 \end{pmatrix}.$$

Then $\mathbf{R}_i$ and $\mathbf{S}_i$ are reflections in hyperplanes (prove$_1$). Moreover, if the **B** part of **A** is absent, then
$$\mathbf{A} = \mathbf{R}_1\mathbf{S}_1\mathbf{R}_2\mathbf{S}_2 \cdots \mathbf{R}_k\mathbf{S}_k,$$
a product of n reflections (prove$_2$). The case where **B** is present is also simply handled (prove$_3$). ∎

EXERCISES

1. Prove that there exists a unique reflection in a flat $\mathcal{F}$.
2. Prove that any rigid motion maps k-flats onto k-flats.
3. Show that the product of two rigid motions
    a) preserves orientation if both rigid motions do;
    b) preserves orientation if both rigid motions reverse orientation;
    c) reverses orientation if one preserves and one reverses.
4. Fill in the details in the proof of 7.1.
5. What are reflections in points and lines in $\mathbb{R}_3$?
6. Finish the proof of 7.2.
7. Finish the proof of 7.4.
8. When will reflection in a k-flat preserve orientation? reverse orientation? why?

... thence must any whatyoulike in the power of empthood be either greater THa<small>N</small> or less <small>TH</small>a<small>N</small> the unitate we have in one or hence shall the vectorious ready-eyes of evertwo circumflicksrent searclhers never film in the elipsities of their gyribouts those fickers which are returnally reprodictive of themselves. Which is unpassible.†

CHAPTER 9 # Symmetric Matrices and Quadratic Forms

In this chapter we first discuss symmetric matrices and their linear transformation counterpart. This type of matrix is of particular importance because (1) it often appears "in nature", (2) it is relatively easy to work with, and (3) it is tied up with other areas of mathematics, such as quadratic forms (that is, quadratic functions). The main theorem about symmetric matrices is that they always have a collection of eigenvectors which forms a basis (this is the Principal Axis Theorem, 1.3). We apply this theorem to the study of quadratic forms and of the general second-degree equation. This chapter also considers positive definite matrices and the nonorthogonal diagonalization of quadratic forms.

1. SYMMETRIC MATRICES AND TRANSFORMATIONS

You will recall that an $n \times n$ matrix **A** is **symmetric** if it is equal to its own transpose: $\mathbf{A}^* = \mathbf{A}$. That is, the ith row of **A** is the same as the ith column (transposed), for $i = 1, \ldots, n$. That is, for each i and j, the i,j-entry a_{ij} of **A** is the same as the j, i-entry a_{ji}:

$$a_{ij} = a_{ji} \quad \text{for} \quad i, j = 1, \ldots, n.$$

As we have seen on many other occasions, it is often more convenient and transparent to work with linear transformations rather than the corresponding matrices. The linear transformation equivalent of "symmetric matrix" takes the following form:

---

† James Joyce, *Finnegan's Wake*, Compass Books, 1959.

A linear transformation **S** from $\mathbb{R}_n$ into $\mathbb{R}_n$ is **symmetric** if
$$S(V) \cdot W = V \cdot S(W) \quad \text{for all} \quad V, W \text{ in } \mathbb{R}_n.$$
We first show how these two notions of "symmetric" correspond.

1.1 (i) If **S** is a symmetric linear transformation on $\mathbb{R}_n$, then, relative to any orthonormal basis, the matrix of **S** is a symmetric matrix.

ii) If **A** is a symmetric $n \times n$ matrix, then $S(V) = AV$ is a symmetric linear transformation.

Proof. (i) Suppose, for convenience, that we take the orthonormal basis to be the usual $\{e_1, \ldots, e_n\}$, so that the matrix of **S** is
$$A = (S(e_1), \ldots, S(e_n)),$$
where $S(e_i) = (a_{1i}, a_{2i}, \ldots, a_{ni})^*$ is the ith column of **A**. We proceed to show that $a_{ij} = a_{ji}$, proving that **A** is symmetric. This follows immediately, since
$$a_{ji} = S(e_i) \cdot e_j = e_i \cdot S(e_j) = a_{ij} \tag{1}$$
(see Exercise 2a).

ii) If **A** is a symmetric matrix then $a_{ij} = a_{ji}$, and if $S(V) = AV$, then, as for (1),
$$S(e_i) \cdot e_j = a_{ji} \quad \text{and} \quad e_i \cdot S(e_j) = a_{ij}.$$
Hence $S(e_i) \cdot e_j = e_i \cdot S(e_j)$. The symmetric condition then follows for all **V** and **W**, by the distributivity of the dot product (Exercise 2b). ∎

One convenient feature of symmetric matrices and mappings is that their eigenvalues are all real:

1.2 If **A** is a symmetric matrix in $\mathbb{R}_{n \times n}$, then the eigenvalues of **A** are real.

Proof. (for which it is necessary to go to $\mathbb{C}_n$). Define a linear transformation on $\mathbb{C}_n$ by
$$S(V) = AV, \quad V \text{ in } \mathbb{C}_n.$$
Let λ be an eigenvector of **A**; we know that λ is in $\mathbb{C}$, and we must show that actually λ is real. We do this by proving that $\bar\lambda = \lambda$. Let **V** be an eigenvector in $\mathbb{C}_n$ corresponding to λ, so
$$S(V) = AV = \lambda V.$$
For any vector **W** in $\mathbb{C}_n$, denote by $\bar{W}$ the vector whose entries are the conjugates of those of **W**:
$$\bar{W} = (\bar{w}_1, \ldots, \bar{w}_n)^* \quad \text{if} \quad W = (w_1, \ldots, w_n)^*.$$
Since **A** has real entries, it follows that
$$A\bar{V} = \bar\lambda \bar{V}, \quad \text{that is,} \quad S(\bar{V}) = \bar\lambda \bar{V} \tag{2}$$

(see Exercise 3). Now, $S(V) \cdot \bar{V} = \lambda(V \cdot \bar{V})$ but also by the symmetry of S,
$$S(V) \cdot \bar{V} = V \cdot S(\bar{V}) = V \cdot \bar{\lambda}\bar{V} = \bar{\lambda}(V \cdot \bar{V}).$$
Since $V \neq 0$, $V \cdot \bar{V} \neq 0$, so $\lambda = \bar{\lambda}$, and λ is real. ∎

The following is of considerably greater interest; the meaning of its name will become clear in Section 2.

1.3 The principal axis theorem. If S is a symmetric linear transformation on a finite-dimensional vector space $\mathcal{V}$ over $\mathbb{R}$, then there exists an orthogonal basis of $\mathcal{V}$ consisting of eigenvectors of S. Thus, relative to this orthogonal basis the matrix of S is of the form

$$\begin{pmatrix} \lambda_1 & & 0 \\ & \cdot & \\ & & \cdot \\ & & & \cdot \\ 0 & & & \lambda_n \end{pmatrix},$$

where $\lambda_1, \ldots, \lambda_n$ are the eigenvalues of S with multiple roots listed according to multiplicity.

Proof. By induction on the dimension n of $\mathcal{V}$. If $\mathcal{V}$ is one-dimensional, the theorem is surely true (Exercise 4a), so suppose it holds for all vector spaces of dimensions less than n. Confronted with a space $\mathcal{V}$ of dimension n, we choose an eigenvalue λ of S (necessarily real) with corresponding eigenvector $V \neq 0$. Let

$$\mathcal{W} = V^\perp = \{W \text{ in } \mathcal{V} \mid W \cdot V = 0\}.$$

Then $\mathcal{W}$ is a vector space of dimension $n - 1$ (why$_1$?). Moreover,

$$S(W) \cdot V = W \cdot S(V) = W \cdot \lambda V = \lambda W \cdot V = 0$$

for W in $\mathcal{W}$, which says that $S(W)$ is in $\mathcal{W}$ if W is. Hence, we may consider S as a linear transformation on $\mathcal{W}$. It will be symmetric there, so, by the induction hypothesis, there will be an orthogonal basis $\{W_1, \ldots, W_{n-1}\}$ of $\mathcal{W}$ consisting of eigenvectors of S. Then $\{V, W_1, \ldots, W_{n-1}\}$ will be an orthogonal basis for $\mathcal{V}$ consisting of eigenvectors of S (why$_2$?). ∎

In applying the Principal Axis Theorem, one usually takes the following steps:

I. Find the eigenvalues of S. In low-dimensional cases, one can take the characteristic polynomial route. In higher-dimensional cases, one applies an approximative procedure, such as the Jacobi Method (see Supplement 2 to this section).

II. Find the subspace of eigenvectors corresponding to each eigenvalue λ (one has to toss in the zero vector to get a subspace, since we've excluded zero as an eigenvector). That is, find all solutions to $S(V) = \lambda V$ for each eigenvalue λ.

III. Take an orthogonal basis for each of these subspaces. They will often be one-dimensional, in which case this is trivial; if a subspace is not one-dimensional, take any convenient orthogonal basis.†

These bases strung together will be the desired basis for the entire space. The subspaces corresponding to different eigenvalues will be orthogonal, so the basis will be too (the orthogonality of the subspaces, while strongly suggested by the Principal Axis Theorem, can be proved directly from 2.5 of Chapter 6; Exercise 5).

If the symmetric transformation happens to be on $\mathbb{R}_n$, so that it is $S(V) = AV$ for some symmetric matrix A, then suppose that the orthogonal basis is $\{V_1, \ldots, V_n\}$; in fact, suppose this is orthonormal. Let $P = (V_1, \ldots, V_n)$ be the matrix with these basis vectors as columns. Then P is an orthogonal matrix (why?). Moreover,

$$AP = A(V_1, \ldots, V_n) = (AV_1, \ldots, AV_n)$$

$$= (\lambda_1 V_1, \ldots, \lambda_n V_n) = P \begin{pmatrix} \lambda_1 & & 0 \\ & \cdot & \\ & & \cdot \\ 0 & & \lambda_n \end{pmatrix}.$$

This yields the following:

1.4 Matrix form of the Principal Axis Theorem. Given an $n \times n$ symmetric matrix A, there is an orthogonal matrix P for which

$$P^{-1}AP = \begin{pmatrix} \lambda_1 & & 0 \\ & \cdot & \\ & & \cdot \\ 0 & & \lambda_n \end{pmatrix} = P^*AP,$$

where λ_i are the eigenvalues of A. In fact, P is a matrix whose columns are an orthonormal set of eigenvectors corresponding to $\lambda_1, \ldots, \lambda_n$ respectively.

Example 1.
Let

$$A = \begin{pmatrix} 5 & -1 & -1 \\ -1 & 3 & 1 \\ -1 & 1 & 3 \end{pmatrix},$$

whose characteristic polynomial is

$$-x^3 + 11x^2 - 36x + 36 = (2 - x)(3 - x)(6 - x).$$

---

† If λ is an eigenvalue of multiplicity m (i.e. an m-fold root of the characteristic polynomial), then it is easy to see that the dimension of this subspace is m (Exercise 9).

1 / Symmetric matrices and transformations

Hence, the eigenvalues are $\lambda_1 = 2$, $\lambda_2 = 3$, $\lambda_3 = 6$, and corresponding eigenvectors are easily calculated to be:

For
$$\lambda_1 = 2, \quad V_1 = (0, 1, -1)^*,$$
$$\lambda_2 = 3, \quad V_2 = (1, 1, 1)^*,$$
$$\lambda_3 = 6, \quad V_3 = (2, -1, -1)^*.$$

When normalized these give the columns of **P**:

$$\mathbf{P} = \begin{pmatrix} 0 & 1/\sqrt{3} & 2/\sqrt{6} \\ 1/\sqrt{2} & 1/\sqrt{3} & -1/\sqrt{6} \\ -1/\sqrt{2} & 1/\sqrt{3} & -1/\sqrt{6} \end{pmatrix}, \quad \mathbf{AP} = \mathbf{P}\begin{pmatrix} 2 & 0 & 0 \\ 0 & 3 & 0 \\ 0 & 0 & 6 \end{pmatrix};$$

and thus

$$\mathbf{A} = \mathbf{P}\begin{pmatrix} 2 & 0 & 0 \\ 0 & 3 & 0 \\ 0 & 0 & 6 \end{pmatrix}\mathbf{P}^{-1} \quad \text{or} \quad \mathbf{P}^{-1}\mathbf{AP} = \begin{pmatrix} 2 & 0 & 0 \\ 0 & 3 & 0 \\ 0 & 0 & 6 \end{pmatrix}.$$

Example 2.

$$\mathbf{A} = \begin{pmatrix} 0 & 1 & 1 \\ 1 & 0 & 1 \\ 1 & 1 & 0 \end{pmatrix},$$

whose characteristic polynomial is $-x^3 + 3x + 2 = (x + 1)^2(2 - x)$. Hence the eigenvalues are $-1, -1, 2$.

For $\lambda = -1$ we get the two linearly independent eigenvectors,

$$V_1 = (-1, 1, 0)^*, \quad V_2 = (-1, 0, 1)^*.$$

Using the Gram-Schmidt process on these, we get the orthogonal basis for the subspace belonging to $\lambda = -1$:

$$V_1' = \left(-\frac{1}{\sqrt{2}}, \frac{1}{\sqrt{2}}, 0\right), \quad V_2' = \left(\frac{1}{\sqrt{6}}, \frac{1}{\sqrt{6}}, -\frac{2}{\sqrt{6}}\right).$$

Corresponding to $\lambda = 2$, we get the normalized vector $\left(\frac{1}{\sqrt{3}}, \frac{1}{\sqrt{3}}, \frac{1}{\sqrt{3}}\right)^*$. Hence,

$$\mathbf{P} = \begin{pmatrix} -1/\sqrt{2} & 1/\sqrt{6} & 1/\sqrt{3} \\ 1/\sqrt{2} & 1/\sqrt{6} & 1/\sqrt{3} \\ 0 & -2/\sqrt{6} & 1/\sqrt{3} \end{pmatrix},$$

and

$$\mathbf{A} = \mathbf{P}\begin{pmatrix} -1 & 0 & 0 \\ 0 & -1 & 0 \\ 0 & 0 & 2 \end{pmatrix}\mathbf{P}^{-1} \quad \text{or} \quad \mathbf{P}^{-1}\mathbf{AP} = \begin{pmatrix} -1 & 0 & 0 \\ 0 & -1 & 0 \\ 0 & 0 & 2 \end{pmatrix}.$$

We know that orthogonal matrices represent orthogonal transformations in $\mathbb{R}_n$, so that, in the language of Chapters 4 and 8, we can rephrase 1.4 as:

Given a symmetric matrix **A**, there is an orthogonal change of basis in $\mathbb{R}_n$ relative to which the transformation represented by **A** has a diagonal matrix.

In many respects the antithesis of symmetric matrices are skew-symmetric matrices (those for which $\mathbf{B}^* = -\mathbf{B}$). This will be brought out more clearly in Chapter 10, Section 5. For now we note that applying the steps in the proof of 1.2 to skew-symmetric matrices proves that

All the eigenvalues of a skew-symmetric matrix are pure imaginary numbers. (3)

(See Exercise 6.) Hence, there is no possibility of obtaining a real matrix **P** for which $\mathbf{P}^{-1}\mathbf{BP}$ is diagonal if **B** is skew-symmetric (and $\mathbf{B} \neq \mathbf{0}$).

EXERCISES

1. Find the diagonal form of the following symmetric matrices; also find an orthogonal **P** for which **P*AP** has this diagonal form. Save your results for future use.

   a) $\begin{pmatrix} -1 & 3 & 1 \\ 3 & -1 & 1 \\ 1 & 1 & 1 \end{pmatrix}$ b) $\begin{pmatrix} 5 & 1 & -3 \\ 1 & 5 & -3 \\ -3 & -3 & 9 \end{pmatrix}$ c) $\begin{pmatrix} 1 & 5 & -3 \\ 5 & 1 & -3 \\ -3 & -3 & 9 \end{pmatrix}$

   d) $\begin{pmatrix} 17 & 6 \\ 6 & 8 \end{pmatrix}$ e) $\begin{pmatrix} 1 & 2 & -2 \\ 2 & 4 & -4 \\ -2 & -4 & 4 \end{pmatrix}$ f) $\begin{pmatrix} 1 & -2 & -2 \\ -2 & 1 & -2 \\ -2 & -2 & 1 \end{pmatrix}$

2. (a) Prove Equation (1). (b) Finish the proof of 1.1(ii).
3. Prove Equation (2).
4. (a) Prove 1.3 for $n = 1$. (b) Answer the questions in the proof of 1.3.
5. Let **A** be a symmetric matrix with eigenvalues λ and μ, $\lambda \neq \mu$, with corresponding eigenvectors **V** and **W** respectively. Prove that **V** and **W** are orthogonal.
6. Prove Statement (3).
7. Prove that the eigenvalues of a 2×2 symmetric matrix are real, by looking at its characteristic polynomial.
8. Find the diagonal form of the following symmetric matrices; also find an orthogonal matrix **P** for which $\mathbf{P}^{-1}\mathbf{AP}$ has this diagonal form:

   a) $\mathbf{A} = \begin{pmatrix} 2 & 0 & 0 & 1 \\ 0 & 2 & 1 & 0 \\ 0 & 1 & 2 & 0 \\ 1 & 0 & 0 & 2 \end{pmatrix}$ b) $\mathbf{A} = \begin{pmatrix} 0 & 1 & 1 & 1 \\ 1 & 0 & 1 & 1 \\ 1 & 1 & 0 & 1 \\ 1 & 1 & 1 & 0 \end{pmatrix}$

   c) $\mathbf{A} = (a_{ij})$, where $a_{ii} = 0$, $a_{ij} = 1$ if $i \neq j$; $i, j = 1, \ldots, n$.

9. Prove that, if λ is an eigenvalue of the symmetric linear transformation **S** of multiplicity m, then the dimension of
$$\mathcal{V}_\lambda = \{\mathbf{V} \mid \mathbf{S}(\mathbf{V}) = \lambda \mathbf{V}\}$$
is m.

Supplement 1: Hermitian Matrices

We saw in the supplement to Section 5, Chapter 8, that the complex analog of an orthogonal matrix ($\mathbf{P}^* = \mathbf{P}^{-1}$) is a unitary matrix ($\bar{\mathbf{P}}^* = \mathbf{P}^{-1}$), where $\bar{\mathbf{P}}$ is the matrix whose entries are the conjugates of those of **P**). Similarly, the complex analog of a symmetric matrix ($\mathbf{A}^* = \mathbf{A}$) is one that satisfies

$$\bar{\mathbf{A}}^* = \mathbf{A}, \quad \mathbf{A} \text{ in } \mathbb{C}_{n \times n},$$

and is called a **Hermitian matrix**. For example,

$$\begin{pmatrix} 1 & 2-i \\ 2+i & 0 \end{pmatrix}$$

is a Hermitian matrix. In terms of its elements, a matrix is Hermitian iff whenever the i,j-entry is $a + ib$ (a, b real), then the j, i-entry is $a - ib$. In particular, the diagonal entries of a Hermitian matrix must be real.

You should check that the results (and proofs!) of this section hold for Hermitian matrices:

1.5 A Hermitian matrix has real eigenvalues.

1.6 If **A** is Hermitian, then there is a unitary matrix **P** for which $\bar{\mathbf{P}}^* \mathbf{AP}$ is diagonal.

The *complex* symmetric matrices ($\mathbf{A}^* = \mathbf{A}$, **A** in $\mathbb{C}_{n \times n}$) do not necessarily enjoy these same properties. For example, the matrix

$$\mathbf{A} = \begin{pmatrix} 2i & 1 \\ 1 & 0 \end{pmatrix}$$

has only one eigenvector and cannot therefore be similar to a diagonal matrix.

EXERCISES

1. If **A** is an Hermitian matrix, show that
$$\mathbf{A} = \mathbf{B} + i\mathbf{C},$$
where **B** is real and symmetric and **C** is real and skew-symmetric.

2. (Reduction of the eigenvalue problem for an $n \times n$ Hermitian matrix to that of a real $2n \times 2n$ symmetric matrix.) Show that if $\mathbf{A} = \mathbf{B} + i\mathbf{C}$ is a complex Hermitian matrix, decomposed as in Exercise 1, with (real) eigenvalues $\lambda_1, \ldots, \lambda_n$, then the

real symmetric matrix

$$\mathcal{A} = \begin{pmatrix} \mathbf{B} & -\mathbf{C} \\ \mathbf{C} & \mathbf{B} \end{pmatrix}$$

has eigenvalues $\lambda_1, \ldots, \lambda_n$; and that two linearly independent eigenvectors of $\mathcal{A}$, corresponding to λ, may be chosen in the form

$$\begin{pmatrix} \mathbf{V}_j \\ \mathbf{W}_j \end{pmatrix} \quad \text{and} \quad \begin{pmatrix} \mathbf{V}_j \\ -\mathbf{W}_j \end{pmatrix},$$

where $\mathbf{V}_j + i\mathbf{W}_j$ is an eigenvector of $\mathbf{A}$ corresponding to the eigenvalue λ_j.

Supplement 2: The Jacobi Method

We herein outline a technique which was originally invented by Jacobi in 1846 to prove that the eigenvalues of a real symmetric matrix are real (1.2). After more than 100 years it was revived by von Neumann as a computer method for finding the eigenvalues and eigenvectors of such matrices. It is one of the most commonly used techniques.

The basic notion is quite simple. Given an $n \times n$ real symmetric matrix $\mathbf{A}$, you perform successive rotations of planes in $\mathbb{R}_n$ so as to reduce the total size of the off-diagonal elements of $\mathbf{A}$. (The next section will make clear why one might think of such a technique.)

Let

$$\mathbf{A} = \begin{matrix} \\ i \\ j \\ \\ \end{matrix} \begin{pmatrix} a_{11} & \cdots & \cdots & a_{1n} \\ \vdots & \ddots & a_{ij} & \vdots \\ \vdots & a_{ij} & \ddots & \vdots \\ a_{1n} & \cdots & \cdots & a_{nn} \end{pmatrix}$$

be symmetric, with a_{ij} its off-diagonal element which is largest in absolute value; let

$$\mathbf{P} = \begin{pmatrix} 1 & & & & & & \\ & \ddots & & & & & \\ & & \cos\theta & \cdots & \sin\theta & & \\ & & \vdots & \ddots & \vdots & & \\ & & -\sin\theta & \cdots & \cos\theta & & \\ & & & & & 1 & \\ & & & & & & 1 \end{pmatrix},$$

which we hereafter call an **elementary orthogonal matrix** (check that $\mathbf{P}^*\mathbf{P} = \mathbf{I}$). We shall see that it is possible to choose an angle of rotation θ so that the matrix $\mathbf{P}^*\mathbf{AP}$ (which is both symmetric and similar to $\mathbf{A}$) has a 0 in its i, j-position. More importantly, for this judicious choice of θ,

the sum of the off-diagonal entries of $\mathbf{P}^*\mathbf{AP}$ =
(the sum of the off-diagonal entries of $\mathbf{A}$) $- 2a_{ij}^2$.

Thus, we have moved from $\mathbf{A}$ to a matrix which is closer to a diagonal matrix by an amount $2a_{ij}^2$.

If we proceed next to do in the largest off-diagonal entry of $\mathbf{P}^*\mathbf{AP}$, we may, unfortunately, introduce a nonzero term to the i, j-position. Thus, one may never end up with an actual diagonal matrix similar to $\mathbf{A}$, although one can make the sum of the squares of the off-diagonal entries less than a preassigned tolerance level by repeated applications (Exercise 3).

Let us first consider an example. Suppose that

$$\mathbf{A} = \begin{pmatrix} a_{11} & a_{12} & a_{13} \\ a_{12} & a_{22} & a_{23} \\ a_{13} & a_{23} & a_{33} \end{pmatrix} \quad \text{and} \quad \mathbf{P} = \begin{pmatrix} \cos\theta & 0 & \sin\theta \\ 0 & 1 & 0 \\ -\sin\theta & 0 & \cos\theta \end{pmatrix},$$

so that we shall attempt to do in the 1,3-entry a_{13} of $\mathbf{A}$. Then

$$\mathbf{P}^*\mathbf{AP} = \begin{pmatrix} a_{11}c^2 - 2a_{13}sc + a_{33}s^2 & a_{12}c - a_{23}s & sc(a_{11} - a_{33}) + a_{13}(c^2 - s^2) \\ \cdots & a_{22} & a_{12}s + a_{23}c \\ \cdots & \cdots & a_{11}s^2 + 2a_{13}sc + a_{33}c^2 \end{pmatrix}$$

(symmetric elements mercifully suppressed, $c = \cos\theta$, $s = \sin\theta$). If $a_{13} \neq 0$, we can obtain a 0 in the 1,3-entry of this latter matrix by taking

$$\frac{a_{33} - a_{11}}{a_{13}} = \frac{\cos^2\theta - \sin^2\theta}{\sin\theta\cos\theta} = \frac{\cos 2\theta}{\frac{1}{2}\sin 2\theta},$$

so that some fancy trigonometry yields

$$\cotan 2\theta = \frac{a_{33} - a_{11}}{2a_{13}}. \tag{1}$$

What is the sum of the squares of the off-diagonal entries in $\mathbf{P}^*\mathbf{AP}$? By symmetry, this will be

$$2[(a_{12}\cos\theta - a_{23}\sin\theta)^2 + (a_{12}\sin\theta + a_{23}\cos\theta)^2] = 2[a_{12}^2 + a_{23}^2],$$

which is less than the corresponding sum for the matrix $\mathbf{A}$ by an amount $2a_{13}^2$.

In general, given an $n \times n$ symmetric matrix $\mathbf{A}$, to find the elementary orthogonal matrix $\mathbf{P}$ for which $\mathbf{P}^*\mathbf{AP}$ has a 0 in the i, j-position, one obtains in place of (1) that

$$\cotan 2\theta = \frac{a_{jj} - a_{ii}}{2a_{ij}}.$$

It is, of course, not necessary to solve directly for θ. It is a simple exercise in trigonometry to calculate $\sin \theta$ and $\cos \theta$. If we put $d = \cos 2\theta$, we find

$$d = (a_{jj} - a_{ii})/\sqrt{(a_{jj} - a_{ii})^2 + 4a_{ij}^2},$$

and hence

$$\sin \theta = \pm\sqrt{\tfrac{1}{2}(1-d)}, \qquad \cos \theta = \sqrt{\tfrac{1}{2}(1+d)};$$

here the plus sign must be taken if $a_{ij} > 0$ and the minus sign must be taken if $a_{ij} < 0$.

It is easy to calculate b_{ii} and b_{jj} (the new diagonal entries). In fact

$$b_{ii} = \tfrac{1}{2}[a_{ii} + a_{jj} - \sqrt{(a_{jj} - a_{ii})^2 + 4a_{ij}^2}],$$
$$b_{jj} = \tfrac{1}{2}[a_{ii} + a_{jj} + \sqrt{(a_{jj} - a_{ii})^2 + 4a_{ij}^2}].$$

If

$$\mathbf{P^*AP} = \begin{pmatrix} b_{11} & \cdots & b_{1i} & \cdots & b_{1j} & \cdots & b_{1n} \\ \vdots & & \vdots & & \vdots & & \vdots \\ b_{i1} & \cdots & b_{ii} & \cdots & b_{ij} & \cdots & b_{in} \\ \vdots & & \vdots & & \vdots & & \vdots \\ b_{j1} & \cdots & b_{ji} & \cdots & b_{jj} & \cdots & b_{jn} \\ \vdots & & \vdots & & \vdots & & \vdots \\ b_{n1} & \cdots & b_{ni} & \cdots & b_{nj} & \cdots & b_{nn} \end{pmatrix},$$

then the above choice of $\sin \theta$ will make $b_{ij} = 0$. The other entries are:

For the ith row and column ($k \neq i, j$),

$$b_{ik} = a_{ik} \cos \theta - a_{jk} \sin \theta;$$

for the jth row and column ($k \neq i, j$),

$$b_{jk} = a_{ik} \sin \theta + a_{jk} \cos \theta;$$

on the diagonal of the ith row and column,

$$b_{ii} = a_{ii} \cos^2 \theta + a_{jj} \sin^2 \theta - 2a_{ij} \sin \theta \cos \theta;$$

on the diagonal of the jth row and column,

$$b_{jj} = a_{ii} \sin^2 \theta + a_{jj} \cos^2 \theta + 2a_{ij} \sin \theta \cos \theta;$$

otherwise ($k, l \neq i, j$), $b_{kl} = a_{kl}$.

To apply the Jacobi Method one chooses a tolerance level ϵ and successively forms
$$\mathbf{A}_1 = \mathbf{P}_1^*\mathbf{A}\mathbf{P}_1, \qquad \mathbf{A}_2 = \mathbf{P}_2^*\mathbf{A}_1\mathbf{P}_2, \qquad \ldots, \qquad \mathbf{A}_k = \mathbf{P}_k^*\mathbf{A}_{k-1}\mathbf{P}_k.$$
When the sum of the squares of the off-diagonal elements of $\mathbf{A}_k$ is less than your preassigned ϵ, then you take the diagonal elements of $\mathbf{A}_k$ as the approximate eigenvalues of $\mathbf{A}$; since $\mathbf{A}$ and $\mathbf{A}_k$ are similar, their eigenvalues are the same.

The eigenvectors will then be approximately the columns of $\mathbf{P} = \mathbf{P}_1\mathbf{P}_2 \cdots \mathbf{P}_k$. As a worthwhile check, you can compare $\mathbf{A}_k$ with $\mathbf{P}^*\mathbf{A}\mathbf{P}$. It would be a valuable exercise for you to write up a flow diagram for the algorithm.

EXERCISES

1. (a)–(f) Try the Jacobi Method on some of the matrices in Exercise 1 of Section 1.
2. Prove that, for the general case, the sum of the squares of the off-diagonal entries is as promised.
3. Prove that given any $\epsilon > 0$, the Jacobi Method ultimately provides you with a matrix $\mathbf{A}_k$, the sum of the squares of whose off-diagonal entries is less than ϵ.

REFERENCES

Ralston, A., and Wilf, H. S. (eds.), *Mathematical Methods for Digital Computers*. Vol. 1, Chapter 7. New York: Wiley, 1964.

Wilkinson, J. H., *The Algebraic Eigenvalue Problem*. Oxford: Clarendon Press, 1965.

2. QUADRATIC FORMS

In analytic geometry and calculus you have encountered the following function:
$$Q(x, y) = ax^2 + 2bxy + cy^2, \tag{1}$$
a specific type of "quadratic form". We can write the form (1) as a matrix expression
$$(x, y)\begin{pmatrix} a & b \\ b & c \end{pmatrix}\begin{pmatrix} x \\ y \end{pmatrix}, \qquad \text{or } \mathbf{X}^*\mathbf{A}\mathbf{X}, \quad \text{where } \mathbf{X} = \begin{pmatrix} x \\ y \end{pmatrix} \quad \text{and} \quad \mathbf{A} = \begin{pmatrix} a & b \\ b & c \end{pmatrix}. \tag{2}$$
Since $\mathbf{A}$ is a symmetric matrix, we know from the Principal Axis Theorem that there is an orthogonal matrix $\mathbf{P}$ for which
$$\mathbf{P}^*\mathbf{A}\mathbf{P} = \begin{pmatrix} \lambda_1 & 0 \\ 0 & \lambda_2 \end{pmatrix},$$

where λ_i are the eigenvalues of **A**. Suppose that

$$\mathbf{P} = \begin{pmatrix} p_{11} & p_{12} \\ p_{21} & p_{22} \end{pmatrix},$$

so that

$$\mathbf{P}^* = \mathbf{P}^{-1} = \begin{pmatrix} p_{11} & p_{21} \\ p_{12} & p_{22} \end{pmatrix}.$$

Define

$$\begin{aligned} x' &= p_{11}x + p_{21}y, \\ y' &= p_{12}x + p_{22}y, \end{aligned} \quad \text{or} \quad \mathbf{X}' = \begin{pmatrix} x' \\ y' \end{pmatrix} = \mathbf{P}^*\mathbf{X}, \tag{3}$$

Then $\mathbf{X} = \mathbf{P}\mathbf{P}^*\mathbf{X} = \mathbf{P}\mathbf{X}'$, so that

$$\begin{aligned} x &= p_{11}x' + p_{12}y', \\ y &= p_{21}x' + p_{22}y', \end{aligned} \quad \text{or} \quad \mathbf{X} = \mathbf{P}\mathbf{X}'. \tag{4}$$

Moreover,

$$Q(x, y) = \mathbf{X}^*\mathbf{A}\mathbf{X} = (\mathbf{X}'^*\mathbf{P}^*)\mathbf{A}(\mathbf{P}\mathbf{X}') = \mathbf{X}'^* (\mathbf{P}^*\mathbf{A}\mathbf{P}) \mathbf{X}'$$

$$= (x', y') \begin{pmatrix} \lambda_1 & 0 \\ 0 & \lambda_2 \end{pmatrix} \begin{pmatrix} x' \\ y' \end{pmatrix} = \lambda_1 x'^2 + \lambda_2 y'^2. \tag{5}$$

Thus, by performing the orthogonal transformation (3) (which is, as we know, a linear rigid motion of $\mathbb{R}_2$), we transform the original quadratic form (1) into the simplified form (5). We could proceed to analyze (5) and then change back by (4), this being particularly easy to do since $\mathbf{P}^* = \mathbf{P}^{-1}$.

Example 1. The quadratic form $4x^2 + 24xy + 11y^2$ has matrix

$$\mathbf{A} = \begin{pmatrix} 4 & 12 \\ 12 & 11 \end{pmatrix}.$$

The characteristic polynomial of **A** is

$$x^2 - 15x - 100 = (x - 20)(x + 5),$$

so its eigenvalues are 20, -5, and its eigenvectors are:

$(3, 4)^*, \quad \lambda = 20,$

$(4, -3)^*, \quad \lambda = -5.$

Hence

$$\mathbf{P} = \begin{pmatrix} \frac{3}{5} & -\frac{4}{5} \\ \frac{4}{5} & \frac{3}{5} \end{pmatrix} \quad \text{(a rotation)}.$$

Check:

$$\mathbf{AP} = \mathbf{P}\begin{pmatrix} 20 & 0 \\ 0 & -5 \end{pmatrix}, \quad \mathbf{P}^{-1}\mathbf{AP} = \begin{pmatrix} 20 & 0 \\ 0 & -5 \end{pmatrix}, \quad \mathbf{A} = \mathbf{P}\begin{pmatrix} 20 & 0 \\ 0 & -5 \end{pmatrix}\mathbf{P}^{-1}.$$

2 / Quadratic forms

The change of variables simplifying the form to $20x'^2 - 5y'^2$ is

$$X = PX', \qquad X' = P*X,$$

or

$$x = \tfrac{3}{5}x' - \tfrac{4}{5}y', \qquad x' = \tfrac{3}{5}x + \tfrac{4}{5}y,$$
$$y = \tfrac{4}{5}x' + \tfrac{3}{5}y', \qquad y' = -\tfrac{4}{5}x + \tfrac{3}{5}y.$$

The above is not particularly awe-inspiring for quadratic forms in two variables. However, the technique may be carried over directly to quadratic forms in n variables, where it becomes a method of considerable power.

In general a **quadratic form** is any polynomial in the variables $x_1, \ldots, x_n$, in which each term is of combined degree two in the variables:

$$a_{11}x_1^2 + a_{12}x_1x_2 + \cdots + a_{1n}x_1x_n$$
$$+ a_{21}x_2x_1 + a_{22}x_2^2 + \cdots + a_{2n}x_2x_n + \cdots$$
$$+ a_{n1}x_nx_1 + \cdots + a_{nn}x_n^2. \tag{6}$$

Given the matrix

$$A = \begin{pmatrix} a_{11} & \cdots & a_{1n} \\ \vdots & & \vdots \\ a_{n1} & \cdots & a_{nn} \end{pmatrix} \text{ in } \mathbb{R}_{n\times n}, \quad \text{and vector } X = \begin{pmatrix} x_1 \\ \vdots \\ x_n \end{pmatrix}$$

in the variables $x_1, \ldots, x_n$, you can quickly see that (6) is the same as

$$X*AX. \tag{7}$$

We shall of course assume that our n variables $x_1, \ldots, x_n$ commute ($x_ix_j = x_jx_i$), so we could combine terms in (6) and obtain for $i \neq j$, $(a_{ij} + a_{ji})x_ix_j$. Let us make the substitution

$$a'_{ij} = (a_{ij} + a_{ji})/2,$$

so that $a'_{ij} = a'_{ji}$. If we then "uncombine" terms, we find that (6) is exactly the same as

$$a'_{11}x_1^2 + a'_{12}x_1x_2 + \cdots + a'_{1n}x_1x_n$$
$$+ a'_{21}x_2x_1 + a'_{22}x_2^2 + \cdots + a'_{2n}x_2x_n + \cdots$$
$$+ a'_{n1}x_nx_1 + \cdots + a'_{nn}x_n^2, \tag{6}'$$

and we now have $a'_{ij} = a'_{ji}$. This, in turn, is the same as

$$X*A'X \tag{7}$$

with A' symmetric! Thus,

any quadratic form may be written as $X*A'X$ with A' a symmetric matrix.

Hereafter, we shall assume that our quadratic forms are so written, and dispense with the primes.

Example 2

$$5x_1^2 + x_1x_2 + 2x_1x_3,$$
$$-3x_2x_1 + 3x_2^2 + 2x_2x_3,$$
$$-4x_3x_1 \quad\quad + 3x_3^2,$$

has the symmetrized form

$$5x_1^2 - x_1x_2 - x_1x_3,$$
$$-x_1x_2 + 3x_2^2 + x_2x_3,$$
$$-x_3x_1 + x_2x_3 + 3x_3^2.$$

In matrix form, this is

$$(x_1, x_2, x_3)\begin{pmatrix} 5 & -1 & -1 \\ -1 & 3 & 1 \\ -1 & 1 & 3 \end{pmatrix}\begin{pmatrix} x_1 \\ x_2 \\ x_3 \end{pmatrix}.$$

If $\mathbf{P}$ is any invertible matrix and we introduce new variables by

$$\mathbf{X} = \mathbf{P}\mathbf{X}',$$

the quadratic form becomes

$$\mathbf{X}^*\mathbf{A}\mathbf{X} = \mathbf{X}'^*\mathbf{P}^*\mathbf{A}\mathbf{P}\mathbf{X}'$$
$$= \mathbf{X}'^*\mathbf{A}'\mathbf{X}' \quad \text{where } \mathbf{A}' = \mathbf{P}^*\mathbf{A}\mathbf{P}.$$

Since it is easily verified that $\mathbf{A}'$ is also symmetric, ($\mathbf{A}'^* = \mathbf{P}^*\mathbf{A}^*\mathbf{P} = \mathbf{P}^*\mathbf{A}\mathbf{P} = \mathbf{A}'$), we have transformed our quadratic form by the introduction of these new variables into another quadratic form. In particular, if $\mathbf{P}$ is orthogonal (that is, $\mathbf{P}^* = \mathbf{P}^{-1}$), we get $\mathbf{A}' = \mathbf{P}^{-1}\mathbf{A}\mathbf{P}$. In this case, we also have $\mathbf{X}' = \mathbf{P}^*\mathbf{X}$, along with $\mathbf{X} = \mathbf{P}\mathbf{X}'$. Now we know that there exists an orthogonal matrix $\mathbf{P}$ for which

$$\mathbf{A}' = \mathbf{P}^*\mathbf{A}\mathbf{P}$$

is a diagonal matrix. If we let x_i' be the ith entry of the vector $\mathbf{X}' = \mathbf{P}^*\mathbf{X}$, Eq. (7) means that our original quadratic form has become utter simplicity:

$$\lambda_1(x_1')^2 + \lambda_2(x_2')^2 + \cdots + \lambda_n(x_n')^2, \tag{8}$$

where $\lambda_1, \ldots \lambda_n$, are the eigenvalues of $\mathbf{A}$. We have proved a theorem:

2.1 Given the quadratic form $\mathbf{X}^*\mathbf{A}\mathbf{X}$ ($\mathbf{A}$ symmetric), let $\mathbf{P}$ be an orthogonal matrix for which $\mathbf{A}' = \mathbf{P}^*\mathbf{A}\mathbf{P}$ is diagonal. Then for the change of variables $\mathbf{X}' = \mathbf{P}^*\mathbf{X}$, the quadratic form $\mathbf{X}'^*\mathbf{A}'\mathbf{X}' = \mathbf{X}^*\mathbf{A}\mathbf{X}$.

In terms of coordinates, given any quadratic form $\mathbf{X}^*\mathbf{A}\mathbf{X}$, there exists an orthogonal change of variables which reduces the form to

$$\lambda_1(x_1')^2 + \cdots + \lambda_n(x_n')^2,$$

where $\lambda_1, \ldots, \lambda_n$ are the eigenvalues of $\mathbf{A}$, each listed according to its multiplicity.

We know techniques for finding $\mathbf{P}$ (See p. 418).

Example 3. Continuing with the quadratic form in Example 2, we saw, in Example 1 of the previous section that

$$\mathbf{P*AP} = \begin{pmatrix} 2 & 0 & 0 \\ 0 & 3 & 0 \\ 0 & 0 & 6 \end{pmatrix} \quad \text{for} \quad \mathbf{P} = \begin{pmatrix} 0 & 1/\sqrt{3} & 2/\sqrt{6} \\ 1/\sqrt{2} & 1/\sqrt{3} & -1/\sqrt{6} \\ -1/\sqrt{2} & 1/\sqrt{3} & -1/\sqrt{6} \end{pmatrix}.$$

Hence, the quadratic form in Example 2 is the same as

$$2(x_1')^2 + 3(x_2')^2 + 6(x_3')^2 \quad \text{where} \quad x_1' = (x_2 - x_3)/\sqrt{2},$$
$$x_2' = (x_1 + x_2 + x_3)/\sqrt{3},$$
$$x_3' = (2x_1 - x_2 - x_3)/\sqrt{6},$$

the latter being found from $\mathbf{X}' = \mathbf{P*X}$. By substituting in the given values of x_i', you can check that these forms are indeed the same.

The set of all points in $\mathbb{R}_n$ satisfying

$$Q(x_1, \ldots, x_n) = c, \quad Q \text{ a quadratic form}, \quad c \text{ a constant}, \tag{9}$$

is called a **quadratic surface**. Our previous theory allows us to make simplifications in the expression (9) for a given quadric surface, simply by changing orthonormal bases for $\mathbb{R}_n$. The eigenvectors of the matrix of the form are called the **principal axes** of the quadric surface. The Principal Axis Theorem says that if we normalize these principal axes, then they form an orthonormal basis for $\mathbb{R}_n$, relative to which the equation for the quadric surface becomes

$$\lambda_1 x_1'^2 + \cdots + \lambda_n x_n'^2 = c.$$

As you can see, this is a very powerful technique for studying quadric surfaces.

Example 4. Given the quadratic surface

$$5x^2 + 3y^2 + 3z^2 - 2xy - 2xz + 2yz = 6,$$

you immediately recognize the quadratic form involved as being the subject of several previous examples. Hence, if we make the basis change

$$x' = (y - z)/\sqrt{2},$$
$$y' = (x + y + z)/\sqrt{3},$$
$$z' = (2x - y - z)/\sqrt{6},$$

the equation becomes

$$2x'^2 + 3y'^2 + 6z'^2 = 6, \tag{10}$$

a much tamer creature than the form with which we started out.

To discuss (10) further, it is convenient to let

$$x' = 0, \quad y' = 0, \quad \text{and} \quad z' = 0,$$

successively, in order to see how the quadric surface intersects these three planes. One obtains an ellipse in each plane, so it comes as no surprise to find that the corresponding egg-shaped quadric surface is called an ellipsoid.

Thus, by making a rotation (or rotation followed by reflection), one can reduce the equation of a quadric surface to a much more tractable form.

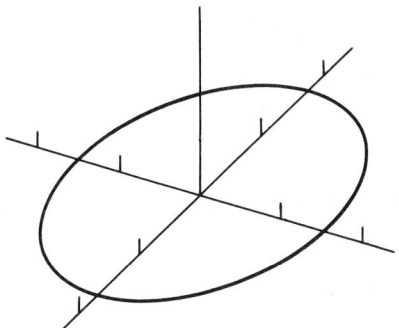

The intersection of the quadric surface and the plane $Z' = 0$

Fig. 1

EXERCISES

1. a) Using the methods we have developed, find a basis for $\mathbb{R}_2$ relative to which the conic section
$$4x^2 + 24xy + 11y^2 = 20$$
has no "cross-product" term. Identify the conic.
   b) Draw a picture, showing the conic relative to the simplifying basis, and also showing the original basis. How exactly did you get from one basis to the other?

2. Reduce the following quadratic forms to diagonal form. Save your work for later use, including the orthogonal matrix **P**.
   a) $-x^2 - y^2 + z^2 + 6xy + 2xz + 2yz$
   b) $5x^2 + 5y^2 + 9z^2 + 2xy - 6xz - 6yz$
   c) $x^2 + y^2 + 9z^2 + 10xy - 6xz - 6yz$
   d) $17x^2 + 12xy + 8y^2$
   e) $x^2 + 4y^2 + 4z^2 + 4xy - 4xz - 8yz$
   f) $x^2 + y^2 + z^2 - 4xy - 4yz - 4xz$

3. Discuss the quadric surfaces obtained by equating the respective quadratic forms in Exercise 2 to the following constants:

   a) -12 b) 144 c) 48 d) 80 e) 9 f) -12

   Draw a sketch of each in $\mathbb{R}_3$, relative to the principal axes, and put in the old axes as well.

3.* THE GENERAL SECOND-DEGREE EQUATION

In this section we shall apply our ability to simplify quadratic forms, to discuss the locus of points satisfying the general second-degree equation

$$\sum_{i,j=1}^{n} a_{ij}x_ix_j + \sum_{i=1}^{n} 2b_ix_i + c = 0.$$

These loci are a generalization of the conic sections which you have studied in the plane. We should point out that a generalization in the other direction, raising the degree of the equation, is hopeless. The quadratic-form theory has no convenient generalization to "cubic forms" and there exists no simple method for simplifying equations of degree greater than two.

We shall restrict ourselves to the general second-degree equation in three variables, since the techniques for dealing with any number of variables is the same (and pleasing pictures can be drawn for our case). Our problem then is to determine the locus of points in $\mathbb{R}_3$ satisfying the equation

$$a_{11}x^2 + a_{22}y^2 + a_{33}z^2 + 2a_{12}xy + 2a_{13}xz + 2a_{23}yz + 2b_1x \\ + 2b_2y + 2b_3z + c = 0. \quad (1)$$

Step I. We first make an orthogonal transformation to reduce the quadratic form occurring in (1) to diagonal form. The matrix of this form is symmetric:

$$\mathbf{A} = \begin{pmatrix} a_{11} & a_{12} & a_{13} \\ a_{12} & a_{22} & a_{23} \\ a_{13} & a_{23} & a_{33} \end{pmatrix};$$

and we saw in 2.1 that we can find an orthogonal matrix $\mathbf{P}$ for which

$$\mathbf{AP} = \mathbf{P}\begin{pmatrix} a_1 & 0 & 0 \\ 0 & a_2 & 0 \\ 0 & 0 & a_3 \end{pmatrix}, \quad \mathbf{P}^{-1} = \mathbf{P}^*.$$

So

$$\mathbf{P}^{-1}\mathbf{AP} = \begin{pmatrix} a_1 & 0 & 0 \\ 0 & a_2 & 0 \\ 0 & 0 & a_3 \end{pmatrix}.$$

In fact a_1, a_2, a_3 are the eigenvalues of $\mathbf{A}$ and the columns of $\mathbf{P}$ are the corresponding normalized eigenvectors. If we make the orthogonal transformation

$$(x', y', z')^* = \mathbf{P}^*(x, y, z)^* \quad \text{or} \quad (x, y, z)^* = \mathbf{P}(x', y', z')^*,$$

we transform (1) to the form

$$a_1x'^2 + a_2y'^2 + a_3z'^2 + 2b_1'x' + 2b_2'y' + 2b_3'z' + c = 0. \quad (2)$$

Example. Let us take the equation

$$-x^2 - y^2 + z^2 + 6xy + 2xz + 2yz - 12x + 4y - 10z - 11 = 0. \quad (3)$$

Here
$$A = \begin{pmatrix} -1 & 3 & 1 \\ 3 & -1 & 1 \\ 1 & 1 & 1 \end{pmatrix},$$

and we find that it has the eigenvalues 3, −4, 0, with corresponding eigenvectors

$$(1, 1, 1)^*, \quad (1, -1, 0)^*, \quad (1, 1, -2)^*,$$

respectively. Hence,

$$P = \begin{pmatrix} 1/\sqrt{3} & 1/\sqrt{2} & 1/\sqrt{6} \\ 1/\sqrt{3} & -1/\sqrt{2} & 1/\sqrt{6} \\ 1/\sqrt{3} & 0 & -2/\sqrt{6} \end{pmatrix}.$$

So

$$AP = P \begin{pmatrix} 3 & 0 & 0 \\ 0 & -4 & 0 \\ 0 & 0 & 0 \end{pmatrix},$$

and

$$P^{-1}AP = \begin{pmatrix} 3 & 0 & 0 \\ 0 & -4 & 0 \\ 0 & 0 & 0 \end{pmatrix} \quad \text{or} \quad A = P \begin{pmatrix} 3 & 0 & 0 \\ 0 & -4 & 0 \\ 0 & 0 & 0 \end{pmatrix} P^{-1};$$

then

$$x' = \frac{1}{\sqrt{3}}(x + y + z), \qquad x = \frac{1}{\sqrt{3}}x' + \frac{1}{\sqrt{2}}y' + \frac{1}{\sqrt{6}}z',$$

$$y' = \frac{1}{\sqrt{2}}(x - y), \qquad y = \frac{1}{\sqrt{3}}x' - \frac{1}{\sqrt{2}}y' + \frac{1}{\sqrt{6}}z',$$

$$z' = \frac{1}{\sqrt{6}}(x + y - 2z), \qquad z = \frac{1}{\sqrt{3}}x' \qquad\qquad - \frac{2}{\sqrt{6}}z'.$$

and when we substitute these in (3) we get

$$3x'^2 - 4y'^2 - 6\sqrt{3}\,x' - 8\sqrt{2}\,y' + 2\sqrt{6}\,z' - 11 = 0. \tag{4}$$

Step II. Having removed the "cross-product" terms, we next remove the linear terms whenever possible. This is easily done by completing the squares. For example, if $a_1 \neq 0$, then

$$a_1 x'^2 + 2b_1' x' = a_1(x' + b_1'/a_1)^2 - b_1'^2/a_1.$$

By setting $x'' = x' + b_1'/a_1$, the translation $(x', y', z') \to (x'', y', z')$ will then change the equation to one which has no term in x''.

Note however that, if $a_1 = 0$, then we cannot effect this change, and we may be stuck with the x' term,† although we then have no x''^2 term.

---

† This should come as no surprise; consider the parabola $y^2 - x = 0$, in which the x term is an essential part.

3 / The general second-degree equation

By doing this whenever possible and performing the corresponding translations,
$$x'' = x' + k_1, \qquad y'' = x' + k_2, \qquad z'' = z' + k_3,$$
we can reduce (2) to one of the three following forms:

1. $a_1 x''^2 + a_2 y''^2 + a_3 z''^2 + c' = 0$,   rank of $\mathbf{A} = 3$, $a_1, a_2, a_3 \neq 0$;
2. $a_1 x''^2 + a_2 y''^2 + 2b_3' z'' + c' = 0$,   rank of $\mathbf{A} = 2$, $a_1, a_2 \neq 0$;
3. $a_1 x''^2 + 2b_2' y'' + 2b_3' z'' + c' = 0$,   rank of $\mathbf{A} = 1$, $a_1 \neq 0$.

The rank of $\mathbf{A}$ enters into this, since
$$\text{rank } \mathbf{A} = \text{rank } \mathbf{P^*AP}$$
$$= \text{rank of diagonal form of } \mathbf{A}$$
$$= \text{number of nonzero eigenvalues } a_i.$$

Example (cont'd). Here the rank of $\mathbf{A}$ is 2 and we write our equation in the form
$$3(x' - \sqrt{3})^2 - 9 - 4(y' + \sqrt{2})^2 + 8 + 2\sqrt{6}\, z' - 11 = 0,$$
or
$$3x''^2 - 4y''^2 + 2\sqrt{6}\, z' - 12 = 0,$$
where
$$x'' = x' - \sqrt{3}, \qquad y'' = y' + \sqrt{2}.$$
by writing the last two terms in the form $2\sqrt{6}(z' - \sqrt{6})$ and putting $z'' = z' - \sqrt{6}$, we get the final result
$$3x''^2 - 4y''^2 + 2\sqrt{6}\, z'' = 0. \tag{5}$$
This has therefore been achieved by making the translation
$$x'' = x' - \sqrt{3}, \qquad y'' = y' + \sqrt{2}, \qquad z'' = z' - \sqrt{6}.$$

It is now easy to picture the surface given by the equation. Slicing the surface with a plane perpendicular to the x''-axis (that is, a plane $x'' = a$), we have in (5),
$$4y''^2 - 2\sqrt{6}\, z'' = 3a^2,$$
a parabola. Cutting it with a plane perpendicular to the y''-axis (that is, $y'' = b$), we get another parabola,
$$3x''^2 + 2\sqrt{6}\, z'' = 4b^2.$$
Finally, its cross section with respect to the plane $z'' = c$ is the hyperbola
$$3x''^2 - 4y''^2 = -2\sqrt{6}\, c.$$
The entire locus is called a "hyperbolic paraboloid". (See Fig. 2 p. 434).

We now make a complete listing of all possibilities. The cases are grouped first according to the rank of the matrix $\mathbf{A}$ of the quadratic form, then according to whether the constant term is or is not zero, and finally according to the relationship of the signs of the coefficients.

1. Rank $\mathbf{A} = 3$. Then $\mathbf{A}$ has no zero eigenvalues and we obtain
$$a_1 x^2 + a_2 y^2 + a_3 z^2 + c = 0,$$
eliminating all first-degree terms.

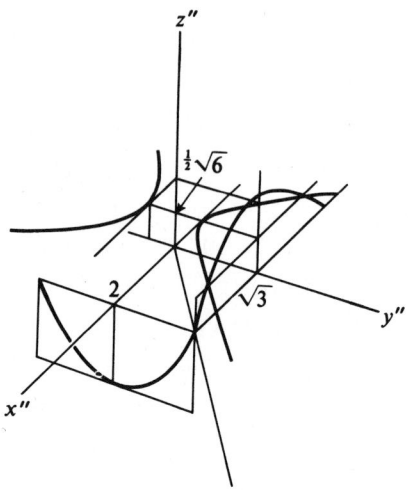

Fig. 2

1.1 $c \neq 0$. Depending on the signs of the coefficients, we get four subcases which can be written in the following standard forms:

1.1(a)

$$\frac{x^2}{a^2} + \frac{y^2}{b^2} + \frac{z^2}{c^2} = 1.$$

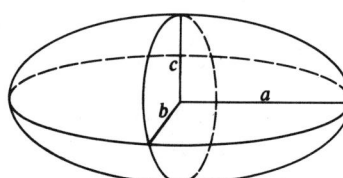

Fig. 3

1.1(b)

$$\frac{x^2}{a^2} + \frac{y^2}{b^2} + \frac{z^2}{c^2} + 1 = 0.$$

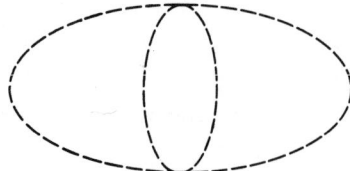

Fig. 4

3 / The general second-degree equation   435

1.1(c)
$$\frac{x^2}{a^2} + \frac{y^2}{b^2} - \frac{z^2}{c^2} = 1.$$

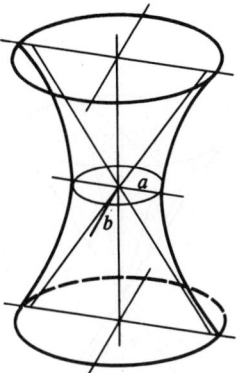

Fig. 5

1.1(d)
$$\frac{x^2}{a^2} + \frac{y^2}{b^2} - \frac{z^2}{c^2} = -1.$$

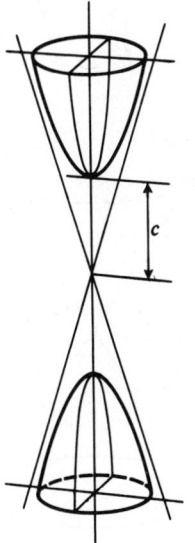

Fig. 6

1.2 $c = 0$. Two sub-cases are:

1.2(a)
$$\frac{x^2}{a^2} + \frac{y^2}{b^2} - \frac{z^2}{c^2} = 0.$$

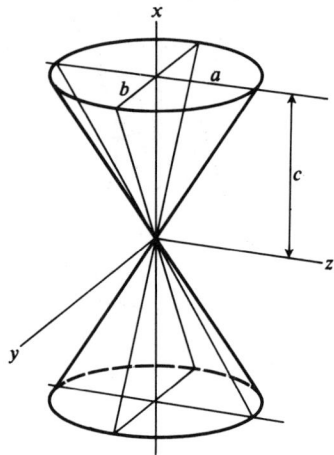

Fig. 7

1.2(b)
$$\frac{x^2}{a^2} + \frac{y^2}{b^2} + \frac{z^2}{c^2} = 0.$$

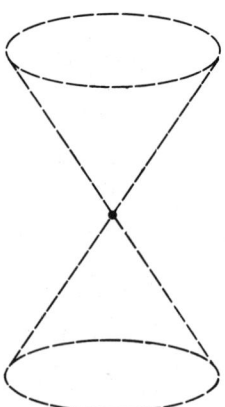

Fig. 8

2. Rank $A = 2$, so there is one zero eigenvalue (yielding one term of degree 1),

$$a_1 x^2 + a_2 y^2 + 2b_3 z + c = 0.$$

2.1 $b_3 \neq 0$. Then we can make $c = 0$. Two subcases are:

2.1(a)

$$\frac{x^2}{a^2} + \frac{y^2}{b^2} = 2cz.$$

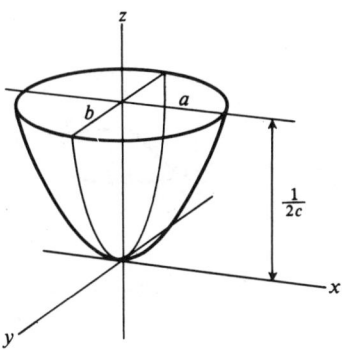

Fig. 9

2.1(b)

$$\frac{x^2}{a^2} - \frac{y^2}{b^2} = 2cz.$$

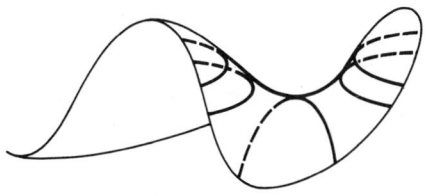

Fig. 10

2.2 $b_3 = 0$, $c \neq 0$. There are three subcases.

2.2(a)
$$\frac{x^2}{a^2} + \frac{y^2}{b^2} = 1.$$

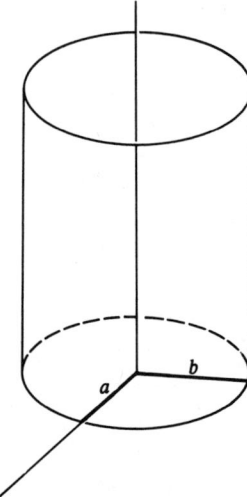

Fig. 11

2.2(b)
$$\frac{x^2}{a^2} + \frac{y^2}{b^2} = -1.$$

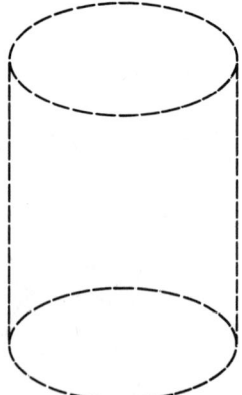

Fig. 12

2.2(c)
$$\frac{x^2}{a^2} - \frac{y^2}{b^2} = 1.$$

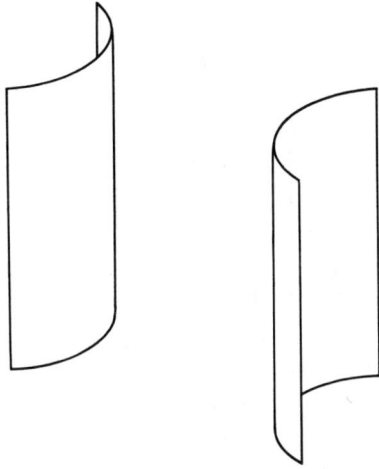

Fig. 13

2.3 $b_3 = 0$, $c = 0$. There are two subcases:

2.3(a)
$$\frac{x^2}{a^2} - \frac{y^2}{b^2} = 0.$$

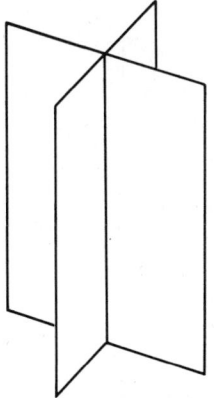

Fig. 14

2.3(b)
$$\frac{x^2}{a^2} + \frac{y^2}{b^2} = 0.$$

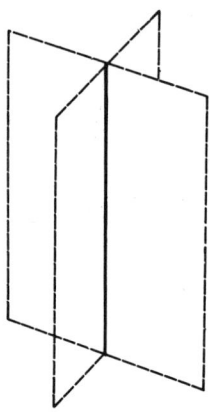

Fig. 15

3. rank $\mathbf{A} = 1$ (two zero eigenvalues giving two terms of degree 1):
$$a_1 x^2 + 2b_2 y + 2b_3 z + c = 0.$$
Here we can make $b_3 = 0$ by rotating about the x-axis. The equation is then
$$a_1 x^2 + 2b_2 y + c = 0,$$
and we get the following possibilities:

3.1 $b_2 \neq 0$. Then we can make $c = 0$:
$$x^2 - 2by = 0.$$

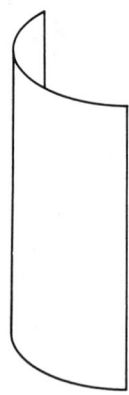

Fig. 16

3.2 $b_2 = 0$, $c \neq 0$. There are two subcases:

3.2(a)
$$x^2 - a^2 = 0.$$

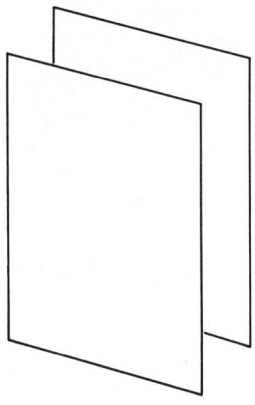

Fig. 17

3.2(b)
$$x^2 + a^2 = 0.$$

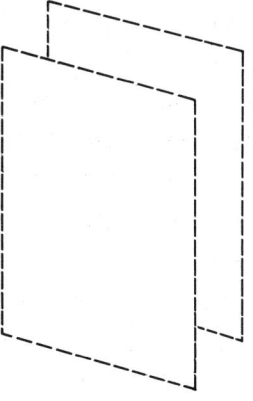

Fig. 18

3.3 $b_2 = 0, c = 0$:
$$x^2 = 0.$$

Fig. 19

There is thus a total of 17 cases, 5 of which yield essentially no real locus.

EXERCISES

In problems 1 through 17, reduce each equation to its standard form and identify the result.

1. $-x^2 - y^2 + z^2 + 6xy + 2xz + 2yz - 14x + 2y - 6z - 11 = 0$
2. $-x^2 - y^2 + z^2 + 6xy + 2xz + 2yz - 14x + 2y - 6z + 1 = 0$
3. $5x^2 + 5y^2 + 9z^2 + 2xy - 6xz - 6yz - 22x - 38y + 42z + 53 = 0$
4. $5x^2 + 5y^2 + 9z^2 + 2xy - 6xz - 6yz - 22x - 38y + 42z + 125 = 0$
5. $5x^2 + 5y^2 + 9z^2 + 2xy - 6xz - 6yz - 22x - 38y + 42z + 89 = 0$
6. $x^2 + y^2 + 9z^2 + 10xy - 6xz - 6yz - 38x - 22y + 42z + 37 = 0$
7. $x^2 + y^2 + 9z^2 + 10xy - 6xz - 6yz - 38x - 22y + 42z + 109 = 0$
8. $x^2 + y^2 + 9z^2 + 10xy - 6xz - 6yz - 38x - 22y + 42z + 73 = 0$
9. $17x^2 + 12xy + 8y^2 + 60x + 80y - 5z + 210 = 0$
10. $17x^2 + 12xy + 8y^2 + 60x + 80y + 180 = 0$
11. $17x^2 + 12xy + 8y^2 + 60x + 80y + 220 = 0$
12. $17x^2 + 12xy + 8y^2 + 60x + 80y + 200 = 0$
13. $x^2 + 4y^2 + 4z^2 + 4xy - 4xz - 8yz - 9y - 9z = 0$
14. $x^2 + 4y^2 + 4z^2 + 4xy - 4xz - 8yz - 4 = 0$
15. $x^2 + 4y^2 + 4z^2 + 4xy - 4xz - 8yz + 4 = 0$
16. $x^2 + 4y^2 + 4z^2 + 4xy - 4xz - 8yz = 0$
17. $x^2 + y^2 + z^2 - 4xy - 4yz - 4xz - 3 = 0$

18. In which of the 17 cases can the surface be a surface of revolution? How do the eigenvalues of **A** relate to this question?
19. Study the general equation of the second degree in x, y by the methods here explained. Describe the loci in the plane that result.
20. Study the general equation of the second degree in w, x, y, and z, and give a classification of those whose quadratic form is of rank 4.

4. POSITIVE DEFINITE MATRICES

Positive definite matrices arise in several important contexts (for example, in giving conditions that a real-valued function of several variables have a minimum at a certain point). There are surprising anologies between positive definite matrices and positive real numbers, and interesting relations between positive definite and orthogonal matrices (see Chapter 10, Section 5).

In this section we define positive definite matrices, and give several theoretical and several practical tests for positive definiteness.

A matrix **A** in $\mathbb{R}_{n \times n}$ is **positive definite** if (i) **A** is symmetric and (ii) $\mathbf{X}^*\mathbf{AX} > 0$ for every nonzero **X** in $\mathbb{R}_n$ (note that $\mathbf{X}^*\mathbf{AX}$ is a real number—a one-by-one matrix; alternatively, it is the value of a quadratic form at the point $\mathbf{X}^* = (x_1, \ldots, x_n)$; so (ii) makes sense).

Example 1. We shall soon see that the familiar matrix

$$\mathbf{A} = \begin{pmatrix} 5 & -1 & -1 \\ -1 & 3 & 1 \\ -1 & 1 & 3 \end{pmatrix}$$

is positive definite. It is surely symmetric. Taking a test value of $\mathbf{X} = (-1, 1, 0)^*$, we have

$$\mathbf{AX} = (-6, 4, 2)^*,$$

so that $\mathbf{X}^*\mathbf{AX} = 10 > 0$. Of course, this does not show that $\mathbf{X}^*\mathbf{AX} > 0$ for *all* $\mathbf{X} \neq \mathbf{0}$. If we try an arbitrary **X**, we obtain

$$\mathbf{X}^*\mathbf{AX} = 5x_1^2 + 3x_2^2 + 3x_3^2 - 2x_1x_2 - 2x_1x_3 + 2x_2x_3, \tag{1}$$

a real-valued function of three variables with which we are even less equipped to deal than with symmetric matrices (in this context).

You recognize (1) as a quadratic form; in fact, one encountered in Section 2. We define such a form to be **positive definite** if it is positive for all nonzero values of **X**. This is the same as saying that the (symmetric) matrix of the form is positive definite.

4.1 A matrix **A** is positive definite iff all of its eigenvalues are positive real numbers.

Proof. By the Principal Axis Theorem, we know that for any symmetric matrix **A** there is an orthogonal **P** for which

$$\mathbf{PAP^*} = \mathbf{D} = \begin{pmatrix} \lambda_1 & & 0 \\ & \cdot & \\ & & \cdot \\ 0 & & \lambda_n \end{pmatrix}.$$

Then

$$\mathbf{X^*AX} \stackrel{\text{why}_1?}{=} \mathbf{(PX)^*(PAP^*)(PX)} = \mathbf{(PX)^*D(PX)}.$$

If $\mathbf{Y} = \mathbf{PX}$, it is clear that $\mathbf{Y^*DY} > 0$ for all **Y** iff each $\lambda_i > 0$ (why$_2$?). But, given any **X**, there is a **Y** in $\mathbb{R}_n$ such that $\mathbf{Y} = \mathbf{PX}$, and conversely (why$_3$?). Hence, $\mathbf{X^*AX} > 0$ for all **X** iff $\mathbf{Y^*DY} > 0$ for all **Y** iff each eigenvalue λ_i of **A** is > 0, and the theorem is proved. ∎

Example 1 (cont'd.). From Example 1, Section 1, we know the eigenvalues of our particular **A** are 2, 3, and 6. Hence **A** is positive definite.

Another theoretical test for positive definiteness which is sometimes useful is

4.2 *The Principal Minors Test*. A symmetric matrix

$$\mathbf{A} = \begin{pmatrix} a_{11} & \cdots & a_{1n} \\ \cdot & & \cdot \\ \cdot & & \cdot \\ a_{n1} & \cdots & a_{nn} \end{pmatrix}$$

is positive definite iff each one of the principal minors

$$\det(a_{11}), \quad \det\begin{pmatrix} a_{11} & a_{12} \\ a_{21} & a_{22} \end{pmatrix}, \ldots, \det\begin{pmatrix} a_{11} & \cdots & a_{1,n-1} \\ \cdot & & \cdot \\ \cdot & & \cdot \\ a_{n-1,1} & & a_{n-1,n-1} \end{pmatrix}, \quad \det \mathbf{A}$$

is positive.

Proof in Section 5. We could give an inductive proof at this time, but a more elegant proof will be a natural byproduct of our researches in the next section (which, we hasten to add, is independent of the rest of the material in this section).

As we know, calculating all the eigenvalues of even a symmetric matrix is a nontrivial task, while computing n determinants will require at least a bit of effort. Consequently, the following result is usually more convenient than our previous theorems.

4.3 *The Pivot Test*. To test whether a symmetric matrix is positive definite, successively pivot on the diagonal entries (that is, pivot on the 1,1-entry of **A**, the 2,2-entry of the resulting matrix, etc.).

A is positive definite iff all the pivot entries thus encountered are positive.

Proof (by induction). This is certainly true for 1×1 matrices, so suppose that it holds for all symmetric $(n-1) \times (n-1)$ matrices. Given an $n \times n$ symmetric matrix

$$\mathbf{A} = \begin{pmatrix} a_{11} & \mathbf{a} \\ \mathbf{a}^* & \mathbf{A}_1 \end{pmatrix},$$

if a_{11} is nonpositive, then **A** cannot be positive definite (try $\mathbf{e}_1^* \mathbf{A} \mathbf{e}_1$). Thus, we may assume that $a_{11} > 0$, and pivot on it. We leave you to verify (Exercise 2) the result:

$$\mathbf{A} = \begin{pmatrix} a_{11} & \mathbf{a} \\ \mathbf{a}^* & \mathbf{A}_1 \end{pmatrix} \to \begin{pmatrix} 1 & a_{11}^{-1}\mathbf{a} \\ 0 & \mathbf{A}_1 - a_{11}^{-1}\mathbf{a}\mathbf{a}^* \end{pmatrix} = \mathbf{PA},$$

where

$$\mathbf{P} = \begin{pmatrix} a_{11}^{-1} & \mathbf{0} \\ -a_{11}^{-1}\mathbf{a}^* & \mathbf{I} \end{pmatrix}.$$

Moreover,

$$\mathbf{PAP}^* = \begin{pmatrix} a_{11}^{-1} & \mathbf{0} \\ \mathbf{0} & \mathbf{A}_1 - a_{11}^{-1}\mathbf{a}\mathbf{a}^* \end{pmatrix}.$$

Now **PAP*** is symmetric (why$_1$?), and since

$$\mathbf{X}(\mathbf{PAP}^*)\mathbf{X}^* = (\mathbf{XP})\mathbf{A}(\mathbf{XP})^*,$$

it follows that **A** is positive definite iff **PAP*** is (why$_2$?). (Compare the proof of 4.1.) Moreover, **PAP*** has $a_{11}^{-1} > 0$ as an eigenvalue. Hence, **A** will be positive definite iff $\mathbf{A}_1 - a_{11}^{-1}\mathbf{a}\mathbf{a}^*$ is positive definite (why$_3$?). But this is an $(n-1) \times (n-1)$ symmetric matrix, to which the induction hypothesis applies. Thus, it is positive definite iff successive pivoting on its diagonal elements yields only positive diagonal elements. Considering how this $(n-1) \times (n-1)$ matrix is situated inside the pivoted-upon matrix **PA**, the result then follows immediately (why$_4$?). ∎

Note that in performing this test, after pivoting on the ith entry one can throw away the ith row and column, since they will have no further bearing on the outcome.

Example 1 again:

$$\begin{pmatrix} \boxed{5} & -1 & -1 \\ -1 & 3 & 1 \\ -1 & 1 & 3 \end{pmatrix} \to \begin{pmatrix} 1 & -\frac{1}{5} & -\frac{1}{5} \\ 0 & \boxed{\frac{14}{5}} & \frac{4}{5} \\ 0 & \frac{4}{5} & \frac{14}{5} \end{pmatrix} \to \begin{pmatrix} 1 & \frac{4}{14} \\ 0 & \boxed{\frac{18}{7}} \end{pmatrix} \to (1)$$

All pivot entries are positive.

There are tests which are not both necessary and sufficient, and which are therefore easier to perform. We give two.

4.4 If any diagonal entry a_{ii} of the symmetric matrix **A** is nonpositive, then **A** cannot be positive definite.

Proof. Calculate $\mathbf{e}_i^* \mathbf{A} \mathbf{e}_i$. ∎

4.5 A symmetric matrix

$$\mathbf{A} = \begin{pmatrix} a_{11} & \cdots & a_{1n} \\ \vdots & & \vdots \\ a_{n1} & \cdots & a_{nn} \end{pmatrix}$$

is positive definite if

$$a_{11} > |a_{12}| + \cdots + |a_{1n}|, \qquad a_{22} > |a_{21}| + |a_{23}| + \cdots + |a_{2n}|, \ldots ,$$
$$a_{nn} > |a_{n1}| + \cdots + |a_{n,n-1}|.$$

Proof. This is reserved for cognoscenti of Gershgorin's Theorem (Supplement 1, Section 1, Chapter 6). ∎

Example 1 (although rather tired) will serve to show how convenient 4.5 can be.

We define a matrix **A** to be **negative definite** if (i) it is symmetric, and (ii) $\mathbf{X}^* \mathbf{A} \mathbf{X} < 0$ for every nonzero **X** in $\mathbb{R}_n$. Since **A** is negative definite iff $-\mathbf{A}$ is positive definite, it follows that all the above theorems hold for negative definite matrices if **A** is replaced by $-\mathbf{A}$.

One must be wary of a too hasty translation of theorems, however. For example, note that

$$\mathbf{A} = \begin{pmatrix} -1 & 0 & 0 & 0 \\ 0 & -2 & 0 & 0 \\ 0 & 0 & -3 & 0 \\ 0 & 0 & 0 & -4 \end{pmatrix}$$

is surely negative definite (e.g., if you don't wish a direct calculation $-\mathbf{A}$ has all positive eigenvalues and by 4.1 is positive definite). However, take care in applying the Principal Minors Test. One cannot simply replace "principal minors all positive" by "principal minors all negative." For det $\mathbf{A} = 24 > 0$, for example. In fact, the principal minors are

$$\det(-1) = -1, \qquad \det \begin{pmatrix} -1 & 0 \\ 0 & -2 \end{pmatrix} = 2,$$

$$\det \begin{pmatrix} -1 & 0 & 0 \\ 0 & -2 & 0 \\ 0 & 0 & -3 \end{pmatrix} = -6, \qquad \det \mathbf{A} = 24.$$

The correct reading of the Principal Minors Test is, of course, that the principal minors of

$$-A = \begin{pmatrix} 1 & & & \\ & 2 & & \\ & & 3 & \\ & & & 4 \end{pmatrix}$$

be all positive, which is certainly the case.

A symmetric matrix for which $X^*AX \geq 0$ for all X in $\mathbb{R}_n$ is called *positive semidefinite*. (It may seem as though we're trying to milk a single concept for all it's worth. We're not. There are actual situations which demand the consideration of each of our "definite" cases).

The proof of 4.1 can be easily amended (Exercise 3d) to prove that

4.6 A symmetric matrix is positive semidefinite iff its eigenvalues are all nonnegative.

Unfortunately, the Principal Minors Test does *not* become "A is positive semidefinite iff all the principal minors in 4.2 are nonnegative". For example,

$$A = \begin{pmatrix} 0 & 0 \\ 0 & -1 \end{pmatrix}$$

has nonnegative principal minors, but A is *"negative* semidefinite". However, 4.5 goes through for positive semidefinite matrices by simply changing $>$ to $\geq$ (Exercise 3c).

Our next result is more for entertainment than practicality:

4.7 (i) A symmetric matrix A is positive definite iff the coefficients of its characteristic polynomial alternate in sign:

$$\det(A - xI) = a_n x^n - a_{n-1} x^{n-1} + \cdots + (-1)^{n-1} a_1 x + (-1)^n a_0, \quad (1)$$

with a's either all positive or all negative.

ii) A is positive semidefinite iff the coefficients alternate in sign down to the first zero, after which all are zero.

iii) A is negative definite iff all coefficients have the same sign.

Proof. By 4.1, we see that this is really a result on the sign of the coefficients of a polynomial vs. the sign of the roots.

i) 1. If all the roots are positive then it is easy to prove by induction that the coefficients alternate in sign (Exercise 4).

2. Conversely, if the characteristic polynomial (1) has coefficients which alternate in sign, consider its value for a negative value of x, say, $x = -c, c > 0$.

Then for each of the terms in (1) we have
$$(-1)^{n-k}a_k x^k = (-1)^{n-k}a_k(-c)^k = (-1)^{n-k}(-1)^k a_k c^k = (-1)^n a_k c^k;$$
since $c > 0$ all the terms in the summation in (1) are of the same sign. Hence, the sum is not zero. Hence, c can't be a root of (1). Since all the roots are real (why?), this proves the converse.
(ii) and (iii) are for you (Exercise 4). ∎

The following two results will be of use in the next chapter. The first is an alternative way of looking at positive definiteness.

4.8 An $n \times n$ symmetric matrix **A** is positive definite iff $\mathbf{AV} \cdot \mathbf{V} > 0$ for all $\mathbf{V} \neq \mathbf{0}$ in $\mathbb{R}_n$. (It is positive semi-definite iff $\mathbf{AV} \cdot \mathbf{V} \geq 0$ for all **V**.)

Proof. Just note that $\mathbf{AV} \cdot \mathbf{V} = (\mathbf{AV})^*\mathbf{V} = \mathbf{V}^*\mathbf{A}^*\mathbf{V} = \mathbf{V}^*\mathbf{AV}$. ∎

The next theorem concerns **A*A** when **A** is not necessarily square. (Note that **A*A** is always both square and symmetric.)

4.9 Let **A** be an arbitrary $m \times n$ matrix. Then
i) **A*A** is always positive semi-definite;
ii) $\mathbf{A}^*\mathbf{AV} = \mathbf{0}$ iff $\mathbf{AV} = \mathbf{0}$;
iii) the rank of **A*A** is the same as the rank of **A**.

Proof. (i)
$$(\mathbf{A}^*\mathbf{AV}) \cdot \mathbf{V} = (\mathbf{A}^*\mathbf{AV})^*\mathbf{V} = \mathbf{V}^*\mathbf{A}^*\mathbf{AV}$$
$$= (\mathbf{AV}) \cdot (\mathbf{AV}) = |\mathbf{AV}|^2 \geq 0,$$
so **A*A** is positive semidefinite by 4.8.

ii) If $\mathbf{AV} = \mathbf{0}$, surely $\mathbf{A}^*\mathbf{AV} = \mathbf{0}$. Conversely, if $\mathbf{A}^*\mathbf{AV} = \mathbf{0}$, then
$$0 = \mathbf{V}^*(\mathbf{A}^*\mathbf{AV}) = (\mathbf{AV})^*(\mathbf{AV}) = (\mathbf{AV}) \cdot (\mathbf{AV}) = |\mathbf{AV}|^2,$$
so $\mathbf{AV} = \mathbf{0}$.

iii) If **A** is $m \times n$, then rank $\mathbf{A} = n - k$, where k is the dimension of the kernel of $\mathbf{T}(\mathbf{V}) = \mathbf{AV}$. By (ii), k is also the dimension of the kernel of $\mathbf{T}'(\mathbf{V}) = \mathbf{A}^*\mathbf{AV}$. Since **A*A** is $n \times n$, its rank is therefore also $n - k$. ∎

It is sometimes necessary to consider the problem of reducing two quadratic forms simultaneously to diagonal form by means of the same invertible change of variables (for example, in studying the intersection of two quadratic surfaces). Let **A** and **B** be the symmetric $n \times n$ matrices of these forms. If we assume that **B** is positive definite, we can show that this simultaneous diagonalization can be done (although perhaps not by an orthogonal change of basis). We first find **P** so that $\mathbf{B} = \mathbf{P}^*\mathbf{P}$, where **P** is invertible (Exercise 5). Then take $\mathbf{A}' = \mathbf{P}^{-1*}\mathbf{AP}^{-1}$ (which is symmetric) and diagonalize it:

$$\mathbf{A}' = \mathbf{U}^*\mathbf{DU}, \quad \text{where } \mathbf{U} \text{ is orthogonal, } \mathbf{D} \text{ is diagonal,}$$

$$D = \begin{pmatrix} \lambda_1 & & 0 \\ & \cdot & \\ & & \cdot \\ 0 & & \lambda_n \end{pmatrix}, \quad \text{where } \lambda_i \text{ are the eigenvalues of } \mathbf{A}'.$$

Then $\quad \mathbf{A} = \mathbf{P}^*\mathbf{A}'\mathbf{P} = (\mathbf{UP})^*\mathbf{D}(\mathbf{UP}) = \mathbf{Q}^*\mathbf{DQ}, \quad$ where $\mathbf{Q} = \mathbf{UP}$,

and $\quad \mathbf{B} = \mathbf{P}^*\mathbf{P} = (\mathbf{UP})^*(\mathbf{UP}) = \mathbf{Q}^*\mathbf{IQ}$,

so that **A** and **B** have been simultaneously diagonalized by the invertible change of variables whose matrix is **Q**. (**Q** will not usually be orthogonal in this situation.)

The diagonal elements $\lambda_1, \ldots, \lambda_n$, being the eigenvalues of $\mathbf{A}'$, are the roots of

$$\det (\mathbf{A}' - \lambda \mathbf{I}) = 0.$$

Since
$$\mathbf{A}' - \lambda \mathbf{I} = \mathbf{P}^{*-1}(\mathbf{A} - \lambda \mathbf{P}^*\mathbf{P})\mathbf{P}^{-1}$$
$$= \mathbf{P}^{*-1}(\mathbf{A} - \lambda \mathbf{B})\mathbf{P}^{-1},$$

we get
$$\det (\mathbf{A}' - \lambda \mathbf{I}) = \det (\mathbf{A} - \lambda \mathbf{B}).$$

Consequently the required diagonal elements can be found immediately as the roots of $\det (\mathbf{A} - \lambda \mathbf{B}) = 0$. This incidentally proves the following result, which is useful in discussing extreme values of functions of several variables.

4.10 If **B** is positive definite and **A** is symmetric, then the roots of

$$\det (\mathbf{A} - \lambda \mathbf{B}) = 0$$

are all real.

It is clear that this result also holds if **B** is *negative* definite (Exercise 6). It is also easy to prove that the same result holds if we assume that **A** is definite (positive or negative) instead of **B**. For $n = 2$ we get the following corollary (which is also helpful in studying extrema):

4.11 If $\mathbf{A} = \begin{pmatrix} a & b \\ b & c \end{pmatrix}$, and $\mathbf{B} = \begin{pmatrix} a' & b' \\ b' & c' \end{pmatrix}$ where $ac - b^2 \geq 0$ or $a'c' - b'^2 \geq 0$

(or both), then the roots of

$$\det \begin{pmatrix} a - \lambda a' & b - \lambda b' \\ b - \lambda b' & c - \lambda c' \end{pmatrix} = (ac - b^2) - \lambda(a'c + ac' - 2bb') + \lambda^2(a'c' - b'^2) = 0$$

are real.

EXERCISES

1. Are the following matrices positive definite? semidefinite? negative definite? semidefinite? Give reasons.

   a) $\begin{pmatrix} 4 & 12 \\ 12 & 11 \end{pmatrix}$ b) $\begin{pmatrix} 17 & 6 \\ 6 & 8 \end{pmatrix}$ c) $\begin{pmatrix} -1 & 3 & 1 \\ 3 & -1 & 1 \\ 1 & 1 & 1 \end{pmatrix}$

   d) $\begin{pmatrix} 5 & 1 & -3 \\ 1 & 5 & -3 \\ -3 & -3 & 9 \end{pmatrix}$ e) $\begin{pmatrix} 0 & 1 & 2 \\ 1 & 0 & -3 \\ 2 & -3 & 0 \end{pmatrix}$ f) $\begin{pmatrix} 1 & 2 & -2 \\ 2 & 4 & -4 \\ -2 & -4 & 4 \end{pmatrix}$

2. a) In 4.3 show that pivoting results in the indicated **PA**.
   b) Answer the questions in the proof.
3. (a) Prove 4.4. (b) Prove 4.5. (c) Revise 4.5 for positive semidefinite matrices.
   (d) Prove 4.6.
4. (a) Prove 4.7(i). (b) (ii). (c) (iii).
5. Prove that **A** is positive semidefinite iff $\mathbf{A} = \mathbf{B^*B}$ for some square matrix **B**; **B** is invertible iff **A** is positive definite.
6. Prove 4.10 (a) when **B** is negative definite, and (b) when **A** is positive definite.

5.* NONORTHOGONAL DIAGONALIZATION OF QUADRATIC FORMS

Our discussion of the diagonalization of quadratic forms in Section 2 involved the diagonalization of the associated symmetric matrix. This usually takes some effort. There is an easier method of diagonalizing forms which involves only simple steps to determine an appropriate linear change of variables (although you do lose some information in the process).

Given the quadratic form

$$Q(x_1, \ldots, x_n) = a_{11}x_1^2 + 2a_{12}x_1x_2 + \cdots + 2a_{1n}x_1x_n + \cdots + a_{nn}x_n^2$$

$$= \sum_{i,j=1}^{n} a_{ij}x_ix_j,$$

we proceed as follows:

If $a_{11} \neq 0$, then

$$Q(x_1, \ldots, x_n) = a_{11}\left(x_1 + \frac{a_{12}}{a_{11}}x_2 + \cdots + \frac{a_{1n}}{a_{11}}x_n\right)^2$$

$$- a_{11}\left(\frac{a_{12}}{a_{11}}x_2 + \cdots + \frac{a_{1n}}{a_{11}}x_n\right)^2 + \sum_{\substack{i \neq 1 \\ j \neq 1}}^{n} a_{ij}x_ix_j$$

$$= a_{11}y_1^2 + Q_1(x_2, \ldots, x_n),$$

where

$$y_1 = x_1 + \frac{a_{12}}{a_{11}}x_2 + \cdots + \frac{a_{1n}}{a_{11}}x_n$$

and Q_1 is what's left over:

$$Q_1(x_2, \ldots, x_n) = \sum_{\substack{i \neq 1 \\ j \neq 1}} b_{ij} x_i x_j,$$

where

$$b_{ij} = a_{ij} - \frac{a_{1i} a_{1j}}{a_{11}}.$$

If $a_{11} = 0$ and some $a_{ii} \neq 0$, we renumber the variables so that x_i becomes x_1' and proceed as above.

If all $a_{ii} = 0$ and some $a_{ij} \neq 0$, we renumber so that $a_{12} \neq 0$ (for convenience) and then put

$$x_1 = y_1 + y_2, \qquad x_2 = y_1 - y_2,$$

leaving the other variables unchanged. We get

$$Q(x_1, \ldots, x_n) = 2a_{12} y_1^2 - 2a_{12} y_2^2 + \cdots + a_{nn} x_n^2,$$

and since the first term is the only one in y_1^2, we can proceed as before.

If all $a_{ij} = 0$, there is nothing to do.

Thus, in all cases (of interest) we get

$$Q(x_1, \ldots, x_n) = a_1 y_1^2 + Q_1(y_2, \ldots, y_n), \qquad \text{where } a_1 \neq 0,$$

by an invertible linear change of variables. Apply the same process to Q_1 and continue until there is nothing left. We finally obtain

$$Q(x_1, \ldots, x_n) = a_1 x_1'^2 + \cdots + a_r x_r'^2, \qquad (1)$$

where $r \leq n$ and all $a_i \neq 0$. Here $\mathbf{X} = \mathbf{PX}'$ for some linear transformation $\mathbf{P}$ (not necessarily orthogonal) and

$$Q(x_1, \ldots, x_n) = \mathbf{X}'^* \mathbf{A}' \mathbf{X}',$$

where

$$\mathbf{A}' = \mathbf{P}^* \mathbf{A} \mathbf{P} = \begin{pmatrix} a_1' & & & & 0 \\ & \cdot & & & \\ & & \cdot & & \\ & & & a_r' & \\ & & & & 0 \\ & & & & & \cdot \\ 0 & & & & & & 0 \end{pmatrix}$$

Here $r = $ rank of $\mathbf{A}'$, and since $\mathbf{A}$ is equivalent to $\mathbf{A}'$, r is also the rank of $\mathbf{A}$.

We next observe that we can even proceed further and replace each a_i in (1) by $+1$ or -1:

If
$$a_i > 0, \quad \text{let } x_i'' = \sqrt{a_i}\, x_i';$$
if
$$a_i < 0, \quad \text{let } x_i'' = \sqrt{-a_i}\, x_i'.$$
Then
$$Q(x_1, \ldots, x_n) = \epsilon_1 x_1''^2 + \cdots + \epsilon_n x_r''^2,$$
where $\epsilon_i = \pm 1$. Renumber the x_i'' to obtain the final form
$$Q(x_1, \ldots, x_n) = x_1''^2 + x_2''^2 + \cdots + x_k''^2 - x_{k+1}''^2 - \cdots - x_r''^2, \qquad (2)$$
so that k is the number of terms with $\epsilon = +1$. It is easy to see (Exercise 2) that

$$\left. \begin{array}{l} \text{the form } Q(x_1, \ldots, x_n) \text{ is positive semi-definite if } k = r; \\ \text{the form } Q(x_1, \ldots, x_n) \text{ is positive definite if } k = r = n; \\ \text{the form } Q(x_1, \ldots, x_n) \text{ is negative semi-definite if } k = 0; \\ \text{the form } Q(x_1, \ldots, x_n) \text{ is negative definite if } k = 0 \text{ and } r = n. \end{array} \right\} \qquad (3)$$

In diagonalizing a quadratic form by means of an orthogonal transformation as in Section 2, the coefficients $\lambda_1, \ldots, \lambda_r$ of the resulting form
$$\lambda_1 x_1''^2 + \cdots + \lambda_r x_r''^2$$
are unique (up to permutation), since they are the eigenvalues of the matrix **A** of the form. However, if we allow the more general types of substitutions above, the coefficients are *not* unique (for example, we went from (1) to (2) and could by similar means obtain a great variety of different coefficients). However, we at least preserve the rank of the associated matrix (as noted above). Moreover,

5.1 Sylvester's Law of Inertia. *If a quadratic form $Q(x_1, \ldots, x_n)$ is changed to a diagonal form*
$$Q'(y_1, \ldots, y_n) = a_1 y_1^2 + \cdots + a_r y_r^2,$$
by an invertible linear transformation of the variables, then the number of positive and negative coefficients of the diagonal form is always the same.

Proof. Let us change the coefficients to ± 1 via the transformation used in going from (1) to (2) (which certainly preserves the signs of coefficients). Thus, we may assume that
$$Q'(y_1, \ldots, y_n) = y_1^2 + \cdots + y_k^2 - y_{k+1}^2 - \cdots - y_r^2.$$
Suppose also that an invertible linear transformation of the variables in Q also yields the form
$$Q''(z_1, \ldots, z_n) = z_1^2 + \cdots + z_l^2 - z_{l+1}^2 - \cdots - z_r^2.$$

(after changing coefficients to ± 1 again). As we observed above, the rank r of the matrices of both forms will be the same. We must show that $k = l$. By symmetry it is enough to show that the assumption that $k < l$ leads to a contradiction.

If $k < l$, then, equating forms Q' and Q'', we obtain

$$y_1^2 + \cdots + y_k^2 + z_{l+1}^2 + \cdots + z_r^2 = z_1^2 + \cdots + z_l^2 + y_{k+1}^2 + \cdots + y_r^2,$$
$$k < l. \quad (4)$$

Moreover, we have expressions for y's in terms of z's:

$$y_i = \sum_{j=1}^{n} b_{ij} z_j, \quad i = 1, \ldots, n, \quad (5)$$

since we can get from the x's to both by means of an invertible linear transformation (elaborate$_1$). Now take

$$z_{l+1}, z_{l+2}, \ldots, z_n \quad \text{all } 0.$$

Then consider the system of homogeneous equations which results from the first k equations in (5) if we take $y_1, \ldots, y_k = 0$:

$$0 = \sum_{j=1}^{l} b_{1j} z_j, \ldots, 0 = \sum_{j=1}^{l} b_{kj} z_j.$$

Since we have k equations in $l > k$ unknowns, there is a solution $z_1, \ldots, z_l$, not all 0 (elaborate$_2$). In summary, we have

$$y_1 = 0, \ldots, y_k = 0, \quad z_{l+1} = 0, \ldots, z_r = 0;$$
$$\text{but } z_1, \ldots, z_l \text{ not all 0.}$$

Hence, in (4) we have

$$0 = y_1^2 + \cdots + y_k^2 + z_{l+1}^2 + \cdots + z_r^2$$
$$= z_1^2 + \cdots + z_l^2 + y_{k+1}^2 + \cdots + y_r^2,$$

the last of which cannot be 0. This contradiction establishes the result. ∎

In the previous section we defined a quadratic form Q to be positive definite if the associated matrix $\mathbf{A}$ of the form was positive definite, that is, if $\mathbf{X}^* \mathbf{A} \mathbf{X} > 0$ for all nonzero $\mathbf{X}$ in $\mathbb{R}_n$. This is the same thing as saying $Q(x_1, \ldots, x_n) > 0$ for all $\mathbf{X} = (x_1, \ldots, x_n)^* \neq \mathbf{0}$.

5.2 The quadratic form $Q(x_1, \ldots, x_n)$ is positive definite iff for the associated matrix $\mathbf{A}$,

$$a_{11} > 0, \quad \det \begin{pmatrix} a_{11} & a_{12} \\ a_{21} & a_{22} \end{pmatrix} > 0, \ldots, \det \begin{pmatrix} a_{11} & \cdots & a_{1n} \\ \vdots & & \vdots \\ a_{n1} & \cdots & a_{nn} \end{pmatrix} > 0.$$

(Since Q is positive definite iff $\mathbf{A}$ is, this is the same as the Principal Minors Test; the following is our postponed proof of 4.2.)

Proof. First, three paragraphs of preliminaries. Suppose Q is positive definite. Then $Q(1, 0, \ldots, 0) = a_{11} > 0$. The first step of the diagonalization procedure at the first of this section can thus be applied:

$$Q(x_1, \ldots, x_n) = a_{11}y_1^2 + Q_1(x_2, \ldots, x_n), \tag{6}$$

where

and

$$y_1 = x_1 + (a_{12}/a_{11})x_2 + \cdots + (a_{1n}/a_{11})x_n,$$

where

$$Q_1(x_2, \ldots, x_n) = \sum_{i \neq 1, j \neq 1}^n b_{ij}x_ix_j,$$

$$b_{ij} = a_{ij} - \frac{a_{1i}a_{1j}}{a_{11}}.$$

Then Q_1 is positive definite. For, take any $x_2, \ldots, x_n$ not all 0, and then choose x_1 so that $y_1 = 0$. Then

$$Q_1(x_2, \ldots, x_n) = Q(x_1, \ldots, x_n) > 0.$$

Conversely, if $a_{11} > 0$ and Q_1 is positive definite in (6), then Q is positive definite: Certainly $Q(x_1, \ldots, x_n) \geq 0$ and in fact $Q(x_1, \ldots, x_n) = 0$ only if $y_1 = 0$ and $x_2 = \cdots = x_n = 0$; but then $x_1 = 0$, too, so Q is positive definite.

We next note what the principal minors are in terms of the substitution used in obtaining (6):

$$a_{11} = a_{11}, \quad \det \begin{pmatrix} a_{11} & a_{12} \\ a_{21} & a_{22} \end{pmatrix} = \det \begin{pmatrix} a_{11} & 0 \\ 0 & b_{22} \end{pmatrix}, \quad \det \begin{pmatrix} a_{11} & a_{12} & a_{13} \\ a_{21} & a_{22} & a_{23} \\ a_{31} & a_{32} & a_{33} \end{pmatrix}$$

$$= \det \begin{pmatrix} a_{11} & 0 & 0 \\ 0 & b_{22} & b_{23} \\ 0 & b_{32} & b_{33} \end{pmatrix}, \ldots, \det \begin{pmatrix} a_{11} & \cdots & a_{1n} \\ \vdots & & \vdots \\ a_{n1} & \cdots & a_{nn} \end{pmatrix} = \det \begin{pmatrix} a_{11} & 0 & \cdots & 0 \\ 0 & b_{22} & \cdots & b_{2n} \\ \vdots & \vdots & & \vdots \\ 0 & b_{n2} & \cdots & b_{nn} \end{pmatrix}$$

(why$_1$?). We are finally ready to prove the theorem. We proceed by induction. It is certainly true for $n = 1$, so assume it for $n - 1$. Then

1. If the principal minors of $\mathbf{A}$ are all positive then also

$$b_{22} > 0, \det \begin{pmatrix} b_{22} & b_{23} \\ b_{32} & b_{33} \end{pmatrix} > 0, \ldots, \det \begin{pmatrix} b_{22} & \cdots & b_{2n} \\ \vdots & & \vdots \\ b_{n2} & \cdots & b_{nn} \end{pmatrix} > 0$$

why$_2$?). Hence, by the inductive assumption, Q_1 will be positive definite (and $a_{11} > 0$), so the above remark shows that Q is positive definite too.

2. Conversely, if Q is positive definite then we saw above that $a_{11} > 0$, and Q_1 in (6) is positive definite, too. Hence

$$b_{22} > 0, \quad \det \begin{pmatrix} b_{22} & b_{23} \\ b_{32} & b_{33} \end{pmatrix} > 0, \ldots, \det \begin{pmatrix} b_{22} & \cdots & b_{2n} \\ \vdots & & \vdots \\ b_{n2} & \cdots & b_{nn} \end{pmatrix} > 0.$$

Hence, the principal minors of **A** are positive (why$_3$?). ∎

EXERCISES

1. Show that the linear change of variables $\mathbf{X} = \mathbf{PX}'$ used in obtaining (1) is invertible.
2. Prove (3).
3. Elaborate on the indicated points in the proof of Sylvester's Law.
4. Describe the diagonalization procedure in terms of pivoting on the matrix **A** of the form Q.
5. Answer the questions in the proof of 5.2.
6. Show that if Q is positive definite, then for the associated matrix **A**, deleting any collection of rows and the corresponding columns from **A** yields a matrix with positive determinant. [*Suggestion:* Renumber.]
7. Perform a nonorthogonal reduction to diagonal form on the quadratic forms:
   a) $-x^2 - y^2 + z^2 + 6xy + 2xz + 2yz$
   b) $5x^2 + 5y^2 + 9z^2 + 2xy - 6xz - 6yz$
   c) $x^2 + y^2 + 9z^2 + 10xy - 6xz - 6yz$
   d) $17x^2 + 12xy + 8y^2$
   e) $x^2 + 4y^2 + 4z^2 + 4xy - 4xz - 8yz$
   f) $x^2 + y^2 + z^2 - 4xy - 4yz - 4xz$

   Compare the results to those of Section 2, Exercise 2.

Quarrellary. The logos of somewome to that base anything, when most characteristically mantissa minus, comes to nullum in the endth†

CHAPTER 10 Matrix Calculus

This chapter begins with a discussion of norms of vectors and matrices. This notion allows us to talk about convergence of sequences and series of vectors and matrices; power series of square matrices are of particular interest.

We apply this material to the complex numbers, which we show can be identified both algebraically and analytically with a certain subset of 2×2 real matrices. In particular, some amazing properties of the complex exponential function follow directly from our matrix work.

In the final section we demonstrate some incredible analogies between the complex number system and the set of $n \times n$ real matrices.

1. VECTOR NORMS

You will recall that in our work on euclidean spaces (Chapter 8), we defined the norm of a vector $\mathbf{V} = (v_1, \ldots, v_n)$ in $\mathbb{R}_n$ to be

$$|\mathbf{V}| = \sqrt{\mathbf{V} \cdot \mathbf{V}} = \sqrt{v_1^2 + \cdots + v_n^2}.$$

We will now call this the **euclidean norm**, to distinguish it from the others we are about to encounter. In general, a **norm** $|\mathbf{V}|$ on $\mathcal{V}$ will be any mapping from $\mathcal{V}$ into the reals $\mathbb{R}$ which has the following properties:

i) For all $\mathbf{V}$ in $\mathbb{R}_n$,

$$|\mathbf{V}| \geq 0; \quad |\mathbf{V}| = 0 \quad \text{iff} \quad \mathbf{V} = \mathbf{0}.$$

---

† James Joyce, *Finnegan's Wake*, Compass Books, 1959.

ii) For any **V** and **W** in $\mathcal{V}$,

$$|V + W| \leq |V| + |W| \quad \text{(Triangle inequality)}.$$

iii) For any **V** in $\mathbb{R}_n$ and scalar c in $\mathbb{R}$,

$$|cV| = |c|\,|V|.$$

Intuitively, the norm of a vector represents its length. A **normed vector space** is a vector space $\mathcal{V}$ together with some specific norm $|V|$ on $\mathcal{V}$.

In some contexts it is more convenient to use norms other than the euclidean norm. Another norm in $\mathbb{R}_n$ is obtained by defining

$$|V|_T = |(v_1, \ldots, v_n)|_T = |v_1| + \cdots + |v_n|. \tag{1}$$

This is sometimes called the **taxicab norm**, since taxis essentially use this norm in $\mathbb{R}_2$ to compute your distance from origin to destination.

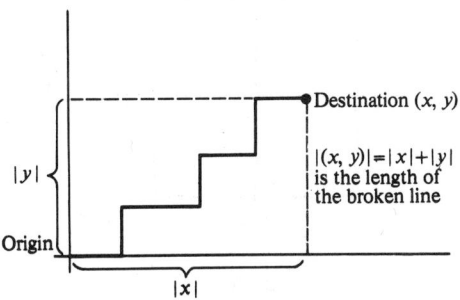

Fig. 1

Another example of a norm in $\mathbb{R}_n$ is

$$|V|_B = |(v_1, \ldots, v_n)|_B = \max\{|v_1|, \ldots, |v_n|\}. \tag{2}$$

This is often called the **box norm**, for reasons to be explained in a moment.

Given any norm $|V|$ in $\mathcal{V}$, one particularly interesting class of points **X** are those with norm 1. The set S of such points,

$$S = \{X \text{ in } \mathcal{V} \mid |X| = 1\},$$

is called the **unit sphere**. We are used to thinking of such a creature as looking like this:

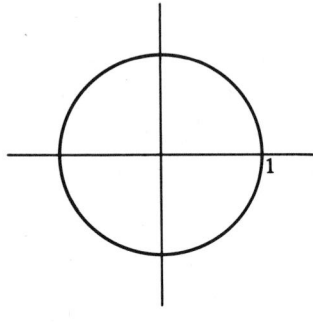

Fig. 2

However, with other norms it will have a different appearance. For example:

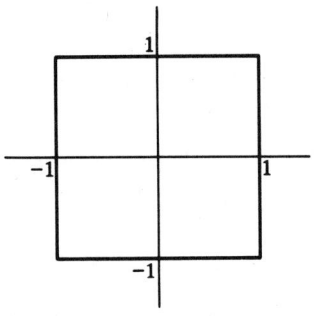

Fig. 3

You may take some comfort in the fact that in any normed vector space over $\mathbb{R}$, the unit ball (the unit sphere plus its interior) is a convex set, at least (definition in Exercise 11).

We have the following pleasant relations among the norms we have thus far mentioned when it comes to $\mathbb{R}_n$:

1.1 *Comparison of norms:* For any **V** in $\mathbb{R}_n$,

box norm **V** $\leq$ euclidean norm **V** $\leq$ taxicab norm **V**.

That is, if $\mathbf{V} = (v_1, \ldots, v_n)$, then

$$\max |v_i| \leq \sqrt{\sum_{i=1}^{n} v_i^2} \leq \sum_{i=1}^{n} |v_i|.$$

Proof. Exercise 12. ∎

Given any norm $|V|$ on a vector space, you can immediately begin to do calculus. Intuitively, this is because the norm is essentially a measure of distance. Being a nonnegative real number, you can talk about making the norm of something arbitrarily small, and hence of convergence. We illustrate this below.

By an **infinite sequence** $V_1, V_2, \ldots, V_k, \ldots$ of points in $\mathbb{R}_n$, we shall mean simply an infinite list of vectors. Given the infinite sequence of vectors†

$$V_1 = (v_1^{(1)}, \ldots, v_n^{(1)}), \qquad V_2 = (v_1^{(2)}, \ldots, v_n^{(2)}), \\ \ldots, \quad V_k = (v_1^{(k)}, \ldots, v_n^{(k)}), \quad \ldots, \tag{3}$$

it is clear that this is the same as the collection of n infinite sequences of real numbers

$$\{v_1^{(1)}, v_1^{(2)}, \ldots, v_1^{(k)}, \ldots\}, \quad \{v_2^{(1)}, v_2^{(2)}, \ldots, v_2^{(k)}, \ldots\}, \\ \ldots, \quad \{v_n^{(1)}, v_n^{(2)}, \ldots, v_n^{(k)}, \ldots\}. \tag{3}'$$

Thus, the study of infinite sequences of vectors in $\mathbb{R}_n$ should contain little that is new.

However, we do have a choice as to how we wish to define convergence of infinite sequences of vectors:

I. The infinite sequence $\{V_k\}$ in (3) converges **elementwise** to $V = (v_1, \ldots, v_n)$ if the n sequences of real numbers

$$\{v_i^{(k)}\}, \quad i = 1, \ldots, n,$$

converge to $v_i, i = 1, \ldots, n$ respectively.

II. The infinite sequence $\{V_k\}$ converges **in** (a given) **norm** to V if the infinite sequence of real numbers

$$|V_1 - V|, \quad |V_2 - V|, \ldots, |V_k - V|, \ldots$$

converges to zero, where $|V|$ is the given norm on $\mathbb{R}_n$.

The norm conditions will ensure that the usual arithmetic of convergent sequences will hold, as we see below.

Example 1. The sequence of vectors in $\mathbb{R}_3$,

$$V_k = \left(1, (2k+3)/k, \sum_{j=1}^{k} 1/k!\right)$$

---

† More precisely, an **infinite sequence** is a function f from the positive integers into $\mathbb{R}_n$. We are often more concerned with the range of values $V_1 = f(1), V_2 = f(2), \ldots$ than with function itself.

converges elementwise to the vector $\mathbf{V} = (1, 2, e)$, as is apparent from the constituent real sequences. The sequence also converges to $\mathbf{V}$ in, say, the taxicab norm, since

$$|\mathbf{V}_k - \mathbf{V}| = \left|\left(1, (2k+3)/k, \sum_{j=1}^{k} 1/k!\right) - (1, 2, e)\right|_T$$

$$= |1 - 1| + |(2k+3)/k - 2| + \left|\sum_{j=1}^{k} 1/k! - e\right|$$

converges to 0.

One may similarly define continuity of vector-valued functions of a real variable (or use sequences to define these concepts, as you saw was possible in Exercise 14, Chapter 8, Section 1, if you saw Exercise 14, Chapter 8, Section 1). You even have your choice of defining continuity elementwise or in norm, and in any norm you please! This multiplicity of possibilities may cause some consternation. For if the unit sphere can be so varied in appearance under different norms, you might suppose that the convergence of sequences and the continuity of functions might vary from norm to norm. Fortunately, this is not the case. We shall show this for limits of sequences and repress our desire to discuss continuity of vector functions at this time.

In order to handle this kind of problem it is convenient to prove the following, which relates any norm to the taxicab norm.

1.2 Let $|\mathbf{V}|$ denote any norm in $\mathbb{R}_n$. Then there exist positive numbers m and M so that, if $\mathbf{V} = v_1\mathbf{e}_1 + \cdots + v_n\mathbf{e}_n$ is any vector in $\mathbb{R}_n$, then

$$m|\mathbf{V}|_T \leq |\mathbf{V}| \leq M|\mathbf{V}|_T,$$

i.e., $m(|v_1| + \cdots + |v_n|) \leq |\mathbf{V}| \leq M(|v_1| + \cdots + |v_n|)$.

Proof (one step of which requires the calculus of functions of several variables). Let M be the largest of the norms

$$|\mathbf{e}_1|, |\mathbf{e}_2|, \ldots, |\mathbf{e}_n|,$$

so that $|\mathbf{e}_i| \leq M$ for $i = 1, 2, \ldots, n$. Then

$$|\mathbf{V}| \leq |v_1||\mathbf{e}_1| + \cdots + |v_n||\mathbf{e}_n|$$

$$\leq M(|v_1| + |v_2| + \cdots + |v_n|),$$

which proves one part of the inequality.

To prove the other part, we need the fact that $|\mathbf{V}|$ is a continuous function of $\mathbf{V}$. That is, if $\mathbf{f}(\mathbf{X}) = |\mathbf{X}|$, then the function $\mathbf{f}: \mathbb{R}_n \to \mathbb{R}$ is continuous in norm: that is, for any $\mathbf{X}_0$ in $\mathbb{R}_n$, given any $\epsilon > 0$ there is a δ for which

$$|\mathbf{f}(\mathbf{X}) - \mathbf{f}(\mathbf{X}_0)| < \epsilon$$

whenever $|\mathbf{X} - \mathbf{X}_0| < \delta$. By the triangle inequality,
$$\big||\mathbf{Y}| - |\mathbf{Z}|\big| \leq |\mathbf{Y} - \mathbf{Z}|$$
(see Exercise 14). Hence,
$$|f(\mathbf{X}) - f(\mathbf{X}_0)| = \big||\mathbf{X}| - |\mathbf{X}_0|\big| \leq |\mathbf{X} - \mathbf{X}_0|.$$
We can therefore make the former $< \epsilon$ by making the latter $< \epsilon$. Hence the norm function is continuous.

We apply this as follows. Since $|\mathbf{X}|$ is continuous, it has a positive minimum value m on the taxicab unit sphere: $|\mathbf{X}| \geq m > 0$ for every $\mathbf{X}$ of taxicab norm 1. As one sees in the study of several variables, this is a consequence of the fact that the unit sphere is a closed and bounded set.

We must finally show that this m has the required property. If $\mathbf{V} \neq \mathbf{0}$ we can write
$$\mathbf{V} = (|v_1| + \cdots + |v_n|) \frac{\mathbf{V}}{|v_1| + \cdots + |v_n|},$$
and since
$$\frac{\mathbf{V}}{|v_1| + \cdots + |v_n|}$$
is on the taxicab unit sphere, we have
$$\frac{|\mathbf{V}|}{|v_1| + \cdots + |v_n|} \geq m.$$
Hence
$$|\mathbf{V}| = (|v_1| + \cdots + |v_n|) \frac{|\mathbf{V}|}{|v_1| + \cdots + |v_n|} \geq m(|v_1| + \cdots + |v_n|).$$
If $\mathbf{V} = \mathbf{0}$, this inequality is true also; hence the left-hand part of our inequality has been proved. ∎

1.3 *Invariance of convergence.* The infinite sequence of vectors $\{\mathbf{V}_k\}$ in $\mathbb{R}_n$ converges to $\mathbf{V}$ in norm (any norm) iff it converges to $\mathbf{V}$ elementwise. Consequently, convergence is independent of the norm.

Proof. 1. Assume that $\mathbf{V}_k$ converges to $\mathbf{V}$ elementwise. Thus, given $\epsilon > 0$, there is an n so that if $k > n$, then
$$|v_1^{(k)} - v_1| < \epsilon, \qquad \ldots, \qquad |v_n^{(k)} - v_n| < \epsilon,$$
in terms of the coordinates of the $\mathbf{V}_k$ and $\mathbf{V}$. Given any norm $|\mathbf{V}|$, by 1.2, there is an M so that
$$|\mathbf{V}_k - \mathbf{V}| \leq M(|v_1^{(k)} - v_1| + \cdots + |v_1^{(k)} - v_1|)$$
$$\leq Mn\epsilon,$$

whenever $k > n$. Thus, $|V_k - V|$ can be made as small as desired by making k sufficiently large, and V_k converges in norm.

2. Conversely, if V_k converges to V in the norm $|V|$, with associated $m > 0$ in 1.2, then one can quickly show that the sequence $V_k - V$ converges in norm to 0. Let

$$W_k = V_k - V = (w_1^{(k)}, \ldots, w_n^{(k)}).$$

Then

$$m(|w_1^{(k)}| + \cdots + |w_n^{(k)}|) \leq |W_k|.$$

We can thus make $|w_i^{(k)}| < \epsilon$ by making $|W_k| < m\epsilon$. Since

$$w_i^{(k)} = v_i^{(k)} - v_i,$$

it follows that $v_i^{(k)}$ converges to v_i; that is, that V_k converges to V elementwise. ∎

To celebrate this invariance of convergence, we shall use the notation

$$\lim_{k \to \infty} V_k = V$$

to indicate that V_k converges to V, for any type of convergence.

But you may ask, why all the convergences and norms if one is as good as the other? There are two reasons:

1) *Convenience*: Some notions of convergence and norms are more natural or more convenient to use in some contexts than in others (*vide* the taxicab norm in the above theorem).

2) *Power*: We introduce a matrix norm in the next section which is a particularly good indicator of convergence of matrix infinite series, but it is somewhat hard to compute (so you might want to use some other norms as a quick check as to the possibility of convergence or divergence).

1.4 *Arithmetic of sequences.* If

$$\lim_{k \to \infty} V_k = V \quad \text{and} \quad \lim_{k \to \infty} W_k = W$$

(all sequences in $\mathbb{R}_n$), then

i) $\lim_{k \to \infty}(V_k + W_k) = V + W;$

ii) $\lim_{k \to \infty} cV_k = cV \quad$ for c in $\mathbb{R}$;

iii) $\lim_{k \to \infty} V_k \cdot W_k = V \cdot W.$

Proof. Exercise 4. ∎

It should now be clear how one handles infinite series of vectors in $\mathbb{R}_n$. For the infinite series

$$\mathbf{V}_1 + \mathbf{V}_2 + \cdots + \mathbf{V}_k + \cdots = \sum_{k=1}^{\infty} \mathbf{V}_k \quad (\text{or just } \sum \mathbf{V}_k),$$

we construct the sequence of partial sums

$$\mathbf{S}_1 = \mathbf{V}_1, \quad \mathbf{S}_2 = \mathbf{V}_1 + \mathbf{V}_2, \ldots, \mathbf{S}_k = \mathbf{V}_1 + \cdots + \mathbf{V}_k;$$

and then define the infinite series to be **convergent to** the sum $\mathbf{S}$ iff the sequence $\mathbf{S}_1, \ldots, \mathbf{S}_k$, has the limit $\mathbf{S}$. As we have just seen, this limit can be taken elementwise or in terms of any norm that may be convenient.

It is natural to make the following definition of absolute convergence:

The series $\sum \mathbf{V}_k$, where $\mathbf{V}_k = (v_1^{(k)}, \ldots, v_n^{(k)})$, is called **absolutely convergent** if each of the series

$$\sum_{k=1}^{\infty} |v_i^{(k)}|, \quad i = 1, 2, \ldots, n$$

converges.

From 1.3 and the appropriate property of convergence of real series we get the well-known fact that

an absolutely convergent series is convergent.

We also get the following useful result:

1.5 The series $\sum \mathbf{V}_k$ converges absolutely iff $\sum |\mathbf{V}_k|$ converges, with $|\mathbf{V}|$ any norm.

Proof. 1. If $\sum \mathbf{V}_k$ converges absolutely and $\mathbf{V} = (v_1^{(k)}, \ldots, v_n^{(k)})$, then each of the series $\sum_{k=1}^{\infty} |v_i^{(k)}|$ converges, and hence the series

$$\sum_{k=1}^{\infty} (|v_1^{(k)}| + \cdots + |v_n^{(k)}|)$$

converges. Using 1.2 we conclude that $\sum |\mathbf{V}_k|$ converges (by the comparison test for series with positive terms).

2. If $\sum_{k=1}^{\infty} |\mathbf{V}_k|$ converges, we again use 1.2 to conclude that

$$|v_i^{(k)}| \leq (1/m) |\mathbf{V}_k| \quad \text{and hence} \quad \sum |v_i^{(k)}| \quad \text{converges}$$

for each $i = 1, 2, \ldots, n$. ∎

Note in particular that 1.5 says

If $\sum |\mathbf{V}_k|$ converges, then $\sum \mathbf{V}_k$ converges.

Project. Although our work in this chapter officially deals with the vector space $\mathbb{R}_n$ over $\mathbb{R}$, much of the material can be carried over to $\mathbb{C}_n$ over $\mathbb{C}$ with only slight modification. This extension is often important as well as charming. Throughout this chapter we shall have a sequence (finite) of honors projects in which you are asked to extend results to $\mathbb{C}_n$.

In Supplement 3 to Section 2, Chapter 7, we saw that the appropriate dot product to use on $\mathbb{C}_n$ is defined by

$$\langle \mathbf{Z}, \mathbf{W} \rangle = \overline{(z_1, \ldots, z_n)} \cdot (w_1, \ldots, w_n) = \bar{z}_1 w_1 + \cdots + \bar{z}_n w_n,$$

with z_i, w_i in $\mathbb{C}$, and where $\bar{}$ denotes complex conjugation. We use this hereafter on $\mathbb{C}_n$.
a) Show that the complex euclidean norm $|V| = \langle \mathbf{V}, \mathbf{V} \rangle^{1/2}$ is a norm on $\mathbb{C}_n$.
b) Show that the box norm and the taxicab norm go over to $\mathbb{C}_n$ as norms if you replace "absolute value of a real number" by "modulus of a complex number".
c) In drawing accurate pictures, we are usually limited to $\mathbb{C}_1$ which we can represent as $\mathbb{R}_2$,

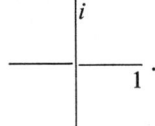

Draw pictures for the various unit spheres in $\mathbb{C}_1$. Can you do anything at all in $\mathbb{C}_2$?
d) Prove 1.1 for $\mathbb{C}_n$.
e) Define convergence of sequences in $\mathbb{C}_n$ and prove 1.3 for $\mathbb{C}_n$ (as much as we did for $\mathbb{R}_n$, anyway).
f) Prove 1.4 for $\mathbb{C}_n$.

EXERCISES

1. (a) Prove that the taxicab norm is a norm. Draw its unit sphere in $\mathbb{R}_2$. (b) Prove that the box norm is a norm.
2. Find a norm in $\mathbb{R}_2$ relative to which the unit sphere is (a) a rectangle of height 2 and width 3, centered at the origin; (b) the ellipse $x^2 + 4y^2 = 4$. In each case prove that you do have a norm.
3. Given any norm $|V|$ in a space $\mathcal{V}$, one can immediately define a *distance function* (or *metric*) on $\mathbb{R}_n$:
   the **distance** from X to Y is $|X - Y|$.
   Prove that the distance function so obtained has the properties
   a) $|X - Y| \geq 0$; $\quad |X - Y| = 0$ iff $X = Y$;
   b) $|X - Y| = |Y - X|$;
   c) $|X - Y| \leq |X - Z| + |Y - Z|$.
4. Prove 1.4.
5. Prove that if $\mathbf{X}$ is a nonzero element of $\mathbb{R}_n$, then $X/|X|$ is on the unit sphere
   $$\{Y \text{ in } \mathbb{R}_n \mid |Y| = 1\}.$$
6. Describe the effect of the following linear transformations on the unit sphere of the euclidean norm:

   a) $\mathbf{A} = \begin{pmatrix} 2 & 0 \\ 0 & 1 \end{pmatrix}$, \qquad b) $\mathbf{B} = \begin{pmatrix} 0 & 2 \\ 1 & 0 \end{pmatrix}$

c) $C = \begin{pmatrix} 2 & 2 \\ 1 & 1 \end{pmatrix}$, \hspace{2cm} d) $D = \begin{pmatrix} 2 & 2 \\ 1 & 1 \end{pmatrix}$

[*Suggestion:* Cherchez la eigenvalue.] Save your results for future endeavors.

7. Do Excercise 6 with respect to the box norm.
8. Is $f(x, y) = |x|$ a norm on $\mathbb{R}_2$? Prove or disprove.
9. a) Prove that if $|V|$ is a norm on $\mathcal{V}$ and c a fixed positive scalar, then $|X|_c = c|X|$ is a norm on $\mathbb{R}_n$.
   b) If $|V|_1$ and $|V|_2$ are two norms on $\mathcal{V}$ prove that $|V|_3 = |V|_1 + |V|_2$ is a norm.
   c) Let

   $|\ |_B$ = box norm, \hspace{1cm} $|\ |_T$ = taxicab norm, \hspace{1cm} $|\ |_E$ = euclidean norm.

   In $\mathbb{R}_2$, describe the unit circle for $|\ |_B + |\ |_T$; for $|\ |_B + |\ |_E$.

10. Recall that the dual space $\mathcal{V}^*$ to a vector space over $\mathbb{R}$ is the set of all linear transformations from $\mathcal{V}$ into $\mathbb{R}$. (See Chapter 3, Section 5.)
    a) Show that if $|V|$ is a norm on $\mathcal{V}$, then for $\mathbf{f}$ in $\mathcal{V}^*$ the definition

    $$||\mathbf{f}|| = \max_{|X|=1} |\mathbf{f}(X)|$$

    (Norm in $\mathcal{V}$) \hspace{2cm} (Absolute value of real number)

    yields a norm on $\mathcal{V}^*$.
    For $\mathcal{V} = \mathbb{R}_2$, $\mathcal{V}^*$ is also two-dimensional with basis $\mathbf{f}_1, \mathbf{f}_2$ satisfying

    $$\mathbf{f}_i(\mathbf{e}_j) = \delta_{ij}.$$

    Use this basis below.
    b) If $|V|$ is the euclidean norm in $\mathbb{R}_2$, draw the unit circle in $\mathbb{R}_2^*$ with respect to the norm $||\mathbf{f}||$.
    c) Same for $|V|$ = box norm.

11. Prove that for any norm $|V|$ on a vector space $\mathcal{V}$ over $\mathbb{R}$, the unit ball

    $$\mathcal{B} = \{X \mid |X| \le 1\}$$

    is convex, that is, if X and Y are any points in $\mathcal{B}$, then any point on the segment between X and Y is also in $\mathcal{B}$, i.e.,

    $$\lambda X + (1 - \lambda)Y$$

    is in $\mathcal{B}$ for any λ satisfying $0 \le \lambda \le 1$.

12. Prove 1.1 (a) for $n = 2$; (b) for arbitrary n. (c) Under what circumstances will equality hold?
13. If $|X|$ is any norm on $\mathbb{R}_n$, prove that

    $$||A| - |B|| \le |A - B|.$$

14. If $w_1, \ldots, w_n$ are all positive, then show that
$$|(v_1, \ldots, v_n)| = \sqrt{w_1 v_1^2 + \cdots + w_n v_n^2}$$
is a norm on $\mathbb{R}_n$. This is the "weighted norm" of Chapter 8, Section 4.

15. Let **A** be symmetric ($n \times n$) and positive definite. Prove
   a) $(V^*AV)^{1/2}$ is a norm on $\mathbb{R}_n$.
   b) $\langle X, Y \rangle = Y^*AX$ is an inner product (see Section 2, Supplement 1, in Chapter 8), and $\langle X, X \rangle^{1/2}$ is the norm described in (a).

2. MATRIX NORMS

Recall that the set $\mathbb{R}_{m \times n}$ of all $m \times n$ matrices with entries in $\mathbb{R}$ forms a vector space over $\mathbb{R}$ with respect to the usual definitions of addition and scalar multiplication. The previous section shows that we have a wide choice of norms which we could inflict on $\mathbb{R}_{m \times n}$. For example, if

$$A = \begin{pmatrix} a_{11} & \cdots & a_{1n} \\ \vdots & & \vdots \\ a_{m1} & \cdots & a_{mn} \end{pmatrix},$$

the euclidean norm of **A** would be

$$\sqrt{\sum a_{ij}^2}.$$

This is a perfectly decent norm, but it does not take into account any of the multiplicative or linear transformation properties of a matrix.

Recall that for **A** in $\mathbb{R}_{m \times n}$ and **X** in $\mathbb{R}_n$, the mapping $X \to AX$ of $\mathbb{R}_n$ into $\mathbb{R}_m$ is a linear transformation. Suppose that we have a norm $|X|$ on $\mathbb{R}_n$. Then **A**, as a linear transformation, is completely determined by its effect on the unit sphere

$$S = \{X \text{ in } \mathbb{R}_n \mid |X| = 1\}.$$

For, if **Y** is any nonzero element in $\mathbb{R}_n$, then $X = Y/|Y|$ has norm 1 (Section 1, Exercise 5). Hence, if we know **AX**, then

$$AY = A(|Y| X) = |Y| AX,$$

and we also know **AY**. A useful indication of the "size" of a matrix might therefore be the maximum value of $A|X|$ on the unit sphere. (This maximum actually exists, as one sees from a theorem in several-variable calculus.) This will in fact turn out to be a particularly useful norm.

We define the **matrix norm** (relating to fixed norms in $\mathbb{R}_m$ and $\mathbb{R}_n$) of a matrix **A** in $\mathbb{R}_{m \times n}$ to be

$$\|A\| = \max_{|X|=1} |AX|$$

(Some norm in $\mathbb{R}_n$) ⟶   ⟵ (Some norm in $\mathbb{R}_m$)

This terminology is not standard. Some people call the euclidean norm the matrix norm. Others would call $\|A\|$ the "sup norm" (pronounced "soup norm") from using "supremum" rather than "maximum" in its definition.

Example 1. Let
$$A = \begin{pmatrix} 2 & 0 \\ 0 & 1 \end{pmatrix},$$
and take the euclidean norm in $\mathbb{R}_2$ both as domain for A and as range of A. To see what A does to X on the unit circle, suppose $x^2 + y^2 = 1$. Then
$$A\begin{pmatrix} x \\ y \end{pmatrix} = \begin{pmatrix} x_0 \\ y_0 \end{pmatrix},$$
where $x_0 = 2x$, $y_0 = y$. Thus, $x_0^2 = 4x^2$, and $y_0^2 = y^2$, so (x_0, y_0) is on the ellipse
$$\tfrac{1}{4}x_0^2 + y_0^2 = 1,$$

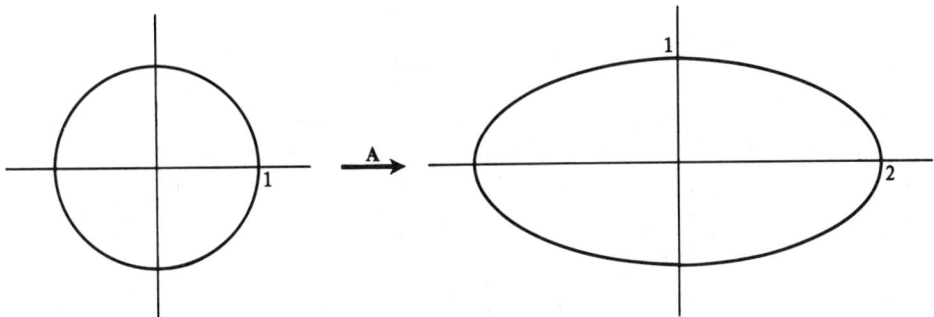

Fig. 4

and hence $\|A\| = 2$. This suggests that the matrix norm may be related to the eigenvalues of a matrix.

Example 2. Now let us take the norm in $\mathbb{R}_2$ as domain of A to be the box norm, and the norm in $\mathbb{R}_2$ as range of A to be the euclidean norm. It is easy to see that:

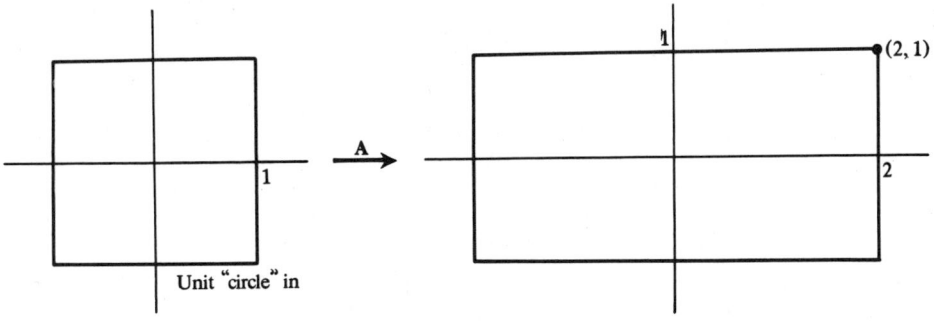

Fig. 5

When we go to measure distances by the euclidean norm on the image of the unit "circle", we find that

$$\|A\| = \max_{|X|_B = 1} |AX| = \sqrt{5}$$

(Euclidean norm)

Thus, the choice of norms in the vector spaces $\mathbb{R}_m$ and $\mathbb{R}_n$ does make a difference:

$\|A\|$ *is dependent on the particular norms in* $\mathbb{R}_m$ *and* $\mathbb{R}_n$.

2.1 Properties of the Matrix Norm. i) $\|A\|$ exists for any A in $\mathbb{R}_{m \times n}$;
ii) $\|A\|$ is a norm:

$\|A\| \geq 0;\quad \|A\| = 0\quad \text{iff}\ A = 0;\quad \|A + B\| \leq \|A\| + \|B\|;\quad \|cA\| = |c|\,\|A\|;$

iii) $|AX| \leq \|A\|\,|X|$ for any X in $\mathbb{R}_n$;

(Norm in $\mathbb{R}_m$)   (Norm in $\mathbb{R}_n$).

iv) If B is in $\mathbb{R}_{n \times l}$, then $\|AB\| \leq \|A\|\,\|B\|$.

Proof (One portion of which is assumed from several-variable calculus). (i) depends upon the theorem about continuous functions of several variables, which we also needed in 1.2.

The first two parts of (ii) are exercise (1). To prove the remaining part of (ii) you need only note that

$$|(A + B)(X)| = |AX + BX| \leq |AX| + |BX|$$

by the appropriate property of the norm in $\mathbb{R}_m$. Since the real-valued function $|(A + B)X|$ is less than or equal to the real-valued function $|AX| + |BX|$ for any X (especially those X with $|X| = 1$), we have

$$\max_{|X|=1} |(A + B)X| \leq \max_{|X|=1} |AX| + \max_{|X|=1} |BX|,$$

which was to be proved.

iii) Given $X \neq 0$ in $\mathbb{R}_n$, let $X_1 = X/|X|$ be the corresponding point on the unit sphere. Then

$$|AX_1| \overset{(\text{why}_1?)}{\leq} \max_{|V|=1} |AV| = \|A\|.$$

Also,

$$|AX_1| = |A(X/|X|)| \overset{(\text{why}_2?)}{=} |AX|\,1/|X|.$$

Hence

$$|AX|\,1/|X| \leq \|A\|\qquad (\text{why}_3?).$$

Hence $|AX| \leq |X|\,\|A\|$.

iv) $|ABX| \leq \|A\||BX| \leq \|A\|\|B\||X|$ by two applications of (iii). Hence $\|AB\| = \max_{|X|=1} |ABX| \leq \|A\|\|B\| \cdot 1$. ∎

You are given an opportunity to test the results of 2.1 in Exercises 2, 3, and 4. It is property (iv) of 2.1,

$$\|AB\| \leq \|A\|\|B\|$$

that is the important feature of matrix norms which we shall use in the next section. This is often called the **submultiplicative property**.

We turn to what is perhaps the most obvious matrix norm, that obtained by using the euclidean norm in both $\mathbb{R}_m$ and $\mathbb{R}_n$. We shall call this the **spectral norm** of A (since it is obtainable as an eigenvalue of a matrix related to A, and the set of eigenvalues is often called the *spectrum* of a matrix). Thus

$$\text{spectral norm of } A = \max_{|X|=1} |AX|$$

Euclidean norm

Although the spectral norm is a very natural norm to consider, it is relatively difficult to compute (see Supplement 1 to this section). However, it does stand in a fixed relationship to various vector norms of the matrix, something which cannot be said for the more computable row and column norms of Supplement 2.

2.2 For any A in $\mathbb{R}_{m \times n}$, the spectral norm of A is less than or equal to the euclidean norm of A.

Proof. For convenience take

$$A = \begin{pmatrix} a_{11} & a_{12} & a_{13} \\ a_{21} & a_{22} & a_{23} \end{pmatrix}.$$

We recall the Schwarz Inequality:

$$|V \cdot W| \leq |V||W| \quad \text{or} \quad \left|\sum_1^n v_i w_i\right| \leq \sqrt{\sum_1^n v_i^2}\sqrt{\sum_1^n w_i^2}.$$

For any $X = (x, y, z)$ in $\mathbb{R}_3$,

$$AX = \begin{pmatrix} a_{11}x + a_{12}y + a_{13}z \\ a_{21}x + a_{22}y + a_{23}z \end{pmatrix},$$

so

$$|AX|^2 = (a_{11}x + a_{12}y + a_{13}z)^2 + (a_{21}x + a_{22}y + a_{23}z)^2$$

$$\overset{\text{2 Schwarz}}{\leq} (a_{11}^2 + a_{12}^2 + a_{13}^2)(x^2 + y^2 + z^2) + (a_{21}^2 + a_{22}^2 + a_{23}^2)(x^2 + y^2 + z^2)$$

$$= |A|^2 |X|^2, \tag{1}$$

where $|A|$ is the euclidean norm. In particular, if X is on the unit sphere in $\mathbb{R}_3$, then

$$x^2 + y^2 + z^2 = 1;$$

and, taking the square root of both sides of (1), we obtain
$$|AX| \leq |A|.$$
Since this is true for all **X** on the unit sphere, we obtain
$$\max_{|X|=1} |AX| = ||A|| \leq |A|.$$

Note that (1) shows that for any **X**, $|AX| \leq |A||X|$.

2.3 *Comparison of some norms of a matrix*. For any **A** in $\mathbb{R}_{m \times n}$,
$$|a_{ij}| \leq \underset{\text{(spectral)}}{||A||} \leq \underset{\text{(euclidean)}}{|A|} \leq \Sigma |a_{ij}|, \qquad \text{where } i = 1, \ldots, m,$$
$$j = 1, \ldots, n.$$

Thus,

box norm **A** $\leq$ spectral norm **A** $\leq$ euclidean norm **A** $\leq$ taxicab norm **A**.

Proof. This follows at once from 2.2, 1.1, and, for the first inequality, from the fact that if
$$\mathbf{A} = (\mathbf{A}_1, \ldots, \mathbf{A}_n),$$
then setting $\mathbf{X} = \mathbf{e}_j$, we can show that
$$||\mathbf{A}|| \geq |\mathbf{A}\mathbf{e}_j| = |\mathbf{A}_j| \geq |a_{ij}|. \quad \blacksquare$$

We shall investigate further properties of the spectral norm in Supplement 1.

EXERCISES

1. a) Prove the first two parts of (ii) in 2.1.
   b) Answer the questions in the proof of part (iii).
2. Let |**X**| be the euclidean norm and
$$\mathbf{A} = \begin{pmatrix} 2 & 0 \\ 0 & 1 \end{pmatrix},$$
as in Example 1. (a) Verify 2.1(iii) for
$$\mathbf{X}_1 = \begin{pmatrix} 3 \\ 0 \end{pmatrix} \quad \text{and} \quad \mathbf{X}_2 = \begin{pmatrix} 3 \\ 3 \end{pmatrix}.$$
Locate both vectors in the picture in Example 1. (b) Same as (a), with the norms now as in Example 2.
3. Let
$$\mathbf{B} = \begin{pmatrix} 0 & 2 \\ 1 & 0 \end{pmatrix}$$
and |**V**| be the euclidean norm in $\mathbb{R}_2$.

a) Calculate $\|\mathbf{B}\|$ with respect to the euclidean norm.
b) Let $\mathbf{A}$ be as in Exercise 2. Verify 2.1(ii).
c) Verify 2.1 (iv).

4. Let $\mathbf{A}$ and $\mathbf{B}$ be as in (2) and (3) and consider

$$\mathbf{B}: \underset{\text{Box norm}}{\mathbb{R}_2} \to \underset{\text{Eucl. norm}}{\mathbb{R}_2}, \quad \mathbf{A}: \underset{\text{Eucl. norm}}{\mathbb{R}_2} \to \underset{\text{Eucl. norm}}{\mathbb{R}_2}$$

Redo parts (a) and (c) of Exercise 3 under these conditions.

5. Prove that $|\mathbf{AB}| \leq |\mathbf{A}| |\mathbf{B}|$, $|\mathbf{A}|$ the euclidean norm, (a) for 2×2 matrices; (b) for arbitrary compatibly sized matrices.
6. Is $|\mathbf{AB}| \leq |\mathbf{A}| |\mathbf{B}|$ for the box norm?
7. Is $|\mathbf{AB}| \leq |\mathbf{A}| |\mathbf{B}|$ for the taxicab norm?
8. With $\mathbf{I}_n$ the $n \times n$ identity matrix, calculate its box norm, spectral norm, euclidean norm, and taxicab norm.

Project. Define the matrix norm for $\mathbf{A}$ in $\mathbb{C}_{m \times n}$ and prove 2.1 in this context.

Supplement 1 to Section 2. The Spectral Norm

For the purpose of this supplement, we shall denote by $|\mathbf{A}|$ the euclidean norm of $\mathbf{A}$ and by $\|\mathbf{A}\|$ the spectral norm of $\mathbf{A}$. Thus,

$$\|\mathbf{A}\| = \max_{|\mathbf{X}|=1} |\mathbf{AX}| \qquad (1)$$

Euclidean norm in $\mathbb{R}_n$ ——————↑   └—————— Euclidean norm in $\mathbb{R}_m$

The next theorem gives a (more or less) convenient method for calculating $\|\mathbf{A}\|$. Recall from Chapter 9, Section 4, that the eigenvalues of $\mathbf{A}^*\mathbf{A}$ are nonnegative for any $\mathbf{A}$.

2.4 The spectral norm in terms of an eigenvalue. For any $\mathbf{A}$ in $\mathbb{R}_{m \times n}$, let λ be the largest eigenvalue of $\mathbf{A}^*\mathbf{A}$. Then $\|\mathbf{A}\| = \sqrt{\lambda}$.

Proof. Whether or not $\mathbf{A}$ is square, $\mathbf{A}^*\mathbf{A}$ is (why$_1$?), and so is allowed to have eigenvalues. Moreover, $\mathbf{A}^*\mathbf{A}$ is symmetric (why$_2$?), so it not only has real eigenvalues (positive in this case), but there exists an orthogonal matrix $\mathbf{U}$, $\mathbf{U}^* = \mathbf{U}^{-1}$, for which

$$\mathbf{U}^*(\mathbf{A}^*\mathbf{A})\mathbf{U} = \mathbf{D} = \begin{pmatrix} \lambda_1 & & 0 \\ & \cdot & \\ & & \cdot \\ 0 & & \lambda_n \end{pmatrix}, \qquad (2)$$

with the λ_i the eigenvalues of $\mathbf{A}^*\mathbf{A}$. Suppose that λ_1 is the largest.

From (2) we obtain $\mathbf{A}^*\mathbf{A} = \mathbf{UDU}^*$, so

$$|\mathbf{AX}|^2 = \mathbf{AX} \cdot \mathbf{AX} \overset{(\text{why}_3?)}{=} (\mathbf{AX})^*\mathbf{AX} \overset{(\text{why}_4?)}{=} \mathbf{X}^*(\mathbf{A}^*\mathbf{A})\mathbf{X} = \mathbf{X}^*(\mathbf{UDU}^*)\mathbf{X}. \qquad (3)$$

Let $\mathbf{Y} = \mathbf{U}^*\mathbf{X} = (y_1, \ldots, y_n)^*$. Then (3) says

$$|\mathbf{AX}|^2 \overset{(\text{why}_5?)}{=} \mathbf{Y}^*\mathbf{DY} \overset{(\text{why}_6?)}{=} \sum_{i=1}^{n} y_i^2 \lambda_i. \qquad (4)$$

If $\mathbf{X}$ is on the unit sphere, then $\mathbf{Y} = \mathbf{U}^*\mathbf{X}$ is also (why$_9$?) so

$$\sum_{1}^{n} y_i^2 = 1.$$

Taking square roots in (4), we obtain $|\mathbf{AX}| \leq \sqrt{\lambda_1}$. Hence

$$\|\mathbf{A}\| \leq \sqrt{\lambda_1}.$$

We need only show that there is some $\mathbf{X}$ with $|\mathbf{X}| = 1$ and for which $|\mathbf{AX}| = \sqrt{\lambda_1}$. Let $\mathbf{X}_1$ be an eigenvector of $\mathbf{A}^*\mathbf{A}$ corresponding to λ_1 and make $|\mathbf{X}_1| = 1$ (why$_{10}$ is this possible?). Then

$$|\mathbf{AX}_1|^2 = \mathbf{X}_1^*(\mathbf{A}^*\mathbf{A})\mathbf{X}_1 = \lambda_1 \mathbf{X}_1^*\mathbf{X}_1 = \lambda_1.$$

Hence $\|\mathbf{A}\| = \sqrt{\lambda_1}$. ∎

Example 1. If

$$\mathbf{A} = \begin{pmatrix} 2 & 1 \\ 2 & 1 \\ 2 & 1 \end{pmatrix}, \quad \text{then} \quad \mathbf{A}^*\mathbf{A} = \begin{pmatrix} 12 & 6 \\ 6 & 3 \end{pmatrix},$$

whose characteristic polynomial is $x(x - 15)$. The maximum root of this is 15, so that $\|\mathbf{A}\| = \sqrt{15}$.

2.5 The Spectral Norm of a Symmetric Matrix. If $\mathbf{A}$ is symmetric, then $\|\mathbf{A}\| = |\lambda|$, with λ its eigenvalue of largest absolute value.

Proof. There is an orthogonal $\mathbf{U}$ and a diagonal $\mathbf{D}$ for which $\mathbf{U}^*\mathbf{DU} = \mathbf{A}$. Hence

$$\mathbf{A}^*\mathbf{A} = \mathbf{U}^*\mathbf{D}^2\mathbf{U},$$

and we can use 2.4. ∎

Our next theorem indicates a possible alternative definition of $\|\mathbf{A}\|$ in terms of the dot product.

2.6 The Spectral Norm and the Dot Product. For $\mathbf{A}$ in $\mathbb{R}_{m \times n}$,

$$\|\mathbf{A}\| = \max_{\substack{|\mathbf{X}|=1 \ (\mathbf{X} \text{ in } \mathbb{R}_n). \\ |\mathbf{Y}|=1 \ (\mathbf{Y} \text{ in } \mathbb{R}_m).}} |\mathbf{AX} \cdot \mathbf{Y}|,$$

Proof. Recall that the angle between $\mathbf{AX}$ and $\mathbf{Y}$ is given by

$$\cos \theta = \mathbf{AX} \cdot \mathbf{Y} / |\mathbf{AX}||\mathbf{Y}|.$$

Hence,

$$|\mathbf{AX} \cdot \mathbf{Y}| / |\mathbf{X}||\mathbf{Y}| = |\cos \theta| \, |\mathbf{AX}|/|\mathbf{X}|.$$

Thus, when $|X| = 1 = |Y|$,
$$|AX \cdot Y| = |\cos \theta| \, |AX|.$$
Hence
$$\|A\| = \max_{|X|=1} (|AX| \, |\cos \theta|) = \max_{\substack{|X|=1 \\ |Y|=1}} |AX \cdot Y|,$$
since Y can always be chosen in the same direction as AX. ∎

Example 2. A picture should clarify both the meaning of this theorem and its proof. If
$$A = \begin{pmatrix} 2 & 2 & 2 \\ 1 & 1 & 1 \end{pmatrix},$$
then

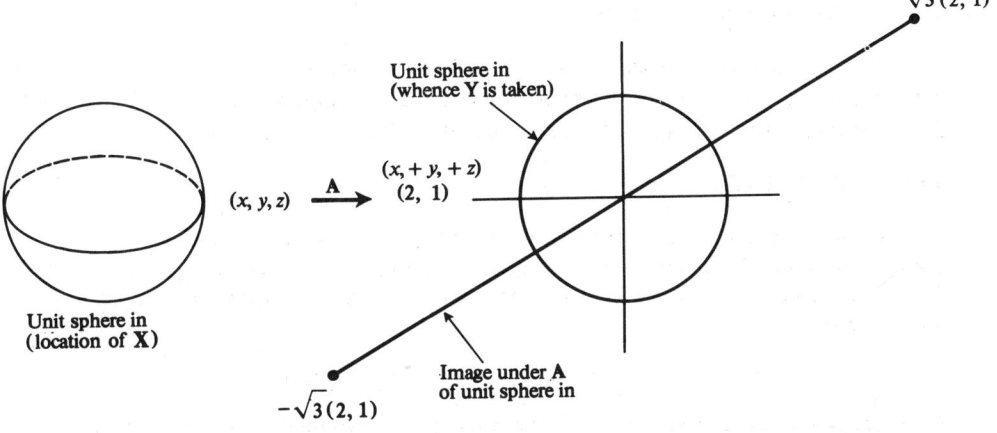

Fig. 6

the image of the unit sphere is shown in the picture. Now $\|A\|$ is the maximum length of a vector on this image ($= \sqrt{15}$). This is also the maximum of $|AX \cdot Y|$ for $|X| = 1$, $|Y| = 1$, since $|AX \cdot Y| = |AX| \, |Y| \, |\cos \theta|$ and the maximum of $|\cos \theta|$ is 1.

2.7 *The spectral norm of a matrix and its transpose.* For A in $\mathbb{R}_{m \times n}$,
$$\|A\| = \|A^*\|.$$
Proof. $AX \cdot Y = (AX)^*Y = X^*A^*Y = X \cdot (A^*Y)$. Hence,
$$|AX \cdot Y| = |X \cdot A^*Y| \stackrel{(\text{why}_1?)}{\leq} |X| \, \|A^*Y\| \stackrel{(\text{why?}_2)}{\leq} \|A^*\| \, |X| \, |Y|.$$
Hence $\|A\| = \max_{|X|=1=|Y|} |AX \cdot Y| \leq \|A\|^*$. If we reverse the roles of A and A^* in the above, we obtain that
$$\|A\| \leq \|A\|.$$
Hence, $\|A^*\| = \|A\|$. ∎

Example 3. If
$$A = \begin{pmatrix} 2 & 2 & 2 \\ 1 & 1 & 1 \end{pmatrix},$$
(so that A is as in Example 2), then
$$A^*A = \begin{pmatrix} 5 & 5 & 5 \\ 5 & 5 & 5 \\ 5 & 5 & 5 \end{pmatrix},$$
whose characteristic polynomial is $-x^2(x-15)$. Hence
$$\|A\| = \sqrt{15},$$
which is also $\|A^*\|$ as seen before.

It is a pleasant consequence of our last theorem that you do not have to remember 2.4 so carefully; whether you look at the eigenvalues of A^*A or AA^*, 2.7 guarantees that the largest eigenvalue of each is the same.

Project. Prove 2.2 for $\mathbb{C}_{m \times n}$.

When it comes to 2.4 onward, we must make a few observations. The natural complex analog of transposition in $\mathbb{R}_{m \times n}$ is conjugate-transposition. Thus, for A in $\mathbb{R}_{m \times n}$, we define $\bar{A}^*$ to be the matrix whose entries are the conjugates of the entries of the transpose of A:
$$\text{if} \quad A = (a_{ij}), \quad \text{then} \quad \bar{A}^* = (\bar{a}_{ji}).$$
(See Chapter 2, Section 4, Ex. 35.) The analog of a symmetric matrix, one for which $\bar{A}^* = A$, is called a *Hermitian matrix* (Chapter 9, Section 1, Supplement 1); the analog of an orthogonal matrix, $\bar{U}^* = U^{-1}$, is called a *unitary matrix* (see the Supplement to Chapter 8, Section 5). One can show, as in Chapter 8, Section 5, that if H is Hermitian, then there is a unitary U for which $\bar{U}^*HU$ is a real diagonal matrix.

With * replaced by "conjugate transpose", "symmetric" replaced by "Hermitian", and dot product replaced by the complex dot product
$$\langle V, W \rangle = \bar{V} \cdot W,$$
show that the proofs to 2.4 through 2.7 carry through as before.

EXERCISES

1. Discuss the possible effects of a matrix in $\mathbb{R}_{2 \times 2}$ on the (euclidean) unit circle. (Consider the matrices A, B, and C in Section 1, Exercise 6, for guidance).
2. (a) Answer the questions in the proof of 2.6. (b) Prove 2.5.

3. Calculate ‖A‖ for the following matrices;

a) $\begin{pmatrix} 0 & 1 \\ 1 & 0 \end{pmatrix}$
b) $\begin{pmatrix} 0 & -1 \\ 1 & 0 \end{pmatrix}$

c) $\begin{pmatrix} 5 & -1 & -1 \\ -1 & 3 & 1 \\ -1 & 1 & 3 \end{pmatrix}$

d) $\begin{pmatrix} 4 & 12 \\ 12 & 11 \end{pmatrix}$

Save your results.

4. Prove that the Euclidean norm of **A** is the square root of the trace of **A*A**.

5. If
$$A = \begin{pmatrix} a & b \\ c & d \end{pmatrix},$$
show that the characteristic polynomial of **A*A** is
$$x^2 - (a^2 + b^2 + c^2 + d^2)x + (ad - bc)^2.$$

6. If
$$A = \begin{pmatrix} a & a' \\ b & b' \\ c & c' \end{pmatrix},$$
show that the characteristic polynomial of **A*A** is
$$x^2 - (a^2 + b^2 + c^2 + a'^2 + b'^2 + c'^2)x + (ab' - a'b)^2$$
$$+ (ac' - a'c)^2 + (bc' - b'c)^2.$$

7. Find the euclidean norm and the spectral norm of

a) $\begin{pmatrix} 2 & -1 \\ 2 & 1 \end{pmatrix}$
b) $\begin{pmatrix} 2 & -1 \\ 3 & 1 \end{pmatrix}$
c) $\begin{pmatrix} \frac{3}{5} & \frac{1}{10} \\ \frac{4}{5} & \frac{9}{5} \end{pmatrix}$.

8. Find the euclidean norm and the spectral norm of

a) $\begin{pmatrix} 2 & -1 \\ 2 & 1 \\ 2 & 1 \end{pmatrix}$
b) $\begin{pmatrix} 1 & 0 \\ 2 & \frac{1}{2} \\ 2 & -\frac{1}{2} \end{pmatrix}$

c) If **A** is as in Exercise 6 and its columns are orthogonal, then the spectral norm equals the length of the column whose length is larger.

9. Suppose that **A** ≠ **0**.
   a) If one of the dimensions of **A** is 1, show that |A| = ‖A‖.
                                              (Euclidean) (Spectral)
   b) If both dimensions of **A** are greater than 1, and **A** ≠ **0**, show that in general
   $$‖A‖ > |A|.$$

Supplement 2. The Row and Column Norms

We now proceed to arm ourselves with two particularly convenient (but non-standard) matrix norms. First of all, in the definition of matrix norm,

$$\|A\| = \max_{|X|=1} |AX| \qquad (1)$$

where $|AX|$ is the Norm in $\mathbb{R}_m$ and $|X|$ is the Norm in $\mathbb{R}_n$,

let us take the box norm in both $\mathbb{R}_m$ and $\mathbb{R}_n$. We call the resulting matrix norm $\|A\|_R$ the **row norm** for the following reason:

2.8 The row norm of a matrix A is the largest of the sums of the absolute values of the entries of the rows of A:

$$\text{row norm } A = \max (|a_{i1}| + |a_{i2}| + \cdots + |a_{in}|) \qquad \text{for } i = 1, \ldots, m.$$

Proof. For convenience, take

$$A = \begin{pmatrix} a_{11} & a_{12} & a_{13} \\ a_{21} & a_{22} & a_{23} \end{pmatrix}, \quad X = \begin{pmatrix} x_1 \\ x_2 \\ x_3 \end{pmatrix}, \quad Y = \begin{pmatrix} y_1 \\ y_2 \end{pmatrix},$$

$$y_1 = a_{11}x_1 + a_{12}x_2 + a_{13}x_3, \qquad y_2 = a_{21}x_1 + a_{22}x_2 + a_{23}x_3;$$

then $|X|_B = 1$ means

$$|x_1| \leq 1, \qquad |x_2| \leq 1, \qquad |x_3| \leq 1$$

with at least one equality. Hence,

$$|y_1| \leq |a_{11}| + |a_{12}| + |a_{13}| \qquad \text{and} \qquad |y_2| \leq |a_{21}| + |a_{22}| + |a_{23}|,$$

where $|\ |$ means ordinary absolute value. Therefore $\|A\|_R \leq$ the larger of these row values. To prove that it is *equal* to the larger of these, suppose that the first is the larger. Then choose

$$x_1 = +1 \text{ or } -1 \qquad \text{so that } a_{11}x_1 = |a_{11}|,$$
$$x_2 = +1 \text{ or } -1 \qquad \text{so that } a_{12}x_2 = |a_{12}|,$$
$$x_3 = +1 \text{ or } -1 \qquad \text{so that } a_{13}x_3 = |a_{13}|.$$

Then $|X|_B = 1$ and $|y_1| = |a_{11}| + |a_{12}| + |a_{13}|$, so that

$$\|A\|_R = |a_{11}| + |a_{12}| + |a_{13}|,$$

as required. ∎

As an example, the row norm of $\begin{pmatrix} 2 & -1 \\ 3 & 1 \end{pmatrix}$ is 4.

For our next norm, we take the taxicab norm for both $\mathbb{R}_m$ and $\mathbb{R}_n$ in (1). The resulting matrix norm $\|A\|_R$ we call the **column norm**, for reasons you can easily guess.

2.9 The column norm of a matrix A is the largest of the sums of the absolute values of the entries of the columns of A:

$$\text{column norm } A = \max(|a_{1j}| + |a_{2j}| + \cdots + |a_{mj}|) \quad \text{for } j = 1, \ldots, n.$$

Proof. Again take

$$A = \begin{pmatrix} a_{11} & a_{12} & a_{13} \\ a_{21} & a_{22} & a_{23} \end{pmatrix},$$

Then $|X|_T = 1$ means $|x_1| + |x_2| + |x_3| = 1$, and

$$|AX|_T = |y_1| + |y_2| = |a_{11}x_1 + a_{12}x_2 + a_{13}x_3| + |a_{21}x_1 + a_{22}x_2 + a_{23}x_3|$$
$$\leq |a_{11}x_1| + |a_{21}x_1| + |a_{12}x_2| + |a_{22}x_2| + |a_{13}x_3| + |a_{23}x_3|$$
$$= |x_1|(|a_{11}| + |a_{21}|) + |x_2|(|a_{12}| + |a_{22}|) + |x_3|(|a_{13}| + |a_{23}|)$$
$$\leq (|x_1|) + |x_2| + |x_3|)(\max |a_{11}| + |a_{21}|, |a_{12}| + |a_{22}|, |a_{13}| + |a_{23}|)$$
$$= \max \text{ of column values}.$$

To prove that $|AX|_T$ is equal to this maximum for some X with $|X|_T = 1$, suppose that the first column yields the maximum. Then take

$$x_1 = 1, \quad x_2 = x_3 = 0,$$

so that $|X|_T = 1$ and

$$|AX|_T = |a_{11}| + |a_{21}|,$$

as required. ∎

Example. If

$$A = \begin{pmatrix} 2 & -1 \\ 3 & 1 \end{pmatrix},$$

the row norm is 4, the column norm is 5, and the euclidean norm is $\sqrt{15}$. The spectral norm turns out to be approximately 3.62 (see Exercise 7 in Supplement 1). Another instructive example is

$$A = \begin{pmatrix} \frac{3}{5} & \frac{1}{10} \\ \frac{4}{5} & \frac{9}{5} \end{pmatrix},$$

for which

$$\text{row norm} = \tfrac{13}{5}, \quad \text{column norm} = \tfrac{19}{10},$$
$$\text{spectral norm} = 2, \quad \text{euclidean norm} = \tfrac{1}{2}\sqrt{17}.$$

MORAL. Although the euclidean and spectral norms have a fixed relation given by 2.2, the others do not. This can sometimes be useful, especially since the row and column norms are so easy to compute.

EXERCISES

1. Find all our above-mentioned norms for

$$\begin{pmatrix} \frac{3}{4} & -\frac{1}{8} \\ \frac{3}{4} & \frac{1}{8} \end{pmatrix}.$$

2. a) Prove that the row norm of a stochastic matrix is 1.
   b) Find all our named norms for

$$\begin{pmatrix} \frac{1}{3} & \frac{2}{3} \\ \frac{1}{2} & \frac{1}{2} \end{pmatrix}.$$

3. INFINITE SERIES OF MATRICES

We have already mastered infinite sequences and series of matrices (in the guise of vectors) in Section 1. An important point developed there was that the convergence of an infinite sequence or series can be decided in terms of any norm; all give the same results. Moreover, elementwise convergence is essentially equivalent to using the box norm.

In Section 2 we developed the notion of matrix norms. We hereafter use the notation $\|A\|$ for matrix norms and $|A|$ for vector norms. In addition to the spectral norm, other attractive matrix norms include the

row norm = maximum of the sums of the absolute values of the entries in the rows of A,

and the

column norm = maximum of the sums of the absolute values of the entries of the columns of A.

(See Supplement 2 to Section 2). All matrix norms share the property that

$$\|AB\| \leq \|A\| \|B\| \qquad \text{(by 2.1)}.$$

3.1 Arithmetic of matrix sequences. i) If

$$\lim_{k \to \infty} A_k = A \quad \text{and} \quad \lim_{k \to \infty} B_k = B,$$

then

$$\lim_{k \to \infty} (A_k + B_k) = A + B.$$

ii) If

$$\lim_{k \to \infty} A_k = A,$$

then

$$\lim_{k \to \infty} BA_k = BA \quad \text{and} \quad \lim_{k \to \infty} A_k C = AC$$

for any compatible B and C.

iii) If
$$\lim_{k \to \infty} A_k = A \quad \text{and} \quad \lim_{k \to \infty} B_k = B,$$
and if multiplication is possible, then
$$\lim_{k \to \infty} A_k B_k = AB.$$

Proof. (i) was in Section 1. You do the rest. It is necessary to use a matrix norm, but the first-year calculus proofs carry over completely. ∎

We shall use
$$\sum_{k=0}^{\infty} A_k = A$$
to denote that the infinite series
$$\sum_{k=0}^{\infty} A_k$$
converges to A.

Another theorem with a new matrix part is

3.2 Arithmetic of matrix series. i) If
$$\sum_{k=0}^{\infty} A_k = A \quad \text{and} \quad \sum_{k=0}^{\infty} A_k = B,$$
then
$$\sum_{k=0}^{\infty} (A_k + B_k) = A + B.$$

ii) If
$$\sum_{k=0}^{\infty} A_k = A$$
and if B and C are compatible matrices, then
$$\sum_{k=0}^{\infty} BA_k = BA \quad \text{and} \quad \sum_{k=0}^{\infty} A_k C = AC.$$

Proof. Yours again. For (ii) a matrix norm is again necessary. ∎

We have also considered the notion of absolute convergence of series. The following will prove useful:

3.3 The comparison theorem. Let $\|A\|$ denote any fixed matrix norm. If $\|A_k\| \leq a_k$ and if $\sum_{k=0}^{\infty} a_k$ converges, then $\sum_{k=0}^{\infty} A_k$ also converges (in fact it converges absolutely).

Proof. This follows immediately from 1.5 and the Comparison Test for (real) infinite series. ∎

The real importance of infinite series of matrices lies in the study of power series, something impossible for us to discuss in the mere vector terms of Section 1. If A is an $n \times n$ matrix, then by a **power series** in A, we shall mean an infinite series of the form

$$\sum_{k=0}^{\infty} c_k A^k,$$

where c_k is a sequence of real numbers.

3.4 Let A be an $n \times n$ matrix and $\sum c_k x^k$ a real power series with radius of convergence R. If $\|A\| < R$ for any matrix norm, then $\sum c_k A^k$ converges.

Proof. Let $a_k = |c^k| \|A\|^k$, so that $\sum a_k$ converges since $\|A\| < R$. Then

$$\|c_k A^k\| = |c_k| \|A^k\| \leq |c_k| \|A\|^k = a_k,$$

and hence the matrix power series converges by the Comparison Theorem. ∎

An important example is the series $I + A + A^2 + \cdots$. Since $1 + x + x^2 + \cdots$ converges for $|x| < 1$ and diverges for $|x| > 1$, we have $R = 1$. Hence

$$I + A + A^2 + \cdots \quad \text{converges} \quad \text{if } \|A\| < 1,$$

for any matrix norm.

Thus, for

$$A = \begin{pmatrix} \frac{3}{4} & -\frac{1}{8} \\ \frac{3}{4} & \frac{1}{8} \end{pmatrix},$$

the row norm is $\frac{7}{8}$, and hence the series converges. The column norm is $\frac{3}{2}$ and the spectral norm is $\frac{3}{4}\sqrt{2} > 1$, so these norms are not good indicators in this particular case.

For

$$A = \begin{pmatrix} \frac{1}{2} & -\frac{1}{4} \\ \frac{3}{4} & \frac{1}{4} \end{pmatrix},$$

the row norm is 1, and the column norm is $\frac{5}{4}$. However, the euclidean norm (while not a matrix norm) is $\frac{1}{4}\sqrt{15} < 1$. Hence

$$\text{spectral norm} \leq \text{euclidean norm} < 1.$$

Consequently, the series converges. In this case, the spectral norm is ≈ 0.905 (see the example at the end of Supplement 2 to Section 2).

If $\sum c_k x^k$ converges for all x (so that we put $R = \infty$), then the matrix power series converges for any square matrix A. An especially important example is the series

$$\sum_{k=0}^{\infty} \frac{1}{k!} A^k,$$

which converges for any A since the exponential series

$$\sum_{k=0}^{\infty} \frac{1}{k!} x^k$$

has $R = \infty$. We call the sum e^A,

$$e^A = I + A + \tfrac{1}{2}A^2 + \cdots = \sum_{k=0}^{\infty} \frac{1}{k!} A^k.$$

Several familiar properties of the exponential function go over to matrices. For example,
if A and B commute ($AB = BA$), then

$$e^{A+B} = e^A e^B. \tag{1}$$

This can be proved by developing properties of the multiplication of series. We choose a more circuitous (and hopefully more entertaining) route in the supplement to this section. Some more easily proved properties:

3.5 i) e^A is always invertible. In fact,

$$(e^A)^{-1} = e^{-A}.$$

ii) $(e^A)^* = e^{A*}$.

Proof. You do it. Feel free to use (1) for (i). ∎

You might stop and calculate e^I, or e^D for any diagonal matrix. We shall return to a more detailed discussion of e^A in a moment. More generally, if

$$f(x) = \sum c_k x^k \quad \text{for} \quad |x| < R,$$

where R is the radius of convergence, then we shall use the notation

$$f(A) = \sum c_k A^k$$

when this series converges; as we have seen, convergence holds if $\|A\| < R$. It is easily seen that if A is a diagonal matrix,

$$A = \begin{pmatrix} \lambda_1 & 0 & \cdots & 0 \\ 0 & \lambda_2 & & \cdot \\ \cdot & & \cdot & \cdot \\ \cdot & & & \cdot \\ 0 & \cdots & \cdots & \lambda_n \end{pmatrix},$$

then

$$f(A) = \begin{pmatrix} f(\lambda_1) & 0 & 0 & \cdots \\ 0 & f(\lambda_2) & 0 & \\ \cdot & & \cdot & \\ \cdot & & & \\ 0 & & & f(\lambda_n) \end{pmatrix}$$

converges if $|\lambda_1| < R, \ldots |\lambda_n| < R$, but diverges if one $|\lambda_i| > R$ (for all λ_i, real and complex). An important result here is the following:

3.6 If **B** is invertible, then $f(x) = \sum c_k x_k$, then
$$f(\mathbf{B}^{-1}\mathbf{AB}) = \mathbf{B}^{-1}f(\mathbf{A})\mathbf{B},$$
meaning that if either side converges, so does the other.

Proof. For you. ∎

Using 3.6 it is easy to evaluate $f(\mathbf{A})$ if **A** is similar to a diagonal matrix (for example, if all n of the eigenvalues are all distinct, or if **A** is symmetric). In fact, if $\mathbf{A} = \mathbf{B}^{-1}\mathbf{DB}$, where

$$\mathbf{D} = \begin{pmatrix} \lambda_1 & 0 & \cdots & 0 \\ 0 & \lambda_2 & & \\ \vdots & & \ddots & \\ 0 & & & \lambda_n \end{pmatrix},$$

then

$$f(\mathbf{A}) = \mathbf{B}^{-1}f(\mathbf{D})\mathbf{B} = \mathbf{B}^{-1} \begin{pmatrix} f(\lambda_1) & 0 & \cdots & 0 \\ 0 & f(\lambda_2) & & \\ \vdots & & \ddots & \\ 0 & & & f(\lambda_n) \end{pmatrix} \mathbf{B}.$$

Now let λ be the eigenvalue of maximum absolute value, so that
$$|\lambda_1| \le |\lambda|, \quad |\lambda_2| \le |\lambda|, \quad \ldots, \quad |\lambda_n| \le |\lambda|.$$
Then it follows that $f(\mathbf{A})$ converges if $|\lambda| < R$ and diverges if $|\lambda| > R$. It will be proved later that this holds for any matrix **A**.

Examples.
$$\mathbf{A} = \begin{pmatrix} \frac{6}{7} & -\frac{1}{7} \\ \frac{6}{7} & \frac{1}{7} \end{pmatrix}$$
has row norm = 1, column norm = $\frac{12}{7}$, and spectral norm = $\frac{6}{7}\sqrt{2} > 1$, so that no conclusion can be drawn about the convergence of
$$\mathbf{I} + \mathbf{A} + \mathbf{A}^2 + \cdots + \mathbf{A}^k + \cdots$$
However, the eigenvalues turn out to be $\frac{3}{7}, \frac{4}{7}$ (check this), so that $\lambda = \frac{4}{7}$, and hence the series converges.
$$\mathbf{A} = \begin{pmatrix} 2 & -\frac{1}{3} \\ 2 & +\frac{1}{3} \end{pmatrix}$$
also has all norms > 1. Here the eigenvalues are 1, $\frac{4}{3}$, so that $\lambda = \frac{4}{3}$, and the series diverges.

To take care of the general case we need the Jordan Form of Chapter 6, Section 5 for **A**. In order to calculate $f(\mathbf{A})$ under these circumstances, we first take the case of a Jordan block and work out what happens to it. In fact

3.7 If

$$\mathbf{J} = \begin{pmatrix} \lambda & 1 & 0 & \cdots & & 0 \\ 0 & \lambda & 1 & \cdots & & 0 \\ 0 & 0 & \lambda & & & \\ \cdot & & \cdot & & & \\ \cdot & & \cdot & & & 1 \\ 0 & & & & & \lambda \end{pmatrix}, \quad \mathbf{J} \; n \times n$$

and $f(x)$ is a power series of radius of convergence R, then $f(\mathbf{J})$ diverges if $|\lambda| > R$. If $|\lambda| < R$, then the series in $\mathbf{J}$ converges and

$$f(\mathbf{J}) = \begin{pmatrix} f(\lambda) & f'(\lambda) & \frac{1}{2!}f''(\lambda) & \cdots & \frac{1}{(n-1)!}f^{(n-1)}(\lambda) \\ 0 & f(\lambda) & f'(\lambda) & \cdots & \frac{1}{(n-2)!}f^{(n-2)}(\lambda) \\ \cdot & & & & \cdot \\ \cdot & & & & \cdot \\ \cdot & & & & f'(\lambda) \\ 0 & & & & f(\lambda) \end{pmatrix},$$

for $|\lambda| < R$.

Proof. We shall prove this for $n = 3$, and leave the general case to the reader. We have

$$\mathbf{J} = \begin{pmatrix} \lambda & 1 & 0 \\ 0 & \lambda & 1 \\ 0 & 0 & \lambda \end{pmatrix} = \lambda \mathbf{I} + \mathbf{N} \quad \text{where} \quad \mathbf{N} = \begin{pmatrix} 0 & 1 & 0 \\ 0 & 0 & 1 \\ 0 & 0 & 0 \end{pmatrix},$$

so that

$$\mathbf{N}^2 = \begin{pmatrix} 0 & 0 & 1 \\ 0 & 0 & 0 \\ 0 & 0 & 0 \end{pmatrix},$$

and $\mathbf{N}^3$ and all higher powers are zero. Since $\lambda \mathbf{I}$ and $\mathbf{N}$ commute,

$$\mathbf{J}^k = \lambda^k \mathbf{I} + k\lambda^{k-1}\mathbf{N} + \frac{1}{2!}k(k-1)\lambda^{k-2}\mathbf{N}^2 \quad \text{for} \quad k \geq 2,$$

and therefore
$$\Sigma c_k \mathbf{J}^k = f(\lambda)\mathbf{I} + f'(\lambda)\mathbf{N} + \frac{1}{2!}f''(\lambda)\mathbf{N}^2.$$
where the series for $f(\lambda), f'(\lambda), f''(\lambda)$ all converge if $|\lambda| < R$. Therefore,
$$f(\mathbf{J}) = \begin{pmatrix} f(\lambda) & f'(\lambda) & \frac{1}{2!}f''(\lambda) \\ 0 & f(\lambda) & f'(\lambda) \\ 0 & 0 & f(\lambda) \end{pmatrix} \quad \text{if} \quad |\lambda| < R.$$

If $|\lambda| > R$, each diagonal term diverges (as do all the others) so that the series diverges. (These things are again true even if λ is complex.) ∎

It should now be clear what one does in general (at least theoretically; recall that there are serious practical difficulties in calculating the Jordan Form of a matrix). First, express $\mathbf{A}$ in terms of its Jordan form,
$$\mathbf{A} = \mathbf{B}^{-1} \begin{pmatrix} \mathbf{J}_1 & 0 & \cdots & 0 \\ 0 & \mathbf{J}_2 & & \\ \vdots & & \ddots & \\ 0 & & & \mathbf{J}_r \end{pmatrix} \mathbf{B}.$$
where $\mathbf{J}_i$ has characteristic value λ_i, $i = 1, 2, \ldots, r$. Then
$$f(\mathbf{A}) = \mathbf{B}^{-1} \begin{pmatrix} f(\mathbf{J}_1) & & 0 \\ & \ddots & \\ 0 & & f(\mathbf{J}_r) \end{pmatrix} \mathbf{B},$$
and this converges if $|\lambda_i| < R$ for all $i = 1, 2, \ldots, r$. The series diverges if one or more of the numbers $|\lambda_i|$ is $> R$. In particular we have proved:

3.8 If λ is the eigenvalue of $\mathbf{A}$ that has maximum absolute value (that is, $|\lambda_i| \le |\lambda|$, for any other eigenvalue λ_i), then the series $\sum c_j \mathbf{A}^j$ converges if $|\lambda| < R$ and diverges if $|\lambda| > R$, where R is the radius of convergence of the power series $\sum c_j x^j$.

From this result we get a curious corollary:

3.9 If $\|\mathbf{A}\|$ is any matrix norm, then $\|\mathbf{A}\| \ge |\lambda|$, where λ is any eigenvalue of $\mathbf{A}$.

Proof. Take λ to be the largest eigenvalue in absolute value. If $\|\mathbf{A}\| < |\lambda|$, take a number R so that $\|\mathbf{A}\| < R < |\lambda|$. Then find a power series $\sum c_k x^k$ whose radius of convergence is R (how can this be done?), and observe that $\sum c_k A^k$ is convergent, since $\|A\| < R$, and divergent since $|\lambda| > R$, so we get a contradiction. ∎

We can now complete our discussion of e^A. In particular, 3.6 tells us that if B is invertible, then
$$Be^A B^{-1} = e^{BAB^{-1}}.$$
Our general discussion shows us that we can then evaluate e^A by obtaining the Jordan form $J = BAB^{-1}$ of A, evaluating e^J, and then calculating $e^A = B^{-1}e^J B$. This of course, is a complicated procedure to carry out in detail and should not be lightly undertaken. The exponential of a Jordan block works out quite nicely. From 3.7, if

$$J = \begin{pmatrix} \lambda & 1 & & & \\ & \ddots & \ddots & & \\ & & \ddots & \ddots & \\ & & & \ddots & 1 \\ & & & & \lambda \end{pmatrix},$$

then

$$e^J = \begin{pmatrix} e^\lambda & \frac{1}{1!}e^\lambda & \frac{1}{2!}e^\lambda & \cdots & \frac{1}{n!}e^\lambda \\ & e^\lambda & \frac{1}{1!}e^\lambda & & \vdots \\ & & \ddots & & \\ & & & & \frac{1}{1!}e^\lambda \\ 0 & & & & e^\lambda \end{pmatrix}$$

Example. If
$$A = \begin{pmatrix} 2 & -\frac{1}{3} \\ 2 & \frac{1}{3} \end{pmatrix},$$
one finds that for
$$B = \begin{pmatrix} -2 & 1 \\ 3 & -1 \end{pmatrix}, \quad B^{-1} = \begin{pmatrix} 1 & 1 \\ 3 & 2 \end{pmatrix}, \quad \text{and} \quad BAB^{-1} = \begin{pmatrix} 1 & 0 \\ 0 & \frac{4}{3} \end{pmatrix}.$$
Hence
$$e^A = e^{B^{-1}\begin{pmatrix} 1 & 0 \\ 0 & 4/3 \end{pmatrix}B} = B^{-1}e^{\begin{pmatrix} 1 & 0 \\ 0 & 4/3 \end{pmatrix}}B$$
$$= \begin{pmatrix} 1 & 1 \\ 3 & 2 \end{pmatrix}\begin{pmatrix} e & 0 \\ 0 & e^{4/3} \end{pmatrix}\begin{pmatrix} -2 & 1 \\ 3 & -1 \end{pmatrix}$$
$$= \begin{pmatrix} -2e + 3e^{4/3} & e - e^{4/3} \\ -6e + 6e^{4/3} & 3e - 2e^{4/3} \end{pmatrix}.$$

The technique for calculating e^A using Jordan Form has its theoretical uses, as we see in the proof of the next theorem.

3.10 *The eigenvalues and eigenvectors of* e^A. λ is an eigenvalue of $\mathbf{A}$ iff e^λ is an eigenvalue of e^A. The eigenvectors of $\mathbf{A}$ and e^A are the same.

Proof. Let

$$\mathbf{BAB}^{-1} = \begin{pmatrix} \lambda_1 & & * \\ & \ddots & \\ 0 & & \lambda_n \end{pmatrix}$$

be in Jordan Form, so that the eigenvalues of $\mathbf{A}$ are $\lambda_1, \ldots, \lambda_n$. Then

$$e^{\mathbf{BAB}^{-1}} = \begin{pmatrix} e^{\lambda_1} & & * \\ & \ddots & \\ 0 & & e^{\lambda_n} \end{pmatrix},$$

so the eigenvalues of e^A are $e^{\lambda_1}, \ldots, e^{\lambda_n}$. (Why$_1$?)

To obtain the result on eigenvectors, suppose that $\mathbf{A}' = \mathbf{BAB}^{-1}$ is in Jordan Form so that in terms of Jordan blocks,

$$\mathbf{A}' = \begin{pmatrix} \mathbf{J}_1 & & & 0 \\ & \mathbf{J}_2 & & \\ & & \ddots & \\ 0 & & & \mathbf{J}_t \end{pmatrix}, \quad e^{\mathbf{A}'} = \begin{pmatrix} e^{\mathbf{J}_1} & & & 0 \\ & e^{\mathbf{J}_2} & & \\ & & \ddots & \\ 0 & & & e^{\mathbf{J}_t} \end{pmatrix}.$$

The eigenvectors of $\mathbf{A}'$ are those vectors $\mathbf{V} = (0 \cdots 0 \ * \ 0 \cdots 0)^*$ with a nonzero entry in a position corresponding to the first row of a Jordan block (why$_2$?). For example, the eigenvectors of

$$\mathbf{A}' = \begin{pmatrix} \lambda & & & & \\ & \mu & 1 & 0 & \\ & 0 & \mu & 1 & \\ & 0 & 0 & \mu & \\ & & & & \nu & 1 \\ & & & & 0 & \nu \end{pmatrix} \quad \text{are} \quad \begin{pmatrix} 1 \\ 0 \\ 0 \\ 0 \\ 0 \\ 0 \end{pmatrix}, \begin{pmatrix} 0 \\ 1 \\ 0 \\ 0 \\ 0 \\ 0 \end{pmatrix}, \text{ and } \begin{pmatrix} 0 \\ 0 \\ 0 \\ 0 \\ 1 \\ 0 \end{pmatrix}$$

(Why$_3$?). Similarly, the eigenvectors of $e^{\mathbf{A}'}$ are vectors with nonzero entries in exactly the same places (why$_4$?). Hence the eigenvectors of $e^{\mathbf{A}'}$ and $\mathbf{A}'$ are the same. We need only note that $\mathbf{V}$ is an eigenvector of $\mathbf{A}'$ iff $\mathbf{B}^{-1}\mathbf{V}$ is an eigenvector of $\mathbf{A}$ (why$_5$?). ∎

3.11 The determinant of e^A is $e^{\text{trace } A}$.

Proof. Exercise 8. ∎

EXERCISES

1. Calculate e^A for the following matrices:

   a) $\begin{pmatrix} 0 & 1 \\ 1 & 0 \end{pmatrix}$

   b) $\begin{pmatrix} 0 & -1 \\ 1 & 0 \end{pmatrix}$

   c) $\begin{pmatrix} 5 & -1 & -1 \\ -1 & 3 & 1 \\ -1 & 1 & 3 \end{pmatrix}$ (Compare Example in Section 1, Chapter 9),

   d) $\begin{pmatrix} 4 & 12 \\ 12 & 11 \end{pmatrix}$ (Compare Example in Section 2, Chapter 9).

2. (a) Prove 3.1; (b) 3.2.
3. (a) Prove 3.5; (b) 3.6.
4. Prove that if ΣA_k converges, then
$$\lim_{k \to \infty} A_k = 0.$$

5. If $\|A\| < 1$ for any matrix norm, prove that
$$\lim_{k \to \infty} A^k = 0.$$

6. Prove that if $\|A\| < 1$ for any matrix norm, then
$$\sum_{k=1}^{\infty} A^k = (I - A)^{-1}.$$

7. Prove 3.7 in general.
8. (a) Answer the questions in the proof of 3.10. (b) Prove 3.11.
9. Which of the matrices in (1) have a geometric series?

H10. If $\|A\| < 1$ for some matrix norm, show that
$$A - A^2/2 + A^3/3 - \cdots + (-1)^{k+1} A^k/k + \cdots$$
converges to a matrix $\log(A + I)$ which satisfies
$$e^{\log(A+I)} = A + I.$$

11. Prove that for
$$A = \begin{pmatrix} 0 & 1 \\ 0 & 0 \end{pmatrix} \quad \text{and} \quad B = \begin{pmatrix} -1 & 0 \\ 0 & 0 \end{pmatrix},$$
$$e^{A+B} \neq e^A e^B.$$

12. Calculate e^A for the following matrices

a) $\begin{pmatrix} 11 & 5 \\ -20 & 11 \end{pmatrix}$
b) $\begin{pmatrix} 3 & -2 & 1 \\ 2 & 1 & -1 \\ 2 & -6 & 4 \end{pmatrix}$
c) $\begin{pmatrix} 3 & 2 & 2 \\ -2 & 1 & -6 \\ 1 & -1 & 4 \end{pmatrix}$

d) $\begin{pmatrix} 0 & 1 \\ -6 & -5 \end{pmatrix}$
e) $\begin{pmatrix} 1 & 0 & -1 \\ 1 & 2 & 1 \\ 2 & 2 & 3 \end{pmatrix}$.

13. Calculate

$$\log(I + A) - A + \frac{A^2}{2} + \frac{A^3}{3} - \cdots$$

for

$$A = \begin{pmatrix} \frac{3}{4} & -\frac{1}{8} \\ \frac{3}{4} & \frac{1}{8} \end{pmatrix},$$

and verify that $e^{\log(I+A)} = I + A$.

14. Prove that for a symmetric matrix, both row and column norm are $\geq$ the spectral norm.

15. Prove that for a stochastic matrix, the eigenvalue of maximum absolute value is 1.

16. Let **P** be a stochastic matrix with the property that 1 is an eigenvalue which is a single root of the characteristic polynomial, and that the absolute values of all other eigenvalues are <1. If $(p_1, p_2, \ldots, p_n) = \mathbf{p}$ is a probability vector which is the eigenvector corresponding to the eigenvalue 1, prove that

$$\lim_{k \to \infty} \mathbf{P}^k = \begin{pmatrix} p_1 & p_2 & \cdots & p_n \\ p_1 & p_2 & \cdots & p_n \\ \cdot & \cdot & & \cdot \\ \cdot & \cdot & & \cdot \\ \cdot & \cdot & & \cdot \\ p_1 & p_2 & \cdots & p_n \end{pmatrix}.$$

17. Let **P** be a stochastic matrix for a Markov chain with m absorbing states, and assume it is of the form

$$\mathbf{P} = \left(\begin{array}{cccc|ccc} 1 & 0 & \cdots & 0 & 0 & \cdots & 0 \\ 0 & 1 & \cdots & 0 & 0 & \cdots & 0 \\ & & \cdot & & & & \\ & & \cdot & & & & \\ 0 & 0 & \cdots & 1 & 0 & \cdots & 0 \\ \hline & \mathbf{B} & & & & \mathbf{Q} & \end{array} \right),$$

where **Q** is $(n - m) \times (n - m)$, and **B** is $(n - m) \times m$ (see Exercise 4 in Chapter 2, Section 5d, Markov Chains). Make the assumption that each row of **B** has at least one nonzero entry. Prove

$$\lim_{k=\infty} \mathbf{P}^k = \begin{pmatrix} 1 & 0 & \cdots & 0 & 0 & \cdots & 0 \\ 0 & 1 & \cdots & 0 & & & \\ \vdots & & & & & & \\ 0 & 0 & \cdots & 1 & 0 & \cdots & 0 \\ \hline & & & & 0 & \cdots & 0 \\ & & \mathbf{NB} & & \vdots & & \\ & & & & 0 & \cdots & 0 \end{pmatrix},$$

where $\mathbf{N} = \mathbf{I} + \mathbf{Q} + \mathbf{Q}^2 + \cdots = (\mathbf{I} - \mathbf{Q})^{-1}$. Give an interpretation of this result. For example, find $\lim_{k \to \infty} \mathbf{P}^k$ if

$$\mathbf{P} = \begin{pmatrix} 1 & 0 & 0 & 0 \\ 0 & 1 & 0 & 0 \\ \hline 0 & \frac{1}{3} & \frac{1}{3} & \frac{1}{3} \\ \frac{1}{4} & \frac{1}{4} & \frac{1}{4} & \frac{1}{4} \end{pmatrix}.$$

Supplement to Section 3: e^{A+B}

We give the outline of a method for proving: if $\mathbf{A}$ and $\mathbf{B}$ commute, then

$$e^{\mathbf{A}+\mathbf{B}} = e^{\mathbf{A}} e^{\mathbf{B}}. \tag{1}$$

The reader should supply missing steps and proofs.

We first quote a result on real functions which is a special case of a well-known theorem on uniformly convergent sequences of functions. You should try to prove this special case directly.

3.12 Let $u_k(t)$ be real functions, $k = 1, 2, \ldots$, with $u'_k(t)$ existing and continuous for $a \leq t \leq b$. Suppose that $\sum M_k$ is a convergent series of real numbers whose terms are bounds for the u'_k:

$$|u'_k(t)| \leq M_k \quad k = 1, 2, \ldots.$$

Under these circumstances, if $\sum u_k(t)$ converges to a function $f(t)$, then $\sum u'_k(t)$ converges, $f'(t)$ exists, and

$$f'(t) = \sum u'_k(t) \quad \text{for} \quad a \leq t \leq b.$$

It is easy to obtain a matrix form of this. To do so, we introduce the notion of the derivative of a vector-valued function of a real variable t. If

$$\mathbf{U}(t) = (u_1(t), \ldots, u_n(t)),$$

with $u_1(t), \ldots, u_n(t)$ real functions which are differentiable at t_0, then we define the **derivative of U at** t_0 to be

$$\mathbf{U}'(t_0) = (u'_1(t_0), \ldots, u'_n(t_0)).$$

Since matrices may be considered as vectors, this immediately yields the notion of the derivative of a matrix-valued function of a real variable: just differentiate componentwise.

3.13 Let $\mathbf{U}_k(t)$ be a sequence of vector-valued functions in $\mathbb{R}_n$ whose derivatives $\mathbf{U}'_k(t)$ exist and have continuous components for $a \leq t \leq b$. Suppose that for some norm,
$$|\mathbf{U}'_k(t)| < M_k,$$
where $\sum M_k$ is convergent.

If $\sum \mathbf{U}_k(t)$ converges to a function $\mathbf{F}(t)$, then $\sum \mathbf{U}'_k(t)$ converges, $\mathbf{F}'(t)$ exists, and
$$\mathbf{F}'(t) = \sum \mathbf{U}'_k(t).$$

We now turn to matrices.

3.14 Let $\mathbf{A}(t)$ and $\mathbf{B}(t)$ be matrix functions of t, and suppose that $\mathbf{A}'(t)$ and $\mathbf{B}'(t)$ exist on some interval. Then on this interval

i) $\dfrac{d}{dt}(\mathbf{A} + \mathbf{B}) = \mathbf{A}' + \mathbf{B}'.$

ii) $\dfrac{d}{dt}(\mathbf{AB}) = \mathbf{A}'\mathbf{B}' + \mathbf{A}'\mathbf{B}.$

You should next prove that for commuting matrices $\mathbf{A}$ and $\mathbf{B}$ we have
$$\mathbf{A}(\mathbf{B} + t\mathbf{A})^k = (\mathbf{B} + t\mathbf{A})^k \mathbf{A},$$
From this and 3.14 we see:

3.15 Let $\mathbf{A}$ and $\mathbf{B}$ be constant $n \times n$ matrices, and suppose that $\mathbf{A}$ and $\mathbf{B}$ commute. Then
$$\frac{d}{dt}(\mathbf{B} + t\mathbf{A})^k = k\mathbf{A}(\mathbf{B} + t\mathbf{A})^{k-1}.$$

We next prove a matrix generalization of a well-known fact:

3.16 Let $\mathbf{A}$ and $\mathbf{B}$ be commuting constant matrices. Then
$$\frac{d}{dt} e^{\mathbf{B}+t\mathbf{A}} = \mathbf{A} e^{\mathbf{B}+t\mathbf{A}}.$$

Proof. $e^{\mathbf{B}+t\mathbf{A}} = \sum \dfrac{1}{k!}(\mathbf{B} + t\mathbf{A})^k$. Take $|t| < R$ for some R, so that for some fixed matrix norm,
$$|\mathbf{B} + t\mathbf{A}| < |\mathbf{B}| + R|\mathbf{A}|.$$
Let $M_k = \dfrac{1}{(k-1)!} |\mathbf{A}| (|\mathbf{B}| + R|\mathbf{A}|)^{k-1}$, so that $\sum M_k$ converges. Now use the

fact that

$$\frac{d}{dt}\left[\frac{1}{k!}(\mathbf{B}+t\mathbf{A})^k\right] = \frac{1}{(k-1)!}\mathbf{A}(\mathbf{B}+t\mathbf{A})^{k-1},$$

together with previous results, to show that $\dfrac{d}{dt}e^{\mathbf{B}+t\mathbf{A}}$ exists and is equal to $\mathbf{A}e^{\mathbf{B}+t\mathbf{A}}$. ∎

3.17 If $\mathbf{A}$ and $\mathbf{B}$ commute, then $e^{-\mathbf{A}}e^{\mathbf{B}+\mathbf{A}} = e^{\mathbf{B}}$.

Proof. Let $\mathbf{f}(t) = e^{-t\mathbf{A}}e^{\mathbf{B}+t\mathbf{A}}$, so that $\mathbf{f}(0) = e^{\mathbf{B}}$. Then justify the fact that

$$\begin{aligned}\mathbf{f}'(t) &= e^{-t\mathbf{A}}\mathbf{A}e^{\mathbf{B}+t\mathbf{A}} + (-\mathbf{A})e^{-t\mathbf{A}}e^{\mathbf{B}+t\mathbf{A}} \\ &= \mathbf{A}e^{-t\mathbf{A}}e^{\mathbf{B}+t\mathbf{A}} - \mathbf{A}e^{-t\mathbf{A}}e^{\mathbf{B}+t\mathbf{A}} = \mathbf{0}.\end{aligned}$$

From this it follows (why?) that $\mathbf{f}(t)$ is constant, so that in particular, $\mathbf{f}(0) = \mathbf{f}(1)$, which proves the statement (why?). ∎

From 3.17 it immediately follows (how?) that $e^{-\mathbf{A}}e^{\mathbf{A}} = \mathbf{I}$, and one can then easily prove (how?) that for commuting matrices,

$$e^{\mathbf{A}}e^{\mathbf{B}} = e^{\mathbf{A}+\mathbf{B}}.$$

4. THE COMPLEX NUMBERS

In this section we shall use our newfound knowledge of infinite series of matrices to prove results about complex numbers. In the next section we use this newfound knowledge of complex numbers to develop analogies between complex numbers and $n \times n$ matrices.

In Exercise 13, Section 4, Chapter 2, we saw that there was a great similarity between complex numbers $z = 1a + ib$ and 2×2 real matrices of the form

$$\begin{pmatrix} a & -b \\ b & a \end{pmatrix}.$$

For example, if we write

$$\mathbf{1} \text{ for } \begin{pmatrix} 1 & 0 \\ 0 & 1 \end{pmatrix} \quad \text{and} \quad \mathbf{i} \text{ for } \begin{pmatrix} 0 & -1 \\ 1 & 0 \end{pmatrix}.$$

then $\mathbf{i}^2 = -\mathbf{1}$. We shall further pursue these similarities in a moment, but we first consider the set of all such 2×2 matrices by itself.

Let $\mathcal{C}$ denote the set of all 2×2 real matrices of the form

$$\begin{pmatrix} a & -b \\ b & a \end{pmatrix} = a\mathbf{1} + b\mathbf{i}.$$

It is easy to check that $\mathcal{C}$ is a subspace of $\mathbb{R}_{2\times 2}$, and that it is closed under multiplication (it is, in fact, an algebra, as defined in Section 4, Chapter 2). Since it inherits norms from $\mathbb{R}_{2\times 2}$, $\mathcal{C}$ is thus a normed vector space, all ready for calculus. We can therefore talk about sequences and series in $\mathcal{C}$.

4.1 $\mathcal{C}$ is closed under convergence. (i) If $\{Z_1, Z_2, \ldots\}$ is a convergent infinite sequence in $\mathbb{R}_{2\times 2}$ and if each Z_k is in $\mathcal{C}$, then the limit is also in $\mathcal{C}$.

ii) If $\sum_{k=0}^{\infty} Z_k$ is a convergent infinite series in $\mathbb{R}_{2\times 2}$ and if each Z_k is in $\mathcal{C}$, then so is the sum of the series.

Proof. (i) Let

$$Z_k = \begin{pmatrix} a_k & -b_k \\ b_k & a_k \end{pmatrix},$$

and use the fact that convergence is elementwise. (ii) Use (i). ∎

Since $\mathcal{C}$ is closed under multiplication, we can consider power series in $\mathcal{C}$. Our results about matrix power series show that

$$\sum_{k=0}^{\infty} c_k Z^k$$

will converge whenever $\|Z\| < R$, where $\|Z\|$ is any matrix norm and R is the radius of convergence of the real series

$$\sum_{k=0}^{\infty} c_k x^k.$$

In particular, we can consider the exponential series

$$\sum_{k=0}^{\infty} \frac{1}{k!} Z^k, \quad Z \text{ in } \mathcal{C}$$

which we know converges for any Z. First, let us note that for $\mathbf{i} = \begin{pmatrix} 0 & -1 \\ 1 & 0 \end{pmatrix}$,

$$\mathbf{i}^2 = -\mathbf{1} = \begin{pmatrix} -1 & 0 \\ 0 & -1 \end{pmatrix}, \quad \mathbf{i}^3 = -\mathbf{i} = \begin{pmatrix} 0 & 1 \\ -1 & 0 \end{pmatrix}, \quad \mathbf{i}^4 = \mathbf{1} = \begin{pmatrix} 1 & 0 \\ 0 & 1 \end{pmatrix},$$

so

$$\mathbf{i}^{2k} = (-1)^k \mathbf{1}, \qquad \mathbf{i}^{2k+1} = (-1)^k \mathbf{i}, \tag{1}$$

where $k = 0, 1, 2, \ldots$. Hence, for

$$\mathbf{i}b = \begin{pmatrix} 0 & -b \\ b & 0 \end{pmatrix},$$

we have

$$\frac{(\mathbf{i}b)^{2k}}{(2k)!} = \frac{(-1)^k b^{2k}}{(2k)!} \mathbf{1} = \begin{pmatrix} (-1)^k b^{2k}/(2k)! & 0 \\ 0 & (-1)^k b^{2k}/(2k)! \end{pmatrix},$$

$$\frac{(\mathbf{i}b)^{2k+1}}{(2k+1)!} = \frac{(-1)^k b^{2k+1}}{(2k+1)!} \mathbf{i} = \begin{pmatrix} 0 & -(-1)^k b^{2k+1}/(2k+1)! \\ (-1)^k b^{2k+1}/(2k+1)! & 0 \end{pmatrix}.$$

Hence,

$$e^{ib} = \sum_{k=0}^{\infty} (ib)^k/k!$$

$$= \begin{pmatrix} \sum_{k=0}^{\infty}(-1)^k b^{2k}/(2k)! & -\sum_{k=0}^{\infty}(-1)^k b^{2k+1}/(2k+1)! \\ \sum_{k=0}^{\infty}(-1)^k b^{2k+1}/(2k+1)! & \sum_{k=0}^{\infty}(-1)^k b^{2k}/(2k)! \end{pmatrix}.$$

You undoubtedly recognize the real series occurring in this matrix:

$$e^{ib} = \begin{pmatrix} \cos b & -\sin b \\ \sin b & \cos b \end{pmatrix} = \mathbf{1}\cos b + \mathbf{i} \sin b. \qquad (2)$$

Comparing this to Section 5, Chapter 8, we see that e^{ib} is an orthogonal matrix with determinant 1, a counterclockwise rotation through an angle b in the plane $\mathbb{R}_2$. The steps are reversible, and we find that any counterclockwise rotation through an angle b in $\mathbb{R}_2$ (or orthogonal matrix with determinant 1) is representable in the form e^{ib}.

To carry things a bit further, if $\mathbf{Z} = \mathbf{1}a + ib$, then $\mathbf{1}a$ and ib commute, so

$$e^{\mathbf{1}a+ib} = e^{\mathbf{1}a}e^{ib}.$$

Now

$$e^{\mathbf{1}a} = e^{\begin{pmatrix} a & 0 \\ 0 & a \end{pmatrix}} = \begin{pmatrix} e^a & 0 \\ 0 & e^a \end{pmatrix} = \mathbf{1}e^a.$$

Thus

$$e^{\mathbf{Z}} = e^{\mathbf{1}a+ib} = e^{\mathbf{1}a}e^{ib} = e^a(\mathbf{1}\cos b + \mathbf{i}\sin b). \qquad (3)$$

On the other hand, given any nonzero

$$\mathbf{W} = \mathbf{1}c + id = \begin{pmatrix} c & -d \\ d & c \end{pmatrix},$$

$\mathbf{W}/\|\mathbf{W}\|$ is an orthogonal matrix of determinant 1, where

$$\|\mathbf{W}\| = \sqrt{c^2 + d^2}$$

is the spectral norm (Exercise 4). Hence

$$\mathbf{W}/\|\mathbf{W}\| = e^{ib}$$

for some b, so $\mathbf{W} = \|\mathbf{W}\| e^{ib}$. Hence, we have the following:

Polar Decomposition. Any $\mathbf{W}$ in $\mathbb{C}$ is of the form

$$\mathbf{W} = re^{ib} \quad \text{where} \quad r \geq 0$$

(in fact, $r = \|\mathbf{W}\|$).
We may identify r with

$$\mathbf{1}r = \begin{pmatrix} r & 0 \\ 0 & r \end{pmatrix}$$

in the polar decomposition. As a linear transformation this stretches everything an amount r (≥ 0). Thus, it follows from the polar decomposition that, considering complex numbers as linear transformations acting in $\mathbb{R}_2$, each complex number can be represented as a stretching followed by a rotation. We now demonstrate that it is reasonable to identify complex numbers with 2×2 matrices in this fashion.

You will recall that the field of complex numbers $\mathbb{C}$ is, first of all, a two-dimensional vector space over $\mathbb{R}$ with basis $\{1, i\}$. Furthermore, for $1a + ib$ and $1c + id$ in $\mathbb{C}$, there is a

Multiplication: $(1a + ib)(1c + id) = 1(ac - bd) + i(ad + bc)$,

Conjugation: $\overline{1a + ib} = 1a - ib$,

Modulus: $|z| = \sqrt{\bar{z}z}$, where $z = 1a + ib$.

This modulus is easily seen to be a norm (in fact, which norm?). We shall now use $a, b, c, \ldots$ to denote real numbers,

$$z = 1a + ib, \qquad w = 1c + id, \qquad \ldots$$

for complex numbers in $\mathbb{C}$ and

$$\mathbf{Z} = \begin{pmatrix} a & -b \\ b & a \end{pmatrix}, \qquad \mathbf{W} = \begin{pmatrix} c & -d \\ d & c \end{pmatrix}, \qquad \ldots$$

for 2×2 matrices in $\mathbb{R}_{2\times 2}$. We investigate the natural correspondence $z \to \mathbf{Z}$ between $\mathbb{C}$ and $\mathcal{C}$. To begin, note that

$$1 \to \mathbf{1} \qquad \text{and} \qquad i \to \mathbf{i}.$$

4.2 $\mathbb{C}$ and $\mathcal{C}$ are essentially the same, both algebraically and analytically.

i) The correspondence $z \to \mathbf{Z}$ is a vector space isomorphism between $\mathbb{C}$ and $\mathcal{C}$.

ii) The correspondence preserves multiplication; i.e., if $z \to \mathbf{Z}$ and $w \to \mathbf{W}$, then $zw \to \mathbf{ZW}$.

iii) Conjugates in $\mathbb{C}$ correspond to transposes in $\mathcal{C}$; i.e., if $z \to \mathbf{Z}$, then $\bar{z} \to \mathbf{Z}^*$.

iv) The modulus in $\mathbb{C}$ equals the spectral norm in $\mathcal{C}$; i.e., if $z \to \mathbf{Z}$, then $|z| = \|\mathbf{Z}\|$.

v) Both the correspondence $z \to \mathbf{Z}$ and its inverse $\mathbf{Z} \to z$ are continuous.†

Proof. Everything is straightforward (Exercise 1). ∎

---

† For the first correspondence, this means that for any w in $\mathcal{C}$ and any $\epsilon > 0$, there is a $\delta > 0$ for which $\|\mathbf{Z} - \mathbf{W}\| < \epsilon$ whenever $|z - w| < \delta$.

4.3 Limits in $\mathbb{C}$ and in $\mathcal{C}$. (i) A sequence $\{z_k\}$ converges to z in $\mathbb{C}$ iff the sequence of images $\{\mathbf{Z}_k\}$ converges to the image $\mathbf{Z}$ of the limit in $\mathcal{C}$.

ii) A series $\sum_{k=0}^{\infty} z_k$ converges to z in $\mathbb{C}$ iff the series of images $\sum_{k=0}^{\infty} \mathbf{Z}_k$ converges to the image $\mathbf{Z}$ of the limit in $\mathcal{C}$.

Proof. Yours again. (See Exercise 2.) ∎

4.3 assures us that any properties of series in $\mathbb{C}$ follow directly from the corresponding properties in $\mathcal{C}$. We may therefore immediately translate our results just obtained for $\mathcal{C}$ into results for $\mathbb{C}$. For example, we see that if the power series $\sum c_k x^k$ has radius of convergence R and if $|z| < R$, then both $\sum c_k \mathbf{Z}^k$ and $\sum k c_k \mathbf{Z}^{k-1}$ converge. This result was used in Section 3. In particular, the infinite series

$$\sum_{k=0}^{\infty} z^k / k!$$

converges for all z in $\mathbb{C}$, and we call its sum $\mathbf{e}^z$. From (2),

Euler's Formula.

$$e^{ib} = \cos b + i \sin b,$$

for b real. We now adopt the usual convention of writing a for $1a$. Then from (3),

$$e^{a+ib} = e^a (\cos b + i \sin b).$$

From the Polar Decomposition in $\mathcal{C}$,

Polar Decomposition in $\mathbb{C}$. Any w in $\mathbb{C}$ is of the form

$$w = re^{ib}, \qquad r \geq 0, \quad r = |w|.$$

In particular, note that Euler's Formula gives us the fantastic relation

$$e^{\pi i} = -1.$$

We know from elementary calculus that any real $r > 0$ is of the form $r = e^a$, a some real number, and the function $a \to e^a$ is a one-to-one mapping from the reals onto the positive reals. There are some differences in the complex case: for *any* complex $w \neq 0$, $w = e^z$ for some z (from the polar decomposition and Euler's formula). We pay for this ability to represent all nonzero complex numbers in this form by the fact that

$$z \to e^z \text{ is not one to one; in fact, } e^{z+2n\pi i} = e^z \text{ for any integer } n. \tag{4}$$

If you recall that the real logarithm function is defined as the inverse of the real exponential function, you see that (4) bodes ill for the complex logarithm function.

In fact there is not just one complex logarithm function, but a whole host of them: given any function log such that $\log e^z = z$ for complex z, then the function $\log z + 2n\pi i$ is another such function for any integer n.

We close with the following observations. Let us traverse the unit circle on some complex logarithm function. Thus, we are considering z of the form $z = e^{i\theta}$; so one pleasant choice of logarithm function would be $\log z = i\theta$:

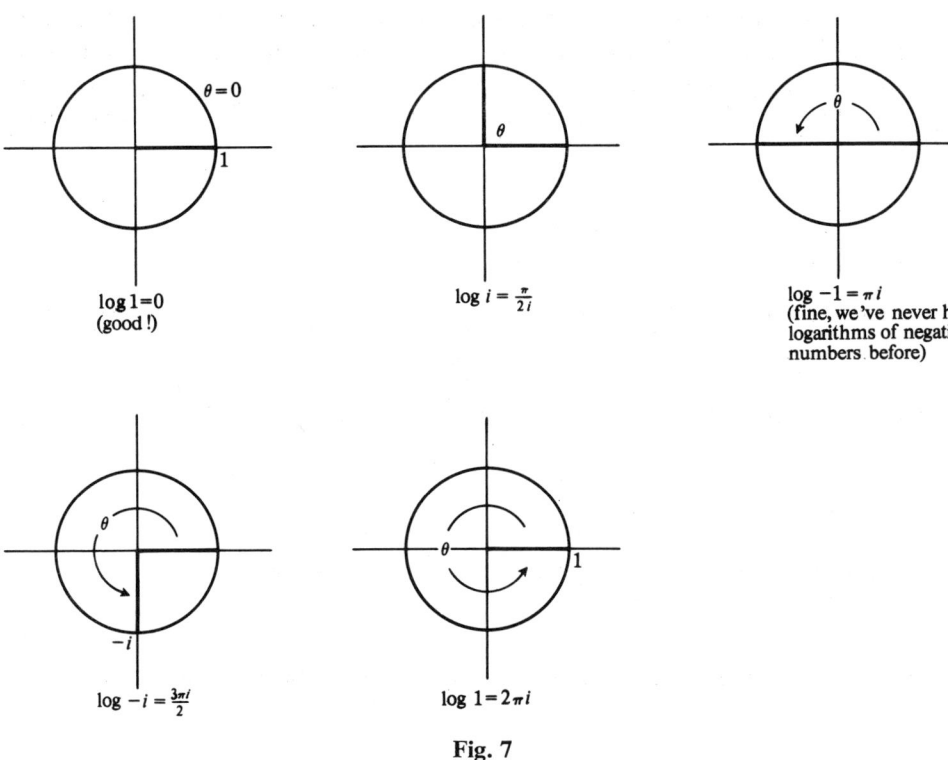

Fig. 7

Now we suppose that you would like a complex logarithm function which is continuous: small changes in z should not result in large jumps in $\log z$. If you insist on this, however, we have seen that this will result in $\log 1 = 0$ and also $\log 1 = 2\pi i$. For a way out of this mess, we suggest you consult a course in functions of a complex variable.

Supplement: Quaternions Again

Let us make our usual effort to extend things from $\mathbb{R}$ to $\mathbb{C}$. In this section, we would therefore consider special matrices in $\mathbb{C}_{2\times 2}$ and try to turn them into a field.

4 / The complex numbers

Our first guess might be to take all matrices of the form

$$\begin{pmatrix} z & -w \\ w & z \end{pmatrix}.$$

However, several of the field axioms for multiplication (Chapter 0, Section 6) would no longer hold; for example,

$$\begin{pmatrix} 1 & -i \\ i & 1 \end{pmatrix}$$

has no inverse (did you recognize that one row is a multiple of the other?).

We can conjure up a set of matrices in $\mathbb{C}_{2\times 2}$ which (if not zero) do have inverses and which satisfy all of the field axioms except commutativity of multiplication. To see how to repair the absence of inverses, let us note that

$$\det \begin{pmatrix} z & -w \\ w & z \end{pmatrix} = z^2 + w^2,$$

and this determinant can be 0 for $z, w \neq 0$, as we saw above. If we wished to avoid this, we would have to make the determinant something like $z\bar{z} + w\bar{w}$, a sum of two nonnegative reals which is 0 iff $z = w = 0$. From whence cometh such a determinant? One appropriate matrix is of the form

$$\begin{pmatrix} z & -w \\ \bar{w} & \bar{z} \end{pmatrix}.$$

We call the set

$$Q = \left\{ \mathbf{q} \text{ in } \mathbb{C}_{2\times 2} \mid \mathbf{q} = \begin{pmatrix} z & -w \\ \bar{w} & \bar{z} \end{pmatrix} \right\}$$

the **quaternions**. We have just shown that any nonzero quaternion is invertible. You show that Q satisfies all the other field axioms except commutativity. (Compare Chapter 8, Supplement to Section 6).

It is convenient to write the matrix

$$\begin{pmatrix} z & 0 \\ 0 & \bar{z} \end{pmatrix}$$

as $\mathbf{1}z$ and the matrix

$$\begin{pmatrix} 0 & -w \\ \bar{w} & 0 \end{pmatrix}$$

as $\mathbf{j}\bar{w}$. Note that these are symbolic and *not* matrix multiplications; for example,

$$\mathbf{1}z \neq \begin{pmatrix} 1 & 0 \\ 0 & 1 \end{pmatrix} z.$$

This immediately gives the decomposition $\mathbf{q} = \mathbf{1}z + \mathbf{j}\bar{w}$. By calculating the product of the two matrices you should show that for

$$\mathbf{q} = \mathbf{1}z + \mathbf{j}\bar{w} \quad \text{and} \quad \mathbf{q}' = \mathbf{1}z' + \mathbf{j}\bar{w}',$$

we have
$$\mathbf{q}\mathbf{q}' = \mathbf{1}(z\bar{z}' - \bar{w}w') + \mathbf{j}(wz' + \bar{z}w'),$$
and compare this with complex multiplication. This shows in particular that
$$(\mathbf{j}1)(\mathbf{1}z) = \mathbf{j}z$$
while
$$(\mathbf{1}z)(\mathbf{j}1) = \mathbf{j}\bar{z},$$
so that $\mathbf{j} = \mathbf{j}1$ does not commute with matrices $\mathbf{1}z$ for z nonreal.

You should notice that Q is not naturally a vector space over $\mathbb{C}$; the scalar multiplication does not make sense:
$$z\begin{pmatrix} 1 & 0 \\ 0 & 1 \end{pmatrix} = \begin{pmatrix} z & 0 \\ 0 & z \end{pmatrix},$$
which is not in Q, in general. (You can make it into a vector space by redefining your scalar multiplication
$$z\begin{pmatrix} 1 & 0 \\ 0 & 1 \end{pmatrix} \stackrel{\text{def}}{=} \begin{pmatrix} z & 0 \\ 0 & \bar{z} \end{pmatrix},$$
if you wish; this substantiates one's feelings that Q is essentially two-dimensional over $\mathbb{C}$, but doesn't do much else.) However, Q is very naturally a vector space over $\mathbb{R}$. One very good basis is
$$\mathbf{1} = \begin{pmatrix} 1 & 0 \\ 0 & 1 \end{pmatrix}, \quad \mathbf{i} = \begin{pmatrix} i & 0 \\ 0 & -i \end{pmatrix}, \quad \mathbf{j} = \begin{pmatrix} 0 & -1 \\ 1 & 0 \end{pmatrix}, \quad \text{and} \quad \mathbf{k} = \begin{pmatrix} 0 & i \\ i & 0 \end{pmatrix}.$$

An an exercise, you should construct a multiplication table for these basis vectors (Be sure to calculate both $\mathbf{XY}$ and $\mathbf{YX}$ for each pair, and also $\mathbf{X}^2$.) You should also muse about vector products in 3-space (again see the Supplement to Section 6 Chapter 8). You might further calculate the product of two elements of Q from the point of view of this basis:
$$(a_1\mathbf{1} + b_1\mathbf{i} + c_1\mathbf{j} + d_1\mathbf{k})(a_2\mathbf{1} + b_2\mathbf{i} + c_2\mathbf{j} + d_2\mathbf{k}) = ?$$

Q has many properties similar to $\mathbb{C}$ (after the next section you'll believe that everything does). For example, Q has a conjugation:
$$\text{For} \quad \mathbf{q} = \mathbf{1}z + \mathbf{j}w, \qquad \bar{\mathbf{q}} = \mathbf{1}\bar{z} - \mathbf{j}w.$$
You should check that this does satisfy $\bar{\bar{\mathbf{q}}} = \mathbf{q}$ and $\overline{\mathbf{q}_1\mathbf{q}_2} = \bar{\mathbf{q}}_2\bar{\mathbf{q}}_1$; that if
$$\mathbf{q} = a\mathbf{1} + b\mathbf{i} + c\mathbf{j} + d\mathbf{k},$$
then
$$\bar{\mathbf{q}} = a\mathbf{1} - b\mathbf{i} - c\mathbf{j} - d\mathbf{k},$$
and that in terms of matrices, $^-$ is the same as conjugate transpose.

Furthermore, this conjugation yields a norm:
$$|q| = \sqrt{q\bar{q}}.$$
You should prove this, and see what this amounts to for the three expressions for q:

$$q = 1z + j\bar{w}, \qquad q = a\mathbf{1} + b\mathbf{i} + c\mathbf{j} + d\mathbf{k}, \quad \text{and} \quad q = \begin{pmatrix} z & -w \\ \bar{w} & \bar{z} \end{pmatrix}.$$

In particular, you should notice that $|q|$ is both the square root of the determinant of q as a matrix, and the matrix norm of q as a matrix.

We finally caution you that if you attempt to take a set of matrices in $Q_{2\times 2}$ and turn them into something resembling a field, the situation is much worse. In fact, if you wish to salvage invertibility of nonzero elements and a norm, then you have to give up associativity of multiplication, which seems a lot to ask in an elementary mathematics course (for future reference, the creatures you do get are called Cayley Numbers).

EXERCISES

1. (a) through (e) Prove 4.2(i) through (v).
2. (a) Prove 4.1. (b) Prove 4.3.
3. Prove (1).
4. If
$$\mathbf{W} = \begin{pmatrix} c & -d \\ d & c \end{pmatrix},$$
prove that $\mathbf{W}/\|\mathbf{W}\|$ is an orthogonal matrix, where $\|\mathbf{W}\|$ is the spectral norm.
5. Prove (4).
6. Find all z for which $w = e^z$ for some fixed w.
7. Show that in the polar decomposition $a + ib = re^{i\theta}$, (r, θ) is the same as the polar coordinates of the point (a, b) in $\mathbb{R}_2$. Make a geometrical interpretation of this.
8. Find the eigenvalues of
$$\mathbf{Z} = \begin{pmatrix} a & -b \\ b & a \end{pmatrix}.$$

5. ANALOGIES BETWEEN $\mathbb{C}$ AND $\mathbb{R}_{n\times n}$

In this section we shall develop some breathtaking similarities between $n \times n$ real matrices and complex numbers. We shall denote real numbers by a, b, c, etc., complex numbers by z, w, etc., and $n \times n$ real matrices by script letters $\mathcal{A}$, $\mathcal{B}, \ldots, \mathcal{Z}$.

It is no surprise to note that both $\mathbb{C}$ and $\mathbb{R}_{n \times n}$ have an addition and multiplication with many similar properties, although we must remind ourselves in the following that the multiplication in $\mathbb{R}_{n \times n}$ is *not* commutative for $n > 1$. In the more interesting analogies which follow, you will note that the properties of $\mathbb{R}$ which are mentioned correspond exactly to the matrix properties when $\mathbb{C}$ is considered as the appropriate set of 2×2 real matrices $\mathcal{C}$, as in the previous section.

Our analogies (some of which we have previously encountered) are based upon the following corresponding concepts:

Complex number $z \leftrightarrow n \times n$ real matrix $\mathcal{Z}$
Complex conjugation $\bar{z} \leftrightarrow$ matrix transpose $\mathcal{Z}^*$
Real number $a \leftrightarrow$ symmetric matrix $\mathcal{A}$
Pure imaginary number $ib \leftrightarrow$ skew symmetric matrix $\mathcal{JB}$

(We use $\mathcal{JB}$ for notation even though there may be no separate matrix $\mathcal{J}$ corresponding to i; see below.)

Analogies

1. Conjugation is a one-to-one mapping of $\mathbb{C}$ onto $\mathbb{C}$, which satisfies
$$\overline{zw} = \bar{w}\bar{z} \quad \text{and} \quad \bar{\bar{z}} = z.$$
Transposition is a one-to-one mapping or $\mathbb{R}_{n \times n}$ onto $\mathbb{R}_{n \times n}$, which satisfies
$$(\mathcal{ZW})^* = \mathcal{W}^*\mathcal{Z}^* \quad \text{and} \quad \mathcal{Z}^{**} = \mathcal{Z} \quad (\text{where } * = \text{tranpose}).$$

2. z is real iff $\bar{z} = z$
   $\mathcal{Z}$ is symmetric iff $\mathcal{Z}^* = \mathcal{Z}$.

3. z is pure imaginary iff $\bar{z} = -z$.
   $\mathcal{Z}$ is skew-symmetric iff $\mathcal{Z}^* = -\mathcal{Z}$.

4. Any $z = a + ib$, a real, ib pure imaginary.
   Any $\mathcal{Z} = \mathcal{A} + \mathcal{JB}$, $\mathcal{A}$ symmetric, $\mathcal{JB}$ skew-symmetric.

To prove this last fact about $\mathbb{R}_{n \times n}$, take any $\mathcal{Z}$ and let
$$\mathcal{A} = \tfrac{1}{2}(\mathcal{Z} + \mathcal{Z}^*), \quad \mathcal{JB} = \tfrac{1}{2}(\mathcal{Z} - \mathcal{Z}^*).$$

Note: if n is odd, there is no separate matrix $\mathcal{J}$ in $\mathbb{R}_{n \times n}$ for which $\mathcal{J}^* = -\mathcal{J}$ and $\mathcal{J}^2 = -\mathcal{J}$ (Exercise 4), so we use $\mathcal{JB}$ as one symbol to denote a skew-symmetric matrix in this case. If $n = 2m$ is even, then there are several such $\mathcal{J}$'s (for example,
$$\mathcal{J} = \begin{pmatrix} 0 & -\mathbf{I}_m \\ \mathbf{I}_m & 0 \end{pmatrix}.$$

In general, noncommutativity enters in and
$$\mathcal{A}(\mathcal{JB}) \neq (\mathcal{JB})\mathcal{A}.$$

Also, two symmetric matrices will not in general commute (or their product be symmetric), and similarly for skew-symmetric matrices.

Recall that a symmetric matrix $\mathcal{R}$ is *positive definite* if $\mathbf{V}^*\mathcal{R}\mathbf{V} > 0$ for all vectors $\mathbf{V} \neq \mathbf{0}$. If $\mathcal{R}$ is positive definite, then there is an $\mathcal{X}$ for which $\mathcal{X}^{-1}\mathcal{R}\mathcal{X}$ is a diagonal matrix with all positive entries, which is about as close to being a positive real number as you might expect a matrix to get. Similarly, we called a symmetric matrix $\mathcal{R}$ *positive semidefinite* if $\mathbf{V}^*\mathcal{R}\mathbf{V} \geq 0$ for all $\mathbf{V}$. We now develop analogies based upon the corresponding notions:

positive real number ⟷ positive definite (symmetric) matrix,
nonnegative real number ⟷ positive semidefinite (symmetric) matrix.

5. $\bar{z}z$ is a nonnegative real number for any z.
$\mathcal{Z}^*\mathcal{Z}$ is a positive semidefinite matrix for any $\mathcal{Z}$ (see Chapter 9, Section 4).

6. For any nonnegative real r there is a nonnegative real $\sqrt{r}$ for which $(\sqrt{r})^2 = r$.

For any positive semidefinite matrix $\mathcal{R}$ there is a positive semidefinite $\sqrt{\mathcal{R}}$ for which $(\sqrt{\mathcal{R}})^2 = \mathcal{R}$.

To prove this last, note that since $\mathcal{R}$ is symmetric, there exists an orthogonal matrix $\mathcal{U}$ such that

$$\mathcal{U}^{-1}\mathcal{R}\mathcal{U} = \mathcal{D} = \begin{pmatrix} \lambda_1 & & 0 \\ & \ddots & \\ 0 & & \lambda_n \end{pmatrix},$$

where λ_i are the eigenvalues of $\mathcal{R}$. Since each $\lambda_i \geq 0$, we can form

$$\sqrt{\mathcal{D}} = \begin{pmatrix} \sqrt{\lambda_1} & & 0 \\ & \ddots & \\ 0 & & \sqrt{\lambda_n} \end{pmatrix}$$

and define $\mathcal{S} = \mathcal{U}\sqrt{\mathcal{D}}\mathcal{U}^{-1}$. Then

$$\mathcal{S}^2 = (\mathcal{U}\sqrt{\mathcal{D}}\mathcal{U}^{-1})(\mathcal{U}\sqrt{\mathcal{D}}\mathcal{U}^{-1}) = \mathcal{U}(\sqrt{\mathcal{D}})^2\mathcal{U}^{-1}$$
$$= \mathcal{U}\mathcal{D}\mathcal{U}^{-1} = \mathcal{R}.$$

We leave it for you to check that $\mathcal{S}$ is symmetric (Exercise 5).

Example.

$$\mathcal{R} = \begin{pmatrix} 5 & 1 & 3 \\ 1 & 5 & 3 \\ 3 & 3 & 3 \end{pmatrix}$$

is positive semidefinite since

$$\mathcal{R} = \mathcal{X} \begin{pmatrix} 9 & 0 & 0 \\ 0 & 4 & 0 \\ 0 & 0 & 0 \end{pmatrix} \mathcal{X}^{-1},$$

with

$$\mathcal{X} = \begin{pmatrix} 1/\sqrt{3} & 1/\sqrt{2} & 1/\sqrt{6} \\ 1/\sqrt{3} & -1/\sqrt{2} & 1/\sqrt{6} \\ 1/\sqrt{3} & 0 & -2/\sqrt{6} \end{pmatrix} \quad \text{orthogonal.}$$

Hence

$$\sqrt{\mathcal{R}} = \mathcal{X} \begin{pmatrix} 3 & 0 & 0 \\ 0 & 2 & 0 \\ 0 & 0 & 0 \end{pmatrix} \mathcal{X}^{-1} = \begin{pmatrix} 2 & 0 & 1 \\ 0 & 2 & 1 \\ 1 & 1 & 1 \end{pmatrix}.$$

You should check that $\sqrt{\mathcal{R}}^2 = \mathcal{R}$ in this case.

7. $r \to e^r$ is a one-to-one mapping of the reals onto the positive reals.
$\mathcal{R} \to e^{\mathcal{R}}$ is a one-to-one mapping of symmetric matrices onto the set of positive definite symmetric matrices.

Since $(e^{\mathcal{R}})^* = e^{\mathcal{R}*}$, $e^{\mathcal{R}}$ is symmetric if $\mathcal{R}$ is, and in fact is positive definite (why?). To show that $\mathcal{R} \to e^{\mathcal{R}}$ is onto, take any positive definite $\mathcal{S}$. Since $\mathcal{S}$ is symmetric, there is an orthogonal $\mathcal{U}$ for which

$$\mathcal{U}^{-1}\mathcal{S}\mathcal{U} = \mathcal{D} = \begin{pmatrix} \lambda_1 & & 0 \\ & \ddots & \\ 0 & & \lambda_n \end{pmatrix}.$$

Since $\mathcal{S}$ is positive definite, $\lambda_i > 0$ and we can take $\log \lambda_i$ (good old real logarithms). Let

$$\mathcal{L} = \begin{pmatrix} \log \lambda_1 & & 0 \\ & \ddots & \\ 0 & & \log \lambda_n \end{pmatrix}.$$

Then $e^{\mathcal{L}} = \mathcal{D}$, so

$$\mathcal{S} = \mathcal{U}\mathcal{D}\mathcal{U}^{-1} = \mathcal{U}e^{\mathcal{L}}\mathcal{U}^{-1} = e^{\mathcal{U}\mathcal{L}\mathcal{U}^{-1}}.$$

You may show that $\mathcal{U}\mathcal{L}\mathcal{U}^{-1} = \mathcal{R}$ is symmetrical. Thus $\mathcal{T} = e^{\mathcal{R}}$ and the mapping is onto.

It is one-to-one: Let $\mathcal{R}$ be a symmetric matrix with eigenvalues $\lambda_1, \ldots, \lambda_n$ corresponding to eigenvectors $V_1, \ldots, V_n$, and suppose that $e^{\mathcal{R}} = e^{\mathcal{R}'}$ for some other symmetric matrix $\mathcal{R}'$. We shall show that $\mathcal{R}$ and $\mathcal{R}'$ have the same eigenvectors and that these correspond to the same eigenvalues. It will follow that $\mathcal{R} = \mathcal{R}'$ (why?). Suppose that $\mathcal{R}'V' = \lambda'V'$ for some λ', V'. Since $\{V_1, \ldots, V_n\}$ forms a basis for $\mathbb{R}_n$,

$$V' = \sum c_i V_i.$$

Now

$$e^{\mathcal{R}'}V' = e^{\lambda'}V' = \sum c_i e^{\lambda'} V_i,$$

and also

$$e^{\mathcal{R}'}V' = e^{\mathcal{R}}V' = \sum c_i e^{\mathcal{R}} V_i = \sum c_i e^{\lambda_i} V_i.$$

We thus have two expressions for $e^{\mathcal{R}'}V'$ in terms of the same basis. We conclude that if $c_i \neq 0$,

$$e^{\lambda_i} = e^{\lambda'}.$$

Thus, if $c_i \neq 0$,

$$\lambda_i = \lambda'.$$

But

$$\mathcal{R}V' = \sum c_i \mathcal{R} V_i = \sum c_i \lambda_i V_i = \sum c_i \lambda' V_i = \lambda' \sum c_i V_i$$
$$= \lambda' V'.$$

So any eigenvector of $\mathcal{R}'$ is an eigenvector of $\mathcal{R}$ with the same eigenvalue. Reversing the roles of $\mathcal{R}$ and $\mathcal{R}'$, we see that they have exactly the same eigenvectors and eigenvalues, and are thus the same. Hence $\mathcal{R} \to e^{\mathcal{R}}$ is one-to-one.

For our final pair of corresponding notions we take:

complex unit (= root of 1) $\leftrightarrow$ proper orthogonal matrix.

($\mathcal{U}$ is **proper orthogonal** means that $\mathcal{U}$ is orthogonal and $\det \mathcal{U} = +1$; properness is not needed in (8) and (10).)

8. z is a complex unit iff $\bar{z}z = 1$.
   $\mathcal{Z}$ is an orthogonal matrix iff $\mathcal{Z}*\mathcal{Z} = \mathbf{I}$.

9. z is a complex unit iff $z = e^{ib}$, ib pure imaginary.
   $\mathcal{Z}$ is proper orthogonal iff $\mathcal{Z} = e^{\mathcal{JB}}$, $\mathcal{JB}$ skew-symmetric.

To prove this, note that if $\mathcal{JB}$ is skew-symmetric, then

$$(e^{\mathcal{JB}})* = e^{\mathcal{JB}*} = e^{-\mathcal{JB}} = (e^{\mathcal{JB}})^{-1},$$

so $e^{\mathcal{JB}}$ is an orthogonal matrix. Moreover,

$$\det e^{\mathcal{JB}} = e^{\text{trace } \mathcal{JB}} = e^0 = 1.$$

Conversely, if $\mathcal{U}$ is proper orthogonal, then there is a matrix $\mathcal{X}$ for which $\mathcal{X}^{-1}\mathcal{U}\mathcal{X}$ is composed of blocks of the form

$$\begin{pmatrix} \cos\theta_i & -\sin\theta_i \\ \sin\theta_i & \cos\theta_i \end{pmatrix} \quad \text{or} \quad (1)$$

running down the diagonal (Chapter 8, 6.1). We leave it to you to obtain

$$\mathcal{X}^{-1}\mathcal{U}\mathcal{X} = e^{\jmath\mathcal{B}'}$$

and thence to finish the proof (Exercise 7).

10. Every complex unit $\neq -1$ can be written as

$$(1 - ib)/(1 + ib),$$

ib pure imaginary.

Every orthogonal matrix with no eigenvalue -1 can be written as

$$(\mathbf{I} - \jmath\mathcal{B})(\mathbf{I} + \jmath\mathcal{B})^{-1},$$

$\jmath\mathcal{B}$ skew-symmetric.

Recall that this was the basis for our parametrization of orthogonal matrices (Chapter 8, Section 5, Supplement 2).

11. *Polar decomposition*

Any $z = ur$, $r \geq 0$, u complex unit.

Any $\mathcal{Z} = \mathcal{U}\mathcal{R}$, $\mathcal{R}$ positive semidefinite, $\mathcal{U}$ orthogonal.

(Note that if $\mathcal{Z} = \mathcal{U}\mathcal{R}$, then also $\mathcal{Z} = \mathcal{R}'\mathcal{U}$ where $\mathcal{R}' = \mathcal{U}\mathcal{R}\mathcal{U}^{-1}$ is positive semidefinite.)

We first consider the case where $\mathcal{Z}$ is invertible. Given such a $\mathcal{Z}$ we aim for $\mathcal{R}$: let $\mathcal{R} = \sqrt{\mathcal{Z}^*\mathcal{Z}}$, which exists by (6). Then $\mathcal{R}$ is also invertible and

$$\mathcal{R}^*\mathcal{R} = \mathcal{R}^2 = \mathcal{Z}^*\mathcal{Z}.$$

We show that $\mathcal{U} = \mathcal{Z}\mathcal{R}^{-1}$ is orthogonal:

$$\mathcal{U}^* = \mathcal{R}^{-1}\mathcal{Z}^*,$$

so that $\mathcal{U}^*\mathcal{U} = \mathcal{R}^{-1}\mathcal{Z}^*\mathcal{Z}\mathcal{R}^{-1} = \mathcal{R}^{-1}\mathcal{R}^2\mathcal{R}^{-1} = \mathbf{I}$. Hence $\mathcal{Z} = \mathcal{U}\mathcal{R}$, as required.

Somewhat more effort is involved if $\mathcal{Z}$ is not invertible, since $\mathcal{R} = \sqrt{\mathcal{Z}^*\mathcal{Z}}$ will not be invertible either, and we must modify the proof. Let $\mathcal{Z}$ have rank r; then $\mathcal{Z}^*\mathcal{Z}\mathbf{V} = \mathbf{0}$ iff $\mathcal{Z}\mathbf{V} = \mathbf{0}$ and $\mathcal{Z}^*\mathcal{Z}$ also has rank r (see Chapter 9, 4.9). We again put $\mathcal{R} = \sqrt{\mathcal{Z}^*\mathcal{Z}}$; then $\mathcal{R}$ has rank r, too. Since $\mathcal{R}$ is symmetric, it is of the form

$$\mathcal{R} = \mathcal{S}\mathcal{D}\mathcal{S}^{-1}$$

where $\mathcal{S}$ is orthogonal and $\mathcal{D}$ is the diagonal matrix

$$\mathcal{D} = \begin{pmatrix} \lambda_1 & 0 & & & & & \\ 0 & \lambda_2 & & & & & \\ & & \ddots & & & & \\ & & & \lambda_r & & & \\ & & & & 0 & & \\ & & & & & \ddots & \\ & & & & & & 0 \end{pmatrix}, \quad \text{with } \lambda_i > 0.$$

In fact, $\mathcal{R}^*\mathcal{R} = \mathcal{S}\mathcal{D}^2\mathcal{S}^{-1}$.

Let $\mathbf{W}_1, \ldots, \mathbf{W}_n$ be the columns of $\mathcal{S}$; these form an orthogonal basis of $\mathbb{R}_n$ and are eigenvectors of $\mathcal{R}$ corresponding to the eigenvalues listed in $\mathcal{D}$. So we have

$$\mathcal{R}\mathbf{W}_i = \lambda_i \mathbf{W}_i, \quad i = 1, 2, \ldots, r,$$
$$\mathcal{R}\mathbf{W}_i = \mathbf{0}, \quad i = r+1, \ldots, n.$$

Now put

$$\mathbf{U}'_i = \frac{1}{\lambda_i} \mathcal{R}\mathbf{W}_i, \quad i = 1, 2, \ldots, r.$$

We shall show that

$$\mathbf{U}'_i \cdot \mathbf{U}'_j = \delta_{ij};$$

that is, $\mathbf{U}'_1, \ldots, \mathbf{U}'_r$ is an orthonormal basis of the space spanned by these vectors. In fact,

$$\mathbf{U}'_i \cdot \mathbf{U}'_j = \frac{1}{\lambda_i \lambda_j} (\mathcal{R}\mathbf{W}_i)^*(\mathcal{R}\mathbf{W}_j)$$
$$= \frac{1}{\lambda_i \lambda_j} \mathbf{W}_i^* (\mathcal{R}^*\mathcal{R}) \mathbf{W}_j$$
$$= \frac{1}{\lambda_i \lambda_j} \mathbf{W}_i^* (\mathcal{R}^*\mathcal{R}) \mathbf{W}_j$$
$$= \frac{1}{\lambda_i \lambda_j} (\mathcal{R}\mathbf{W}_i) \cdot (\mathcal{R}\mathbf{W}_j)$$
$$= \mathbf{W}_i \cdot \mathbf{W}_j$$
$$= \delta_{ij}.$$

Next we extend $\mathbf{U}'_1, \ldots, \mathbf{U}'_r$ to an orthonormal basis of $\mathbb{R}_n$:

$$\mathbf{U}'_1, \mathbf{U}'_2, \ldots, \mathbf{U}'_n.$$

Now set up a linear transformation $\mathbf{T}$ by putting

$$\mathbf{T}(\mathbf{W}_i) = \mathbf{U}'_i, \qquad i = 1, 2, \ldots, n;$$

$\mathbf{T}$ is then orthogonal. We shall show that for any vector $\mathbf{V}$ in $\mathbb{R}_n$,

$$\mathbf{T}(\mathcal{R}\mathbf{V}) = \mathfrak{Z}\mathbf{V}.$$

In fact, let

$$\mathbf{V} = a_1\mathbf{W}_1 + \cdots + a_r\mathbf{W}_r + \cdots + a_n\mathbf{W}_n;$$

then

$$\mathcal{R}\mathbf{V} = a_1\lambda_1\mathbf{W}_1 + \cdots + a_r\lambda_r\mathbf{W}_r$$

and

$$\mathbf{T}(\mathcal{R}\mathbf{V}) = a_1\lambda_1\mathbf{U}'_1 + \cdots + a_r\lambda_r\mathbf{U}'_r$$
$$= a_1\mathfrak{Z}\mathbf{W}_1 + \cdots + a_r\mathfrak{Z}\mathbf{W}_r.$$

On the other hand,

$$\mathfrak{Z}\mathbf{V} = a_1\mathfrak{Z}\mathbf{W}_1 + \cdots + a_r\mathfrak{Z}\mathbf{W}_r,$$

since $\mathbf{W}_{r+1}, \ldots, \mathbf{W}_n$ are annihilated by $\mathcal{R}$ and therefore by $\mathfrak{Z}$. Consequently,

$$\mathbf{T}(\mathcal{R}\mathbf{V}) = \mathfrak{Z}\mathbf{V}.$$

Now let $\mathcal{U}$ be the matrix of $\mathbf{T}$ relative to the standard basis. Then, for any $\mathbf{V}$ in $\mathbb{R}_n$,

$$\mathfrak{Z}\mathbf{V} = \mathcal{U}(\mathcal{R}\mathbf{V}) = (\mathcal{U}\mathcal{R})\mathbf{V},$$

and applying this to $\mathbf{e}_1, \ldots, \mathbf{e}_n$ in turn, we get $\mathfrak{Z} = \mathcal{U}\mathcal{R}$. Since $\mathcal{U}$ is orthogonal, this gives the required expression for $\mathfrak{Z}$. ∎

The proof just given will, of course, also apply (with obvious simplifications) in the case where $\mathfrak{Z}$ is invertible ($r = n$).

The matrix $\mathcal{U}$ is easily found if one notes that

$$\mathcal{U} = \mathcal{U}'\mathcal{T}^{-1},$$

where $\mathcal{U}'$ has the columns $\mathcal{U}'_1, \ldots, \mathcal{U}'_n$ that appeared above. We can write $\mathfrak{Z}$ in the form

$$\mathfrak{Z} = \mathcal{U}\mathcal{R} = \mathcal{U}\mathcal{T}\mathcal{D}\mathcal{T}^{-1} = \mathcal{U}'\mathcal{D}\mathcal{U}''\mathcal{R} \text{ where } \mathcal{U}' = \mathcal{U}\mathcal{T}, \mathcal{U}'' = \mathcal{T}^{-1}.$$

This proves the following

Diagonal Decomposition. Any $n \times n$ matrix $\mathfrak{Z}$ can be written in the form $\mathfrak{Z} = \mathfrak{U}'\mathfrak{D}\mathfrak{U}''$, where $\mathfrak{U}'$, $\mathfrak{U}''$ are orthogonal and $\mathfrak{D}$ is the diagonal matrix

$$\mathfrak{D} = \begin{pmatrix} \lambda_1 & 0 & & & & \\ 0 & \lambda_2 & & & & \\ & & \ddots & & & \\ & & & \lambda_r & & \\ & & & & 0 & \\ & & & & & \ddots \\ & & & & & & 0 \end{pmatrix}, \quad \lambda_i > 0.$$

(There is no interesting analogy in the complex numbers since the situation is much simpler there.) In this symmetrical form our theorem has a simple geometric interpretation. In fact it states that any linear transformation, given by the matrix $\mathfrak{Z}$, can be factored into a linear rigid motion (given by $\mathfrak{U}''$) followed by a stretching in n mutually orthogonal directions (given by $\mathfrak{D}$), which is finally followed by another linear rigid motion (given by $\mathfrak{U}'$).

Note also, as a remembrance of things past, that the largest of the numbers $\lambda_1, \ldots, \lambda_r$ is the spectral norm of $\mathfrak{Z}$.

If $\mathfrak{Z}$ has a determinant which is nonnegative, then $\det \mathfrak{U} = 1$. This fact is obvious if $\det \mathfrak{Z} > 0$ and it can be shown (Exercise 9) that we can make $\det \mathfrak{U} = +1$ in the case where $\det \mathfrak{Z} = 0$. Hence by (9) we can, in this case, restate (11) as follows:

12. Any z can be written in the form $z = re^{ib}$, where $r \geq 0$ and ib is pure imaginary.
    If $\det \mathfrak{Z} \geq 0$, then $\mathfrak{Z}$ is of the form $\mathfrak{Z} = \mathfrak{R}e^{\mathfrak{J}\mathfrak{B}}$, where $\mathfrak{R}$ is positive semidefinite and $\mathfrak{J}\mathfrak{B}$ is skew-symmetric.

We illustrate the Polar Decomposition and the Diagonal Decomposition Theorems with some specific examples.

Example 1.

$$\mathfrak{Z} = \begin{pmatrix} \frac{3}{5} & \frac{1}{10} \\ \frac{4}{5} & \frac{9}{5} \end{pmatrix},$$

where $\mathfrak{Z}$ is invertible.

$$\mathfrak{Z}^*\mathfrak{Z} = \begin{pmatrix} 1 & \frac{3}{2} \\ \frac{3}{2} & \frac{13}{4} \end{pmatrix}$$

$$= \mathfrak{S} \begin{pmatrix} 4 & 0 \\ 0 & \frac{1}{4} \end{pmatrix} \mathfrak{S}^{-1}, \quad \mathfrak{S} = \begin{pmatrix} 1/\sqrt{5} & -2/\sqrt{5} \\ 2/\sqrt{5} & 1/\sqrt{5} \end{pmatrix}.$$

Therefore
$$\mathcal{R} = \mathcal{S}\begin{pmatrix} 2 & 0 \\ 0 & \frac{1}{2} \end{pmatrix}\mathcal{S}^{-1}$$
$$= \begin{pmatrix} \frac{4}{5} & \frac{3}{5} \\ \frac{3}{5} & \frac{17}{10} \end{pmatrix}, \quad \mathcal{R}^{-1} = \begin{pmatrix} \frac{17}{10} & -\frac{3}{5} \\ -\frac{3}{5} & \frac{4}{5} \end{pmatrix}.$$

Put
$$\mathcal{U} = 3\mathcal{R}^{-1} = \begin{pmatrix} \frac{24}{25} & -\frac{7}{25} \\ \frac{7}{25} & \frac{24}{25} \end{pmatrix},$$

which is orthogonal, as it should be. Then $3 = \mathcal{U}\mathcal{R}$. This can be written $3 = \mathcal{R}'\mathcal{U}$, where
$$\mathcal{R}' = \mathcal{U}\mathcal{R}\mathcal{U}^{-1} = \begin{pmatrix} \frac{137}{250} & \frac{33}{125} \\ \frac{33}{125} & \frac{244}{125} \end{pmatrix}.$$

We also have
$$3 = \mathcal{U}\mathcal{S}\begin{pmatrix} 2 & 0 \\ 0 & \frac{1}{2} \end{pmatrix}\mathcal{S}^{-1}$$
$$= \mathcal{U}'\begin{pmatrix} 2 & 0 \\ 0 & \frac{1}{2} \end{pmatrix}\mathcal{U}'' \quad \text{(Diagonal Decomposition)},$$

where
$$\mathcal{U}' = \mathcal{U}\mathcal{S} = \begin{pmatrix} 2/5\sqrt{5} & -11/5\sqrt{5} \\ 11/5\sqrt{5} & 2/5\sqrt{5} \end{pmatrix}$$

and
$$\mathcal{U}'' = \mathcal{S}^* = \begin{pmatrix} 1/\sqrt{5} & 2/\sqrt{5} \\ -2/\sqrt{5} & 1/\sqrt{5} \end{pmatrix}.$$

Example 2. For
$$3 = \begin{pmatrix} 2 & \frac{4}{5} & \frac{7}{5} \\ 1 & -\frac{3}{5} & \frac{1}{5} \\ 0 & 2 & 1 \end{pmatrix},$$

$r = 2$. For
$$3*3 = \begin{pmatrix} 5 & 1 & 3 \\ 1 & 5 & 3 \\ 3 & 3 & 3 \end{pmatrix}$$

has eigenvalues 9, 4, 0, and can be written as
$$\mathcal{S}\begin{pmatrix} 9 & 0 & 0 \\ 0 & 4 & 0 \\ 0 & 0 & 0 \end{pmatrix}\mathcal{S}^{-1}, \quad \mathcal{S} = \begin{pmatrix} 1/\sqrt{3} & 1/\sqrt{2} & 1/\sqrt{6} \\ 1/\sqrt{3} & -1/\sqrt{2} & 1/\sqrt{6} \\ 1/\sqrt{3} & 0 & -2/\sqrt{6} \end{pmatrix},$$

where W_1, W_2, W_3 are the columns of $\mathcal{S}$. Therefore
$$\mathcal{R} = \mathcal{S}\begin{pmatrix} 3 & 0 & 0 \\ 0 & 2 & 0 \\ 0 & 0 & 0 \end{pmatrix}\mathcal{S}^{-1} = \begin{pmatrix} 2 & 0 & 1 \\ 0 & 2 & 1 \\ 1 & 1 & 1 \end{pmatrix}$$

(check that $\mathcal{R}^2 = \mathfrak{z}*\mathfrak{z}$). Then
$$\mathcal{U}_1' = \tfrac{1}{3}\mathfrak{z}W_1 = (7/5\sqrt{3},\ 1/5\sqrt{3},\ 1/\sqrt{3})*,$$
$$\mathcal{U}_2' = \tfrac{1}{3}\mathfrak{z}W_2 = (3/5\sqrt{2},\ 4/5\sqrt{2},\ -1/\sqrt{2})*.$$

(Check that $\mathcal{U}_1'$, $\mathcal{U}_2'$ form an orthogonal basis for the subspace.) We find
$$\mathcal{U}_3' = (-1/\sqrt{6},\ 2/\sqrt{6},\ 1/\sqrt{6})*.$$

Hence
$$T(W_1) = \mathcal{U}_1', \qquad T(W_2) = \mathcal{U}_2', \qquad T(W_3) = \mathcal{U}_3'$$
and
$$\mathcal{U}' = \begin{pmatrix} 7/5\sqrt{3} & 3/5\sqrt{2} & -1/\sqrt{6} \\ 1/5\sqrt{3} & 4/5\sqrt{2} & 2/\sqrt{6} \\ 1/\sqrt{3} & -1/\sqrt{2} & 1/\sqrt{6} \end{pmatrix}.$$

The matrix of **T** with respect to the standard basis is
$$\mathcal{U} = \mathcal{U}'\mathfrak{I}^* = \begin{pmatrix} \tfrac{3}{5} & 0 & \tfrac{4}{5} \\ \tfrac{4}{5} & 0 & -\tfrac{3}{5} \\ 0 & 1 & 0 \end{pmatrix},$$

and we have $\mathfrak{z} = \mathcal{U}\mathcal{R}$. Also,
$$\mathfrak{z} = \mathcal{U}' \begin{pmatrix} 3 & 0 & 0 \\ 0 & 2 & 0 \\ 0 & 0 & 0 \end{pmatrix} \mathcal{U}''$$

where $\mathcal{U}'' = \mathfrak{I}^*$.

It should be clear from the proof just given that the orthogonal matrix $\mathcal{U}$ in $\mathfrak{z} = \mathcal{U}\mathcal{R}$ will not be unique if the kernel of $\mathfrak{z}$ is too large. However

For any *invertible* $\mathfrak{z}$, the decomposition $\mathfrak{z} = \mathcal{U}\mathcal{R}$ is unique;
if $\det \mathfrak{z} > 0$, so that $\mathfrak{z} = \mathcal{R}e^{\mathfrak{I}\mathcal{B}}$, then $e^{\mathfrak{I}\mathcal{B}}$ will be unique but $\mathfrak{I}\mathcal{B}$ is not.

To prove this suppose that
$$\mathfrak{z} = \mathcal{U}\mathcal{R} \qquad \text{and} \qquad \mathfrak{z} = \mathcal{U}'\mathcal{R}'.$$
Since
$$\mathfrak{z}*\mathfrak{z} = \mathcal{R}*\mathcal{U}*\mathcal{U}\mathcal{R} = \mathcal{R}^2 \qquad \text{and} \qquad \mathfrak{z}*\mathfrak{z} = \mathcal{R}'*\mathcal{U}'*\mathcal{U}'\mathcal{R}' = \mathcal{R}'^2,$$
we have $\mathcal{R}^2 = \mathcal{R}'^2$, so that $\mathcal{R}' = \mathcal{R}$ (why?). Since $\mathcal{R}$ is invertible it then follows that $\mathcal{U}' = \mathcal{U}$. (In Exercise 10 you are invited to discuss the situation if $\mathcal{R}$ is not invertible.)

Project. Obtain the same analogies (1) through (12) between $\mathbb{C}$ and $\mathbb{C}_{n \times n}$ by making the following substitutions)

$$\mathbb{R}_{n \times n} \leftrightarrow \mathbb{C}_{n \times n}$$
Transpose $\leftrightarrow$ Conjugate transpose
Symmetric $\leftrightarrow$ Hermitian ($A* = A$)
Skew-symmetric $\leftrightarrow$ Skew-Hermitian ($A* = -A$)
Positive definite symmetric $\leftrightarrow$ Positive definite Hermitian (same definition)
Orthogonal $\leftrightarrow$ Unitary.

For $\mathbb{C}_n$ there is always an $\mathfrak{J}$ for which

$$\mathfrak{J}^* = -\mathfrak{J} \quad \text{and} \quad \mathfrak{J}^2 = -I \text{ (namely, } iI\text{)}.$$

EXERCISES

1. Prove that the indicated decomposition in (4) actually works.
2. a) Decompose
$$\mathfrak{z} = \begin{pmatrix} a & c \\ b & d \end{pmatrix}$$
in $\mathbb{R}_{2\times 2}$ á là (4).
   b) If $\mathfrak{z}$ is in $\mathbb{C}$, what does this decomposition amount to?
3. How many 4×4 matrices $\mathfrak{J}$ can you find for which
$$\mathfrak{J}^2 = -I \quad \text{and} \quad \mathfrak{J}^* = -\mathfrak{J}?$$
*4. Prove that there is no matrix $\mathfrak{J}$ in $\mathbb{R}_{n\times n}$, n odd, for which
$$\mathfrak{J}^2 = -I \quad \text{and} \quad \mathfrak{J}^* = -\mathfrak{J}.$$

   [*Suggestions:* The characteristic polynomial of $\mathfrak{J}$ must have a real root. Study an eigenvector corresponding to this root.]

5. (a) Prove that the $\mathcal{S} = \sqrt{\mathcal{R}}$ in the proof of (6) is symmetric. (b) Prove that the $\mathcal{R}$ in the proof of (7) is symmetric.
6. Why does $\mathcal{R}' = \mathcal{R}$ in the proof of (7)?
7. Finish the proof of (9).
H8. If $\mathfrak{z}$ is an invertible matrix in $\mathbb{R}_{n\times n}$ and has all its eigenvalues positive, show that $\mathfrak{z} = e^{\mathcal{W}}$ for some $\mathcal{W}$ in $\mathbb{R}_{n\times n}$, (a) for $n = 2$; (b) for $n = 3$; (c) for arbitrary n. [*Suggestion:* Use the Jordan form of $\mathfrak{z}$ and consider each Jordan block separately; note that, for example,
$$\begin{pmatrix} \lambda & 1 \\ 0 & \lambda \end{pmatrix} = \begin{pmatrix} \lambda & 0 \\ 0 & \lambda \end{pmatrix} \begin{pmatrix} 1 & \frac{1}{\lambda} \\ 0 & 1 \end{pmatrix},$$
and each matrix of the right-hand side is of the form $e^{\mathcal{W}}$, where $\mathcal{W}$ is easy to find).

9. Prove that if $\det \mathfrak{z} = 0$, then one can choose a $\mathcal{U}$ in the polar decomposition $\mathfrak{z} = \mathcal{U}\mathcal{R}$ for which
$$\det \mathcal{U} = +1.$$

10. If $\mathfrak{z}$ is not invertible, discuss the uniqueness of the $\mathcal{R}$ and the $\mathcal{U}$ in the polar decomposition.

Answers to Selected Exercises

CHAPTER 0

Section 1

3. **W** − **V** is to be interpreted as **W** + (−**V**). Use the associative law.

Supplement 1

3. There are three, represented by (say) 0, 1, 2.
6. No you can't.

Supplement 2

4. e is represented by 0.
7. [*Hint:* Use 1.3.]

Section 2

4. $3\mathbf{V} + 2\mathbf{W} = (7, -5, 7, 6)$

Section 3

8. $f(x) = e^{-3x}$ is one such function; there are many others.
9. Closure fails for both operations.
12. To prove that, if $p(x) = 0$ on S, then $a_i = 0$ for $i = 0, 1, 2, \ldots, n$, you may quote a theorem from elementary algebra.

Section 4

2. An example is $f(1) = 1, f(n) = n - 1$ for $n \geq 2$, the domain being the set of all positive integers.
3. b) f is one-to-one if whenever two ordered pairs (s_1, t) and (s_2, t) occur in the collection which constitutes f, then we must have $s_1 = s_2$.

Section 5

1. a) gf assigns to each element s in S the element $g(t)$ where $t = f(s)$. Since f is a function, t is unique and since g is a function, $g(t)$ is unique. Hence gf is a function with domain S.
   b) [*Hint:* gf is the collection of ordered pairs (s, u) such that t exists where (s, t) belongs to f and (t, u) belongs to g.]

Section 6

2. [*Hint:* Use 1.2, 1.3.]
3. Use 6.4.
5. For example, 2 does not have a multiplicative inverse.
6. a) The value of the *function* is zero for any x in the field while the *polynomial* is not the zero polynomial.

CHAPTER 1

Section 1

9. Axiom (ii) is not satisfied.
10. [*Hint:* It is not necessary to find these functions.]
11. c) Any nonnegative multiple.

Section 2

1. There are five ways of doing this if each group is to contain no more than two elements. An example is $((V + W) + X) + Y$.
5. a) Either use 2.7 or imitate its proof.
6. Use induction on n.

Section 3

4. There is only one.
5. [*Hint:* If the subspace contains a nonzero element prove that it is $\mathbb{R}$ itself.]
8. b) $\mathcal{W}$ consists of the vectors from the origin to points in a certain plane.
10. c) and d) Quote theorems from elementary calculus to prove these.
    e) You need the theorem which states that if $\sum_1^\infty a_n$ converges, then $\lim_{n \to \infty} a_n = 0$.
    f) To show that $C_1 \subset C_0$ is a proper inclusion quote an example where $\lim a_n = 0$ but $\sum_1^\infty a_n$ does not converge.

Supplement to Section 3

2. a) If $\mathfrak{Z}$ is a subspace containing $\mathfrak{X}$ and $\mathfrak{Y}$, then $\mathfrak{Z}$ contains all vectors of the form $X + Y$, where X is in $\mathfrak{X}$ and Y is in $\mathfrak{Y}$. Hence $\mathfrak{Z}$ contains $\mathfrak{X} + \mathfrak{Y}$.

Section 4

1. b) Show directly that any vector in $\mathbb{R}_4$ can be expressed as a linear combination of the four vectors.
   c) The dimension of $\mathbb{R}$ over $\mathbb{R}$ is one.
4. a) The third vector is twice the first plus the second.
   b) The vectors are linearly independent.
5. a) [*Hint:* Show how to express the vector **0** as a linear combination of any set of vectors.]
11. One basis is the set $1, x, x^2, \ldots, x^{n-1}$.
14. b) Use 4.7.
    c) Use 4.9.
16. First prove that S' is infinite-dimensional.
17. a) Let $\alpha = \{V_1, \ldots, V_n\}$, $\beta = \{W_1, \ldots, W_m\}$. Then $\alpha \cup \beta = \{V_1, \ldots, V_n, W_1, \ldots, W_m\}$ spans $\mathfrak{X} \oplus \mathfrak{Y}$. To prove that this set is linearly independent assume that $a_1, \ldots, a_n, b_1, \ldots, b_m$ exist so that $a_1V_1 + \cdots + a_nV_n + b_1W_1 + \cdots + b_mW_m = 0$. This gives $a_1V_1 + \cdots + a_nV_n = -b_1W_1 - \cdots - b_mW_m$ whence it follows that

$$a_1W_1 + \cdots + a_nV_n = 0 = b_1W_1 + \cdots + b_mW_m,$$

since **0** is the only vector in both $\mathfrak{X}$ and $\mathfrak{Y}$. Hence finally $a_1 = a_2 = \cdots = a_n = 0$ and $b_1 = b_2 = \cdots = b_m = 0$.
18. To show that $\mathfrak{X} + \mathfrak{Y}$ has finite dimension note that it is spanned by the union of the bases for $\mathfrak{X}$ and $\mathfrak{Y}$ respectively, as in 17(a).
20. a) Note that the dimension of $\mathfrak{X} + \mathfrak{Y}$ is ≤ 4 and use the result of 19.

Supplement to Section 4

4. a) $(1, i)$ is a basis for $\mathbb{C}$ over $\mathbb{R}$ so that the dimension of $\mathbb{C}$ over $\mathbb{R}$ is 2.
   b) The dimension of $\mathbb{C}$ over $\mathbb{C}$ is 1.
5. a) If S is not linearly independent, distinct primes $p_1, \ldots, p_k$ and nonzero integers $a_1, \ldots, a_k$ must exist so that

$$a_1 \log p_1 + \cdots + a_k \log p_k = 0.$$

   This gives

$$p_1^{a_1} p_2^{a_2} \cdots p_k^{a_k} = 1.$$

   Now show that this contradicts the following theorem from arithmetic: Every integer $n > 1$ can be factored *uniquely* into a product of prime powers $n = p_1^{b_1} \cdots p_r^{b_r}$ where $b_i > 0$.
   b) $\mathbb{R}$ is infinite dimensional over $\mathbb{Q}$ while the dimension of $\mathbb{R}$ over $\mathbb{R}$ is one.

CHAPTER 2

Section 1

1. See Exercise 4(a) in Section 4, Chapter 1.
2. The dimension is 3.
4. The dimension is 2.
6. Use Exercise 19 in Supplement to Section 4 in Chapter 1.

Section 2

1. A basis is $(1, 0, 1), (0, 1, -2)$.
4. Dimension of row space is 3.
5. In Exercise 1, $\{V_1, V_2\}$ is a basis. In Exercise 3, the first three rows form a basis of the row space.

Section 3

1. a) $\begin{pmatrix} 1 & 0 & 0 \\ 0 & 1 & 0 \\ 0 & 0 & 1 \end{pmatrix}$

2. b) $AB' = \begin{pmatrix} 0 & \frac{1}{3} & 2 \\ 3 & -1 & 0 \\ 4 & -\frac{4}{3} & 2 \\ 5 & 0 & 10 \end{pmatrix}$

a) $BA = \begin{pmatrix} 0 & 1 & 2 \\ 1 & -1 & 0 \\ 4 & -4 & 2 \\ 5 & 0 & 10 \end{pmatrix}$

4. a) If A is $m \times n$ and B is $n \times m$.

5. $VW = (-3)$, $WV = \begin{pmatrix} 3 & 6 & 0 \\ -1 & -2 & 0 \\ 4 & 8 & 0 \end{pmatrix}$

9. a) $A + B = \begin{pmatrix} 2 & 2 \\ 4 & 1 \end{pmatrix}$

b) $AB = \begin{pmatrix} 3 & 3 \\ 0 & 0 \end{pmatrix}$, $BA = \begin{pmatrix} 3 & 0 \\ 6 & 0 \end{pmatrix}$

10. d) The $l \times n$ matrix each of whose rows is the sum of the rows of A.
19. $\tau_+ = \frac{1}{2}\tau_1 + \frac{1}{2}i\tau_2$
20. $\sigma_1\sigma_2 = i\sigma_3$

21. $(\lambda_2 + \lambda_3)\lambda_5 = \begin{pmatrix} 0 & 0 & -1 \\ 0 & 0 & -i \\ 0 & 0 & 0 \end{pmatrix}$

Section 4

1. a) $A^2 = 4\begin{pmatrix} 0 & 0 & 1 & 0 & 0 \\ 0 & 0 & 0 & 1 & 0 \\ 0 & 0 & 0 & 0 & 1 \\ 0 & 0 & 0 & 0 & 0 \\ 0 & 0 & 0 & 0 & 0 \end{pmatrix}$

2. a) $B^2 = 4\begin{pmatrix} 0 & 0 & 1 & 0 \\ 0 & 0 & 0 & 1 \\ 1 & 0 & 0 & 0 \\ 0 & 1 & 0 & 0 \end{pmatrix}$

3. b) [*Hint:* Take **A** to be a matrix unit E_{ij} for $i, j = 1, 2, \ldots, n$. See also Exercise 4, which follows.]

6. a) This can be done by using elementary matrices.

12. See also Exercise 20, which follows.

13. c) $i = \begin{pmatrix} 0 & -1 \\ 1 & 0 \end{pmatrix}$

17. See the hint for Exercise 3(b).

20. b) To prove that A^{-1} cannot exist if $ad - bc = 0$, assume that it does and get a contradiction from the fact that $AB = 0$.

21. $A^k = \begin{pmatrix} 1 & kh \\ 0 & 1 \end{pmatrix}$; use mathematical induction.

23. $E_5 E_4 E_3 E_2 E_1 A = \begin{pmatrix} 1 & 0 & \frac{17}{2} \\ 0 & 1 & -\frac{7}{2} \\ 0 & 0 & 0 \end{pmatrix}$

$E_5 E_4 E_3 E_2 E_1 = \begin{pmatrix} -2 & \frac{3}{2} & 0 \\ 1 & -\frac{1}{2} & 0 \\ -3 & 2 & 1 \end{pmatrix}$

24. [*Hint:* Many of the matrices used in Exercise 23 can be used here.]

25. $A^{-1} = 3I - 4A + A^2$

26. d) Use Exercise 20(b).

33. $A^{-1} = \begin{pmatrix} \frac{2}{9} & -\frac{1}{5} & \frac{4}{45} \\ \frac{2}{9} & 0 & -\frac{1}{9} \\ \frac{1}{9} & \frac{2}{5} & \frac{2}{45} \end{pmatrix}$

Section 5b

2. a) a total of 6833   b) a total of 5166
   c) same as the original distribution

5. $M_1^{-1} = \begin{pmatrix} 0 & 2 & 0 \\ 0 & 0 & 3 \\ \frac{1}{6} & 0 & 0 \end{pmatrix}$, $M_2^{-1} = \begin{pmatrix} 0 & 2 & 0 \\ 0 & 0 & 3 \\ \frac{1}{3} & 0 & -1 \end{pmatrix}$

Section 5d

1. b) If $\mathbf{pu} = 1$ and $\mathbf{Pu} = \mathbf{u}$, then $\mathbf{pPu} = 1$.
   c) If $\mathbf{Pu} = \mathbf{u}$ and $\mathbf{Qu} = \mathbf{u}$, then $(\mathbf{PQ})\mathbf{u} = \mathbf{u}$.

2. c) $(\frac{1}{7}, \frac{35}{56}, \frac{13}{56})$

3. a) $\mathbf{P} = \begin{array}{c} \\ \mathbf{v} \\ \mathbf{c} \end{array}\!\!\begin{array}{c} \mathbf{v} \quad \mathbf{c} \\ \begin{pmatrix} \frac{1}{2} & \frac{1}{2} \\ 1 & 0 \end{pmatrix} \end{array}$
   b) $\mathbf{e}_2 \mathbf{P} = \mathbf{e}_1$
   c) $\mathbf{e}_2 \mathbf{P}^2 = (\frac{1}{2}, \frac{1}{2})$
   d) $(\frac{1}{2}, \frac{1}{2})\mathbf{P}^2 = (\frac{5}{8}, \frac{3}{8})$
   f) $\bar{\mathbf{p}} = (\frac{2}{3}, \frac{1}{3})$
   h) $\mathbf{P}^m = \begin{pmatrix} \frac{2}{3} + \frac{1}{3}(-\frac{1}{2})^m & \frac{1}{3} - \frac{1}{3}(-\frac{1}{2})^m \\ \frac{2}{3} - \frac{2}{3}(-\frac{1}{2})^m & \frac{1}{3} + \frac{2}{3}(-\frac{1}{2})^m \end{pmatrix}$

4. b) $\mathbf{P} = \left(\begin{array}{c|c} \mathbf{I}_m & \mathbf{0} \\ \hline \mathbf{B} & \mathbf{Q} \end{array}\right)$ where $\mathbf{I}_m$ is the $m \times m$ identity, $\mathbf{Q}$ is $(n-m) \times (n-m)$, $\mathbf{B}$ is $(n-m) \times n$.
   c) $\mathbf{P}^k = \left(\begin{array}{c|c} \mathbf{I}_m & \mathbf{0} \\ \hline \mathbf{B}_k & \mathbf{Q}^k \end{array}\right)$, $\mathbf{B}_k = (\mathbf{I}_{n-m} + \mathbf{Q} + \cdots + \mathbf{Q}^{k-1})\mathbf{B}$.

Section 5e

7. c) $n!$
8. $\mathbf{W}^2 = \mathbf{I}$, $\mathbf{E}^2 = \mathbf{C}$, $\mathbf{WC} = \mathbf{D}$
11. commutative for $n = 1$ and 2 only

13. $\mathbf{W} = \begin{pmatrix} 0 & 1 & 0 & 0 \\ 0 & 0 & 1 & 0 \\ 0 & 0 & 0 & 1 \\ 1 & 0 & 0 & 0 \end{pmatrix}$, $\mathbf{C} = \mathbf{I}$

14. a) $\mathbf{CW}$ gives the clan of the son's wife.
    b) $\mathbf{W}^2$ gives the clan of the husband's brother-in-law's wife.
    c) $\mathbf{C}^2$ gives the clan of the father's grandson.
    d) $(\mathbf{C}^2)\mathbf{C} = \mathbf{C}(\mathbf{C}^2)$

15. a) Each clan swaps wives with another clan.
16. a) yes   b) yes   c) yes   d) 2
17. c) no

CHAPTER 3

Section 1

1. c) $\mathbf{T}(a, b) = (b, a)$ is a reflection in the line $y = x$; $\mathbf{T}(a, b) = (-a, -b)$ is a rotation about the origin through an angle π.
   d) affine, linear iff $a = b = 0$

h) linear
i) linear
j) not linear (nor affine)
k) homogeneous but not additive
l) linear
m) not a function
2. a) $2T(0) = T(0)$ by homogeneity, hence $T(0) = 0$.
5. b) rotation about the origin through a right angle
   d) the rotation in (b) followed by a 2:1 stretching in the direction of the y-axis
6. c) Determine the image $P'(x', y')$ of $P(x, y)$ by noting that PP' is perpendicular to the line $ax = by$ and the midpoint of PP' is on that line. The result is given by the matrix
$$A = \begin{pmatrix} \dfrac{b^2 - a^2}{a^2 + b^2} & \dfrac{2ab}{a^2 + b^2} \\ \dfrac{2ab}{a^2 + b^2} & \dfrac{a^2 - b^2}{a^2 + b^2} \end{pmatrix}.$$
   d) $\begin{pmatrix} \cos\theta & -\sin\theta \\ \sin\theta & \cos\theta \end{pmatrix}$
7. a) $A = (a, b)$.
   b) h, i cannot be obtained from matrices.

Section 2

3. a) $(T + 2S)(x, y) = (x, 3x + 7y)$
   b) The image space is $\mathbb{R}_2$.
   c) The inverse is $(T + 2S)^{-1}(x, y) = (x, -\frac{3}{7}x + \frac{1}{7}y)$.
4. a) $TS(x, y) = (-5x - 3y, 4y)$
   c) There is an inverse.
5. a) $ST(x, y) = (-2x + 3y, 6x + y)$.
   b) The image space is $\mathbb{R}_2$.
   c) There is an inverse.
6. a) $(T^2 - 4T + 5I)(x, y) = (0, 0)$
7. $(TS)(x, y) = (4y, -3x - y, -4x - 8y)$
   $U(TS)(x, y) = (-3x + 3y, -4x - 8y, 3x + 5y, -7x - 9y)$
   $(UT)(x, y) = (3x, x - 3y, -x + 2y, 3x - 4y)$
8. [Hint: Set $x' = ax + by$, $y' = cx + dy$ and solve for x, y if possible—to obtain a unique solution.]
9. a) T_1, T_3 are shears; T_2, T_4 are stretchings, possibly combined with a reflection.
   b) A shear leaves the area unchanged; a stretching multiplies the area by the multiplier of the stretch.
   [Hint: You can begin by deriving (or quoting) the fact that the area of the given parallelogram is $|xy' - x'y|$.]

Section 3

1. a) $T(x, y, z) = (y, 2x, -z)$.
   d) $\begin{pmatrix} 0 & 1 & 0 \\ 2 & 0 & 0 \\ 0 & 0 & -3 \end{pmatrix}$

2. a) The matrices of **S, T,** and **S + T** are

$$\begin{pmatrix} 0 & 1 \\ 1 & 2 \\ -1 & 1 \end{pmatrix}, \begin{pmatrix} 1 & 1 \\ 2 & -1 \\ 1 & -3 \end{pmatrix}, \begin{pmatrix} 1 & 2 \\ 3 & 1 \\ 0 & -2 \end{pmatrix},$$

respectively.

3. Matrix of **U** is $\begin{pmatrix} 1 & 1 & 0 \\ 0 & 0 & 1 \\ 1 & -1 & 0 \\ 0 & 1 & 1 \end{pmatrix}.$

Matrix of **UT** is $\begin{pmatrix} 3 & 0 \\ 1 & -3 \\ -1 & 2 \\ 3 & -4 \end{pmatrix}.$

4. Matrix of **S** is $\begin{pmatrix} -1 & 1 \\ 1 & 3 \end{pmatrix}.$

Matrix of **TS** is $\begin{pmatrix} 0 & 4 \\ -3 & -1 \\ -4 & -8 \end{pmatrix}.$

Matrix of **U(TS)** is $\begin{pmatrix} -3 & 3 \\ -4 & -8 \\ 3 & 5 \\ -7 & -9 \end{pmatrix}$ = matrix of **(UT)S**.

6. Put $V = e_1, e_2, \ldots, e_n$.

Section 4

1. a) Kernel is (0, 0); invertible.
   b) A basis of the kernel is $(1, -1)$; not invertible.
   c) Invertible.
   e) A basis of the kernel is $(5, -4, 3)$.
   f) A basis of the kernel is $(3, 0, -1), (0, 3, -2)$.
2. a) Transformation is onto.
   b) Image space has basis (1, 2).
   c) Image space has basis $(1, 2, 1), (1, -1, -3)$.
   e) Transformation is onto.
   f) Image space has basis (1, 1).
3. a) If a, b are not both zero the dimension of the kernel is 1 and so is the dimension of the image space.
   b) Dimension of the image space is 3, hence the dimension of the kernel is 2.
   c) Dimension of the kernel is zero, hence the dimension of the image space is 3.
4. b) The dimension of the kernel is 1.
8. Use the dimension theorem.
13. (a) and (d) have an inverse, (b) and (c) do not.

CHAPTER 4

Section 1

1. $x = 1 - 2z, y = 2 + 3z$
2. $(x, y, z) = (1, -2, 1)t$, t arbitrary
3. $x = 1 + 2z, y = -3 - 7z$
4. no solutions
5. $x_1 = \frac{1}{3} - 2x_3$, $x_2 = \frac{7}{15} - x_3$, $x_4 = \frac{4}{15}$
7. Has unique solution.
8. $(2, 1, -1, 0, 0)t$, t arbitrary.
9. Use the result of Exercise 8.
10. $x_1 = \frac{35}{26}y_1 - \frac{19}{26}y_2 - \frac{29}{26}y_3 - \frac{17}{2}x_3$
    $x_2 = -\frac{1}{2}y_1 + \frac{1}{2}y_2 + \frac{1}{2}y_3 + \frac{7}{2}x_3$
    $x_4 = \frac{3}{13}y_1 - \frac{2}{13}y_2 - \frac{1}{13}y_3$
11. Condition is $17y_1 - 7y_2 - 2y_3 = 0$.
12. Conditions are $2y_1 - 7y_2 + y_3 = 0$, $y_1 - 3y_2 - y_4 = 0$.
13. Conditions are $2y_1 + y_2 - y_3 = 0$, $y_2 - 3y_3 + 2y_4 = 0$.
14. If $a_1 \neq 0$, then $(a_2, -a_1, 0, \ldots, 0)$, $(a_3, 0, -a_1, 0, \ldots, 0), \ldots, (a_n, 0, \ldots, 0, -a_1)$ is a basis.

Section 2

1. $\begin{pmatrix} 1 & 0 & 2 \\ 0 & 1 & -3 \end{pmatrix}$ has rank 2; kernel has basis $(-2, 3, 1)^*$; image space is $\mathbb{R}_2$; left kernel is $\mathbf{0}$.

2. $\begin{pmatrix} 1 & 1 \\ 1 & 2 \\ 1 & 3 \end{pmatrix}$ has rank 2; kernel is $\mathbf{0}$; image space has basis $(1, 1, 1)^*$, $(1, 2, 3)^*$; left kernel has basis $(1, -2, 1)$.

3. a) Rank $= 2$ b) Kernel has basis $(-2, 7, 1)^*$.
   c) First two columns are a basis of image space.
   d) Left kernel has basis $(-2, -1, 1)$.
4. a) Rank $= 3$ b) Kernel is $\mathbf{0}$.   d) Left kernel has basis $(-2, -1, 1, 0)$.
5. Use the result of Exercise 4 by noting that the matrices are transposes of each other.
6. a) Rank $= 2$ b) Kernel has basis $(17, -7, -2)^*$.
   d) Left kernel is $\mathbf{0}$ (cf. Exercise 10 in Section 1).
7. Transpose of matrix in Exercise 6 (cf. Exercise 11 in Section 1).
8. c) Kernel has basis $(-2, -2, 1)^*$.
   d) Left kernel has basis $(2, -7, 1, 0)$, $(-1, 3, 0, 1)$ (cf. Exercise 12 in Section 1).
9. a) Rank $= 1$
10. a) Rank $= 2$
11. a) Rank $= 2$ c) Kernel has basis $(4, 2, 1, 0)^*$, $(-10, -3, 0, 1)^*$ (cf. Exercise 13 in Section 1).
12. Use 2.5.

14. Basis of intersection is $(1, -1, 0, 2)^*$, $(3, -1, -1, 3)^*$.
15. Use 2.5.
17. $A^{-1} = \begin{pmatrix} -1 & 1 & -1 \\ -3 & 2 & -1 \\ 3 & -2 & 2 \end{pmatrix}$.
18. No inverse (rank = 2) (cf. Exercises 3 and 4 in Section 1).
19. Inverse exists (cf. Exercise 7 in Section 1).
20. No inverse (rank = 2) (cf. Exercise 13 in Section 1).

Section 3

19. a) (1) $(0, 0)$ (2) no solution   (3) $(1, 1)$
    b) Kernel has basis $(1, 0, 1, 0)^*$.
    c) Kernel is $(0, 0, 0)^*$.
    d) Kernel has basis $(2, 1, -1)^*$.
    e) Kernel has basis $(2, -7, 1, 0)^*$, $(-1, 3, 0, 1)^*$.
20. a) Left kernel has basis $(-2, 13, 7)$.

25. a) $\begin{pmatrix} 2 & -1 \\ -1 & 1 \end{pmatrix} \begin{pmatrix} 1 & 1 & 1 \\ 1 & 2 & 3 \end{pmatrix} \begin{pmatrix} 1 & 0 & 1 \\ 0 & 1 & -2 \\ 0 & 0 & 1 \end{pmatrix} = \begin{pmatrix} 1 & 0 & 0 \\ 0 & 1 & 0 \end{pmatrix}$

    b) $P = \begin{pmatrix} \frac{35}{26} & -\frac{19}{26} & -\frac{29}{26} \\ -\frac{1}{2} & \frac{1}{2} & \frac{1}{2} \\ \frac{3}{13} & -\frac{2}{13} & -\frac{1}{13} \end{pmatrix}$, $Q = \begin{pmatrix} 1 & 0 & 0 & -\frac{17}{2} \\ 0 & 1 & 0 & \frac{7}{2} \\ 0 & 0 & 0 & 1 \\ 0 & 0 & 1 & 0 \end{pmatrix}$ gives

$PAQ = \begin{pmatrix} 1 & 0 & 0 & 0 \\ 0 & 1 & 0 & 0 \\ 0 & 0 & 1 & 0 \end{pmatrix}$.

    e) $P = \begin{pmatrix} 1 & 0 & 0 \\ -2 & 1 & 0 \\ 1 & 0 & 1 \end{pmatrix}$, $Q = \begin{pmatrix} 1 & -2 & -1 & -3 \\ 0 & 1 & 0 & 0 \\ 0 & 0 & 1 & 0 \\ 0 & 0 & 0 & 1 \end{pmatrix}$ gives $PAQ = \begin{pmatrix} 1 & 0 & 0 & 0 \\ 0 & 0 & 0 & 0 \\ 0 & 0 & 0 & 0 \end{pmatrix}$.

26. $A^{-1} = QP$
27. Use 3.4.

Section 4

1. b) $P = \begin{pmatrix} -1 & 1 & -1 \\ -3 & 2 & -1 \\ 3 & -2 & 2 \end{pmatrix}$.
    c) $X = (1, -1, 2)^*$ gives $X' = (-4, -7, 9)^*$.
    d) $X = (1, 1, -2)^*$

2. $Q = \begin{pmatrix} 2 & 3 \\ -1 & 1 \end{pmatrix}$

3. a) $\mathbf{PA} = \begin{pmatrix} 1 & -1 \\ -2 & 2 \\ 0 & 1 \end{pmatrix}$

b) $\mathbf{AQ} = \begin{pmatrix} 5 & 5 \\ 1 & 4 \\ -7 & -3 \end{pmatrix}$

c) $\mathbf{PAQ} = \begin{pmatrix} 3 & 2 \\ -6 & -4 \\ -1 & 1 \end{pmatrix}$

CHAPTER 5

Section 1

4. See Exercise 20 in Section 4 of Chapter 2.

Section 2

1. a) 26 b) 12 c) $(x - 3)^4$
   d) -5 e) 26
2. a) Use 2.4 b) -4 c) -2^9

Section 4

8. a) $\operatorname{adj} \mathbf{A} = \begin{pmatrix} 11 & 6 & 2 \\ 5 & -3 & -1 \\ 9 & 3 & -6 \end{pmatrix}$

11. c) $-x^3 + 8x^2 - 15x$
14. b) Use Exercise 13.
    c) Use 4.6.

Section 5

2. Use the fact that an *even* permutation followed by an *odd* permutation gives an *odd* permutation.

CHAPTER 6

Section 1

2. $\mathbf{A}$ has 0 as an eigenvalue iff there exists a nonzero vector $\mathbf{V}$ such that $\mathbf{AV} = 0$.
4. a) $x^2 - (a + d)x + (ad - bc)$
   b) $(3 - x)^4$
   c) $(2 - x)^2(3 - x)$
   d) $-x^3 + 8x^2 - 15x$
   f) $(2 - x)(3 - x)(6 - x)$

5 and 6. b) $x = 3$; eigenvectors $(1, 0, 0, 0)^*$, $(0, 0, 0, 1)^*$
  c) $x = 2$; eigenvector $(1, 0, 0)^*$
     $x = 3$; eigenvector $(0, 0, 1)^*$
  d) $x = 0$; eigenvector $(1, 5, 7)^*$
     $x = 3$; eigenvector $(2, 1, 2)^*$
     $x = 5$; eigenvector $(1, 0, 2)^*$
  e) $x = 0$; eigenvector $(2, -2, -1)^*$
     $x = 3$; eigenvector $(2, 1, -1)^*$
     $x = 5$; eigenvector $(1, 4, -3)^*$
  f) $x = 2$; eigenvector $(0, -1, 1)^*$
     $x = 3$; eigenvector $(1, 1, 1)^*$
     $x = 6$; eigenvector $(-2, 1, 1)^*$.

Section 2

1. a) Matrix is $\begin{pmatrix} -3 & 2 \\ 1 & -2 \end{pmatrix}$ which has eigenvalues $-1, -4$. Corresponding eigenvectors are $(1, 1)^*$, $(-2, 1)^*$. Hence the new basis is $e'_1 = e_1 + e_2$, $e'_2 = -2e_1 + e_2$ and $T(e'_1) = -e'_1$, $T(e'_2) = -4e'_2$.
  b) Matrix is that of Exercise 4(d) in Section 1. New basis is $e'_1 = e_1 + 5e_2 + 7e_3$, $e'_2 = 2e_1 + e_2 + 2e_3$, $e'_3 = e_1 + 2e_3$ and $T(e'_1) = 0$, $T(e'_2) = 3e'_2$, $T(e'_3) = 5e'_3$.
  c) Matrix is that of Exercise 4(e) in Section 1. New basis is $V'_1 = 2V_1 - 2V_2 - V_3$, $V'_2 = 2V_1 + V_2 - V_3$, $V'_3 = V_1 + 4V_2 - 3V_3$, and $T(V'_1) = 0$, $T(V'_2) = 3V'_2$, $T(V'_3) = 5V'_3$.

2. a) The characteristic polynomial of A^{-1} is $(ad - bc)x^2 - (a + d)x + 1$. Show directly that the roots of this polynomial are the reciprocals of those of the characteristic polynomial of A.

5. A is stochastic iff its entries are nonnegative and $AU = U$ where $U = (1, 1, \ldots, 1)^*$.

Supplement to Section 2

2. b) The eigenvalues of P are $1, \frac{1}{5}, \frac{3}{10}$.
8. [*Hint:* Apply Bernoulli's method to the equation $x^2 - x - 1 = 0$.]

Section 3

9. This problem is easier if you use 4.1 and/or 4.3 and 4.4 in Section 4.
10. *Hint:* The required characteristic polynomial is

$$\det(A^{-1} - xI) = \frac{1}{\det A} \det A(A^{-1} - xI) = \frac{1}{\det A} \det(I - xA).$$

11. If A is invertible, use the trace of adj A.

Section 4

3. d) $P = \begin{pmatrix} 1 & 2 & 1 \\ 5 & 1 & 0 \\ 7 & 2 & 2 \end{pmatrix}$, $D = \begin{pmatrix} 0 & 0 & 0 \\ 0 & 3 & 0 \\ 0 & 0 & 5 \end{pmatrix}$

e) $\mathbf{P} = \begin{pmatrix} 2 & 2 & 1 \\ -2 & 1 & 4 \\ -1 & -1 & -3 \end{pmatrix}$, $\mathbf{D}$ as in (d)

f) $\mathbf{P} = \begin{pmatrix} 0 & 1 & -2 \\ -1 & 1 & 1 \\ 1 & 1 & 1 \end{pmatrix}$, $\mathbf{D} = \begin{pmatrix} 2 & 0 & 0 \\ 0 & 3 & 0 \\ 0 & 0 & 6 \end{pmatrix}$

6. $\mathbf{P} = \begin{pmatrix} 1 & 1 & 2 \\ 1 & 0 & 3 \\ 0 & -1 & 0 \end{pmatrix}$, $\mathbf{D} = \begin{pmatrix} 1 & 0 & 0 \\ 0 & 1 & 0 \\ 0 & 0 & -1 \end{pmatrix}$

10. $\mathbf{P} = \begin{pmatrix} 1 & 1 \\ 0 & 1 \end{pmatrix}$

11. a) $\mathbf{P} = \begin{pmatrix} 1 & 1 & 1 \\ 1 & \omega & \omega^2 \\ 1 & \omega^2 & \omega \end{pmatrix}$, $\mathbf{D} = \begin{pmatrix} 1 & 0 & 0 \\ 0 & \omega & 0 \\ 0 & 0 & \omega^2 \end{pmatrix}$, where $\omega^3 = 1$, $\omega \neq 1$

b) $\mathbf{P} = \begin{pmatrix} 1 & 1 & 1 & 1 \\ 1 & i & -1 & -i \\ 1 & -1 & 1 & -1 \\ 1 & -i & -1 & i \end{pmatrix}$, $\mathbf{D} = \begin{pmatrix} 1 & 0 & 0 & 0 \\ 0 & i & 0 & 0 \\ 0 & 0 & -1 & 0 \\ 0 & 0 & 0 & -i \end{pmatrix}$

c) $\mathbf{P}$ as in (a), $\mathbf{D} = \begin{pmatrix} 0 & 0 & 0 \\ 0 & \omega - \omega^2 & 0 \\ 0 & 0 & \omega^2 - \omega \end{pmatrix}$

Section 5

12. b) Iff, for each eigenvalue λ_i, the dimension of the kernel of $\mathbf{A} - \lambda_i \mathbf{I}$ is one.

13. $\mathbf{P}^k = \begin{pmatrix} \frac{1}{7} & \frac{5}{8} & \frac{13}{56} \\ \frac{1}{7} & \frac{5}{8} & \frac{13}{56} \\ \frac{1}{7} & \frac{5}{8} & \frac{13}{56} \end{pmatrix} + (\tfrac{1}{5})^k \begin{pmatrix} 0 & -\frac{5}{8} & \frac{5}{8} \\ 0 & \frac{3}{8} & -\frac{3}{8} \\ 0 & -\frac{5}{8} & \frac{5}{8} \end{pmatrix} + (\tfrac{3}{10})^k \begin{pmatrix} \frac{6}{7} & 0 & -\frac{6}{7} \\ -\frac{1}{7} & 0 & \frac{1}{7} \\ -\frac{1}{7} & 0 & \frac{1}{7} \end{pmatrix}$

14. Use the fact that, if $\mathbf{A}$ is a permutation matrix, there exists an integer N so that $\mathbf{A}^N = \mathbf{I}$.

CHAPTER 7

Section 1

5. a) $\{(1, 2) + k(1, 0)\}$
   c) $\{(1, 2) + k(1, 1)\}$
7. $\mathbf{P} = (0, 0, -10)$, basis for $\mathcal{V}$: $(1, 0, 2)$, $(0, 1, 3)$

Section 2

2. $x = 1 + t_1 + 2t_2$
   $y = -1 + 2t_1 + t_2$
   $z = 2 + 2t_1 - t_2$

Answers to selected exercises / Chapter 8

3. a) $x = 1 + t, y = 2$
   c) $x = 1 + t, y = 2 + t$
5. a) $y = 0$ for $\lambda = 3$, which gives the point $(-1, 0)$.
   b) $x = 0$ for $\lambda = -\frac{1}{2}$
6. a) the points between **P** and **Q** on the line
7. $x = \lambda_1 + 2\lambda_2 + 3\lambda_3$
   $y = -\lambda_1 + \lambda_2$
   $z = 2\lambda_1 + 4\lambda_2 + \lambda_3$
   where $\lambda_1 + \lambda_2 + \lambda_3 = 1$
9. $x_1 = 1 + 6t_1 + 6t_2$
   $x_2 = 8t_1 - 3t_2$
   $x_3 = -2 + 11t_1$
   $x_4 = 3 + 2t_1 + 2t_2$

Section 3

1. b) The possibilities are: no intersection, one point, $\mathcal{F}_1$ lies entirely in $\mathcal{F}_2$.
2. $4x - 5y + 3z = 15$
4. a) $3x_1 - 3x_2 - 2x_3 + 2x_4 = 3$
   b) the equation in (a) and $x_1 - 3x_2 - x_3 = 5$
5. $x_1 - 3x_4 = -8, 2x_2 - 2x_3 + 3x_4 = 13$
6. Use Exercise 2 above.
7. a) the line $(x_1, x_2, x_3, x_4) = (1, 2, 0, 3) + t(30, 16, 31, 10)$
8. $x + 2y = 5$ and $x + z = 4$
14. Use 2.5 in Chapter 4.
15. Assume that the flats have a point **P** in common and get a contradiction by proving that $\mathbf{P} + \mathcal{W}$ is contained in $\mathbf{P} + \mathcal{V}$ (if $\mathcal{W}$ is a subspace of $\mathcal{V}$).
16. Represent the hyperplanes as solution sets of linear equations.
18. Use the parametric form of the straight line.
19. If the lines $\mathbf{P} + t_1\mathbf{V}, \mathbf{Q} + t_2\mathbf{W}$ are not parallel and lie in the same plane, use **V**, **W** as a basis for the direction space of that plane. Then write the plane in the form $\mathbf{P} + t_1\mathbf{V} + t_2\mathbf{W}$.

CHAPTER 8

Section 1

3. $\mathbf{PQ} = (6, 8, 11, 2)$, length $= 15$, cosine of the angle $= \dfrac{4}{\sqrt{14}\sqrt{219}}$

4. a) Lengths are $3, \sqrt{11}, \sqrt{6}$.

   b) $\cos A = \dfrac{2}{3\sqrt{6}}$, $\cos B = \dfrac{7}{3\sqrt{11}}$, $\cos C = \dfrac{4}{\sqrt{66}}$

   c) $\sin A = \dfrac{5}{3\sqrt{3}}$

   d) Find $\sin(A + B)$ and $\cos(A + B)$ and compare to $\sin C$ and $\cos C$.

Chapter 8 / Answers to selected exercises   525

9. a) To prove (ii) write $\mathbf{V} \circ \mathbf{V}$ as the sum of two squares.
   b) Length of $(0, 1)$ is $\sqrt{2}$.
   c) Angle between $(0, 1)$ and $(1, 0)$ is $\pi/4$.
10. a) To prove (ii) you need the continuity of f.
    b) Distance between x and x^2 is $1/\sqrt{30}$.
13. b) [*Hint:* If $|\mathbf{V} + x\mathbf{W}|^2 = 0$ then $\mathbf{V} = -x\mathbf{W}$ so that $\mathbf{V}, \mathbf{W}$ are linearly dependent.]
14. a) $\lim_{\mathbf{V} \to \mathbf{V}_0} f(\mathbf{V}) = c$ (a real number) means: for any positive ϵ there exists a number δ so that, if $0 < |\mathbf{V} - \mathbf{V}_0| < \delta$, then $|f(\mathbf{V}) - c| < \epsilon$.
    g) The Schwarz inequality can be used here.

Section 2

5. a) $\mathbf{U}, \mathbf{V}', \mathbf{W}'$,
   where
   $$\mathbf{V}' = \mathbf{V} - 3\mathbf{U} = (6, -3, 0, 2)$$
   $$\mathbf{W}' = \mathbf{W} - 3\mathbf{U} - \tfrac{1}{7}\mathbf{V}' = (-\tfrac{6}{7}, -\tfrac{4}{7}, 1, \tfrac{12}{7})$$
   b) $(\tfrac{1}{3}, \tfrac{2}{3}, \tfrac{2}{3}, 0)$
      $(\tfrac{6}{7}, -\tfrac{3}{7}, 0, \tfrac{2}{7})$
      $$\left(-\frac{6}{7\sqrt{5}}, -\frac{4}{7\sqrt{5}}, \frac{1}{\sqrt{5}}, \frac{12}{7\sqrt{5}}\right)$$
   c) Add $(4, 26, -28, 27)$ to (a).

Section 3

2. a) $\mathbf{V} = (\tfrac{3}{2}, \tfrac{3}{2})$, $\mathbf{W} = (-\tfrac{1}{2}, \tfrac{1}{2})$ minimum distance $= |\mathbf{W}| = 1/\sqrt{2}$
   b) Minimize $(1 - x)^2 + (2 - x)^2$; the minimum value occurs for $x = \tfrac{3}{2}$, as it should.
3. $2x + y + 3z = 9$.
5. $3/\sqrt{14}$.
6. Basis of $\mathcal{V}^\perp$ is $(1, 2, 2)$.
7. a) Let $\mathbf{P}_0 = (x_0, y_0)$. We must find the point $\mathbf{P}(x, y)$ on the line so that $\mathbf{P}_0\mathbf{P}$ is in $\mathcal{V}^\perp$. Since $\mathcal{V}^\perp$ has basis (a, b), this gives $\mathbf{P}_0 - \mathbf{P} = t(a, b)$ and hence $t(a^2 + b^2) = (a, b) \cdot (\mathbf{P}_0 - \mathbf{P}) = ax_0 + by_0 - (ax + by) = ax_0 + by_0 + c$.
   The required distance is then $|\mathbf{P}_0 - \mathbf{P}| = \dfrac{|ax_0 + by_0 + c|}{\sqrt{a^2 + b^2}}$.
7. b) $\dfrac{|ax_0 + by_0 + cz_0 + d|}{\sqrt{a^2 + b^2 + c^2}}$.

16. Assume lines not parallel so that $\mathbf{V}_1, \mathbf{V}_2$ form a basis of a subspace $\mathcal{V}$. Express $\mathbf{P}_1 - \mathbf{P}_2$ as $\mathbf{P}_1 - \mathbf{P}_2 = a_1\mathbf{V}_1 + a_2\mathbf{V}_2 + \mathbf{V}$ where $\mathbf{V}$ is in $\mathcal{V}^\perp$. Then $(\mathbf{P}_1 - a_1\mathbf{V}_1) - (\mathbf{P}_2 + a_2\mathbf{V}_2)$ is $\mathcal{V}^\perp$ so that the vector from $\mathbf{P}_1 - a_1\mathbf{V}_1$ on $\mathcal{L}_1$ to $\mathbf{P}_2 + a_2\mathbf{V}_2$ on $\mathcal{L}_2$ is orthogonal to $\mathcal{L}_1$ and $\mathcal{L}_2$. Its length is the required distance. We can find a_1, a_2 from the equations
$$a_1(\mathbf{V}_1 \cdot \mathbf{V}_1) + a_2(\mathbf{V}_1 \cdot \mathbf{V}_2) = \mathbf{V}_1 \cdot (\mathbf{P}_1 - \mathbf{P}_2)$$
$$a_1(\mathbf{V}_1 \cdot \mathbf{V}_2) + a_2(\mathbf{V}_2 \cdot \mathbf{V}_2) = \mathbf{V}_2 \cdot (\mathbf{P}_1 - \mathbf{P}_2)$$

and then write $|V|^2 = |P_1 - P_2|^2 + |a_1V_1 + a_2V_2|^2$. In the case $n = 3$ we can let V_3 be a basis of $\mathcal{V}^\perp$ so that $V = a_3V_3$ where $a_3(V_3 \cdot V_3) = V_3 \cdot (P_1 - P_2)$. Since $|V| = |a_3| \, |V_3|$ we get $|V| = \dfrac{|V_3 \cdot (P_1 - P_2)|}{|V_3|}$.

We can take $V_3 = V_1 \times V_2$ if we wish (see Supplement to Section 6 of this chapter).

Section 4

1. $A = \begin{pmatrix} 1 & x_1 & y_1 \\ 1 & x_2 & y_2 \\ & \vdots & \\ 1 & x_n & y_n \end{pmatrix}, \ Y = \begin{pmatrix} z_1 \\ \vdots \\ z_n \end{pmatrix}, \ X = \begin{pmatrix} a \\ b \\ c \end{pmatrix},$

$$A^*A = \begin{pmatrix} n & \sum x_i & \sum y_i \\ \sum x_i & \sum x_i^2 & \sum x_i y_i \\ \sum y_i & \sum x_i y_i & \sum y_i^2 \end{pmatrix}, \ A^*Y = \begin{pmatrix} \sum z_i \\ \sum x_i z_i \\ \sum y_i z_i \end{pmatrix}$$

A is of rank 3 unless the points (x_i, y_i) are on the same line.

6. Put $A' = W'A$, $Y' = W'Y$, where

$$W' = \begin{pmatrix} \sqrt{w_1} & 0 & \cdots & 0 \\ 0 & \sqrt{w_2} & & 0 \\ \vdots & & & \vdots \\ & & \cdots & \sqrt{w_n} \end{pmatrix}$$

and show that the weighted norm problem transforms into the ordinary problem with matrix A' and vector Y'. Then use the known results for the ordinary problem.

7. [*Hint:* If A is of rank <3 there exists a nonzero vector $Z = (a, b, c)^*$ so that $AZ = 0$. Show that this contradicts a theorem about quadratic equations.]

15. Let V_1, V_2 be the orthogonal basis of $\mathcal{S}$. Then $V_0 = x_1V_1 + x_2V_2$ where
$$x_1(V_1 \cdot V_1) = (V_1 \cdot Y), \qquad x_2(V_2 \cdot V_2) = (V_2 \cdot Y),$$
$$|Y - V_0|^2 = |Y|^2 - (V_1 \cdot Y)^2 - (V_2 \cdot Y)^2.$$

Section 5

4. $\begin{pmatrix} -1 & 0 \\ 0 & -1 \end{pmatrix} = \begin{pmatrix} \cos \pi & -\sin \pi \\ \sin \pi & \cos \pi \end{pmatrix}$

5. Use Exercise 1.

6. The matrix is the first of the matrices in 5.4, since $T(e_1) = e_1 \cos \theta + e_2 \sin \theta$ and
$$T(e_2) = e_1 \cos\left(\theta + \frac{\pi}{2}\right) + e_2 \sin\left(\theta + \frac{\pi}{2}\right)$$
$$= -e_1 \sin \theta + e_2 \cos \theta.$$

8. a) If $A^{-1} = A^*$ and $B^{-1} = B^*$, then $(AB)^{-1} = B^*A^* = (AB)^*$.

10. $A = \begin{pmatrix} \cos\theta & -\sin\theta & 0 \\ \sin\theta & \cos\theta & 0 \\ 0 & 0 & 1 \end{pmatrix}$.

Section 6

1. a) rotation about the point $(1, 7)$
   b) glide reflection in the line $y = \frac{1}{2}x - \frac{7}{4}$, amount of glide $(\frac{8}{5}, \frac{4}{5})$
   c) reflection in the line $y = \frac{1}{2}x - \frac{5}{2}$
   d) rotary reflection
   e) glide reflection
   f) reflection in the plane $y - 2z = 5$
   g) rotary reflection
   h) screw displacement
   i) rotation about the line $(0, \frac{9}{2}, 0) + t(1, 2, 2)$ through an angle θ with $\cos\theta = \frac{3}{5}$

7. a) Rotation about z-axis has matrix $\begin{pmatrix} \cos\theta & -\sin\theta & 0 \\ \sin\theta & \cos\theta & 0 \\ 0 & 0 & 1 \end{pmatrix}$.

   Rotation about y-axis has matrix $\begin{pmatrix} \cos\varphi & 0 & -\sin\varphi \\ 0 & 1 & 0 \\ \sin\varphi & 0 & \cos\varphi \end{pmatrix}$.

   Product of these is $\begin{pmatrix} \cos\varphi\cos\theta & -\cos\varphi\sin\theta & -\sin\varphi \\ \sin\theta & \cos\theta & 0 \\ \sin\varphi\cos\theta & -\sin\varphi\sin\theta & \cos\varphi \end{pmatrix}$

   b) If ψ is the angle of rotation for the product, then $\cos^2\frac{\psi}{2} = \cos^2\frac{\theta}{2}\cos^2\frac{\varphi}{2}$.

9. b) rotation, if $x_0 + 2y_0 + 2z_0 = 0$ about the line $(\frac{5}{8}x_0, -z_0, 0) + t(1, 2, 2)$

CHAPTER 9

Section 1

1. in (a), (b), (c), (f): $P = \begin{pmatrix} 1/\sqrt{3} & 1/\sqrt{2} & 1/\sqrt{6} \\ 1/\sqrt{3} & -1/\sqrt{2} & 1/\sqrt{6} \\ 1/\sqrt{3} & 0 & -2/\sqrt{6} \end{pmatrix}$

   a) Diagonal elements are $3, -4, 0$.
   b) Diagonal elements are $3, +4, 12$.
   c) Diagonal elements are $3, -4, 12$.

   d) $P = \begin{pmatrix} 1/\sqrt{5} & 2/\sqrt{5} \\ -2/\sqrt{5} & 1/\sqrt{5} \end{pmatrix}$, $P^*AP = \begin{pmatrix} 5 & 0 \\ 0 & 20 \end{pmatrix}$

e) $\mathbf{P} = \begin{pmatrix} \frac{1}{3} & 0 & 4/3\sqrt{2} \\ \frac{2}{3} & 1/\sqrt{2} & -1/3\sqrt{2} \\ -\frac{2}{3} & 1/\sqrt{2} & 1/3\sqrt{2} \end{pmatrix}$, $\mathbf{P*AP} = \begin{pmatrix} 9 & 0 & 0 \\ 0 & 0 & 0 \\ 0 & 0 & 0 \end{pmatrix}$

f) Diagonal elements are $-3, 3, 3$.

7. Show that the discriminant of the characteristic polynomial is positive or zero.

8. a) Diagonal elements are $1, 1, 3, 3$.

b) $\mathbf{P} = \begin{pmatrix} \frac{1}{2} & +1/\sqrt{2} & 1/\sqrt{6} & 1/\sqrt{12} \\ \frac{1}{2} & -1/\sqrt{2} & 1/\sqrt{6} & 1/\sqrt{12} \\ \frac{1}{2} & 0 & -2/\sqrt{6} & 1/\sqrt{12} \\ \frac{1}{2} & 0 & 0 & -3/\sqrt{12} \end{pmatrix}$.

Diagonal elements are $3, -1, -1, -1$.

Section 2

1. a) Use the results of Example 1. New basis is $\mathbf{e}'_1 = (\frac{3}{5}, \frac{4}{5}), \mathbf{e}'_2 = (-\frac{4}{5}, \frac{3}{5})$. Equation becomes $x'^2 - \frac{1}{4}y'^2 = 1$; the curve is a hyperbola.
2. All are obtainable from Exercise 1 in Section 1.
   a) $3x'^2 - 4y'^2$
3. a) $3x'^2 - 4y'^2 = -12$ gives a hyperbolic cylinder.
   b) $3x'^2 + 4y'^2 + 12z'^2 = 144$ (ellipsoid)
   d) $x'^2 + 4y'^2 = 16$ (ellipse) in $\mathbb{R}_2$; ellipitcal cylinder in $\mathbb{R}_3$).
   e) two parallel planes.
   f) hyperboloid of two sheets, of revolution

Section 3

1. $3x''^2 - 4y''^2 = 12$; new origin is $(0, 2, 1)$.
2. two intersecting planes.
3. $3x''^2 + 4y''^2 + 12z''^2 = 36$; new origin is $(1, 3, -1)$.
5. Locus is the point $(1, 3, -1)$.
6. $3x''^2 - 4y''^2 + 12z''^2 = 36$, hyperboloid of one sheet
7. hyperboloid of two sheets
8. cone
9. $5x''^2 + 20y''^2 = 5z''$; new origin is $(0, -5, 2)$, elliptic paraboloid.
10. elliptic cylinder
13. $x'^2 = \sqrt{2}\, y'$, parabolic cylinder
17. hyperboloid of revolution of one sheet

Section 4

1. a) none
   b) positive definite
   c) none
   f) positive semidefinite, not positive definite

5. If **A** is positive semidefinite, then **P** (invertible) exists so that $\mathbf{A} = \mathbf{P}^*\mathbf{D}^2\mathbf{P}$ where the diagonal matrix **D** has *real* elements. Then show that we can take $\mathbf{B} = \mathbf{DP}$. Note that **B** is invertible iff **D** has no zero elements in the diagonal.

6. b) [*Hint:* Write det $(\mathbf{A} - \lambda\mathbf{B}) = (-\lambda)^n$ det $(\mathbf{B} - \frac{1}{\lambda}\mathbf{A})$.]

Section 5

7. a) $-y_1^2 + 8y_2^2$, where $y_1 = x - 3y - z$, $y_2 = y + \frac{1}{2}z$.

CHAPTER 10

Section 1

2. a) $|\mathbf{V}| = \max(\frac{2}{3}|x|, |y|)$
6. a) $x^2 + 4y^2 = 4$ (ellipse)
   c) the line segment from $(-2\sqrt{2}, -\sqrt{2})$ to $(2\sqrt{2}, \sqrt{2})$
7. c) the line segment from $(-4, -2)$ to $(4, 2)$
10. b) unit circle in $\mathbb{R}_2^*$
    c) taxicab norm in $\mathbb{R}_2^*$
11. [*Hint:* Remember that $0 \leq \lambda \leq 1$.]
12. [*Hint:* Use the triangle inequality in $\mathbb{R}_1$.]
13. Apply (ii) to $\mathbf{V} = \mathbf{A} - \mathbf{B}$, $\mathbf{W} = \mathbf{B}$.
15. Use Exercise 5 in Section 4 of Chapter 9.

Section 2

3. a) $\|\mathbf{B}\| = 2$.
   b) $\|\mathbf{A} + \mathbf{B}\| = \sqrt{10}$ (see Exercise 6(c) in Section 1)
5. Use the Schwarz inequality generously.
6. No

Supplement 1 to Section 2

3. c) 6.
   d) 20.
6. Supplement to Section 6 in Chapter 8 (especially (vi)) is useful here.
7. The spectral norms are: (a) $2\sqrt{2}$, (b) $\sqrt{\frac{1}{2}(15 + 5\sqrt{5})} \approx 3.62$, (c) 2.
8. b) Spectral norm is 3.

Supplement 2 to Section 2

1. Row norm $= \frac{7}{8}$; column norm $= \frac{3}{2}$; spectral norm $= \frac{3}{4}\sqrt{2}$; euclidean norm $= \frac{1}{8}\sqrt{74}$.

Section 3

1. a) Can be done directly; $e^\mathbf{A} = (\cosh 1)\mathbf{I} + (\sinh 1)\mathbf{A}$.
   b) $(\cos 1)\mathbf{I} + (\sin 1)\mathbf{A}$

6. The series converges; put $N = I + A + A^2 + \cdots$ and get $AN = A + A^2 + \cdots = N - I$ so that $(I - A)N = I$.

12. a) Write the matrix in the form $11I + \begin{pmatrix} 0 & 5 \\ -20 & 0 \end{pmatrix}$ and you can proceed directly to obtain $e^{11} \begin{pmatrix} \cos 10 & \frac{1}{2}\sin 10 \\ -2\sin 10 & \cos 10 \end{pmatrix}$.

13. The series converges since the row norm of **A** is <1.

$$\log(I + A) = \begin{pmatrix} 1 & 1 \\ 2 & 3 \end{pmatrix} \begin{pmatrix} \log \frac{3}{2} & 0 \\ 0 & \log \frac{11}{8} \end{pmatrix} \begin{pmatrix} 3 & -1 \\ -2 & 1 \end{pmatrix}$$

14. Use 3.9 in this section and 2.5 in Supplement 1 to Section 2.

16. Use the Jordan form.

17. First show that the row norm of **Q** is <1 so that $\lim_{k \to \infty} Q^k = 0$ and

$$\lim_{k \to \infty}(I + Q + \cdots + Q^{k-1}) = (I - Q)^{-1} = N.$$

Then use the answer given for Exercise 4 in Section 5d of Chapter 2 (Markov chains).

In the example, $N = \begin{pmatrix} \frac{9}{5} & \frac{4}{5} \\ \frac{3}{5} & \frac{8}{5} \end{pmatrix}$, $NB = \begin{pmatrix} \frac{1}{5} & \frac{4}{5} \\ \frac{2}{5} & \frac{3}{5} \end{pmatrix}$.

If we start in State 3 the probability is $\frac{1}{5}$ that we end in State 1 and is $\frac{4}{5}$ that we end in State 2 (and is 0 that we end in State 3 or 4). Similarly, if we start in State 4.

Index

absolute convergence, 463
adjoint, 257
affine equations, 331
affine subspace, 327
affine transformation, 148
algebra, an, 104 f.n.
alternating function, 236, 240
angle between **V** and **W**, 349
augmented matrix, 190
axiom system, 5
axiomatic method, 4–9

basis, 57
basis algorithm, 82
Bernoulli's method, 287 (ex. 6)
bijection, 29
Blythe, Mrs. Dorothy, 346 f.n.
box norm, 457

$\mathbb{C}_{m \times n}$, definition, 103
$\mathbb{C}_n$, definition, 86
Cauchy-Schwarz inequality, 347
Cauchy's polynomial root theorem, 278
Cayley numbers, 499

Cayley-Hamilton theorem, 291
   proof of, 305
change of basis matrix, 221
characteristic polynomial, 270
characteristic value, vector, 268
clique, 125
coefficient of correlation, 379
coefficient matrix, 189
cofactor, 254
collinear, 332
column norm, 477
column rank, 76, 174
column space, 76, 174
commutative group, 20–21
   definition, 17, 20
companion matrix, 277
complex numbers $\mathbb{C}$, 41, 117 (ex. 13), 491–496
complex dot product, 361
complex euclidean space, 361
complex inner product space, 362
complex matrices, 103
compression in a line, 150
computers, 87

consistency, 9
convergence
  of sequences
    elementwise, 459
    in norm, 459
  of series, 463
Cramer's Rule, 252
cross product, 402
cyclic basis, 311
cyclic nilpotent, 311

derivative, 489
determinant, 235 ff.
determinant algorithm, 246
determinant rank, 260
diagonal decomposition, 507
dimension, 62
dimension theorem, 172
direct sum, 54
direction angles, 350
  cosines, 350
direction space
  of a flat, 327
  of a line, 325
  of a plane, 326
distance between points, 348, 464
distance from a point to a subspace, 372
distance from a point to a flat, 379 (ex. 13)
domain, 28
dominant diagonal theorem, 182
dot product, 345
dual basis, 183
dual mapping, 185
dual space, 182

eigenvalue, 268, 269
eigenvalue problem, 268
eigenvector, 268, 269
elementary matrices, 107–108
elementary orthogonal matrix, 423
elementary row operations, 77
epimorphism, 29
equivalence relations, 18–20
equivalent matrices, 217
equivalent systems of equations, 188
euclidean norm, 456
euclidean space, 345, 358

Euler's formula, 495
Euler's identity, 405
exponential of a matrix, 481

$\mathbb{F}_{m \times n}$, definition, 103
$\mathbb{F}_n$, definition, 86
Fibonacci sequence, 288
field, 34
flat, 327
flow diagram, 85, 87, 88
function
  composition, 31
  equality, 29
  informal definition, 27
  precise definition, 30
  real
    addition, 24
    scalar multiplication, 25
functor
  additive contravariant, 160
  additive covariant, 160

Gershgorin's Theorem, 276
glide reflection, 392
Gram-Schmidt process, 356
group, 21

Hermitian matrix, 421
homogeneity, 104
homogeneous system of equations, 197
homomorphism, 154 f.n.
hyperbolic rotation, 150
hyperplane, 327

identity function, 33, 145
identity matrix, 104
i,j-entry, 79
image of a linear transformation, 156
infinite dimensional, 62
infinite sequence in $\mathbb{R}_n$, 459
injection, 29
inner product, 346, 358
inverse function, 32
inverse of a linear transformation, 157
invertible matrix, 106
invertibility, criterion for, 174
isometric isomorphism, 359, 380

isometry, 389
isomorphic vector spaces, 69
isomorphism, 29, 70
   natural, 184

Jacobi Method, 422
Jordan block, 305
Jordan Form, 300

kernel, 170
kernel algorithm, 214
kinship systems, 140
Kronecker delta, 116

latent root, 268
   vector, 268
least squares, 373 ff.
left kernel, 209
Legendre polynomials, 360
length of vector, 348
line, 324
   two point formula for, 331
linear algebra
   overview, 2–3
   uses and importance, 3–4
linear combination, 50
linear functional, 182
linear independence, computational test for, 85
linear transformation(s)
   composition of, 154
   definition, 145
   over $\mathbb{C}$, 152
   over a field, 153
   scalar multiplication of, 153
   sum of, 153
linear variety, 327
linearly dependent, 59
linearly independent, 58
logarithm matrix, 487, (ex. 10)

main diagonal, 114
Markov chain, 133
matrices, partitioned, 96 ff.
matrix, 75
   addition, 90

   equality, 90
   scalar multiplication, 90
matrix multiplication, 91, 92
   computational check, 101 (ex. 11), 102 (ex. 12)
matrix norm, 466
matrix unit, 100
matrix with respect to bases, 220
metric, 464 (ex. 3)
minor, 256
models, mathematical, 9–12
modulus of complex number, 275, 494
monomorphims, 29
morra, game of, 130
multilinear, 235, 240
multiplicity of eigenvalue, 300

negative definite, 446
nilpotent, 303
nonsingular, 106
nontrivial equation, 337
norm, 346, 456
normal form of equation of plane, 368
normal vector to a hyperplane, 367
normed vector space, 457
nullity, 173

one-to-one, 29
   correspondence, 29
onto, 29
orientation
   preserving, 395, 399, 406
   reversing, 395, 399, 406
orthogonal, 353
orthogonal basis, 355
orthogonal complement, 362
orthogonal group, 386 (ex. 8)
orthogonal linear transformation, 379
orthogonal matrix, 383
   and parameters, 387
orthogonal projection, 364
orthogonal vectors, 353
orthonormal basis, 355

parallel
   flats, 344
   lines, 344

parameter(s), 329, 331
parametric equations
    of a line, 329
    of a plane, 331
permutation, 138
    even or odd, 266
permutation matrix, 138, 263
perpendicular, 353
    from a subspace to a point, 364
perpendicular vectors, 353
product theorem for determinants, 248
pivot, 79
pivot test, 444
plane, 326
    three-point formula, 333
point, 322
polar decomposition, 493
    of complex numbers, 495
    of matrices, 504
polynomials, function *vs.* formal symbol, 26
positive definite matrix, 443
    quadratic form, 443
positive semidefinite, 447
power method, 284 ff.
power series, 480
principal axis, 429
principal axis theorem, 417
    matrix form, 418
principal minors test, 444
probability vector, 44
projection, 178
projection
    orthogonal, 181
    of vector on a subspace, 364
projection on a line, 150
projection of **V** on **W**, 353
proper orthogonal, 503
proper value, vector, 268

quadratic form, 427
quadratic surface, 429
quaternions, 119 (ex. 26), 404, 496 f.

$\mathbb{R}_{m \times n}$, definition, 91
$\mathbb{R}_n$, definition, 22
range, 28

rank, 173
real numbers, differences from other fields, 36
reflection about a line, 391
reflection in a flat, 411
reflections in $\mathbb{R}_n$, 409
regular, 106
rigid motion, 389
ring, 105 f.n.
rotary reflection, 398
rotation about a point, 391
rotations and quaternions, 404–406
row addition matrix, 108
row-echelon form, 192
row-equivalent matrices, 213
row interchange matrix, 107
row multiplier matrix, 108
row norm, 476
row rank, 76
row space, 76

scalar product, 346
scalars, 40
Schwarz inequality, 347
screw displacement, 400
shear, 150
similar matrices, 293
singular (= non-nonsingular), 106
skew compression in a line, 150
skew reflection, 149
skew-symmetric matrix, 119 (ex. 27), 387, 420
sociometric matrix, 121
solution algorithm, 193
    transposed form, 216
span, 50
sparse matrix, 124
special orthogonal group, 386 (ex. 8)
spectral norm, 469
square root, matrix, 501
stochastic matrix, 134
strategy, in game theory, 131, 132
submultiplicative, 469
subspace(s)
    definition, 49
    direct sum of, 54
    intersection of, 54
    sum of, 53

subspace spanned by S (= subspace generated by S), 51
surjection, 29
Sylvester's Law of Inertia, 452
symmetric linear transformation, 416
symmetric matrix, 113, 415, Chapter 9

taxicab norm, 457
trace, 290
translate, 323, 389
transpose, 111
triangle inequality, 348

unit sphere, 458
unit vector, 349
unitary space, 362
unitary transformation, 386

van der Monde determinant, 259
vector, 40
vector from $\mathbf{P}$ to $\mathbf{Q}$, 322
vector product, 403
vector space
   over the complex numbers, 44
   over other fields, 44
vector space over the reals
   definition, 39
vectors, geometrical, 16–18

weighted norm, 377
well-defined, 22 (ex. 4), 30

zero divisor, 105
zero function, 145